Coulson and Richardson's
Chemical Engineering

Coulson & Richardson's Chemical Engineering Series

Chemical Engineering, Volume 1A, Seventh edition
Fluid Flow: Fundamentals and Applications
Raj Chhabra and V. Shankar

Chemical Engineering, Volume 1B, Seventh edition
Heat and Mass Transfer: Fundamentals and Applications
Raj Chhabra and V. Shankar

Chemical Engineering, Volume 2A, Sixth edition
Particulate Systems and Particle Technology
Raj Chhabra and Madivala G. Basavaraj

Chemical Engineering, Volume 2B, Sixth edition
Separation Processes
A. K. Ray

Chemical Engineering, Volume 3A, Fourth edition
Chemical and Biochemical Reactors and Reaction Engineering
R. Ravi, R. Vinu and S. N. Gummadi

Chemical Engineering, Volume 3B, Fourth edition
Process Control
Sohrab Rohani

Chemical Engineering
Solutions to the Problems in Volume 1
J. R. Backhurst, J. H. Harker and J. F. Richardson

Chemical Engineering
Solutions to the Problems in Volumes 2 and 3
J. R. Backhurst, J. H. Harker and J. F. Richardson

Chemical Engineering, Volume 6, Third edition
Chemical Engineering Design
R. K. Sinnott

Coulson and Richardson's Chemical Engineering

Volume 2A: Particulate Systems and Particle Technology

Sixth Edition

Volume Editors

Raj Chhabra

Department of Chemical Engineering, Indian Institute of Technology Ropar, Rupnagar, Punjab, India

Madivala G. Basavaraj

Department of Chemical Engineering, Indian Institute of Technology Madras, Chennai, India

Editor in Chief

Raj Chhabra

Butterworth-Heinemann
An imprint of Elsevier

Butterworth-Heinemann is an imprint of Elsevier
The Boulevard, Langford Lane, Kidlington, Oxford OX5 1GB, United Kingdom
50 Hampshire Street, 5th Floor, Cambridge, MA 02139, United States

Library of Congress Cataloging-in-Publication Data
A catalog record for this book is available from the Library of Congress

British Library Cataloguing-in-Publication Data
A catalogue record for this book is available from the British Library

ISBN: 978-0-08-101098-3

For information on all Butterworth-Heinemann publications
visit our website at https://www.elsevier.com/books-and-journals

Working together
to grow libraries in
developing countries

www.elsevier.com • www.bookaid.org

Publisher: Susan Dennis
Acquisition Editor: Anita Koch
Editorial Project Manager: Ali Afzal-Khan
Production Project Manager: Prem Kumar Kaliamoorthi
Cover Designer: Victoria Pearson

Typeset by SPi Global, India

Contents

Contributors

Madivala G. Basavaraj   Department of Chemical Engineering, Indian Institute of Technology Madras, Chennai, India

Jayanta Chakraborty   Department of Chemical Engineering, Indian Institute of Technology Kharagpur, Kharagpur, India

Raj Chhabra   Department of Chemical Engineering, Indian Institute of Technology Ropar, Rupnagar, Punjab, India

Abhijit Deshpande   Department of Chemical Engineering, Indian Institute of Technology Madras, Chennai, India

Kanjarla Anand Krishna   Department of Metallurgical and Materials Engineering, Indian Institute of Technology Madras, Chennai, India

Anand Prakash   Department of Chemical and Biochemical Engineering, University of Western Ontario, London, ON, Canada

V. Shankar   Department of Chemical Engineering, Indian Institute of Technology Kanpur, Kanpur, India

Anurag Tripathi   Department of Chemical Engineering, Indian Institute of Technology Kanpur, Kanpur, India

About Professor Coulson

John Coulson, who died on 6 January 1990 at the age of 79, came from a family with close involvement with education. Both he and his twin brother Charles (renowned physicist and mathematician), who predeceased him, became professors. John did his undergraduate studies at Cambridge and then moved to Imperial College where he took the postgraduate course in chemical engineering—the normal way to qualify at that time—and then carried out research on the flow of fluids through packed beds. He became an Assistant Lecturer at Imperial College and, after wartime service in the Royal Ordnance Factories, returned as Lecturer and was subsequently promoted to a Readership. At Imperial College, he initially had to run the final year of the undergraduate course almost single-handed: a very demanding assignment. During this period, he collaborated with Sir Frederick (Ned) Warner to write a model design exercise for the I. Chem. E. Home Paper on 'The Manufacture of Nitrotoluene'. He published research papers on heat transfer and evaporation, on distillation, and on liquid extraction, and coauthored this textbook of Chemical Engineering. He did valiant work for the Institution of Chemical Engineers, which awarded him its Davis medal in 1973, and was also a member of the Advisory Board for what was then a new Pergamon journal, *Chemical Engineering Science*.

In 1954, he was appointed to the newly established Chair at Newcastle-upon-Tyne, where Chemical Engineering became a separate Department and independent of Mechanical Engineering of which it was formerly part, and remained there until his retirement in 1975. He took a period of secondment to Heriot Watt University where, following the splitting of the joint Department of Chemical Engineering with Edinburgh, he acted as adviser and de facto Head of Department. The Scottish university awarded him an Honorary DSc in 1973.

John's first wife Dora sadly died in 1961; they had two sons, Anthony and Simon. He remarried in 1965 and is survived by Christine.

JFR

About Professor Richardson

Professor John Francis Richardson, "Jack" to all who knew him, was born at Palmers Green, North London, on 29 July 1920, and attended the Dame Alice Owens School in Islington. Subsequently, after studying chemical engineering at Imperial College, he embarked on research into the suppression of burning liquids and fires. This early work contributed much to our understanding of the extinguishing properties of foams, carbon dioxide, and halogenated hydrocarbons, and he spent much time during the war years on large-scale fire control experiments in Manchester and at the Llandarcy Refinery in South Wales. At the end of the war, Jack returned to Imperial College as a lecturer where he focused on research in the broad area of multiphase fluid mechanics, especially sedimentation and fluidization, and the two-phase flow of gasses and liquids in pipes. This laid the foundation for the design of industrial processes like catalytic crackers and led to a long-lasting collaboration with the Nuclear Research Laboratories at Harwell. This work also led to the publication of the famous paper, now common knowledge, the so-called Richardson-Zaki equation, which was selected as the week's citation classic (*Current Contents*, 12 February 1979)!

After a brief spell with Boake Roberts in East London, where he worked on the development of novel processes for creating flavours and fragrances, he was appointed professor of chemical engineering at what was then the University College of Swansea, (now University of Swansea) in 1960. He remained there until his retirement in 1987 and thereafter continued as an emeritus professor until his death on 4 January 2011.

Throughout his career, his major thrust was on the wellbeing of the discipline of chemical engineering. In the early years of his teaching duties at Imperial College, he and his colleague John Coulson recognized the lack of satisfactory textbooks available in the field of chemical engineering. They set about rectifying the situation, and this is how the now well-known Coulson-Richardson series of books on chemical engineering was born. The fact that this series of books (six volumes) is as relevant today as it was at the time of their first appearance is a testimony to the foresight of John Coulson and Jack Richardson.

Throughout his entire career spanning almost 40 years, Jack contributed significantly to all facets of professional life, teaching, research in multiphase fluid mechanics, and service to the Institution of Chemical Engineers (IChem E, UK). His professional work and long-standing public service were well recognized. Jack was president of IChem E during the period from 1975 to 1976; he was named a Fellow of the Royal Academy of Engineering in 1978 and was awarded an OBE in 1981.

In his spare time, Jack and his wife Joan were keen dancers, having been founding members of the Society of International Folk Dancing, and they also shared a love of hill walking.

Raj Chhabra

Preface to the Sixth Edition

The fifth edition of this title appeared in 2002, and obviously, there is a gap of more than 15 years between the fifth edition and the sixth edition in your hands now. The fact that this title has never been out of print is a testimony to its timelessness and "evergreen" character, both in terms of content and style. The sole reason for this unusually long gap between the two editions is the fact that Jack Richardson passed away in 2011, and, therefore, the publishers needed to be sure whether it was a worthwhile project to continue. The question was easily answered in the affirmative by numerous independent reviews and by the continuous feedback from the users of this volume—students, teachers, and working professionals—from all over the world. Having established that there was a definite need for this title, the next step was to identify individuals who would have the inclination to carry forward the legacy of Coulson and Richardson. Indeed, we feel privileged to have been entrusted with this onerous task.

The basic philosophy and the objectives of this edition remain the same as articulated so very well by the previous authors of the fifth edition, except for the fact that it has been split into two subvolumes, 2A and 2B. In essence, this volume, 2A, continues to concentrate on the fundamentals of particulate processing operations, as applied to wide-ranging industrial settings. In particular, consideration is given to the characterisation of particles, both at the individual level as well as in bulk, in terms of their size, shape, texture, packing characteristics in the context of storage, and flow in bins and hoppers, etc. Similarly, the processes used to effect size reduction and enlargement of particles and the segregation and mixing phenomena in dry solids are also introduced in this volume. The next set of topics includes the all-too-familiar operations of the flow and heat transfer in packed and fluidised beds and in the settling of concentrated slurries and suspensions. These ideas, in turn, are applied to the classification of solids based on their size and/or density differences using the gravity or centrifugal flow fields. The entire volume has been reviewed while keeping in mind the feedback received from the readers and reviewers. Wherever needed, both contents and presentation have been improved by reorganising the existing material for easier understanding, or new material has been added to provide updated and more reliable information. Apart from the general revision, the specific changes made in this edition are summarized below:

(i) A new Chapter 1 is introduced, not only to bring out the truly interdisciplinary character of this field, but also to illustrate the broad range of applications of particulate matter cutting across various industrial settings.

(ii) The material in the old version of Chapter 1 has been redistributed in three chapters, Chapters 2–4, to improve the clarity of presentation for easier understanding of the numerous concepts involved here in the characterisation of particulates.

(iii) In Chapter 5, a new section on the application of population balance modelling approach to size reduction has been added.

(iv) In addition, three new chapters have been added in this edition. Chapter 13 deals with the behaviour of colloidal and nanoparticles in solutions. Chapter 14 is concerned with the health and process hazards associated with the handling and processing of dry powders. Chapter 15 deals with recent developments and novel aspects of particle technology.

Most of these changes are based on one of the volume editors—Raj Chhabra's—extensive conversations and discussion with Jack Richardson over a period of 30 years (from 1981 onward). It is also appropriate to mention here that it is beyond the expertise of a single individual to undertake the revision of the entire volume. Therefore, we are grateful to our many friends who have volunteered to shoulder the responsibility of revising various chapters in this edition. These deserve to be singled out as follows: Abhijit Deshpande and Anand Krishna Kanjarla (Chapter 1), Anurag Tripathi (Chapters 2 and 3), Jayanta Chakraborty (Chapter 5), Anand Prakash (Chapter 9), and V. Shankar (Chapter 15). Very little editorial changes have been made to the chapters written by these individuals.

We are grateful to the many other individuals who have facilitated the publication of the sixth edition. Over the past 2 years, it has been a wonderful experience working with the staff at Butterworth-Heinemann (Elsevier). Each one of them has been extremely helpful, and some of these individuals deserve a mention here. First and foremost, we are grateful to Anita Koch for commissioning the new edition of this volume. She not only patiently answered our endless queries but also came to our rescue on several occasions. Similarly, Joshua Bayliss and Ali Afzal-Khan went far beyond their call of duty to see this project through. Finally, Prem Kumar Kaliamoorthi assembled the numerous fragments in different forms and formats—ranging from handwritten notes to latex files—into the finished product in your hands.

We end this preface with an appeal to our readers to please let us know as and when you spot errors or inconsistencies so that these can be rectified at the earliest opportunity.

Raj Chhabra
Madivala G. Basavaraj

Rupnagar and Chennai, March 2019

Preface to the Fifth Edition

It is now 47 years since Volume 2 was first published in 1955, and during the intervening time the profession of chemical engineering has grown to maturity in the United Kingdom, and worldwide; the Institution of Chemical Engineers, for instance, has moved on from its 33rd to its 80th year of existence. No longer are the heavy chemical and petroleum-based industries the main fields of industrial applications of the discipline, but chemical engineering has now penetrated into areas, such as pharmaceuticals, health care, foodstuffs, and biotechnology, where the general level of sophistication of the products is much greater, and the scale of production is often much smaller, though the unit value of the products is generally much higher. This change has led to a move away from large-scale continuous plants to smaller scale batch processing, often in multipurpose plants. Furthermore, there is an increased emphasis on product purity, and the need for more refined separation technology, especially in the pharmaceutical industry where it is often necessary to carry out the difficult separation of stereo-isomers, one of which may have the desired therapeutic properties while the other is extremely malignant. Many of these large molecules are fragile and are liable to be broken down by the harsh solvents commonly used in the manufacture of bulk chemicals. The general principles involved in processing these more specialised materials are essentially the same as in bulk chemical manufacture, but special care must often be taken to ensure that processing conditions are mild.

One big change over the years in the chemical and processing industries is the emphasis on designing products with properties that are specified, often in precise detail, by the customer. Chemical composition is often of relatively little importance provided that the product has the desired attributes. Hence *product design*, a multidisciplinary activity, has become a necessary precursor to *process design*.

Although undergraduate courses now generally take into account these new requirements, the basic principles of chemical engineering remain largely unchanged and this is particularly the case with the two main topics of Volume 2, *Particle Mechanics* and *Separation Processes*. In preparing this new edition, the authors have faced a typical engineering situation where a compromise has to be reached on size. The knowledgebase has increased to such an extent that many of the individual chapters appear to merit expansion into separate books. At the same time, as far as students and those from other disciplines are concerned, there is still a need for an integrated concise treatment in which there is a consistency of approach across the board and, most importantly, a degree of uniformity in the use of symbols. It has to be remembered that the learning capacity of students is certainly no greater than it was in the past, and a book of manageable proportions is still needed.

The advice that academic staffs worldwide have given in relation to revising the book has been that the layout should be retained substantially unchanged—*better the devil we know, with all his faults!* With this in mind the basic structure has been maintained. However, the old Chapter 8 on *Gas Cleaning*, which probably did not merit a chapter on its own, has been incorporated into Chapter 1, where it sits comfortably beside other topics involving the separation of solid particles from fluids. This has left Chapter 8 free to accommodate *Membrane Separations* (formerly Chapter 20) which then follows on logically from *Filtration* in Chapter 7. The new Chapter 20 then provides an opportunity to look to the future, and to introduce the topics of *Product Design* and the *Use of Intensified Fields* (particularly centrifugal in place of gravitational) and *miniaturisation*, with all the advantages of reduced hold-up, leading to a reduction in the amount of out-of-specification material produced during the changeover between products in the case multipurpose plants, and in improved safety where the materials have potentially hazardous properties.

Other significant changes are the replacement of the existing chapter on *Crystallisation* by an entirely new chapter written with expert guidance from Professor J.W. Mullin, the author of the standard textbook on that topic. The other chapters have all been updated and additional Examples and Solutions incorporated in the text. Several additional Problems have been added at the end, and solutions are available in the Solutions Manual, and now on the Butterworth-Heinemann website.

We are, as usual, indebted to both reviewers and readers for their suggestions and for pointing out errors in earlier editions. These have all been taken into account. Please keep it up in future! We aim to be error-free but are not always as successful as we would like to be! Unfortunately, the new edition is somewhat longer than the previous one, almost inevitably so with the great expansion in the amount of information available. Whenever in the past we have cut out material which we have regarded as being out-of-date, there is inevitably somebody who writes to say that he now has to keep both the old and the new editions because he finds that something which he had always found particularly useful in the past no longer appears in the revised edition. It seems that you cannot win, but we keep trying!

J.F. Richardson
J.H. Harker

Preface to the Fourth Edition

Details of the current restructuring of this Chemical Engineering Series, coinciding with the publication of the Fourth Edition of Volumes 1 and 2 and to be followed by new editions of the other volumes, have been set out in the Preface to the Fourth Edition of Volume 1. The revision involves the inclusion in Volume 1 of material on non-Newtonian flow (previously in Volume 3) and the transference from Volume 2 to Volume 1 of *Pneumatic and Hydraulic Conveying* and *Liquid Mixing*. In addition, Volume 6, written by Mr. R.K. Sinnott, which first appeared in 1983, nearly 30 years after the first volume, acquires some of the design-orientated material from Volume 2, particularly that related to the hydraulics of packed and plate columns.

The new subtitle of Volume 2, *Particle Technology and Separation Processes*, reflects both the emphasis of the new edition and the current importance of these two topics in Chemical Engineering. *Particle Technology* covers the basic properties of systems of particles and their preparation by comminution (Chapters 1 and 2). Subsequent chapters deal with the interaction between fluids and particles, under conditions ranging from those applicable to single isolated particles, to systems of particles freely suspended in fluids, as in sedimentation and fluidisation; and to packed beds and columns where particles are held in a fixed configuration relative to one another. The behaviour of particles in both gravitational and centrifugal fields is also covered. It will be noted that *Centrifugal Separations* are now brought together again in a single chapter, as in the original scheme of the first two editions, because the dispersal of the material between other chapters in the Third Edition was considered to be not entirely satisfactory.

Fluid–solid separation processes are discussed in the earlier chapters under the headings of Sedimentation, Filtration, Gas Cleaning, and Centrifugal Separations. The remaining separations involve applications of mass-transfer processes, in the presence of solid particles in leaching (solid–liquid extraction), drying, and crystallisation. In distillation, gas absorption and liquid–liquid extraction, interactions occur between two fluid streams with mass transfer taking place across a phase boundary. Usually these operations are carried out as continuous countercurrent flow processes, either stagewise (as in a plate-column) or with differential contacting (as in a packed column). There is a case therefore for a generalised treatment of countercurrent contacting processes with each of the individual operations, such as Distillation, treated as particular cases. Although this approach has considerable merit, both conceptually and in terms of economy of space, it has not been adopted here, because the authors' experience of teaching suggests that the student more readily grasps the principles involved, by considering each topic in turn, provided of course that the teacher makes a serious attempt to emphasise the common features.

The new edition concludes with four chapters which are newcomers to Volume 2, each written by a specialist author from the Chemical Engineering Department at Swansea—

Adsorption and Ion Exchange (Chapters 17 and 18)
(topics previously covered in Volume 3)
by J.H. Bowen
Chromatographic Separations (Chapter 19)
by J.R. Conder
and
Membrane Separations (Chapter 20)
by W.R. Bowen

These techniques are of particular interest in that they provide a means of separating molecular species which are difficult to separate by other techniques and which may be present in very low concentrations. Such species include large molecules, submicrometre size particles, stereo-isomers, and the products from bioreactors (Volume 3). The separations can be *highly specific* and may depend on molecular size and shape, and the configuration of the constituent chemical groups of the molecules.

Again I would express our deep sense of loss on the death of our colleague, Professor John Coulson, in January 1990. His two former colleagues at Newcastle, Dr. John Backhurst and the Reverend Dr. John Harker, have played a substantial part in the preparation of this new edition both by updating the sections originally attributable to him, and by obtaining new illustrations and descriptions of industrial equipment.

Finally, may I again thank our readers who, in the past, have made such helpful suggestions and have drawn to our attention errors, many of which would never have been spotted by the authors. Would they please continue their good work!

J.F. Richardson
Swansea

Note to Fourth Edition—Revised Impression 1993

In this reprint corrections and minor revisions have been incorporated. The principal changes are as follows:

(1) Addition of an account of the construction and operation of the Szego Grinding Mill (Chapter 2).

(2) Inclusion of the Yoshioka method for the design of thickeners (Chapter 5).

(3) Incorporation of Geldart's classification of powders in relation to fluidisation characteristics (Chapter 6).

(4) The substitution of a more logical approach to filtration of slurries yielding compressible cakes and redefinition of the specific resistance (Chapter 7).

(5) Revision of the nomenclature for the underflow streams of washing thickeners to bring it into line with that used for other stagewise processes, including distillation and absorption (Chapter 10).

(6) A small addition to the selection of dryers and the inclusion of Examples (Chapter 16).

Preface to the 1983 Reprint of the Third Edition

In this volume, there is an account of the basic theory underlying the various unit operations, and typical items of equipment are described. The equipment items are the essential components of a complete chemical plant, and the way in which such a plant is designed is the subject of Volume 6 of the series which has just appeared. The new volume includes material on flowsheeting, heat and material balances, piping, mechanical construction, and costing. It completes the Series and forms an introduction to the very broad subject of Chemical Engineering Design.

Preface to Third Edition

In producing a third edition, we have taken the opportunity, not only of updating the material but also of expressing the values of all the physical properties and characteristics of the systems in the SI System of units, as has already been done in Volumes 1 and 3. The SI system, which is described in detail in Volume 1, is widely adopted in Europe and is now gaining support elsewhere in the world. However, because some readers will still be more familiar with the British system, based on the foot, pound, and second, the old units have been retained as alternatives wherever this can be done without causing confusion.

The material has, to some extent, been re-arranged and the first chapter now relates to the characteristics of particles and their behaviour in bulk, the blending of solids, and classification according to size or composition of material. The following chapters describe the behaviour of particles moving in a fluid and the effects of both gravitational and centrifugal forces and of the interactions between neighbouring particles. The old chapter on centrifuges has now been eliminated and the material dispersed into the appropriate parts of other chapters. Important applications which are considered include flow in granular beds and packed columns, fluidisation, transport of suspended particles, filtration, and gas cleaning. An example of the updating which has been carried out is the addition of a short section on fluidised bed combustion, potentially the most important commercial application of the technique of fluidisation. In addition, we have included an entirely new section on flocculation, which has been prepared for us by Dr. D.J.A. Williams of University College, Swansea, to whom we are much indebted.

Mass transfer operations play a dominant role in chemical processing and this is reflected in the continued attention given to the operations of solid–liquid extraction, distillation, gas absorption, and liquid–liquid extraction. The last of these subjects, together with material on liquid–liquid mixing, is now dealt within a single chapter on liquid–liquid systems, the remainder of the material which appeared in the former chapter on mixing having been included earlier under the heading of solids blending. The volume concludes with chapters on evaporation, crystallisation, and drying.

Volumes 1–3 form an integrated series with the fundamentals of fluid flow, heat transfer, and mass transfer in the first volume, the physical operations of chemical engineering in this, the second volume, and in the third volume, the basis of chemical and biochemical reactor design, some of the physical operations which are now gaining in importance and the underlying theory of both process control and computation. The solutions to the problems listed in Volumes 1 and 2 are now available as Volumes 4 and 5, respectively. Furthermore, an additional volume in the series is in course of preparation and will provide an introduction to chemical engineering design and indicate how the principles enunciated in the earlier volumes can be translated into chemical plant.

We welcome the collaboration of J.R. Backhurst and J.H. Harker as co-authors in the preparation of this edition, following their assistance in the editing of the latest edition of Volume 1 and their authorship of Volumes 4 and 5. We also look forward to the appearance of R.K. Sinnott's volume on chemical engineering design.

Preface to Second Edition

This text deals with the physical operations used in the chemical and allied industries. These operations are conveniently designated 'unit operations' to indicate that each single operation, such as filtration, is used in a wide range of industries, and frequently under varying conditions of temperature and pressure.

Since the publication of the first edition in 1955 there has been a substantial increase in the relevant technical literature but the majority of developments have originated in research work in government and university laboratories rather than in industrial companies. As a result, correlations based on laboratory data have not always been adequately confirmed on the industrial scale. However, the section on absorption towers contains data obtained on industrial equipment and most of the expressions used in the chapters on distillation and evaporation are based on results from industrial practice.

In carrying out this revision we have made substantial alteration to Chapters 1, 5–7, 12, 13, and 15[1] and have taken the opportunity of presenting the volume paged separately from Volume 1. The revision has been possible only as the result of the kind cooperation and help of Professor J.D. Thornton (Chapter 12), Mr. J. Porter (Chapter 13), Mr. K.E. Peet (Chapter 10), and Dr. B. Waldie (Chapter 1), all of the University at Newcastle, and Dr. N. Dombrowski of the University of Leeds (Chapter 15). We want in particular to express our appreciation of the considerable amount of work carried out by Mr. D.G. Peacock of the School of Pharmacy, University of London. He has not only checked through the entire revision but has made numerous additions to many chapters and has overhauled the index.

We should like to thank the companies that have kindly provided illustrations of their equipment and also the many readers of the previous edition who have made useful comments and helpful suggestions.

Chemical engineering is no longer confined to purely physical processes and the unit operations, and a number of important new topics, including reactor design, automatic control of plants, biochemical engineering, and the use of computers for both process design and

---

[1] N.B. Chapter numbers are altered in the current (third) edition.

control of chemical plant will be covered in a forthcoming Volume 3 which is in course of preparation.

Chemical engineering has grown in complexity and stature since the first edition of the text, and we hope that the new edition will prove of value to the new generation of university students as well as forming a helpful reference book for those working in industry.

In presenting this new edition we wish to express our gratitude to Pergamon Press who have taken considerable trouble in coping with the technical details.

J.M. Coulson
J.F. Richardson

Preface to First Edition

In presenting Volume 2 of *Chemical Engineering*, it has been our intention to cover what we believe to be the more important unit operations used in the chemical and process industries. These unit operations, which are mainly physical in nature, have been classified, as far as possible, according to the underlying mechanism of the transfer operation. In only a few cases is it possible to give design procedures when a chemical reaction takes place in addition to a physical process. This difficulty arises from the fact that, when we try to design such units as absorption towers in which there is a chemical reaction, we are not yet in a position to offer a thoroughly rigorous method of solution. We have not given an account of the transportation of materials in such equipment as belt conveyors or bucket elevators, which we feel lie more distinctly in the field of mechanical engineering.

In presenting a good deal of information in this book, we have been much indebted to facilities made available to us by Professor Newitt, in whose department we have been working for many years. The reader will find a number of gaps, and a number of principles which are as yet not thoroughly developed. Chemical engineering is a field in which there is still much research to be done, and, if this work will in any way stimulate activities in this direction, we shall feel very much rewarded. It is hoped that the form of presentation will be found useful in indicating the kind of information which has been made available by research workers up to the present day. Chemical engineering is in its infancy, and we must not suppose that the approach presented here must necessarily be looked upon as correct in the years to come. One of the advantages of this subject is that its boundaries are not sharply defined.

Finally, we should like to thank the following friends for valuable comments and suggestions: Mr. G.H. Anderson, Mr. R.W. Corben, Mr. W.J. De Coursey, Dr. M. Guter, Dr. L.L. Katan, Dr. R. Lessing, Dr. D.J. Rasbash, Dr. H. Sawistowski, Dr. W. Smith, Mr. D. Train, Mr. M.E.O'K. Trowbridge, Mr. F.E. Warner, and Dr. W.N. Zaki.

Introduction

Welcome to the next generation of Coulson-Richardson series of books on *Chemical Engineering*. I would like to convey to you all my feelings about this project which have evolved over the past 30 years, and are based on numerous conversations with Jack Richardson himself (from 1981 onward until his death in 2011) and with some of the other contributors to previous editions including Tony Wardle, Ray Sinnott, Bill Wilkinson, and John Smith. So, what follows here is the essence of these interactions combined with what the independent (solicited and unsolicited) reviewers had to say about this series of books on several occasions.

The Coulson-Richardson series of books has served academia, students, and working professionals extremely well since their first publication more than 50 years ago. This is a testimony to their robustness and, to some extent, their timelessness. I have often heard much praise from different parts of the world for these volumes both for their informal and user-friendly yet authoritative style and for their extensive coverage. Therefore, there is a strong case for continuing with its present style and pedagogical approach.

On the other hand, advances in our discipline in terms of new applications (energy, bio, microfluidics, nanoscale engineering, smart materials, new control strategies, and reactor configurations, for instance) are occurring so rapidly and in such a significant manner that it would be naive, even detrimental, to ignore them. Therefore, while we have tried to retain the basic structure of this series, the contents have been thoroughly revised. Wherever the need was felt, the material has been updated, revised, and expanded as deemed appropriate. Therefore, the reader, whether a student, a researcher, or a working professional, should feel confident that what is in the book is the most up-to-date, accurate, and reliable piece of information on the topic he/she is interested in.

Obviously, this is a massive undertaking which cannot be managed by a single individual. Therefore, we now have a team of volume editors responsible for each volume having the individual chapters written by experts in some cases. I am most grateful to all of them for having joined us in the endeavour. Furthermore, based on extensive deliberations and feedback from a large number of individuals, some structural changes were deemed to be appropriate as detailed here. Due to their size, each volume has been split into two subvolumes as follows:

Volume 1A: Fluid Flow

Volume 1B: Heat and Mass Transfer

Volume 2A: Particulate Technology and Processing

Volume 2B: Separation Processes

Volume 3A: Chemical Reactors

Volume 3B: Process Control

Undoubtedly, the success of a project of such a vast scope and magnitude hinges on the cooperation and assistance of many individuals. In this regard, we have been extremely fortunate in working with some of the outstanding individuals at Butterworth-Heinemann, a few of whom deserve to be singled out: Jonathan Simpson, Fiona Geraghty, Anita Koch, Maria Convey, Ashlie Jackman, Joshua Bayliss, and Afzal Ali, who have taken personal interest in this project and have come to our rescue whenever needed, going far beyond the call of duty.

Finally, this series has had a glorious past, but I sincerely hope that its future will be even brighter by presenting the best possible books to the global chemical engineering community for the next 50 years, if not longer. I sincerely hope that the new edition of this series will meet (if not exceed) your expectations! Last, a request to the readers: please continue to do good work by letting me know if (no, not "if," but "when") you spot a mistake so that these can be corrected at the first opportunity.

Raj Chhabra

Editor-in-Chief

Rupnagar, March 2019

CHAPTER 1

Introduction

Solid materials and their dispersion in fluids are the subject matter of this book. An individual unit of dispersed solid material is referred to as a particle. These solid-fluid material systems are referred to as particulate systems, particulate materials or particulates. Powders (concentrated solid particles in air), slurries (solid particles dispersed in liquid), and dispersions (fine solid particles dispersed in liquid) are few other terms commonly used to describe solid-fluid material systems. The inherent multiphasic nature of these systems can lead to a set of properties that can be solid-like and another set that can be fluid-like. Additionally, they exhibit several unique phenomena that are difficult to explain based on the knowledge of single-phase systems. Therefore, the engineering and science of particles and particulates (as collection of particles) offers great challenges in terms of design and understanding.

Because the particle behaviour is very crucial in determining the response of particulate systems, several key features of particles are introduced in this chapter. We begin by reviewing the material science and mechanical properties of solids. In the following sections, we outline some of the key characteristics of particulate systems. Finally, several past, current, and future examples in industrial applications, Fig. 1.1, and consumer applications, Fig. 1.2, are described. Several important features of particulate systems in these applications are highlighted.

1.1 Material Science of Solids
1.1.1 Bonding in Solids

Solids are made of aggregates of atoms held together by interatomic forces. The geometric arrangement of these atoms and the forces that bind them together to form cohesive solids that have significant influence on the properties of these materials. The fact that atoms prefer to stay in solid form instead of individual (free) atoms implies that the overall potential energy of a solid is lower than that of the free atoms.

It is worth revising the nature of the interatomic energies and the consequent forces. Consider two neutral atoms that are far apart. There exists attractive energy between these atoms due to the electrostatic attraction between the nucleus and the electron clouds of the individual atoms. This energy, which is of long range nature, depends weakly on the distance between the atoms, denoted by r, and is typically formulated as $U_A = -\frac{a}{r^m}$ where a and m are constants.

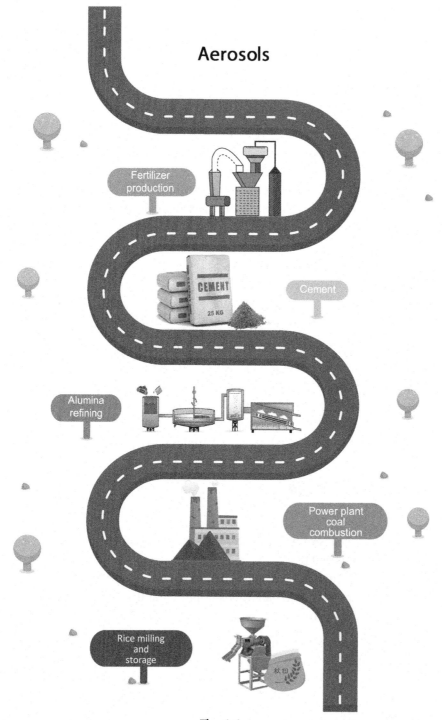

Fig. 1.1
Examples of particulate systems in industrial applications.

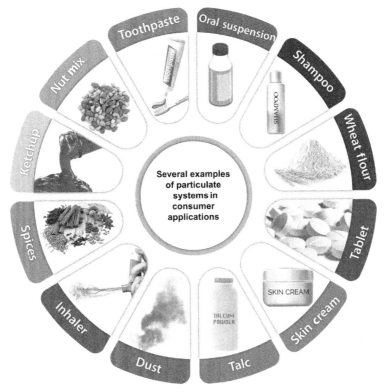

Fig. 1.2
Examples of particulate systems in consumer applications.

Convention dictates that attractive energy is negative. When the atoms get too close to each other, there is a repulsive energy arising from the nucleus-nucleus and electron-electron repulsions. The repulsive energy, which is of short range, due to Pauli's exclusion principle, is idealised as $U_R = \frac{b}{r^n}$. The net interatomic potential is then treated as the sum of the attractive and the repulsive energies as, $U = U_A + U_R = -\frac{a}{r^m} + \frac{b}{r^n}$.

The most commonly assumed values for m and n are 6 and 12 and this form of interatomic interaction/potential is referred to as Lennard-Jones potential. One can gain insights into many material properties with a knowledge of the potential curves. For instance, the equilibrium separation (r_0 in Fig. 1.3) between two atoms can readily be obtained by knowing that, at that point, the net force acting on the atom is zero. Thus, the value of F can be obtained by negative derivative of the potential curve $F = -dU/dr$. Similarly, the depth of the potential well is an indication of the amount of energy needed to separate the two atoms, which, in turn, decides the melting point of the material. Other properties such as coefficient of thermal expansion or elastic modulus can also be derived knowing the interatomic interactions. It is worth noting here that the above description is an idealisation of interatomic interactions where there is no explicit participation of electrons.

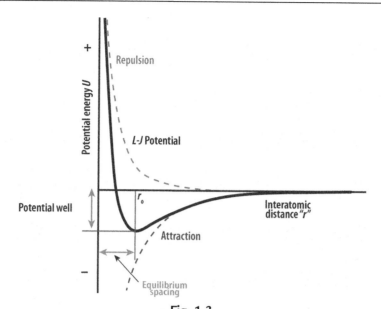

Fig. 1.3

Schematic of Lennard-Jones potential showing the variation of potential energy with distance between two atoms.

1.1.2 Classification of Solids

The chemical bonds between the atoms can be categorised as primary bonds and secondary bonds. Primary bonds, also referred to as strong bonds, are a consequence of the directional interaction of atoms either through the transfer or sharing of (valence) electrons. Secondary bonds, also referred to as weak bonds, are due to indirect interaction between are molecules/atoms.

Ionic bonds are formed due to complete transfer of one or more electrons from an electropositive atom (cation) to an electronegative(anion) atom. The electrostatic attraction between oppositely charged ions is typically long range and is nondirectional. The bond strength is high which makes ionic compounds mechanically hard. The bond strength of ionic bonds can range from 600 to 1500 kJ/mol.

Covalent bonds are formed between two atoms with similar electronegativities by sharing of their valence electrons. The electron cloud between the atoms is now localised between the atoms, resulting in high directionality of these bonds. Covalent bonds are also very strong and, therefore, result in compounds with high melting point. The bond strength of covalent bonds can range from 300 to 1500 kJ/mol.

Metallic bonds are typically seen between electropositive atoms wherein the outer most electrons of these atoms are delocalised across many atoms. Metallic bonds are typically nondirectional in nature. The bond strength of metallic bonds can range from 50 to 1000 kJ/mol.

On the other hand, the van der Waals bonds are weak electrostatic attractions between polarised atoms or molecules. The bond strength can range from 10 to 40 kJ/mol. Hydrogen bonds are similar to van der Waals bonds, except that these are caused by induced dipoles in molecules containing hydrogen. Hydrogen bonds are slightly stronger than the van der Waals bonds, in the range of 10–50 kJ/mol.

It is important to note here that materials seldom have bonding of only one kind. There is always a combination of different bonds. For example, the Si–C bond in silicon carbide has about 11% ionic bonding and 89% covalent bonding characteristics.[1]

Based on the geometric arrangement of atoms, materials can be classified into crystalline or amorphous materials. Materials in which there is a regular 3D arrangement of atoms located at points related by translational symmetry are often referred to as crystalline and the underlying periodic arrangement as the crystal structure. The smallest repeating atomic arrangement is called a unit cell. As shown in the Fig. 1.4, the unit cell is a parallelepiped with edges given by a_1, a_2, a_3 and the angles between them as $\alpha_{12}, \alpha_{23}, \alpha_{31}$. The periodic arrangement typically extends over distances much larger than the interatomic distances. Hence, crystalline materials are considered to have long-range order with translational symmetry. There are 14 different ways in which identical atoms can be arranged in 3D space. These are often referred to as the Bravais lattices. They are grouped into seven crystal systems, the details of which are shown in Fig. 1.4. Crystal structures of some of the common engineering materials are shown in Table 1.1. A key characteristic of crystals is a sharp melting temperature T_m, at which transition from ordered solid to a disordered liquid state is observed.

Amorphous/noncrystalline materials lack regular arrangement of atoms over large atomic distances, that is, they do not possess long-range order. One important consequence of the lack of long-range order is the absence of a sharp melting point. Instead, amorphous materials have what is commonly referred to as glass transition temperature T_g. Below this temperature, materials are generally hard and brittle, whilst above this temperature they are soft and flexible (Fig. 1.5).

Solid materials are typically classified into three major groups: metals, ceramics, and polymers. Apart from this broad classification, one also often comes across two other categories: composites and biomaterials.

Metals and alloys are inorganic materials composed of one or more metallic elements. They are good conductors of heat and electricity. Most metallic systems are crystalline with some exceptions, such as metallic glasses. They also exhibit good ductility, which gives them ability to be shaped. Pure metals are seldom used in many applications. Most metals are used as alloys, formed by mixing two or more metals. Alloying allows improvement on the properties of pure metals.

Ceramics are inorganic, nonmetallic compounds. Most ceramics are oxides, but the term is also used for nitrides, carbides, etc. The bonding in ceramics can be either ionic or covalent.

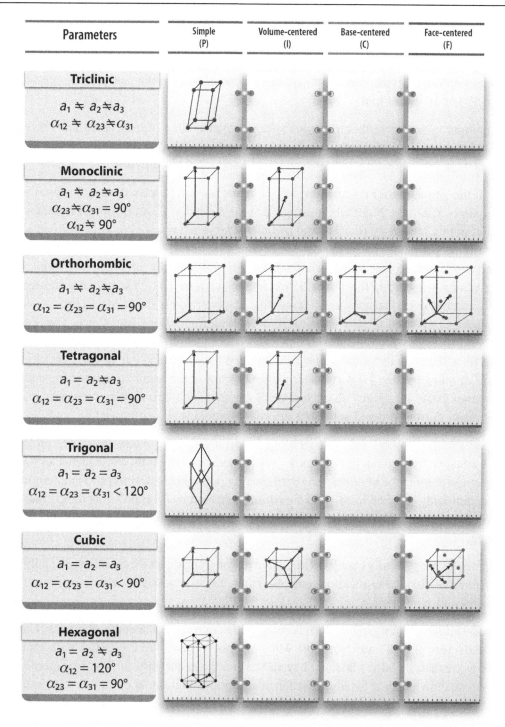

Parameters	Simple (P)	Volume-centered (I)	Base-centered (C)	Face-centered (F)

Triclinic

$a_1 \neq a_2 \neq a_3$
$\alpha_{12} \neq \alpha_{23} \neq \alpha_{31}$

Monoclinic

$a_1 \neq a_2 \neq a_3$
$\alpha_{23} \neq \alpha_{31} = 90°$
$\alpha_{12} \neq 90°$

Orthorhombic

$a_1 \neq a_2 \neq a_3$
$\alpha_{12} = \alpha_{23} = \alpha_{31} = 90°$

Tetragonal

$a_1 = a_2 \neq a_3$
$\alpha_{12} = \alpha_{23} = \alpha_{31} = 90°$

Trigonal

$a_1 = a_2 = a_3$
$\alpha_{12} = \alpha_{23} = \alpha_{31} < 120°$

Cubic

$a_1 = a_2 = a_3$
$\alpha_{12} = \alpha_{23} = \alpha_{31} < 90°$

Hexagonal

$a_1 = a_2 \neq a_3$
$\alpha_{12} = 120°$
$\alpha_{23} = \alpha_{31} = 90°$

Fig. 1.4

14 three-dimensional Bravais lattices and 7 crystal systems from which they are derived.

Table 1.1 **Examples of crystal structures of some engineering materials**[2]

Material Class	Material	Crystal Structure
Metals	Aluminium	Face-centred cubic
	Iron	Body-centred cubic
	Iron (above 923°C)	Face-centred cubic
	Titanium	Hexagonal close-packed
Ceramics	Alumina Al_2O_3	Hexagonal close-packed
	Silicon carbide SiC (zinc blende)	Face-centred cubic
	Silicon carbide SiC (wurtzite)	Hexagonal close-packed
Polymers	Polyethylene	Orthorhombic
	Nylon 66	Triclinic
	α Isotactic polypropylene	Monoclinic
	β Isotactic polypropylene	Hexagonal

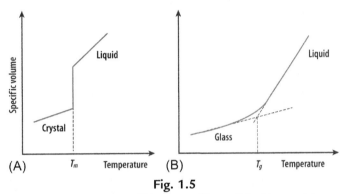

Fig. 1.5

Specific volume versus temperature curves: (A) a crystalline solid with a melting point T_m; (B) a glass, with a glass transition temperature T_g.

Ceramics are generally strong and chemically inert. Ionic ceramics are typically compounds of metals with nonmetals such as MgO, Al_2O_3, ZrO_2, whereas covalent ceramics are compounds of two nonmetals (e.g. SiO_2, SiC). Stone and clay porcelain are natural ceramics. High-performance ceramics are synthesised; they include alumina (Al_2O_3), silica (SiO_2), other oxides, such as TiO_2, ZrO_2, carbides WC, TiC, SiC, and BC; nitrides Si_3N_4, TiN, and BN; and borides such as TiB_2. Glasses fall under the category of ceramics. They exhibit high melting point, high compressive strength, and are often brittle.

Polymers are generally organic materials made of long chain or network of small repeating carbon containing molecules called monomers. Typically, covalent bonding is seen in polymers along with weak van der Waals and hydrogen bonds. Polymers can be either thermoplastic (those that can be melted and processed repeatedly), thermosets (can be formed only once and cannot be remelted), and elastomers or rubbers that can be stretched considerably and return to their original size rapidly when the force is removed. Polymers are also commonly refereed to as plastics. Plastics are sometimes defined as polymers that can be easily formed at low temperatures. Most polymers are noncrystalline, with some exhibiting a mixture of both crystalline and amorphous nature. They generally have low density and are typically electrical insulators. Natural polymers include cellulose, starch, and rubbers, whilst examples of the manmade or engineering polymers are polyethylene, nylon, and polytetrafluoroethylene.

1.1.3 Response of Solids to Stress

Rigid and deformable

A rigid solid material implies that it does not deform. For example, in fluidisation, we commonly assume that particles are subjected to forces due to phenomena such as drag and collision. However, we consider that particles do not deform due to application of such forces, and, therefore, the particles are assumed to be rigid. On the other hand, particles undergo deformation and eventual failure when subjected to size reduction operations. In these cases, particles are considered to be deformable. The validity of this assumption of rigid behaviour of materials would depend on the loading condition as well as the material properties.

Small and large deformations

As discussed in Section 1.1.2, for both ordered or disordered solids, the positions of atoms and molecules are fixed. When a force is applied on a solid material, it can lead to relative displacement of the atoms and molecules with respect to each other. This relative displacement results in deformation of the material on a bulk scale. When a small force is applied, small deformation is expected whilst large force leads to large deformation. Moreover, the deformation would also depend on interatomic and intermolecular interactions, or in other words, on the material properties. The deformation response of a material is characterised in terms of stress (=force per unit area) and strain (=relative displacement).

When strain is small, a linear relationship between stress and strain is generally observed. This is termed the elastic response of solids. In the elastic region, if the stress is reduced to zero, the stain also approaches zero. In other words, when the solid is deformed in the elastic region, it regains the original configuration upon the removal of the stress.

When a solid material is subjected to larger strains, it can exhibit three possible behaviours, as shown in Fig. 1.6. When the material is brittle, it breaks or fails when the strain is progressively increased. Another possibility is that the material shows a yielding and ductile behaviour. In the yielding region, the stress increase is no longer proportional to the strain. Beyond the yield, the material deforms, even though the stress does not increase significantly. Moreover, if the stress is removed, the material retains a residual strain. Therefore, beyond the yield stress, the material deformation is plastic. Due to such plastic deformation, the ductile materials deform significantly before they fail. The third type of behaviour shown

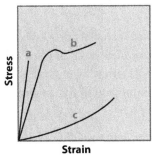

Fig. 1.6

Stress as a function of strain for representative solid materials: (a) brittle material, (b) yielding and ductile material; (c) rubbery material.

in Fig. 1.6 is of rubbery materials. These materials deform too much larger extent, and with increasing strain, higher and higher stress is required at larger deformation.

The stress-strain response can be quantified based on a few mechanical properties. In the elastic region, the proportionality constant describing the linear stress-strain relation is called the modulus or stiffness. The stress at which yielding occurs is called the yield stress. The stress at which the material breaks or fails is termed its ultimate strength. Ceramics, which are mostly brittle materials, have generally high moduli and lower strengths. On the other hand, ductile metals have relatively lower moduli and higher strengths. Polymers, also ductile, have lower moduli and strengths when compared to metals. Toughness can be quantified as the area enclosed by the stress-strain curve, and it is a measure of the energy required to fracture or break the material. Ductile materials exhibit higher toughness than the brittle materials (Fig. 1.7).

Mechanical characterisation of solids is carried out in tensile, compressive or flexural loading. In particle technology, the compressive mechanical response of solids is relevant for size reduction. Similarly, attrition during particle handling, storage, and operation is also affected by this response.

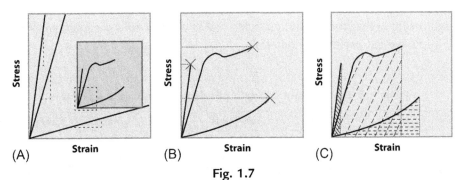

Fig. 1.7

Mechanical properties of solid materials. (A) Modulus: the slope of stress versus strain; (B) strength: stress at failure; (C) toughness: energy of deformation.

Failure

Failure, breakage, or fracture of solid materials proceeds in several stages. A processed material generally consists of defects such as crystalline defects and voids. When a stress is applied, the defects can cause the formation of microcracks due to stress concentration. These microcracks propagate and combine to form a macrocrack, which then leads to the failure of the material.

In the case of brittle materials, defects lead to rapid generation and propagation of microcracks and macrocracks, and eventually to catastrophic failure. In the case of ductile metals, migration of crystal defects leads to dissipation of energy and plastic deformation. For polymeric materials, conformational changes and orientation of molecules lead to the dissipation of energy, and, therefore, resulting in large plastic deformation. Due to these mechanisms, size reduction is easier (or less energy intensive) for ceramics than for metals. Size reduction for polymers is most difficult due to their ductility or plastic deformation.

Failure of solid materials implies the creation of new solid interface whilst breaking ionic, metallic, or covalent bonds. Therefore, energy involved in fracture depends on surface energy and bond energies. Since metals and polymers can dissipate energy without breakage of bonds, size reduction is more difficult.

When a solid material is processed, defects are always present, and they are distributed within the material. Of the large number of defects, microcracks start at few locations in the material. At fewer locations, microcracks combine to form a macrocrack. At a specific location, macrocracks grow and lead to failure. Therefore, fracture of solids is inherently a statistical process. Materials fail because defects can never be eliminated, and generally, the greater the number of defects, the easier it would be for a few of the defects to lead to failure. Hence, a smaller sample size of the same material shows more strength and toughness (endures more stress before failure) because of fewer defects. Therefore, failure is more and more difficult with the smaller particle sizes.

The compressive strength of silica particles of 4–25 mm size-range varies from 10 to 150 MPa,[3] with strength being higher for smaller particles. Table 1.2 provides compressive strength values for a few typical particulate materials.

Table 1.2 Compressive strength of selected particulate systems

Material	Size (mm)	Compressive Strength (MPa)
Glass	4–25	10–150[3]
Salt	0.7–4	2–40[4]
Sandstone	Bulk	55[5]
Sandstone	16–45	150–1600[3]
Dolomite	16–45	300–2000[3]

1.2 Particulate Solids

1.2.1 Characteristics of Particles

As we discussed earlier, particulate solids consist of different material types and of different atomic and molecular arrangements. Consequently, a diverse set of interactions are involved when we consider them. The mechanical properties of the particulate materials are also of varied nature, as described in Section 1.1.3. The diversity is also apparent when we consider the following factors that are very influential in determining the behaviour of particulate solids.

- *Size.* Particle size is one of the defining characteristics of particulate solids. The nature of interactions as well as the underlying physical mechanisms involved in a particulate process can vary depending on the particle size. Therefore, the same material can be used in different applications based on the particle size. Fig. 1.8 shows a few example characteristics, and their dependence on the particle size. As is evident from the figure, particulate solids generally have a distribution of particle sizes. Hence, to understand their behaviour, it is not only important to characterise the average size, but also the nature of distribution. Therefore, size reduction and size enlargement processes are used to arrive at an appropriate size distribution for a given application.

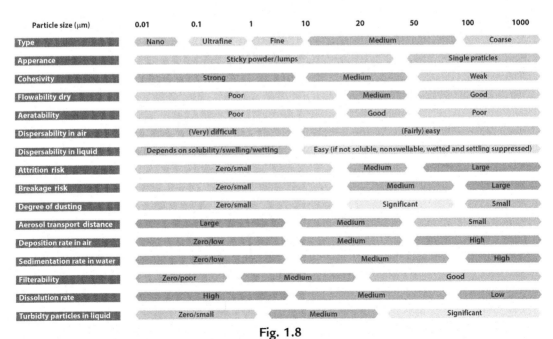

Fig. 1.8
Influence of particle size on example characteristics of particulate solids.

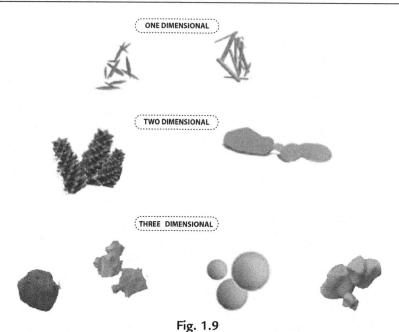

Fig. 1.9

Typical shapes of particles. *(Redrawn based on information in Nouri A, Sola A. Metal particle shape: a practical perspective.* Metal Powder Rep *2018. 10.1016/j.mprp.2018.04.001.)*

- *Shape.* The shape of particles is another important factor that influences the behaviour of particulate solids. It is helpful to classify shapes based on the dimensionality involved, as shown in Fig. 1.9. Particles can be 1D, for example, being rod-like in shape, 2D, such as platelets and flakes, or 3D, such as spheres and irregular shapes. An industrial powder may also contain different shapes, for example, a mixture of rod-like and irregular-shaped particles.

- *Thermal properties.* In several particulate operations, heat transfer and thermal processing are involved. For such operations, thermal properties of particles are very important. Table 1.3 shows melting temperatures, thermal conductivities, specific heats, and glass transition temperatures of a few solid materials relevant to particulate systems. It should be noted that the thermal property of a particulate system is different from the properties of bulk solid.

- *Surface properties.* Surface energy is a thermodynamic property of a material and is crucial in determining the behaviour of particulate solids in several applications. Some values of solid surface energies are listed in Table 1.4. However, there are other aspects pertaining to surfaces that also influence the properties and behaviour of a particulate system. These aspects include roughness, surface functionalisation, adsorbed species, and surface charges. For handling, storage, and transport of particulate solids, the friction between particles as well as the friction between solid substrates (such as pipe wall, vessel surface, etc.) and particles are also important. Angle of repose, which is an indication of flowability of particulate solids, for a few materials is given in Table 1.5.

Table 1.3 Thermal properties of selected materials

Material	Melting Temperature (°C)	Thermal Conductivity (W/mK)	Specific Heat (J/kgK)	Glass Transition Temperature (°C)
Aluminium[6]	660	247	897	
Iron[6]	538	80.4	449	
Al_2O_3[6]	2050	30	873	
SiC[6]	2400	90	873	
Styrene-butadiene rubber		0.21[7]	1970[8]	−60
Polycarbonate[6]	265	0.20	840	150
Polystyrene[6]	220	0.13	1170	80
Rice[9]		0.09		1200
Sorghum[9]		0.12	1400	

Table 1.4 Surface energies of selected solids

Material	Surface Energy (mJ/m^2)
Silver	1086[10]
Tungsten	2765[10]
Polystyrene	42[11]
Polytetrafluoroethylene	19[11]
SiO_2	287[11]
Fe_2O_3	1357[11]

Table 1.5 Angle of repose for selected particulate solids

Material	Angle of Repose (degrees)
Aluminium powder	60[12]
Iron ore	36[13]
Wheat	27[14]
Wheat flour	45[14]
Dry sand	34[14]
Wet sand	45[14]

- *Mixtures.* The diversity of industrial particulate systems is also due to them being mixtures of materials, sizes, and shapes. It is relatively easy for gases to mix. Liquids, on the other hand, can be miscible or immiscible. Particulates, on the other hand, as a rule, tend to segregate.[15]

Figs 1.10–1.13 provide a few examples of commercial particulate systems. Depending on the material system and the application, engineers tend to highlight different properties as part of specifications or datasheets. As described in this section, size is the most important characteristic described in such datasheets. Additional information about physical, chemical, thermal, surface, and mechanical properties is given depending on the target applications of particulate system.

PRODUCT NAME		SA0125	SA0103	SA0109	SA0118	SA0180
Particle Diameter	um	22 ~ 28	18 ~ 24	2.5 ~ 4.5	2.5 ~ 4.5	30 ~ 40
Specific Surface Area	m²/g	0.22	0.16	0.65	0.60	0.20
Al_2O_3	%	99.8	99.9	99.8	99.8	99.8
SiO_2	ppm	780	400	800	800	780
Fe_2O_3	ppm	140	76	155	155	140
CaO_2	ppm	60	30	75	75	60
MgO_2	ppm	10	15	22	22	10
Na_2O	ppm	1100	560	620	620	1100
K_2O	ppm	22	27	31	31	22
Upper Sieve Size	um	75	125	15	32	75

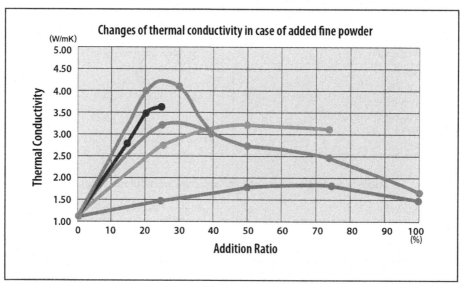

Fig. 1.10

Examples of datasheets for particulate systems: for alumina. By adding fine powder, particle size distribution of a sample can be manipulated to control thermal conductivity. In the graph, data for 5 representative samples are shown to indicate the range of thermal conductivity possible. *(Extracted from datasheet at www.industrialpowder.com.)*

1.2.2 Agglomeration and Sintering

Particulate solids are subjected to variety of unit operations before they are transformed into final products. The diversity of particulate systems used today arises from the variety of raw materials and constituents used for eclectic end uses that range from food and pharmaceutical products to advanced engineering applications. One of the common

Part Number	Chemistry, %			Particle Size Distribution, US Mesh					Other Physical		
	BN	O₂	B₂O₃	mV	D10	D50	D90	Max	US Sieve, 95%	Tap Density, g/cc	Surface Area m²/g
PHPP325	94.0	6.0	1.0	7	0.5	2	20	53	-325	0.55	25
PHPP325B	98.0	2.0	1.0	6	1	4	14	37	-325	0.6	60
MCFP	96.0	3.0	2.0	12	2	12	30	60	-325	0.7	14
PSHP325	99.5	0.5	0.3	12	2	12	30	60	-325	0.6	15
PSHP605	99.5	0.4	0.02	6	4	6	11	22	-400	0.4	7
PCPS302	98.8	1.2	0.1	2	1	2	4	11	NA	0.2	15
PCPS308	99.5	0.5	0.02	10	4	8	18	44	NA	0.5	4
PCPS3012	99.6	0.4	0.02	13	5	12	22	52	NA	0.5	3.5
PCPS3016	99.6	0.4	0.02	18	7	16	31	74	NA	0.6	2
PCPS330	99.8	0.2	0.02	31	11	30	49	103	NA	0.6	1

General Properties

Appearance		White
Crystal Structure		Hexagonal
Apparent Density	gm / cc	2.2
Refractive Index		1.74
Coefficient of Friction		< 0.3
Dielecric Constant		3 - 4
Thermal Conductivity	W/mK	30 - 130

Fig. 1.11

Examples of datasheets for particulate systems: for boron nitride. *(Extracted from datasheet at www.bn.saint-gobain.com.)*

operations performed is agglomeration and aggregation of particles. Agglomeration is a particle size-enlargement technique in which the primary particles are joined together, either by short range, relatively weak forces between the particles themselves or through externally added agents. The original primary particles in an agglomerate are still visible (Fig. 1.14). Typically, agglomerates have interconnected pores, independent of the presence or absence of pores in the primary particles. Depending on the particle size and the nature of the surface, one can expect a certain degree of natural agglomeration amongst fine particles. However, more common are the engineered agglomerates, in which there is better control over the characteristics of agglomerates. Forming agglomerates from primary particles significantly improves the flowability and reduces dust hazards. Agglomeration also enhances wettability and dispersibility. Agglomeration is now a routine process applied across a wide range of industries, including chemical, food, pharmaceutical, and mineral industries. Sometimes, the term "granulation" is used interchangeably with agglomeration.

Sintering is another technique to transform primary powders into dense engineering components through application of heat and pressure, either independently or together, without melting the powders to the point of liquefaction. Most inorganic powders, that is, metallic and ceramic powders, are subjected to sintering process in their processing. A few polymeric particulates are also subjected to sintering. Powders are seldom sintered in their loose state.

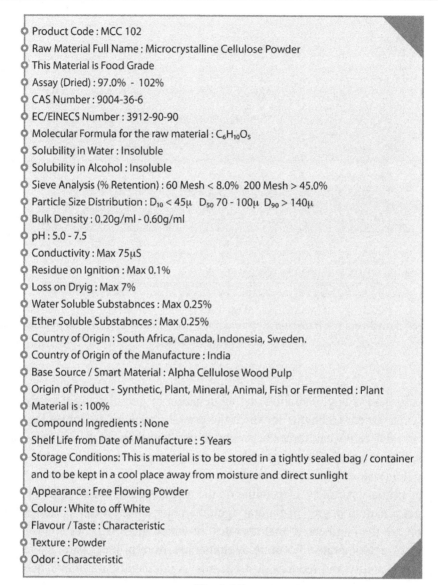

Product Code : MCC 102
Raw Material Full Name : Microcrystalline Cellulose Powder
This Material is Food Grade
Assay (Dried) : 97.0% - 102%
CAS Number : 9004-36-6
EC/EINECS Number : 3912-90-90
Molecular Formula for the raw material : $C_6H_{10}O_5$
Solubility in Water : Insoluble
Solubility in Alcohol : Insoluble
Sieve Analysis (% Retention) : 60 Mesh < 8.0% 200 Mesh > 45.0%
Particle Size Distribution : $D_{10} < 45\mu$ D_{50} 70 - 100μ $D_{90} > 140\mu$
Bulk Density : 0.20g/ml - 0.60g/ml
pH : 5.0 - 7.5
Conductivity : Max 75μS
Residue on Ignition : Max 0.1%
Loss on Dryig : Max 7%
Water Soluble Substabnces : Max 0.25%
Ether Soluble Substabnces : Max 0.25%
Country of Origin : South Africa, Canada, Indonesia, Sweden.
Country of Origin of the Manufacture : India
Base Source / Smart Material : Alpha Cellulose Wood Pulp
Origin of Product - Synthetic, Plant, Mineral, Animal, Fish or Fermented : Plant
Material is : 100%
Compound Ingredients : None
Shelf Life from Date of Manufacture : 5 Years
Storage Conditions: This is material is to be stored in a tightly sealed bag / container
and to be kept in a cool place away from moisture and direct sunlight
Appearance : Free Flowing Powder
Colour : White to off White
Flavour / Taste : Characteristic
Texture : Powder
Odor : Characteristic

Fig. 1.12

Examples of datasheets for particulate systems: for microcrystalline cellulose. *(Extracted from datasheet at www.lfatabletpresses.com.)*

First, a green compact is made by pressing the raw or blended powders into the desired shape of the final product. Often, small quantities of binders are added to the raw powders to impart enough strength to the green compacts that they do not crumble and are easy to handle between the compaction and sintering stages. Water-soluble polymers and polymers that evaporate without any residue are generally preferred as binding agents. The powder compact is then kept

HuberCal® Ground Calcium Carbonate - Typical Properties

HuberCal® Product Grade	Median Particle Size (microns)	Surface Area BET (m²/g)	Bulk Density Loose (g/cc)	Bulk Density Tamped (g/cc)	Screen Residue 325 Mesh (%)	Physical Form
950	3.6*	1.7	0.75	1.1	0.005	Powder
850	4*	1.7	0.72	1.2	0.005	Powder
500	6*	1.2	0.85	1.3	0.10	Powder
250	12**	1.1	1.07	1.5	30	Powder
150	20**	0.9	1.15	1.6	35	Powder

Fig. 1.13

Examples of datasheets for particulate systems: for calcium carbonate. *(Extracted from datasheet at hubermaterials.com.)*

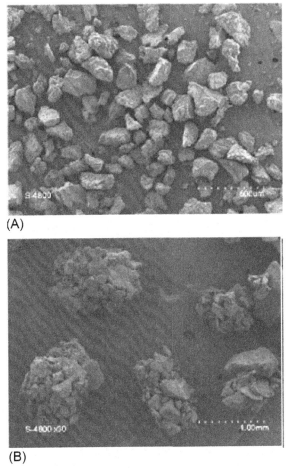

(A)

(B)

Fig. 1.14

(A) Primary particles and (B) agglomerates in semolina.[16]

at constant temperature for a duration of time. During the sintering process, a combination of heat, local surface curvature gradients, and the resultant internal pressure differences enable material transport to the contact regions between the particles, leading to strong chemical bonds between them, and filling of voids results in densification of compacts. Various steps commonly employed during metal powder processing are shown in Fig. 1.15.

1.3 Diverse Applications of Particulate Systems

1.3.1 Examples From Industrial and Consumer Applications

Particulate systems are involved in all stages in a chemical industry, namely, raw materials handling, intermediates processing, and product packaging. Due to various operations involved at various stages, particulate systems have to be stored, transported, mixed, dissolved, produced, and separated.

As mentioned earlier, the performance of powders depends very strongly on their size and shape, and distribution. For example, the size and shape of cement and starch powder are shown in Table 1.6. Given the irregular shapes of such systems, analysis and estimation are quite often carried out with equivalent sizes and shapes. On the other hand, consideration of varied sizes and complex shapes makes it very difficult to design the particulate system. For example, to achieve effective performance and durability of cement, there is a need to design optimum combinations of particles sizes and shapes.[17]

Catalyst particles and pellets (catalyst and support) in various sizes and shapes are used in chemical processes. In addition to their handling and use in industries, particulate processes are involved in the synthesis of these catalyst systems. For example, one of the most commonly used catalysts, zeolites, can be mined from natural resources, or synthesised using sol-gel and hydrothermal processes.[20]

Starch, a product extracted from renewable agricultural resources, can have different particle sizes depending on the source.[21] For example, rice starch particles can be on the order ~1 μm; corn starch particles are generally larger and on the order ~10 μm. Potato starch particles tend to be even larger, and can be as large as 100 μm.

Food products contain a mixture of particles of different sizes, shapes, and mechanical properties. They are, therefore, inherently heterogeneous particulate systems. Breakfast cereals are one common example in which size and shape vary considerably. In addition to their processing, the particle properties are also important in oral processing, that is, the chewing behaviour and the breakdown of the food particles.[22] Generally, each food product has particulate systems with a typical range of particle sizes, as shown in Fig. 1.16.

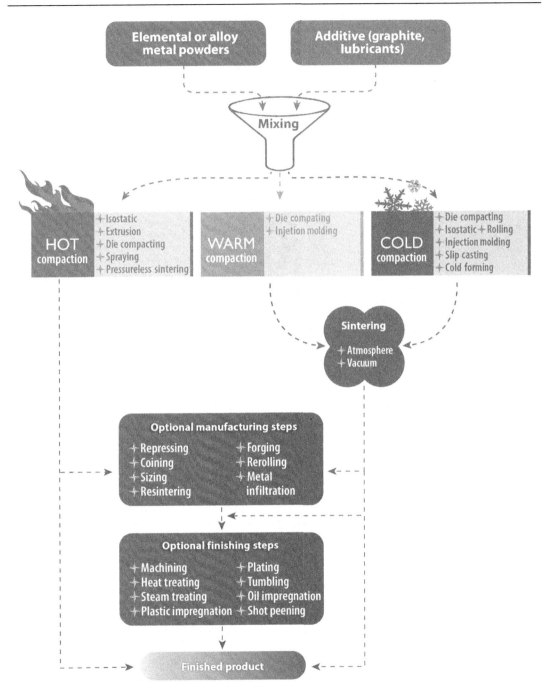

Fig. 1.15

Stages of typical metal powder processing route. *(Redrawn based on information from https://www.mpif. org/IntroPM/process.asp?linkid=2.)*

Table 1.6 Examples of particulate systems with irregular shapes

Material	Particle Shape	Important Consideration
Cement	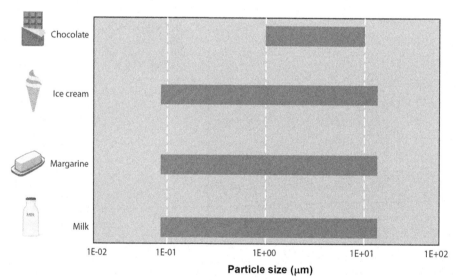	Effect of shape on hydration and setting[18]
Starch		Optimum size and shape to serve as fat replacement in processed foods[19]

Fig. 1.16
Examples of food products with their particle size range.[23]

In recent years, there has been an increased emphasis on efficient chemical processing, the impact on energy and environment, and the use of natural ingredients in health. The following list shows a few recent examples of operations in which particulate processing is important:

- Process intensification in solids handling, which can be achieved by microwave-assisted drying.[24]

- Photocatalysts in nanometre size range, which can be synthesised using aerosol synthesis.[25]
- CO_2 capture and sequestration, which can be achieved using flow and adsorption in particulate adsorbents such as oxides, coal.[26,27]
- Herbal plant use in medicine, phytomedicines, which can be enabled by producing powders suitable for bulk handling and processing.[28,29]

Large quantities of solids are handled and processed in chemical industries, which include fertilisers, soda ash, sand, cement, alumina, polyethylene, and polypropylene. In these industries, particulate systems are processed in a wide variety of forms from fine powders to pellets.[30]

With process intensification, microscopic systems are being used for chemical production. Microreactors are now being considered for various fine, pharmaceutical, specialty chemicals, and natural products.[31] Evolving appropriate set of strategies for handling and processing of particulate solids in these microscopic systems is also a challenge.[31]

For agricultural grains, drying, storage, handling, and size reduction are very important operations. Some of the important properties involved in these operations are the size, grain weight, shape, bulk density, porosity, surface area, coefficient of friction, angle of repose, thermal conductivity, heat capacity, moisture diffusivity, and equilibrium moisture content.[9] Sphericity, which can be used to quantify nonspherical shapes, varies greatly for grains. For reference, sphericity of spherical particles is 1. Nonspherical grains such as rice and wheat have very low sphericity, ~0.4, whilst grains such as soybean or mung have sphericity of ~0.9.[9] Due to their shapes and sizes, their aerodynamic properties are significantly different. For example, the terminal velocity of a grain in air can differ from ~4 m/s for sesame to ~13 m/s for blackeye peas.[9]

In addition to these large areas of industrial applications, particulate systems are relevant in the following examples:

- *Personal care products.* Powders, pastes, and dispersions are very often the form of various products such as makeup powders, toilet soaps, shower preparations, skin, hair, and oral and nail care products.[32] Particulates are often functional agents, or they serve as delivery systems. For example, a cosmetic ingredient can be delivered using molecules adsorbed on a particle.
  In modern personal care products, the number and potency of functional materials has been increasing, and, therefore, controlled delivery of these agents has become crucial.[33] Particulate systems are an integral part of the delivery systems, as rheological modifiers such as clay and acrylics, and as delivery vehicles such as crosslinked gel particles.[33]
- *Minerals.* Because mineral processing involves recovery of valuable minerals from ores, it relies on various particulate operations such as comminution (blast fragmentation, crushing, grinding), size separation (screening, cycloning), materials handling (conveying,

pumping, piping), solid-liquid separations (thickening, filtering, drying), storage between unit operations, tailings management, and disposal.[34] Size and shape of particles at different stages, mineral exposure/liberation, and particle fracture characteristics are very important characteristics for designing efficient operations.[34] Fig. 1.17 shows an example of size and shape of pyrite ore particles. For the same mineral, different particle sizes are used in different applications. For example, Fig. 1.19 shows the particle size and applications for Silica.

- *Pharmaceutical and biomedical applications.* Pharmaceutical products involve particulate systems as active drug ingredients as well as excipients (carriers) and processing/performance aids. For example, inhalation drug products use an aerosol of an active ingredient and a carrier.[35] The characteristics of active drugs in their solid state are very important in determining their performance. These characteristics are influenced by crystallinity, size, and shape of the particles.[36] The behaviour of particulate solids is crucial in the design and development of pharmaceutical dosage forms. This is due to the influence of powder properties on the formulation, manufacturability, dissolution, and the bioperformance of the dosage.[37] The sphericity of active ingredients can vary from 0.4 to 1, although most of them are likely to have a sphericity ~0.8.[37]

Improvement in the quality of pharmaceutical products can be obtained by increasing the mechanical resistance of particles, providing controlled release of active ingredients,

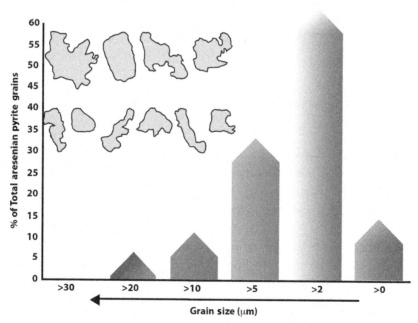

Fig. 1.17

Size distribution of arsenic-containing pyrite ore, the *inset* highlights shape of particles. *(Redrawn based on data from Malhotra D, Taylor PR, Spiller E, LeVier M, editors. Recent advances in mineral processing plant design. Society for Mining, Metallurgy, and Exploration; 2009.)*

providing protection from microorganisms and other factors such as excessive heat, moisture, and light exposure.[38] To achieve this, particle granulation and coating processes are being increasingly used in the pharmaceutical industry.[38]

The pharmaceutical industry has historically relied upon batch production, as opposed to the continuous processing. However, to harness the advantages of continuous processing, many batch-wise operations such as high-shear granulation, drum coating, and drum blending can be converted into continuous operations.[39] Alternately, such batch-wise operations could be replaced by a suitable continuous operations.[39]

- *Particles in additive manufacturing.* Additive manufacturing (AM), or 3D printing, is a generic term used to describe the manufacturing process in which a component is built layer upon layer. The process typically involves consolidating a thin homogeneous layer of powder, either by sintering or by melting, using lasers or electron beam sources. The powders used as raw materials in AM processes are often custom produced, as the powders used in traditional powder metallurgical processes may not always work for AM processes. The ability of the particles to flow smoothly from the feedstock container onto the substrate and to pack tightly are important in AM and are dictated by the particle morphology.[40] Highly spherical particles, known to improve the packing density, are generally preferred. Furthermore, spherical particles are also known to flow smoothly compared to irregular-shaped particles. Apart from the shape, the particle size and their distribution also play a significant role. A tightly controlled Gaussian distribution of particle sizes is generally recommended with cut-off for the largest and smallest particle sizes, see Fig. 1.18.[41] Having fine particles along with large particles can potentially improve the packing density but would adversely affect the flowability due to their tendency to agglomerate. Thermal properties of powders, such as conductivity, are also important because there is complex heat transfer between the built component and the substrate. As mentioned earlier in this chapter, the thermal conductivity of powders is significantly less than that of their bulk counterpart. The conductivity in case of powder beds also depends on the packing density and the particle size distribution. In densely packed powder beds, there is less trapped gas, which leads to overall improved conductivity. It is also worth noting here that conductivity also alters the sintering kinetics.[42]

- *Particles in composite materials.* Composites are formed by dispersion of a solid in another solid or a viscous matrix, and they exhibit performance which is a combination of its constituents. Dispersed solids can be of various sizes and shapes. For example, a 1D dispersed solid can be whisker, chopped fibre, or continuous fibre. Metal matrix composite refers to ceramic or metal dispersed in metals, for example, silicon carbide particles in aluminium. Fibre-reinforced plastics are composites of generally glass or carbon fibres dispersed in plastic materials. Examples of these are chopped glass fibres in polypropylene and clay in nylon. Ceramic materials are also used as matrices to form ceramic matrix composites, for example, carbon fibres in alumina.

Behaviour of particulate systems during composite processing and their role in determining their overall performance are important factors to be considered. As described in Fig. 1.9,

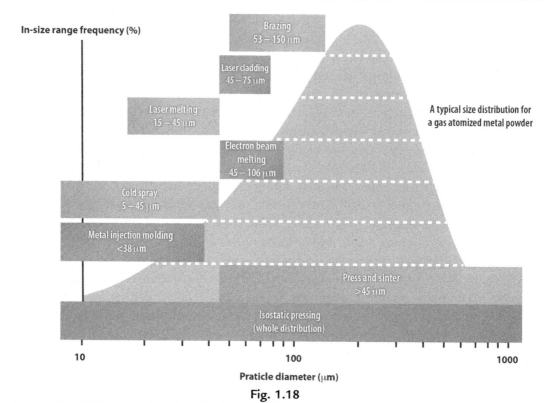

Fig. 1.18

Particles in additive manufacturing. *(Redrawn based on DeNigris J. Taking control of metal powder properties: exploring the benefits of real-time particle sizing.* Metal Powder Rep *2018. 10.1016/j.mprp.2018.03.001.)*

dispersed phases of various shapes such as spheres, platelets, short, long, and continuous fibres are used in composite materials. The size of these fillers or reinforcement can be in the broad range of a few nanometres to millimetres, and much larger for continuous fillers. Whilst discontinuous fillers are easy to disperse in matrices, and composite processing is relatively easy, property improvement, such as reinforcement of mechanical properties, is limited. On the other hand, continuous fillers such as fibres lead to significant improvement in properties, but these composites are relatively difficult to process.

1.3.2 Particles in Environment and the Resultant Pollution

Given the wide-ranging applications of particulate systems, they form part of industrial as well as nonindustrial effluents. They can lead to air, water, and soil pollution and can have long-term impacts on the ecology, climate, environment, and public health.[43–45]

Particles in air can be classified based on origin, such as anthropogenic, biological, or geogenic, or by source, such as combustion products, industrial emission, or vehicular emission.[43] The

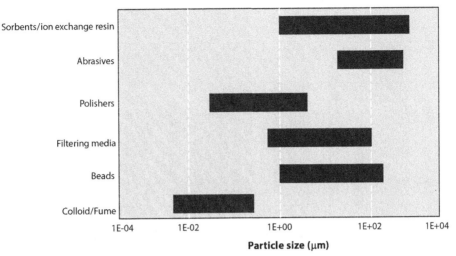

Fig. 1.19

Silica: different size ranges for different applications.[23]

size ranges of particles in the atmosphere, aerosols, and some possible sources from which aerosols arise are shown in Fig. 1.20. Aerosols arising from biological sources, bioaerosols, are also important and span the overall range from a few nanometres to submillimetres. Typical shapes of these bioaerosol particles are shown in Fig. 1.21. Given the rich diversity of biological sources, very interesting shapes of bioaerosols abound. The figure also shows an example of bioaerosol coated with salt, and images of sea salt and mineral dust particles in atmosphere.

Particulate solids can also exist as suspended matter in water and can be a significant source of water pollution. In addition to the direct impact of these particles, it is also important to consider and understand interactions between other inorganic and organic pollutants, and the particulate systems in water.

Extensive use and disposal of polymeric materials over the decades has led to large deposits in the environment. The degradation process of plastics, rubbers, and other polymeric materials is extremely slow. However, over time, due to solar radiation and other physical, chemical, or biological factors, these materials can break into smaller fragments, with sizes as small as few micrometres.[47] The extent and the impact of these microscopic polymeric particles, called microplastics, on aquatic systems has become a matter of concern and investigation. These unintended particulate systems can have significant ecological as well as environmental consequences.[48,49]

Soil contamination in urban and rural environment can be due to industrial effluents, mining, traffic routes, pipeline transport, and dust deposition. Given the scale and the land-use pattern, soil remediation and rehabilitation are extremely challenging.[50] Large-scale operations, which include handling and separation processes, process particulate soil in these efforts.[50]

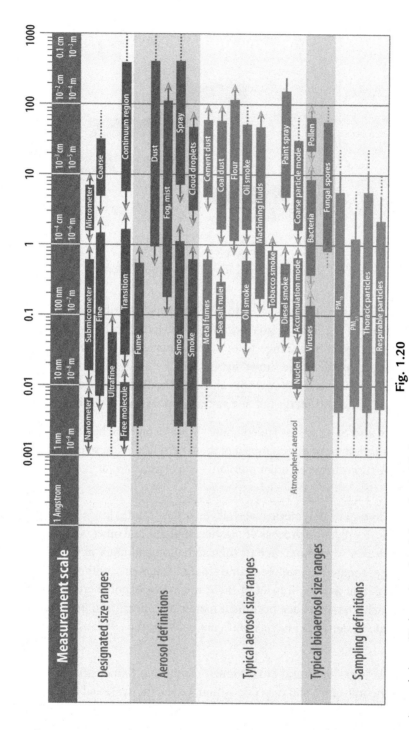

Fig. 1.20

Aerosol sizes. *(Redrawn based on information in Hinds WC. Aerosol technology: properties, behavior, and measurement of airborne particles. Wiley; 1999.)*

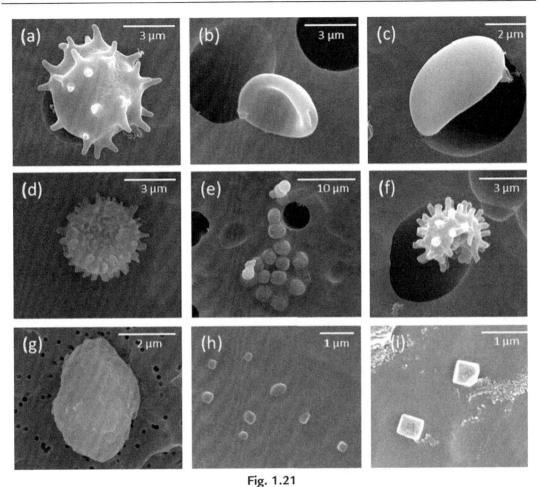

Fig. 1.21

Bioaerosol shapes, image taken from the article by Valsan et al.[46], licensed under CC Attribution 3.0: (A)–(D) Different types of spores; (E) Spore coated with salt particles; (F) Spore; (G) Mineral dust; (H) and (I) Sea salt.

1.3.3 Energy Sector: Coal and Petroleum

In the petroleum industry, particulate systems are ubiquitous. They may be a consequence as a side product, or they may be utilised as an aiding system for better product recovery. Therefore, they are present in all aspects of petroleum production process, in reservoirs themselves, during recovery of oil from oilwells, and during surface processing operations. The presence or absence of dispersed particles can determine both the economic and technical successes of the industrial process concerned.[51] Drilling suspensions and their rheological properties are important considerations in oil extraction from rocks, and Fig. 1.22 shows examples of particle ranges involved in these fluids.

Coal is processed in several sizes, and coal combustion leads to particles of different sizes. This is shown in Fig. 1.23 for a few coal-related particulate systems.

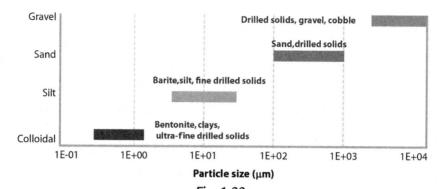

Fig. 1.22
Example of drilling fluid particulate systems, with different ranges of particle size.[52]

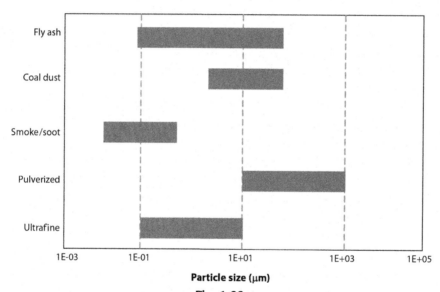

Fig. 1.23
Example of material systems with the relevant particle size.[23]

1.3.4 Particle Systems of the Future

The understanding and control of the behaviour of a single particle, as well as of the particulate systems, has improved tremendously over the years. With this, specially targeted particles and their collective systems are being envisioned for future applications. Following are some examples of particulate systems of the future, where the particles are being designed for unique features and properties:

• *Self-propelling particles.* Particles that can consume energy and move or exert mechanical forces, are called active particles and are inspired by microorganisms. These self-propelling particles can undergo motion due to catalytic reaction, or stimuli such as optical, magnetic field, ultrasound, electricity, and thermophoresis.[53] Such particles, which are being considered for their applications as nano- and micromotors, can also be propelled with the help of bacteria and enzymatic reactions.[53]

- *Functional and responsive particles.* Biological cells involve selective physical and chemical transformations in response to the environment, and they also undergo shape transformations based on the requirements.[54] A futuristic goal is to make stimuli-responsive particles capable of signalling, recognising, and self-assembly, so as to enable multidimensional 3D constructs.[54] Such functional particulate systems may help in futuristic low-carbon and sustainable society through control of materials, their assembly, and the creation of the specific structures.[55]

  Particles, specifically due to their size-dependent properties, can serve as smart biomaterials because their biodistribution can be modulated by changing the size of the particles.[56] Additionally, particles and their responsive nature can be utilised for controlled release. The particles are very promising not only because of the intrinsic properties of the component elements and the surface modification, but also because of the nanosized effects of the particles.[56]

- *Janus particles.* Janus particles are a class of particles that incorporate two or more surface regions with different properties, and the regions are distributed asymmetrically in each particle.[57] They are being considered for applications in diverse fields such as surfactants, sensors, drug delivery, displays, coatings, self-propelled particles, or probes.[57] Janus particles have been prepared with different shapes of particles, such as rods, spheres, and discs.

1.4 Scope and Objectives of This Book

The foregoing discussion not only clearly brings out the highly interdisciplinary nature of this subject, but also demonstrates wide-ranging industrial settings wherein particulate material is encountered including their production, or transformation into high-value products, or separation from fluid streams and from a mixture of solid components. At the other end of the spectrum is the design of particulate products in various sectors of prespecified attributes in terms of their size, shape, morphology, texture, etc. in order to impart and retain their functional properties, like magnetic, or electronic or optical, or simply mechanical properties, etc. Naturally, it is not possible to cover in a single volume such a wide range of topics. Therefore, some choices need to be made right at the start. Bearing in mind the objectives and contents of the previous editions of this work, the topics included here are what might be loosely referred to as the unit operations of particulate processing. In particular, consideration is given here to the characterisation of individual particles as well as that in bulk (storage and flow), segregation and mixing in solid-solid systems, and size changes (reduction and enlargement). The next set of topics included here rely on the fluid-particle interactions when there is a relative motion between the two phases, such as that encountered in estimating the drag or settling velocity of a single particle in fluids. This knowledge can be used in the context of classification of particles based on differences in size and/or density. This discussion is subsequently extended to the so-called concentrated fluid-solid systems with significant particle-particle interactions in the context of flow in packed beds, fluidised beds, and sedimentation of concentrated slurries and suspensions. Some of the ideas developed thus far are then employed to develop some understanding of liquid filtration and centrifugal solid-fluid separations. The relevance of green processing of solids in the context of particulate product design and process intensification are also covered briefly in this volume. The important field of the behaviour of colloidal and nanoparticles in solutions is covered in Chapter 13. The final chapter provides a brief introduction to the area of health and process hazards associated with the handling and processing of dry powders. Each chapter also provides an extensive list of the currently available resources for an interested reader to turn to for more detailed in-depth coverage of these topics.

References

1. Smith WF, Hashemi J, Prakash R. *Material science and engineering: (in SI units)*. India: Tata McGraw-Hill Education; 2013.
2. Mitchell BS. *An introduction to materials engineering and science: for chemical and materials engineers*. John Wiley & Sons, Inc.; 2004.
3. Yu Z, Lei W-S, Zhai J. A synchronized statistical characterization of size dependence and random variation of breakage strength of individual brittle particles. *Powder Technol* 2017;**317**:329–38.
4. Rozenblat Y, Portnikov D, Levy A, Kalman H, Aman S, Tomas J. Strength distribution of particles under compression. *Powder Technol* 2011;**208**:215–24.
5. Armaghani DJ, Amin MFM, Yagiz S, Faradonbeh RS, Abdullah RA. Prediction of the uniaxial compressive strength of sandstone using various modeling techniques. *Int J Rock Mech Min Sci* 2016;**85**:174–86.
6. Callister WD, Rethwisch DG. *Materials science and engineering: an introduction*. Wiley; 2009.
7. Saxena NS, Pradeep P, Mathew G, Thomas S, Gustafsson M, Gustafsson SE. Thermal conductivity of styrene butadiene rubber compounds with natural rubber prophylactics waste as filler. *Eur Polym J* 1999;**35**:1687–93.
8. Nah C, Park JH, Cho CT, Chang Y-W, Kaang S. Specific heats of rubber compounds. *J Appl Polym Sci* 1999;**72**:1513–22.
9. Chakraverty A, Mujumdar AS, Raghavan GSV, Ramaswamy HS. *Handbook of postharvest technology*. New York: Marcel Dekker; 2003.
10. Tyson WR, Miller WA. Surface free energies of solid metals: estimation from liquid surface tension measurements. *Surf Sci* 2000;**62**:267–76.
11. Cognard P. *Adhesives and sealants: general knowledge, application techniques, new curing techniques*. 1st ed. vol. 2. Elsevier; 2006.
12. Jallo LJ, Schoenitz M, Dreizin EL, Dave RN, Johnson CE. The effect of surface modification of aluminum powder on its flowability, combustion and reactivity. *Powder Technol* 2010;**204**:63–70.
13. Li C, Honeyands T, O'Dea D, Moreno-Atanasio R. The angle of repose and size segregation of iron ore granules: DEM analysis and experimental investigation. *Powder Technol* 2017;**320**:257–72.
14. Al-Hashemi HMB, Al-Amoudi OSB. A review on the angle of repose of granular materials. *Powder Technol* 2018;**330**:397–417.
15. Ottino JM, Khakhar DV. Mixing and segregation of granular materials. *Ann Rev Fluid Mech* 2000;**32**:55.

16. Hafsa I, Mandato S, Ruiz T, Schuck P, Jeantet R, Mejean S, et al. Impact of the agglomeration process on structure and functional properties of the agglomerates based on the durum wheat semolina. *J Food Eng* 2015;**145**:25–36.

17. Provis JL, Duxson P, van Deventer JSJ. The role of particle technology in developing sustainable construction materials. *Adv Powder Technol* 2010;**21**:2–7.

18. Bullard JW, Garboczi EJ. A model investigation of the influence of particle shape on Portland cement hydration. *Cem Conc Res* 2006;**36**:1007–15.

19. Liu K, Stieger M. van der Linden E, van de Velde F. Tribological properties of rice starch in liquid and semi-solid food model systems. *Food Hydrocoll* 2016;**58**:184–93.

20. Abdullahi T, Harun Z, Othman MHD. A review on sustainable synthesis of zeolite from kaolinite resources via hydrothermal process. *Adv Powder Technol* 2017;**28**:1827–40.

21. Ossen MS, Otome IS, Akenaka MT, Sobe SI, Akajima MN, Kadome HO. Effect of particle size of different crop starches and their flours on pasting properties. *Jpn J Food Eng* 2011;**12**:29–35.

22. Kim EHJ, Jakobsen VB, Wilson AJ, Waters IR, Motoi L, Hedderley DI, et al. Oral processing of mixtures of food particles. *J Texture Stud* 2017;**46**:487–98.

23. Merkus HG. *Particle size measurements: fundamentals, practice, quality.* Springer Dodrecht; 2009.

24. Wang H, Mustaffar A, Phan AN, Zivkovic V, Reay D, RLaw KB. A review of process intensification applied to solids handling. *Chem Eng Process Process Intensif* 2017;**118**:78–107.

25. Akurati KK, Vital A, Klotz UE, Bommer B, Graule T, Winterer M. Synthesis of non-aggregated Titania nanoparticles in atmospheric pressure diffusion flames. *Powder Technol* 2006;**165**:73–82.

26. Ghadirian E, Abbasian J, Arastoopour H. Three-dimensional CFD simulation of an MgO-based sorbent regeneration reactor in a carbon capture process. *Powder Technol* 2017;**318**:314–20.

27. Yang H, Xu J, Peng S, Nie W, Geng J, Zhang C. Large-scale physical modelling of carbon dioxide injection and gas flow in coal matrix. *Powder Technol* 2016;**294**:449–53.

28. Gallo L, Llabot JM, Allemandi D, Bucalá V, Pina J. Influence of spray-drying operating conditions on *Rhamnus purshiana* (cáscara sagrada) extract powder physical properties. *Powder Technol* 2011;**208**:205–14.

29. Battista CAD, Constenla D, Rigo MVR, Pina J. Process analysis and global optimization for the microencapsulation of phytosterols by spray drying. *Powder Technol* 2017;**321**:55–65.

30. McGlinchey D. *Bulk solids handling equipment selection and operation.* Wiley - Blackwell; 2008.

31. Hartman RL. Managing solids in microreactors for the upstream continuous processing of fine chemicals. *Org Process Res Dev* 2012;**16**:870–87.

32. Costa R, Santos L. Delivery systems for cosmetics—from manufacturing to the skin of natural antioxidants. *Powder Technol* 2017;**322**:402–16.

33. Rosen MR. *Delivery system handbook for personal care and cosmetic products.* Norwich, NY, USA: William Andrew; 2005.

34. Malhotra D, Taylor PR, Spiller E, LeVier M. *Recent advances in mineral processing plant design.* Society for Mining, Metallurgy, and Exploration; 2009.

35. Nagao LM, Lyapustina S, Munos MK, Capizzi MD. Aspects of particle science and regulation in pharmaceutical inhalation drug products. *Cryst Growth Des* 2005;**5**:2261–7.

36. Calvo NL, Maggio RM, Kaufman TS. Characterization of pharmaceutically relevant materials at the solid state employing chemometrics methods. *J Pharm Biomed Anal* 2018;**147**:538–64.

37. Al-Hashemi HMB, Al-Amoudi OSB. What is the "typical" particle shape of active pharmaceutical ingredients? *Powder Technol* 2017;**313**:1–8.

38. Abdullahi T, Harun Z, Othman MHD. Monitoring and control of coating and granulation processes in fluidized beds—a review. *Adv Powder Technol* 2014;**25**:195–210.

39. Tezyk M, Milanowski B, Ernst A, Lulek J. Recent progress in continuous and semi-continuous processing of solid oral dosage forms: a review. *Drug Dev Ind Pharm* 2016;**42**:1195–214.

40. Slotwinski JA, Garboczi EJ. Metrology needs for metal additive manufacturing powders. *JOM* 2015;**67** (3):538–43.

41. DeNigris J. Taking control of metal powder properties: exploring the benefits of real-time particle sizing. *Metal Powder Rep* 2018. https://doi.org/10.1016/j.mprp.2018.03.001.

42. Slotwinski JA, Garboczi EJ, Stutzman PE, Ferraris CF, Watson SS, Peltz MA. Characterization of metal powders used for additive manufacturing. *J Res Natl Inst Stand Technol* 2014;**119**(3):460–93.

43. Englert N. Fine particles and human health—a review of epidemiological studies. *Toxicol Lett* 2004;**149**:235–42.

44. Fuzzi S, Baltensperger U, Carslaw K, Decesari S, van der Gon HD, Facchini MC, et al. Particulate matter, air quality and climate: lessons learned and future needs. *Atmos Chem Phys* 2015;**15**:213–39.

45. Grantz DA, Garner JHB, Johnson DW. Ecological effects of particulate matter. *Environ Int* 2003;**29**:213–39.

46. Valsan AE, Ravikrishna R, Biju CV, Pohlker C, Despres VR, Huffman JA, et al. Fluorescent biological aerosol particle measurements at a tropical high-altitude site in southern India during the southwest monsoon season. *Atmos Chem Phys* 2016;**16**:9805–30.

47. Yu Y, Zhou D, Li Z, Zhu C. Advancement and challenges of microplastic pollution in the aquatic environment: a review. *Water Air Soil Pollut* 2018;**229**:140.

48. Anbumani S, Kakkar P. Ecotoxicological effects of microplastics on biota: a review. *Environ Sci Pollut Res* 2018;**25**:14373–96.

49. Conkle JL, Valle CDBD, Turner JW. Are we underestimating microplastic contamination in aquatic environments? *Environ Manag* 2018;**61**:1–8.

50. Meuser H. *Soil remediation and rehabilitation.* Dordrecht: Springer; 2013.

51. Schramm LL. *Suspensions: fundamentals and applications in the petroleum industry.* Netherlands: Springer; 1996.

52. ASME. *Drilling fluids processing handbook.* Elsevier; 2005. https://www.sciencedirect.com/book/978075 0677752/drilling-fluids-processing-handbook.

53. Guix M, Weiz SM, Schmidt OG, Medina-Sánchez M. Self-propelled micro/nanoparticle motors. *Part Syst Charact* 2018;**35**:1700382.

54. Lu C, Urban MW. Stimuli-responsive polymer nano-science: shape anisotropy, responsiveness, applications. *Prog Polym Sci* 2018;**78**:24–46.

55. Mancic L, Nikolic M, Gomez L, Rabanal ME, Milosevic O. The processing of optically active functional hierarchical nanoparticles. *Adv Powder Technol* 2017;**28**:3–22.

56. Tanaka K, Chujo Y. Design of functionalized nanoparticles for the applications in nanobiotechnology. *Adv Powder Technol* 2014;**25**:101–13.

57. Poggi E, Gohy J-F. Janus particles: from synthesis to application. *Colloid Polym Sci* 2017;**295**:2083–108.

CHAPTER 2

Particulate Solids

2.1 Introduction

In Vol. 1 (A & B), the behaviour of liquids and gases is considered, with particular reference to their flow properties and their heat and mass transfer characteristics. Once a fluid's composition, temperature, and pressure are specified, then its relevant physical properties, such as density, viscosity, thermal conductivity, and molecular diffusivity are defined. In this and the next chapter, attention is given to the properties and behaviours of systems containing bulk solids. Such systems are generally more complicated, not only because of the complex geometrical arrangements that are possible, but also because of the basic problem of defining the physical state of the material.

The key characteristics of an individual particle include composition, size, shape, density, and hardness. Composition determines properties such as density and conductivity if the particle is uniform. However, in many cases the particle is porous or it may consist of a continuous matrix in which small particles of a second material are distributed. Particle size is important because it affects properties such as the surface per unit volume and the rate at which a particle will settle in a fluid. A particle shape may be regular, such as spherical or cubic, or it may be irregular like a piece of broken glass. Regular shapes can be defined by mathematical equations, while irregular shapes and properties of irregular particles are usually found by comparison to specific characteristics of a regularly shaped particle.

Large quantities of particles are handled on an industrial scale and it is frequently required to define the system as a whole. Thus, in place of particle size, it is necessary to know the distribution of particle sizes in the mixture and to be able to define a mean size that can represent the behaviour of the entire particulate mass. Particle shape has strong influence on the flow and packing behaviour of the material. Characterisation of the particle shape by quantitative means is preferred. The measurement of all relevant properties of the material depends crucially on the sample. Hence, appropriate sampling of the particulate solids is essential for obtaining reliable and reproducible information about the material.

2.2 Particle Characterisation

2.2.1 Single Particles

The simplest shape of a particle is the sphere. Because of its symmetry, the particle looks exactly the same from whatever direction it is viewed, and behaves in the same manner in fluid, regardless of its orientation. No other particle shape has this characteristic. However, perfect spheres are rarely found and, generally, a typical particle may not be of any regular known shape. In fact, irregular-shaped particles are common. They are typically observed in nature and regularly handled and processed in many industries. Frequently, the size of a particle of an irregular shape is defined in terms of the size of an equivalent sphere, although the particle is represented by a sphere of a different size according to the property that is selected. Some of the important sizes of equivalent spheres are that the spheres have the same:

(a) volume as the particle.
(b) surface area as the particle.
(c) surface area per unit volume (i.e. specific surface area) as the particle.
(d) area as the particle when projected onto a plane perpendicular to its direction of motion.
(e) projected area as the particle, as viewed from above, when lying in its position of maximum stability such as on a microscope slide, for example.
(f) ability to pass through the same size of square aperture as the particle, such as on a screen for example.
(g) same settling velocity as the particle in a specified fluid.

Several definitions depend on the measurement of a particle in a particular orientation. Thus, Feret's statistical diameter is the mean distance between two parallel lines that are tangential to the particle in an arbitrarily fixed direction, irrespective of the orientation of each particle coming up for inspection. Similarly, the Martin diameter is the length of the chord that bisects the projected area of the particle. Both these diameters are shown in Fig. 2.1 for two different orientations of the same particle. While these measurements are orientation dependent for a specific particle, it will probably be unimportant when the measurements are done for a sufficiently large number of particles (assuming all possible orientations of the particles are equally probable). Depending on the process of interest, the relevant particle size is typically chosen as the method to define the particle size. A measure of particle shape that is frequently used is the sphericity, ψ, defined as:

$$\psi = \frac{\text{surface area of sphere of same volume as particle}}{\text{surface area of particle}} \tag{2.1}$$

Another simple method to indicate shape is to use the factor by which the cube of the size of the particle must be multiplied to give the volume. In this case the particle size is usually defined by method (e) above. A more detailed discussion about particle shape indicators is provided in Section 2.2.5.

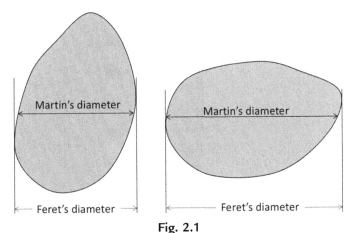

Fig. 2.1
Feret's and Martin's diameter for two different orientations of a particle.

Other properties of the particle that may be important are particle density, which can be crystalline, amorphous, or porous, and the properties of its surface, which includes roughness and presence of adsorbed films. Particle hardness could also be important if the particle is subject to heavy loading and/or impact.

2.2.2 Measurement of Particle Size

Measurement of particle size and of particle size distribution is a highly specialised topic, and considerable skill is needed to make accurate measurements and interpretation. Specialised texts such as those by Allen[1, 2] are recommended for details of the experimental techniques in common use.

No attempt is made here to give a detailed account or critical assessment of the various methods of measuring particle size, which can be seen in Fig. 2.2 to cover a range of 10^7 in linear dimension, or 10^{21} in volume. Only a brief account of some of the principal methods of measurement is given here, and for further details it is necessary to refer to one of the specialist texts on particle size measurement. An outstanding example of such a text is the two-volume monograph by Allen[1, 2] with Herdan[3] providing additional information. Both the size range in the sample and the particle shape may be as important, or even more so, than a single characteristic linear dimension, which at best can represent only one single property of an individual particle or of an assembly of particles. The ability to make accurate and reliable measurements of particle size is achieved only after many years of practical experience.

Before a size analysis can be carried out, it is necessary to collect a representative sample of the solids, and then reduce this to the quantity, which is required for the chosen method of analysis. Again, Allen's work provides information on how this is best carried out. Samples will

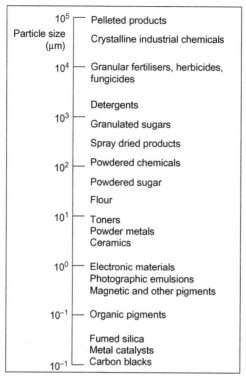

Fig. 2.2
Sizes of typical powder products.[1]

generally need to be taken from the bulk of the powder, whether this is in a static heap, in the form of an airborne dust, in a flowing or falling stream, or on a conveyor belt. In each case, the precautions that need to be taken to obtain a representative sample are different. A brief discussion about the sampling method is given in Section 2.3.

A wide range of measuring techniques is available for both single particles and systems of particles, ranging from visual methods utilising microscopy and image analysis to stream or field scanning methods such as electronic particle counters and laser diffractions. In practice, each method is applicable to a finite range of sizes and gives a particular equivalent size, dependent on the nature of the method. The principles of some of the chief methods are now considered together with an indication of the size range to which they are applicable.

Sieving (>50 μm)

Sieve analysis may be carried out using a nest of sieves, each lower sieve being of smaller aperture size. Generally, sieve series is arranged so that the ratio of aperture sizes on consecutive sieves is 2, $2^{1/2}$, or $2^{1/4}$ according to the closeness of sizing that is required. The sieves may either be hand-shaken or mounted on a vibrator, which should be designed to give a degree of vertical movement and horizontal vibration. Whether or not a particle passes through an aperture depends not only upon its size, but also on the probability that it will be presented at the required orientation at the surface of the screen. The sizing is based purely on the linear dimensions of the particle and the lower limit of size is determined by two principal factors. First, the proportion of free space on the screen surface becomes small as the size of the aperture is reduced. Second, the attractive forces between particles become larger at small particle sizes, and consequently particles tend to stick together and block the screen. Sieves are available in a number of standard series. There are several standard series of screens and the sizes of the openings are determined by the thickness of the wire used. In the United Kingdom, British Standard (B.S.)[4] screens are made in sizes from 300-mesh and higher, although they are too fragile for some work. The Institute of Mining and Metallurgy (IMM)[5] screens are more robust, with the thickness of the wire approximately equal to the size of the apertures. The Tyler series, which is standard in the United States, is intermediate, between the two British series. Details of the three series of screens[4] are given in Table 2.1, together with the American Society for Testing Materials (ASTM) series.[6]

The efficiency of screening is defined as the ratio of the mass of material that passes the screen, to that which is capable of passing through the screen. This will differ according to the size of the material. It may be assumed that the rate of passage of particles of a given size through the screen is proportional to the number or mass of particles of that size on the screen at any instant. Thus, if w is the mass of particles of a particular size on the screen at a time t, then:

$$\frac{\mathrm{d}w}{\mathrm{d}t} = -kw \tag{2.2}$$

where k is a constant for a given size and shape of particle and for a given screen.

Thus, the mass of particles $(w_1 - w_2)$ passing the screen in time t is given by:

$$\ln \frac{w_2}{w_1} = -kt$$

or:

$$w_2 = w_1 e^{-kt} \tag{2.3}$$

If the screen contains a large proportion of material that is slightly larger than the maximum size of particle that will pass, its capacity is considerably reduced. Screening is generally continued either for a predetermined time or until the rate of screening falls off to a certain fixed value.

Screening may be carried out with either wet or dry material. In wet screening, material is washed evenly over the screen and clogging is prevented. In addition, small particles are washed off the surface of large ones. The obvious disadvantage is that it may be necessary to dry the material afterwards. With dry screening, the material is sometimes brushed lightly over the

Table 2.1 Standard sieve sizes

British Fine Mesh (BSS 410)[4]			IMM[5]			U.S. Tyler[6]			U.S. ASTM[5]		
Sieve No.	Nominal Aperture		Sieve No.	Nominal Aperture		Sieve No.	Nominal Aperture		Sieve No.	Nominal Aperture	
	in.	μm		in.	μm		in.	μm		in.	μm
						325	0.0017	43	325	0.0017	44
						270	0.0021	53	270	0.0021	53
300	0.0021	53				250	0.0024	61	230	0.0024	61
240	0.0026	66	200	0.0025	63	200	0.0029	74	200	0.0029	74
200	0.0030	76							170	0.0034	88
170	0.0035	89	150	0.0033	84	170	0.0035	89			
150	0.0041	104				150	0.0041	104	140	0.0041	104
120	0.0049	124	120	0.0042	107	115	0.0049	125	120	0.0049	125
100	0.0060	152	100	0.0050	127	100	0.0058	147	100	0.0059	150
			90	0.0055	139	80	0.0069	175	80	0.0070	177
85	0.0070	178	80	0.0062	157	65	0.0082	208	70	0.0083	210
			70	0.0071	180				60	0.0098	250
72	0.0083	211	60	0.0083	211	60	0.0097	246	50	0.0117	297
60	0.0099	251							45	0.0138	350
52	0.0116	295	50	0.0100	254	48	0.0116	295	40	0.0165	420
			40	0.0125	347	42	0.0133	351	35	0.0197	500
44	0.0139	353				35	0.0164	417	30	0.0232	590
36	0.0166	422	30	0.0166	422	32	0.0195	495			
30	0.0197	500				28	0.0232	589			
25	0.0236	600									
22	0.0275	699	20	0.0250	635	24	0.0276	701	25	0.0280	710
18	0.0336	853	16	0.0312	792	20	0.0328	833	20	0.0331	840
16	0.0395	1003				16	0.0390	991	18	0.0394	1000
14	0.0474	1204	12	0.0416	1056	14	0.0460	1168	16	0.0469	1190
12	0.0553	1405	10	0.0500	1270	12	0.0550	1397			
10	0.0660	1676	8	0.0620	1574	10	0.0650	1651	14	0.0555	1410
8	0.0810	2057				9	0.0780	1981	12	0.0661	1680
7	0.0949	2411				8	0.0930	2362	10	0.0787	2000
6	0.1107	2812	5	0.1000	2540	7	0.1100	2794	8	0.0937	2380
5	0.1320	3353				6	0.1310	3327			
						5	0.1560	3962	7	0.1110	2839
						4	0.1850	4699			
									6	0.1320	3360
									5	0.1570	4000
									4	0.1870	4760

screen to form a thin, even sheet. It is important that any agitation is not so vigorous that size reduction occurs because screens are usually quite fragile and are easily damaged by rough treatment. In general, the larger and more abrasive the solids, the more robust the screen should be.

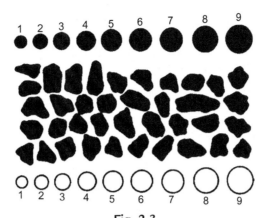

Fig. 2.3
Particle profiles and comparison circles.

Microscopic analysis (1–100 μm)

Microscopic examination permits measurement of the projected area of the particle and also enables an assessment to be made of its two-dimensional shape. In general, the third dimension cannot be determined except when using special stereomicroscopes. The apparent size of a particle is compared with that of circles engraved on a graticule in the eyepiece as shown in Fig. 2.3. Automatic methods of scanning have been developed. By using the electron microscope,[7] the lower limit of size can be reduced to about 0.001 μm.

Sedimentation and elutriation methods (>1 μm)

These methods depend on the fact that the terminal falling velocity of a particle in a fluid increases with size. There are two main sedimentation methods. With the pipette method, samples are abstracted from the settling suspension at a fixed horizontal level at intervals of time. Each sample contains a representative sample of the suspension, with the exception of particles larger than a critical size, all of which settle below the level of the sampling point. The most commonly used equipment, the Andreasen pipette, is described by Allen.[1, 2] In the second method, which involves the use of the sedimentation balance, particles settle on an immersed balance pan, which is continuously weighed. The largest particles are deposited preferentially and consequently the rate of increase of weight falls off progressively as the particles settle out.

Sedimentation analyses must be carried out at concentrations that are sufficiently low for interactive effects between particles to be negligible. In this case, their terminal falling velocities can be taken as equal to those of isolated particles. Careful temperature control (preferably to ± 0.1 K) is necessary to suppress convection currents. The lower limit of particle size is set by the increasing importance of Brownian motion for progressively smaller particles. It is possible however, to replace gravitational forces by centrifugal forces, which reduces the lower size limit to about 0.05 μm.

The elutriation method is a reverse sedimentation process in which the particles are dispersed in an upward flowing stream of fluid. All particles with terminal falling velocities less than the upward velocity of the fluid will be carried away. A complete size analysis can be obtained by using successively higher fluid velocities. Fig. 2.4 shows the standard elutriator (BS 893)[8] for particles with settling velocities between 7 and 70 mm/s.

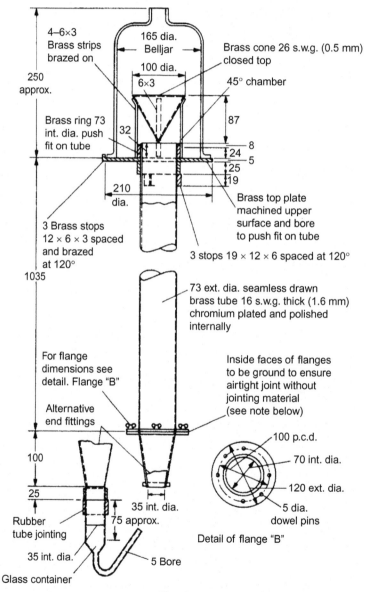

Fig. 2.4
Standard elutriator with 70-mm tube (all dimensions in mm).[8]

Permeability methods (>1 μm)

These methods depend on the fact that at low flow rates, the flow rate through a packed bed is directly proportional to the pressure difference, and the proportionality constant being proportional to the square of the specific surface (surface to volume ratio) of the powder. From this method it is possible to obtain the diameter of the sphere with the same specific surface as the powder. The reliability of the method is dependent upon the care with which the sample of powder is packed. Further details are given in Chapter 7.

Electronic particle counters

A suspension of particles in an electrolyte is drawn through a small orifice with an electrode positioned on either side. A constant electrical current supply is connected to the electrodes. The electrolyte within the orifice constitutes the main resistive component of the circuit. As particles enter the orifice, they displace an equivalent volume of the electrolyte, thereby producing a change in the electrical resistance of the circuit, the magnitude of which is related to the displaced volume. The consequent voltage pulse across the electrodes is fed to a multi-channel analyser. The distribution of pulses arising from the passage of many thousands of particles is then processed to provide a particle (volume) size distribution.

The main disadvantage of the method is that the suspension medium must be so highly conducting that its ionic strength may be such that surface-active additives may be required to maintain colloidal stability of fine particle suspensions as discussed in Chapter 8. The technique is suitable for the analysis of the nonconducting particles, and for the conducting particles when electrical double layers confer a suitable degree of electrical insulation. This is also discussed in Chapter 8.

By using orifices of various diameters, different particle size ranges may be examined, and the resulting data may then be combined to provide size distributions extending over a large proportion of the submillimeter size range. The prior removal from the suspension of particles of sizes upwards of about 60% of the orifice diameter helps to prevent problems associated with the blocking of the orifice. The *Coulter Counter* and the *Elzone Analyser* work on this principle.

Laser diffraction analysers

These instruments[9] exploit the radial light scattering distribution functions of particles. A suspension of particles is held in, or usually passed across the path of a collimated beam of laser light, and the radially scattered light is collected by an array of photodetectors positioned perpendicular to the optical axis. The scattered light distribution is sampled and processed using appropriate scattering models to provide a particle size distribution. The method is applicable to the analysis of a range of different particles in a variety of media. Consequently, it is possible to examine the aggregation phenomena as discussed in Chapter 8, and to monitor particle size for online control of process streams. Instruments are available that provide particle size information over the range 0.1–600 μm. Light scattered from particles smaller than 1 μm is strongly influenced by their optical properties and care is required in data processing and interpretation.

The scattering models employed in data processing methods invariably involve the assumption of particle sphericity. Size data obtained from the analysis of suspensions of asymmetrical particles using laser diffraction tend to be more ambiguous than those obtained by electronic particle counting, where the solid volumes of the particles are detected.

X-ray or photo-sedimentometers

Information on particle size may be obtained from the sedimentation of particles in dilute suspensions. The use of pipette techniques can be tedious and care is required to ensure that measurements are sufficiently precise. Instruments such as X-ray or photo-sedimentometers serve to automate this method in a nonintrusive manner. The attenuation of a narrow collimated beam of radiation passing horizontally through a sample of suspension is related to the mass of solid material present in the path of the beam. This attenuation can be monitored at a fixed height in the suspension, or can be monitored as the beam is raised at a known rate. This latter procedure reduces the time required to obtain sufficient data from which the particle size distribution can be calculated. This technique is limited to the analysis of particles whose settling behaviour follows Stokes' law, as discussed in Chapter 6, and to conditions where any diffusive motion of particles is negligible.

Submicron particle sizing

Particles of a size of <2 μm are of particular interest in process engineering applications because of their large specific surface and colloidal properties, as discussed in Chapter 8. The diffusive velocities of such particles are significant in comparison with their settling velocities. Provided that the particles scatter light, dynamic light scattering techniques, such as photon correlation spectroscopy (PCS), may be used to provide information about particle diffusion.

In the PCS technique, a quiescent particle suspension behaves as an array of mobile scattering centres over which the coherence of an incident laser light beam is preserved. The frequency of the light intensity fluctuations at a point outside the incident light path is related to the time taken for a particle to diffuse a distance equivalent to the wavelength of the incident light. The dynamic light signal at such a point is sampled and correlated with itself at different time intervals using a digital correlator and associated computer software. The relationship of the (so-called) autocorrelation function to the time intervals is processed to provide estimates of an average particle size and variance (polydispersity index). Analysis of the signals at different scattering angles enables more detailed information to be obtained about the size distribution of this fine, and usually problematical, end of the size spectrum. The technique allows fine particles to be examined in a liquid environment so that estimates can be made of their effective hydrodynamic sizes. This is not possible using other techniques.

If fluid motion is uniform in the illuminated region of the suspension, then similar information may also be extracted by analysis of laser light scattering from particles undergoing electrophoretic motion. This is migratory motion in an electric field that is superimposed on the electrophoretic motion. Instrumentation and data processing techniques for systems employing dynamic light scattering for the examination of fine particle motion are under development.

Refer to the monograph by Xu[10] for a detailed account of the particle characterisation using light scattering methods.

Digital image processing

Advances in digital image capturing and image processing make measurement of particle sizes relatively easy. Commercial imaging setups utilising the image processing techniques and providing reliable information about the particle sizes are also available. With the advent of image processing software (such as ImageJ, SigmaScan Pro, Matlab Image Processing Toolbox, etc.) and enhanced computational power, the analysis of the samples for particle size can be done in a fast, reliable and automated manner, and various possible parameters associated with the 2D contour of the imaged particle (silhouette) can be estimated. The subjectivity and difficulty associated with the techniques of the size measurement such as the one mentioned in Fig. 2.3 are greatly reduced by the digital image analysis method.

The digital image of the particles is taken by a camera or microscope and accurate results require the dispersal of the particles in a non-touching fashion on a background having good colour contrast. Conversion of the colour image (Fig. 2.5A) to a greyscale image followed by thresholding is typically done to obtain a binary image having only pure black and white regions (Fig. 2.5B). These binary images are used to identify various parameters of interest of the particle contour. Typically, the projected area (A_p) of the particle is obtained as the sum of the pixel area in calibrated units (e.g. μm^2 or mm^2) within the boundary of the particle (shown as the white region in Fig. 2.5B).

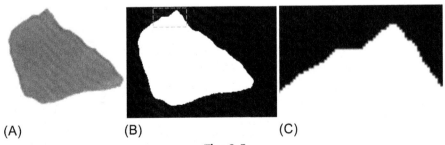

(A)   (B)   (C)

Fig. 2.5

Digitisation of the particle image: (A) original image, (B) binary image obtained after thresholding, and (C) zigzag nature of the particle boundary.

Often, instead of reporting this area, the equivalent diameter $D_p = (4A_p/\pi)^{1/2}$ of the circle having area equal to the projected area of particle is reported. Measurements for multiple particles can be done simultaneously and the particle size distributions can be easily obtained using this technique. Other parameters such as perimeter (P), major axis (L), minor axis (B) and different Feret lengths can also be measured. Fig. 2.5C shows the zoomed-in view of the rectangular region of the image in part (B). The zigzag border between the particle and the background due to the digitisation of the image using square pixels is evident. Such representation of the particle contour leads to the loss of the information at the boundary and introduces errors in the measurement. At lower image resolutions, the measurements are often in error because of inaccurate representation of the particle contour.[11] Good resolution of the particle image is needed for reasonable estimates of the particle contour properties. While higher resolutions offer more detailed information about the particle, the cost of information storage and processing time increases and a trade-off is typically needed. For higher image resolutions, the area overestimations and underestimations cancel each other out leading to accurate measurements for the projected area and projected diameter.[12] However, higher resolutions are not necessarily better for a few other properties, such as the perimeter. Also, the algorithms used in the image analysis software seem to significantly influence the measurements, particularly that of the perimeter.[13] The use of image processing techniques for characterisation of the shape of the particles is common and is discussed in Section 2.2.5.

2.2.3 Particle Size Distribution

Most particulate systems of practical interest consist of particles of a wide range of sizes and it is necessary to give a quantitative indication of the mean size and of the spread of sizes. The results of a size analysis can most conveniently be represented by means of a *cumulative mass fraction curve*, in which the proportion of particles (x) smaller than a certain size (d) is plotted against that size (d). When the particle sizes are determined by image analysis, the results are represented by a *cumulative number fraction curve*, from which the mass fraction curve can be obtained. In most practical determinations of particle size, the size analysis is obtained by a series of steps. Each step represents the proportion of particles lying within a certain small range of size. From these results a cumulative size distribution can be built up and this can then be approximated by a smooth curve provided that the size intervals are sufficiently small. A typical curve for size distribution on a cumulative basis is shown in Fig. 2.6. This curve rises from zero to unity over the range from the smallest to the largest particle size present in the sample. The cumulative size distribution curve is often used to calculate the particle size d_α below which $\alpha\%$ of distribution lies. Using this approach, d_{50} is the size for which half of the distribution lies below this size. Similarly, d_{90} (d_{10}) represents the particle size for which 90% (10%) of the material size lies below this size. Often, d_{90} and d_{10} are used to represent the size of the coarsest and the finest particles in the distribution.

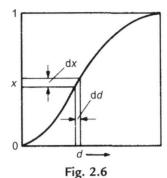

Fig. 2.6
Size distribution curve—cumulative basis.

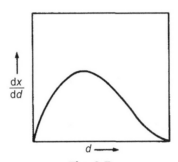

Fig. 2.7
Size distribution curve—frequency basis.

The distribution of particle sizes can be seen more readily by plotting a *size frequency curve*, Fig. 2.7, in which the slope (dx/dd) of the cumulative curve (Fig. 2.6) is plotted against particle size (d). The most frequently occurring size is then shown by the maximum of the curve. For naturally occurring materials, the curve will generally have a single peak. For mixtures of particles, there may be as many peaks as components in the mixture. Again, if the particles are formed by crushing larger particles, the curve may have two peaks, one characteristic of the material and the other characteristic of the equipment.

2.2.4 Mean Particle Size

The expression of the particle size of a powder in terms of a single linear dimension is often reported. For coarse particles, Bond[14, 15] has somewhat arbitrarily chosen the size of the opening through which 80% of the material will pass. This size d_{80} is a useful rough comparative measure for the size of material that has been through a crusher. A more common practice of reporting particle sizes, especially those obtained from image analysis, is to report d_{50} which will indicate the mean particle size along with the values of d_{10} and d_{90} to denote the range of the particle sizes of the mixture.

A mean size will describe only one particular characteristic of the powder and it is important to decide what that characteristic is before the mean is calculated. Thus, it may be desirable to define the size of the particle so that its mass, surface, or length is the mean value for all the particles in the system. In the following discussion, it is assumed that each of the particles has the same shape.

Considering unit mass of particles consisting of n_1 particles of characteristic dimension d_1, constituting a mass fraction x_1, n_2 particles of size d_2, and so on, then:

$$x_1 = n_1 k_1 d_1^3 \rho_s \tag{2.4}$$

and:

$$\sum x_1 = 1 = \rho_s k_1 \sum \left(n_1 d_1^3 \right) \tag{2.5}$$

Thus:

$$n_1 = \frac{1}{\rho_s k_1} \frac{x_1}{d_1^3} \tag{2.6}$$

If the size distribution can be represented by a continuous function, then:

$$dx = \rho_s k_1 d^3 dn$$

or:

$$\frac{dx}{dn} = \rho_s k_1 d^3 \tag{2.7}$$

and:

$$\int_0^1 dx = 1 = \rho_s k_1 \int d^3 dn \tag{2.8}$$

where ρ_s is the density of the particles and k_1 is a constant whose value depends on the shape of the particle.

Mean sizes based on volume

The mean abscissa in Fig. 2.6 is defined as the *volume mean diameter d_v*, or as the *mass mean diameter*, where:

$$d_v = \frac{\int_0^1 d\, dx}{\int_0^1 dx} = \int_0^1 d\, dx. \tag{2.9}$$

Expressing this relation in finite difference form, then:

$$d_v = \frac{\Sigma(d_1 x_1)}{\Sigma x_1} = \Sigma(x_1 d_1) \tag{2.10}$$

which, in terms of particle numbers, rather than mass fractions gives:

$$d_v = \frac{\rho_s k_1 \Sigma(n_1 d_1^4)}{\rho_s k_1 \Sigma(n_1 d_1^3)} = \frac{\Sigma(n_1 d_1^4)}{\Sigma(n_1 d_1^3)} \tag{2.11}$$

Another mean size based on volume is the *mean volume diameter* d'_v. If all the particles are of diameter d'_v, then the total volume of particles is the same as in the mixture.

Thus:

$$k_1 d'^3_v \Sigma n_1 = \Sigma(k_1 n_1 d_1^3)$$

or:

$$d'_v = \sqrt[3]{\left(\frac{\Sigma(n_1 d_1^3)}{\Sigma n_1}\right)} \tag{2.12}$$

Substituting from Eq. (2.6) gives:

$$d'_v = \sqrt[3]{\left(\frac{\Sigma x_1}{\Sigma(x_1/d_1^3)}\right)} = \sqrt[3]{\left(\frac{1}{\Sigma(x_1/d_1^3)}\right)} \tag{2.13}$$

Mean sizes based on surface

In Fig. 2.6, if instead of fraction of total mass, the surface in each fraction is plotted against size, then a similar curve is obtained, although the mean abscissa d_s is then the *surface mean diameter*.

Thus:

$$d_s = \frac{\Sigma[(n_1 d_1) S_1]}{\Sigma(n_1 S_1)} = \frac{\Sigma(n_1 k_2 d_1^3)}{\Sigma(n_1 k_2 d_1^2)} = \frac{\Sigma(n_1 d_1^3)}{\Sigma(n_1 d_1^2)} \tag{2.14}$$

where $S_1 = k_2 d_1^2$, and k_2 is a constant whose value depends on particle shape. d_s is also known as the *Sauter mean diameter* and is the diameter of the particle with the same specific surface as the powder.

Substituting for n_1 from Eq. (2.6) gives:

$$d_s = \frac{\Sigma x_1}{\Sigma\left(\dfrac{x_1}{d_1}\right)} = \frac{1}{\Sigma\left(\dfrac{x_1}{d_1}\right)} \tag{2.15}$$

The *mean surface diameter* is defined as the size of particle d'_s, which is such that if all the particles are of this size, the total surface will be the same as in the mixture.

Thus:

$$k_2 d'^2_s \sum n_1 = \sum\left(k_2 n_1 d_1^2\right)$$

or:

$$d'_s = \sqrt{\left(\frac{\sum\left(n_1 d_1^2\right)}{\sum n_1}\right)} \tag{2.16}$$

Substituting for n_1 gives:

$$d'_s = \sqrt{\left(\frac{\sum\left(x_1/d_1\right)}{\sum\left(x_1/d_1^3\right)}\right)} \tag{2.17}$$

Mean dimensions based on length

A *length mean diameter* may be defined as:

$$d_l = \frac{\Sigma[(n_1 d_1)d_1]}{\Sigma(n_1 d_1)} = \frac{\Sigma\left(n_1 d_1^2\right)}{\Sigma(n_1 d_1)} = \frac{\Sigma\left(\dfrac{x_1}{d_1}\right)}{\Sigma\left(\dfrac{x_1}{d_1^2}\right)} \tag{2.18}$$

A *mean length diameter* or arithmetic mean diameter may also be defined by:

$$d'_l \Sigma n_1 = \Sigma(n_1 d_1)$$

$$d'_l = \frac{\Sigma(n_1 d_1)}{\Sigma n_1} = \frac{\Sigma\left(\dfrac{x_1}{d_1^2}\right)}{\Sigma\left(\dfrac{x_1}{d_1^3}\right)} \tag{2.19}$$

Example 2.1

The size analysis of a powdered material on a mass basis is represented by a straight line from 0% mass at 1 µm particle size to 100% mass at 101 µm particle size as shown in Fig. 2.8. Calculate the surface mean diameter of the particles constituting the system.

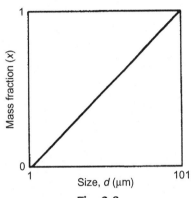

Fig. 2.8
Size analysis of powder.

Solution

From Eq. (2.15), the surface mean diameter is given by:

$$d_s = \frac{1}{\Sigma(x_1/d_1)}$$

Because the size analysis is represented by the continuous curve:

$$d = 100x + 1$$

then:

$$d_s = \frac{1}{\displaystyle\int_0^1 \frac{dx}{d}}$$

$$= \frac{1}{\displaystyle\int_0^1 \frac{dx}{100x + 1}}$$

$$= (100/\ln 101)$$

$$= \underline{\underline{21.7\,\mu m}}$$

Example 2.2

The equations giving the number distribution curve for a powdered material are $dn/dd = d$ for the size range 0–10 µm and $dn/dd = 100{,}000/d^4$ for the size range 10–100 µm where d is in µm. Sketch the number, surface and mass distribution curves and calculate the surface mean diameter for the powder. Explain briefly how the data required for the construction of these curves may be obtained experimentally.

Solution

Note: The equations for the number distributions are valid only for d expressed in µm.

For the range, $d = 0$–10 µm, $dn/dd = d$.

On integration:

$$n = 0.5d^2 + c_1 \qquad \qquad (i)$$

where c_1 is the constant of integration.

For the range, $d = 10$–100 µm, $dn/dd = 10^5 d^{-4}$.

On integration:

$$n = c_2 - \left(0.33 \times 10^5\, d^{-3}\right) \qquad \qquad (ii)$$

where c_2 is the constant of integration.

When $d = 0$, $n = 0$, and from (i): $c_1 = 0$

When $d = 10$ µm, in (i): $n = (0.5 \times 10^2) = 50$

In (ii): $50 = c_2 - (0.33 \times 10^5 \times 10^{-3})$, and $c_2 = 83.0$.

Thus for $d = 0$–10 µm: $n = 0.5d^2$ and for $d = 10$–100 µm: $n = 83.0 - (0.33 \times 10^5 d^{-3})$

Using these equations, the following values of n are obtained:

d (µm)	n	d (µm)	n
0	0	10	50.0
2.5	3.1	25	80.9
5.0	12.5	50	82.7
7.5	28.1	75	82.9
10.0	50.0	100	83.0

and these data are plotted in Fig. 2.9.

From this plot, values of d are obtained for various values of n and hence n_1 and d_1 are obtained for each increment of n. Values of $n_1 d_1^2$ and $n_1 d_1^3$ are calculated and the totals obtained. The surface area of the particles in the increment is then given by:

$$s_1 = n_1 d_1^2 \Big/ \sum n_1 d_1^2$$

and s is then found as $\Sigma\, s_1$. Similarly, the mass of the particles, $x = \Sigma\, x_1$ where:

$$x_1 = n_1 d_1^3 \Big/ \sum n_1 d_1^3$$

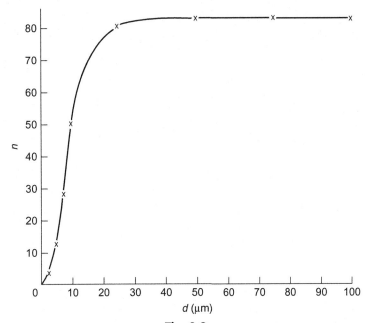

Fig. 2.9
Plot of data for Example 2.2.

The results are:

n	d	n_1	d_1	$n_1d_1^2$	$n_1d_1^3$	s_1	S	x_1	x
0	0								
		20	3.1	192	596	0.014	0.014	0.001	0.001
20	6.2								
		20	7.6	1155	8780	0.085	0.099	0.02	0.022
40	9.0								
		10	9.5	903	8573	0.066	0.165	0.020	0.042
50	10.0								
		10	10.7	1145	12,250	0.084	0.249	0.029	0.071
60	11.4								
		5	11.75	690	8111	0.057	0.300	0.019	0.090
65	12.1								
		5	12.85	826	10,609	0.061	0.0361	0.025	0.115
70	13.6								
		2	14.15	400	5666	0.029	0.390	0.013	0.128
72	14.7								
		2	15.35	471	7234	0.035	0.425	0.017	0. 45
74	16.0								
		2	16.75	561	9399	0.041	0.466	0.022	0.167
76	17.5								
		2	18.6	692	12,870	0.057	0.517	0.030	0.197

78	19.7								
		2	21.2	890	18,877	0.065	0.582	0.044	0.241
80	22.7								
		1	24.1	581	14,000	0.043	0.625	0.033	0.274
81	25.5								
		1	28.5	812	23,150	0.060	0.685	0.055	0.329
82	31.5								
		1	65.75	4323	284,240	0.316	1.000	0.670	1.000
83	100								
				13,641	424,355				

Values of s and x are plotted as functions of d in Fig. 2.10.

The surface mean diameter,

$$d_s = \Sigma\left(n_1 d_1^3\right)/\Sigma\left(n_1 d_1^2\right) = 1/\Sigma(x_1 d_1)$$
$$= \int d^3 \, dn / \int d^2 \, dn \quad \text{(Eqs 2.14 and 2.15)}$$

For $0 < d < 10\,\mu\text{m}$, $dn = d \, dd$

For $10 < d < 100\,\mu\text{m}$, $dn = 10^5 d^{-4} \, dd$

$$\therefore d_s = \left(\int_0^{10} d^4 \, dd + \int_{10}^{100} 10^5 d^{-1} \, dd\right) / \left(\int_0^{10} d^3 \, dd + \int_{10}^{100} 10^5 d^{-2} \, dd\right)$$
$$= \left([d^5/5]_0^{10} + 10^5[\ln d]_{10}^{100}\right) / \left([d^4/4]_0^{10} + 10^5[-d^{-1}]_{10}^{100}\right)$$
$$= \left(2 \times 10^4 + 2.303 \times 10^5\right) / \left(2.5 \times 10^3 + 9 \times 10^3\right) = \underline{\underline{21.8\,\mu\text{m}}}$$

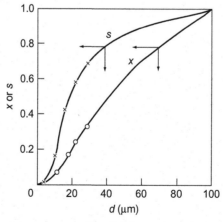

Fig. 2.10
Calculated data for Example 2.2.

The size range of a material is determined by sieving for relatively large particles and by sedimentation methods for particles that are too fine for sieving.

2.2.5 Shape Descriptors

The shape of the particles is known to strongly influence the properties of the material. For example, the densest possible packing fraction of the same size spheres is $\varphi = 0.74$, whereas cubical shape particles can be arranged to get the maximum solids fraction equal to one. Solids volume fraction in a packed bed is strongly influenced by the shape of the particles, which in turn affects the pressure drop through the bed. With increasing departure from the spherical shape of grains, the angle of repose and internal angle of friction was found to increase and the bulk density and flowability were found to decrease by various researchers.[16–18]

Since the shape of the particles affects the bulk material properties significantly, describing the shape of the particle is an important aspect of particle characterisation. Historically, the shape of an irregular shape particle is obtained by its comparison with regular shapes such as sphere or cuboid in 3-D (circle or rectangle in 2-D). The irregular shape of the particle, in these cases, is described either in terms of the dimensions of the known shape or their ratios. Particles with a regular shape can be accurately described by using one or more length scale associated with its shape. Thus, a sphere/cube can be accurately described by only one length (radius/side), a cylinder/cone requires two lengths (radius and height) for accurate description, while a cuboid/ellipsoid requires three lengths for their accurate description (three side/axis lengths). However, an irregular shape particle cannot be unambiguously defined by using only one or a few lengths. Because of this, the particle size is typically expressed in terms of some equivalent diameter. Some of the popular ones were mentioned in Section 2.2.1. While all these different ways of characterising the size of the particle would be identical for a spherical particle, that is not the case for irregular particles. The measured particle size using two different methods will not be identical and these differences in size measurements obtained from different techniques are also utilised to quantify the shape factor of the particles. In this case, the shape factor serves as a proportionality factor for size measurement by two different techniques, such as sedimentation and sieve analysis or image processing and sieve analysis.

Surface-volume conversion shape factors are often used based on the empirical observations, as reported by Herdan,[3] that for various grades of particle size of the same material, the ratio of the mean surface area to square of the mean particle size remains sensibly constant. Similarly, the ratio of the mean particle volume to cube of the mean particle size is also reported to remain sensibly constant. These ratios are commonly used to define the surface and volume shape factors and, for spherical particles, these ratios are equal to π and $\pi/6$, respectively. The values for non-spherical particles will be higher than the values for sphere because the spherical shape has the smallest surface and volume as compared to other shapes.

Dimensionless geometric shape descriptors are the more common way to characterise the particle shape and categorisation in 1-D, 2-D or 3-D shape descriptors based on the associated methods and variables has been suggested.[19, 20] 1-D shape descriptors, such as elongation (length to width ratio) and flatness (width to thickness ratio) are based on the particle lengths in three dimensions; these can be easily measured by a ruler and have been traditionally used by researchers.[21, 22] On the other hand, 3-D shape descriptors, other than sphericity, are not commonly used because they require the full three-dimensional measurement of the particle surface using methods such as 3-D laser scanning or scanning electron microscope microcomputed tomography (SEM micro-CT). A study by Bagheri et al.[20] indicates that while correlating the 3-D particle parameters and shape descriptors (surface area, volume and sphericity) with 1-D parameters is possible, correlations obtained using 2-D parameters should be preferred due to smaller errors. 2-D image parameters are most commonly obtained by digital image processing and are discussed below.

Shape characterisation using digital image processing

Digital image processing methods have removed the subjectivity and difficulty associated with the earlier techniques of the shape measurement, which were primarily based on shape comparison from charts, e.g. those by Krumbein.[23] The digitised image contains all the information about the particle contour in the projected plane and the shape descriptors are calculated as dimensionless ratios of the particle contour parameters. As mentioned in Section 2.2.2, the resolution of the image and the analysing software play important roles in accurate calculation of the particle contour properties. Hence, the shape descriptors obtained from the image analysis are significantly affected by the image resolution and analysis software.

The projected area A_p is one important contour parameter that is calculated as the sum of the pixel area within the boundary of the particle.[24, 25] As mentioned in Section 2.2.2, projected diameter D_p is often reported instead of the projected area. Perimeter is typically obtained as the total length of the boundary region pixels and Duris et al.[13] have mentioned that different softwares use different methods to calculate the perimeter. In addition, the major axis (L) and minor axis (B) of an ellipse having the same second moment as that of the particle are also easily calculated using the available softwares. The horizontal and vertical Feret lengths (F_h and F_v) can be obtained from the rectangular bounding box of the region. Custom written programs can be used to rotate the particle image by small angles to obtain other length scales associated with the particle silhouette, such as minimum (F_{min}), maximum (F_{max}), and average (F_{avg}) Feret length of the particle over all orientations.

Due to the large number of candidates that can serve as the measure of particle sizes, the number of various possible shape descriptors is enormous. For example, considering only the nine measurements with the dimensions of length mentioned above (i.e. five Feret lengths, two elliptical axis lengths, P and D_p), leads to a total of 36 shape descriptors (when no distinction is made between inverse combinations), by considering the ratios of these length scales. It is important to note that the number of length scales associated with the particle contour mentioned above is by no means exhaustive and many more such length scales (and area scales) are not only possible, but also in use in the literature (e.g. Hentschel and Page[26] mention eight Feret lengths, eight chord lengths, two lengths associated with the minimum bounding rectangle, Martin's diameter, perimeter and area). The large number of particle contour parameters that can be used for calculating the shape descriptors is one of the major difficulties of shape characterisation. The task is further complicated due to the lack of a common terminology and notations, which seems to be the case not only across different disciplines, but also within a particular discipline and different shape factors are defined and used. Due to these issues, identifying the most appropriate shape descriptor is not straightforward and no list can claim to be exhaustive. Table 2.2 lists some of the popular shape descriptors obtained from image analysis that are commonly used.

A survey of the available literature shows the lack of agreement on the usage of shape parameters; and it is not clear which shape descriptor is the best in a given situation. Hence, it should be kept in mind that the shape descriptors listed in Table 2.2 are not exhaustive. For example, elongation is commonly used to describe the shape factor of particles in the comminution process (particle size reduced by breakage) and is defined as $(F_{min}/F_{min,90}) -1$, where $F_{min,90}$ is the Feret diameter perpendicular to the F_{min}.[27] Sozer[28] has summarised more than ten ways in which the angularity of the particles has been calculated by researchers and more than thirty shape descriptors that have been used to quantify the particle shape at different levels.

By considering six different powders, spanning a wide range of particle shapes, and performing cluster analysis, Hentschel and Page[26] identified five key (out of 17 considered) length scales and used them to define 10 different shape descriptors. The authors conclude that all the shape descriptors were correlated with either aspect ratio (AR) or the root of the form factor ($\sqrt{FF}$), suggesting that minimum set of shape descriptors to characterise the shape of the powders produced by commercial methods must include these two shape descriptors. Before generalising this result to the shape descriptors listed in Table 2.2, it should be noted that while the elongation and compactness were found to be correlated with the AR and $\sqrt{FF}$, the last three shape descriptors listed in Table 2.2 were not considered in the study. While combining multiple shape descriptors to obtain a single unifying shape descriptor has been tried,[18] it is better to retain the separate values of the different shape descriptors and they should be reported individually.

It must be emphasised that the resolution of the image plays an important role in the calculation of particle properties. It was shown that while the projected diameter, major axis and minor axis are not significantly affected by changes in resolution, the measurement of the perimeter is more sensitive to image resolution.[13] This is because the pixelated boundary (see Fig. 2.5C)

Table 2.2 Some common shape descriptors obtained from image analysis

Expression	Shape Descriptor Name	Notes/Comments
$\dfrac{F_{MBB}}{F_{MBB,90}}$ (or $\tfrac{F_{MBB,90}}{F_{MBB}}$)	Elongation Chunkiness	F_{MBB} and $F_{MBB,90}$ are the sides of the minimum bounding rectangle
$\dfrac{F_{min}}{F_{max}}$ (or $\tfrac{F_{max}}{F_{min}}$)	Aspect ratio	$AR = F_{min}/F_{max}$ is defined by ISO 9276-6:2008 F_{min} and F_{max} are minimum and maximum Feret lengths of the particle over all orientations
$\dfrac{P^2}{4\pi A_p}$ (or $\tfrac{4\pi A_p}{P^2}$) $\dfrac{\pi D_p}{P}$	Circularity Sphericity (in 2D) Form Factor (FF) Root of form factor ($\sqrt{FF}$)	P is perimeter, A_p is projected area of the particle, and D_p is the diameter of circle having area A_p Expression in second line is the square root of the first expression
$\dfrac{4A_p}{\pi F_{max}^2}$ $\dfrac{D_p}{F_{max}}$	Roundness Compactness	A_p is the projected area of the particle, D_p is the diameter of circle having area A_p, and F_{max} is maximum Feret length Expression in second line is the square root of the first expression
$\sqrt{\dfrac{D_{inscr}}{D_{circum}}}$	Inscribed circle sphericity Riley circularity Irregularity parameter	D_{inscr} is the diameter of the maximum inscribed circle and D_{circum} is the diameter of the minimum circumscribed circle
$\dfrac{P_{convex}}{P}$	Convexity	P_{convex} is the convex hull perimeter (i.e. stretched length of a rubber band surrounding the particle) and P is the perimeter. Serves as a measure of particle roughness
$\dfrac{A_p}{A_{convex}}$	Solidity	A_{convex} is the convex hull area (i.e. area bounded by a rubber band surrounding the particle) and A is the projected area. Serves as a measure of overall particle concavity

of the particles becomes more and more zigzag with the increasing resolution and different software programs use different algorithms to estimate the length of this zigzag line, which is reported as the perimeter. Thus, the circularity and convexity measurements will be most sensitive to the image analysis software used as compared to all other shape descriptors listed in Table 2.2. Based on the available literature,[13, 29] it appears that the projected particle diameter to the pixel size ratio should be greater than twenty for reliable measurements of the relevant particle properties and the shape descriptors.

Mathematically more sophisticated and complex ways of characterising the particle shape have also been proposed in the literature.[30–34] However, due to the complexity involved in the analysis, these approaches have not yet gained wide acceptance, at least in industry. One such approach is based on the Fourier analysis of the geometric radial signature waveforms of the two-dimensional profile of irregular shape particles. The radial distance of the boundary from the centroid is measured as a function of the angle from a reference line. This signature waveform is then analysed using Fourier analysis and the first few (typically five) harmonics are used to get the overall shape of the particles whereas high-frequency harmonics provide information about the texture.[30, 31] The second approach makes use of fractal geometry to obtain an index of ruggedness or boundary texture of irregular particles. In this approach, the extended irregularities on the boundary of an irregularly shaped particle are likened to a convoluted line, and the fractal dimension of this line is calculated to characterise the textured profiles of the particle. The analysis by Hentschel and Page[26] suggests that the inverse of the fractal dimension is correlated with the root of the form factor (FF). More details can be found in the works of Mandelbrot,[32] Flook,[33] and Berube et al.[34]

2.2.6 Particle Density and Hardness

In addition to size and shape, the particle density and porosity can also play an important role in many applications. Particle density is defined as the ratio of the mass of the particles to the volume of the particles. While the measurement of the mass is relatively straightforward, the measurement of particle volume is not easy due to their irregular shape and the presence of open or closed pores in the particle. The particle volume measurement is typically done in three different ways, leading to *true*, *apparent*, and *effective* density of the particles.

When the particle volume measurements consider only the *true volume* occupied by the solid and exclude the volume of the open and closed pores of the particles, the measured density is called the *true density* of the particles, which is the same as the density of the solid material.

The *apparent density* accounts for the total *apparent volume* of the particle, which includes both the volume occupied by the solid and that of the closed pores; the volume of the open pores is not included. The apparent density is measured by liquid or gas pycnometry, which are essentially liquid or gas displacement methods that are commonly used in industry.

The *effective density* of the particle accounts for the *effective volume* of the particles, as would be seen by a gas flowing past the particle. It includes the true volume occupied by the solid as well as the volume of the open pores and the closed pores. Due to these different ways of accounting for the particle volume, the true density of the particle is the largest and the effective density of the particle is the smallest among the three particle densities. For nonporous particles, the effective and true densities will be identical. The particle densities are also reported in terms of the specific gravity by dividing the particle density with that of water. It is important to note that these particle densities are different from the bulk density of the material, which depends on the relative packing of the particles with respect to each other and is discussed in Chapter 3.

Particle hardness is another important property that determines the rate of wear of the equipment if the particles are harder than the equipment material of construction. Particles of significantly lower hardness, on the other hand, will be subject to attrition and breakage while handling and processing, which can not only influence the process and product quality but may even cause a dust explosion. Cahn and Lifshin[35] report that the term 'hardness' has meaning only in terms of specific tests, which range from a simple, practical and commonly used semiquantitative Mohs scratch test to sophisticated macro- or micro-indentation hardness tests. The Mohs hardness is measured on a scale of 1 (very soft, corresponding to talc that can be powdered by a finger) to 10 (hardest, corresponding to diamond). A material of a given Mohs hardness can only scratch materials of lower Mohs hardness and the intervals between different scales are not of the same magnitude.

Indentation hardness tests involve penetrating an indenter of a specific shape made of a particular material over the surface of interest. For example, a spherical steel ball is used in the Brinell hardness tester whereas a square-based diamond pyramid is used in the Vickers hardness tester. A hardness number is calculated using the load and the surface area of the impression left by the indenter using appropriate formulae. While conversion tables are available for some specific materials to obtain relationships among the major types of tests, these are limited in their applicability and may lead to substantial errors if used outside the stated scope.[35]

2.3 Sampling of Solids

In a bulk material handling industry, a typical amount of the particulate solid processed may range from a few hundred kilograms to a few hundred tons per day. The measurement of the material properties, however, is done using a miniscule amount of the material. Reliable measurement of any relevant property of the particulate solids (e.g. particle size distribution, shape descriptors, bulk density, angle of repose, etc.) thus requires that the material used in the measurement is truly representative of the material. A poor, nonrepresentative sample may lead to erroneous conclusions about the material, which can have many undesired consequences. In this section, only the key points to keep in mind while sampling bulk solids are highlighted, and the interested reader is referred to the works of Allen,[1] Crosby and Patel,[36] and Smith.[37]

The error in sampling can result either from the primary sampling itself or from the subsequent division of the sample, and hence care must be taken to avoid both types of errors. Even in an ideal random mixture, the quantitative distribution will be subject to random fluctuations. This statistical error will be present even in ideal sampling, and hence, cannot be eliminated. However, this statistical error can be estimated using mathematical techniques. It can also be reduced by increasing the mass of the total physical sample and by collecting several random increments from the lot and combining them to form the sample.

Another major source of error in sampling the bulk solids can occur because of the segregation of the material. Free-flowing, noncohesive particulate solids, typically segregate based on the particle size and density, and hence, the distribution of the particles depends upon the flow history. The segregation of bulk solids occurs due to the relative motion of the grains with respect to each other and is commonly observed in many situations, for example, while feeding the material into containers or on the belt conveyors. Segregation can also occur due to the shocks and vibrations while the material is transported from one place to another (see Section 3.10). Because the bulk solids typically have a distribution of particle sizes, segregation is inevitable for free-flowing, noncohesive solids, and it becomes worse as the particle size distribution becomes wider. For example, a conical heap of bulk solids contains a larger fraction of fines in the centre while coarse grains concentrate to the periphery. Similarly, grain containers transported via road typically have coarser grains near the free surface due to the vibrations and shocks during the road trip. Due to these reasons, static sampling in such cases will lead to sampling errors and hence should be avoided. As a general rule, if possible, dynamic sampling is to be preferred over static sampling.

2.3.1 Static Sampling

Cohesive, nonflowing powders and sticky/moist material show weak segregation. Therefore, such materials may be passed through a mixer or shaken in a container using different modes of shaking; static sampling may be sufficient to provide a representative sample for laboratory tests. Static sampling consists of three basic methods: scooping, thieving, and coning and quartering. Scooping is essentially taking the sample out from the surface of heaped material using a scoop while thieving makes use of a device having one or more sampling chambers and the material is collected from the bulk. The operator can control the opening and closing of the sampling chambers. Another static sampling method used for reducing the amount of the material for sampling is the coning and quartering method. In this method, the material is poured in the form of a conical heap and is flattened at the top. This flattened heap is cut into four equal parts, and one of the parts is used for further analysis. This process is repeated until the desired amount of the material is obtained (see Fig. 2.11). The skill of the operator is an important factor for accurate sampling using this method as the symmetry of the heap during the flattening and cutting is critical for reducing sampling errors. Both scoop sampling and coning and quartering are commonly used for sample reduction, primarily due to the negligible cost associated with the methods. However, as reported by Allen and Khan,[38] the associated errors with static sampling are significantly higher as compared to the dynamic methods (see Table 2.3).

Because only a small fraction of the heap is accessible in static sampling methods, the basic premise of the good sampling that every particle must have an identical chance of being included in the sample is not fulfilled. In such cases, satisfactory sampling can be achieved

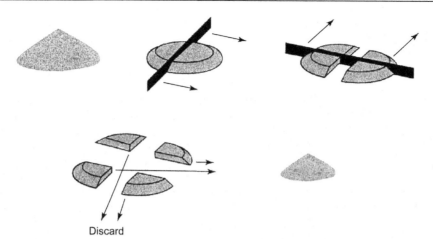

Discard

Fig. 2.11

Coning and quartering method. *Reproduced from Gerlach RW, Dobb DE, Rabb GA, Nocerino JM. Gy sampling theory in environmental studies:1. Assessing soil splitting protocols.* Chemometrics *2002;***16***:321–8.*

Table 2.3 Standard deviation for different sampling methods (Allen and Khan[38])

Sampling Method	Standard Deviation (%)
Random variation for theoretically perfect sampler	0.076
Rotary sample divider	0.125
Chute splitter	1.01
Table splitting	2.09
Scoop sampling	5.14
Coning and quartering	6.81

either during the transfer of the heaped material by the conveyor belt or during loading/unloading. If the loading/unloading operation is carried out using grabs/shovels, each grab should be treated as a defined lot. These lots should be numbered sequentially, and then random sampling of these numbers should be done to get the sampling lot of the material. In case of the loading/unloading by a constant speed conveyor belt, the sampling lots should be identified with time and a random selection is to be done for taking samples. The material from the conveyor should be collected from a short length of the belt covering the entire width.

2.3.2 Dynamic Sampling

The first golden rule for sampling bulk solids is that they should be sampled when in motion. Only in scenarios, where dynamic sampling is not possible, should static sampling be used. Because static sampling for free-flowing, noncohesive bulk solids leads to sampling errors due to segregation, dynamic sampling techniques should be used for such materials. The second golden rule of sampling is that the entire stream of the material should be preferably sampled for short durations of time instead of sampling only part of the stream for a long duration. Conforming to this rule, when samples are obtained from a falling stream, multiple increments should be obtained for short durations of time by a cutter whose length should be enough to collect the material from the entire stream. The depth of the cutter should be large enough to avoid complete filling and width should be greater than three times the size of the largest particle present in the stream. The cutter should ideally move with a constant speed and should have parallel edges at the openings, which should be large enough to avoid bridging/jamming. While the linear traversing cutters (cutter moves in a straight line through the stream) are the most accurate samplers, rotational traversing cutters (cutter moves in an arc through the stream) and stationary cutters (cutter does not move and cuts only a part of the stream) are often used in industry because of their low cost and simplicity of design and operation.

Typically, the amount of sample collected from the stream is large and needs to be reduced by using table sampling, chute splitting or a rotary sample divider. Usually, the table sampling is done over an inclined plane that has a series of prisms and holes in different rows. These are used to direct and discard the part of the stream and retain only a fraction of the stream as the sample which is collected at the end. One such table splitter (16:1 sample splitter, courtesy Sepor Inc.) is shown in Fig. 2.12.

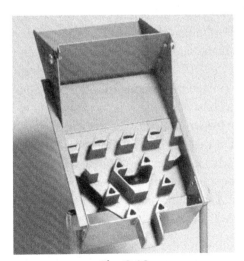

Fig. 2.12
Table sample splitter. *Used with permission from Sepor Inc., Wilmington, CA, USA.*

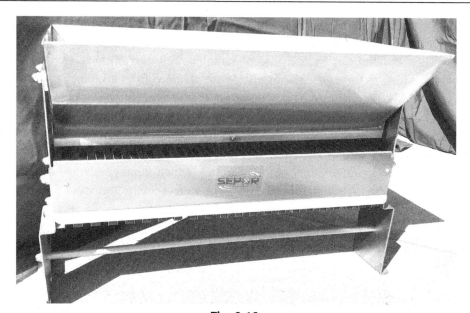

Fig. 2.13
Chute splitter. *Used with permission from Sepor Inc., Wilmington, CA, USA.*

Because the parts of the stream are removed sequentially, errors compound at each separation leading to loss of accuracy. In addition, the reliability over a uniformly distributed initial feed and lack of complete mixing after each separation further contribute to the relatively less accurate sample splitting. A chute splitter, also known as a Jones riffle sampler, is shown in Fig. 2.13 (image provided by Sepor Inc.) This sample splitter has a V-shaped box that consists of a series of identical chutes of rectangular cross-sections which alternately feed containers on both sides of the equipment. The width of the chute should be greater than three times the largest particle size in the sample. Satisfactory sample division can be obtained by this method when used with great care. The rotary sample divider (or spinning riffle), an example of which is shown in Fig. 2.14 (image provided by Sepor Inc.), consists of multiple collection chambers that are rotated at a constant speed while the material to be sampled is poured either from a hopper or a vibratory feeder at a constant flow rate. In order to obtain reliable and accurate results, it is important that the hopper should be filled in a way that no or little segregation occurs (Section 3.10), and the mass flow of the material occurs through the hopper (see Section 3.6). Rotary sample divider is the most robust method of sample reduction that conforms to the golden rules of sampling and is considered to be more accurate than the chute splitting and table splitting (see Table 2.3).

In brief, the importance of following the golden rules of sampling cannot be overemphasised and Gerlach et al.[39] have shown that a rotary sample divider made of paper cones in a laboratory environment outperforms both the coning and quartering and grab sampling by scoops; it was found to be comparable to the commercially available chute splitters.

Fig. 2.14

Rotary sample splitter. *Used with permission from Sepor Inc., Wilmington, CA, USA.*

2.4 Nomenclature

		Units in SI System	Dimensions in **M, L, T**
AR	Aspect ratio	–	–
A_{convex}	Convex hull area	m^2	**L^2**
A_P	Projected area of the particle	m^2	**L^2**
B	Minor axis	m	**L**
D_{circum}	Diameter of minimum circumscribed circle	m	**L**
D_{inscr}	Diameter of maximum inscribed circle	m	**L**
D_P	Diameter of circle having area A_P	m	**L**
d	Particle size, diameter of sphere or characteristic dimension of particle	m	**L**
d_l	Length mean diameter	m	**L**
d_s	Surface mean diameter (Sauter mean)	m	**L**
d_v	Volume mean diameter	m	**L**
d_{10}	Size below which 10% of the size distribution lies	m	**L**

d_{50}	Size below which 50% of the size distribution lies	m	**L**
d_{80}	Size below which 80% of the size distribution lies	m	**L**
d_{90}	Size below which 90% of the size distribution lies	m	**L**
d'_l	Mean length diameter	m	**L**
d'_s	Mean surface diameter	m	**L**
FF	Form factor	–	–
$F_{MBB,}$ $F_{MBB,90}$	Side of minimum bounding rectangle	m	**L**
F_{avg}	Average Feret length of the particle	m	**L**
F_h	Horizontal Feret length	m	**L**
F_{min}	Minimum Feret length of the particle	m	**L**
F_{max}	Maximum Feret length of the particle	m	**L**
F_v	Vertical Feret length	m	**L**
k	Constant in Eq. (2.2)	s^{-1}	T^{-1}
k_1	Constant depending on particle shape (volume/d^3)	–	–
k_2	Constant depending on particle shape (Area/d^2)	–	–
L	Major axis length of particle	m	**L**
n	Number of particles per unit mass	kg^{-1}	M^{-1}
n_1	Number of particles of characteristic dimension d_1, constituting a mass fraction x_1	–	–
P	Perimeter	m	**L**
P_{convex}	Convex hull perimeter	m	**L**
t	Time	s	**T**
w	Mass of particles on screen at time t	kg	**M**
x	Proportion of particles smaller than size d	–	–
ρ_s	Density of solid or particle	kg/m^3	ML^{-3}
ψ	Sphericity of particle	–	–
φ	Packing fraction	–	–

References

1. Allen T. *Particle size measurement, volume-1, Powder sampling and particle size measurement*. London: Chapman and Hall; 1997.
2. Allen T. *Particle size measurement, volume-2, Area and pore size determination*. London: Chapman and Hall; 1997.
3. Herdan G. *Small particle statistics*. London: Butterworths Scientific; 1960.
4. BS410. *Specification for test sieves*. London: British Standards Institution; 1962.
5. Dallavalle JM. *Micromeritics: the technology of fine particles*. New York: Pitman Publishing Corporation; 1948.
6. Rose JW, Cooper JR. Technical data on fuel, In: *British National Committee, World energy conference, London*; 1977.

7. Green RSB, Murphy PJ, Posner AM, Quirk JP. Preparation techniques for electron microscopic examination of colloidal particles. *Clay Clay Miner* 1974;**22**:185.
8. BS893. *The method of testing dust extraction plant and the emission of solids from chimneys of electric power stations.* London: British Standards Institution; 1940.
9. Bohren CF, Huffman DR. *Absorption and scattering of light by small particles.* New York: Wiley & Sons; 1983.
10. Xu R. *Particle characterization: light scattering methods.* New York: Kluwer Academic; 2002.
11. Schafer M. Digital optics: some remarks on the accuracy of particle image analysis. *Part Part Syst Charact* 2002;**19**:158–68.
12. Zeidan M, Jia X, Williams RA. Errors implicit in digital particle characterisation. *Chem Eng Sci* 2007;**62**:1905–14.
13. Đuris M, Arsenijevic Z, Jacimovski D, Kaluderovic Radoicic T. Optimal pixel resolution for sand particles size and shape analysis. *Powder Technol* 2016;**302**:177–86.
14. Bond FC. Some recent advances in grinding theory and practice. *Br Chem Eng* 1963;**8**:631–4.
15. Bond FC. Costs of process equipment. *Chem Eng Albany* 1964;**71**:134.
16. Ridgway K, Rupp R. The effect of particle shape on powder properties. *J Pharm Pharmacol* 1969;**21**:30S–39S.
17. Wong LW, Pilpel N. The effect of particle shape on the mechanical properties of powders. *Int J Pharm* 1990;**59**:145–54.
18. Podczeck F, Miah Y. The influence of particle size and shape on the angle of internal friction and the flow factor of unlubricated and lubricated powders. *Int J Pharm* 1996;**144**:187–94.
19. Blott SJ, Pye K. Particle shape: a review and new methods of characterization and classification. *Sedimentology* 2008;**55**:31–63.
20. Bagheri GH, Bonadonna C, Manzella I, Vonlanthen P. On the characterization of size and shape of irregular particles. *Powder Technol* 2015;**270**:141–53.
21. Heywood H. Symposium on particle size analysis. *Inst Chem Eng* 1947;**14**:.
22. Kaye BH, Clark GG, Liu Y. Characterizing the structure of abrasive fine particles. *Part Part Syst Charact* 1992;**9**:1–8.
23. Krumbein WC. Measurement and geological significance of shape and roundness of sedimentary particles. *Sediment Petrol* 1941;**11**:64–72.
24. Olson E. Particle shape factors and their use in image analysis—part 1: theory. *GXP Compliance* 2011;**15**:85–96.
25. Olson E. Particle shape factors and their use in image analysis—part II: practical applications. *GXP Compliance* 2011;**15**:.
26. Hentschel ML, Page NW. Selection of descriptors for particle shape characterization. *Part Part Syst Charact* 2003;**20**:25–38.
27. Hogg R, Turek ML, Kaya E. The role of particle shape in size analysis and the evaluation of comminution processes. *Part Sci Technol* 2004;**22**:355–66.
28. Sozer ZB. *Two-dimensional characterization of topographies of geomaterial particles and surfaces.* (Ph.D. dissertation), Atlanta, GA: Georgia Institute of Technology; 2005.
29. Kroner S, Domenech Carbo MT. Determination of minimum pixel resolution for shape analysis: proposal of a new data validation method for computerized images. *Powder Technol* 2013;**245**:297–313.
30. Wang L, Wang X, Mohammad L, Abadie C. Unified method to quantify aggregate shape angularity and texture using Fourier analysis. *J Mater Civ Eng* 2005;**17**(5):498–504.
31. Zhanwei Y, Fuguo L, Peng Z, Bo C. Description of shape characteristics through Fourier and wavelet analysis. *Chin J Aeronaut* 2014;**27**(1):160–8.
32. Mandelbrot BP. *Fractals, forms, chance and dimension.* San Francisco, CA: WH Freeman; 1977.
33. Flook AG. The use of dilation logic on the quantimet to achieve fractal dimension characterisation of textured and structured profiles. *Powder Technol* 1978;**21**:295–8.
34. Berube D, Jebrak M. High precision boundary fractal analysis for shape characterization. *Comput Geosci* 1999;**25**:1059–71.
35. Cahn RW, Lifshin E. *Concise Encyclopaedia of materials characterization.* Oxford: Pergamon Press; 1993.

36. Crosby NT, Patel I. *General principles of good sampling practice*. The Royal Society of Chemistry; 1995.
37. Smith PL. *A primer for sampling solids, liquids, and gases based on the seven sampling errors of Pierre Gy*. Philadelphia, PA: Society for Industrial and Applied Mathematics; 2001.
38. Allen T, Khan AA. Critical evaluation of powder sampling procedures. *Chem Eng* 1970;**238**:108–12.
39. Gerlach RW, Dobb DE, Rabb GA, Nocerino JM. Gy sampling theory in environmental studies: 1. Assessing soil splitting protocols. *Chemometrics* 2002;**16**:321–8.

Further Reading

Beddow JK. *Particle characterization in technology, volume-1 application and micro-analysis*. Boca Raton, FL: CRC Press; 1984.

Beddow JK. *Particle characterization in technology, volume-2 morphological analysis*. Boca Raton, FL: CRC Press; 1984.

Bohren CF, Huffman DR. *Absorption and scattering of light by small particles*. New York: Wiley & Sons; 1983.

Dallavalle JM. *Micromeritics: the technology of fine particles*. New York: Pitman Publishing Corporation; 1948.

ISO9276-6:2008(E). *Representation of results of particle size analysis*. Geneva: International Organization for Standardisation; 2008.

Mcglinchy D. *Characterisation of bulk solids*. Oxford: Blackwell; 2005.

Rhodes M. *Introduction to particle technology*. Chichester: Wiley; 1998.

Rumpf H. *Particle technology*. London: Chapman and Hall; 1990.

Smith PL. *A primer for sampling solids, liquids, and gases based on the seven sampling errors of Pierre Gy*. Philadelphia, PA: Society for Industrial and Applied Mathematics; 2001.

CHAPTER 3

Particulate Solids in Bulk: Storage and Flow

3.1 Scope and General Characteristics

Particulate solids or bulk solids are an assembly of discrete solid particles interspersed with a fluid, generally, air. They are distinguished from the suspensions and fluidised beds because the constituent particles in particulate/bulk solids are typically in substantial contact with neighbouring particles. While the interaction between the fluid and the particles may have a considerable effect on the behaviour of the bulk material, in the case of air as the interstitial fluid, the influence of the fluid properties becomes negligible for grain sizes larger than a few hundred microns. The cohesive/adhesive forces also become less important compared to the particle weight beyond this size, and the behaviour of the granular material is determined almost entirely by the normal and frictional contact forces between the individual grains. Bulk solids with particle sizes $>100\,\mu m$ are typically referred to as *granular solid*. For size beyond 5 mm, the term *broken solid* is commonly used. Even such an oversimplified system, excluding the interaction with interstitial fluid and cohesion/adhesion effects, poses significant challenges for theoretical understanding and modelling, and is an active area of research.

Particulate solids in the size range of 0.1–$100\,\mu m$ are typically referred to as powders. Further subdivisions in terms of fine powders (10–$100\,\mu m$), superfine powders (1–$10\,\mu m$), and hyperfine/ultrafine powders (0.1–$1\,\mu m$) are also common. At these scales, the cohesion/adhesion due to van-der-Waals forces, humidity, and the air drag become important and cannot be ignored. In general, the properties of solids in bulk are a function of the properties of the individual particles, including their shapes, sizes, size distribution, and of the way the particles interact with one another. Thus in addition to the size, shape, density, hardness, and surface roughness of the particles, small variations in moisture and prior loading/compaction history can lead to significant changes in the bulk behaviour. Establishing a cause and effect relationship between bulk and particle level properties is an extremely difficult task due to the complexities involved.[1]

Due to their poor flowability, fine and cohesive bulk solids often show much more complex behaviours, such as time consolidation and caking. The scope of the discussion in this chapter will be primarily limited to relatively free-flowing, noncohesive bulk solids with appropriate observations about the applicability or limitations for cohesive material. For these materials, theoretical analysis for some simple problems is possible and will be demonstrated by taking examples available in the literature.

Coulson and Richardson's Chemical Engineering. https://doi.org/10.1016/B978-0-08-101098-3.00003-2

3.1.1 Frictional Nature of Bulk Materials

Interparticle forces in the bulk material are frictional in nature, and the magnitude of the frictional force between the grains can sustain any value between zero and a limiting value. This limiting value depends upon the normal force between the grains. It is reached when the contacting particles are on the verge of sliding with respect to each other and has interesting macroscopic consequences. Two examples illustrating the effect of the limiting value of the friction force are explained below to emphasise the difference in the behaviour of the bulk solids, as opposed to fluids in common scenarios.

Example 3.1

Stress saturation in a bin containing bulk solids: Janssen (1895) measured the stress at the bottom of a silo filled with corn and showed that with the pouring of the corn in the silo, stress at the bottom does not increase, but saturates to a maximum value. Following Janssen's approach of force balance on a slice of the material and accounting for the sidewall friction force, show that the normal stress tends to saturate as the distance from the free surface increases.

Solution

Since the original analysis by Janssen had some inconsistencies, following Nedderman,[2] we provide the refined Janssen-Walker-Nedderman solution for stress variation in a container of constant cross section area, A, and perimeter, P. Consider a granular medium of bulk density ρ_b in a container of constant cross section area A as shown in Fig. 3.1A. Let $\bar{\sigma}_{zz} = \int \sigma_{zz} dA/A$ be the mean axial stress over the cross-section. Similarly, let $\bar{\tau}_w$ and $\bar{\sigma}_w$ be the values of the wall shear stress and wall normal stress, averaged over the perimeter, P, respectively. Assume that the material at the wall is on the verge of sliding downward, so that the shear force due to friction at the wall is directed upward. For a slice of the material of thickness dz located at distance z from the free surface, the force balance in the z direction leads to (Fig. 3.1B)

$$\bar{\sigma}_{zz}A\big|_z - \bar{\sigma}_{zz}A\big|_{z+dz} + \rho_b(Adz)g - \bar{\tau}_w P dz = 0. \tag{3.1}$$

Since the material is on the verge of sliding, friction will be fully mobilised at the walls, and assume $\bar{\tau}_w = \mu_w \bar{\sigma}_w$, where μ_w is the friction coefficient between the material and the container wall. Further assuming that the wall-normal stress is proportional to the axial stress at the wall, i.e., $\bar{\sigma}_w = K \sigma_{zz,w}$, where the axial stress at the wall is related to the mean axial stress by a Walker's distribution factor, $\mathfrak{D}$, so that $\sigma_{zz,w} = \mathfrak{D}\bar{\sigma}_{zz}$. With these assumptions, the force balance equation becomes

$$\frac{d\bar{\sigma}_{zz}}{dz} + \frac{(P\mu_w K\mathfrak{D})}{A}\bar{\sigma}_{zz} = \rho_b g. \tag{3.2}$$

Assuming that both K and $\mathfrak{D}$ are constants, along with the boundary condition that the axial stress at the free surface is zero, i.e., $\bar{\sigma}_{zz}\big|_{z=0} = 0$, the above equation can be integrated to obtain the variation of the axial stress (shown in Fig. 3.1C)

$$\bar{\sigma}_{zz} = \rho_b g \lambda \left(1 - e^{-\frac{z}{\lambda}}\right), \tag{3.3}$$

where $\lambda = \frac{A}{\mu_w K \mathfrak{D} P}$ is the characteristic length. Using the hydraulic mean diameter $D_h = 4A/P$ (equal to the diameter for a cylindrical container), the characteristic length becomes $\lambda = \frac{D_h}{4\mu_w K \mathfrak{D}}$. Since both K and $\mathfrak{D} \sim O(1)$,[2] for a typical value of $\mu_w \sim 0.5$, this characteristic length turns out to be

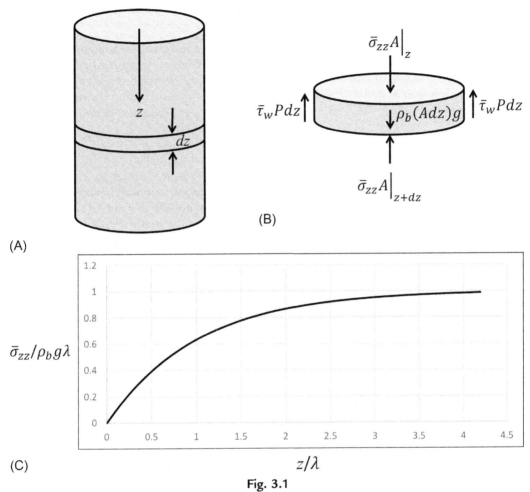

Fig. 3.1

Stress saturation in a container filled with bulk granular material. (A) The z-axis points downward with $z=0$ being the free surface. (B) Forces acting on a slice of the material. (C) Variation of the mean axial stress with the distance from the top surface.

comparable to the hydraulic diameter, i.e. $\lambda/D_h \sim O(1)$. The axial stress (or the pressure) at the base saturates to a value of $\bar{\sigma}_{zz} \approx \rho_b g \lambda$ far away from the surface. The mobilisation of the wall friction leads to a pressure screening at the base since part of the material weight is supported by frictional side walls. The axial stress reaches 95% of the saturated value for $z/\lambda = 3$, i.e. within a distance of three hydraulic diameters from the free surface. In contrast, due to their inability to sustain shear stresses, liquids exhibit linear variation of the pressure with depth. This is because the last term in the LHS of Eq. (3.1) (and second term containing μ_w in LHS of Eq. 3.2) is absent for the case of liquids, and hence, the pressure varies linearly with depth. Due to this linear pressure variation, higher thickness of the container at the base is needed in the containers used for storing liquids. No such requirement exists for grain storage silos, and hence, silos with larger H/D ratio are not uncommon in industry, whereas the smaller aspect ratio cylinders are more commonly used for liquid storage.

Example 3.2

Free surface shape of a granular solid in a rotating bucket[3,4]: As a second example, consider a granular material in a cylinder that is spinning about its axis. It is well known that for a liquid in a rotating cylinder, the free surface takes the shape of a parabola. However, the shape of the free surface of the granular material at a given rotation speed is not unique, and depends on the initial shape of the surface. This can be understood from the analysis given below.

Solution

Consider a cylinder with a radius R filled up to a height H with a granular solid, e.g., sand (Fig. 3.2A). Assume the free surface is initially flat when the cylinder is not rotating. The cylinder is then rotated about its axis, gradually increasing the angular speed, eventually attaining a steady angular velocity Ω. As the cylinder rotates, the surface takes the shape of a curved surface at equilibrium. Consider an element of the material at the flattened surface at a distance r from the axis. Owing to the symmetry along the axis, the variations in the azimuthal direction are neglected and the problem is solved in a two-dimensional r-z plane. In a reference frame attached to the rotating material, the centrifugal force in the positive r direction will be balanced by the frictional force F_f in the negative r direction, i.e., $mr\Omega^2 = F_f$. From Coulomb's law of friction $F_f \leq \mu mg$, where μ is the friction coefficient of the material. Hence for $r \leq r_c = \mu g / \Omega^2$, the centrifugal force will not be sufficient to overcome the friction force and the material will retain the flat shape.

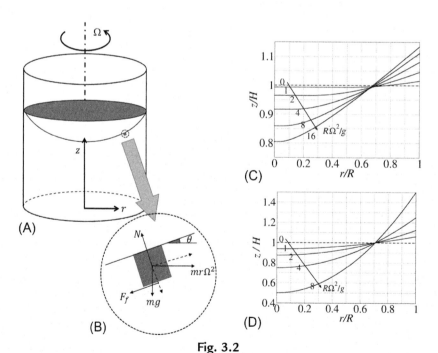

Fig. 3.2
(A) Granular material in a rotating bucket, (B) free body diagram on a material element at the surface, (C) free surface height for the case of sand with $\mu = 0.5$. Note that the surface retains its flat shape near the axis. (D) Parabolic free surface shape obtained in case of liquids.

For $r>r_c$, the free body diagram of the material element is shown in Fig. 3.2B. A force balance along the normal and tangential direction on an element at the free surface gives,

$$F_f = m\left(r\Omega^2\cos\theta - g\sin\theta\right), \tag{3.4}$$

$$N = m\left(r\Omega^2\sin\theta + g\cos\theta\right), \tag{3.5}$$

where $\tan\theta = \frac{dz}{dr}$ is the local slope of the surface at a radial distance r from the axis. Since $F_f \leq \mu N$, from Eq. (3.4) and Eq. (3.5), we have

$$\tan\theta = \frac{dz}{dr} \geq \frac{r\Omega^2/g - \mu}{\mu r\Omega^2/g + 1}. \tag{3.6}$$

Since this expression is valid only for $r>r_c=\mu g/\Omega^2$, the slope of the free surface is greater than zero, i.e., $\tan\theta = \frac{dz}{dr} > 0$, beyond the region $r>r_c$. This slope is the minimum permissible value of the slope of the free surface in the rotating bucket. For the limiting condition when the material is on the verge of sliding, the friction force is fully mobilised, and we have $F_f = \mu N$, so that

$$\frac{dz}{dr} = \frac{r\Omega^2/g - \mu}{\mu r\Omega^2/g + 1}. \tag{3.7}$$

From Eq. (3.7), it is clear that as $\Omega \to 0$, the slope of the surface becomes constant and negative, so that the angle from the horizontal is $\theta = -\tan^{-1}(\mu)$, i.e., equal to the angle of internal friction, taking the shape of a conical pile. For large values of $r\Omega^2/g$, the slope tends to $1/\mu$ and the angle of the surface approaches $90-\tan^{-1}(\mu)$. Eq. (3.7) can be integrated with the condition that at $r=r_c$, $z=z_c$, to obtain the equation of the free surface as:

$$z = z_c + \frac{r-r_c}{\mu} - \frac{\mu^2+1}{\mu^2\Omega^2/g}\log\left(\frac{1+\mu r\Omega^2/g}{1+\mu^2}\right). \tag{3.8}$$

The height z_c can be calculated by using mass conservation. Assuming that the bulk density of the material (sand) does not change significantly due to the compression by the centrifugal force in the radial direction, the total volume of the material should be equal to the initial volume of the material, i.e.,

$$\pi r_c^2 z_c + 2\pi \int_{r_c}^{R} z\, r\, dr = \pi R^2 H. \tag{3.9}$$

Fig. 3.2C shows the variation of the free surface height z with the radial coordinate r for sand with friction coefficient $\mu=0.5$ for different values of $R\Omega^2/g$, for an initial fill height of $H=4R$. The parabolic shape profiles for the fluid case are also shown for comparison in Fig. 3.2D. For $R\Omega^2/g$ $=8$, the dip at the centre is $\sim15\%$ of the original height in the sand, in contrast to a dip of $\sim50\%$ in the case of liquid. Notice that the sand surface retains its initial flat shape in the region close to the axis. If the sand is poured in the container using a funnel, the initial shape of the sand in the container will be in the form of a conical heap. In that case, the region near the centre will not be flat.[4]

It is worth emphasising here that both the examples considered above highlight the macroscopic consequence of the limiting value of friction force, which occurs only when the material is on the verge of failure, that is the grains are on the verge of sliding past each other. When this condition is not met, which may be the case in a general static bulk solid, indeterminate value of the friction force will be operative. Due to this reason, the stress distribution in a static bulk material cannot be determined[5] without making simplifying assumptions such as those made in Examples 3.1 and 3.2. Laboratory experiments, as well as simulations, show that to obtain reliable data confirming Janssen's theoretical analysis require the supporting plate at the base of the cylindrical silo to be moved by a small amount so that the assumption of full friction mobilisation at the walls is realised.[6] Similarly, the slope of the surface given by Eq. (3.7) is the value of the slope for the case of full friction mobilisation towards the wall; as per Eq. (3.6), higher slopes are also possible. For the other limiting case, when the fully mobilised friction acts towards the centre, analysis similar to Example 3.2 shows that the slope of the free surface should be less than or equal to the maximum value of $\frac{dz}{dr} \leq \frac{r\Omega^2/g + \mu}{1 - \mu r\Omega^2/g}$. Thus an infinite number of shapes of the free surface are possible for the sand in rotating bucket, the only constraint being the slope of the surface $\frac{dz}{dr}$ must remain within the bounds $\frac{r\Omega^2/g - \mu}{\mu r\Omega^2/g + 1} \leq \frac{dz}{dr} \leq \frac{r\Omega^2/g + \mu}{1 - \mu r\Omega^2/g}$. Similarly, the existence of different bulk densities (Section 3.1.2) and various angles of repose (Section 3.3) in bulk solids can also be attributed to the existence of the range of the frictional forces.

In general, particulate solids present considerably greater problems than fluids in storage, in removal at a controlled rate from storage, and when introduced into vessels or reactors where they become involved in a process. Important operations relating to systems of particles include storage in hoppers, flow through orifices and pipes, and metering of flows. It is frequently necessary to reduce the size of particles, or alternatively, to form them into aggregates or sinters. Although there has recently been a considerable amount of work carried out on the properties and behaviour of solids in bulk, there is still a considerable lack of understanding of all the factors determining their behaviour.

3.1.2 Voidage, Bulk Density and Flowability

One of the most important characteristics of a particulate solid is its voidage, the fraction of the total volume of the bulk solid which is made up of the free space between the particles and is filled with fluid, typically air. The voidage ε can be defined as the ratio of the void volume to the total volume occupied by the bulk solid. A related complementary property to the voidage is the *packing fraction* (or *solids fraction*) ϕ of the bulk solid. It is defined as the ratio of the *effective volume* (see Section 2.2.6) occupied by the solid particles to the total volume occupied by the bulk solid. The total volume is equal to the sum of the *effective volume* occupied by the particles and the volume of the voids. Note that the void volume represents volume *between* the solid particles and does not account for the pore volume *within* the particles. By definition, both the voidage (ε) and the solids fraction (ϕ) are less than unity and are interrelated, i.e. $\phi = 1 - \varepsilon$. To avoid ambiguity in the voidage value, it is better to explicitly state whether the voidage value includes the pore volume inside the particles or not. Clearly, a low voidage corresponds to a high density of packing of the particles. The packing fraction depends not only on the physical properties of the constituent particles, including shape and size distribution, but also on the way in which the particulate mass has been introduced to its particular location. In general, isometric particles, which have approximately the same linear dimension in each of the three principal directions, will pack more densely than long thin particles or plates, in case of random packing of the particles. This will not be the case, for example, if perfectly aligned rectangular bars are filled into a box and will lead to very high values of packing factions. The more rapidly material is poured on to a surface or into a vessel, the more densely it will pack. If it is then subjected to vibration, further consolidation may occur.

Since the packing fraction determines the bulk density of the material, bulk solids do not have a unique bulk density. If ρ_f is the density of the fluid occupying the void space between the particles, and ρ_p is the effective density of the solid particle, then the bulk density of the material is calculated as $\rho_b = \phi\rho_p + (1-\phi)\rho_f$. In case of interstitial fluid being air, $\rho_f \sim 10^{-3}\rho_p$, so that $\rho_b \cong \phi\rho_p$. For cohesionless, monodisperse spherical particles, the value of packing fraction can range from the random loose packing fraction $\phi_{RLP} \sim 0.56$ to random close packing fraction $\phi_{RCP} \sim 0.64$. However, the packing fraction for fine cohesive powders can be much smaller than these values.[7] Packing fraction of a material having a wider-sized distribution will be significantly higher than that having nearly monodisperse particles. For example, the packing fraction of a binary mixture of spherical grains depends upon both the composition and the size ratio; for a given size ratio of 1:0.09 the packing fraction can attain a maximum value of ~ 0.81 for a volume concentration of $\sim 70\%$ large particles.[6] Packing fraction also affects the tendency for agglomeration of the particles, and it critically influences the resistance which the material offers to the percolation of fluid through it—as, for example, in filtration as discussed in Chapter 10.

Measurement of the bulk density is typically done by completely filling a container of known volume until the material overflows, and then levelling it with a straight edge. The weight of the container filled with the material is measured using a balance and the weight of the material is obtained by subtracting the weight of the empty container. Bulk density is calculated as the ratio of the weight of the material to the volume of the container. This value of the bulk density, obtained by pouring the material into the container, is commonly called the 'poured bulk density'. The poured density depends upon the filling mechanism and the height of fall and standards for these have been defined in the relevant ASTM, British, and ISO standards. For practical usage, often the container is either tapped or dropped from a small height several times on a surface to obtain the 'tapped bulk density'. Typically, the tapped bulk density is higher than the poured bulk density. Another, commonly used bulk density for fine powders, is 'aerated bulk density', which is measured by fluidising a known mass of powder in the column and noting the powder volume after slowly reducing the fluidising gas flowrate to zero. For materials which are subjected to significant loading, *compacted bulk density* under a known loading condition is also reported.

The ratio of the tapped bulk density (ρ_{tapped}) to aerated bulk density (ρ_{aerated}) is reported as the often quoted 'Hausner ratio' $H = \rho_{\text{tapped}}/\rho_{\text{aerated}}$,[8] which serves as a qualitative measure of the powder flowability. Use of the poured (or loose) bulk density in place of aerated bulk density while calculating Hausner ratio is also common. These bulk densities are used to define the Carr's compressibility index CI which relates to the Hausner ratio as $\text{CI}(\%) = (1 - 1/H) \times 100$. This index is used to rank the powders in different flowability ranges. Low values of CI indicate good flowing powders and large values of CI suggest poorly flowing cohesive powders.[8] A more common way of numerical characterisation of flowability is in terms of the flow factor, which is the ratio of consolidation stress to unconfined yield stress (the compression strength of the powder in uniaxial loading). A large value of this ratio indicates low yield stress, and hence, better flowability, of the powder.

3.2 Agglomeration

Because it is necessary in processing plant to transfer material from storage to processing equipment, it is important to know how the particulate material will flow. If a significant amount of the material is in the form of particles smaller than 10 μm or if the particles deviate substantially from isometric form, it may be inferred that the flow characteristics will be poor. If the particles tend to agglomerate, poor flow properties may again be expected. Agglomeration arises from interaction between particles, as a result of which they adhere to one another to form clusters. The main mechanisms giving rise to agglomeration are:

(1) *Mechanical interlocking.* This can occur particularly if the particles are long and thin, in which case large masses may become completely interlocked.
(2) *Surface attraction.* Surface forces, including van der Waals' forces, may cause substantial bonds between particles, particularly where particles are very fine (<10 μm), with the result that their surface per unit volume is high. In general, freshly formed surface, such as that resulting from particle fracture, gives rise to high surface forces.
(3) *Plastic welding.* When irregular particles are in contact, the forces between the particles will be borne on extremely small surfaces and the very high pressures developed may give rise to plastic welding.
(4) *Electrostatic attraction.* Particles may become charged as they are fed into equipment and significant electrostatic charges may be built up, particularly on fine solids.
(5) *Effect of moisture.* Moisture may have two effects. Firstly, it will tend to collect near the points of contact between particles and give rise to surface tension effects. Secondly, it may dissolve a little of the solid, which then acts as a bonding agent on subsequent evaporation.
(6) *Temperature fluctuations* bring about changes in particle structure and to greater cohesiveness.

Because interparticle forces in very fine powders make them very difficult to handle, the effective particle size is frequently increased by agglomeration. This topic is discussed in Section 5.4.

3.3 Angle of Repose

One of the commonly measured property of the particulate solids is its *angle of repose,* which often serves as a rapid method of assessing the behaviour of a particulate mass. If a solid is poured from a nozzle on to a plane surface, it will form an approximately conical heap and the angle between the sloping side of the cone and the horizontal is called the angle of repose. When this is determined in this manner, it is sometimes referred to as the *poured angle of repose.* In practice, the heap will not be exactly conical and there will be irregularities in the sloping surface. In addition, there will be a tendency for large particles to roll down from the top and collect at the base, thus giving a greater angle at the top and a smaller angle at the bottom.

In order to avoid the influence of the height of free fall, an approach that is sometimes used (though not easy to follow) is to move the nozzle up slowly so as to keep the distance between the nozzle and heap top constant (Fig. 3.3A). The other approach involves the flow from a fixed height so that the heap keeps rising and eventually blocks the flow of the material (Fig. 3.3B). The poured angle of repose may exhibit slight dependency on the properties of the flat surface, and hence, a plane sheet with a layer of particles from the powder glued on it is also used instead of the plane surface. Alternatively, to minimise the influence of the flat surface, the material can also be poured into a container of suitable size and the poured angle of repose is obtained from the height of the heap and the radius of the cylinder (Fig. 3.3C).

A more reliable way of measuring the angle of repose is the 'drained angle of repose' where the material is drained from a flat bottom container and the angle of the surface of the remaining material from the horizontal is measured. A cuboid shape box can be used as the container and

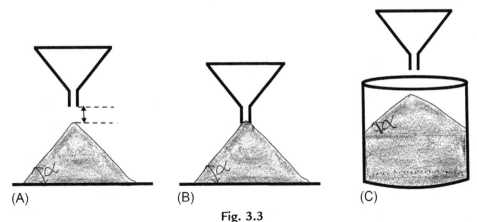

(A)   (B)   (C)

Fig. 3.3
Schematic and definition of poured angle of repose. See the text for details.

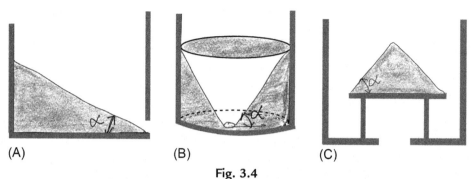

Fig. 3.4

Schematic and definition of drained angle of repose. See the text for details.

the material is drained by removing one side of the box (Fig. 3.4A). Alternatively, the draining of the material can be done from a hole at the centre of a cylindrical container, leaving the material in form of a crater (Fig. 3.4B). It is imperative that the size of the hole should be large enough to ensure the free flow of the material. To eliminate the possibility of the effect of the container walls, another method is also used where the material is poured into a cylinder, which contains a circular tray supported over a hole at the base of the cylinder. The height of the cylinder should be reasonably larger than the platform height so that material can be filled well above the circular platform. The material is then drained by opening the hole at the centre and the height of the heap of the material retained on the circular tray of known radius is used to calculate the drained angle of repose (Fig. 3.4C).

Another angle of repose that is commonly reported is the 'dynamic angle of repose', which is measured as the angle of the free surface from the horizontal in a slowly rotating (~4–6 rpm) horizontal drum. For free-flowing, non-cohesive grains, the free surface shape is fairly linear and the dynamic angle of repose is estimated from the slope of this line (Fig. 3.5A). This angle depends upon the rotating speed of the cylinder and seems to vary linearly with the rpm for rotation speeds in the range of 4–12 rpm. Rotation speeds lower than 3 rpm are typically avoided since the angle of the free surface keeps oscillating; starting from a minimum angle, it increases with the rotation of the cylinder to a maximum angle and then falls back to the minimum angle.

To some extent, the angle of repose depends upon the way the heap is prepared, and various parameters seem to affect the measurement, such as the height of the fall in the case of the poured angle of repose, and the size of the draining hole in the case of the drained angle of repose. Similarly, the dynamic angle of repose measured at the sidewall of a rotating cylinder is found to be few degrees higher than the angle measured at the middle of the cylinder.[9] However, if the standardised procedures are followed, a reasonably consistent value for a given method can be obtained. In general, the values of the angle of the repose measured in a laboratory using different methods show a bit of variation and the differences of few degrees between the methods is not uncommon.

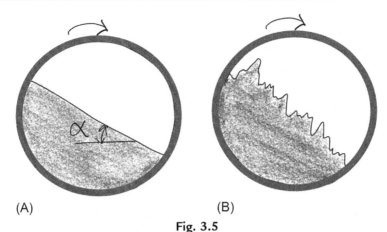

Fig. 3.5

Sketch of the free surface shape in a rotating cylinder for (A) free-flowing coarse material (B) poorly flowing cohesive material.

The angle of repose measurements generally tend to be more reliable for noncohesive, free flowing, and homogeneous materials having a narrow particle size distribution as opposed to the cohesive, poorly flowing, non-homogeneous materials having a broad size distribution. For example, the dynamic angle of repose measurements for a highly cohesive material in a rotating tumbler are not meaningful due to the highly irregular, sawtooth like shape of the free surface (Fig. 3.5B). Generally, material, which contains no particles smaller than 100 µm, has a low angle of repose. Powders with low angles of repose tend to pack rapidly to give a high packing density almost immediately. If the angle of repose is large, a loose structure is formed initially, and the material subsequently consolidates if subjected to vibration. Hence, the angle of repose can also be used as a rough guide for indicating powder flowability. Angles of repose vary from about 20° for very free-flowing solids, to about 60° for very cohesive solids with poor flow characteristics. In extreme cases of highly agglomerated solids, angles of repose up to nearly 90° or even higher can also be encountered.

Sometimes, the drained angle of repose of the powder is used to measure the *angle of friction*. The powder is contained in a two-dimensional bed with transparent walls, and is allowed to flow out through a slot in the centre of the base. A triangular wedge of material in the centre flows out leaving stationary material at the outside. The angle between the cleavage separating stationary and flowing material, and the horizontal is then termed as the angle of friction. Mcglinchey[10] states that while such values of the internal angle of friction can provide a reasonable estimate for ground area requirements for a stockpile or surcharge in a storage vessel, they should not be used to estimate the wall angle required for the converging section of a hopper. As mentioned in Section 3.4.4, shear testers should be used to obtain more reliable values of the angle of friction since they permit evaluation of the coefficient of friction in a variety of loading conditions.

3.4 Resistance to Shear and Tensile Forces

A particulate mass may offer a significant resistance to both shear and tensile forces, and this is specially marked when there is a significant amount of agglomeration. Even in non-agglomerating powders there is some resistance to relative movement between the particles and it is always necessary for the bed to dilate, that is, for the particles to move apart to some extent, before internal movement can take place. The greater the density of packing, the higher will be this resistance to shear and tension.

3.4.1 Coulomb Yield Criterion

Consider a block of consolidated particulate material on which a force in the vertical direction is applied (Fig. 3.6A). While the material is able to sustain small amounts of force, when the applied force reaches a critical value, the material breaks into two parts, which slide past each other (Fig. 3.6B). The plane along which the material breaks is known as the failure, yield, or slip plane. The value of the yield shear stress τ_y (tangential force per unit area) on the yield plane is found to be independent of the extent or the rate of the deformation and depends only on the normal stress σ (normal force per unit area). For many materials, a linear relationship between the shear and normal stresses exists and such materials are referred to as the ideal Coulomb material. The *Coulomb yield criterion* for such materials can be written as

$$|\tau_y| = \mu\sigma + c \tag{3.10}$$

where μ is the friction coefficient of the material and c is the cohesion of the material. The friction coefficient is often reported in terms of the angle of internal friction $\delta = \tan^{-1}(\mu)$. The values of μ and c are obtained experimentally and typical values of the coefficients of friction (angle of internal friction) lie in the range of about 0.3 ($\delta \sim 17°$) for smooth spherical particles to 1.5($\delta \sim 56°$) for rough, angular particles. The cohesion c is negligible for many coarse, noncohesive materials and can reach values of up to $50\,kN/m^2$ for stiff clay.[2] Cohesion is

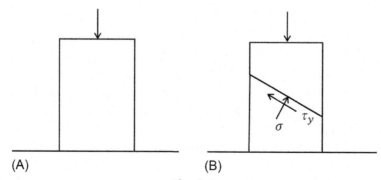

Fig. 3.6
Failure of a block of ideal Coulomb material under (A) vertical loading leads to (B) failure along a slip plane.

sensitive to the moisture content of the material and keeps varying frequently due to the change in weather conditions.

This simplest model describing the response of the bulk solids to stress is known as the 'ideal Coulomb model'. Just as the Newtonian fluid model serves as a foundation for the study of viscous flows, this oversimplified model serves as a starting point for studying bulk solids.[2] Ideal coulomb materials, by definition, do not permit any motion along the failure plane for $|\tau| < |\tau_y|$. As soon as the shear stress reaches the value of $|\tau| = |\tau_y| = \mu\sigma + c$, a slip plane is formed, and the material is said to be in a state of incipient failure. If the plane of failure is known, the Coulomb yield criterion is able to predict the value of the stress along the plane for which the material will undergo failure or yielding. Consider, for example, the formation of a heap by pouring a dry granular material. Since the failure plane in this case is along the surface of the heap, the maximum permitted angle for the heap will be $\tan^{-1}(\mu)$ so that the yield criterion is fulfilled along the heap surface.

However, the Coulomb yield criterion does not provide any information about the orientation of the plane of failure in case of Fig. 3.6B. The analysis of the material stress state using Mohr circle given in the next section helps us overcome this difficulty. Note that the ideal Coulomb model does not say anything about the extent of yielding, which could be infinitesimal in the case of cracks or very large in the case of steady deformation. Values of $|\tau| > |\tau_y|$ are not permitted in this model. Many of the real-life scenarios, however, deal with the flow of particulate solids, which are beyond the scope of the Coulomb model. Flow of bulk solids, as they flow through orifice/hopper, is discussed in Sections 3.5 and 3.6.

3.4.2 Concept of Stress and Mohr Circle

By definition, a classical fluid cannot sustain shear stress and will have the same normal stress in all directions at any point in stationary condition (or in rigid body motion). Bulk solids, however, can sustain shear stress (up to a critical value), and can have different normal stresses in different directions. Since the stress can be understood as force per unit area acting on a plane, it depends on the force vector as well as on the orientation of the area on which the force is acting. Hence, the material element for a given force loading will have different stresses depending upon the area under consideration. Consider an element of a bulk solid in equilibrium under the influence of the forces acting on it, shown in Fig. 3.7. A vector force $\vec{F}^{(1)}$, having components $F_x^{(1)}$ and $F_y^{(1)}$, acts on plane ABCD and a vector force $\vec{F}^{(2)}$ having components $F_x^{(2)}$ and $F_y^{(2)}$ acts on plane ADEH; equal and opposite forces act on the planes EFGH and BCFG so that the material element is in equilibrium. For the sake of simplicity, the forces in the z-direction are not considered. The normal stress on the plane ABCD (area A_x) normal to the x axis is denoted as $\sigma_{xx} = \lim_{A_x \to 0} \left(F_x^{(1)}/A_x \right)$, whereas the normal stress on the plane ADEH (area A_y) normal to the y-axis is denoted as $\sigma_{yy} = \lim_{A_y \to 0} \left(F_y^{(2)}/A_y \right)$. Similarly,

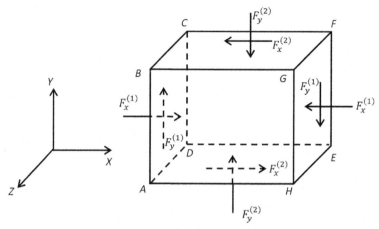

Fig. 3.7
Bulk solids element in equilibrium under the influence of forces acting on it.

the shear stress on the plane ABCD is denoted as $\sigma_{xy} = \lim_{A_x \to 0}\left(F_y^{(1)}/A_x\right)$ and that on plane ADEH is denoted as $\sigma_{yx} = \lim_{A_y \to 0}\left(F_y^{(2)}/A_y\right)$.

Note that the force, being a vector quantity, is denoted with one subscript to indicate the direction in which it acts. The stress, however, needs two subscripts to indicate the two directions involved, and is a tensor quantity. The force vector in 2-D (3-D) is represented by its two (three) components, whereas the stress tensor in 2-D (3-D) is represented by its four (nine) components. Our discussion in this chapter will be restricted to the 2-D case with plane stresses.

We follow the notation of the Bird, Stewart and Lightfoot (BSL),[11] where the first subscript denotes the area normal direction and the second subscript indicates the direction of the force. Further, as shown in Fig. 3.8A, the stress σ_{ij} is force in the positive j-direction on a plane of lesser i, in other words, on a plane having outward normal in negative i-direction. Other choices for the pair of directions are also possible and indeed are commonly used; the benefit of this notation being that the often-encountered compressive stresses in the bulk solids are positive (as in Example 3.1).

The stress state of a material in 2-D (Fig. 3.8A) is thus denoted as $[\sigma_{xx}, \sigma_{xy}; \sigma_{yx}, \sigma_{yy}]$. The application of the angular momentum balance suggests the two shear stresses be equal, i.e. $\sigma_{xy} = \sigma_{yx} = \tau$, i.e. the stress tensor be symmetric. Due to the symmetric nature of the stress tensor, we use the symbol τ to denote the shear stress and drop both the subscripts. Further, one of the repeated subscripts in the normal stress is dropped and the stress tensor in 2D is represented as $[\sigma_x, \tau; \tau, \sigma_y]$ as shown in Fig. 3.8B, which is equivalent to the stress state shown in Fig. 3.8A.

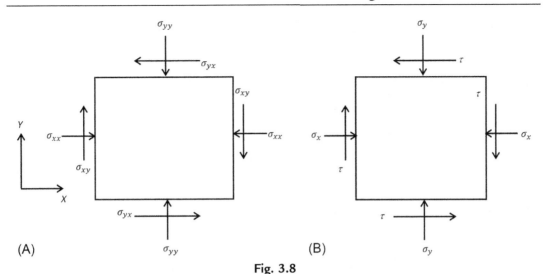

Fig. 3.8

(A) Stress components on a two-dimensional element according to the sign convention of BSL.[11]
(B) Simplified stress notation after dropping the repeated index for normal stress.

Note that our choice of coordinate system (i.e. x and y directions in 2D) is arbitrary and other directions for the coordinate axes are also equally valid choices. Moreover, mother nature and the material need not have any regard for our arbitrary imposed choice of the axes; the gravitational force vector on an object does not depend on the choice of the axes. However, our choice of the axes does affect the components of the force vector, which change with the rotation of the axes and some particular choices of the axes are advantageous since they simplify the analysis involved in solving the problem. In the same fashion, the material under the influence of the vertical load in Fig. 3.6 will fail the same way, no matter what set of axes we prefer to use. However, the components of the stress tensor will change with the rotation of the axes and some particular choices will simplify the analysis. For example, the failure analysis of the ideal Coulomb material would be simpler, if we choose our axes in the directions parallel and normal to the failure plane. However, the direction of the failure plane is not known *a-priori*. In addition, the stress components along the desired axes also need to be known.

Consider a material under the influence of various loads shown in Fig. 3.9A. Assume that the stress state in XY coordinates $[\sigma_x, \tau; \tau, \sigma_y]$ at a point P is known. In other words, if the material is cut along the planes shown by broken lines in Fig. 3.9A, the normal and shear stress on the plane will be as shown in the Fig. 3.9A. Following the above-mentioned notation, the stress state at point P, denoted by a differential rectangular element of length dx and width dy, is shown in Fig. 3.9B. Since the stress state depends upon the orientation of the area normal vector, the value of the normal and shear stresses will be different along the planes (shown by broken lines) in Fig. 3.9C. For this choice of coordinate axes, let the stress components at the point P be $[\sigma_x', \tau'; \tau', \sigma_y']$ as shown in Fig. 3.9D. Given the components of the stress tensor for a

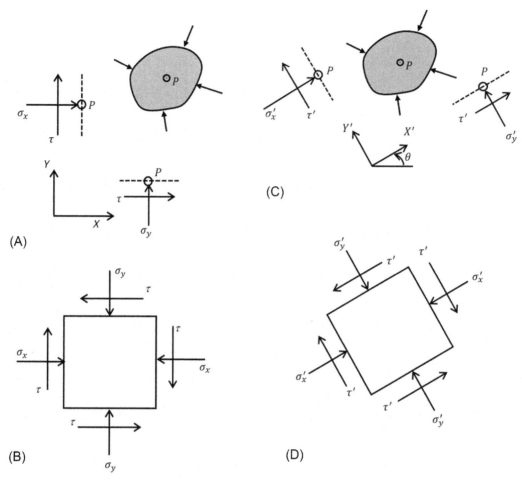

Fig. 3.9

Stress at any point *P* in the material depends on the orientation of the plane. See the text for details.

particular choice of the coordinate axes (*XY* axes in this case), the Mohr circle analysis is used to obtain the components of the stress tensor $[\sigma_x', \tau'; \tau', \sigma_y']$ for any arbitrary choice of the coordinate axes (*X'Y'* in this case). Thus our objective is to find the stress components at the point *P* $[\sigma_x', \tau'; \tau', \sigma_y']$ in the *X'Y'* coordinate axes that is rotated by an angle θ in the counterclockwise direction, given the stress components $[\sigma_x, \tau; \tau, \sigma_y]$.

The stress in the rotated frame can be obtained by writing a force balance on a triangular block *ABC* (Fig. 3.10) obtained from cutting the material shown in Fig. 3.9B along a plane so that the normal on the plane *AC* points in the *X'* direction. The stresses on the faces of the triangular element are shown in Fig. 3.10. Force balance on the element *ABC* in *X'* direction gives

$$(\sigma_x')(1.AC) = (\sigma_y \sin\theta + \tau\cos\theta)(1.BC) + (\sigma_x\cos\theta + \tau\sin\theta)(1.AB) \qquad (3.11)$$

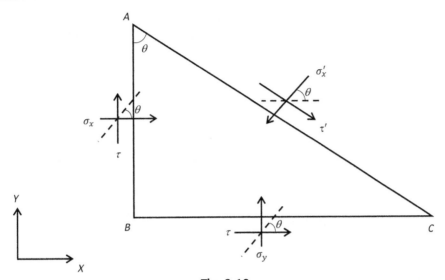

Fig. 3.10

Force balance on a triangular element to obtain the stresses in rotated frame.

Here we have assumed that the thickness of the block in the z direction is unity. Let θ be the angle between AB & AC. Using the fact that $AB = AC\cos\theta$ and $BC = AC\sin\theta$, the above equation simplifies to give the value of the normal stress $\sigma_x' = \sigma_x\cos^2\theta + \sigma_y\sin^2\theta + 2\tau\sin\theta\cos\theta$. Similarly, a force balance on the element ABC in Y' direction gives

$$(\tau')(1.AC) = (\sigma_y\cos\theta - \tau\sin\theta)(1.BC) + (\tau\cos\theta - \sigma_x\sin\theta)(1.AB) \tag{3.12}$$

Eqs (3.11) and (3.12) can be rearranged to obtain the expressions,

$$\sigma_x' = \frac{(\sigma_x + \sigma_y)}{2} + \frac{(\sigma_x - \sigma_y)}{2}\cos 2\theta + \tau\sin 2\theta, \tag{3.13}$$

$$\tau' = -\frac{(\sigma_x - \sigma_y)}{2}\sin 2\theta + \tau\cos 2\theta. \tag{3.14}$$

Complete determination of the stress state requires the knowledge of σ_y', which can be obtained by doing a similar analysis on a block whose normal points in the Y' direction. Alternatively, since the expression given by Eq. (3.13) is valid for general θ, the expression for σ_y' can be obtained by replacing θ by $90 + \theta$, giving

$$\sigma_y' = \frac{(\sigma_x + \sigma_y)}{2} - \frac{(\sigma_x - \sigma_y)}{2}\cos 2\theta - \tau\sin 2\theta \tag{3.15}$$

Eqs (3.13)–(3.15) represent the stress state in the coordinate system rotated by an angle θ with respect to the given coordinate system, where positive θ is measured in counterclockwise direction. It can be shown that the locus of the points $(\sigma_x', -\tau')$ and (σ_y', τ') in the $\sigma - \tau$ plane is a circle centred at $(\frac{\sigma_x + \sigma_y}{2}, 0)$ having a radius $r = \sqrt{\frac{(\sigma_x - \sigma_y)^2}{4} + \tau^2}$. Consequently, the points corresponding to the stress in the rotated axes can be represented graphically on a circle, often referred to as Mohr's circle.

The stress state in the given coordinates is represented by a pair of points A_1 and B_1 on σ-τ plane. Point A_1 is represented by $(\sigma_x, -\tau)$ and point B_1 is represented as (σ_y, τ). The line A_1B_1 joining these two points is one of the diameters of the Mohr circle as shown in Fig. 3.11A with point $C(\frac{\sigma_x + \sigma_y}{2}, 0)$ being the centre of the circle. The stress on a material element corresponding to pair A_1B_1 is shown in Fig. 3.11B. The stress state in a frame rotated by an angle θ in counterclockwise direction can be obtained by rotating the line A_1B_1 by an angle 2θ in counterclockwise direction about centre C so that the new stress state is represented by a pair of points A_2B_2. The coordinates of A_2 give the value of $(\sigma_x', -\tau')$ and that of point B_2 give the value of (σ_y', τ'). The stress on a material element in this rotated frame is then represented by Fig. 3.11C. As evident from the Fig. 3.11A, the centre of the Mohr circle is at $(\frac{\sigma_x + \sigma_y}{2}, 0)$ and the

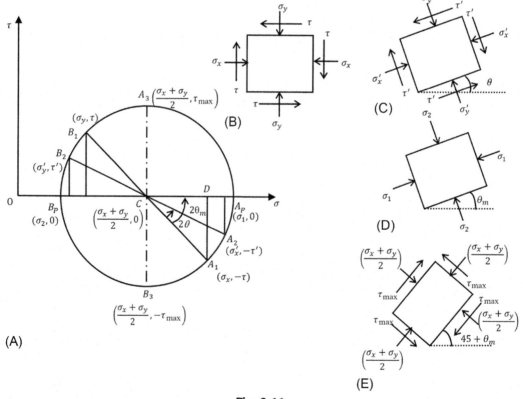

Fig. 3.11
Stress representation using Mohr circle. See the text for details.

radius is $r = A_1C = \sqrt{CD^2 + A_1D^2}$, i.e. $r = \sqrt{\frac{(\sigma_x - \sigma_y)^2}{4} + \tau^2}$. As the angle between the given coordinates and rotated coordinates increases from zero (in counterclockwise direction), the value of the shear stress decreases ($\tau' < \tau$). One of the normal stress increases, while the other one decreases, with increase in θ. At $\theta = \theta_m$, normal stresses achieve their maximum and minimum values and the shear stress becomes zero, i.e. the material element experiences only normal stresses in this orientation, which occurs at an angle $\theta_m = \frac{1}{2}\tan^{-1}\left(\frac{A_1D}{CD}\right)$, i.e.

$\theta_m = \frac{1}{2}\tan^{-1}\left(\frac{2\tau}{\sigma_x - \sigma_y}\right)$. The two normal stresses σ_1 and σ_2 are called the principal stresses; the larger value σ_1 is known as the major principal stress and the smaller value σ_2 is known as the minor principal stress (see Fig. 3.11D). From Fig. 3.11A, the value of the major principal stress

is obtained as $OA_P = OC + CA_P = \left(\frac{\sigma_x + \sigma_y}{2}\right) + r$, i.e. $\sigma_1 = \left(\frac{\sigma_x + \sigma_y}{2}\right) + \sqrt{\frac{(\sigma_x - \sigma_y)^2}{4} + \tau^2}$. The minor

principal stress is given as $\sigma_2 = \left(\frac{\sigma_x + \sigma_y}{2}\right) - \sqrt{\frac{(\sigma_x - \sigma_y)^2}{4} + \tau^2}$ and the maximum possible value of the

shear stress $\tau_{max} = r$, i.e. $\tau_{max} = \sqrt{\frac{(\sigma_x - \sigma_y)^2}{4} + \tau^2}$.

Example 3.3

Stress state of a material element is given as $\sigma_x = 11$ kPa, $\sigma_y = 5$ kPa and $\tau = 4$ kPa. Find the major and minor principal stresses as well as the maximum possible value of the shear stress acting on the material element. At what angle of the plane will these stresses be acting?

Solution

The expressions given above can be used to directly obtain the values of σ_1, σ_2 and, τ_{max}. However, we prefer to use the construction of the Mohr circle to obtain these so that the application and strength of this geometrical approach can be illustrated.

Step 1: We proceed with indicating the stress state in the $\sigma - \tau$ plane using points A_1 (11,−4) and B_1 (5,4). Join the points A_1B_1 as shown in Fig. 3.12A.

Step 2: Draw a circle with A_1B_1 as the diameter (Fig. 3.12B). Centre of the Mohr circle is point C (8, 0).

Step 3: Obtain the coordinates of A_P and B_P which represent the major and minor principal stresses. From Fig. 3.12B, the radius of the Mohr circle is $CA_1 = \sqrt{(11 - 8)^2 + 4^2} = 5$. The major principal stress $\sigma_1 = OA_P = OC + CA_P = OC + CA_1$, i.e. $\sigma_1 = 8 + 5 = 13$ kPa. The minor principal stress $\sigma_2 = OB_P$. Since $OB_P = OC - CB_P = OC - CA_1$, $\sigma_2 = 8 - 5 = 3$ kPa. Thus the coordinates of A_P and B_P are (13, 0) and (3, 0), respectively. The stress state on the material element is shown in Fig. 3.12B.

Step 4: The angle between the axis of given stress state A and the major principal stress is θ_m, which can be calculated from $\triangle A_1CD$. Since the angle on the Mohr circle is twice the angle between the axis, $\angle A_1CD = 2\theta_m$, using $\tan 2\theta_m = A_1D/CD$, i.e. $\tan 2\theta_m = \frac{4}{\sqrt{5^2 - 4^2}} = \frac{4}{3} \Rightarrow \theta_m = \frac{1}{2}\tan^{-1}(4/3) = 26.6°$. Thus a plane whose normal is inclined at an

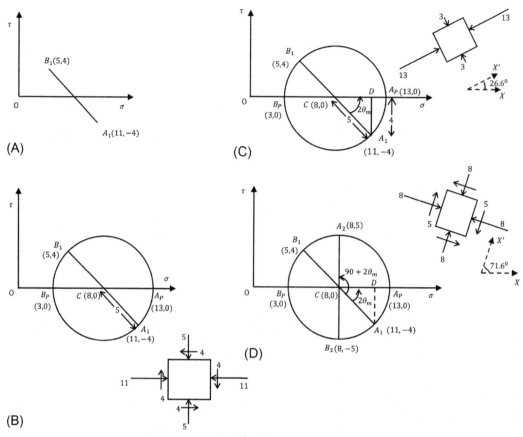

Fig. 3.12
Construction of Mohr circle for Example 3.3. See text for details.

angle of 26.6° (in the counterclockwise direction) from the X-axis, experiences only normal stress of 13 kPa and does not experience any shear stress. Similarly, a plane having a normal incline at an angle of 26.6° (in the counterclockwise direction) from the Y-axis experiences only normal stress of 3 kPa and does not experience any shear stress.

Step 5: The maximum shear stress corresponds to the top most points of the circle, i.e. points A and B whose coordinates can be determined as (8,5) and (8,−5). The normal stress at these points is $\sigma_x' = \sigma_y' = 8$ kPa and the shear stress $\tau' = \tau_{max} = 5$ kPa. This value of shear stress is observed on a plane whose normal is inclined at an angle of
$\frac{1}{2}\angle A_1CA_2 = \frac{1}{2}(90 + 2\theta_m) = 45 + \theta_m = 45 + 26.6 = 71.6°$ from the X-axis. The same value of the shear stress also occurs on a plane inclined at 71.6° from the Y-axis. Note that the coordinates of point A_2, is $(\sigma_x', -\tau)$ as per the notation, and hence, $\tau = -5$, which means that the shear stress on the plane is negative as per our convention, which is indicated in the schematic next to point A_2.

It is important to remember that the stress state in the Mohr circle representation is denoted by a pair of points at diametrically opposite ends, and hence, a rotation of 180° on the Mohr circle, which corresponds to a physical rotation of 90° of the coordinate axes, does not change the stress state. Using the stress state of the previous example, let us plot the schematic stress diagram on the material element after a rotation of 180°, so that points A_1 (11,−4) and B_1 (5,4), shown in Fig. 3.13A are interchanged and represented by the pair A_2 (5,4) and B_2 (11,−4) as shown in Fig. 3.13B.

Fig. 3.13B shows the schematic of stress on an element using the rotated axes X' and Y' by 90° in counterclockwise direction. Point $A_2(\sigma_x', -\tau')=(5,4)$ now corresponds to normal stress in X' direction to be 5 kPa and the shear stress to be −4 kPa. By convention, shear stress on a plane ab with outward normal in negative X' direction will be positive in positive Y' direction. Thus shear stress of −4 kPa points in the negative Y' direction. Similarly, point $B_2(\sigma_y', \tau')=(11, -4)$ corresponds to compressive normal stress of 11 kPa and shear stress of −4 kPa. The outward normal on plane ad (Fig. 3.13B) points in positive Y' direction, hence positive value of shear stress points in negative X' direction and negative values of the shear stress points in positive X' direction. Comparing the stress on the material element in Fig. 3.13A and B shows that both the cases correspond to identical stresses on the material element, as expected.

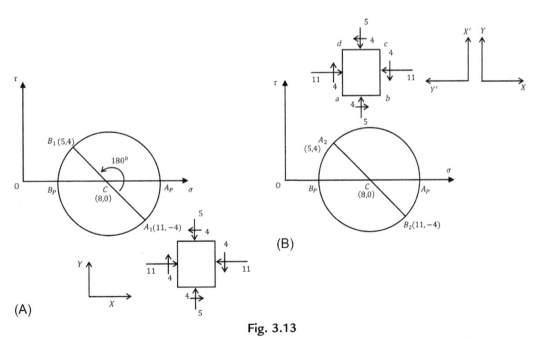

Fig. 3.13
Same pair of points represent identical stress on the material element. See the text for details.

Example 3.4

Consider the stress state given in Example 3.3. At what orientations of the axes, will the stress state of the material element be given as $\sigma_x' = 12$ kPa, $\sigma_y' = 4$ kPa and $\tau' = 3$ kPa? At what orientations of the axes, the shear stress becomes -3 kPa?

Solution

The stress state of Example 3.3 is represented by points A_1 and B_1 in Fig. 3.14. The stress state given in this example corresponds to points A_2 $(12, -3)$ and B_2 $(4, 3)$. The angle between the lines A_1C and A_2C is $2\theta_1$ where θ_1 corresponds to the angle by which the X-axis needs to be rotated in counterclockwise direction. The numerical value of θ_1 can be easily calculated to be equal to $[\tan^{-1}(4/3) - \tan^{-1}(3/4)]/2 = 8.13°$. The stress state on the material element is shown next to point A_2 in Fig. 3.14. Note that the rotation of line A_1B_1 around the centre C in the counterclockwise direction reduces the magnitude of the shear stress. At points A_PB_P, the shear stress becomes zero. Further rotation leads to a change in the direction of the shear stress and it also becomes negative. The shear stress value of -3 kPa corresponds to point A_3 $(12, 3)$ and B_3 $(4, -3)$. Let θ_2 be the angle between the given and rotated axes. The angle between CA_1 and CA_3 is $2\theta_2 = \angle A_1CA_3 = \angle A_1CA_2 + \angle A_2CA_3 = 2\theta_1 + \angle A_2CA_3$. Since $\frac{1}{2}\angle A_2CA_3 = \angle A_3CA_P$ is equal to $\tan^{-1}(3/4) = 36.86°$, using the value of θ_1, we get $\theta_2 = 45°$.

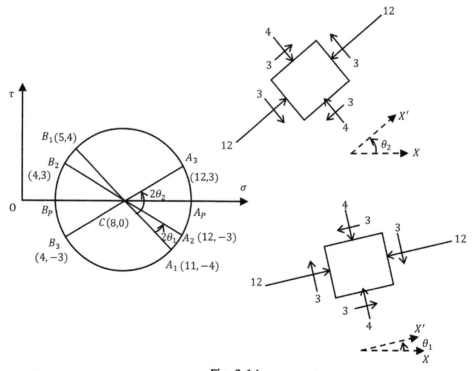

Fig. 3.14
Mohr circle and stress states corresponding for Example 3.4. See the text for details.

To plot the stress on the material element, recall that the coordinates of point A_3 $(\sigma_x', -\tau')$ and B_3 (σ_y', τ), we have $\sigma_x' = 12$, $\tau' = -3$, $\sigma_y' = 4$. Note that the positive shear stress points in the positive Y' direction and the negative shear stress points in the negative Y' direction on the left plane (i.e. on the plane of lesser x). Hence, the shear stress of 3 kPa points in the negative Y' direction on the left plane, as shown in the schematic next to point A_3 in Fig. 3.14.

To summarise, given the stress state at a point in a particular coordinate system, the Mohr circle analysis allows us to obtain the values of the normal and shear stress at that point in the material along all possible planes passing through the point. Major and minor principal stresses are the largest and smallest possible values of the normal stress that occur at an angle θ_m and $90 + \theta_m$, respectively and the shear stress becomes zero at these orientations of the plane. The maximum value of the shear stress occurs at an angle of 45° from the major principal direction.

3.4.3 Mohr-Coulomb Yield Criterion

The Coulomb yield criterion, given by Eq. (3.10), is represented by a pair of straight lines on the σ-τ plane. Consider the yield line represented by the line EF_1 shown in Fig. 3.15. The negative intercept OE on the σ axis indicates that the cohesion constant c has a nonzero value. This yield criterion, coupled with the Mohr circle analysis of the stress state, can be used to obtain the failure plane orientation. Note that the yield criterion is fulfilled

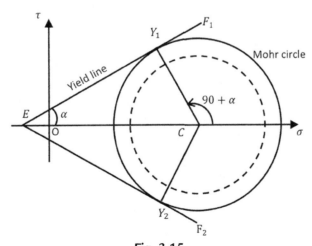

Fig. 3.15
Mohr-Coulomb yield criterion.

only at the points that fall on the yield line; for all points lying below EF_1, the value of the shear stress is less than the yield stress. Let the stress state at a point in the material be represented by the Mohr circle shown by broken line in Fig. 3.15. Since all possible combinations of the normal and shear stresses (present along different planes passing through the point) lie on the Mohr circle, for all of these combinations, the shear stress remains below the yield stress, and hence, the yield criterion is never met along any orientation of the plane. The material in this stress state is able to sustain the stresses without yielding or failing.

However, if the stress state is changed so that the Mohr circle becomes large enough and touches the yield line at Y_1, there exists a plane at which the yield criterion is fulfilled. The material will fail along that plane. Since the yield criterion involves the magnitude of the shear stress, the yield line is not unique and a pair of yield lines exist (Fig. 3.15), and the Mohr circle touches these lines at points Y_1 and Y_2. These points correspond to the planes oriented at angles of $\pm(45+\alpha/2)$ degrees from the major principal axis. On planes other than these two yielding planes, the yield criterion is not satisfied, and hence, the ideal Coulomb material of a perfectly isotropic nature cannot fail along any plane other than these two yield planes. In practice, particulate solids do show some anisotropy and inhomogeneity and hence the failure surfaces may not be the perfect failure planes as predicted by the preceding analysis.

Example 3.5

Dry, noncohesive sand with a Coulomb friction coefficient of $\mu=\tan(20°)$ is subject to a biaxial compression test starting from a stress state $\sigma_x=\sigma_y=\sigma_0$ and $\tau=0$. What is the value σ_y that the sample can sustain before failing if σ_x is held fixed at σ_0?

Solution

Since the material is noncohesive, the yield line passes through the origin and has a slope of $20°$. The stress state at the beginning is represented by coinciding points $A(\sigma_x,0)$ and $B(\sigma_y,0)$ so that the Mohr circle is reduced to a point, i.e. a circle of radius zero centred at point A in Fig. 3.16. As the value of σ_y increases, the point B $(\sigma_y, 0)$ moves to B_1 and the corresponding Mohr circle is shown by the dashed line. Further increase in σ_y makes the Mohr circle even larger (passing through B_2, shown by the dotted line). When the value of σ_y reaches the maximum possible value of $\sigma_{y,max}$, the Mohr circle passes through point B_3 and touches the yield line at point Y_1. Line OY_1 is then tangent to the circle at point Y_1. In $\triangle OCY_1$, $\sin\alpha=CY_1/OC$, since $CY_1=r=\left(\frac{\sigma_y^{max}-\sigma_0}{2}\right)$ and $OC=\frac{\sigma_0+\sigma_y^{max}}{2}$, using $\alpha=20°$, we have $\sin 20° =\frac{\sigma_y^{max}-\sigma_0}{\sigma_y^{max}+\sigma_0}$, i.e. $\sigma_y^{max}=\left(\frac{1+\sin\alpha}{1-\sin\alpha}\right)\sigma_0\approx 2.039\sigma_0$.

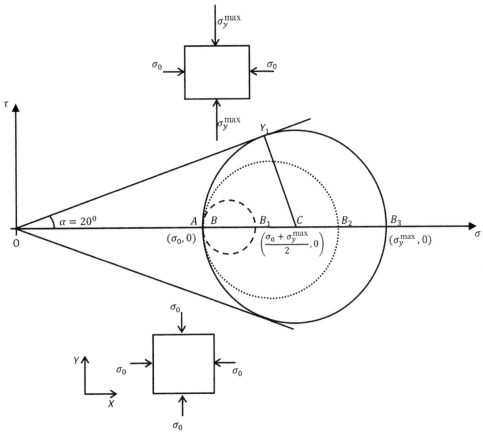

Fig. 3.16

Determination of maximum permissible value of the normal stress in biaxial compression test problem, Example 3.5.

It is important to note that one can also choose to reduce the value of σ_y so that point B moves to the left of point A, passing through points B_4, B_5, and eventually B_6 so that the Mohr circle at the time of yielding could be as shown by the circle represented by the solid line in Fig. 3.17. An analysis as before then gives $\sigma_y^{min} = \left(\frac{1-\sin\alpha}{1+\sin\alpha}\right)\sigma_0 \approx 0.49\sigma_0$. Thus, the sample will yield only when $\sigma_y < \sigma_y^{min}$ or $\sigma_y > \sigma_y^{max}$ and will not yield for $\sigma_y^{min} < \sigma_y < \sigma_y^{max}$.

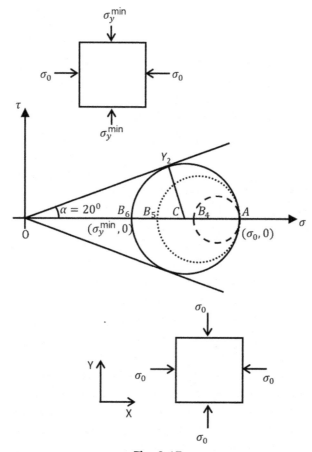

Fig. 3.17

Determination of the minimum permissible value of the normal stress in biaxial compression test, Example 3.5.

3.4.4 Measurement of the Coefficient of Friction and Cohesion

The resistance of a particulate mass to shear may be measured in a shear cell such as that described by Jenike et al.[12,13] The powder is contained in a shallow cylindrical cell (with a vertical axis), which is split horizontally. The lower half of the cell is fixed and the upper half is subjected to a shear force, which is applied slowly and continuously measured until the upper half moves with a constant speed and the shear stress reaches a steady state value. The shearing is carried out for a range of normal loads and the value of the steady state shear stress is plotted with the applied normal stress to obtain the locus of the yield point. For Jenike shear testers (or translational shear testers), the value of the shear displacement is limited to twice the thickness of the cell wall, and the applied shear and normal forces are not uniform.[14] Ring (or annular) shear testers, containing the bottom ring and upper lid of annular shapes, are rotational shear boxes that permit relatively much larger shear displacements and are less prone to operator-dependent errors. Introduced initially by Walker[15] in 1960s, these testers have been continuously improvised[14] and automated as well as semi-automated versions are available, leading to easier and faster measurements. Powder rheometers have also been developed in last decade, which measure the torque or force needed to move/rotate a linear/helical blade through the powder after appropriately conditioning the sample to reduce the variability in the measurements due to sample preparation.[16] These seem to provide satisfactory results for various industrial powders.[17]

The locus of the yield points (or the yield locus), obtained from these devices will be nearly linear for coarse, noncohesive/slightly cohesive materials. A linear fit to the yield locus data (Eq. 3.10) is used to obtain the slope of the yield line, which determines the coefficient of friction μ (from which the angle of internal friction δ is calculated). The intercept on the shear stress axis is used to determine the cohesion constant c. However, deviations from the linear yield line behaviour of ideal Coulomb material are commonly observed for fine, cohesive materials. The intercept of the yield locus on the normal stress axis represents the tensile strength of the powder and the intercept on the shear stress is called the cohesion of the powder. In addition to the indirect measurement of the tensile strength using the data from the shear testers, some methods of direct measurement of the tensile strength have been also developed by researchers.[18,19] For both direct and indirect measurements, the tensile strength has been shown to depend on the degree of compaction of the powder. In general, the yield locus is sensitive to the initial packing of the material, and hence, in practice, a family of the yield loci is obtained depending upon the initial packing, as well as the consolidation history of the sample. Time consolidation, where the properties of the particulate solids keep changing with time due to storage, is also observed for many powders. Refer to the book by Schulze[14] for a detailed discussion about these aspects.

While the coefficient of friction determines the yield (i.e. incipient flow) of the material in bulk, the motion of bulk solids along the surface wall of the silo or bin is determined by the coefficient of wall friction μ_w. For a given wall normal stress (σ_w), the wall friction coefficient determines the minimum shear stress (τ_w) required to obtain a continuous motion of the bulk solid along the wall and is defined as the ratio of the wall shear stress to wall normal stress, $\mu_w = \tau_w/\sigma_w$. The angle of the wall friction is calculated $\delta_w = \tan^{-1}(\mu_w)$. The wall friction coefficient is measured by the Jenike (or ring) shear tester by replacing one of the cell (or ring) containing the bulk solid by a sample surface of the wall material and following the procedure of determining the yield locus mentioned above. The slope of this wall yield locus equals the wall friction coefficient for linear yield locus. In the case of normal stress dependent wall friction, such as that for highly adhesive materials, yield locus is curved and a decreasing trend of the wall friction with the increasing normal stress is commonly observed since the adhesion forces do not increase proportionally with the normal load. The intercept of the yield locus on the shear stress axis determines the adhesion of the particulate solid with the wall surface and represents the value of the shear stress that can be supported in case of zero normal stress, indicating that such a particulate material can adhere to the vertical walls of the silo/bin due to large adhesive forces.

The magnitude of the shear and tensile strengths of the powder has a considerable effect on the way in which the powder will flow, and particularly on the way it will discharge from a storage hopper through an outlet nozzle. Detailed measurements of the flow functions at different times need to be carried out to obtain the flow factors for the design of a silo in such cases. For more details, refer to the book by Schulze.[14] For noncohesive, relatively free-flowing granular materials, this dependence on the shear and tensile strengths is rather weak and correlations such as those described in Sections 3.5 and 3.6 prove handy.

3.4.5 Beyond Mohr-Coulomb Failure

From the above discussion, it should be evident that the Mohr-Coulomb yield criterion discussed in Section 3.4.3 is the simplest possible description of failure of the bulk solids. More advanced models dealing with the yielding of granular materials in the slow flow regime are critically reviewed by Rao and Nott.[7] Once the material yields, the stresses acting on it can lead to the flow of the bulk material. Research in the past few decades has enhanced our understanding of dry, cohesionless granular materials significantly and it may, in turn, pave the way for understanding the behaviour of cohesive particulate solids. For dry granular materials, steady flows are observed beyond the yield stress and it is observed that the shear stress remains proportional to the normal stress. The ratio of the shear stress to pressure (average value of the normal stresses), called the effective friction coefficient, depends upon a nondimensional parameter called the inertial number which depends upon both the shear rate and the pressure, in addition to the grain size and material density. Since the granular materials behave like solids before the failure and flow like liquids through orifice beyond the yield, they show properties resembling solids, as well as that of liquids. The text book by Andreotti, Forterre, and Pouliquen[6] is an excellent reference highlighting this intermediate or dual nature of bulk solids. As per the current understanding, granular materials can be modelled as a viscoplastic fluid with a pressure dependent yield stress and an apparent viscosity that depends on both the shear rate and the pressure. It is important to keep in mind that most of the phenomena observed in bulk solids still lack explanations from the first principles and are understood by means of empirical relations, including the most basic Coulomb's law of friction.

3.5 Flow of Solids Through Orifices

The discharge rate of solid particles is usually controlled by the size of the orifice or the aperture at the base of the hopper, though sometimes screw feeders or rotating table feeders may be incorporated to encourage an even flowrate. The flow of solids through an orifice depends on the ability of the particles to dilate in the region of the aperture. Discussion in this section is restricted to the flow through orifice in flat bottom containers and flow through conical/wedge-shaped hoppers is discussed in the following section.

In contrast to the fluids, the rate of discharge of bulk solids through the outlet orifice is nearly independent of the depth of solids in the hopper, provided it exceeds few times the diameter of the hopper.[2,20] For cylindrical bins having the diameter larger than twice of the orifice diameter, the flowrate becomes independent of the height for fill heights greater than twice the orifice diameter. This height independence of the flowrate is sometimes explained using the height independence of the pressure at the base. However, the weakness of this argument becomes obvious as soon as one recalls that the saturated pressure depends on the silo diameter (via the characteristic length), whereas no dependence on the silo diameter is observed for the mass flowrate through the orifice. Recent studies show that for granular flow through an aperture, the pressure and flowrate are independent of each other.[21,22]

Wieghardt[23] reports that Hagen in 18th century had suggested that for a circular orifice at the bottom of a silo, the mass flowrate rate varies as the orifice diameter is raised to the power 2.5. In his experiments, Beverloo[20] noticed slight variations in the exponent for different granular materials and proposed that the mass flowrate is proportional to the *effective diameter* of the orifice, raised to the power 2.5. The effective diameter is the actual orifice diameter less a correction, which is equal to few times the particle diameter. The Beverloo correlation for granular flow of nearly monodispersed grains through a circular orifice is given as

$$\dot{m} = c\rho_b g^{0.5} D_{eff}^{2.5} \tag{3.16}$$

where $\dot{m}$ is the mass flowrate, ρ_b is the bulk density of the material, g is the acceleration due to gravity, $D_{eff} = D - kd$ is the effective diameter of the orifice.

In the above expression, D is the diameter of the orifice, d is the diameter of the solid particles and the value of the proportionality constant, also called Beverloo constant, is $c \sim 0.58$. The parameter k lies within the range of 1–3. While the use of 'flowing density' has been suggested by some researchers, Beverloo used the bulk density ρ_b. Use of the bulk density eliminates the challenges and errors associated with the measurement of the flowing density, which is calculated by dividing the mass flowrate by the volumetric flowrate of the material. The latter is calculated as the product of the cross-section area and the rate of the change of the level of the material at the walls.

Dependence of the mass flowrate on the diameter in Eq. (3.16) can be understood by dimensional arguments. For orifice sizes much lager compared to grains size ($D \gg d$) and given the height independence of the flowrate, the only relevant length scale in the problem is the orifice size D. Dimensional constraints then ensure that $\dot{m} \propto \rho_b g^{0.5} D^{2.5}$ since the velocity scale is $\sqrt{gD}$ and the orifice cross-section area is $\pi D^2/4$. Assuming that the flowrate through the orifice can be calculated by only considering the motion of particles from a hemispherical free-fall dome, Hilton and Cleary[24] derived the Beverloo correlation for flow through a circular orifice, (see Example 3.6). It is important to note that the value of the Beverloo constant depends on the frictional properties of the material, and the reported values for this proportionality constant typically lie in range of 0.55–0.65. However, for nonspherical grains, values smaller than 0.55 are often reported.[25,26] Nedderman[2] suggests that Beverloo's correlation should be used only for particle sizes in the range $0.4\,\text{mm} < d < D/6$. For $d < 0.4\,\text{mm}$, the air drag may affect the flowrate, whereas intermittent flow is often observed if $d > D/6$.

Example 3.6

Derivation of Beverloo correlation: Adopted from Hilton and Cleary[24]: Consider a granular material flowing out of a circular orifice of diameter D at the base of a cylinder. Assuming that grains fall freely under the influence of gravity from a hemispherical arch, obtain the Beverloo correlation for flowrate through an orifice.

Solution

In the region far away from the orifice, the material flows slowly with a high packing fraction and the frictional contacts between the grains dominate. In the region close to the outlet, the collisions and frictional contacts between the grains become less important and the flow transforms to rapid, dilute flow. The boundary between the slow-flowing region and the fast-flowing region is referred to as the 'free-fall arch'. Hilton and Cleary[24] proposed the hemispherical shape of the free-fall arch spanning the orifice (Fig. 3.18A). The velocity of the grains in the slow flowing region is assumed to be negligible. Once the grains reach the hemispherical arch, they fall freely under the influence of gravity. Thus all the particles flowing through an annular region of thickness dr at the orifice (Fig. 3.18B) will have the velocity $v(r) = \sqrt{2gh(r)}$, due to the free fall through a distance $h(r)$ under gravity. The mass flowrate of grains flowing through the annulus is $d\dot{m} = \rho v(r)2\pi r\,dr$. Total mass flowrate through the orifice

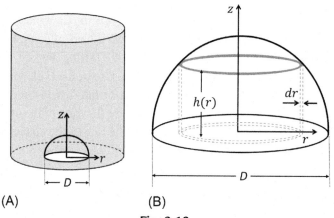

(A)   (B)

Fig. 3.18

(A) Cylinder with a circular orifice of diameter D at the base. A hemispherical 'free-fall arch' spans the orifice. (B) The grains with zero initial velocity fall freely under the influence of gravity starting from the free-fall arch.

can be obtained by integrating the mass flowrate over the entire orifice cross section, i.e. $\dot{m} = \int_0^{\frac{D}{2}} \rho v(r) 2\pi r dr$. Since the height $h(r) = \sqrt{D^2/4 - r^2}$, the total mass flowrate $\dot{m} = 2\sqrt{2g}\,\pi\rho \int_0^{\frac{D}{2}} (D^2/4 - r^2)^{\frac{1}{4}} r dr$. The above integral can be easily solved to obtain the mass flowrate through circular orifice as $\dot{m} = \frac{\pi}{5}\rho g^{0.5} D^{2.5}$. This simple analysis gives the value of the Beverloo constant as $c = \pi/5 = 0.628$, which is well within the empirical range of values.

Using the experimental value of the Beverloo constant, the average velocity of the grains falling out of the orifice can be obtained as $\dot{m}/\left(\rho_b \pi D_{eff}^2/4\right) = 0.74\sqrt{gD_{eff}}$, indicating that the relevant velocity scale can be obtained using the acceleration due to gravity and the diameter of the orifice. For a noncircular orifice, modifications to the Beverloo correlation has been proposed.[2] A velocity scale of $\sqrt{gD_h}$, based on the hydraulic mean diameter, is used in the free-fall region so that the expression for mass flowrate through a noncircular orifice then takes the form:

$$\dot{m} = c'\rho_b A_{eff}\sqrt{gD_h} \tag{3.17}$$

with $c' = 0.74$. In this expression, A_{eff} is the effective area (after accounting for the correction factor kd) and $D_h = 4A_{eff}/P_{eff}$ is the hydraulic diameter, with A_{eff} and P_{eff} being the area and perimeter of the orifice, respectively, after the effective orifice size correction using the length scale kd. Thus for instance, for rectangular slots of length $L(=L_{eff}+kd)$ and width $B(=B_{eff}+kd)$ with $L \gg B$, the Eq. (3.17) reduces to the following form, which can be used for calculating the mass flowrate:

$$\dot{m} = c''\rho_b L_{eff}\sqrt{gB_{eff}^3} \tag{3.18}$$

with $c'' = 1.04$. Other forms of such correlations for noncircular orifices, and for nonspherical particles, have been proposed by Al-Din and Gunn.[27] In general, the flowrate for nonspherical particles will be smaller than that of the spherical particles under otherwise identical conditions.

Example 3.7

For equal orifice area of different shapes, flowrate through which orifice shape will be minimum: circle, semi-circle, square, equilateral triangle and rectangle?

Solution

Let us assume that the orifice size is much larger than the grain size, so that $A_{eff} \cong A$. Using Eq. (3.17), i.e. $\dot{m} = c'\rho_b A_{eff}\sqrt{gD_h}$, for same orifice area, $\dot{m} \propto \sqrt{D_h} \propto 1/\sqrt{P}$, where P is the perimeter of the orifice. One can easily show that, for a given area, $P_{circle} < P_{square} < P_{semi\ circle} < P_{rectangle} < P_{triangle}$. Hence, $\dot{m}_{circle} > \dot{m}_{square} > \dot{m}_{semi\ circle} > \dot{m}_{rectangle} > \dot{m}_{triangle}$. This is consistent with the observations of Kotchanova[28] who studied the effect of orifice shapes on the flowrate of food grains from flat-bottom bins and found the highest discharge rates for circular orifices followed by square, semi-circular and rectangular orifices.

The origin of the effective diameter has been attributed to the finite size of the grains as compared to the orifice size. Brown and Richards[5] explained the presence of term kd in Eq. (3.16) by suggesting an 'empty annulus' region of width $d/2$ near the boundary of the orifice in which no particle centre can approach. However, this does not explain the observed values of $k > 1$. The recent study of Mankoc et al.[29] shows that the value of k is not unique for a given shape and can vary from one to three depending on the range of the orifice sizes considered for fitting the data. Based on simulations spanning nearly 100-fold variation in the orifice sizes, the authors conclude that the 'empty annulus' concept is not entirely satisfactory to explain the simulation observations. The mass flowrate for a much wider range of the orifice size can be obtained by applying a correction factor, X, to the mass flowrate obtained using the Beverloo correlation with $k = 1$, i.e. $\dot{m} = cX\rho_b g^{0.5}(D-d)^{2.5}$ where $X = \left(1 - 0.5e^{-0.051\left(\frac{D}{d}-1\right)}\right)$. The correction factor X accounts for the effect of the density variation of the material near the outlet of the orifice and it also seems to be applicable for orifice sizes, which permit only intermittent flows, provided that the mass flowrate is calculated by considering the time of flow duration between the two jamming events. For orifice to particle size ratios of $D/d = 10$, 20, 50 and 100, the values of the correction factor are given as $X = 0.68$, 0.81, 0.96 and 0.997, respectively. This variation of the density along the orifice has been observed in experiments as well,[30] indicating that the assumption of the uniform bulk density across the orifice cross section in Example 3.6 is not correct.

3.5.1 Does the Free-Fall Arch Exist?

The success of the Beverloo correlation in predicting the mass flowrate through a circular orifice suggests that the average grain velocity is proportional to the square root of the orifice size. This observation leads to the hypothesis of the existence of a free-fall arch, which has been historically invoked to explain this velocity scale. However, the shape of the free-fall arch is not very well agreed upon. Traditionally, the free-fall arch shape has been assumed to be a portion of a spherical section.[2] Hilton and Cleary[24] showed that hemispherical shape of the free-fall arch is able to predict the Beverloo constant within a reasonable agreement. By measuring the outlet velocity at different locations across the orifice, recent studies[30,31] show that assuming a parabolic shape of the free-fall arch seems more reasonable than the hemispherical shape. The shapes of the three types of arches used in literature are shown in Fig. 3.19 (upper curves) along with the velocity profiles (lower curves).

However, it is now generally agreed upon that the proportionality of the average velocity at the exit to the square root of the orifice size stems from the dimensional constraints and recent simulation studies[30,31] investigating the existence of the free-fall arch have questioned the validity of this widely used assumption of the free-fall arch. Though the existence of a 'region above the outlet where certain dynamical features determine the velocity profile' has been

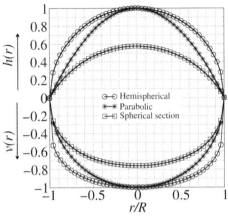

Fig. 3.19

Various shapes of the free-fall arch used for explaining the success of the Beverloo correlation. The curves in the upper half show the free-fall arch height $h(r)$. Curves in the lower half show the corresponding velocity profiles across the orifice.

confirmed in the simulation studies,[31] the observations in this region are not consistent with the naive historical picture of the free-fall arch. The free-fall arch description assumes that once the particles reach the free-fall arch, they lose all their kinetic energy and start to fall freely under gravity. According to this description, the particle accelerations are equal to g in the free-fall arch region and become zero beyond that (as assumed in Example 3.6). However, detailed measurements of the particle accelerations near the orifice[31] show a near linear variation of the acceleration of the grains along the centre is observed at the centreline. The height of this region is comparable to the orifice diameter and the acceleration beyond this region is small. The particle accelerations approach the acceleration due to gravity only very close at the exit, occasionally even showing values slightly higher than g, suggesting that the compressive pressure is significant even at the exit.

In addition, the stress along the centreline is continuous throughout the silo and no stress discontinuity exists near the region above the outlet. These findings contradict the hypothesis of the stress discontinuity at the free-fall arch, which has been suggested as one of the possible resolves for theoretical descriptions. Though a parabolic arch of height equal to the orifice radius is observed, it does not correspond to the beginning of the region of free fall. Instead, this arch corresponds to the locations where the fluctuations of the grains become maximum.

The deficiencies of the free-fall arch description can be more clearly understood by considering the flow through orifices that are located at the sidewalls of the silo as opposed to the bottom wall. As mentioned in Section 3.5.2, the mass flow rate through a lateral orifice is also found to be proportional to $g^{1/2}D^{5/2}$, indicating that the average velocity through lateral orifice is also proportional to $(gD)^{1/2}$. However, the exit velocity from the lateral orifice is not parallel to the gravitational acceleration, and hence, the idea of the free-fall arch cannot explain the flowrate through the lateral orifice at the sidewalls.

3.5.2 Flow Through Sidewall Orifice

The studies investigating the flow of bulk solids through an orifice typically consider flow through a horizontal orifice located at the bottom of the silo. Studies investigating flow of coarse grains through the lateral orifices on the vertical sidewalls of the silo show important differences compared to the flow through orifice at the bottom of the silo.[25,32,33] While the wall thickness of the container has no influence on the flow through horizontal orifices at the bottom of the silo, its effect on the flowrate through the vertical orifice located at the sidewalls of the silo can rarely be ignored. In fact, for wall thickness greater than a minimum value, the flowrate through a lateral orifice goes to zero, leading to the complete stoppage of the flow. Consider a granular material filled in a container of wall thickness w (Fig. 3.20A). The flowrate through the vertical orifice depends strongly on the angle of repose of the flowing material and is given as (for the case $D \gg d$)

$$\dot{m} = c_\perp \rho_b g^{0.5} D^{2.5} (\theta - \alpha) \tag{3.19}$$

where $\theta = \tan^{-1}(D/w)$ is a characteristic angle of the orifice, α is the angle of repose of the material, D is the orifice size and d is the grain size (see Fig. 3.20B). As the thickness of the wall increases from zero, the value of θ decreases from the value of $\pi/2$, leading to the reduction in the mass flowrate with the container wall thickness. This is due to the formation of a stagnant

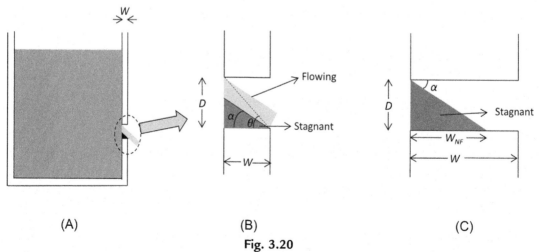

Fig. 3.20

(A) Granular flow through a lateral orifice depends upon the container thickness W. (B) Flowing and stagnant regions near a square shape orifice of size D. (C) No flow can be sustained through the lateral orifice if the container thickness W is larger than W_{NF}. Figure adopted from ref. 25.

region across the thickness of the container that leads to a reduction in the orifice area available for flow. When the container wall thickness becomes equal to $w_{NF} = D/\tan\alpha$, the value of θ becomes equal to angle of repose α and the flowrate becomes zero, i.e. complete stoppage of the flow occurs for container wall thickness $w > w_{NF}$ as shown in Fig. 3.20C. In other words, for a given container thickness, there exists a minimum size of lateral orifice below which no flow is possible.

From the above discussion, it is clear that attaching a pipe through an orifice at the sidewall will reduce the flow and can even lead to complete stoppage of the flow. However, attaching a pipe through the hole at the bottom of the silo will have a countereffect. It has been found that the attachment of a discharge pipe of the same diameter as the orifice immediately beneath it increases the flowrate, particularly of fine solids. Thus in one case, with a pipe with a length to diameter ratio of 50, the discharge rate of a fine sand could be increased by 50%, and that of a coarse sand by 15%. The increased flowrate can be attributed to the build-up of a pressure gradient due to the air, which adds to gravity within the silo and opposes the gravity in the pipe.[7] While the effect of the air drag on the material while flowing through the orifice is small for coarse grains, the effect on the fine particulate solids is significant and can either add or oppose the flow due to gravity. The latter case leads to slower flows, increased flowrate of the powder shall be observed in the earlier case. Another method of increasing the discharge rate of fine particles is to fluidise the particles in the neighbourhood of the orifice by the injection of air; more details about fluidisation are discussed in Chapter 9.

3.6 Flow of Solids in Hoppers

Solids may be stored in heaps or in sacks, although subsequent handling problems may be serious with large-scale operations. Frequently, solids are stored in hoppers, which are usually circular or rectangular in cross-section, with conical or tapering sections at the bottom. The hopper is filled at the top, and it should be noted that if there is an appreciable size distribution of the particles, some segregation may occur during filling since the larger particles tend to roll to the outside of the piles in the hopper.

Discharge from the hopper takes place through an aperture at the bottom of the cone, and difficulties are commonly experienced in obtaining a regular, or sometimes any, flow. Commonly experienced types of behaviour are shown in Fig. 3.21 taken from the work of Weigand.[34] Bridging of particles may take place and sometimes stable arches, as shown in Fig. 3.21(A), may form inside the hopper, and although these can usually be broken down by vibrators attached to the walls, problems of persistent blockage are not unknown. Such blockages are more common for hoppers containing fine, cohesive particulate solids. Jamming of relatively freely flowing grains may also occur due to an increase in the moisture content (e.g. due to rains), eventually leading to flow blockage. Such jammed arches may be broken by hammering the hopper near the exit, and hence, the sight of the hammered hopper wall near the

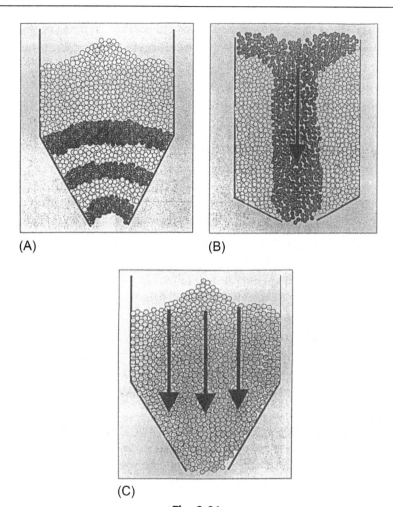

(A)   (B)

(C)

Fig. 3.21
(A) Formation of stable arches leads to jamming and blockage. (B) Funnel flow leading to 'piping' or
'rat-holing'. (C) Mass flow.

exit is not rare. A further problem, which is commonly encountered, is that of "funnel flow", as
shown in Fig. 3.21(B), in which the central core of material is discharged leaving a stagnant
surrounding mass of solids, known as the 'dead storage'. As a result, some solids may be
retained for long periods in the hopper and may deteriorate. Since funnel flow is not predictable,
it leads to segregation issues that may cause flooding or uncontrolled discharge for fine bulk
solids of low permeability. To avoid these issues, pneumatic discharge aids (inflatable aeration
pads/nozzles, air cannon, etc.) or mechanical discharge aids (vibrators or agitators) are
commonly used. More details are in the text by Schulze[14] and Woodcock and Mason.[35]

Ideally, "mass flow", as shown in Fig. 3.21(C), is required, which means the solids are in
plug flow and move downward *en masse* in both parallel and tapered segments of the hopper.

The residence time of all particles in the hopper will then be similar. Complete discharge of the material is ensured in mass flow at a predictable rate. The nature of the surface of the hopper is important, and smooth surfaces at steep angles give improved discharge characteristics. Monel metal cladding of steel is frequently used for this purpose. In general, tall, thin hoppers give better flow characteristics than short, wide ones, and the use of long small-angle conical sections at the base is advantageous since it reduces the chances of blockage and increases the possibility of mass flow.

3.6.1 Flowrate of Solids Through Hoppers

Rose and Tanaka[2,7] found that the flowrate through steep conical hoppers depends on two angles: (a) the hopper half angle θ_w and (b) the angle β that the boundary of stagnant and flowing regions makes with the vertical axis measured in a cylindrical bin. Both these angles are shown in Fig. 3.22. For small hopper wall angles, $\theta_w \leq \beta$, the flowrate was found to be proportional to $(\tan\beta/\tan\theta_w)^{0.35}$. Combining these results with the Beverloo correlation, Nedderman[2] suggested that flow through converging hoppers can be estimated reasonably well using the Beverloo correlation for flowrate through the orifice by accounting for the proportionality factor proposed by Rose and Tanaka as a correction factor. Thus the flowrate through a converging hopper of a particular orifice shape can be calculated by the 'Rose-Tanaka-Beverloo-Nedderman' (RTBN) correlation.[7]

$$\dot{m}_{\text{hopper}} = \dot{m}_{\text{orifice}}(\tan\beta/\tan\theta_w)^{0.35}, \quad \theta_w \leq \beta \qquad (3.20a)$$

$$\dot{m}_{\text{hopper}} = \dot{m}_{\text{orifice}}, \quad \theta_w > \beta \qquad (3.20b)$$

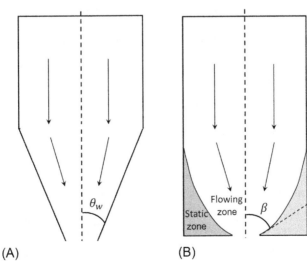

Fig. 3.22
Schematic diagram showing angles used in Eq. (3.20). See text for details.

where $\dot{m}_{orifice}$ is given by Eq. (3.17). Rao and Nott[7] show that the predictions from RTBN correlation are in agreement with the experimental observations for glass beads in size range 0.85–1.25 mm. The measured values of angle β for glass beads, sand, rice and coal lie within the range of 34°–58°; in the absence of data, a value of $\beta = 45°$ is suggested as a reasonable choice by Nedderman.[2]

3.6.2 Control of Solids Flowrate Through Hoppers

The mass flow rate correlations, such as those given by Eq. (3.20), describe the maximum permissible flow rate through the hopper outlet. Often, bulk solids are required to be discharged at a controlled flowrate and necessitate some kind of a feeder installed at the hopper outlet. While the discharge aids help by preventing the flow obstruction by the arch formation or ratholing (as discussed in Section 3.6), the feeders, help control the flowrate of the bulk solids flowing out of the hoppers/bins. Various types of feeders are used to control the rate of feed of solids. Some common examples include belt or apron feeders, screw feeders, rotating tables or vibrating feeders, such as magnetically vibrated troughs. The volumetric flow rates may be controlled by regulating the speeds of rotation of star feeders or rotary vane valves. To ensure a uniform flow from the entire outlet of the hopper, the feeder and hopper need to be designed as one integral unit. More details about the various kinds of feeders are available in the book by Schulze[14] and Woodcock and Mason.[35]

3.7 Measurement of Flowrate and Level of Bulk Solids

The flowrate of solids can be measured by weighing the mass of the material either as they leave the hopper or as they are conveyed over the belt. In the former case, the hopper itself is supported on load cells so that a continuous record of the mass of the contents may be obtained as a function of time. Typically, 100–400 readings can be collected per hour and this frequency can be changed as per the requirement of the process.[36] Alternatively, a belt-type feeder can be used in which the mass of material on the belt is continuously measured using a belt scale or load cell. The product of the weight of the material on unit length of the belt and the belt speed provides the flowrate. These methods provide direct measurement of the mass flowrate from the hopper, and hence, are more reliable and accurate for powder flow measurements. The mass of material on the belt can also be measured by a nuclear densitometer, which measures the degree of absorption of gamma rays transmitted vertically through the solids on the belt. Another method for measuring the flow rate of bulk solids utilises an impulsive force flowmeter in which a solids stream impacts on a sensing plate orientated at an angle to the vertical. The components of the resulting force are measured either by a load cell attached to the plate or by using impulse-momentum relations, from which the mass of the stream can be calculated.[36]

The problems associated with the measurement and control of the flowrate of solids are much more complicated than those in the corresponding situation with liquids. The flow characteristics will depend, not only on particle size (or size range) and shape, but also on how densely the particles are packed. In addition, surface and electrical properties and moisture content all exert a strong influence on the flow behaviour, and the combined effect of these factors is almost impossible to predict in advance. It is therefore desirable to carry out a preliminary qualitative assessment before making a selection of the most appropriate technique for controlling and measuring the flowrate for any particular application.

A detailed description of the various methods of measuring solids flowrates has been given by Liptak[37] and Table 3.1, taken from this work, gives a summary of the principal types of devices in common use. Alternatively, the flowrate of the bulk solid can be measured by measuring the change in the level of the solids in the hopper. The product of the level change,

Table 3.1 Different types of solids flowmeter[37]

Type of Meter	Flowrate (kg/s)	Accuracy (Percent FSD[a] Over 10:1 Range)	Type of Material
Gravimetric belt	<25 (or <0.3 m³/s)	±0.5R	Dependent on mechanism used to feed belt. Vertical gate feeder suitable for non-fluidised materials with particle size <3 mm
Belt with nucleonic sensor	<25 (or <0.3 m³/s)	±0.5 to ±1	Preferred when material is difficult to handle, e.g. corrosive, hot, dusty, or abrasive. Accuracy greatly improved when particle size, bulk density, and moisture content are constant, when belt load is 70%–100% of maximum
Vertical gravimetric	Limited capacity	±0.5 for 5:1 turndown. ±1.0 for 20:1 turndown	Dry and free-flowing powders with particle size <2.5 mm
Loss-in-weight	Depends on size of hopper	±1.0R	Liquids, slurries, and solids
Dual hopper	0.13–40	±0.5R	Free-flowing bulk solids
Impulse	0.4–400	±1 to ±2	Free-flowing powders. Granules/pellets <13 mm in size
Volumetric	<0.3 m³/s	±2 to ±4	Solids of uniform size

[a]FSD = full-scale deflection.

the hopper cross-section area, and the bulk density (measured separately in similar loading conditions), can be used to estimate the mass flowrate.

Measuring the level of the bulk solid may be important for some continuous operations, and the level may be monitored either continuously or in a discrete fashion. Such level measurements can be done, for example, using transducers covered by flexible diaphragms flush with the walls of the hopper. The diaphragm responds to the presence of the solids, and thus indicates whether there are solids present at a particular level. Measurement of electrical resistance or capacitance between a vertically inserted wire electrode and the vessel is also used to obtain the information about the solid level. Microwave or ultrasonic wave level meters operate by emitting pulsated waves over a powder bed. Part of the wave gets reflected by the bed and the reflection time of the wave is measured by a sensor. The elapsed time is used to calculate the level of solids in the hopper by accounting for the wave travel speed. Microwave level meters are less sensitive to the dust and water vapour in the container, and hence, are generally preferred over the ultrasonic level meters. Optical methods utilising high-power laser to scan the surface profile of the powder bed are also commonly used to measure the level of the solids in bins/hoppers. Discrete level meters, also known as level switches, are used to maintain the level of the powder between an upper and a lower limit in the hopper. More details about the measurement of flow rate and level of the bulk solids can be found in Ref. 36 along with the information about a few other properties such as temperature and moisture content.

3.8 Conveying of Solids

The variety of requirements in connection with the conveying of solids has led to the development of a wide range of equipment. This includes:

(a) *Gravity chutes* – solids fall down the chutes under the action of gravity.

(b) *Air slides* – particles, which are maintained partially suspended in a channel by the upward flow of air through a porous distributor, flow at a small angle to the horizontal.

(c) *Belt conveyors* – solids are conveyed horizontally, or at small angles (usually <15°) to the horizontal, on a continuous moving belt.

(d) *Screw conveyors* – solids are moved along a pipe or channel by a rotating helical impeller, as in a screw lift elevator.

(e) *Bucket elevators* – particles are carried upwards in buckets attached to a continuously moving vertical belt, as illustrated in Fig. 3.23.

(f) *Vibrating conveyors* – particles are subjected to an asymmetric vibration and travel in a series of steps over a table. During the forward stroke of the table the particles are carried forward in contact with it, but the acceleration in the reverse stroke is so high that the table slips under the particles. With fine powders, vibration of sufficient intensity results in a fluid-like behaviour.

(g) *Pneumatic/hydraulic conveying installations* – particles are transported in a stream of air/water. The operation is described in detail in Volume 1A, Chapter 5.

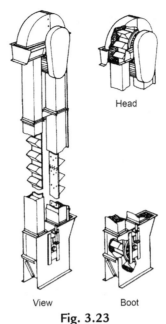

Head

View   Boot

Fig. 3.23

Bucket elevator. *Used with permission from Satake Europe Limited.*

The book by Woodcock and Mason[35] gives a detailed account of gravity chutes, air assisted gravity conveying and pneumatic conveying. Details of the belt conveyors, bucket conveyors, screw conveyors, vibrating conveyors along with few other types of conveyors are also discussed in the text. Another reference for detailed description of various types of conveyors is the bulk materials handling handbook by Fruchtbaum.[38]

3.9 Mixing of Bulk Solids

Mixing or blending of bulk solids is needed in many industrial processes. For example, powders of different sizes often need to be mixed in pharmaceutical industry. Other common examples include the food (breakfast cereals, for instance) and beverages industry, the cement and construction materials industry, and the agricultural industry. Appropriate mixing of the different components is crucial for the desired product quality. Poorly mixed material in any intermediate stage may adversely affect further processing of the material. However, mixing of bulk solids differs from the mixing of liquids in many significant ways. Due to thermal fluctuations of the liquid molecules, good mixing can be achieved in the case of miscible liquids by thermal diffusion only. Thus, a mixture of water and alcohol, given enough time, will mix on their own due to diffusion. However, thermal fluctuations rarely play a role in particulate solids, and hence, a mixture of two powders will never mix on their own. Due to highly dissipative nature of these materials, an external source of energy is always required to induce relative motion between the grains for any mixing to occur. However, the relative motion of grains with respect to each other does not lead to mixing only, it can also cause segregation (discussed in Section 3.10) or unmixing, and is a source of trouble in many industries. A fundamental understanding about many aspects of bulk solids flow and mixing behaviour continues as a topic of active research. Generalised standards for mixing of bulk solids that are applicable for a diverse range of mixers and applications are difficult, and relation between operation variables of mixing and mixing quality indices are hard to derive.[16]

Lacey[39,40] suggests that in the mixing of solid particles, the following three mechanisms may be involved:

(a) Convective mixing, in which groups of particles are moved from one position to another.
(b) Diffusion mixing, where the particles are distributed over a freshly developed interface.
(c) Shear mixing, where slipping zones are formed and particle exchange between the layers in these zones leads to mixing.

These mechanisms operate to varying extents in different kinds of mixers and with different kinds of particles. A trough mixer with a ribbon spiral involves almost pure convective mixing, and a simple barrel-mixer mainly involves a form of diffusion mixing. In case of dry, free-flowing granular materials, convective mixing is the major mixing mechanism since diffusion and shear mixing lead to segregation.[41]

The mixing of pastes is discussed in the section on Non-Newtonian Technology in Volume 1A, Chapter 7.

3.9.1 Degree of Mixing

It is difficult to quantify the degree of mixing, although any index related to the properties of the required mix should be easy to measure, and should be suitable for a variety of different mixers. When dealing with solid particles, the statistical variation in composition among samples withdrawn at any time from a mix is commonly used as a measure of the degree of mixing. A collection of representative samples is a crucial step for determining the degree of mixing. Golden rules for sampling, discussed in Chapter 2, must be followed while collecting the samples. Rhodes[41] suggests that the size of the sample taken for quantitative determination of mixing should be decided by the 'scale of scrutiny'[42] imposed by the end use of the material. Dankwerts[42] states that the scale of scrutiny means, 'the maximum size of regions of segregation in mixture, which would cause it to be regarded as imperfectly mixed'. Thus the scale of scrutiny for a nutrient powder is the mass of nutrient contained in the spoon used by the consumer, and that of a pharmaceutical drug will be the amount used in the tablet/capsule.[41]

The level of scrutiny based on the end usage ensures that the differences in any two tablets (or spoons of the powder) are not significant. The standard deviation s (the square root of the mean of the squares of the individual deviations) or the variance s^2 is generally used. A particulate material cannot attain the perfect mixing that is possible with two fluids, and the best that can be obtained is a degree of randomness in which two similar particles may well be side by side. No amount of mixing will lead to the formation of a uniform mosaic as shown in Fig. 3.24, but only to a condition, such as shown in Fig. 3.25, where there is an overall uniformity but not point uniformity. For a random mix of uniform particles distinguishable, for instance, only by colour, Lacey[39,40] determined:

Particulate solids

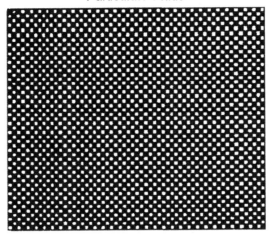

Fig. 3.24
Uniform mosaic.

Fig. 3.25
Overall but not point uniformity in mix.

$$s_r^2 = \frac{p(1-p)}{n} \tag{3.21}$$

where s_r^2 is the variance for the mixture, p is the overall proportion of particles of one colour, and n is the number of particles in each sample.

This equation illustrates the importance of the size of the sample in relation to the size of the particles. In an incompletely randomised material, s^2 will be greater, and in a completely unmixed system, indicated by the suffix 0, it may be shown that:

$$s_0^2 = p(1-p) \tag{3.22}$$

which is independent of the number of particles in the sample. Only a definite number of samples can, in practice, be taken from a mixture, and hence, s will be subject to random errors. This analysis was extended to systems containing particles of different sizes by Buslik.[43]

When a material is partly mixed, then the degree of mixing may be represented by some term b, and several methods have been suggested for expressing b in terms of measurable quantities.[36] If s is obtained from examination of a large number of samples then, as suggested by Lacey, b may be defined as being equal to s_r/s, or $(s_0 - s)/(s_0 - s_r)$, where, as before, s_0 is the value of s for the unmixed material. This form of expression is useful in that $b=0$ for an unmixed material and one for a completely randomised material where $s=s_r$. If s^2 is used instead of s, then this expression may be modified to give:

$$b = \frac{\left(s_0^2 - s^2\right)}{\left(s_0^2 - s_r^2\right)} \tag{3.23}$$

or:

$$1 - b = \frac{\left(s^2 - s_r^2\right)}{\left(s_0^2 - s_r^2\right)} \tag{3.24}$$

For diffusive mixing, b will be independent of the sample size provided the sample is small. In the case of convective mixing, when groups of particles are randomly distributed, each group behaves as a unit containing n_g particles. As mixing proceeds, n_g becomes smaller. The number of groups will then be n_p/n_g, where n_p is the number of particles in each sample. Applying Eq. (3.21):

$$s^2 = \frac{p(1-p)}{n_p/n_g} = n_g s_r^2$$

which gives:

$$1 - b = \frac{\left(n_g s_r^2 - s_r^2\right)}{\left(n_p s_r^2 - s_r^2\right)} = \frac{\left(n_g - 1\right)}{\left(n_p - 1\right)} \tag{3.25}$$

Thus with convective mixing, $1 - b$ depends on the size of the sample.

Example 3.8

The performance of a solids mixer was assessed by calculating the variance occurring in the mass fraction of a component among a selection of samples withdrawn from the mixture. The quality was tested at intervals of 30 s and the data obtained are:

Sample variance (−)	0.025	0.006	0.015	0.018	0.019
Mixing time (s)	30	60	90	120	150

If the component analysed represents 20% of the mixture by mass and each of the samples removed contains approximately 100 particles, comment on the quality of the mixture produced and present the data in graphical form showing the variation of the mixing index with time.

Solution

For a completely unmixed system:

$$s_0^2 = p(1-p) = 0.20(1-0.20) = 0.16 \quad (\text{Eq.3.22})$$

For a completely random mixture:

$$s_r^2 = p(1-p)/n = 0.20(1-0.20)/100 = 0.0016 \quad (\text{Eq.3.21})$$

The degree of mixing b is given by Eq. (3.23) as: $b = (s_0^2 - s^2)/(s_0^2 - s_r^2)$. In this case, $b = (0.16 - s^2)/(0.16 - 0.0016) = 1.01 - 6.313s^2$. The calculated data are therefore:

$t(s)$	30	60	90	120	150
s^2	0.025	0.006	0.015	0.018	0.019
b	0.852	0.972	0.915	0.896	0.890

These data are plotted in Fig. 3.26 from which it is clear that the degree of mixing is a maximum at $t = 60$ s.

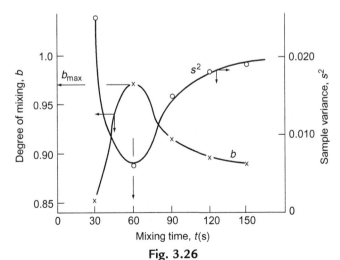

Fig. 3.26

Example 3.8: Degree of mixing as a function of mixing time.

It is important to note that although the degree of mixing proposed by Lacey is one of the most popular mixing index, a variety of other mixing indices have been proposed in literature, especially for usage in particle based simulation methods.[44] However, many of these indices are not useful for characterising mixing in practical applications due to the requirement of particle level details. Kaye[45] mentions that despite significant knowledge gained about the mixing of powders in academic institutes, its impact on the industrial powder mixing operations has been minimal. A large number of factors interact during the operation of a powder mixing system and the complex interactions between these varied factors lead researchers to investigate the powder mixing in the context of 'chaos', which is a separate branch of mechanics altogether.[45] Some popular methods to assess granular mixing in a noninvasive manner for industrial processes is recently reviewed by Nadeem and Heindel.[46]

3.9.2 Rate of Mixing

Expressions for the rate of mixing may be developed for any one of the possible mechanisms. Since mixing involves obtaining an equilibrium condition of uniform randomness, the relation between b and time might be expected to take the general form:

$$b = 1 - e^{-ct} \tag{3.26}$$

where c is a constant depending on the nature of the particles and the physical action of the mixer.

Considering a cylindrical vessel (Fig. 3.27) in which one material **A** is poured into the bottom and the second **B** on top, then the boundary surface between the two materials is a minimum. The process of mixing consists of making some of **A** enter the space occupied by **B**, and some of **B** enter the lower section originally filled by **A**. This may be considered as the diffusion of **A** across the initial boundary into **B**, and of **B** into **A**. This process will continue until there is a

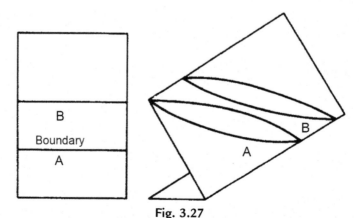

Fig. 3.27
Initial arrangement in a mixing test.

maximum degree of dispersion, and a maximum interfacial area between the two materials. This type of process is somewhat akin to that of diffusion, and tentative use may be made of the relationship given by Fick's law. This law may be applied as follows.

If a is the area of the interface per unit volume of the mix, and a_m the maximum interfacial surface per unit volume that can be obtained, then:

$$\frac{da}{dt} = c(a_m - a) \tag{3.27}$$

and:

$$a = a_m(1 - e^{-ct}) \tag{3.28}$$

It is assumed that, after any time t, a number of samples are removed from the mix, and examined to see how many contain both components. If a sample contains both components, it will contain an element of the interfacial surface. If Y is the fraction of the samples containing the two materials in approximately the proportion in the whole mix, then:

$$Y = 1 - e^{-ct} \tag{3.29}$$

Coulson and Maitra[47] examined the mixing of a number of pairs of materials in a simple drum mixer, and expressed their results as a plot of $\ln(100/X)$ against t, where X is the percentage of the samples that is unmixed and $Y = 1 - (X/100)$. It was shown that the value of constant c depends on:

(a) the total volume of the material,
(b) the inclination of the drum,
(c) the speed of rotation of the drum,
(d) the particle size of each component,
(e) the density of each component,
(f) the relative volume of each component.

While the precise values of c are only applicable to the particular mixer under examination, they do bring out the effect of these variables. Thus Fig. 3.28 shows the effect of speed of rotation, where the best results are obtained when the mixture is not just taken round by centrifugal action. If fine particles are put in at the bottom and coarse on the top, then no mixing occurs on rotation and the coarse material remains on top. If the coarse particles are put at the bottom and the fine on top, then on rotation mixing occurs to an appreciable extent, although on further rotation Y falls, and the coarse particles settle out on the top. This is shown in Fig. 3.29, which shows a maximum degree of mixing. Again, with particles of the same size but differing density, the denser ones migrate to the bottom and the lighter ones to the top. Thus if the lighter particles are put in at the bottom and the heavier at the top, rotation of the drum will give improved mixing to a given value, after which the heavier particles will settle to the bottom.

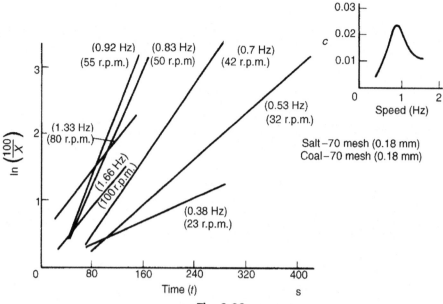

Fig. 3.28

Effect of speed on rate of mixing.

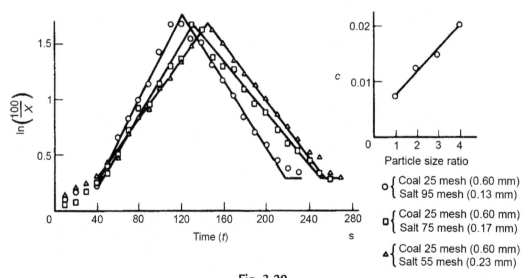

Fig. 3.29

Mixing and subsequent separation of solid particles.

This kind of analysis of simple mixers was made by Blumberg and Maritz,[48] although the results of their experiments are little more than qualitative.

Brothman et al.[49] have given an alternative theory for the process of mixing of solid particles, which has been described as a shuffling process. It is proposed that, as mixing takes place, there will be an increasing chance that a sample of a given size intercepts more than one part of the interface, and hence, the number of mixed samples will increase less rapidly than the surface. This is further described by Herdan.[50] The application of some of these ideas to continuous mixing has been attempted by Danckwerts.[42,51]

A variety of mixers are commonly used in industry. Tumbling mixers are essentially closed vessels of different shapes (as shown in Fig. 3.30) that are rotated about their axis at speeds around 50%–80% of the critical rotation speed for centrifuging.[16] These mixers are typically filled up to 50%–60% of their volume and dominant mechanism of segregation is diffusive mixing in these mixers.

Horizontal trough or vertical screw mixers promote primarily convective mixing (though shear and diffusive mixing is also present). Ribbon blenders, supporting a pair of counter-rotating helical blades (ribbons) on a shaft are an example of a continuously operating horizontal trough mixer. In this mixer, one of the ribbon moves the material slowly in one direction while the other ribbon moves the material faster in the opposite direction.[16] The gap between the vessel walls and the ribbon in these mixers are typically small, and hence, particle damage may occur in some cases. Nauta mixer is an example of vertical screw mixer in which an Archimedean

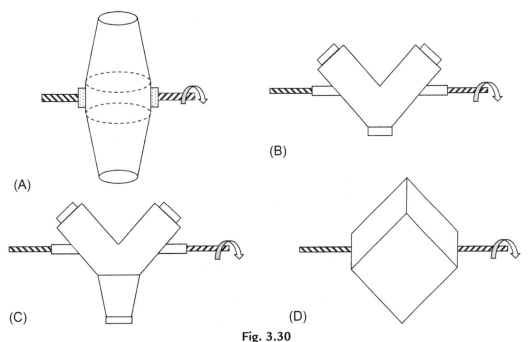

Fig. 3.30
Tumbling mixers of different shapes. (A) Double cone shape, (B) V shape, (C) Y-shape, and (D) cube shape.

screw, orbiting along the periphery of the vessel lifts the material from the base of conical vessel and moves it upwards along the vessel wall.[45] In this mixer, three types of convective motion occur: motion around the screw, motion around the wall of the vessel and convective current from top to bottom of the chamber. In addition to the orbital screw, central screws are also used in mixers.[16] High shear mixers are commonly used for highly agglomerated cohesive powders in which rotating blades at high velocity are used to break the agglomerates and improve mixing. More details about the mixing of cohesive powders can be found in Harnby.[52]

Two different types of industrial mixers are illustrated in Figs 3.31 and 3.32. The Oblicone blender, which is available in sizes from 0.3 to 3 m, is used where ease of cleaning and 100% discharge is essential such as in the pharmaceutical, food or metal powder industries. The Buss kneader shown in Fig. 3.32 is a continuous mixing or kneading machine in which the characteristic feature is that a reciprocating motion is superimposed on the rotation of the kneading screw, resulting in the interaction of specially profiled screw flights with rows of kneading teeth in the casing. The motion of the screw causes the kneading teeth to pass between the flights of the screw at each stroke backwards and forwards. In this manner, a positive exchange of material occurs both in axial and radial directions, resulting in a homogeneous distribution of all components in a short casing length and with a short residence time. For a detailed description of various types of powder mixers, see Ref. 53.

Fig. 3.31
The Oblicone blender. *Used with permission from Baker Perkins Ltd.*

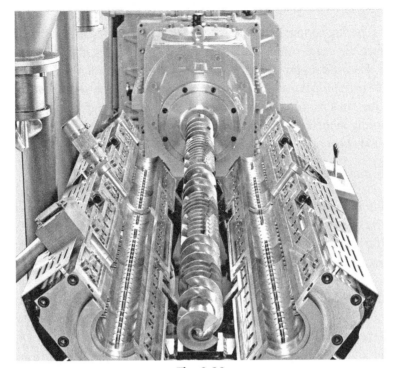

Fig. 3.32
The Buss kneader. *Used with permission from Buss AG, Pratteln, Switzerland.*

3.10 Segregation in Bulk Solids

Constituting grains of a bulk solid always exhibit differences in terms of particle properties, difference in size being the most common. Particle size distributions spanning a wide size range (sometimes up to an order of magnitude) are not uncommon in industrial applications. In addition, when different bulk solids or powders are mixed together, differences in particle densities, particle shape, particle surface roughness, particle hardness etc. are also present. The differences in any one or more such particle properties lead to one of the most fundamental problem of bulk solids handling: segregation.

Segregation is defined as the opposite of mixing, where particles of a particular type concentrate in one region, giving rise to significant variation in the concentration of the components in different regions. Due to this reason, segregation is detrimental to process and product quality. Consider an example from the iron and steel making industry where iron ore feed (a mixture of lump iron ore, sinter, and pellet) and coke is charged into the blast furnace. Hot air is passed through the bottom of the furnace, which burns the coke and melts the iron ore to form a molten slag, which is then processed further to make iron and steel. During the charging of the material in the furnace, coke and iron ore feed are distributed by means of a rotating chute in alternating layers of coke and iron ore. This layer by layer loading is done to avoid the segregation of coke and the ore. In case a combined feed of coke and iron ore is used, the ensuing segregation of coke and iron ore may lead to burning of the coke without melting the iron ore feed, posing serious challenges for the further processing. Hence, it is often desirable to reduce the segregation, and if possible, avoid it altogether to ensure the smooth operation of the downstream processes. In addition, the end product quality itself may be compromised due to segregation as it can lead to large variations between two different samples of the same product. In a pharmaceutical industry, this can be manifested in the form of tablets significantly varying in their chemical compositions, including the possibility of the overdose of drug components beyond the safe limits.

3.10.1 Causes and Consequences of Segregation

While segregation is observed due to differences in density,[54–56] shape,[57] surface roughness,[58] inelasticity,[59] etc., size segregation is one of the most commonly observed forms of segregation due to the ubiquity of the particle size variation. It has been widely studied by researchers for years.[60–67] For example, consider the case of when a granular material having grains of two different sizes (size ratio 2:1) is poured in a cylinder to form a heap. Fig. 3.33 shows the isometric view and top view of such a heap produced by EDEM® academic software. The larger particles concentrate towards the periphery of the heap and the fine ones concentrate towards the centre of the heap. Concentration of the small and large particles varies with the location of the heap. Due to this 'sifting segregation', a material sample collected from the surface of the heap will not be a representative sample of the average concentration of the heap and will have more than an average amount of coarse grains.

Another example is shown in Fig. 3.34 where a mixture of two different size grains is filled in a rotating tumbler. Initially, the particles are segregated in two portions of the cylinder. After a

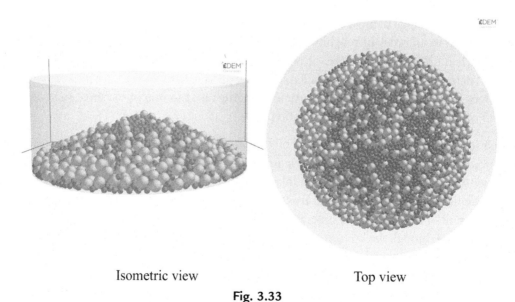

Isometric view   Top view

Fig. 3.33

'Sifting segregation' during heap formation (images produced from simulations using EDEM®
Academic software).

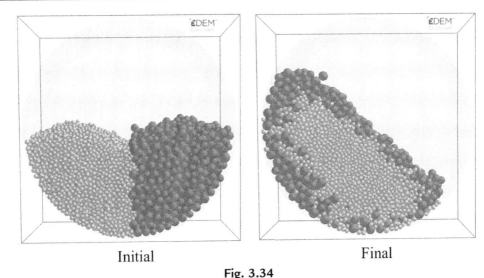

Initial	Final

Fig. 3.34

Radial size segregation in a rotating cylinder (images produced from simulations using EDEM®
Academic software).

few rotations of the cylinder, the mixture gets segregated in such a way that the large particles
concentrate towards the periphery and the smaller size grains concentrate near the centre of the
cylinder. This final state of segregated material is obtained even if one starts from an initial
condition where the particles are well mixed. Thus a rotating tumbler ends up un-mixing the
components of a free-flowing granular material. Due to this reason, rotating tumblers
(including those shown in Fig. 3.30) are typically avoided to mix free-flowing granular
materials. Since cohesive materials show much less segregation compared to the cohesionless,
free-flowing materials, use of tumbler does not lead to intense segregation in case of fine,
cohesive particulate solids.

Bulk solids are transported using belt conveyors or by means of road transport. Vibrations
during the motion of the belt/vehicle often lead to the coarse grains to concentrate on the top.
This effect is often referred to as the 'Brazil nut effect'[61] since the Brazilian nuts, being larger in
size, usually end up at the top surface during the transportation of the nut containers. Fig. 3.35
shows a mixture of grains of two different sizes (size ratio 2:1) in a container that is vibrated in
vertical direction. The vibration of the container causes the large particles to concentrate
at the free surface while the small size particles concentrate near the base.

Note that the segregation pattern obtained in the vibrated bed example shown in Fig. 3.35 is
sensitive to strength of the vibration and segregation pattern different from the one shown in
Fig. 3.35 are also observed.[68] In general, segregation is more dominant for freely flowing
granular materials and slightly cohesive materials and decreases for more cohesive materials
comprised of grains smaller than $500\,\mu m$[65] due to increase in the cohesion owing to
van-der-Waals, electrostatic, and surface tension forces as compared to gravity force.[52]
However, for fine grains, the effect of air drag also starts to play important role and can lead to
segregation.[14,53]

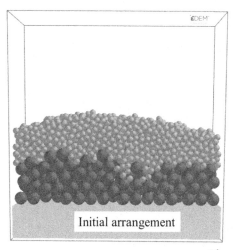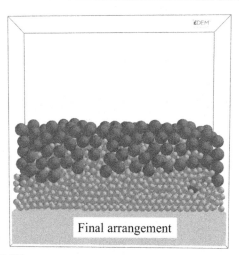

Fig. 3.35
Size segregation in a vertically vibrated bed (images produced from simulations using EDEM®
Academic software).

While increasing the size ratio seems to increase the segregation of free-flowing grains,[69] experiments in rotating tumblers[70] and computer simulations[71] show that scaled concentration profiles seem to be nearly independent of the size of the large particle. Tang and Puri[69] mention that segregation of a mixture of non-spherical and spherical grains differing in size is easier compared to a mixture of two different size grains having nearly spherical shape. However, a mixture of non-spherical grains of two different sizes is harder to segregate compared to mixture of spherical grains of two different sizes due to lower flowability of the mixture of non-spherical grains. Johanson[72] mentions that segregation is closely related to the flowability of the mixture, and though segregation in cohesive mixtures is substantially reduced, a quantitative relationship between these two is not available.[69]

Rhodes[41] mentions that segregation due to density difference is comparatively less important than segregation due to size difference in many situations, except in the case of gas fluidisation where density plays a much more important role compared to size. For the examples shown in Figs 3.33–3.35, segregation can also occur if grains differ only in their densities. In that case, the grains of lower density concentrate near the free surface while the large density grains concentrate in regions away from the free surface. Thus density difference can either enhance or reduce the segregation due to size difference depending upon whether the combined effect of the two factors is cooperating (larger grains have low density) or competitive (larger grains have high density).[71,73]

3.10.2 Mechanisms of Segregation

More than 10 different mechanisms of segregation have been suggested by researchers.[67,69] However, some of these proposed mechanisms can be combined together[69] and only few dominant ones are considered to be primarily responsible for segregation. Depending upon the direction of the particle movement, the differences in the particle properties can lead to either horizontal, i.e. side-by-side segregation or vertical, i.e. top-to-bottom segregation. The key mechanisms of size segregation are discussed below.

Percolation of small particles, also referred to as *sieving segregation,* is one of the commonly occurring mechanism of segregation that is operative during hopper flows (when material flow below the bin-to-hopper transition point), transport of material through belt conveyors and transfer chutes as well as during the formation of heaps. When a mixture of grains of different sizes undergoes shear or vibration, rearrangements in the particle packings lead to formation of gaps between the particles. The size and distribution of these gaps/holes keeps fluctuating with time. The grains can fall through these temporary holes due to the presence of gravity (Fig. 3.36) and the smaller particles have a higher probability to fall through a hole compared to the large grains which require a rather large size opening to fall through. This leads to a preferential percolation of the small particles through the 'fluctuating kinetic sieve'[74] of the temporary holes. The preferential percolation of the small particles is balanced by an upward flux of the large and small particles through 'squeeze expulsion'.[74] Percolation of small particles typically leads to top-to-bottom[69] segregation where the large size grains concentrate near the free surface. In the case when the material is flowing over an inclined surface (e.g. a heap), this mechanism of segregation, compounded by trajectory segregation, leads to the often observed 'sifting segregation'. Note that although sifting segregation observed in a heap (shown in Fig. 3.33) has been qualitatively described using other mechanisms,[75] recent research[76,77] has shown that quantitative prediction of the segregation in heap as well as rotating cylinder geometry is possible by correlating the percolation velocity of the particles with the local shear rate and the local concentration.

Trajectory segregation typically leads to side-by-side segregation[69] and usually occurs 'when the bulk solids moves with a horizontal velocity component through a gas',[14] for example, in flight or free-fall after the exit from a transfer chute. The effect of the air drag on the motion of the grains starts to affect the grain trajectory for sizes below $100\,\mu m$ and becomes significant for particle size below $10\,\mu m$. Fine particles cannot move with a large speed relative to the gas, and hence, are typically carried away by the air. For example, when the material exits from an inclined chute, the drag on the fine particles rather than inertia dominates and hence, different size grains have different trajectories after the exit. This leads to trajectory segregation shown in Fig. 3.37A.

Note that percolation segregation will occur during the chute flow provided shearing of layers over each other happens due to the presence of the base (and not just sliding of the material as a whole over the chute base). The shear in the layer will lead to top-to-bottom segregation with small particles concentrating towards the base and large particles concentrating towards the free surface (as in Fig. 3.36). Due to the presence of the velocity gradient along the layer thickness, the velocity of the larger grains near the free surface will be higher and the velocity of the smaller grains near the base will be lower. This difference in the exit velocities will also lead

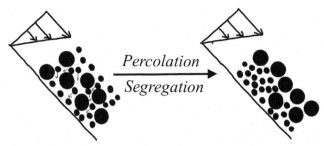

Fig. 3.36

Percolation mechanism of size segregation over an inclined surface leads to large particles near the free surface.

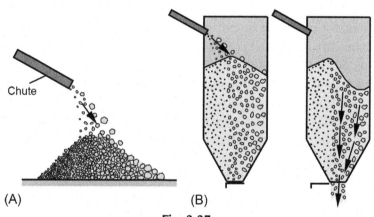

(A)   (B)

Fig. 3.37

(A) Trajectory segregation in a pile after discharge from a chute. (B) Side-by-side segregation during hopper filling may lead to eccentric flow during discharge. *Reproduced from Schulze D.* Powders and bulk solids: behavior, characterization, storage and flow. *Springer; 2008.*

segregation similar to the trajectory segregation shown in Fig. 3.37A. This type of segregation, however, occurs due to the combined effect of percolation and advection and can be observed even for coarse grains moving at moderate velocities so that the effect of the air drag is not important. Similarly, the percolation of the small particles over the inclined surface of the heap leads to rise of large grains near the free surface, which are then carried over to the periphery of the heap due to the higher speed, giving rise to 'sifting segregation' shown in Fig. 3.33.

In case of material having significant concentration of fine particles, the trajectory segregation effects are compounded due to the percolation-advection, giving rise to side-by-side segregation. Trajectory segregation influences the filling and subsequent discharge from a hopper as the material filled in the hopper shows side-by-side segregation (Fig. 3.37B). In case the fine particles have poor flowability compared to the coarse grains, eccentric discharge from the hopper may be seen as shown in Fig. 3.37B.

Segregation due to elutriation/fluidisation occurs when a powder having significant concentration of fines (smaller than 50 μm) is filled into a hopper or container.[41,69] Since the settling velocities of the fine particles in the air are very small (typically less than a few cm/s), the upward velocity of the displaced air (due to the pouring of the bulk material) will be typically larger than the settling velocity of the fine grains. Hence, the fines may remain suspended in the air longer than the time required for the coarse grains to settle in the hopper. The fines, carried by the air, then drop out and slowly settle on the material away from the centre as the air velocity decreases with the distance away from the hopper centre and becomes zero at the wall due to no-slip condition. Thus the centre of the hopper will contain relatively large size grains and the region away from the centre will be rich in fines (Fig. 3.38). The larger the height of the free fall of the material in the hopper, the more severe the intensity of the elutriation segregation.

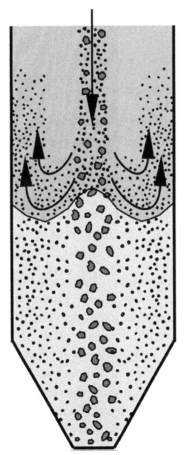

Fig. 3.38
Elutriation segregation while filling a hopper. *Reproduced from Schulze D. Powders and bulk solids: behavior, characterization, storage and flow. Springer; 2008.*

The poor flowability of the fines may lead to funnel flow (Fig. 3.21). Since the fine grains have a higher tendency to stick to each other and to the wall, a layer of very fine grains often sticks to the wall due to elutriation segregation, and falls on the material beyond a minimum thickness. If the grains are fed into the hopper by means of pneumatic conveying, the segregation effects get significantly enhanced due to higher velocity of the incoming air. Possibility of vertical segregation also exists due to the fluidisation of the layer of the fine grains due to the air escape from the bed.[14]

Before introducing possible ways to reduce segregation in the next section, it is important to reiterate that although a large number of mechanisms have been suggested for segregation by researchers, only a few have been addressed here. For example, as many as seven different mechanisms of segregation (including the formation of convection rolls) have been identified in a vertically vibrated bed of grains and the percolation mechanism mentioned above is just one such mechanism that dominates at low vibrations.[68] In addition, it appears that despite the large number of mechanisms proposed in literature, explanation of segregation patterns in an unconventional geometry, such as a horizontally vibrating macro-ratchet, requires invoking of an additional mechanism.[78]

3.10.3 Segregation Reduction Techniques

It should be clear by now that the segregation of the bulk solids occurs due to many factors and can have significant impact on the subsequent flow and handling of the material. Due to this reason, segregation in bulk solids is an active field of research attracting researchers from different disciplines. Schulze[14] highlights that 'quantitative prediction of segregation to obtain the distribution of the fines and coarse grains across the cross section is not possible due to several factors involved'. However, an understanding of the possible causes and mechanisms of segregation do offer some insights to minimise segregation, which are discussed below.

Since segregation typically occurs due to the difference in material properties, improving these properties is one possible way of reducing the segregation. Realising that size difference is the most important factor that affects segregation, narrowing of the particle size distribution can reduce the segregation significantly. However, the effect of changing the particle size distribution on the behaviour of the material needs to be well assessed. Particulate materials with a narrow size distribution often have lower packing fractions and compressive/tensile strength compared to the material having a wider size distribution. In case of a powder which is to be compressed in the form of a tablet, poor packing and low strength may be a cause of concern.[69] Huge variations in particle size can be avoided in two ways. One approach is to increase the size of the fines by granulation to remove the tail of the fines in the particle size distribution curve. Alternatively, large size grains can be crushed (e.g. by a crusher or in a mill) or broken (e.g. using high velocity impact). In most industrial operations, a minimum and a maximum permissible size of the grains is often prescribed as a cutoff to avoid unwanted issues in handling and processing.

In case the components of the mixtures differ in density, choosing a particular size ratio of larger to smaller grains (larger grain size of the material having higher grain density) can help prevent segregation by exploiting the competing effect of density and size segregation in some flow and handling scenarios.[71,73] Similarly, grains of highly irregular shape or long aspect ratios should be avoided, if possible.[69] Use of crushing/milling or agglomeration to obtain grains showing relatively less variation in shape may also be helpful. Since segregation is more intense in case of freely flowing bulk solids, reducing the size below a hundred microns also reduces the possibility of segregation, although it comes at a cost of reducing the flowability of the material, which may be significant.[52] Addition of small amounts of liquids or increase in the moisture can also reduce the flowability, and hence, reduce segregation. Care must be taken, however, to ensure that it does not lead to unwanted problems such as ratholing or arching in case of flow through a hopper.

In addition, to the generic methods discussed above, unique methods depending upon the specific scenario are discussed by Tang and Puri[69] and Schulze.[14] These include optimisation of the material filling process in hoppers/containers and remixing of the segregated components during the discharging process. While filling the material in a hopper through a central feed, formation of a large heap leads to sifting segregation (see Fig. 3.33). In addition, elutriation segregation may occur due to presence of fines and large free-fall height. To avoid this, various types of distributors and inserts (such as dispersion baffles) are used to distribute the feed over a much larger cross-sectional area, and reduce the height of the heap formed (see Fig. 3.39A). Alternatively, feeding of the material can be done in the form of layers

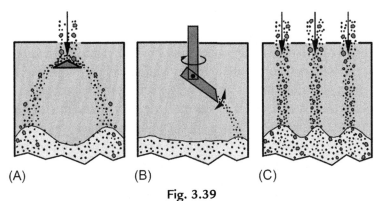

(A)   (B)   (C)

Fig. 3.39

Segregation reduction during hopper filling. (A) Use of dispersion baffle. (B) Rotating chute with adjustable inclination. (C) Multiple feeding points to obtain distributed feeding over the entire cross section. *Reproduced from Schulze D.* Powders and bulk solids: behavior, characterization, storage and flow. *Springer; 2008.*

by using a rotating chute with an inclination that can be adjusted to cover most of the cross-sectional area (see Fig. 3.39B). Note that this arrangement happens to be much more complex and expensive, and hence, is typically used only in highly specific applications such as loading of a blast furnace in the iron and steel industry. Another way to reduce the segregation is to include multiple feeding points (Fig. 3.39C) or a branched pipe distributor so segregation due to the formation of a large heap is avoided.

Due to the unpredictability of their efficiency and higher cost of the above solutions, Schulze[14] suggests that hoppers and bins should be designed in such a way that discharge from the hopper ensures remixing of the segregated material. In case of horizontal or side-by-side segregation, the mass flow happens to be a good solution. As mentioned before, the material flows through the entire cross-sectional area towards the exit in mass flow and gets remixed during this movement. The requirement of the steep walls for mass flow often increases the space requirement as well as the financial cost. Hence, inserts of various types (conical insert, cone-in-cone insert, bullet insert)[69] or discharge tubes[14] are used in hoppers. Segregation due to the effect of air resistance can be minimised by directing the material upward on a deflector plate or by pouring it over a helical slide. More details can be found in the book by Schulze.[14] Ordered (or interactive) mixing, in which fine grains (e.g. $<5\,\mu m$) are adsorbed on to large size 'carrier' grains by carefully controlling the particle sizes and interparticle forces, is often used in pharmaceutical industry to achieve mixture of little variance and minimise segregation.[41] More details about the ordered mixing can be found in the work of Harnby.[52]

In the end, it is worth emphasising that even for the simplest case of free-flowing, noncohesive, coarse grains, both the rheology and the segregation are yet to be completely understood and are an active area of research. It has been shown that even for this simplest case, the rheology and the segregation are intercoupled with each other.[54,71] In other words, the segregation seems to affect the rheology, which in turn affects the segregation. Since the theoretical understanding of the bulk solids' rheology, as well as segregation in various flow regimes is still being developed, quantitative predictions using continuum simulations are still a long way from being formed. Discrete element method simulations,[79] which model the motion of each individual grain due to the effect of various forces, are currently being used by researchers to explore and understand the behaviour of bulk solids in various scenarios. A number of opensource (e.g. Ligghts,[80] Yade,[81] MercuryCG[82]) as well as commercial packages (e.g. EDEM,[83] Rocky,[84] PFC3D[85]) are available for simulating the behaviour of bulk solids. Figs 3.33–3.35 that show the examples of segregation in heap, rotating cylinder, and vibrated bed are produced using the EDEM® Academic software. Through these simulations, a detailed understanding of the process can be obtained. Choice of appropriate simulation parameters and calibration of the simulation parameters, however, needs to be carefully done to obtain quantitative comparison with experiments.

3.11 Nomenclature

		Units in SI System	Dimensions in **M, L, T**
A	Cross-sectional area of container in Example 3.1	m²	L^2
A_{eff}	Effective area of the orifice in Eq. (3.17)	m²	L^2
a	Interfacial area per unit volume between constituents	m²/m³	L^{-1}
a_m	Maximum interfacial area per unit volume	m²/m³	L^{-1}
B	Breadth of the rectangular orifice	m	L
B	Angle between cone wall and horizontal	–	–
b	Degree of mixing	–	–
CI	Carr's compressibility index	–	–
c	Cohesion constant of material	N/m²	$ML^{-1}T^{-2}$
c	Beverloo constant in Eq. (3.16)	–	–
c	Coefficient of t (Eqs. 3.26, 3.28, 3.29)	s⁻¹	T^{-1}
c', c''	Modified Beverloo constants in Eqs (3.17) and (3.18)	–	–
D	Diameter of silo in Example 3.1	m	L
D_{eff}	Effective diameter of the orifice in Eq. (3.16)	m	L
D_h	Hydraulic mean diameter in Eq. (3.17)	m	L
d	Diameter of the solid particle	m	L
F	Force	N	MLT^{-2}
F_f	Frictional force	N	MLT^{-2}
g	Acceleration due to gravity	m/s²	LT^{-2}
H	Hausner ratio	–	–
H	Height of material in Example 3.1	m	L
K	Proportionality constant in Example 3.1	–	–
k	Shape coefficient in Eq. (3.16)	–	–
L	Length of rectangle	m	L
m	Mass of material in Example 3.2	kg	M
$\dot{m}$	Mass flowrate	kg/s	MT^{-1}
N	Normal force in Example 3.2	N	MLT^{-2}
n	No. of particles in each sample	–	–
n_g	Number of particles in a group	–	–
P	Perimeter in Example 3.1	m	L
P	Perimeter of the orifice in Example 3.7	m	L
P_{eff}	Perimeter of the orifice in Eq. (3.17)	m	L
p	Overall proportion of particles of one colour	–	–
R	Radius of cylinder in Example 3.2	m	L

r_c	Radial distance till the surface retains initial flat surface shape in Example 3.2	m	**L**
s	Standard deviation	–	–
s_0	Standard deviation of unmixed material	–	–
s_r	Standard deviation of random mixture	–	–
s_r^2	Variance for the mixture	–	–
t	Time	s	**T**
V	Velocity of particle	m/s	$\mathbf{LT^{-1}}$
w	Container wall thickness in flow through horizontal orifice	m	**L**
X	Correction factor in Example 3.7	–	–
X	Percentage of samples in which particles are not mixed	–	–
Y	Fraction of samples in which particles are mixed	–	–
z_c	Height of the flat surface at the centre in Example 3.2	m	**L**
α	Angle of repose	–	–
β	Angle between vertical hopper axis and the boundary of stagnant and flowing regions	–	–
$\mathfrak{D}$	Walker's distribution factor	–	–
δ	Angle of internal friction	–	–
δ_w	Angle of wall friction		
ε	Voidage	–	–
λ	Characteristic length in Example 3.1	m	**L**
μ	Friction coefficient in Example 3.2	–	–
μ	Friction coefficient of the material in Eq. (3.10)	–	–
μ_w	Friction coefficient between the material and container wall in Example 3.1	–	–
Ω	Angular velocity in Example 3.2	s^{-1}	$\mathbf{T^{-1}}$
ϕ	Solid fraction	–	–
ϕ_{RLP}	Random loose packing fraction	–	–
ϕ_{RCP}	Random close packing fraction	–	–
$\rho_{aerated}$	Aerated bulk density	kg/m^3	$\mathbf{ML^{-3}}$
ρ_b	Bulk density of material	kg/m^3	$\mathbf{ML^{-3}}$
ρ_f	Density of the fluid	kg/m^3	$\mathbf{ML^{-3}}$
ρ_p	Effective density of the solid	kg/m^3	$\mathbf{ML^{-3}}$
ρ_{tapped}	Tapped bulk density	kg/m^3	$\mathbf{ML^{-3}}$
σ	Normal stress	N/m^2	$\mathbf{ML^{-1}T^{-2}}$
σ_1	Major principal stress in Example 3.3	N/m^2	$\mathbf{ML^{-1}T^{-2}}$

σ_2	Minor principal stress in Example 3.3	N/m^2	$\mathbf{ML^{-1}\,T^{-2}}$
$\bar{\sigma}_w$	Wall normal stress in Example 3.1	N/m^2	$\mathbf{ML^{-1}\,T^{-2}}$
σ_w	Wall normal stress	N/m^2	$\mathbf{ML^{-1}\,T^{-2}}$
$\bar{\sigma}_{zz}$	Mean axial stress in Example 3.1	N/m^2	$\mathbf{ML^{-1}\,T^{-2}}$
$\sigma_{zz,w}$	Axial stress at wall	N/m^2	$\mathbf{ML^{-1}\,T^{-2}}$
τ	Shear stress	N/m^2	$\mathbf{ML^{-1}\,T^{-2}}$
$\tau_{\max}$	Maximum possible value of shear stress	N/m^2	$\mathbf{ML^{-1}\,T^{-2}}$
$\bar{\tau}_w$	Wall shear stress in Example 3.1	N/m^2	$\mathbf{ML^{-1}\,T^{-2}}$
τ_w	Wall shear stress	N/m^2	$\mathbf{ML^{-1}\,T^{-2}}$
τ_y	Yield shear stress	N/m^2	$\mathbf{ML^{-1}\,T^{-2}}$
θ	Characteristic angle of the orifice in Eq. (3.19)	–	–
θ_w	Hopper half angle	–	–

References

1. Cahn RW, Bever MB. *Concise encyclopaedia of materials characterization*. Pergamon Press; 1993.
2. Nedderman RM. *Statics and kinematics of granular materials*. Cambridge University Press; 1992.
3. Medina A, Luna E, Alvarado R. Axisymmetrical rotation of a sand heap. *Phys Rev E* 1995;**51**:4621–5.
4. Baxter GW, Yeung C. The rotating bucket of sand: experiment and theory. *Chaos* 1999;**9**:631–8.
5. Brown RL, Richards JC. *Principles of powders mechanics: essays on the packing and flow of powders and bulk solids*. Pergamon Press; 1970.
6. Andreotti B, Forterre Y, Pouliquen O. *Granular media: between fluid and solid*. Cambridge University Press; 2013.
7. Rao KK, Nott PR. *An introduction to granular flow*. Cambridge University Press; 2008.
8. Aulton ME, Taylor KMG. Aulton's pharmaceutics: the design and manufacture of medicines. In: *Handbook of behavioral neuroscience*. Vol. 12. Elsevier; 2013.
9. Dury CM, Ristow GH. Boundary effects on the angle of repose in rotating cylinders. *Phys Rev E Stat Phys Plasmas Fluids Relat Interdiscip Topics* 1998;**57**:4491–7.
10. Mcglinchey D. *Characterisation of bulk solids*. CRC Press; 2005.
11. Bird RB, Stewart WE, Lightfoot EN. *Transport phenomena*. John Wiley & Sons, Inc.; 2002
12. Jenike AW, Elsey PJ, Woolley RH. Flow properties of bulk solids. *Proc Am Soc Test Mat* 1960;**60**:1168.
13. Jenike AW. Gravity flow of solids. *Trans Inst Chem Eng* 1962;**40**:264.
14. Schulze D. *Powders and bulk solids: behavior, characterization, storage and flow*. Springer; 2008.
15. Carr JF, Walker DM. An annular shear cell for granular materials. *Powder Technol* 1968;**1**:369–73.
16. Ortega-Rivas E. *Unit operations of particulate solids: theory and practice*. CRC Press, Taylor and Francis Group; 2012.
17. Freeman RE, Cooke JR, Schnneider LCR. Measuring shear properties and normal stresses generated within a rotational shear cell for consolidated and non-consolidated powders. *Powder Technol* 2009;**190**:65–9.
18. Ashton MD, Farley R, Valentin FH. An improved apparatus for measuring the tensile strength of powders. *J Sci Instrum* 1964;**41**:763.
19. Orband JLR, Geldart D. Direct measurement of powder cohesion using a torsional device. *Powder Technol* 1997;**92**:25–33.
20. Beverloo WA, Leniger HA, Van De Velde J. The flow of granular solids through orifices. *Chem Eng Sci* 1961;**15**:260–9.

21. Aguirre MA, Grande JG, Calvo A, Pugnaloni LA, Geminard J-C. Pressure independence of granular flow through an aperture. *Phys Rev Lett* 2010;**104**:1–4.

22. Aguirre MA, Grande JG, Calvo A, Pugnaloni LA, Geminard J-C. Granular flow through an aperture: pressure and flow rate are independent. *Phys Rev E - Stat Nonlinear Soft Matter Phys* 2011;**83**:1–6.

23. Wieghardt K. Experiments in granular flow. *Annu Rev Fluid Mech* 1975;1–2.

24. Hilton JE, Cleary PW. Granular flow during hopper discharge. *Phys Rev E - Stat Nonlinear Soft Matter Phys* 2011;**84**:1–10.

25. Medina A, Cabrera D, Lopez-Villa A, Pliego M. Discharge rates of dry granular material from bins with lateral exit holes. *Powder Technol* 2014;**253**:270–5.

26. Cleary PW, Sawley ML. DEM modelling of industrial granular flows: 3D case studies and the effect of particle shape on hopper discharge. *Appl Math Model* 2002;**26**:89–111.

27. Al-Din N, Gunn DJ. The flow of non-cohesive solids through orifices. *Chem Eng Sci* 1984;**39**:121–7.

28. Kotchanova II. Experimental and theoretical investigations on the discharge of granular materials from bins. *Powder Technol* 1970;**4**:32–7.

29. Mankoc C, et al. The flow rate of granular materials through an orifice. *Granul Matter* 2007;**9**:407–14.

30. Janda A, Zuriguel I, Maza D. Flow rate of particles through apertures obtained from self-similar density and velocity profiles. *Phys Rev Lett* 2012;**108**:1–5.

31. Rubio-Largo SM, Janda A, Maza D, Zuriguel I, Hidalgo RC. Disentangling the free-fall arch paradox in Silo discharge. *Phys Rev Lett* 2015;**114**:1–5.

32. Sheldon HG, Durian DJ. Granular discharge and clogging for tilted hoppers. *Granul Matter* 2010;**12**:579–85.

33. Serrano DA, Medina A, Chavarria GR, Pliego M, Klapp J. Mass flow rate of granular material flowing from tilted bins. *Powder Technol* 2015;**286**:438–43.

34. Weigand J. Viscosity measurements on powders with a new viscometer. *Appl Rheol* 1999;**9**:204.

35. Woodcock CR, Mason JS. *Bulk solids handling: an introduction to the practice and technology.* Blackie Academic & Professional; 1987.

36. Masuda H, Higashitani K, Yoshida H. *Powder technology handbook.* Taylor & Francis; 2006.

37. Liptak BG. *Instrument engineer's handbook—process instrumentation and analysis.* Butterworth-Heinemann; 1995.

38. Fruchtbaum J. *Bulk materials handling handbook.* Van Nostrand Reinhold Inc Company Inc; 1988.

39. Lacey PMC. The mixing of solid particles. *Trans Inst Chem Eng* 1943;**21**:53.

40. Lacey PMC. Developments in the theory of particle mixing. *J Appl Chem* 1954;**4**:257–68.

41. Rhodes M. *Introduction to particle technology.* 2nd ed. John Wiley & Sons Ltd; 2008.

42. Danckwerts PV. The definition and measurement of some characteristics of mixtures. *Appl Sci Res* 1952;**3**:279–96.

43. Buslik D. Mixing and sampling with special reference to multisized granular particles. *Bull Am Soc Test Mat* 1950;**165**:66.

44. Kaye BH. Mixing of powders. In: Fayed ME, Otten L, editors. *Handbook of powder science & technology.* Chapman and Hall; 1997.

45. Wen Y, Liu M, Liu B, Shao Y. Comparative study on the characterization method of particle mixing index using DEM method. *Procedia Eng* 2015;**102**:1630–42.

46. Nadeem H, Heindel TJ. Review of noninvasive methods to characterize granular mixing. *Powder Technol* 2018;**332**:331–50.

47. Coulson JM, Maitra NK. The mixing of solid particles. *Ind Chem* 1950;**26**:55.

48. Blumberg R, Maritz JS. Mixing of solid particles. *Chem Eng Sci* 1953;**2**:240–6.

49. Brothman A, Wollan GN, Feldman SM. New analysis provides formula to solve mixing problems. *Chem Met Eng* 1945;**52**:102.

50. Herdan G. *Small particle statistics.* Butterworths Scientific; 1960.

51. Danckwerts PV. Theory of mixtures and mixing. *Research* 1953;**6**:355.

52. Harnby N. An engineering view of pharmaceutical powder mixing. *Pharm Sci Technol Today* 2000;**3**:303–9.

53. APPIE (Association of Powder Process Industry and Engineering). *Mixing technology for particulate materials.* Nikkan Kogyo Press; 2001.

54. Tripathi A, Khakhar DV. Density difference-driven segregation in a dense granular flow. *J Fluid Mech* 2013;**717**:643–69.

55. Hsiau S-S, Wang P-C, Tai C-H. Convection cells and segregation in a vibrated granular bed. *AIChE J* 2002;**48**:1430–8.

56. Pereira GG, Cleary PW. Radial segregation of multi-component granular media in a rotating tumbler. *Granul Matter* 2013;**15**:705–24.

57. Pollard BL, Henein H. Kinetics of radial segregation of different sized irregular particles in rotary cylinders. *Can Metall Q* 1989;**28**:29–40.

58. Ulrich S, Schroter M, Swinney HL. Influence of friction on granular segregation. *Phys Rev E* 2007;**76**:042301.

59. Brito R, Enriquez H, Godoy S, Soto R. Segregation induced by inelasticity in a vibrofluidized granular mixture. *Phys Rev E - Stat Nonlinear Soft Matter Phys* 2008;**77**:3–8.

60. Drahun JA, Bridgwater J. The mechanisms of free surface segregation. *Powder Technol* 1983;**36**:39–53.

61. Rosato A, Strandburg KJ, Prinz F, Swendsen RH. Why the Brazil nuts are on top: size segregation of particulate matter by shaking. *Phys Rev Lett* 1987;**58**:1038–40.

62. Shinbrot T, Muzzio FJ. Reverse buoyancy in shaken granular beds. *Phys Rev Lett* 1998;**81**:4365–8.

63. Shinbrot T. The Brazil nut effect—in reverse. *Nature* 2004;**429**:352–3.

64. Knight JB, Jaeger HM, Nagel SR. Vibration-induced size separation in granular media: the convection connection. *Phys Rev Lett* 1993;**70**:3728–31.

65. Williams J, Khan M. Mixing and segregation of particulate solids of different particle-size. *Inst Chem Eng* 1973;**269**:19.

66. Fan Y, Jacob KV, Freireich B, Lueptow RM. Segregation of granular materials in bounded heap flow: a review. *Powder Technol* 2017;**312**:67–88.

67. Gray JMNT. Particle segregation in dense granular flows. *Annu Rev Fluid Mech* 2018;**50**:407–33.

68. Schroter M, Ulrich S, Kreft J, Swift JB, Swinney HL. Mechanisms in the size segregation of a binary granular mixture. *Phys Rev E - Stat Nonlinear Soft Matter Phys* 2006;**74**:011307.

69. Tang P, Puri VM. Methods for minimizing segregation: a review. *Part Sci Technol* 2004;**22**:321–37.

70. Hajra SK, Khakhar DV. Radial segregation of ternary granular mixtures in rotating cylinders. *Granul Matter* 2011;**13**:475–86.

71. Tripathi A, Khakhar DV. Rheology of binary granular mixtures in the dense flow regime. *Phys Fluids* 2011;**23**:113302.

72. Johanson JR. Predicting segregation of bimodal particle mixtures using the flow properties of bulk solids. *Pharm Technol* 1996;**8**:38–44.

73. Jain N, Ottino JM, Lueptow RM. Regimes of segregation and mixing in combined size and density granular systems: an experimental study. *Granul Matter* 2005;**7**:69–81.

74. Savage SB, Lun CKK. Particle size segregation in inclined chute flow of dry cohesionless granular solids. *J Fluid Mech* 1988;**189**:311–35.

75. Mosby J, De Silva SR, Enstad GG. Segregation of particulate materials-mechanisms and testers. *Kona Powder Part J* 1996;**14**:31–42.

76. Fan Y, Schlick CP, Umbanhowar PB, Ottino JM, Lueptow RM. Modelling size segregation of granular materials: the roles of segregation, advection and diffusion. *J Fluid Mech* 2014;**741**:252–79.

77. Schlick CP, et al. Modeling segregation of bidisperse granular materials using physical control parameters in the quasi-2D bounded heap. *Am Inst Chem Eng J* 2015;**61**:1524–34.

78. Bhateja A, Sharma I, Singh JK. Segregation physics of a macroscale granular ratchet. *Phys Rev Fluids* 2017;**2**:1–9.

79. Thornton C. *Granular dynamics, contact mechanics and particle system simulations: a DEM study.* Springer; 2015.

80. https://www.cfdem.com/liggghts-open-source-discrete-element-method-particle-simulation-code.

81. https://yade-dem.org/doc/.

82. http://www.mercurydpm.org/.
83. https://www.edemsimulation.com/.
84. https://rocky.esss.co/.
85. https://www.itascacg.com/software/pfc.

Further Reading

Allen T. *Particle size measurement*. 5th ed. Vols. 1 and 2. London: Chapman and Hall; 1997.

Andreotti B, Forterre Y, Pouliquen O. *Granular media: between fluid and solid*. New York: Cambridge University Press; 2013.

Aulton ME, Taylor KMG. Aulton's pharmaceutics: the design and manufacture of medicines. In: *Handbook of behavioral neuroscience*. Vol. 12. Elsevier; 2013.

Brown RL, Richards JC. *Principles of powders mechanics: essays on the packing and flow of powders and bulk solids*. Pergamon Press; 1970.

Fruchtbaum J. *Bulk materials handling handbook*. Springer; 1988.

Herdan G. *Small particle statistics*. 2nd ed. London: Butterworths; 1960.

Masuda H, Higashitani K, Yoshida H. *Powder technology handbook*. Taylor & Francis; 2006.

Mcglinchey D. *Characterisation of bulk solids*. CRC Press; 2005.

Nedderman RM. *Statics and kinematics of granular materials*. Cambridge University Press; 1992.

Rao KK, Nott PR. *An introduction to granular flow*. New York: Cambridge University Press; 2008.

Rhodes M. *Introduction to particle technology*. 2nd ed. Wiley; 1998.

Rumpf H. *Particle technology*. London: Chapman and Hall; 1990.

Schulze D. *Powders and bulk solids: behavior, characterization, storage and flow*. Springer; 2008.

Sterbacek Z, Tausk P. *Mixing in the chemical industry*. (translated from Czech by K. Mayer)Oxford: Pergamon Press; 1965.

Thornton C. *Granular dynamics, contact mechanics and particle system simulations: a DEM study*. Springer; 2015.

Woodcock CR, Mason JS. *Bulk solids handling: an introduction to the practice and technology*. Blackie Academic and Professional; 1987.

CHAPTER 4

Classification of Solid Particles From Liquids and Gases

4.1 Introduction

Multiphase and multicomponent mixtures containing one or more types of particles, liquids, and/or gases are often encountered in particle processing. The word 'classification' refers to the separation of particles from solid–solid, solid–liquid, solid–gas, or other complex mixtures. In general, the separation of particles can be achieved by exploiting the difference in size, shape, physical, or chemical properties of solid particles. In Chapter 3, sieves or screens (that will be discussed further in this chapter) are introduced, wherein the separation is purely based on particle dimensions (or sizes). The use of sieves or screens if carried as 'dry' operation is an example 'solid–solid' separation. In the processing of solid–liquid and solid–gas mixtures, which are often encountered, for example, during the flow and transport of particles, a requirement post such operation is the complete separation of solids from the fluids or the separation of solids into different fractions. Certain multicomponent materials generated in an industry whether it be gaseous or liquid, may contain solid particles which are either valuable or toxic, and need to be separated from a fluid stream. Another common example that involves the separation of solids from gases is vehicular emission. For separating solids from solids, one of the strategies commonly employed is to identify a suitable fluid in which the solid–solid particle mixtures can be dispersed, and then exploit the difference in chemical or physical properties of the solid particles or solid–liquid mixtures to effect classification. This chapter deals with classification of particles from liquids and gases.

4.2 Classification of Solid Particles

4.2.1 Introduction

The problem of separating solid particles according to their physical properties is important for large-scale operations in the mining industry, where it is necessary to separate the valuable constituents in a mineral from the adhering gangue which is usually of a lower density. In this case, it is first necessary to crush the material so that individual particles contain only one constituent. There is a similar problem in coal washing plants in which dirt is separated from the clean coal. The processing industries are usually more concerned with separating a single

Coulson and Richardson's Chemical Engineering. https://doi.org/10.1016/B978-0-08-101098-3.00005-6

material, such as the product from a size-reduction plant, into a number of size fractions, or in obtaining a uniform material for incorporation in a system in which a chemical reaction takes place. As similar problems are involved in separating a mixture into its constituents and into size fractions, the two processes are considered together.

Separation depends on the selection of a process in which the behaviour of the material is influenced to a marked degree by some physical property. Thus if a material is to be separated into various size fractions, a sieving method may be used because this process depends primarily on the size of the particles, though other physical properties such as the shape of the particles and their tendency to agglomerate, may also be involved. Other methods of separation depend on the differences in the behaviour of the particles in a moving fluid. In this case, the size and the density of the particles are the most important factors, and the shape is of secondary importance. Other processes make use of differences in electrical or magnetic properties of the materials or in their surface properties.

In general, large particles are separated into size fractions by means of screens. Small particles that would clog the fine apertures of the screen or make it impractical to collect them by narrowing the opening are separated in a fluid. Fluid separation is commonly used for dividing a mixture of two materials, though magnetic, electrostatic, and froth flotation methods are also used where appropriate.

Considerable advances have been made in developing techniques for size separation in the subsieve range. As discussed by Work[1], the emphasis has been on techniques lending themselves to automatic working. Many of the methods of separation and much of the equipment have been developed for use in the mining and metallurgical industries, as described by Taggart[2].

Most processes, which depend on differences in the behaviour of particles in a stream of fluid, separate materials according to their terminal falling velocities, as reported in Chapter 6, which in turn depends primarily on density and size, and to a lesser extent, on shape. Thus in many cases it is possible to use the method to separate a mixture of two materials into its constituents, or to separate a mixture of particles of the same material into a number of size fractions.

If, for example, it is desired to separate particles of a relatively dense material **A** of density ρ_A from particles of a less dense material **B**, and the size range is large, the terminal falling velocities of the largest particles of **B** of density ρ_B may be greater than those of the smallest particles of **A**, and therefore a complete separation will not be possible. The maximum range of sizes that can be separated is calculated from the ratio of the sizes of the particles of the two materials, which have the same terminal falling velocities. It is shown in Chapter 6 that this condition is given by Eq. (6.32) as:

$$\frac{d_B}{d_A} = \left(\frac{\rho_A - \rho}{\rho_B - \rho}\right)^j \tag{4.1}$$

where $j = 0.5$ for fine particles, where Stokes' law applies, and $j = 1$ for coarse particles where Newton's law applies.

The equation shows that the size range becomes wider with an increase in the density of the separating fluid. When the fluid has the same density as the less-dense material, complete separation is possible, whatever the relative sizes. Although water is the most commonly used fluid, an effective specific gravity that is greater than unity is obtained if *hindered settling* takes place. Frequently, the fluid density is increased artificially by forming a suspension of small particles of a dense solid, such as galena, ferro-silicon, magnetite, sand, or clay, in the water. These suspensions have effective densities of up to about $3500 \, kg/m^3$. Alternatively, zinc chloride or calcium chloride solutions may be used. On the small scale, liquids with a range of densities are produced by mixing benzene and carbon tetrachloride.

If the particles are allowed to settle in the fluid for only a short time, they will not attain their terminal falling velocities, and a better degree of separation can be obtained. A particle of material **A** will have an initial acceleration $g[1 - (\rho/\rho_A)]$, because there is no fluid friction when the relative velocity is zero. Thus the initial velocity is a function of density only, and it is unaffected by size and shape. A small particle of the denser material will therefore always commence settling at a greater rate than a large particle of the less dense material. Theoretically, it should be possible to separate materials completely, irrespective of the size range, provided that the periods of settling are sufficiently short. In practice, the required periods will often be so short that it is impossible to make use of this principle alone, although a better degree of separation may be obtained if the particles are not allowed to become fully accelerated. As the time of settling increases, the larger particles of the less dense material catch up and overtake the small, heavy particles.

Size separation equipment in which particles move in a fluid stream is now considered, noting that most of the plant utilises the difference in the terminal falling velocities of the particles: However, in the hydraulic jig, the particles are allowed to settle for only very brief periods at a time, and this equipment may therefore be used when the size range of the material is large.

Example 4.1

A mixture of quartz and galena of a size range from 0.015 to 0.065 mm is to be separated into two pure fractions using a hindered settling process. What is the minimum apparent density of the fluid that will give this separation? The density of galena is $7500 \, kg/m^3$ and the density of quartz is $2650 \, kg/m^3$.

Solution

Assuming the galena and quartz particles are of similar shapes, then from Eq. (4.1), the required density of fluid when Stokes' law applies is given by:

$$\frac{0.065}{0.015} = \left(\frac{7500 - \rho}{2650 - \rho}\right)^{0.5}$$

and:

$$\rho = 2377 \, \text{kg/m}^3$$

The required density of fluid when Newton's law applies is given by:

$$\frac{0.065}{0.015} = \left(\frac{7500 - \rho}{2650 - \rho}\right)^{1.0}$$

and hence:

$$\rho = 1196 \, \text{kg/m}^3$$

Thus the required density of the fluid is between 1196 and 2377 kg/m³.

4.2.2 Gravity Settling

The settling tank

Material is introduced in the form of suspension into a tank containing a relatively large volume of water moving at a low velocity, as shown in Fig. 4.1. The particles soon enter the slowly moving water and, because the small particles settle at a lower rate, they are carried farther before they reach the bottom of the tank. The very fine particles are carried away in the liquid overflow. Receptacles at various distances from the inlet collect different grades of particles according to their terminal falling velocities, with the particles of high terminal falling velocity collecting near the inlet. The positions at which the particles are collected may be calculated on the assumption that they rapidly reach their terminal falling velocities, and attain the same horizontal velocity as the fluid.

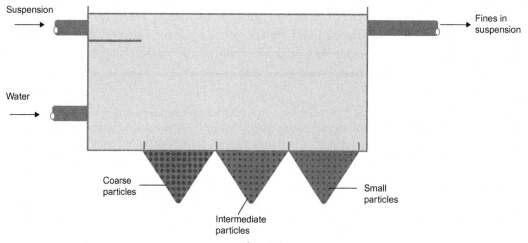

Fig. 4.1
Gravity settling tank.

If the material is introduced in solid form down a chute, the position at which the particles are deposited will be determined by the rate at which they lose the horizontal component of their velocities. The larger particles will therefore be carried farther and the smaller particles will be deposited near the inlet. The position at which the particles are deposited may be calculated using the relations given in Chapter 6.

The elutriator

Material may be separated by means of an elutriator, which consists of a vertical tube where fluid is passed up at a controlled velocity. The particles are introduced, often through a side tube, and the smaller particles are carried over in the fluid stream while the large particles settle against the upward current. Further size fractions may be collected if the overflow from the first tube is passed vertically upwards through a second tube of greater cross section, and any number of such tubes can be arranged in a series.

The Spitzkasten

The Spitzkasten consists of a series of vessels of conical shape arranged in a series. A suspension of the material is fed into the top of the first vessel and the larger particles settle, while the smaller ones are carried over in the liquid overflow and enter the top of a second conical vessel of greater cross-sectional area. The bases of the vessels are fitted with wide diameter outlets, and a stream of water can be introduced near the outlet so that the particles have to settle against a slowly rising stream of liquid. The size of the smallest particle, which is collected in each of the vessels, is influenced by the upward velocity at the bottom outlet of each of the vessels. The size of each successive vessel is increased, partly because the amount of liquid to be handled includes all the water used for classifying in the previous vessels, and partly because it is desired to reduce, in stages, the surface velocity of the fluid flowing from one vessel to the next. The Spitzkasten thus combines the principles used in the settling tank and in the elutriator.

The size of the material collected in each of the units is determined by the rate of feeding of suspension, the upward velocity of the liquid in the vessel, and the diameter of the vessel. The equipment can also be used for separating a mixture of materials into its constituents, provided that the size range is not large. The individual units can be made of wood or sheet metal.

The Hydrosizer, illustrated in Fig. 4.2, works on the same principle, although it has a number of compartments trapezoidal in section. It is suitable for use with materials finer than about four-mesh ($\sim$5 mm), and it works at high concentrations in order to obtain the advantages of hindered settling. Altogether eight sharply classified fractions are produced, ranging from coarse nearest to the feed inlet, to fine from the eighth compartment, at feed rates in the range 2–15 kg/s.

Fig. 4.2

Hydrosizer. *The authors and publishers acknowledge with thanks the assistance given by Dorr-Oliver Co Ltd, Croydon, Surrey for their permission for reproduction of this figure.*

The double cone classifier

This classifier, shown in Fig. 4.3, consists of a conical vessel, with a second hollow cone of greater angle arranged apex downwards inside it, so that there is an annular space of an approximately constant cross-section between the two cones. The bottom portion of the inner cone is cut away and its position relative to the outer cone can be regulated by a screw adjustment. Water is passed in an upward direction, as in the Spitzkasten, and overflows into a launder arranged around the periphery of the outer cone. The material to be separated is fed as suspension to the centre of the inner cone, and the liquid level is maintained slightly higher than the overflow level, so that there is a continuous flow of liquid downwards in the centre cone. The particles are therefore brought into the annular space where they are subjected to a classifying action; the smaller particles are carried away in the overflow, and the larger particles settle against the liquid stream and are taken off at the bottom.

The mechanical classifier

Several forms of classifiers exist in which the material of lower settling velocity is carried away in a liquid overflow, and the material of higher settling velocity is deposited at the bottom of the equipment, and is dragged upwards against the flow of the liquid by some mechanical means. During the course of the raking action, the solids are turned over so that any small particles trapped under the larger ones are brought to the top again.

The mechanical classifier is extensively used where it is necessary to separate a large amount of fines from the oversize material. Arrangement of a number of mechanical classifiers in a series makes it possible for a material to be separated into several size fractions.

In the rake classifier, which consists of a shallow rectangular tank inclined to the horizontal, the feed is introduced in the form of a suspension near the middle of the tank and water for

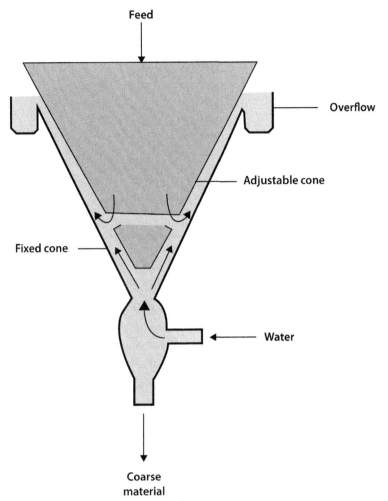

Fig. 4.3

Double cone classifier. *The authors and publishers acknowledge with thanks the assistance given by Denver Process Equipment Ltd, Leatherhead, Surrey for their permission for reproduction of this figure.*

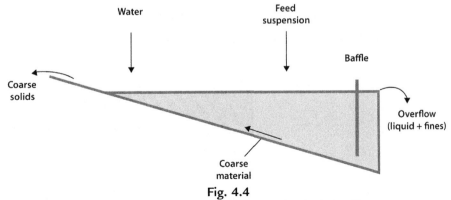

Fig. 4.4

The principle of operation of the mechanical classifier.

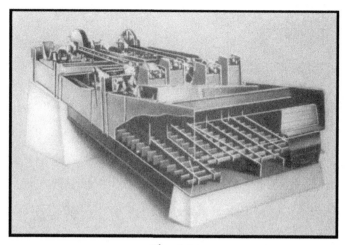

Fig. 4.5

A rake classifier. *The authors and publishers acknowledge with thanks the assistance given by Dorr-Oliver Co Ltd, Croydon, Surrey for their permission for reproduction of this figure.*

classifying is added at the upper end, as shown in Fig. 4.4. The liquid, together with the material of low terminal falling velocity, flows down the tank under a baffle and is then discharged over an overflow weir. The heavy material settles to the bottom and is then dragged upwards by means of a rake. It is thus separated from the liquid and is discharged at the upper end of the tank. Fig. 4.5 shows the Dorr-Oliver rake classifier. Fig. 4.6 shows the Denver classifier, which is similar in action, although a trough that is semi-circular in cross section is used, and the material, which settles to the bottom, is continuously moved to the upper end by means of a rotating helical scraper.

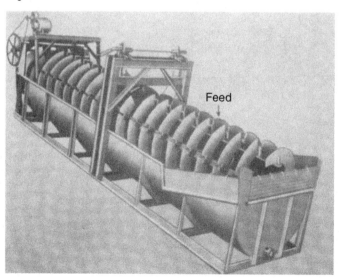

Feed

Fig. 4.6

The Denver classifier. *The authors and publishers acknowledge with thanks the assistance given by Denver Process Equipment Ltd, Leatherhead, Surrey for their permission for reproduction of this figure.*

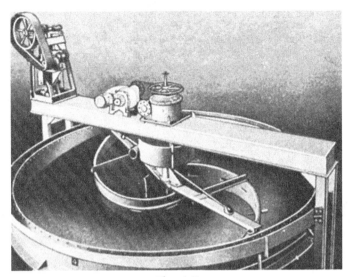

Fig. 4.7
Bowl classifier.

The bowl classifier

The bowl classifier, which is used for fine materials, consists of a shallow bowl with a concave bottom as shown in Fig. 4.7. The suspension is fed into the centre of the bowl near the liquid surface, the liquid and the fine particles are carried in a radial direction and pass into the overflow, which is an open launder running round the periphery of the top of the bowl. The heavier or larger material settles to the bottom and is raked towards the outlet at the centre. The classifier has a large overflow area, and consequently, high volumetric rates of flow of liquid may be used without producing a high linear velocity at the overflow. The action is similar to that of a thickener which is effecting incomplete clarification.

The hydraulic jig

The hydraulic jig operates by allowing material to settle for brief periods so that the particles do not attain their terminal falling velocities, and is therefore suitable for separating materials of a wide size range into their constituents. The material to be separated is fed dry, or more usually in suspension, over a screen and is subjected to a pulsating action by liquid, which is set in oscillation by means of a reciprocating plunger. The particles on the screen constitute a suspension of high concentration, and therefore the advantages of hindered settling are exploited. The jig usually consists of a rectangular-section tank with a tapered bottom, divided into two portions by a vertical baffle. In one section, the plunger operates in a vertical direction; the other incorporates the screen over which the separation is carried out. In addition, a stream of liquid is fed to the jig during the upward stroke.

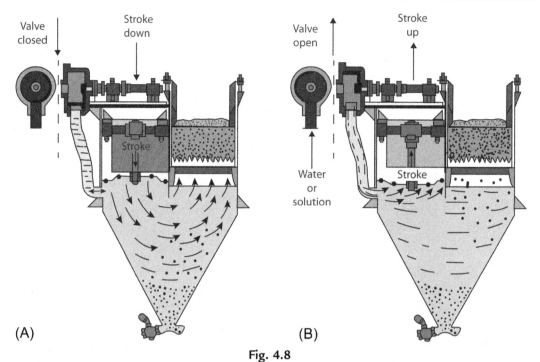

Fig. 4.8

The Denver hydraulic jig: (A) downward stroke, (B) upward stroke. *The authors and publishers acknowledge with thanks the assistance given by Denver Process Equipment Ltd, Leatherhead, Surrey for their permission for reproduction of this figure.*

The particles on the screen are brought into suspension during the downward stroke of the plunger (Fig. 4.8A). As the water passes upwards the bed opens up, starting at the top, and thus tends to rise *en masse*. During the upward stroke, the input of water is adjusted so that there is virtually no flow through the bed (Fig. 4.8B). During this period, differential settling takes place and the denser material tends to collect near the screen and the lighter material above it. After a short time, the material becomes divided into three strata: the bottom layer consisting of the large particles of the heavy material, the next of large particles of the lighter material together with small particles of the heavy material, and the top stratum of small particles of the light material. Larger particles wedge at an earlier stage than smaller ones, and therefore the small particles of the denser material are able to fall through the spaces between the larger particles of the light material. Many of these small particles then fall through the supporting gauze.

Four separate fractions are obtained from the jig and their successful operation depends on the rapid removal of these fractions. The small particles of the heavy material, which have fallen through the gauze, are taken off at the bottom of the tank. The larger particles of each of the materials are retained on the gauze in two layers, the denser material at the bottom and the less dense on top. These two fractions are removed through gates at the side of the jig. The remaining material, consisting of small particles of the lighter material, is carried away in the liquid overflow.

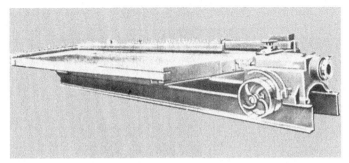

Fig. 4.9

Wilfley riffled table. *The authors and publishers acknowledge with thanks the assistance given by Wilfley Mining Machinery Co Ltd for their permission for reproduction of this figure.*

Riffled tables

The riffled table, of which the Wilfley table shown in Fig. 4.9 is a typical example, consists of a flat table, which is inclined at an angle of about 3° to the horizontal. Running approximately parallel to the top edge is a series of slats, or riffles, as they are termed, about 6 mm in height. The material to be separated is fed to one of the top corners and a reciprocating motion, consisting of a slow movement in the forward direction and a very rapid return stroke, causes it to move across the table. The particles also tend to move downwards under the combined action of gravity and of a stream of water, which is introduced along the top edge, but are opposed by the riffles behind which the smaller particles or the denser material tend to be held. Thus the large particles and the less dense material are carried downwards, and the remainder is carried parallel to the riffles. In many cases, each riffle is tapered, which allows a number of fractions to be collected. Riffled tables can be used for separating materials down to about 50 μm in size, provided that the difference in the densities is large.

4.2.3 Centrifugal Separators

The use of cyclone separators for the removal of suspended dust particles from gases is discussed in Section 4.3. By suitable choice of operating conditions, it is also possible to use centrifugal methods for the classification of solid particles according to their terminal falling velocities. Fig. 4.10 shows a typical air separation unit, which is similar in construction to the double cone classifier. The solids are fed to the bottom of the annular space between the cones, carried upwards in the air stream, and enter the inner cone through a series of ports fitted with adjustable vanes. As depicted in the diagram, the suspension enters approximately tangentially and is therefore subjected to the action of centrifugal force. The coarse solids are thrown outwards against the walls and fall to the bottom under the action of gravity, while the small particles are removed by means of an exhaust fan. This type of separator is widely used for separating the oversize material from the product from a ball mill, and is suitable for materials as fine as 50 μm.

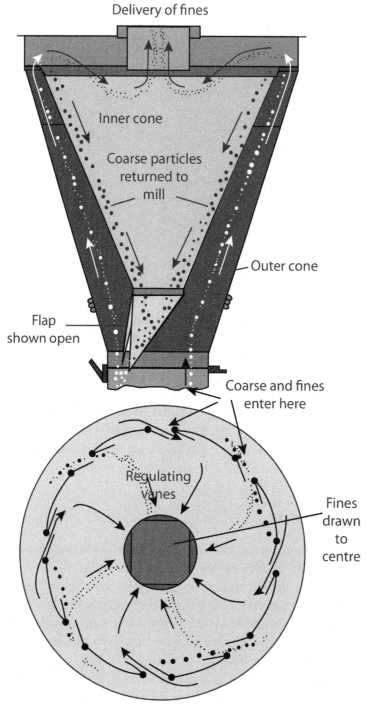

Delivery of fines

Inner cone

Coarse particles
returned to
mill

Outer cone

Flap
shown open

Coarse and fines
enter here

Regulating
vanes

Fines
drawn
to
centre

Fig. 4.10

NEI International Combustion vacuum air separator. *The authors and publishers acknowledge with thanks the assistance given by NEI International Combustion Ltd, Derby for their permission for reproduction of this figure.*

A mechanical air separator is shown in Figs 4.11 and 4.12. The material is introduced at the top through the hollow shaft, and it falls on to the rotating disc, which throws it outwards. Very large particles fall into the inner cone, and the remainder are lifted by the air current produced by the rotating vanes above the disc. Because a rotary motion has been imparted to the air stream, the coarser particles are thrown to the walls of the inner cone and, together with the very large particles, are withdrawn from the bottom. The fine particles remain in suspension, carried down the space between the two cones, and are collected from the bottom of the outer cone. The air returns to the inner cone through a series of deflector vanes, and blows any fines from the surfaces of the larger particles as they fall. The size at which the 'cut' is made is controlled by the number of fixed and rotating blades, and the speed of rotation. The Delta sizer is designed to separate dry powders of mixed particle size within the range 5–200 μm into distinct fractions at

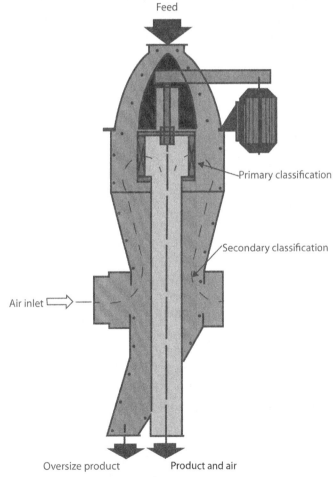

Fig. 4.11

The NEI Delta sizer. *The authors and publishers acknowledge with thanks the assistance given by NEI International Combustion Ltd, Derby for their permission for reproduction of this figure.*

Fig. 4.12

NEI Delta sizer ultrafine classifier. *The authors and publishers acknowledge with thanks the assistance given by NEI International Combustion Ltd, Derby for their permission for reproduction of this figure.*

a high extraction efficiency and low power consumption. The equipment shown in Fig. 4.12 has a feed rate from 2 to 7 kg/s depending upon the desired cut point and the bulk density of the material.

4.2.4 The Hydrocyclone or Liquid Cyclone

In the hydrocyclone, or hydraulic cyclone, which is discussed extensively in the literature,[3–9] separation is effected in the centrifugal field, which is generated as a result of introducing the feed at a high tangential velocity into the separator. The hydrocyclone may be used for:

(a) separating particles (suspended in a liquid of lower density) by size or density, or more generally, by terminal falling velocity;
(b) the removal of suspended solids from a liquid;
(c) separating immiscible liquids of different densities;
(d) dewatering of suspensions to give a more concentrated product;
(e) breaking down liquid–liquid or gas–liquid dispersions; and
(f) the removal of dissolved gases from liquids.

In this section, the general design of the hydrocyclone and its application in the grading of solid particles, or their separation from a liquid, is considered and then the special features required in hydrocyclones required for the separation of immiscible liquids will be addressed. The use of cyclones for separating suspended particles from gases is discussed in Section 4.3.2.

General principles and applications in solids classification

A variety of geometrical designs exists, the most common being of the form shown in Fig. 4.13, with an upper cylindrical portion of between 20 and 300 mm in diameter and a lower conical portion, although larger units are occasionally employed. The fluid is introduced near the top of the cylindrical section, and overflow is removed through a centrally located offtake pipe at the top, usually terminating approximately at the level corresponding to the junction of the cylindrical and tapered portions of the shell. Other configurations include an entirely cylindrical shell, a conical shell with no cylindrical portion, and curved sides to the tapered section. Generally, a long, tapered section is preferred.

As flow patterns are influenced only slightly by gravitational forces, hydrocyclones may be operated with their axes inclined at any angle, including the horizontal, although the removal of the underflow is facilitated with a vertical axis.

Much of the separating power of the hydrocyclone is associated with the interaction of the primary vortex which follows the walls, and the secondary vortex, revolving about a low pressure core, which moves concentrically in a countercurrent direction as shown in Fig. 4.13. The separating force is greatest in this secondary vortex, which causes medium-sized particles to be rejected outwards to join the primary vortex flow in which they are then carried back towards the apex. It is this secondary vortex which exerts the predominant influence in determining the largest size or heaviest particle, which will remain in the overflow stream. The tangential fluid velocity is maximum at a radius roughly equal to that of the overflow discharge pipe or 'vortex finder'.

The flow patterns in the hydrocyclone are complex, and much development work has been necessary to determine the most effective geometry, as theoretical considerations alone will not allow the accurate prediction of the size cut, which will be obtained. A mathematical model has been proposed by Rhodes et al.,[10] and predictions of streamlines from their work are shown in Fig. 4.14. Salcudean and Gartshore[11] have also carried out numerical simulations of the three-dimensional flow in a hydrocyclone, and have used the results to predict cut sizes. Good agreement has been obtained with experimental measurements.

Near the top of the hydrocyclone there will be some short-circuiting of the flow between the inlet and the overflow, although the effects are reduced as a result of the formation of circulating eddies, often referred to as the mantle, which tend to act as a barrier. Within the secondary vortex the pressure is low, and there is a depression in the liquid surface in the region of the axis. Frequently, a gas core is formed, and any gas dispersed in the form of fine bubbles, or coming out of solution, tends to migrate to this core. In pressurised systems, the gas core may be very much reduced in size, and sometimes completely eliminated.

The effectiveness of a hydrocyclone as a separator depends primarily on the liquid flow pattern and, in particular, on the values of the three principal components of velocity (tangential, u_t,

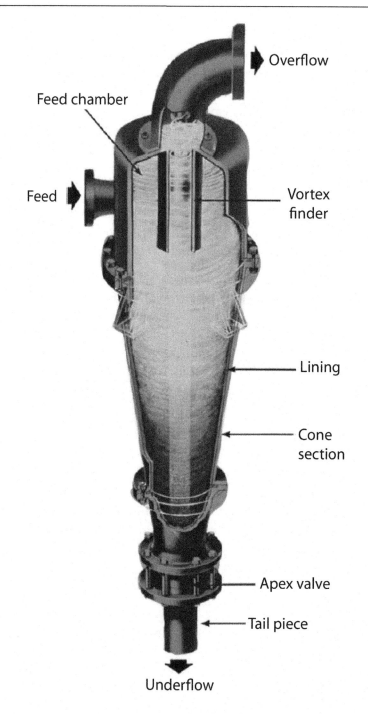

Fig. 4.13

Liquid cyclone or hydrocyclone. *The authors and publishers acknowledge with thanks the assistance given by NEI International Combustion Ltd, Derby for their permission for reproduction of this figure.*

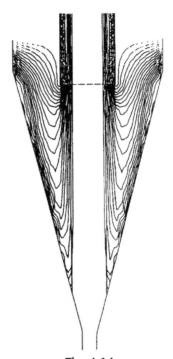

Fig. 4.14
Predicted streamlines in a hydrocyclone.[10]

axial u_a, and radial u_r) throughout the body of the separator. The efficiency with which fractionation takes place is highest at very low concentrations, and therefore there is frequently a compromise between selectivity and throughput in selecting the optimum concentration. Turbulent flow conditions should be avoided wherever possible as they give rise to undesirable mixing patterns which reduce the separating capacity.

Most studies of hydrocyclone performance for particle classification have been carried out at particle concentrations of about 1% by volume. The simplest theory for the classification of particles is based on the concept that particles tend to orbit at the radius at which the centrifugal force is exactly balanced by the fluid friction force on the particles. Thus the orbits will be of increasing radius as the particle size increases. Unfortunately, there is scant information on how the radial velocity component varies with location. In general, a particle will be conveyed in the secondary vortex to the overflow, if its orbital radius is less than the radius of that vortex. Alternatively, if the orbital radius were greater than the diameter of the shell at a particular height, the particle will be deposited on the walls and will be drawn downwards to the bottom outlet.

The general characteristics of vortices have been considered in Volume 1A, Chapter 2, where it has been shown that, in the absence of fluid friction, the relation between tangential fluid velocity u_t and r is as follows:

(a)   in a *forced vortex* (formed, for example, when a rotating member, such as an impeller in a pump or mixer, imparts a constant angular velocity ω to the fluid):

$$u_t/r = \omega = \text{constant} \tag{4.2}$$

(b)   in a *free* (or *natural*) vortex, in which the energy per unit mass of fluid is constant throughout:

$$u_t r = \text{constant} \tag{4.3}$$

Laboratory measurements on hydrocylones have shown that, in the primary vortex, the relation between tangential velocity u_t and radius r, is approximately of the form:

$$u_t r^n = \text{constant} \tag{4.4}$$

where $n = 0.5 - 0.9$.

Thus as might have been expected, the primary vortex in the hydrocyclone is more akin to a free ($n = 1$) than to a forced ($n = -1$) vortex.

Typical profiles of the components u_t, u_r and u_a of the liquid velocity in a hydrocyclone, as given by Kelsall[12] and Svarovsky,[8] are shown in Fig. 4.15. These profiles are quite critically dependent on the geometry of the separator. Entry velocities, corresponding to u_t near the walls at that level, will be up to about 5 m/s.

Qualitatively similar velocity profiles are set up within gas cyclones, discussed in Section 4.3.2, which are extensively used for the removal of suspended solids from gases. In this case, the velocities are generally considerably higher.

For suspended particles, the axial and tangential velocities will differ little from those of the liquid. As already suggested, however, in the radial direction, the particles tend to rotate at the radius at which the centrifugal force is balanced by the drag force of the fluid on the particle. This means that the radius of the particle orbit in a liquid will increase with both particle size and density, or more particularly with terminal falling velocity. As a result, the larger and denser particles will move selectively towards the walls of the hydrocyclone, and will be discharged at the apex of the cone. For a sharp fractionation to be achieved, the residence times of the particles must be sufficiently long for them to attain their equilibrium orbits within the hydrocyclone.[13] Velocity profiles do in practice tend to be flatter than those predicted by Eq. (4.3), especially in large units where turbulence may develop.

Design considerations

Hydrocyclones are 20–500 mm in diameter, with the smaller units giving a much better separation. Typical values of length to diameter ratios range from about 5 to 20. Because of the very high shearing stresses which are set up, flocs will be broken down and the suspension in the

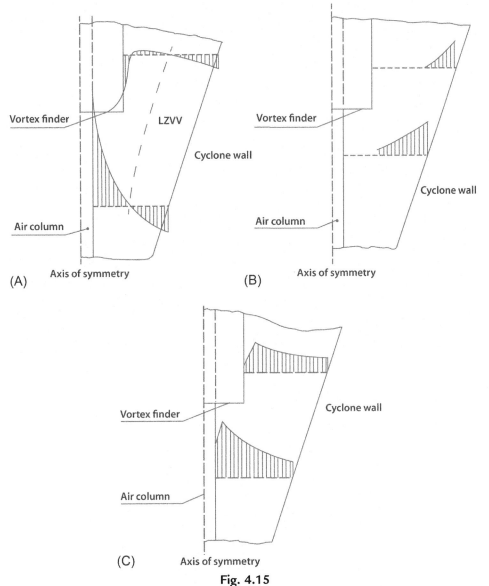

Fig. 4.15
Typical velocity distributions in a hydrocyclone[12]: (A) axial, (B) radial, (C) tangential (*broken line* LZVV is the locus of zero axial velocity).

secondary vortex will be completely deflocculated, irrespective of its condition on entry. Generally, hydrocylones are not effective in removing particles smaller than about 2–3 μm.

Because separating power is greatest in hydrocyclones of small diameter, and the cut size is approximately proportional to the diameter of the cylindrical shell raised to the power of 1.5, it is common practice to operate banks of small hydrocyclones in parallel inside a large containing vessel. Furthermore, this procedure also makes scale-up easier to carry out. With units connected only in parallel (or as single units), however, it is not possible to control the compositions of the overflow and underflow independently, and therefore some form of series–parallel operation is often employed. This is an important consideration when the hydrocyclone is used for thickening a suspension. In this case, there is a requirement to produce an underflow of the appropriate solids concentration and for the overflow to be particle-free. These two conditions cannot usually be satisfied in a single unit. In thickening, the particle concentrations are high and little classification by size occurs.

In general, the performance of the hydrocyclone is improved by increasing the operating pressure, and the principal control variable is the size of the orifice on the underflow discharge. Several theoretical and practical studies have been made in an attempt to present a sound basis for design. These have been described by Bradley[3] and Savrovsky[4, 8] among others, but the description is not entirely satisfying. Also, some design charts and formulae have been given by Zanker.[14] In practice, tests with the actual materials to be used are desirable for the evaluation for the various parameters. Since systems are usually scaled-up by increasing the number of units in parallel, it is seldom necessary to carry out tests on large units.

The optimum design of a hydrocyclone for a given function depends upon reconciling a number of conflicting factors and reference should be made to specialist publications. Because it is simple in construction and has no moving parts, maintenance costs are low. The chief problem arises from the abrasive effect of the solids; materials of construction, such as polyurethane, show less wear than metals and ceramics.

Effect of non-Newtonian fluid properties

There have been very few studies of the effects of non-Newtonian properties on flow patterns in hydrocyclones, although Dyakowski et al.[15] have carried out numerical simulations for power-law fluids, and these have been validated by experimental measurements in which velocity profiles were obtained by laser–Doppler anemometry.

Liquid–liquid separations

For the separation of two liquids, the hydrocyclone is normally operated at a pressure high enough to suppress the formation of the gas core by appropriate throttling of the outlets on the underflow and overflow streams. Hydrocyclones are now used extensively for separating mixtures of oil (as the less dense component) from water. A wide variety of designs has been developed to cope with different proportions of the light and heavy components. Colman[16] described a plant for cleaning up large quantities of sea or river water following an oil spill, the proportion of the light component may be very low (2–3%). A separator with an enlarged cylindrical top section may be fitted with two tangential inlets, and the conical section may be very long with an included angle as small as 1–2°. The gradual taper leads to the formation of a secondary vortex of very small diameter to accommodate the relatively small proportion of the lighter (oil) phase. By using this configuration, it has been possible to recover a high proportion (up to 97%) of the oil in the feed. In this case, the correct setting of the orifice on the underflow discharge is of critical importance.

Liquid–gas separations

Hydrocyclones are used for removing entrained gas bubbles from liquids, and the extracted gas collects in the gas core of the secondary vortex before leaving through the vortex finder. Nebrensky[17] points out that because of the low pressure in the region close to the axis, they will also remove dissolved gases.[17]

Applications

Hydrocyclones are now being used for a wide range of applications and are displacing other types of separation equipment in many areas. They are compact and have low maintenance costs, with no moving parts, and have substantially lower liquid holdups than gravity-driven separators. Since any aggregates tend to break down in the high shear fields, they usually give cleaner separations. However, they are relatively inflexible in that a given unit will operate satisfactorily over only a narrow range of flowrates and particle concentrations. This is not usually a serious drawback as hydrocyclones are usually operated with banks of small units in parallel, and their number can be adjusted to suit the current flowrate.

4.2.5 Sieves or Screens

Sieves or screens are used on a large scale for the separation of particles according to their sizes, and on a small scale for the production of closely graded materials in carrying out size analyses. The method is applicable for particles of a size as small as about 50 μm, but not for very fine materials because of the difficulty of producing accurately woven fine gauze of sufficient strength, and the fact that the screens become clogged. Woven wire cloth is generally used for fine sizes and perforated plates are used for the larger meshes. Some large industrial screens are formed, either from a series of parallel rods or from H-shaped links bolted together, though square or circular openings are usual. Screens may be operated on both a wet or dry basis. With coarse solids, the screen surface may be continuously washed by means of a flowing stream of water which tends to keep the particles apart so the finer particles can be removed from the surface of larger particles and to keep the screen free of adhering materials. Fine screens are normally operated wet, with the solids fed continuously as a suspension. Concentrated suspensions, particularly when flocculated, have highly effective viscosities and frequently exhibit shear-thinning, non-Newtonian characteristics (see Chapter 13, Section 13.9). By maintaining a high cross-flow velocity over the surface of the screen, or by rapid vibration, the apparent viscosity of the suspension may be reduced and the screening rate substantially increased.

The only large screen that is hand operated is the *grizzly*. This has a plane screening surface composed of longitudinal bars up to 3 m long, fixed in a rectangular framework. In some cases, a reciprocating motion is imparted to alternate bars to reduce the risk of clogging. The grizzly is usually inclined at an angle to the horizontal. The greater the angle, the greater is the throughput, although the screening efficiency is reduced. If the grizzly is used for wet screening, a much smaller angle is employed. In some screens, the longitudinal bars are replaced by a perforated plate.

Mechanically operated screens are vibrated by an electromagnetic device as shown in Fig. 4.16, or mechanically as shown in Fig. 4.17. In the former case, the screen itself is vibrated, and in the latter, the whole assembly. Because very rapid accelerations and retardations are produced, the power consumption and the wear on the bearings are high. These screens are sometimes mounted in a multideck fashion with the coarsest screen on top, either horizontally or inclined at angles up to 45°. With the horizontal machine, the vibratory motion fulfils the additional function of moving the particles across the screen.

The screen area which is required for a given operation cannot be predicted without testing the material under similar conditions on a small plant. In particular, the tendency of the material to clog the screening surface can only be determined experimentally.

A very large mechanically operated screen is the *trommel*, shown in Fig. 4.18, which consists of a slowly rotating perforated cylinder with its axis at a slight angle to the horizontal. The material to be screened is fed in at the top and gradually moves down the screen and passes over

Fig. 4.16
Hummer electromagnetic screen. *The authors and publishers acknowledge with thanks the assistance given by Lockers Engineers Ltd, Warrington for their permission for reproduction of this figure.*

Fig. 4.17

Tyrock mechanical screen. *The authors and publishers acknowledge with thanks the assistance given by Lockers Engineers Ltd, Warrington for their permission for reproduction of this figure.*

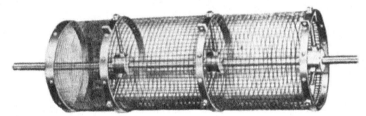

Fig. 4.18

Trommel. *The authors and publishers acknowledge with thanks the assistance given by Lockers Engineers Ltd, Warrington for their permission for reproduction of this figure.*

apertures of gradually increasing size, which means all the material passes over the finest screen. There is therefore a tendency for blockage of the apertures by the large material and for oversize particles to be forced through. Further, the relatively fragile fine screen is subjected to the abrasive action of the large particles. These difficulties are obviated to some extent by arranging the screens in the form of concentric cylinders, with the coarsest in the centre. The disadvantage of all screens of this type is that only a small fraction of the screening area is in use at any time. The speed of rotation of the trommel should not be so high that the material is carried completely around in contact with the screening surface. The lowest speed at which this occurs is known as the critical speed and is analogous to the critical speed of the ball mill, discussed in Chapter 5. Speeds between one-third and a half of the critical speed are usually recommended. In a modified form of the trommel, the screen surfaces are in the form of truncated cones. Such screens are mounted with their axes horizontal, and the material always flows away from the apex of the cone.

Screening of suspensions of very fine particles can present difficulties because of the tendency of the apertures to clog, and thus impede the passage of further particles. Furthermore, if flocculated, the suspensions can exhibit very high apparent viscosities, and flowrates can fall to unacceptably low levels. Many such suspensions exhibit highly shear-thinning non-Newtonian characteristics with their apparent viscosities showing a marked decrease as the velocity of flow is increased (see Volume 1A, Chapter 3). With such materials, the rate of screening may be substantially increased by imparting a high velocity to the suspension in a direction parallel to the screen surface. Such motion has been successfully imparted by imposing a rotational or a reciprocating motion to the screen.

4.2.6 Magnetic Separators

In the magnetic separator, material is passed through the field of an electromagnet, which causes the retention or retardation of the magnetic constituent. It is important that the material should be supplied as a thin sheet so that all the particles are subjected to a field of the same intensity, and so that the free movement of individual particles is not impeded. The two main types of equipment are:

(a) Eliminators, which are used for the removal of small quantities of magnetic material from the charge to the equipment. These are frequently employed, for example, for the removal of stray pieces of scrap iron from the feed to crushing equipment (discussed in Chapter 5). A common type of eliminator is a magnetic pulley incorporated in a belt conveyor so that the non-magnetic material is discharged in the normal manner, and the magnetic material adheres to the belt and falls off from the underside.

(b) Concentrators, which are used for the separation of magnetic ores from the accompanying mineral matter. These may operate with dry or wet feeds and an example of the latter is the Mastermag wet drum separator, the principle of operation of which is shown in Fig. 4.19. An industrial machine is shown in operation in Fig. 4.20. A slurry containing the magnetic component is fed between the rotating magnet drum cover and the casing. The stationary magnet system has several radial poles, which attract the magnetic material to the drum face, and the rotating cover carries the magnetic material from one pole to another, at the same time gyrating the magnetic particles, allowing the non-magnetics to fall back into the slurry mainstream. The clean magnetic product is discharged clear of the slurry tailings. Operations can be co-current or counter-current, and the recovery of magnetic material can be as high as 99.5%.

An example of a concentrator operating on a dry feed is a rotating disc separator. The material is fed continuously as a thin layer beneath a rotating magnetic disc, which picks up the magnetic material in the zone of high magnetic intensity. The captured particles are carried by the disc to the discharge chutes where they are released. The non-magnetic material is then passed to a second magnetic separation zone where secondary separation occurs in the same way, leaving a

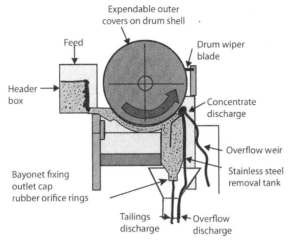

Fig. 4.19

Principle of operation of a wet drum separator. *The authors and publishers acknowledge with thanks the assistance given by Master Magnets Ltd, Birmingham for their permission for reproduction of this figure.*

Fig. 4.20

An industrial scale wet drum separator in action. *The authors and publishers acknowledge with thanks the assistance given by Master Magnets Ltd, Birmingham for their permission for reproduction of this figure.*

clean non-magnetic product to emerge from the discharge end of the machine. A Mastermagnet disc separator is shown in Fig. 4.21.

High gradient magnetic separation

The removal of small quantities of finely dispersed ferromagnetic materials from fine minerals, such as china clay, may be effectively carried out in a high gradient magnetic field. The suspension of mineral is passed through a matrix of ferromagnetic wires, which is magnetised by

Fig. 4.21
Rotating disc separator. *The authors and publishers acknowledge with thanks the assistance given by Master Magnets Ltd, Birmingham for their permission for reproduction of this figure.*

the application of an external magnetic field. The removal of the weakly magnetic particles containing iron may considerably improve the 'brightness' of the mineral, and thereby enhance its value as a coating or filler material for paper, or for use in the manufacture of high-quality porcelain. In cases where the magnetic susceptibility of the contaminating component is too low, adsorption may first be carried out on the surface of a material with the necessary magnetic properties. The magnetic field is generated in the gap between the poles of an electromagnet into which a loose matrix of fine stainless steel wire, usually of voidage of about 0.95, is inserted.

The attractive force on a particle is proportional to its magnetic susceptibility and to the product of the field strength and its gradient, and the fine wire matrix is used to minimise the distance between adjacent magnetised surfaces. The attractive forces which bind the particles must be sufficiently strong to ensure that the particles are not removed by the hydrodynamic drag exerted by the flowing suspension. As the deposit of separated particles builds up, the capture rate progressively diminishes and, at the appropriate stage, the particles are released by reducing the magnetic field strength to zero and flushing out with water. Commercial machines usually have two reciprocating canisters, in one of which particles are being collected from a stream of suspension, and in the other released into a waste stream. The dead time during which the canisters are being exchanged may be as short as 10 s.

Magnetic fields of very high intensity may be obtained by the use of superconducting magnets which operate most effectively at the temperature of liquid helium, and conservation of both gas and 'cold' is therefore of paramount importance. The reciprocating canister system employed in the china clay industry is described by Svarovsky[30] and involves the use of a single superconducting magnet and two canisters. At any time, one is in the magnetic field while the other is withdrawn for cleaning. The whole system needs delicate magnetic balancing so that the two canisters can be moved without the use of large forces and, for this to be the case, the amount of iron in the magnetic field must be maintained constant throughout the transfer process. The superconducting magnet then remains at high field strength, thereby reducing the demand for liquid helium.

Magnetic microorganisms

Microorganisms can play an important role in the removal of certain heavy metal ions from effluent solutions. In the case of uranyl ions, which are paramagnetic, the cells which have adsorbed the ions may be concentrated using a high gradient magnetic separation process. If the ions themselves are not magnetic, it may be possible to precipitate a magnetic deposit on the surfaces of the cells. Some microorganisms incorporate a magnetic component in their cellular structure and are capable of taking up nonmagnetic pollutants and are then themselves recoverable in a magnetic field. Such organisms are referred to being *magnetotactic.*

4.2.7 Electrostatic Separators

Electrostatic separators, in which differences in the electrical properties of the materials are exploited in order to effect a separation, are sometimes used with small quantities of fine material. The solids are fed from a hopper onto a rotating drum, which is either charged or earthed, and an electrode bearing the opposite charge is situated at a small distance from the drum, as shown in Fig. 4.22. The point at which the material leaves the drum determined by the

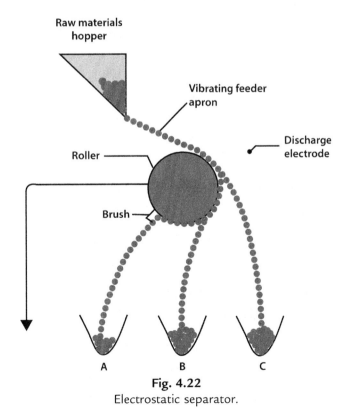

Fig. 4.22
Electrostatic separator.

charge it acquires, and by suitable arrangement of the collecting bins (A, B, C), a sharp classification can be obtained.

4.2.8 Flotation

Froth flotation

The separation of a mixture using flotation methods depends on differences in the surface properties of the materials involved. If the mixture is suspended in an aerated liquid, the gas bubbles will tend to adhere preferentially to one of the constituents, the one which is more difficult to wet by the liquid, and its effective density may be reduced to such an extent that it will rise to the surface. If a suitable frothing agent is added to the liquid, the particles are held in the surface by means of the froth until they can be discharged over a weir. Froth flotation is widely used in the metallurgical industries where, in general, the ore is difficult to wet and the residual particle from earth is readily wetted. Both the theory and practical application of froth flotation are discussed by Clarke and Wilson.[18]

The froth flotation process depends on the existence, or development, of a selective affinity of one of the constituents for the envelopes of the gas bubbles. In general, this affinity must be induced. The reagents which increase the angle of contact between the liquid and one of the materials are known as *promoters* and *collectors*. Promoters are selectively adsorbed on the surface of one material and form a monomolecular layer. The use of excess material destroys the effect. Concentrations of the order of 0.05 kg/ton of solids are usually required. A commonly used promoter is sodium ethyl xanthate:

$$NaS-C(-O-C_2H_5)=S$$

In solution, this material ionises giving positively charged sodium ions and negatively charged xanthate ions. The polar end of the xanthate ion is adsorbed and the new surface of the particles is therefore made up of the nonpolar part of the radical, so that the contact angle with water is increased (the particle becomes hydrophobic). A large number of other materials are used in particular cases, such as other xanthates and diazoaminobenzene. Collectors are the materials which form surface films on the particles. These films are thicker than those produced by promoters, and collectors therefore have to be added in higher concentrations–about 0.5 kg/ton of solids. Pine oil is a commonly used collector, and petroleum compounds are frequently used, though they often form a very greasy froth which is then difficult to disperse.

In many cases, it is necessary to modify the surface of the other constituent so that it does not adsorb the collector or promoter. This is effected by means of materials known as *modifiers,* which are either adsorbed on the surface of the particles or react chemically with the particle surface, and thereby prevent the adsorption of the collector or promoter. Mineral acids, alkalis, and salts are frequently used for this purpose.

An essential requirement of this process is the production of a froth that is of a stability that is sufficient to retain the particles of the one constituent at the bubble surface, so that they can be discharged in the overflow over the weir. On the other hand, the froth should not be so stable that it is difficult to break down at a later stage. The frothing agent should reduce the interfacial tension between the liquid and gas, and the quantities required are less when the frothing agent is adsorbed at the interface. Liquid soaps, soluble oils, cresol, and alcohols, in the range between amyl and octyl alcohols, are frequently used as frothing agents. In many cases, the stability of the foam is increased by adding a stabiliser–usually a mineral oil–which increases the viscosity of the liquid film forming the bubble, and therefore reduces the rate at which liquid can drain from the envelope. Pine oil is sometimes used as a frothing agent. This produces a stable froth and, in addition, acts as a collector.

The reagents which are used in froth flotation are usually specific in their action, and the choice of suitable reagents is usually made as a result of tests on small-scale equipment. Mixtures of more than two components may be separated in stages using different reagents at each stage. Sometimes the behaviour of a system is considerably influenced by change in the pH of the solution, and it is then possible to remove materials at successive stages by progressive alterations in the pH. In general, froth flotation processes may be used for particles of 5–250 μm. The tendency of various materials to respond to froth flotation methods is discussed by Doughty.[19] Hanumanth and Williams[20, 21] have published details of the design and operation of laboratory flotation cells and the effect of froth height on the flotation of china clay.

It is advantageous to generate bubbles of micronsize when the particles to be floated are tiny. The generation of such bubbles is almost impossible in conventional equipment, which relies on mechanical means of breaking down the gas. If air, or another gas, is dissolved under pressure in the suspension before it is introduced into the cell, numerous microbubbles are formed when the pressure is reduced, and these then attach themselves to the hydrophobic particles. Similar effects can be obtained by operating the cells under vacuum, or producing gas bubbles electrolytically. Dissolved air flotation and electro filtration are discussed later.

The Denver DR flotation machine

The Denver DR flotation machine, an example of a typical froth flotation unit used in the mining industry, is illustrated in Fig. 4.23. The pulp is introduced through a feed box and is distributed over the entire width of the first cell. Circulation of the pulp through each cell is such that, as the pulp comes into contact with the impeller, it is subjected to intense agitation and aeration. Low pressure air for this purpose is introduced down the standpipe surrounding the shaft and is thoroughly disseminated throughout the pulp in the form of minute bubbles when it leaves the impeller/diffuser zone, thus assuring maximum contact with the solids, as shown in Fig. 4.23. Each unit is suspended in an essentially open trough and generates a 'ring doughnut' circulation pattern, with the liquid being discharged radially from the impeller, through the diffuser, across the base of the tank, and then rising vertically as it returns to the eye of the

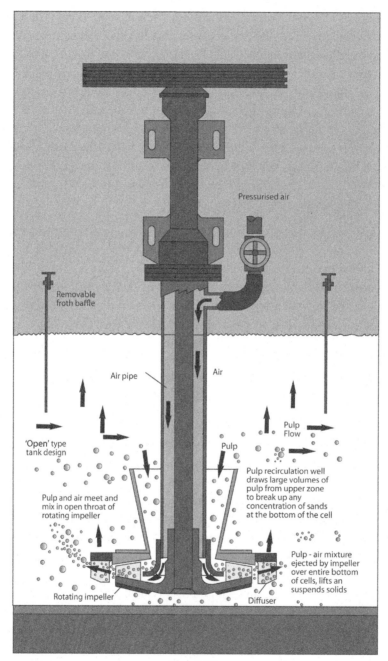

Pressurised air

Removable
froth baffle

Air pipe

Air

'Open' type
tank design

Pulp
Flow

Pulp

Pulp and air meet and
mix in open throat of
rotating impeller

Pulp recirculation well
draws large volumes of
pulp from upper zone
to break up any
concentration of sands
at the bottom of the cell

Pulp - air mixture
ejected by impeller
over entire bottom
of cells, lifts an
suspends solids

Rotating impeller

Diffuser

Fig. 4.23

Typical cross section through a Denver DR flotation machine. *The authors and publishers acknowledge with thanks the assistance given by Denver Process Equipment Ltd, Leatherhead, Surrey for their permission for reproduction of this figure.*

impeller through the recirculation well. This design gives strong vertical flows in the base zone of the tank to suspend coarse solids, and by recirculation through the well, isolates the upper zone which remains relatively quiescent.

Froth baffles are placed between each unit mechanism to prevent migration of froth as the liquid flows along the tank. The liquor level is controlled at the end of each bank section by a combination of weir overflows and dart valves which can be automated. These units operate with a fully flooded impeller. A low-pressure air supply is required to deliver air into the eye of the impeller where it is mixed with the recirculating liquor at the tip of the air bell. Butterfly valves are used to adjust and control the quantity of air delivered into each unit.

Each cell is provided with an individually controlled air valve. Air pressure is between 108 and $124 \, kN/m^2$ (7 and 23 kN/m^2 gauge) depending on the depth and size of the machine and the pulp density. Typical energy requirements for this machine range from $3.1 \, kW/m^3$ of cell volume for a $2.8 \, m^3$ unit to $1.2 \, kW/m^3$ for a $42 \, m^3$ unit.

In the froth flotation cell used for coal washing, illustrated in Fig. 4.24, the suspension contains about 10% of solids, together with the necessary reagents. The liquid flows along the cell bank and passes over a weir, and directly enters the unit via a feed pipe and feed hood. Liquor is

Fig. 4.24
A Denver flotation cell for coal washing. *The authors and publishers acknowledge with thanks the assistance given by Denver Process Equipment Ltd, Leatherhead, Surrey for their permission for reproduction of this figure.*

discharged radially from the impeller, through the diffuser, and is directed along the cell base and recirculated through ports in the feed hood. The zone of maximum turbulence is confined to the base of the tank; a quiescent zone exists in the upper part of the cell. These units induce sufficient air to ensure effective flotation without the need for an external air blower.

Dissolved air flotation

In general, the efficiency with which air bubbles attach themselves to fine particles is considerably improved if the bubbles themselves are tiny. In dissolved air flotation, described in detail by Gochin[22] and by Zabel,[23] the suspension is saturated with air under pressure before it is introduced through a nozzle into the flotation chamber. Because shear rates are much lower than they are in dispersed air flotation, flocculated suspensions may be processed without deflocculation occurring. If the flocs are weak, however, they may be broken down in the nozzle, and it may then be desirable to avoid passing the suspension through the high-pressure system by recycling part of the effluent from the flotation cell. However, in this case higher pressures will be needed to maintain the same released air to suspension ratio.

The suspension must be saturated, or nearly saturated, with air at high pressure in a contacting device, such as a packed column, before it is introduced into the flotation cell. At a pressure of $500 \, kN/m^2$, the solubility of air in water at 291 K is about $0.1 \, m^3 \, air/m^3$ water and, on reducing the pressure to $100 \, kN/m^2$, from Henry's law approximately $0.08 \, m^3 \, air/m^3$ water will be released as a dispersion of fine bubbles in the liquid.

Table 4.1 provides data on the solubility of air in water as a function of temperature and pressure.

Electroflotation

Kuhn et al.[24] described the electroflotation process which represents an interesting development for the treatment of dilute suspensions. Gas bubbles are generated electrolytically within the suspension, and attach themselves to the suspended particles, which then rise to the surface. Because the bubbles are very small, they have a high surface to volume ratio and are therefore very effective in suspensions of fine particles.

The method has been developed particularly for the treatment of dilute industrial effluents, including suspensions and colloids, especially those containing small quantities of organic

Table 4.1 Solubility of air in water and air released (litres free air/m^3 water)

Temperature (K)	Pressure 500 kN/m^2 (5 bar)	Pressure 100 kN/m^2 (1 bar)	Air Released on Reducing Pressure
273	0.145	0.029	0.116
283	0.115	0.023	0.092
293	0.095	0.019	0.076
303	0.080	0.016	0.064

materials. It allows the dilute suspension to be separated into a concentrated slurry and a clear liquid and, at the same time, permits oxidation of unwanted organic materials at the positive electrode.

Typically, the electrodes are made of lead dioxide on a titanium substrate in the form of horizontal perforated plates, usually from 5 to 40 mm apart, depending on the conductivity of the liquid. A potential difference of 5 to 10 V may be applied to achieve current densities of the order of $100 \, A/m^2$. Frequently, the conductivity of the suspension itself is adequate, though it may be necessary to add ionic materials, such as sodium chloride or sulphuric acid. Electrode fouling can usually be prevented by periodically reversing the polarity of the electrodes. Occasionally, consumable iron or aluminium anodes may be used because the ions released into the suspension may then assist flocculation of the suspended solids.

An electroflotation plant usually consists of a steel or concrete tank with a sloping bottom as shown in Fig. 4.25. The liquid depth in the tank typically is about 1 m. Since the flotation process is much faster than sedimentation with fine particles, flotation can be achieved with much shorter retention times, usually about 1 h (3.6 ks)—and the land area required may be only about one-eighth of that for a sedimentation tank.

The floated sludge, which frequently contains 95% of the solids, forms a blanket on the liquid surface and may be continuously removed by means of slowly moving brushes or scrapers mounted across the top of the tank.

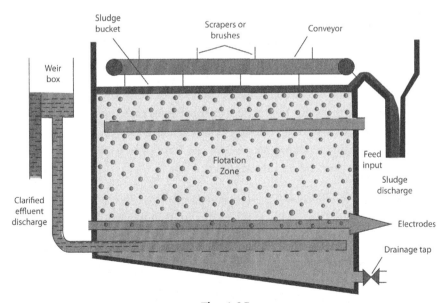

Fig. 4.25
Schematic diagram of an electroflotation plant.

4.3 Separation of Suspended Solid Particles From Fluids

4.3.1 Introduction

Solids are removed from fluids in order to purify the fluid although, in some cases, and particularly with liquids, it is the solid material that is the product. Where solids are to be removed from liquids, a variety of operations are available including:

(a) Sedimentation, in which the solids are allowed to settle by gravity through the liquid they are removed from, usually as a pumpable sludge. Such an operation is discussed in Chapter 8.

(b) Filtration, in which the solids are collected on a medium, such as a porous material or a layer of fine particles, through which the liquid is pumped. The theory, practice, and available equipment for filtration are discussed in Chapter 10.

(c) Centrifugal separation in which the solids are forced to sediment on to the walls of a vessel, which is rotated to provide the centrifugal force. Such an operation forms the basis of the discussion in Chapter 11.

Since the removal of solids from liquids in these ways is discussed elsewhere, this section is concerned only with the removal of suspended particles and liquid mists from gases.

The need to remove suspended dust and mist from a gas arises not only in the treatment of effluent gas from a plant before it is discharged into the atmosphere, but also in processes where solids or liquids are carried over in the vapour or gas stream. For example, in an evaporator it is frequently necessary to eliminate droplets, which become entrained in the vapour, and in a plant involving a fluidised solid and the removal of fine particles. This prevents the loss of material, as well as contamination of the gaseous product. Further, in all pneumatic conveying plants, some form of separator must be provided at the downstream end.

Whereas relatively large particles with settling velocities greater than about 0.3 m/s readily disengage themselves from a gas stream, fine particles tend to follow the same path as the gas and separation is therefore difficult. In practice, dust particles may have an average diameter of about 0.01 mm (10 μm) and a settling velocity of about 0.003 m/s, so that a simple gravity settling vessel would be impracticable because of the long time required for settling and the large size of the separator, for a given throughput of gas.

The main problems involved in the removal of particles from a gas stream have been reviewed by Ashman,[25] Stairmand and Nonhebel,[26] and Swift.[27] The main reasons for removing particles from an effluent gas are:

(a) To maintain the health of operators in the plant and the surrounding population. In general, the main danger arises from the inhaling of dust particles. The most dangerous range of sizes is generally between about 0.5 and 3 μm, as discussed in Chapter 14.

(b) In order to eliminate explosion risks. A number of carbonaceous materials and finely powdered metals give rise to explosive mixtures with air. Also flames can be propagated over large distances.

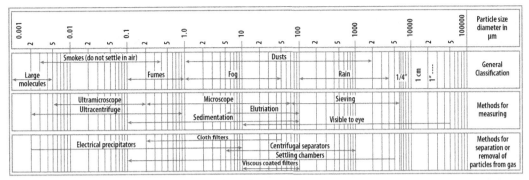

Fig. 4.26
Characteristics of aerosols and separators.[25]

(c)   In order to reduce the loss of valuable materials.

(d)   Because the gas itself may be required for use in a further process such as the blast furnace gas used for firing stoves.

The size range of commercial aerosols and the methods available for determination of particle size and for removing the particles from the gas are shown in Fig. 4.26, which is taken from the work of Ashman.[25] It may be noted that the ranges over which the various equipments operation overlap to some extent. The choice of equipment depends not only on the particle size, but also on such factors as the quantity of gas to be handled, the concentration of the dust or mist, and the physical properties of the particles.

Tables 4.2 and 4.3 provide data which aid the selection of suitable equipment.[26]

Separation equipment may depend on one or more of the following principles, the relative importance of which is sometimes difficult to assess:

Table 4.2 Summary of dust arrester performance[26]

Type of Equipment	Field of Application	Pressure Loss
Settling chambers	Removal of coarse particles, larger than about 100–150 μm	Below 50 N/m²
Scroll collectors, shutter collectors, low-pressure-drop cyclones	Removal of fairly coarse dusts down to about 50–60 μm	Below 250 N/m²
High-efficiency cyclones	Removal of average dusts in the range 10–100 μm	250–1000 N/m²
Wet washers (including spray towers, venturi scrubbers)	Removal of fine dusts down to about 5 μm (or down to sub-micron sizes for the high-pressure-drop type)	250–600 N/m² or more
Bag filters	Removal of fine dusts and fumes, down to about 1 μm or less	100–1000 N/m²
Electrostatic precipitators	Removal of fine dusts and fumes down to 1 μm or less	50–250 N/m²

Table 4.3 Efficiency of dust collectors[a,26]

Dust Collector	Efficiency at 5 μm (%)	Efficiency at 2 μm (%)	Efficiency at 1 μm (%)
Medium-efficiency cyclone	27	14	8
High-efficiency cyclone	73	46	27
Low-pressure-drop cellular cyclone	42	21	13
Tubular cyclone	89	77	40
Irrigated cyclone	87	60	42
Electrostatic precipitator	99	95	86
Irrigated electrostatic precipitator	98	97	92
Fabric filter	99.8	99.5	99
Spray tower	94	87	55
Wet impingement scrubber	97	95	80
Self-induced spray deduster	93	75	40
Disintegrator	98	95	91
Venturi scrubber	99.8	99	97

[a]For dust of density $2700 \, kg/m^3$.

(a) Gravitational settling.
(b) Centrifugal separation.
(c) Inertia or momentum processes.
(d) Filtration.
(e) Electrostatic precipitation.
(f) Washing with a liquid.
(g) Agglomeration of solid particles and coalescence of liquid droplets.

Example 4.2

The collection efficiency of a cyclone is 45% over the size range 0–5 μm, 80% over the size range 5–10 μm, and 96% for particles exceeding 10 μm. Calculate the efficiency of collection for a dust with a mass distribution of 50% 0–5 μm, 30% 5–10 μm and 20% above 10 μm.

Solution

For the collector:

Size (μm)	0–5	5–10	>10
Efficiency (%)	45	80	96

For the dust:

Mass (%)	50	30	20

On the basis of 100 kg dust:

Mass at inlet (kg)	50	30	20
Mass retained (kg)	22.5	24.0	19.2—a total of 65.7 kg

Thus : Overall efficiency $= 100(65.7/100) = \underline{65.7\%}$.

Example 4.3

The size distribution by mass of the dust carried in a gas, together with the efficiency of collection over each size range, is as follows:

Size range (µm)	0–5	5–10	10–20	20–40	40–80	80–160
Mass (%)	10	15	35	20	10	10
Efficiency (%)	20	40	80	90	95	100

Calculate the overall efficiency of the collector, and the percentage by mass of the emitted dust that is smaller than 20 µm in diameter. If the dust burden is $18\,g/m^3$ at entry and the gas flow $0.3\,m^3/s$, calculate the mass flow of dust emitted.

Solution

Taking $1\,m^3$ of gas as the basis of calculation, the following table of data may be completed noting that the inlet dust concentration is $18\,g/m^3$. Thus:

Size range (µm)	0–5	5–10	10–20	20–40	40–80	80–160	Total
Mass in size range (g)	1.8	2.7	6.3	3.6	1.8	1.8	18.0
Mass retained (g)	0.36	1.08	5.04	3.24	1.71	1.80	13.23
Mass emitted (g)	1.44	1.62	1.26	0.36	0.09	0	4.77

Hence:

$$\text{Overall efficiency} = 100\,(13.23/18.0) = \underline{73.5\%}$$

$$\text{Dust} < 20\,\mu m\;\text{emitted} = 100[(1.44 + 1.62 + 1.26)/4.77] = \underline{90.1\%}$$

$$\text{The inlet gas flow} = 0.3\,m^3/s$$

and hence:

$$\text{Mass emitted} = (0.3 \times 4.77) = 1.43\,g/s\;\text{or} : 1.43 \times 10^{-3}\,kg/s\,(0.12\,\text{tonne/day})$$

4.3.2 Gas Cleaning Equipment

The classification given in Section 4.3.1 is now used to describe commercially available equipment.

Gravity separators

If the particles are large, they will settle out of the gas stream if the cross-sectional area is increased. The velocity will then fall so that the eddy currents which are maintaining the particles in suspension are suppressed. In most cases, however, it is necessary to introduce

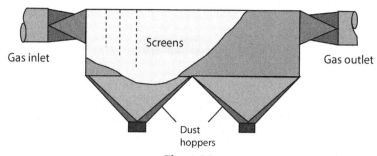

Fig. 4.27
Settling chamber.

baffles or screens as shown in Fig. 4.27, or to force the gas over a series of trays as shown in Fig. 4.28, which depicts a separator used for removing dust from sulphur dioxide produced by the combustion of pyrites. This equipment is suitable when the concentration of particles is high, because it is easily cleaned by opening the doors at the side. Gravity separators are seldom used as they are very bulky and will not remove particles smaller than $50–100\,\mu m$.

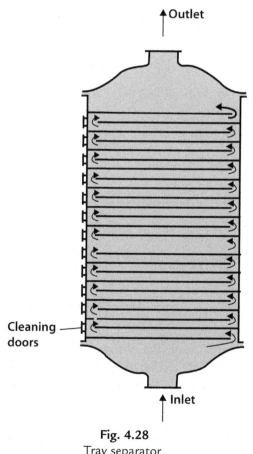

Fig. 4.28
Tray separator.

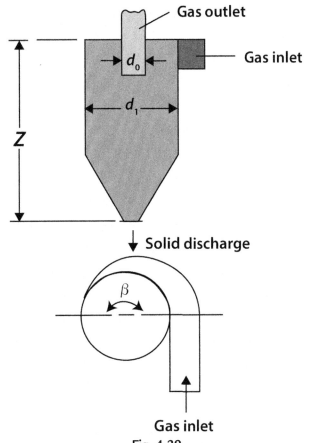

Fig. 4.29
A schematic of a cyclone separator.

Gravity separation is little used since the equipment must be very large in order to reduce the gas velocity to a reasonably low value which allows the finer particles to settle. For example, a settling chamber designed to remove particles with a diameter of $20\,\mu m$ and density $2000\,kg/m^3$ from a gas stream flowing at $10\,m^3/s$ would have a volume of about $3000\,m^3$. Clearly, this large volume imposes a severe limitation, and this type of equipment is normally restricted to small plants as a pre-separator, which reduces the load on a more efficient secondary collector.

Centrifugal separators

The rate of settling of suspended particles in a gas stream may be greatly increased if centrifugal rather than gravitational forces are employed. In the cyclone separator shown in Figs 4.29 and 4.30, the gas is introduced tangentially into a cylindrical vessel at a velocity of about $30\,m/s$ and the clean gas is taken off through a central outlet at the top. The solids are thrown outwards against the cylindrical wall of the vessel, and then move away from the gas inlet and are

Fig. 4.30

Industrial scale cyclone separator. *The authors and publishers acknowledge with thanks the assistance given by AAF Ltd, Cramlington, Northumberland for their permission for reproduction of this figure.*

collected at the conical base. This separator is very effective unless the gas contains a large proportion of particles less than about 10 μm in diameter and is equally effective when used with either dust-laden or mist-laden gases. It is now the most commonly used general purpose separator. The use of cyclones for liquid–solid systems is considered in Section 4.2.4.

Because the rotating motion of the gas in the cyclone separator arises from its tangential entry and no additional energy is imparted within the separator body, a free vortex is established. The energy per unit mass of gas is then independent of its radius of rotation and the velocity distribution in the gas may be calculated approximately by methods discussed in Volume 1A, Chapter 2.

There have been several studies in which the flow patterns within the body of the cyclone separator have been modelled using a Computational Fluid Dynamics (CFD) technique. An example is the work of Slack et al.[28] in which the computed three-dimensional flow fields have been plotted and compared with the results of experimental studies in which a backscatter Laser Doppler Anemometry (LDA) system was used to measure flowfields. Agreement between the computed and experimental results was good. When using very fine grid meshes, the existence of time-dependent vortices was identified. These had the potentiality of adversely affecting the separation efficiency, as well as leading to increased erosion at the walls.

The effects of the dimensions of the separator on its efficiency have been examined experimentally by ter Linden[29] and Stairmand[30, 31] who studied the flow pattern in a glass cyclone separator with the aid of smoke injected into the gas stream. The gas was found to move spirally downwards, gradually approaching the central portion of the separator, and then to rise and leave through the central outlet at the top. The magnitude and direction of the gas velocity have been explored by means of small Pitot tubes, and the pressure distribution has been measured. The tangential component of the velocity of the gas appears to be predominate throughout the whole depth, except within a highly turbulent central core of diameter that is about 0.4 times that of the gas outlet pipe. The radial component of the velocity acts inward, and the axial component is away from the gas inlet near the walls of the separator, but is in the opposite direction in the central core. Pressure measurements indicate a relatively high pressure throughout, except for a region of reduced pressure corresponding to the central core. Any particle is therefore subjected to two opposing forces in the radial direction, the centrifugal force, which tends to throw it to the walls, and the drag of the fluid, which tends to carry the particle away through the gas outlet. Both of these forces are a function of the radius of rotation and of the size of the particles, with the result that particles of different sizes tend to rotate at different radii. As the outward force on the particles increases with the tangential velocity and the inward force increases with the radial component, the separator should be designed to make the tangential velocity as high as possible and the radial velocity low. This is generally effected by introducing the gas stream with a high tangential velocity, with as little shock as possible, and by making the height of the separator large.

The radius at which a particle will rotate within the body of a cyclone corresponds to the position where the net radial force on the particle is zero. The two forces acting are the centrifugal force outwards and the frictional drag of the gas acting inwards.

Considering a spherical particle of diameter d rotating at a radius r, then the centrifugal force is:

$$\frac{mu_t^2}{r} = \frac{\pi d^3 \rho_s u_t^2}{6r} \tag{4.5}$$

where m is the mass of the particle and u_t is the tangential component of the velocity of the gas.

It is assumed here that there is no slip between the gas and the particle in the tangential direction.

If the radial velocity is low, the inward radial force due to friction will, from Chapter 6, be equal to $3\pi\mu du_r$, where μ is the viscosity of the gas, and u_r is the radial component of the velocity of the gas.

The radius r at which the particle will rotate at equilibrium is then given by:

$$\frac{\pi d^3 \rho_s u_t^2}{6r} = 3\pi\mu du_r$$

$$\text{or:} \quad \frac{u_t^2}{r} = \frac{18\mu}{d^2 \rho_s} u_r$$

(4.6)

The free-falling velocity of the particle u_0, when the density of the particle is large compared with that of the gas, is given from Eq. (6.23) as:

$$u_0 = \frac{d^2 g \rho_s}{18\mu}$$

(4.7)

Substituting in Eq. (4.6):

$$\frac{u_t^2}{r} = \frac{u_r}{u_0} g$$

$$\text{or:} \quad u_0 = \frac{u_r}{u_t^2} r g$$

(4.8)

Thus the higher the terminal falling velocity of the particle, the greater is the radius at which it will rotate, and the easier it is to separate. If it is assumed that a particle will be separated provided it tends to rotate outside the central core of diameter $0.4d_0$, the terminal falling velocity of the smallest particle, which will be retained, is found by substituting $r = 0.2d_0$ in Eq. (4.8) to give:

$$u_0 = \frac{u_r}{u_t^2} (0.2d_0) g$$

(4.9)

In order to calculate u_0, it is necessary to evaluate u_r and u_t for the region outside the central core. The radial velocity u_r is found to be approximately constant at a given radius and to be given by the volumetric rate of flow of gas divided by the cylindrical area for flow at the radius r. Thus, if G is the mass flow rate of gas through the separator and ρ is its density, the linear velocity in a radial direction at a distance r from the centre is given by:

$$u_r = \frac{G}{2\pi r Z \rho}$$

(4.10)

where Z is the depth of the separator.

For a free vortex, it is shown in Volume 1A, Chapter 2, that the product of the tangential velocity and the radius of rotation is a constant. Because of fluid friction effects, this relation does not hold exactly in a cyclone separator where it is found experimentally that the tangential velocity is more nearly inversely proportional to the square root of radius, rather than to the radius. This relation appears to hold at all depths in the body of the separator. If u_t is the tangential component of the velocity at a radius r, and u_{t0} is the corresponding value at the circumference of the separator, then:

$$u_t = u_{t0}\sqrt{(d_t/2r)} \tag{4.11}$$

It is found that u_{t0} is approximately equal to the velocity with which the gas stream enters the cyclone separator. If these values for u_r and u_t are now substituted into Eq. (4.9), the terminal falling velocity of the smallest particle, which the separator will retain, is given by:

$$\begin{aligned} u_0 &= \left(\frac{G}{2\pi \times 0.2d_0\rho Z}\right)\left(\frac{2 \times 0.2d_0}{d_t}\right)\left(\frac{1}{u_{t0}^2}\right)0.2d_0 g \\ &= \frac{0.2Gd_0 g}{\pi\rho Zd_t u_{t0}^2} \end{aligned} \tag{4.12}$$

If the cross-sectional area of the inlet is A_i, $G = A_i \rho u_{t0}$ and:

$$u_0 = \frac{0.2A_i^2 d_0\rho g}{\pi Zd_t G} \tag{4.13}$$

A small inlet and outlet therefore result in the separation of smaller particles but, as the pressure drop over the separator varies with the square of the inlet velocity and the square of the outlet velocity,[30] the practical limit is set by the permissible drop in pressure. The depth and diameter of the body should be as large as possible because the former determines the radial component of the gas velocity and the latter controls the tangential component at any radius. In general, the larger the particles, the larger the diameter of the separator should be because the greater is the radius at which they rotate. The larger the diameter, the greater too is the inlet velocity, which can be used without causing turbulence within the separator. The factor which ultimately settles the maximum size is, of course, the cost. Because the separating power is directly related to the throughput of gas, the cyclone separator is not very flexible though its efficiency can be improved at low throughputs by restricting the area of the inlet with a damper and thereby increasing the velocity. Generally, however, it is better to use a number of cyclones in parallel and to keep the load on each approximately the same whatever the total throughput.

Because the vertical component of the velocity in the cyclone is downwards everywhere outside the central core, the particles will rotate at a constant distance from the centre and move continuously downwards until they settle on the conical base of the plant. Continuous removal of the solids is desirable so that the particles do not get entrained again in the gas stream due to

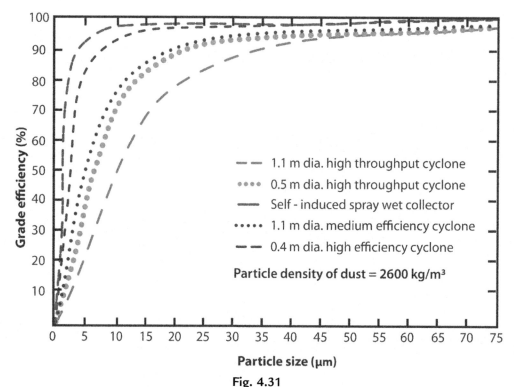

Fig. 4.31

Typical grade efficiency curves for cyclones and a self-induced spray wet collector.[27]

the relatively low pressures in the central core. Entrainment is reduced to a minimum if the separator has a deep conical base of small angle.

Though the sizes of particles which will be retained or lost by the separator can be calculated, it is found in practice that some smaller particles are retained and some larger particles are lost. The small particles, which are retained, have in most cases collided with other particles and adhered to form agglomerates, which behave as large particles. Relatively large particles are lost because of eddy motion within the cyclone separator, and because they tend to bounce off the walls of the cylinder back into the central core of fluid. If the gas contains a fair proportion of large particles, it is desirable to remove these in a preliminary separator before the gas is fed into the cyclone.

The efficiency of the cyclone separator is greater for larger than for smaller particles, and it increases with the throughput until the point is reached where excessive turbulence is created. Fig. 4.31 shows the efficiency of collection plotted against particle size for an experimental separator for which the theoretical 'cut' occurs at about 10 μm. It may be noted that an appreciable quantity of fine material is collected, largely as a result of agglomeration, and that some of the coarse material is lost with the result that a sharp cut is not obtained.

The effect of the arrangement and size of the gas inlet and outlet has been investigated and it has been found that the inlet angle β in Fig. 4.29 should be of the order of 180°. Further, the depth of the inlet pipe should be small, and a square section is generally preferable to a circular one because a greater area is then obtained for a given depth. The outlet pipe should extend downwards well below the inlet in order to prevent short-circuiting.

Various modifications may be made to improve the operation of the cyclone separator in special cases. If there is a large proportion of fine material present, a bag filter may be attached to the clean gas outlet. Alternatively, the smaller particles may be removed by means of a spray of water which is injected into the separator. In some cases, the removal of the solid material is facilitated by running a stream of water down the walls, which also reduces the risk of the particles becoming reentrained in the gas stream. The main difficulty lies in wetting the particles with the liquid.

Because the separation of the solid particles, which have been thrown out to the walls, is dependent on the flow of gas parallel to the axis rather than to the effect of gravity, the cyclone can be mounted in any desired direction. In many cases horizontal cyclone separators are used, and occasionally the separator is fixed at the junction of two mutually perpendicular pipes and the axis is then a quadrant of a circle. The cyclone separator is usually mounted vertically, except where there is a shortage of headroom, because removal of the solids is more readily achieved especially if large particles are present.

A double cyclone separator is sometimes used when the range of the size of particles in the gas stream is large. This consists of two cyclone separators, one inside the other. The gas stream is introduced tangentially into the outer separator and the larger particles are deposited. The partially cleaned gas then passes into the inner separator through tangential openings and the finer particles are deposited there because the separating force in the inner separator is greater than that in the outer cylinder. In Fig. 4.32, a multicyclone is illustrated in which the gas is subjected to further action in a series of tubular units, the number of which can be varied with the throughput. This separator is therefore rather more flexible than the simple cyclone.

Szekely and Carr[32] have studied the heat transfer between the walls of a cyclone and a gas–solid suspension, and have shown that the mechanism of heat transfer is quite different from that occurring in a fluidised bed. There is a high rate of heat transfer directly from the wall to the particles, but the transfer direct to the gas is actually reduced. Overall, the heat transfer rate at the walls is slightly greater than that obtained in the absence of the particles.

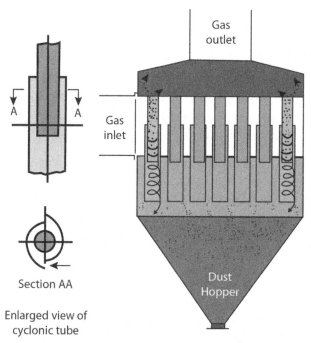

Fig. 4.32
Multi-tube cyclone separator.

Example 4.4

A cyclone separator, 0.3 m in diameter and 1.2 m long, has a circular inlet 75 mm in diameter and an outlet of the same size. If the gas enters at a velocity of 1.5 m/s, at what particle size will the theoretical cut occur?

The viscosity of air is $0.018 \, \text{mN s/m}^2$, the density of air is $1.3 \, \text{kg/m}^3$ and the density of the particles is $2700 \, \text{kg/m}^3$.

Solution

Using the data provided:

cross-sectional area at the gas inlet, $A_i = (\pi/4)(0.075)^2 = 4.42 \times 10^{-3} \, \text{m}^2$

gas outlet diameter, $d_0 = 0.075 \, \text{m}$

gas density, $\rho = 1.30 \, \text{kg/m}^3$

height of separator, $Z = 1.2 \, \text{m}$, separator diameter, $d_t = 0.3 \, \text{m}$.

Thus: mass flow of gas, $G = (1.5 \times 4.42 \times 10^{-3} \times 1.30) = 8.62 \times 10^{-3} \, \text{kg/s}$

The terminal velocity of the smallest particle retained by the separator,

$$u_0 = 0.2A_i^2 d_0 \rho g/(\pi Z d_t G) \text{(Eq. 4.13)}$$

$$u_0 = \left[0.2 \times (4.42 \times 10^{-3})^2 \times 0.075 \times 1.3 \times 9.81\right]/\left[\pi \times 1.2 \times 0.3 \times 8.62 \times 10^{-3}\right]$$

$$= 3.83 \times 10^{-4}\,\text{m/s}$$

Use is now made of Stokes' law (Chapter 6) to find the particle diameter, as follows:

$$u_0 = d^2 g(\rho_s - \rho)/18\mu \ \text{(Eq. 6.24)}$$

$$d = [u_0 \times 18\mu/g(\rho_s - \rho)]^{0.5}$$

$$= \left[(3.83 \times 10^{-4} \times 18 \times 0.018 \times 10^{-3})/(9.81(2700 - 1.30))\right]^{0.5}$$

$$= 2.17 \times 10^{-6}\,\text{m or } \underline{\underline{2.17\,\mu\text{m}}}$$

Inertia or momentum separators

Momentum separators rely on the fact that the momentum of the particles is far greater than that of the gas, so that the particles do not follow the same path as the gas if the direction of motion is suddenly changed. Equipment developed for the separation of particles from mine gases is shown in Fig. 4.33, which reveals that the direction of the gas is changed suddenly at the end of each baffle. Again, the separator shown in Fig. 4.34 consists of a number of vessels—up to about 30 connected in a series—in each of which the gas impinges on a central baffle. The dust drops to the bottom and the velocity of the gas must be arranged so that it is sufficient for effective separation to be made without the danger of reentraining the particles at the bottom of each vessel.

As an alternative to the use of rigid baffles, the separator may be packed with a loose fibrous material. In this case, the separation will be attributable partly to gravitational settling on to the packing, partly to inertia effects and partly to filtration. If the packing is moistened with a viscous liquid, the efficiency is improved because the film of liquid acts as an effective filter and prevents the particles being picked up again in the gas stream. The viscous filter as shown in Fig. 4.35 consists of a series of corrugated plates, mounted in a frame and covered with a non-drying oil. These units are then arranged in banks to give the required area. They are readily cleaned and offer a low resistance to flow. Packs of slag wool offer a higher resistance though they are very effective.

Fig. 4.33
Baffled separator.

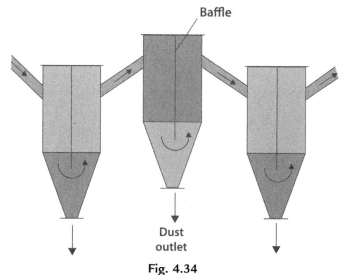

Fig. 4.34
Battery of momentum separators.

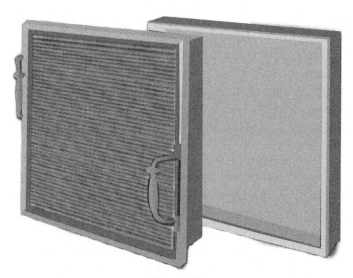

Fig. 4.35
Viscous filter. *The authors and publishers acknowledge with thanks the assistance given by Institution of Mechanical Engineers for their permission for reproduction of this figure.*

Fabric filters

Fabric filters include all types of bag filters in which the filter medium is in the form of a woven or felted textile fabric, which may be arranged as a tube or supported on a suitable framework. As reported by Dorman,[33] this type of filter is capable of removing particles of a size down to 1 μm or less by the use of glass fibre paper or pads. In a normal fabric filter, particles smaller

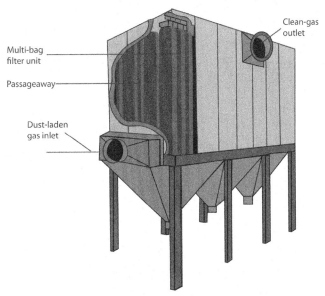

Clean-gas
outlet

Multi-bag
filter unit

Passageaway

Dust-laden
gas inlet

Fig. 4.36

General view of 'bag-house' (low face-velocity type). *The authors and publishers acknowledge with thanks the assistance given by Sturtevant Engineering Co Ltd for their permission for reproduction of this figure.*

than the apertures in the fabric are trapped by impingement on the fine 'hairs', which span the apertures. Typically, the main strands of the material may have a diameter of 500 μm, spaced 100–200 μm apart. The individual textile fibres with a diameter of 5–10 μm crisscross the aperture and form effective impingement targets capable of removing particles of sizes down to 1 μm.

In the course of operation, filtration efficiency will be low until a loose 'floc' builds up on the fabric surface. These flocs act as effective filter for the removal of fine particles. The cloth will require cleaning from time to time to avoid excessive build-up of solids which gives rise to a high pressure drop. The velocity at which the gases pass through the filter must be kept low, typically 0.005–0.03 m/s, in order to avoid compaction of the floc and consequently high pressure drops, or to avoid local breakdown of the filter bed, which would allow large particles to pass through the filter.

There are three main types of bag filter. The simplest, which is shown in Fig. 4.36, consists of a number of elements assembled together in a 'bag-house'. This is the cheapest type of unit and operates with a velocity of about 0.01 m/s across the bag surface.

A more sophisticated and robust version incorporates some form of automatic bag-shaking mechanism which may be operated by mechanical, vibratory, or air-pulsed methods. A heavier fabric allows higher face velocities, up to 0.02 m/s, to be used and permits operation under more difficult conditions than the simpler bag-house type can handle.

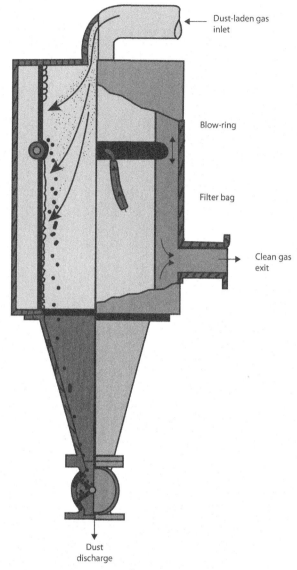

Dust-laden gas
inlet

Blow-ring

Filter bag

Clean gas
exit

Dust
discharge

Fig. 4.37
Reverse-jet filter.[26]

The third type of bag filter is the reverse-jet filter, described by Hersey[34] and illustrated in Figs 4.37 and 4.38. With face velocities of about 0.05 m/s and with the capability of dealing with high dust concentrations at high efficiencies, this type of filter can deal with difficult mixtures in an economic and compact unit. Use of the blow ring enables the cake to be dislodged in a cleaning cycle which takes only a few seconds.

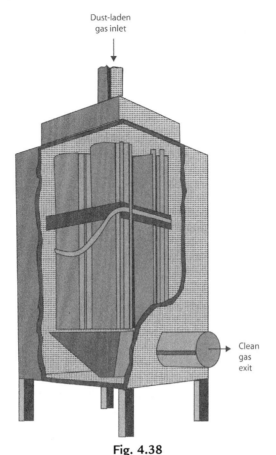

Fig. 4.38
General view of reverse-jet filter (high face-velocity type).[26]

Electrostatic precipitators

Electrostatic precipitators, such as that shown in Fig. 4.39, are capable of collecting very fine particles, $<2\,\mu$m, at high efficiencies. Their capital and operating costs are high, however, electrostatic precipitation should only be considered in place of alternative processes, such as filtration, where the gases are hot or corrosive. Electrostatic precipitators are used extensively in the metallurgical, cement, and electrical power industries. Their main application is in the removal of the fine fly ash formed in the combustion of pulverised coal in power-station boilers. The basic principle of operation is simple. The gas is ionised in passing between a high-voltage electrode and an earthed at grounded electrode. The dust particles become charged and are attracted to the earthed electrode. The precipitated dust is removed from the electrodes mechanically, usually by vibration, or by washing. Wires are normally used as the high-voltage electrode shown in Fig. 4.40 and plates or tubes for the earthed electrode shown in Fig. 4.41. A typical design is shown in Fig. 4.39. A full description of the construction, design, and application of electrostatic precipitators is given by Schneider et al.[35] and by Rose and Wood.[36]

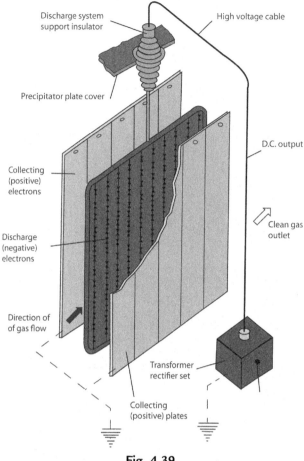

Fig. 4.39
Electrostatic precipitator.

If the gas is passed between two electrodes charged to a potential difference of 10–60 kV, it is subjected to the action of a corona discharge. Ions which are emitted by the smaller electrode—on which the charge density is greater—attach themselves to the particles which are then carried to the larger electrode under the action of the electric field. The smaller electrode is known as the discharge electrode and the larger one, which is usually earthed, as the receiving electrode. Most industrial gases are sufficiently conducting to be readily ionised, the most important conducting gases being carbon dioxide, carbon monoxide, sulphur dioxide, and water vapour. If the conductivity of the gas being handled is low, water vapour can be added.

The gas velocity over the electrodes is usually between about 0.6 and 3 m/s, with an average contact time of about 2 s. The maximum velocity is determined by the maximum distance through which any particle must move in order to reach the receiving electrode, and by the attractive force acting on the particle. This force is given by the product of the charge on the particle and the strength of the electrical field, although calculation of the path of a particle is

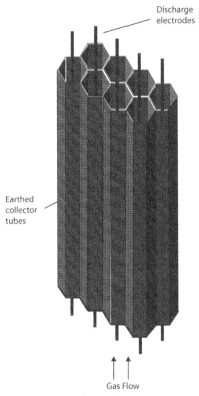

Fig. 4.40
General arrangement of wire-in-tube precipitator.

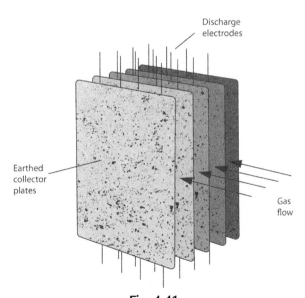

Fig. 4.41
General arrangement of wire-and-plate precipitator.

difficult because it gradually becomes charged as it enters the field, and the force therefore increases during the period of charging. This rate of charging cannot be estimated with any degree of certainty. The particle moves towards the collecting electrode under the action of the accelerating force due to the electrical field and the retarding force of fluid friction. The maximum rate of passage of gas is that which will just allow the most unfavourably located particle to reach the collecting electrode before the gas leaves the precipitator. Collection efficiencies of nearly 100% can be obtained at low gas velocities, although the economic limit is usually about 99%.

Electrostatic precipitators are made in a wide range of sizes and will handle gas flows up to about 50 m³/s. Although they operate more satisfactorily at low temperatures, they can be used up to about 800 K. Pressure drops over the separator are low.

4.3.3 Liquid Washing

If the gas contains an appreciable proportion of fine particles, liquid washing provides an effective method of cleaning which gives a gas of high purity. In the spray column illustrated in Fig. 4.42, the gas flows upwards through a set of primary sprays to the main part of the column

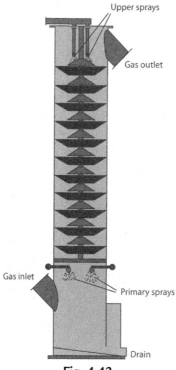

Fig. 4.42
Spray washer.

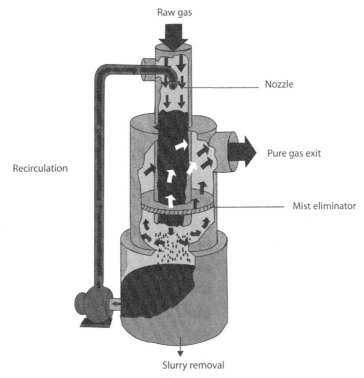

Raw gas

Nozzle

Pure gas exit

Recirculation

Mist eliminator

Slurry removal

Fig. 4.43

The Sulzer cocurrent scrubber. *The authors and publishers acknowledge with thanks the assistance given by Sulzer (UK) Ltd, Farnborough, Hants for their permission for reproduction of this figure.*

where it flows countercurrently to a water spray, which is redistributed at intervals. In some cases packed columns are used for gas washing, though it is generally better to arrange the packing on a series of trays to facilitate cleaning.

A cocurrent washer is shown in Fig. 4.43 where the gas and the water flow down through a packed section before entering a disengagement space. The gas leaves through a mist eliminator and the bulk of the water is recirculated.

Fig. 4.44 shows a wet gas scrubber designed initially for the removal of small quantities of hydrogen sulphide from gas streams, though used for cleaning gases, vapours, and mists from process and ventilation air flows. This works on the absorption principle of dissolving impurities, such as acids and solvents in cleaning liquors like water or chemical solutions. The contaminated air is drawn through a packing zone filled with saddles or any other suitable packings, which are irrigated with the liquor. Due to the design of the saddles and the way they pack, the impurities achieve good contact with the absorbing liquor, and hence, a high cleaning efficiency is obtained. A demister following the packing zone removes any entrained liquid droplets, leaving the exhaust air containing <2% of the original contaminant.

Fig. 4.45 shows a venturi scrubber in which water is injected at the throat and the separation is then carried out in a cyclone separator.

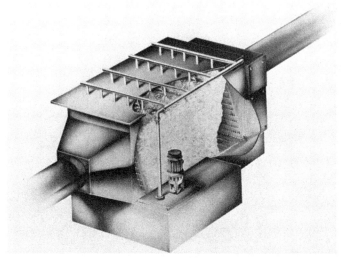

Fig. 4.44

The AAF HS gas scrubber. *The authors and publishers acknowledge with thanks the assistance given by AAF Ltd, Cramlington, Northumberland for their permission for reproduction of this figure.*

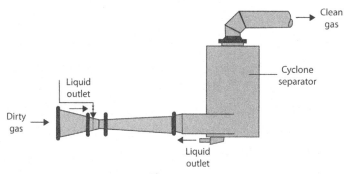

Fig. 4.45

Venturi washer with cyclone separator.

The AAF Kinpactor scrubber, illustrated in Fig. 4.46, operates in the same way and is capable of handling gas flowrates as high as $30\,m^3/s$.

In general, venturi scrubbers work by injecting water into a venturi throat through which contaminated air is passing at 60–100 m/s. The air breaks the water into small droplets, which are then accelerated to the air velocity. Both are then decelerated relatively gently in a tapered section of ducting. The difference between the water and dust densities gives different rates of deceleration, as does the larger size of water droplets, all resulting in collisions between dust particles and droplets. Thus the dust particles are encapsulated in the droplets, which are of a size easily collected by a conventional spray eliminator system. The end result is a very good grade efficiency curve shown in Fig. 4.47, but at the price of a very high pressure drop, from

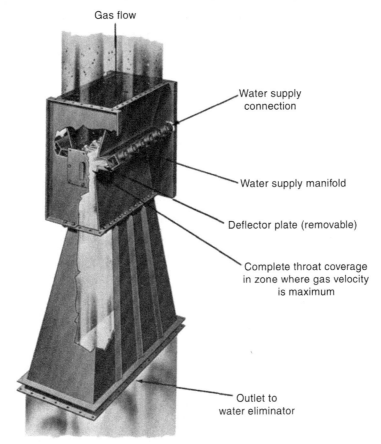

Fig. 4.46

The AAF Kinpactor venture scrubber. *The authors and publishers acknowledge with thanks the assistance given by AAF Ltd, Cramlington, Northumberland for their permission for reproduction of this figure.*

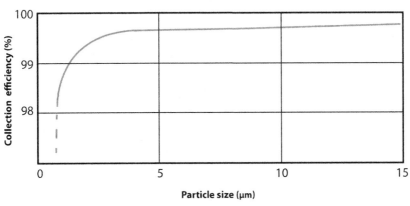

Fig. 4.47

Typical grade efficiency curve for a medium velocity venturi scrubber.[27]

300 mm of water upwards. Water/gas ratios of up to $1 m^3$ water/$1000 m^3$ gas may be used, although commercial venturi scrubbers tend to use considerable flows of water.

The main difficulty in operating any gas washer is the wetting of the particles because it is often necessary for the liquid to displace an adsorbed layer of gas. In some cases, the gas is sprayed with a mixture of water and oil to facilitate wetting and it may then be bubbled through foam. This method of separation cannot be used, of course, where a dry gas is required.

Agglomeration and coalescence

Separation of particles or droplets is often carried out by first achieving an increase in the effective size of the individual particles by causing them to agglomerate or coalesce, and then separating the enlarged particles. A number of methods are available. Thus if the dust-laden or mist-laden gas is brought into contact with a supersaturated vapour, condensation occurs on the particles which act as nuclei. Again if the gas is brought into an ultrasonic field, the vibrational energy of the particles is increased so that they collide and agglomerate. Ultrasonic agglomeration[37, 38] has been successful in causing fine dust particles to form agglomerates, about 10 μm in diameter, which can then be removed in a cyclone separator.

Odour removal

In addition to the removal of solids from gases, the problem of small quantities of compounds causing odour nuisance must frequently be handled. Activated carbon filters are used for this purpose and, in continuous operation, gaseous impurities from the air-stream are adsorbed on the activated carbon granules. The carbon will frequently adsorb contaminants until they reach a level of 20%–30% of the mass of the carbon. When saturation level is reached, the carbon is regenerated by the injection of live steam for a period of 30–60 min. Both the desorbed substances and steam are condensed in the lower part of the unit. In case of solvent recovery some units feature a water separator permitting recycling of the recovered solvent. Adsorption processes are discussed in Volume 2B.

4.4 Recent Developments in Size and Shape Separation of Particles

Due to the increase in use of particles in several fields including biology, material science, and surface engineering, there have been several developments in size and shape-based particle separations in which the use of well-established techniques has been exploited. Some examples of such separation are discussed in this section. In Chapter 15, one of the recent developments, namely, the techniques based on microfluidic separation are discussed.

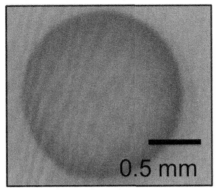

Fig. 4.48

Patterns left by dried coffee spills on solid surfaces—*dark brown colour* at the periphery is due to accumulation of coffee particles at the edge.

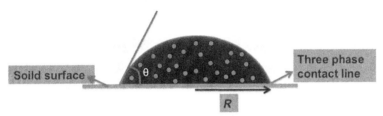

Fig. 4.49

A side-view of an axisymmetric hemispherical sessile characterised by contact angle θ and contact radius R.

4.4.1 Size-Based Particle Separation Using Drying Drops

If a liquid contains two or more particles of different sizes, it is possible to separate particles into pure fractions by evaporating particle-laden drops on solid surfaces.[39] This separation process exploits a phenomenon called 'coffee-ring' effect.[40–46] The name comes from typical patterns that result when the coffee drops that spill on the solid surfaces dry. A closer look at the dried coffee spills reveals the accumulation of coffee-particles as evidenced by the dark brown colour at the edge or periphery of the dried drop shown in Fig. 4.48. The formation as well as suppression of such coffee-ring deposit has been a subject of considerable research for more than two decades.[42,43,45,46] The interest in understanding this phenomenon is due to its widespread occurrence when drops containing colloids, nanoparticles or any dispersed component are dried on solid substrates and relevance to coating, printing, and paint technologies.

Small volume (of the order of few microlitres) sessile drops, that is, the drops that reside on the top of a solid surface, form a axisymmetric spherical cap as soon as the drop is placed on a solid surface, as shown schematically in Fig. 4.49. Initially, the particles in the drop are uniformly distributed throughout as shown in Fig. 4.49. The boundary of the drop where the three phases, solid, liquid, and vapour (typically air) meet is called three-phase contact line, which for a perfect hemispherical drop is a circle. A sessile drop is characterised by a contact angle, θ defined as the angle that a tangent to the three-phase contact makes with the horizontal and the radius (R) of the contact line. The contact angle θ is related to the three interfacial tensions—namely, solid–liquid (γ_{SL}), solid-vapour (γ_{SV}) and liquid-vapour (γ_{LV}) as[47]:

$$\gamma_{LV}\cos\theta = \gamma_{SV} - \gamma_{SL} \tag{4.14}$$

Eq. (4.14) is the well-known Young's equation.

For the coffee-ring deposit to occur: (i) the contact angle of the drop on the solid substrate should be non-zero (i.e. $\theta > 0$) (ii) the continuous phase or the solvent should evaporate (iii) the contact line should be pinned to its initial position during majority of the drying period.[40–46] The contact line pinning occurs on most solid substrates due to physical or chemical surface heterogeneities on the surface. A schematic representation of the variation of the contact radius (solid line) and the height of the drop (dashed line) measured experimentally by capturing the image (side view) of a drying drop is shown in Fig. 4.50. The drawing shows that while the drop height decreases linearly, the radius of the contact line does not change for a significant period, typically until $\sim$0.75–0.85t_f, where t_f is the total drying time of the drop. This is characteristic of a drop drying in constant contact radius (CCR) mode. To ensure CCR drying mode, a radial fluid flow is generated to replace the fluid that evaporates from the contact line. This flow directed from the drop centre towards the drop edge carries the particles to the periphery of the drop and therefore, the accumulation of particles at the three-phase contact line occurs during

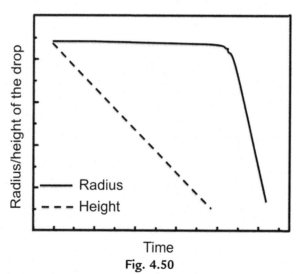

Fig. 4.50

A schematic showing the variation of contact radius (*solid line*) and height of the drop (*dashed line*) during the drying of drops in the constant contact radius (CCR) mode.

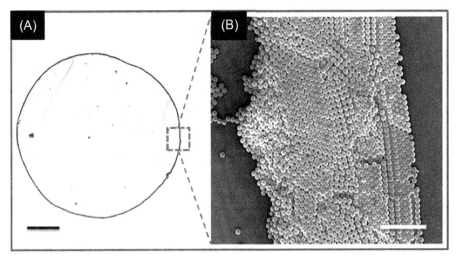

Fig. 4.51

An example of coffee ring formed after drying drops containing polystyrene colloids. The scale bar in the optical microscopy in (A) corresponds to 500 μm and in the scanning electron microscopy image in (B) is 10 μm. *Figure courtsey of M. Mayarani and Dillip K. Satapathy, IIT Madras.*

the CCR period. A 4 μL aqueous dispersion of charge stabilised polystyrene spheres of 1 μm diameter dried on glass substrate leaves a coffee ring as shown in Fig. 4.51A. The microstructural arrangement of polystyrene colloids in Fig. 4.51B reveals a multilayer deposit with particles arranged in square and hexagonal lattices.

When drops containing binary mixture of nanoparticles and microparticles are dried, size dependent sorting of particles occurs in the vicinity of the three-phase contact line.[48] The smaller-sized nanoparticles segregate and accumulate at the edge as these particles can access smaller drop height at the edge of the drop. When liquid contains two or more particles of different sizes, it is possible to separate particles into pure fractions by evaporating particle-laden drops on solid surfaces. Fig. 4.52 depicts the separation of particles of three different sizes by exploiting the coffee-ring phenomena.[39] In Fig. 4.52A, it is evident that the thickness of the droplet is maximum at the drop apex and reduces progressively towards the contact line. Therefore the particles that are carried by the radial flow are conveyed to locations such that the diameter of the particle matches with the local thickness of the drop.[39] Thus the smaller particles are transported closer to the three-phase contact line. Therefore the outermost ring in Fig. 4.52B, i.e. the ring close to the drop edge, consists of the smallest particles, the middle ring consists of particles which are intermediate in size and the innermost ring consists of larger ones leading to particle separation. The formation of three distinct, concentric coffee rings with the green ring containing 40 nm particles, the red ring containing 1 μm and the blue ring containing 2 μm particles captured in the optical fluorescence image shown in Fig. 4.52B illustrates the separation of particles into pure fractions.[39]

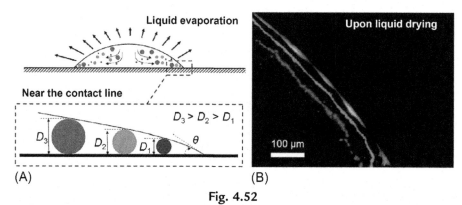

Fig. 4.52

Separation of particles by exploiting coffee-ring phenomena: (A) The configuration of drops containing mixtures of particles of three different sizes and the height gradient close to the drop edge. (B) The formation of three distinct concentric coffee-rings with each ring containing particles of different sizes is illustrated in the optical fluorescence microscopy image. Colour code corresponds to particles of different sizes: 40 nm *(green)*, 1 μm *(red)*, and 2 μm *(blue)*.[39] *A (partially modified) and B (reproduced) with permission from Wong T-S, Chen T-H, Shen X, Ho C-M. Nanochromatography driven by the coffee ring effect. Anal Chem 2011;83(6):1871–3. Copyright (2011) American Chemical Society.*

4.4.2 Separation by Density Gradient Sedimentation

In the mathematical treatment and applications of gravity and centrifugal sedimentation discussed in Chapters 8 and 11 as well as Sections 4.2 and 4.3, the density of the fluid in the column in which the particles move is constant across the column height. In a class of techniques called *density gradient sedimentation*,[49, 50] the separation of particles is achieved by exploiting the motion of particles in a fluid column with density gradient. In such separation processes, the density gradient columns used consist of a solution whose density increases with the column height. Traditionally, this technique has been used for the determination of particle density. Fig. 4.53 shows a particle of mass (m), volume (V) and density (ρ^*) falling in a density gradient column characterised by a linear density gradient (k) with the density ρ at the top. Neglecting the initial inertial effects, the distance of fall x as a function of time is governed by[49]:

$$x = \frac{\rho^* - \rho}{\rho k} - \left[1 - \exp - \left(\frac{V g \rho k}{f} \right) t \right] \tag{4.15}$$

Eq. (4.15) is valid when $f \gg 4mVg\rho k$, where f is the Stokes friction factor assuming that the density gradient solution in the column has the same viscosity and g is the gravitational acceleration. Therefore a particle of unknown density dropped into a density gradient column, ultimately reaches a position in the column such that the density of the particle matches with the density of the solution at that location.[49]

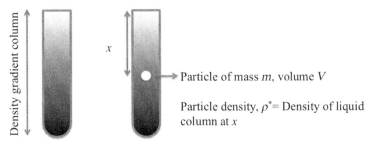

Fig. 4.53
Density gradient sedimentation.

Similar to conventional centrifugal separation, the ability to accelerate particle motion and achieve settling rates much higher than gravity settling has also been exploited in density gradient centrifugation. In here, the separation of particles is carried out by allowing the particles to settle in a fluid column with density gradient under the influence of centrifugal forces. Typically a small quantity of sample containing particles is placed on top of the density gradient column taken in a cylindrical tube and subjected to a centrifugal field. Separation of colloids, nanoparticles, particle aggregates, viruses, cells, proteins, and other biological materials using density gradient centrifugation is an active area of research.[50–54]

Depending on the density gradient of the column and the density of particles that need to be separated, density gradient centrifugation is further classified as (1) isopycnic method and (2) rate zonal method.[49–53] When the density of the particles in a sample is within the density of the gradient column, i.e. if the particle densities lie between the density at the top (lowest density) and the bottom (highest density) of the column gradient, the separation is referred to as 'isopycnic centrifugation'. In this method, as the name implies, the particles in the column sediment to corresponding position in the column such that the density of the particle matches with (equal to) the density of the column at that location. Once the particles reach such an equal density zone, further centrifugation does not have any effect on further sedimentation. In the separation by the rate zonal centrifugation method, the density of objects in the sample that need to be separated is greater than the highest density of the density-gradient liquid column (i.e. density at the bottom of the density gradient column). The particles in the sample separate into different zones depending on their density, size, and morphology.[49–55]

In a recent study, single-walled carbon nanotubes (SWCNTs) of subtle differences in densities have been engineered by using bile salts, and bile salt-surfactant mixtures to encapsulate SWCNTs and separated into pure fractions based on diameter, band gap, and electronic type using density gradient ultracentrifugation.[54] Size-based separation is achieved in spite of remarkably small difference in the diameter (0.7–1.6 nm) of SWCNTs. Density gradient centrifugation is exploited to separate a dispersion consisting of gold nanoparticle (AuNP) clusters consisting of monomers (AuNP1), dimers (AuNP2), trimers (AuNP3), and a small fraction of tetramers (AuNP4) into pure fractions of AuNP2 and AuNP3.[55] The separation is carried out using centrifuging of a density gradient column consisting of dispersion of AuNP mixture at the top, followed by a middle layer of 11% caesium chloride (CsCl) solution and a bottom layer of 62% CsCl solution at 8500 rpm (5800 g) for a period of 20 min. The density gradient column marked A in Fig. 4.54 shows a colour band corresponding to a dispersion consisting of a mixture of 71% monomers, 24% dimers, 3.5% trimers, and 0.5% tetramers (see the bar plot below the columns).[55] The transmission electron microscopy (TEM) labelled a1 in Fig. 4.54 corresponding to the particle mixtures in the band represented by a1 in the density

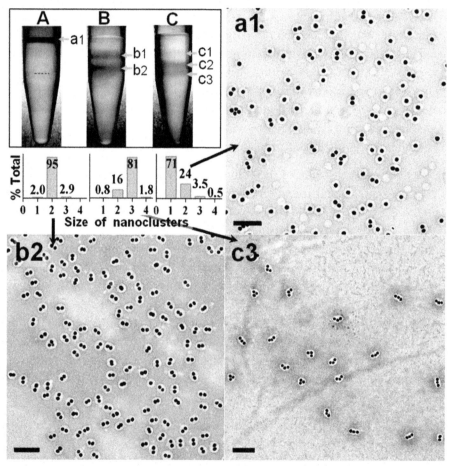

Fig. 4.54
Density gradient sedimentation of gold nanoparticle (AuNP) clusters consisting of monomers (AuNP1), dimers (AuNP2), trimers (AuNP3) and a small fraction of tetramers (AuNP4) into pure fractions of AuNP2 and AuNP3.[55] *Reprinted with permission from Chen G, Wang Y, Tan LH, Yang M, Tan LS, Chen Y, Chen H. High-purity separation of gold nanoparticle dimers and trimers. J Am Chem Soc 2009;**131** (12):4218–9. Copyright (2009) American Chemical Society.*

gradient column marked A. The TEM images in (b2) and (c3) correspond to the zones or bands marked b2 and c3 in the columns B and C. Note that the TEM images in (b2) and (c3), respectively, contain 95% dimers and 81% trimmers indicating the efficiency of separation that can be achieved by this technique.[55]

4.4.3 Shape Dependent Particle Separation

By centrifugation

The use of centrifugation for the separation of particles in a dispersion containing a mixture of nonspherical particles to obtain pure fraction of monodisperse high aspect ratio rods has been demonstrated.[56,57] The TEM in Fig. 4.55A shows the mixture of gold nanoparticles of different sizes and shape in the mother solution that need to be separated.[56–57] Aqueous dispersion of particles shown in (A), subjected to centrifugal sedimentation in a laboratory centrifuge at 5600 g leaves a deposit of particles at the side walls and bottom of the centrifuge tube as depicted in Fig. 4.55B.[56–57] The redispersion of the sediment formed on the sidewall of the centrifuge tube, followed by electron microscopy, showed that this deposit consisted of highly monodisperse gold nanorods. The sediment at the bottom consisted of particles of different shapes—cubes, spheres and lower aspect ratio rods. The shape dependent separation is dictated by the shape dependent effective mass and friction factor for the particle in the dispersion.[56,57]

By exploiting depletion interactions

Depletion attraction driven shape and size-based separation of multicomponent nanoparticles mixtures into pure fractions has been reported.[58, 59] Depletion interaction is class of particle-particle interaction induced by the addition of small molecules into a dispersion of particles. The depletion effects are typically induced by adding polymers, nanoparticles or micelles to a dispersion consisting of particles in the size range from a few nanometres (10^{-9} m) to several microns (10^{-6} m). For depletion effects to occur, the additives should be non-adsorbing (i.e. they should not adsorb on particle surface), and their size must be much smaller than the size of the particles. When the larger particles in the dispersion approach surface-to-surface distance smaller than the diameter of the additive, the gap between the particles is depleted of the additive. This leads to a zone between the particles where the osmotic pressure is zero (as the concentration of the additive in this zone is zero). Therefore the osmotic pressure imbalance in the zone between the particles and that outside the particles leads to attraction between the particles. This effect is discussed further in detail in Chapter 13.

The separation that results when a mixture of cationic surfactants cetyltrimethylammonium bromide (CTAB) and benzyldimethylhexadecylammonium chloride (BDAC) is added to an aqueous dispersion of gold nanoparticles is demonstrated in Fig. 4.56.[58] The attractants in solution form micelles (~6 nm in size) and induces depletion attraction between gold nanoparticles in the solution. The strength of attraction increases with increasing the

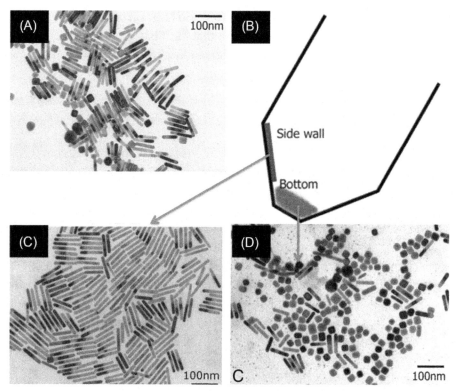

Fig. 4.55

Transmission electron microscopy (TEM) images mixture of gold nanoparticles of different shape shown in (A) sediment at different regions following centrifugation as shown in (B). The TEM images in (C) shows highly monodisperse gold nanorods in the sediment deposited on the side wall of the centrifuge tube and (D) a mixture particles of different shapes—cubes, spheres, lower aspect ratio rods settled at the bottom.[57] *Reprinted with permission from Sharma V, Park K, Srinivasarao M. Shape separation of gold nanorods using centrifugation.* Proc Natl Acad Sci U S A 2009;*106*:4981–5. *Copyright (2009) National Academy of Sciences.*

concentration or number of micelles, i.e., with the increase in surfactant concentration in the solution. The strength of the attraction that varies with the overlap volume (i.e. the volume depleted with micelles) also depends on the shape of the particles. Depletion attraction due to CTAB-BDAC micelles causes the flocculation of gold nanorods, which eventually settle to the bottom of the solution and segregate as pure fraction of rods (Fig. 4.56B).[58] The individual cubes shown in Fig. 4.56C remain in the supernatant. Similar depletion-induced separation is also observed due to the presence of CTAB micelles in the rod-sphere mixtures shown in Fig. 4.56D into pure fraction of spherical particles in the sediment shown in Fig. 4.56E and pure fraction of rods in the supernatant shown in Fig. 4.56F.[58]

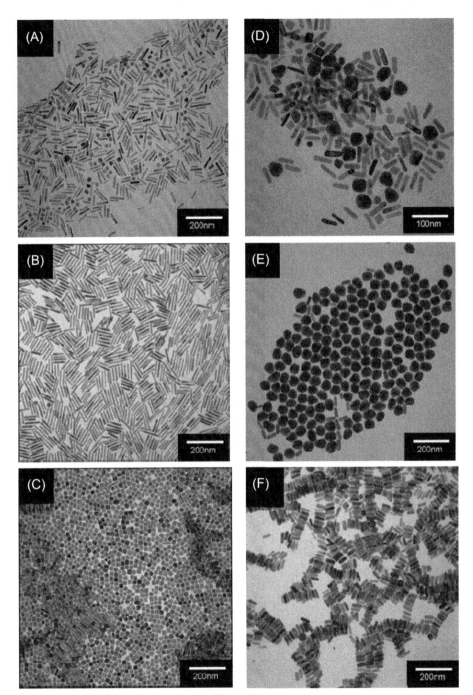

Fig. 4.56

Transmission electron microscopy images in (A) and (D) show gold nanoparticles in a solution before the separation. As a result of the depletion effects induced by the micelles, the particle mixtures shown in (A) separate into nanorods (77 nm length and 11 nm diameter) in the sediment (B) and cubes (20 nm length) in the supernatant (C). In a similar fashion, the particle mixtures shown in (D) separate into sediment of large 60 nm diameter spheres (E) and rods of 60 nm length and 18 nm diameter in the supernatant (F).[58] *Reprinted with permission from Park K, Koerner H, Vaia RA. Depletion-induced shape and size selection of gold nanoparticles.* Nano Lett *2010;**10**(4):1433–9. Copyright (2010) American Chemical Society.*

4.5 Nomenclature

		Units in SI System	Dimensions in M, L, T
A_i	Cross-sectional area of gas inlet to cyclone separator	m^2	L^2
d	Particle size, diameter of sphere or characteristic dimension of particle	m	L
d_A, d_B	Equivalent spherical diameters of particles **A**, **B**	m	L
d_t	Tube diameter *or* diameter of cyclone separator	m	L
d_0	Diameter of outlet of cyclone separator	m	L
G	Mass rate of flow of solids or of fluid	kg/s	MT^{-1}
g	acceleration due to gravity	m/s^2	LT^{-2}
j	Exponent in Eq. (4.1)	–	–
m	Mass of particle	kg	M
n	Exponent in Eq. (4.4)	–	–
r	Radius at which particle rotates or radius in centrifuge bowl or hydrocyclone	m	L
t	Time	s	T
u_a	Axial component of gas in separator	m/s	LT^{-1}
u_r	Radial component of velocity of gas in separator	m/s	LT^{-1}
u_t	Tangential component of gas in separator	m/s	LT^{-1}
u_{t0}	Tangential component of gas velocity at circumference	m/s	LT^{-1}
u_0	Free falling velocity of particle	m/s	LT^{-1}
Z	Total height of separator	m	L
β	Angle between cone wall and horizontal *or* angle of entry into cyclone separator	–	–
γ	Surface or interfacial tension	N/m	MT^{-2}
μ	Viscosity of fluid	Ns/m^2	$ML^{-1}T^{-1}$
ρ	Density of fluid	kg/m^3	ML^{-3}
ρ_A, ρ_B	Density of material **A**, **B**	kg/m^3	ML^{-3}
ρ_s	Density of solid or particle	kg/m^3	ML^{-3}
ω	Angular velocity	s^{-1}	T^{-1}

References

1. Work LT. Annual review–size reduction. *Ind Eng Chem* 1965;**57**(11):97 [21 refs.].
2. Taggart AF. *Handbook of mineral dressing*. Wiley; 1945.

3. Bradley D. *The hydrocyclone*. Pergamon Press; 1965.

4. Svarovsky L, editor. *Solid–liquid separation*. 3rd ed. Butterworths; 1990.

5. Trawinski HF. The application of hydrocyclones as versatile separators in chemical and mineral industries In: Hydrocyclones conference, Cambridge, UK; 1980. BHRA fluid engineering 179.

6. Trawinski HF. Theory, applications and practical operation of hydrocyclones. *Eng Min J* 1976;**115**.

7. Trawinski HF. Practical hydrocyclone operation. *Filtr Sep* 1985;**22**:22.

8. Svarovsky L. *Hydrocyclones*. Holt, Rinehart and Winston; 1984.

9. Mozley R. Selection and operation of high performance hydrocyclones. *Filtr Sep* 1983;**20**:474.

10. Rhodes N, Pericleous KA, Drake SN. The prediction of hydrocyclone performance with a mathematical model. *Solid Liquid Flow* 1989;**1**:35.

11. Salcudean M, He P, Gartshore IS. A numerical simulation of hydrocyclones. *Chem Eng Res Des* 1999;**77** (A5):429.

12. Kelsall DF. A study of the motion of solid particles in a hydraulic cyclone. *Trans Inst Chem Eng* 1957;**30**:87.

13. Rietema K. Performance and design of hydraulic cyclones. *Chem Eng Sci* 1961;**15**:298.

14. Zanker A. Hydrocyclones: dimensions and performance. In: *Separation Techniques Vol 2 Gas Liquid Solid Systems*. McGraw-Hill; 1980. p. 178.

15. Dyakowski T, Hornung G, Williams RA. Simulation of non-Newtonian flow in a hydrocyclone. *Chem Eng Res Des* 1994;**72**(A4):513.

16. Colman DA, Thew MT, Corney DR. Hydrocyclones for oil-water separation, In: International conference on hydrocyclones. Vol. 11. 1980. p. 143–66 BHRA (Cranfield) paper.

17. Nebrensky JR, Morgan GE, Oswald BJ. Cyclone for gas/oil separation. In: International conference on hydrocyclones. Cranfield: BHRA; 1980. Paper 12, pp. 167–78.

18. Clarke AN, Wilson DJ. *Foam flotation, theory and applications*. Marcel Dekker; 1983.

19. Doughty FIC. Floatability of minerals. *Min Mag* 1949;**81**:268.

20. Hanumanth GS, Williams DJA. Design and operation characteristics of an improved laboratory flotation cell. *Miner Eng* 1988;**1**:177.

21. Hanumanth GS, Williams DJA. An experimental study of the effects of froth height on flotation of China clay. *Powder Technol* 1990;**60**:133.

22. Gochin RJ. Flotation. In: Svarovsky L, editor. *Solid-liquid separation*. 3rd ed. Butterworths; 1990. p. 591–613.

23. Zabel T. The advantages of dissolved air flotation for water treatment. *J Am Water Works Assoc* 1985;**77**:42–76.

24. Kuhn AT. The electrochemical treatment of aqueous effluent streams. In: Bockris JO'M, editor. *Electrochemistry of cleaner environments*. Plenum Press; 1972.

25. Ashman R. Control and recovery of dust and fume in industry. *Proc Inst Mech Eng* 1952;**1B**:157.

26. Stairmand CJ. In: Nonhebel G, editor. *Gas purification processes for air pollution control*. 2nd ed. London: Newnes-Butterworths; 1972.

27. Swift P. A user's guide to dust control. *Chem Eng Lond* 1984;**403**:22.

28. Slack MD, Prasad RO, Bakker A, Boysan F. Advances in cyclone venturi using unstructured grids. *Trans Inst Chem Eng* 2000;**78**(Part A):1098.

29. ter Linden AJ. Cyclone dust collectors. *Proc Inst Mech Eng* 1949;**160**:233.

30. Stairmand CJ. Pressure drop in cyclone separators. *Engineering* 1949;**168**:409.

31. Stairmand CJ. The design and performance of modern gas cleaning equipment. *J Inst Fuel* 1956;**29**:58.

32. Szekely J, Carr R. Heat transfer in a cyclone. *Chem Eng Sci* 1966;**21**:1119.

33. Dorman RG, Maggs FAP. Filtration of fine particles and vapours from gases. *Chem Eng Lond* 1976;**314**:671.

34. Hersey Jr HJ. Reverse-jet filters. *Ind Chem Mfr* 1955;**31**:138.

35. Schneider GG, Horzella TI, Cooper J, Striegl PJ. Selecting and specifying electrostatic precipitators. *Chem Eng Albany* 1975;**82**:94.

36. Rose HE, Wood AJ. *An introduction to electrostatic precipitation in theory and practice*. 2nd ed. Constable; 1966.

37. Dallavalle JM. *Micromerities*. 2nd ed. Pitman; 1948.

38. White ST. Inaudible sound: a new tool for air cleaning. *Heat Vent* 1948;**45**:59.

39. Wong T-S, Chen T-H, Shen X, Ho C-M. Nanochromatography driven by the coffee ring effect. *Anal Chem* 2011;**83**:1871–3.

40. Deegan RD, Bakajin O, DuPont TF, Huber G, Nagel SR, Witten TA. Capillary flow as the cause of ring stains from dried liquid drops. *Nature* 1997;**389**:827–9.

41. Deegan RD. Pattern formation in drying drops. *Phys Rev E* 2000;**61**:475–85.

42. Han W, Lin Z. Learning from "coffee rings": ordered structures enabled by controlled evaporative self-assembly. *Angew Chem Int Ed* 2012;**51**:1534–46.

43. Dugyala VR, Daware SV, Basavaraj MG. Shape anisotropic colloids: synthesis, packing behavior, evaporation driven assembly, and their application in emulsion stabilization. *Soft Matter* 2013;**9**:6711–25.

44. Dugyala VR, Basavaraj MG. Control over coffee-ring formation in evaporating liquid drops containing ellipsoids. *Langmuir* 2014;**30**:8680–6.

45. Sefiane K. Patterns from drying drops. *Adv Colloid Interf Sci* 2014;**206**:372–81.

46. Mampallila D, Eralb HB. A review on suppression and utilization of the coffee-ring effect. *Adv Colloid Interf Sci* 2018;**252**:38–54.

47. De Gennes P, Brochard-Wyart F, Quéré D. *Capillarity and wetting phenomena.* New York: Springer; 2004.

48. Monteux C, Lequeux F. Packing and sorting colloids at the contact line of a drying drop. *Langmuir* 2011;**27**:2917–22.

49. Oster G, Yamamoto M. Density gradient techniques. *Chem Rev* 1963;**63**:257–68.

50. Kowalczyk B, Lagzi I, Grzybowski BA. Nanoseparations: strategies for size and/or shape-selective purification of nanoparticles. *Curr Opin Colloid Interface Sci* 2011;**16**:135–48.

51. Brakke MK. Density gradient centrifugation: a new separation technique. *J Am Chem Soc* 1951;**73**:1847–8.

52. Lammers WT. Density-gradient separation of organic and inorganic particles by centrifugation. *Science* 1963;**139**:1298–9.

53. Brakke MK. Zonal separations by density-gradient centrifugation. *Arch Biochem Biophys* 1953;**45**:275–90.

54. Arnold MS, Green AA, Hulvatt JF, Stupp SI, Hersam MC. Sorting carbon nanotubes by electronic structure using density differentiation. *Nat Nanotechnol* 2006;**1**:60–5.

55. Chen G, Wang Y, Tan LH, Yang M, Tan LS, Chen Y, Chen H. High-purity separation of gold nanoparticle dimers and trimers. *J Am Chem Soc* 2009;**131**:4218–9.

56. Sharma V, Park K, Srinivasarao M. Colloidal dispersion of gold nanorods: historical background, optical properties, seed-mediated synthesis, shape separation and self-assembly. *Mater Sci Eng R* 2009;**65**:1–38.

57. Sharma V, Park K, Srinivasarao M. Shape separation of gold nanorods using centrifugation. *Proc Natl Acad Sci U S A* 2009;**106**:4981–5.

58. Park K, Koerner H, Vaia RA. Depletion-induced shape and size selection of gold nanoparticles. *Nano Lett* 2010;**10**:1433–9.

59. Baranov D, Fiore A, Van Huis M, Giannini C, Falqui A, Lafont U, Zandbergen H, Zanella M, Cingolani R, Manna L. Assembly of colloidal semiconductor nanorods in solution by depletion attraction. *Nano Lett* 2010;**10**:743–9.

Further Reading

Allen T. *Particle size measurement.* 5th ed. Vols. 1 and 2. London: Chapman and Hall; 1997.

Beddow JK. *Particle characterization in technology, Vol 1: Application and micro-analysis, Vol 2: Morphological analysis.* Boca Raton, FL: CRC Press; 1984.

Bohren CF, Huffman DR. *Absorption and scattering of light by small particles.* New York: Wiley; 1983.

Dorman RG. *Dust control and air cleaning.* Oxford: Pergamon Press; 1974.

Dullien, FAL. Introduction to industrial gas cleaning. (Academic Press, San Diego, CA, 1989); Dust and fume control, A user guide. 2nd ed. (Institution of Chemical Engineers, 1992).

Flnch JA, Dobby GS. *Column flotation.* Oxford: Pergamon Press; 1990.

Herdan G. *Small particle statistics.* 2nd ed. London: Butterworths; 1960.

Lewis Publications Inc. *Wet scrubbers: a practical handbook.*

Nonhebel G, editor. *Gas purification processes for air pollution control.* 2nd ed. London: Newnes-Butterworths; 1972.

Rhodes M. *Introduction to particle technology.* Wiley; 1998.

Rumpf H. *Particle technology.* London: Chapman and Hall; 1990.

Šterbaček Z, Tausk P. *Mixing in the chemical industry.* (translated from Czech by K. Mayer)Oxford: Pergamon Press; 1965.

Svarovsky L, editor. *Solid–liquid separation.* 3rd ed. Oxford: Butterworths; 1990*Solid–liquid separation.* 4th ed. Oxford: Butterworth-Heinemann; 2000.

White HJ. *Industrial electrostatic precipitation.* London: Addison-Wesley Publ. Co. Inc. and Pergamon Press; 1963.

CHAPTER 5

Particle Size Reduction and Enlargement

5.1 Introduction

Materials are rarely found in the size range required, and it is often necessary either to decrease or to increase the particle size. When, for example, the starting material is too coarse, and possibly in the form of large rocks, and the final product needs to be a fine powder, the particle size will have to be progressively reduced in stages. The most appropriate type of machine at each stage depends not only on the size of the feed and of the product, but also on such properties as compressive strength, brittleness, and stickiness of the particles. For example, the first stage in the process may require the use of a large jaw crusher and the final stage a sand grinder, two machines of very different characters.

At the other end of the spectrum, very fine powders are difficult to handle in some cases. Such fine powders may also give rise to hazardous dust clouds during transport (see chapter 14 in this volume). It may, therefore, be necessary to increase the particle size. Examples of size enlargement processes include granulation for the preparation of fertilisers, and compaction using compressive forces to form the tablets required for the administration of pharmaceuticals.

In this chapter, the two processes of size reduction and size enlargement are considered in Sections 5.2 and 5.4, respectively.

5.2 Size Reduction of Solids

5.2.1 Introduction

In the materials processing industry, size reduction, or *comminution*, is usually carried out in order to increase the surface area. In most unit operations and chemical reactions involving solid particles, the rates of heat transfer, mass transfer, and chemical reactions are directly proportional to the area. Hence, increased surface area helps in such operations. Thus, the rate of combustion of solid particles is proportional to the area presented to the gas, though a number of secondary factors may also be involved. For example, the free flow of a gas may be impeded because of the higher resistance to flow of a bed of small particles. In leaching, not only is the rate of extraction increased by virtue of the increased area of contact between the solvent and the solid, but the distance the solvent has to penetrate into the particles in order to gain access to the more remote pockets of solute is also shortened. This factor is

Coulson and Richardson's Chemical Engineering. https://doi.org/10.1016/B978-0-08-101098-3.00006-8

also important in the drying of porous solids, where reduction in size causes both an increase in contact area and a reduction in the distance the moisture must travel within the particles in order to reach the surface. In this case, the capillary forces acting on the moisture are also affected.

There are a number of other reasons for carrying out size reduction. It may, for example, be necessary to break a material into very small particles in order to separate two constituents, especially where one is dispersed in small isolated pockets. In addition, the properties of a material may be considerably influenced by the particle size; for example, the chemical reactivity of fine particles is greater than that of coarse particles, and the colour and covering power of a pigment are considerably affected by the size of the particles. Also, far more intimate mixing of solids can be achieved if the particle size is relatively small.

5.2.2 Mechanism of Size Reduction

While the mechanism of the process of size reduction is extremely complex, in recent years, a number of attempts have been made at a more detailed analysis of the problem. If a single lump of material is subjected to a sudden impact, it will generally break so as to yield a few relatively large particles and a number of fine particles, with relatively few particles of intermediate size. If the energy in the blow is increased, the larger particles will be of a rather smaller size and more numerous, and, whereas the number of fine particles will be appreciably increased, their size, however, will not be greatly altered. Therefore, it appears that the size of the fine particles is closely connected with the internal structure of the material, and the size of the larger particles is more closely connected with the process by which the size reduction is effected.

This effect is well illustrated by a series of experiments on the grinding of coal in a small mill, carried out by Heywood.[1] The results are shown in Fig. 5.1, in which the distribution of particle size in the product is shown as a function of the number of revolutions of the mill. The initial size distribution shows a single mode corresponding to a relatively coarse size, but as the degree of crushing is gradually increased, this mode progressively decreases in magnitude, and a second mode emerges at a particular size. This process continues until the first mode has completely disappeared. Here, the second mode is characteristic of the material and is known as the *persistent mode*, and the first is known as the *transitory mode*. There appears to be a *grind limit* for a particular material and machine. After some time, there seems to be little change in particle size if grinding is continued, though the particles may show some irreversible plastic deformation, which results in a change in shape rather than in size.

The energy required to effect size reduction is related to the internal structure of the material, and the process consists of two parts: first, opening up any small fissures which are already present, and second, forming a new surface. A material such as coal contains a number of small cracks and tends first to break along these weak spots, and therefore, the large pieces

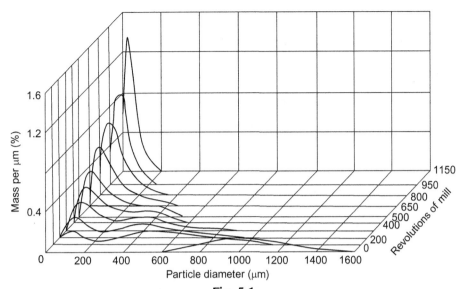

Fig. 5.1

Effect of progressive grinding on size distribution.

are broken up more readily than the small ones. Because a very much greater increase in surface results from crushing a given quantity of fine as opposed to coarse material, fine grinding requires very much more power. Very fine grinding can be impeded by the tendency of some relatively soft materials, including gypsum and some limestones, to form aggregates. These are groups of relatively weakly adhering particles held together by cohesive and van der Waals forces. Materials such as quartz and clinker form agglomerates in which the forces causing adhesion may be chemical in nature, and the bonds are then very much stronger.

In terms of energy utilisation, size reduction is a very inefficient process, and only between 0.1% and 2.0% of the energy supplied to the machine appears as increased surface energy in the solids. The efficiency of the process is very much influenced by the manner in which the load is applied and its magnitude. In addition, the nature of the force exerted is also very important, depending, for example, on whether it is predominantly a compressive, an impact, or a shearing force. If the applied force is insufficient for the elastic limit to be exceeded, and the material is compressed, energy is stored in the particle. When the load is removed, the particle expands again to its original condition without doing useful work. The energy appears as heat, and no size reduction is effected. A somewhat greater force will cause the particle to fracture, however, and in order to obtain the most effective utilisation of energy, the force should be only slightly in excess of the crushing strength of the material. The surface of the particles will generally be of a very irregular nature so that the force is initially taken on the high spots, with the result that very high stresses and temperatures may be set up locally in the material. As soon as a small amount of breakdown of material takes place, the point of application of the force alters. Bemrose and Bridgwater[2] and Hess and Schönert[3] have studied the breakage of single particles. All large lumps of material contain cracks, and size reduction occurs as a result of crack propagation that takes place above a critical parameter, F, where:

$$F = \frac{\tau^2 a}{Y} \tag{5.1}$$

where a is the crack length (m), τ is stress (Pa), and Y is Young's modulus(Pa).

Hess and Schönert[3] suggest that, at lower values of F, elastic deformation occurs without fracture, and the energy input is completely ineffective in achieving size reduction. Fundamental studies of the application of fracture mechanics to particle size reduction have been carried out by Schönert.[4] In essence, an energy balance is applied to the process of crack extension within a particle by equating the loss of energy from the strain field within the particle to the increase in surface energy when the crack propagates. Because of plastic deformation near the tip of the crack, however, the energy requirement is at least 10 times greater, and, in addition, kinetic energy is associated with the sudden acceleration of material as the crack passes through it. Orders of magnitude of the surface fracture energy per unit volume are:

Glass	$1\text{--}10\,\text{J/m}^2$
Plastics	$10\text{--}10^3\,\text{J/m}^2$
Metals	$10^3\text{--}10^5\,\text{J/m}^2$

All of these values are several orders of magnitude higher than the thermodynamic surface energy, which is about $10^{-1}\,\text{J/m}^2$. When a crack is initially present in a material, the stresses near the tip of the crack are considerably greater than those in the bulk of the material. Calculation of the actual value is well nigh impossible, as the crack surfaces are usually steeply curved and rough. The presence of a crack modifies the stress field in its immediate location, with the increase in energy being approximately proportional to $\sqrt{(a/l)}$ where a is the crack length and l is the distance from the crack tip to the point of interest. Changes in crack length are accompanied by modifications in stress distribution in the surrounding material and, hence, in its energy content.

During the course of the size reduction processes, much energy is expended in causing plastic deformation, and this energy may be regarded as a waste as it does not result in fracture. Only part of it is retained in the system as a result of elastic recovery. It is not possible, however, to achieve the stress levels necessary for fracture to occur without first passing through the condition of plastic deformation, and, in this sense, this must be regarded as a necessary state which must be achieved before fracture can possibly occur.

The nature of the flaws in the particles changes with their size. If, as is customary, fine particles are produced by crushing large particles, the weakest flaws will be progressively eliminated as the size is reduced, and thus, small particles tend to be stronger and require more energy for fracture to occur. In addition, as the capacity of the particle for storing energy is

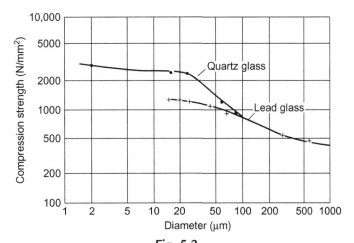

Fig. 5.2

Compressive strength of glass spheres as a function of their diameter.

proportional to its volume ($\propto d^3$), and the energy requirement for propagating geometrically similar cracks is proportional to the surface area ($\propto d^2$), the energy available per unit crack area increases linearly with particle size (d). Thus, breakage will occur at lower levels of stress in large particles. This is illustrated in Fig. 5.2, which shows the results of experimental measurements of the compressive strengths for shearing two types of glass. It may be noted from Fig. 5.2 that, for quartz glass, the compressive strength of $2\,\mu m$ particles is about three times that of $100\,\mu m$ particles.

The exact method by which fracture occurs is not known, although it is suggested by Piret[5] that the compressive force produces small flaws in the material. If the energy concentration exceeds a certain critical value, these flaws will grow rapidly and will generally branch, and the particles will break up. The probability of fracture of a particle in an assembly of particles increases with the number of contact points, up to a number of around 10, although the probability then decreases for further increase in number. The rate of application of the force is important because there is generally a time lag between attainment of maximum load and fracture. Thus, a rather small force will cause fracture provided it is maintained for a sufficient time. This is a phenomenon similar to the ignition lag which is obtained with a combustible gas–oxidant mixture. Here, the interval between introducing the ignition source and the occurrence of ignition is a function of the temperature of the source, and when it is near the minimum ignition temperature, delays of several seconds may be encountered. The greater the rate at which the load is applied, the less effectively is the energy utilised, and the higher is the proportion of fine material which is produced. If the particle shows any viscoelastic behaviour, a high rate of application of the force is needed for fracture to occur. The efficiency of utilisation of energy as supplied by a falling mass has been compared with that of energy applied slowly by means of hydraulic pressure. Up to three or four times more surface can be produced per unit of energy if it is applied by the latter method. Piret[5] suggests that there is a close similarity between the crushing operation and a chemical reaction. In both cases, a critical energy level must be exceeded before the process will start, and in both cases, time is an important variable.

The method of application of the force to the particles may affect the breakage pattern. Prasher[6] suggests that four basic patterns may be identified, though it is sometimes difficult to identify the dominant mode in any given machine. The four basic patterns are:

(a) *Impact*—particle concussion by a single rigid force.
(b) *Compression*—particle disintegration by two rigid forces.
(c) *Shear*—produced by a fluid or by particle–particle interaction.
(d) *Attrition*—arising from particles scraping against one another or against a rigid surface.

5.2.3 Energy Requirement for Size Reduction

Energy requirements

Although it is impossible to estimate accurately the amount of energy required in order to effect a size reduction of a given material, a number of empirical laws have been proposed. The two earliest laws are due to Kick[7] and von Rittinger,[8] and a third law due to Bond[9, 10] has also been proposed. These three laws may all be derived from the basic differential equation:

$$\frac{\mathrm{d}E}{\mathrm{d}L} = -CL^p \tag{5.2}$$

which states that the energy $\mathrm{d}E$ required to effect a small change $\mathrm{d}L$ in the size of unit mass of material is a simple power function of the size. If $p = -2$, then integration gives:

$$E = C\left(\frac{1}{L_2} - \frac{1}{L_1}\right)$$

Writing $C = K_R f_c$, where f_c is the crushing strength of the material, then *Rittinger's law*, first postulated in 1867, is obtained as:

$$E = K_R f_c\left(\frac{1}{L_2} - \frac{1}{L_1}\right) \tag{5.3}$$

Because the surface of unit mass of material is proportional to $1/L$, the interpretation of this law is that the energy required for size reduction is directly proportional to the increase in surface area.

If $p = -1$, then:

$$E = C\,\ln\frac{L_1}{L_2}$$

and, writing $C = K_K f_c$:

$$E = K_K f_c \ln \frac{L_1}{L_2} \tag{5.4}$$

which is known as *Kick's law*. This supposes that the energy required is directly related to the reduction ratio L_1/L_2, which means that the energy required to crush a given amount of material from a 50-mm to a 25-mm size is the same as that required to reduce the size from 12 mm to 6 mm. In Eqs (5.3) and (5.4), K_R and K_K are known respectively as Rittinger's constant and Kick's constant. It may be noted that neither of these constants is dimensionless.

Neither of these two laws permits an accurate calculation of the energy requirements. Rittinger's law is applicable mainly to that part of the process where new surface is being created and holds most accurately for fine grinding where the increase in surface per unit mass of material is large. Kick's law more closely relates to the energy required to effect elastic deformation before fracture occurs, and is more accurate than Rittinger's law for coarse crushing where the amount of surface produced is considerably less.

Bond[9, 10] has suggested a law intermediate between Rittinger's and Kick's laws, by using $p = -3/2$ in Eq. (5.2). Thus:

$$E = 2C \left(\frac{1}{L_2^{1/2}} - \frac{1}{L_1^{1/2}} \right)$$

$$= 2C \sqrt{\left(\frac{1}{L_2} \right) \left(1 - \frac{1}{q^{1/2}} \right)} \tag{5.5}$$

where

$$q = \frac{L_1}{L_2}$$

the reduction ratio. Writing $C = 5E_i$, then:

$$E = E_i \sqrt{\left(\frac{100}{L_2} \right) \left(1 - \frac{1}{q^{1/2}} \right)} \tag{5.6}$$

Bond[9, 10] calls E_i the *work index*, and expresses it as the amount of energy required to reduce the unit mass of material from an infinite particle size to a size L_2 of 100 μm, that is $q = \infty$. The size of material is taken as the size of the square hole through which 80% of the material will pass. Expressions for the work index are given in the original papers[8, 9] for various types of materials and various forms of size reduction equipment.

Austin and Klimpel[11] have reviewed these three laws and their applicability, and Cutting[12] has described laboratory work to assess grindability using rod mill tests.

Example 5.1

A material is crushed in a Blake jaw crusher such that the average size of particle is reduced from 50 to 10 mm with the consumption of energy of 13.0 kW/(kg/s). What would be the consumption of energy needed to crush the same material of an average size of 75 mm to an average size of 25 mm:

(a) assuming Rittinger's law applies?
(b) assuming Kick's law applies?

Which of these results would be regarded as being more reliable and why?

Solution

(a) *Rittinger's law.*

This is given by:

$$E = K_R f_c[(1/L_2) - (1/L_1)] \tag{5.3}$$

Thus:

$$13.0 = K_R f_c[(1/10) - (1/50)]$$

and:

$$K_R f_c = (13.0 \times 50/4) = 162.5 \, \text{kW/(kg mm)}$$

Thus the energy required to crush 75 mm material to 25 mm is:

$$E = 162.5[(1/25) - (1/75)] = \underline{4.33 \, \text{kJ/kg.}}$$

(b) *Kick's law.*

This is given by:

$$E = K_K f_c \, \ln(L_1/L_2) \tag{5.4}$$

Thus:

$$13.0 = K_K f_c \, \ln(50/10)$$

and:

$$K_K f_c = (13.0/1.609) = 8.08 \, \text{kW/(kg/s)}$$

Thus, the energy required to crush 75 mm material to 25 mm is given by:

$$E = 8.08 \, \ln(75/25) = \underline{8.88 \, \text{kJ/kg.}}$$

The size range involved in this case should be considered as that for coarse crushing. Because Kick's law is more accurate for coarse crushing, this would be taken as given the more reliable result.

Energy utilisation

One of the first important investigations into the distribution of the energy fed into a crusher was carried out by Owens,[13] who concluded that energy was utilised as follows:

(a) In producing elastic deformation of the particles before fracture occurs,
(b) In producing inelastic deformation which results in size reduction,
(c) In causing elastic distortion of the equipment,
(d) In friction between particles, and between particles and the machine,
(e) In noise, heat, and vibration in the plant, and
(f) In friction losses in the plant itself.

Owens[13] estimated that only about 10% of the total power is usefully employed.

In an investigation by the US Bureau of Mines,[14] in which a drop weight type of crusher was used, it was found that the increase in surface was directly proportional to the input of energy, and that the rate of application of the load was an important factor.

This conclusion was substantiated in a subsequent investigation of the power consumption in a size reduction process that is reported in three papers by Kwong et al.,[15] Adams et al.,[16] and Johnson et al.[17] A sample of material was crushed by placing it in a cavity in a steel mortar, placing a steel plunger over the sample and dropping a steel ball of known weight on the plunger over the sample from a measured height. Any bouncing of the ball was prevented by three soft aluminium cushion wires under the mortar, and these wires were calibrated so that the energy absorbed by the system could be determined from their deformation. Losses in the plunger and ball were assumed to be proportional to the energy absorbed by the wires, and the energy actually used for size reduction was then obtained as the difference between the energy of the ball on striking the plunger and the energy absorbed. Surfaces were measured by a water or air permeability method or by gas adsorption. The latter method gave a value approximately double that obtained from the former, indicating that, in these experiments, the internal surface was approximately the same as the external surface. The experimental results showed that, provided the new surface did not exceed about $40 \, \text{m}^2/\text{kg}$, the new surface produced was directly proportional to the energy input. For a given energy input, the new surface produced was independent of:

(a) The velocity of impact,
(b) The mass and arrangement of the sample,
(c) The initial particle size, and
(d) The moisture content of the sample.

Between 30% and 50% of the energy of the ball on impact was absorbed by the material, although no indication was obtained of how this was utilised. An extension of the range of the experiments, in which up to $120 \, \text{m}^2$ of new surface was produced per kilogram of material, showed that the linear relationship between energy and new surface no longer held strictly. In further tests in which the crushing was effected slowly using a hydraulic press, it was found, however, that the linear relationship still held for the larger increases in surface.

In order to determine the efficiency of the surface production process, tests were carried out with sodium chloride, and it was found that 90 J was required to produce 1 m² of new surface. As the theoretical value of the surface energy of sodium chloride is only $0.08\,J/m^2$, the efficiency of the process is about 0.1%. Zeleny and Piret[18] have reported calorimetric studies on the crushing of glass and quartz. It was found that a fairly constant energy was required of $77\,J/m^2$ of new surface created, compared with a surface-energy value of $<5\,J/m^2$. In some cases, over 50% of the energy supplied was used to produce plastic deformation of the steel crusher surfaces.

The apparent efficiency of the size reduction operation depends on the type of equipment used. Thus, for instance, a ball mill is rather less efficient than a drop weight type of crusher because of the ineffective collisions that take place in the ball mill.

Further work[5] on the crushing of quartz showed that more surface was created per unit of energy with single particles than with a collection of particles. This appears to be attributable to the fact that the crushing strength of apparently identical particles may vary by a factor as large as 20, and it is necessary to provide a sufficient energy concentration to crush the strongest particle. Some recent developments, including research and mathematical modelling, are described by Prasher.[6]

5.2.4 Methods of Operating Crushers

There are two distinct methods of feeding material to a crusher. The first, known as *free crushing*, involves feeding the material at a comparatively low rate so that the product can readily escape. Its residence time in the machine is, therefore, short, and the production of appreciable quantities of undersize material is avoided. The second method is known as *choke feeding*. In this case, the machine is kept full of material, and discharge of the product is impeded so that the material remains in the crusher for a longer period. This results in a higher degree of crushing, although the capacity of the machine is reduced, and energy consumption is high because of the cushioning action produced by the accumulated product. This method is, therefore, used only when a comparatively small amount of materials is to be crushed, and when it is desired to complete the whole of the size reduction in one operation.

If the plant is operated, as in *choke feeding*, so that the material is passed only once through the equipment, the process is known as *open circuit grinding*. If, on the other hand, the product contains material which is insufficiently crushed, it may be necessary to separate the product and return the oversize material for a second crushing. This system, which is generally to be preferred, is known as *closed circuit grinding*. A flow sheet for a typical closed circuit grinding process, in which a coarse crusher, an intermediate crusher, and a fine grinder are used, is shown in Fig. 5.3. In many plants, the product is continuously removed, either by allowing the material to fall on to a screen or by subjecting it to the action of a stream of fluid, such that the small particles are carried away, and the oversize material falls back to be crushed again.

The desirability of using a number of size reduction units when the particle size is to be considerably reduced arises from the fact that it is not generally economical to effect a large size reduction ratio in a single machine. The equipment used is usually divided into classes, as given in Table 5.1, according to the size of the feed and the product.

A greater size reduction ratio can be obtained in fine crushers than in coarse crushers.

The equipment may also be classified, to some extent, according to the nature of the force which is applied, though, as a number of forces are generally involved, it is a less convenient basis.

Grinding may be carried out either wet or dry, although wet grinding is generally applicable only with low speed mills. The advantages of wet grinding are:

(a)   The power consumption is reduced by about 20%–30%.
(b)   The capacity of the plant is increased.
(c)   The removal of the product is facilitated, and the amount of fines is reduced.
(d)   Dust formation is eliminated.
(e)   The solids are more easily handled.

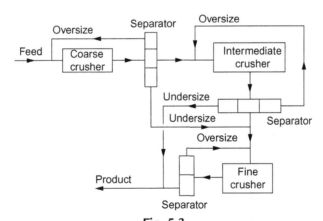

Fig. 5.3
Flow diagram for closed circuit grinding system.

Table 5.1 Classification of size reduction equipment

	Feed Size (mm)	Product Size
Coarse crushers	1500–40	50–5 mm
Intermediate crushers	50–5	5–0.1 mm
Fine crushers	5–2	0.1 mm
Colloid mills	0.2	Down to 0.01 μm

Against this, the wear on the grinding medium is generally about 20% greater, and it may be necessary to dry the product.

The separators in Fig. 5.3 may be either a cyclone type, as typified by the Bradley microsizer, or a mechanical air separator. Cyclone separators, the theory of operation and application of which are fully discussed in Chapter 4, may be used. Alternatively, a *whizzer* type of air separator, such as the NEI air separator shown in Figs 4.10 and 4.11, is often included as an integral part of the mill, as shown in the examples of the NEI pendulum mill in Fig. 5.21. Oversize particles drop down the inner case and are returned directly to the mill, while the fine material is removed as a separate product stream.

5.2.5 Nature of the Material To Be Crushed

The choice of a machine for a given crushing operation is influenced by the nature of the product required and the quantity and size of material to be handled. The more important properties of the feed, apart from its size, are as follows:

Hardness. The hardness of the material affects the power consumption and the wear on the machine. With hard and abrasive materials, it is necessary to use a low-speed machine and to protect the bearings from the abrasive dusts that are produced. Pressure lubrication is recommended. Materials are arranged in order of increasing hardness in the *Mohs* scale, in which the first four items rank as soft and the remainder as hard. The Mohs scale of Hardness is:

1. Talc	5. Apatite	8. Topaz
2. Rock salt or gypsum	6. Felspar	9. Carborundum
3. Calcite	7. Quartz	10. Diamond
4. Fluorspar		

Structure. Normal granular materials such as coal, ores, and rocks can be effectively crushed employing the normal forces of compression, impact, and so on. With fibrous materials, a tearing action is required.

Moisture content. It is found that materials do not flow well if they contain between about 5% and 50% of moisture. Under these conditions, the material tends to cake together in the form of balls. In general, grinding can be carried out satisfactorily outside these limits.

Crushing strength. The power required for crushing is almost directly proportional to the crushing strength of the material.

Friability. The friability of the material is its tendency to fracture during normal handling. In general, a crystalline material will break along well-defined planes, and the power required for crushing will increase as the particle size is reduced.

Stickiness. A sticky material will tend to clog the grinding equipment, and it should, therefore, be ground in a plant that can be cleaned easily.

Soapiness. In general, this is a measure of the coefficient of friction of the surface of the material. If the coefficient of friction is low, the crushing may be more difficult.

Explosive materials must be ground wet or in the presence of an inert atmosphere.

Materials yielding dusts that are harmful to the health must be ground under conditions where the dust is not allowed to escape.

Work[19] has presented a guide to equipment selection based on size and abrasiveness of material.

5.3 Types of Crushing Equipment

The most important coarse, intermediate, and fine crushers may be classified as described in Table 5.2.

The features of these crushers are now considered in detail.

5.3.1 Coarse Crushers

The Stag jaw crusher, shown in Fig. 5.4, has a fixed jaw and a moving jaw pivoted at the top with the crushing faces formed of manganese steel. Because the maximum movement of the jaw is at the bottom, there is little tendency for the machine to clog, though some uncrushed material may fall through and have to be returned to the crusher. The maximum pressure is exerted on the large material, which is introduced at the top. The machine is usually protected so that it is not damaged if lumps of metal inadvertently enter, by making one of the toggle plates in the driving mechanism relatively weak so that, if any large stresses are set up, this is the first part to fail. Easy renewal of the damaged part is then possible.

Table 5.2 Crushing equipment

Coarse Crushers	Intermediate Crushers	Fine Crushers
Stag jaw crusher	Crushing rolls	Buhrstone mill
Dodge jaw crusher	Disc crusher	Roller mill
Gyratory crusher	Edge runner mill	NEI pendulum mill
Other coarse crushers	Hammer mill	Griffin mill
	Single roll crusher	Ring roller mill (Lopulco)
	Pin mill	Ball mill
	Symons disc crusher	Tube mill
		Hardinge mill
		Babcock mill

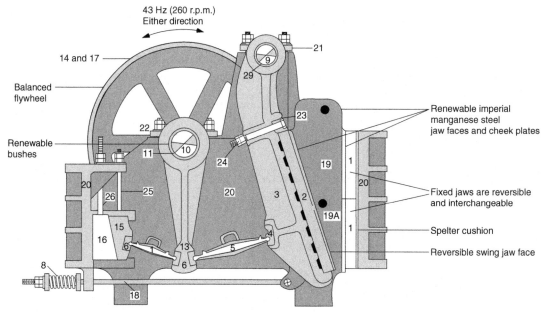

Fig. 5.4

Typical cross-section of Stag jaw crusher. 1: Fixed jaw face, 2: swing jaw face, 3: swing jaw stock, 4: toggle seating, 5: front toggle plate, 6: toggle seating, 7: back toggle plate, 8: springs and cups, 9: swing jaw shaft, 10: eccentric shaft, 11: Pitman bush, 13: Pitman, 14: flywheel grooved for V rope drive, 15: toggle block, 16: wedge block, 17: flywheel, 18: tension rods, 19: cheek plates (top), 19A: cheek plates (bottom), 20: body, 21: swing jaw shaft bearing caps, 22: eccentric shaft bearing caps, 23: wedge for swing jaw face, 24: bolts of wedge, 25: bolts for toggle block, 26: bolts for wedge block, 27: eccentric shaft bearing bush (bottom), 28: eccentric shaft bearing bush (top), 29: swing stock bush. *The author and editor thank Vaba Process Plant Ltd, Rotherham, Yorks for this figure.*

Stag crushers are made with jaw widths varying from about 150 mm to 1 m, and the running speed is about 4 Hz (240 rpm) with the smaller machines running at the higher speeds. The speed of operation should not be so high that a large quantity of fines is produced as a result of material being repeatedly crushed because it cannot escape sufficiently quickly. The angle of nip, the angle between the jaws, is usually about 30 degrees.

Because the crushing action is intermittent, the loading on the machine is uneven, and the crusher, therefore, incorporates a heavy flywheel. The power requirements of the crusher depend upon size and capacity and vary from 7 to about 70 kW, the latter figure corresponding to a feed rate of 10 kg/s.

The Dodge jaw crusher

In the Dodge crusher, shown in Fig. 5.5, the moving jaw is pivoted at the bottom. The minimum movement is, thus, at the bottom, and a more uniform product is obtained, although the crusher is less widely used because of its tendency to choke. The large opening at the top enables it to take very large feed and to effect a large size reduction. This crusher is usually made in smaller

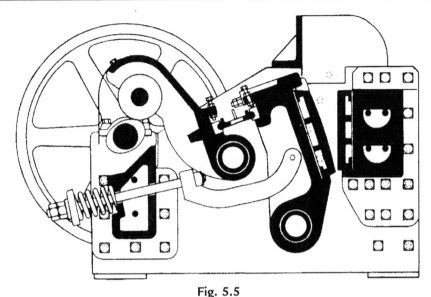

Fig. 5.5
Dodge crusher. *Used with permission from The Weir Group PLC.*

sizes than the Stag crusher, because of the high fluctuating stresses that are produced in the members of the machine.

The gyratory crusher

The gyratory crusher shown in Fig. 5.6 employs a crushing head, in the form of a truncated cone mounted on a shaft, the upper end of which is held in a flexible bearing, while the lower end is driven eccentrically so as to describe a circle. The crushing action takes place around the whole of the cone, and, because the maximum movement is at the bottom, the characteristics of the machine are similar to those of the Stag crusher. As the crusher is continuous in action, the fluctuations in the stresses are smaller than in jaw crushers, and the power consumption is lower. This unit has a large capacity per unit area of grinding surface, particularly if it is used to produce a small size reduction. It does not, however, take such a large size of feed as a jaw crusher, although it gives a rather finer and more uniform product. Because the capital cost is high, the crusher is suitable only where large quantities of material are to be handled.

The jaw crushers and the gyratory crusher all employ a predominantly compressive force.

Other coarse crushers

Friable materials, such as coal, may be broken up without the application of large forces, and, therefore, a less robust plant may be used. A common form of coal breaker consists of a large hollow cylinder with perforated walls. The axis is at a small angle to the horizontal, and the feed is introduced at the top. The cylinder is rotated, and the coal is lifted by means of arms attached to the inner surface and then falls against the cylindrical surface. The coal

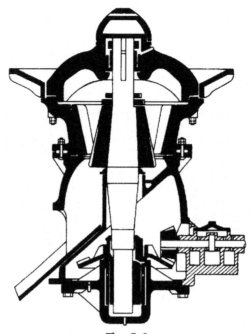

Fig. 5.6
Gyratory crusher.

breaks by impact and passes through the perforations as soon as the size has been sufficiently reduced. This type of equipment is less expensive and has a higher throughput than the jaw or gyratory crusher. Another coarse rotary breaker, the rotary coal breaker, is similar in action to the hammer mill described later and is shown in Fig. 5.7. The crushing action depends upon the transference of kinetic energy from hammers to the material, and these pulverisers are essentially high speed machines with a speed of rotation of about 10 Hz (600 rpm) giving hammer tip velocities of about 40 m/s.

5.3.2 Intermediate Crushers

The edge runner mill

In the edge runner mill, shown in Fig. 5.8, a heavy cast iron or granite wheel, or *muller* as it is called, is mounted on a horizontal shaft which is rotated in a horizontal plane in a heavy pan. Alternatively, the muller remains stationary, and the pan is rotated, and in some cases, the mill incorporates two mullers. Material is fed to the centre of the pan and is worked outwards by the action of the muller, while a scraper continuously removes material that has adhered to the sides of the pan and returns it to the crushing zone. In many models, the outer rim of the bottom of the pan is perforated, so that the product may be removed

Fig. 5.7

Rotary coal breaker. *The author and editor thank Vaba Process Plant Ltd, Rotherham, Yorks for this figure.*

Fig. 5.8

Edge runner mill. *Used with permission from Baker Perkins Ltd.*

continuously as soon as its size has been sufficiently reduced. The mill may be operated wet or dry, and it is used extensively for the grinding of paints, clays, and sticky materials.

The hammer mill

The hammer mill is an impact mill employing a high speed rotating disc, to which are fixed a number of hammer bars, which are swung outwards by centrifugal force. An industrial model is illustrated in Fig. 5.9, and a laboratory model in Fig. 5.10. Material is fed in, either at the top or at the centre, and it is thrown out centrifugally and crushed by being beaten between the hammer bars, or against breaker plates fixed around the periphery of the cylindrical casing. The material is beaten until it is small enough to fall through the screen that forms the lower portion of the casing. Because the hammer bars are hinged, the presence of any hard material does not cause damage to the equipment. The bars are readily replaced when they are worn out. The machine is suitable for the crushing of both brittle and fibrous materials, and, in the latter case, it is usual to employ a screen with cutting edges. The hammer mill is suitable for hard materials, although, because a large amount of fines is produced, it is advisable to employ positive pressure lubrication to the bearings in order to prevent the entry of dust. The size of the product is regulated by the size of the screen and the speed of rotation.

Fig. 5.9
Swing claw hammer mill.

Fig. 5.10
The Raymond laboratory hammer mill. *The author and editor thank NEI International Combustion Ltd, Derby, for this figure.*

A number of similar machines are available, and in some, the hammer bars are rigidly fixed in position. Because a large current of air is produced, the dust must be separated in a cyclone separator or a bag filter.

The pin-type mill

The Alpine pin disc mill, shown in Fig. 5.11, is a form of pin mill and consists of two vertical steel plates with horizontal projections on their near faces. One disc may be stationary, while the other disc is rotated at high speed; sometimes, the two discs may be rotated in opposite directions. The material is gravity fed in through a hopper or air conveyed to the centre of the discs, and is thrown outwards by centrifugal action and broken against of the projections before it is discharged to the outer body of the mill and falls under gravity from the bottom of the casing. Alternatively, the pins may be replaced by swing beaters or plate beaters, depending on the setup and application. The mill gives a fairly uniform fine product with little dust and is extensively used with chemicals, fertilisers, and other materials that are nonabrasive, brittle, or

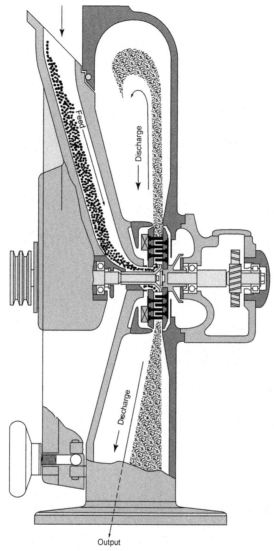

Fig. 5.11
Alpine pin mill with both discs and sets of pins rotating.

crystalline. Control of the size of the product is effected by means of the speed and the spacing of the projections, and a product size of $20\,\mu m$ is readily attainable.

The Alpine universal mill with turbine beater and grinding track shown in Fig. 5.12 is suitable for both brittle and tough materials. The high airflow from the turbine keeps the temperature rise to a minimum.

Fig. 5.12

Alpine Universal mill. *Used with permission from Hosokawa Micron Ltd.*

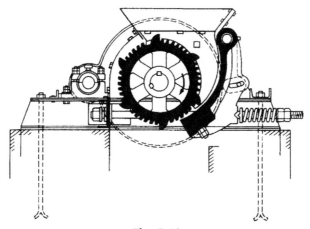

Fig. 5.13

Single roll crusher. *The author and editor thank Vaba Process Plant Ltd, Rotherham, Yorks for this figure.*

The single roll crusher

The single roll crusher, shown in Fig. 5.13, consists of a toothed crushing roll that rotates close to a breaker plate. Fig. 5.14 is an illustration of an industrial model. The material is crushed by compression and shearing between the two surfaces. It is used extensively for crushing coal.

Crushing rolls

Two rolls, one in adjustable bearings, rotate in opposite directions as shown in Fig. 5.15, and the clearance between them can be adjusted according to the size of feed and the required size of product. The machine is protected, by spring loading, against damage from very hard

Fig. 5.14
The Stag single roll crusher. *The author and editor thank Vaba Process Plant Ltd, Rotherham, Yorks for this figure.*

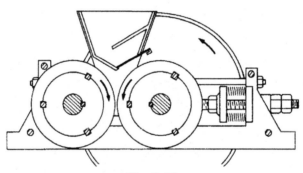

Fig. 5.15
Crushing rolls.

material. Both rolls may be driven, or one directly and the other by friction with the solids. The crushing rolls, which may vary from a few centimetres up to about 1.2 m in diameter, are suitable for effecting a small size reduction ratio, 4:1 in a single operation, and it is, therefore, common to employ a number of pairs of rolls in series, one above the other. Roll shells with either smooth or ridged surfaces are held in place by keys to the main shaft. The capacity is usually between one-tenth and one-third of that calculated on the assumption that a continuous ribbon of the material forms between the rolls.

An idealised system where a spherical or cylindrical particle of radius r_2 is being fed to crushing rolls of radius r_1 is shown in Fig. 5.16. 2α is the angle of nip, the angle between the two common tangents to the particle and each of the rolls, and $2b$ is the distance between the rolls. It may be seen from the geometry of the system that the angle of nip is given by:

$$\cos\alpha = \frac{(r_1 + b)}{(r_2 + r_1)} \tag{5.7}$$

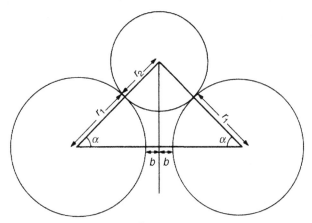

Fig. 5.16
Particle fed to crushing rolls.

For steel rolls, the angle of nip is not greater than about 32 degrees. It is clear that the angle of nip is dependent on the coefficient of friction between the rolls and the material. It can be shown that $\tan(\alpha) = \mu$ (the coefficient of friction).[19a]

Crushing rolls are extensively used for crushing oil seeds and in the gunpowder industry, and they are also suitable for abrasive materials. They are simple in construction and do not give a large percentage of fines.

Example 5.2

If crushing rolls, 1 m in diameter, are set so that the crushing surfaces are 12.5 mm apart and the angle of nip is 31 degrees, what is the maximum size of particle which should be fed to the rolls?

If the actual capacity of the machine is 12% of the theoretical value, calculate the throughput in kg/s when running at 2.0 Hz if the working face of the rolls is 0.4 m long, and the bulk density of the feed is 2500 kg/m³.

Solution

The particle size may be obtained from:

$$\cos \alpha = (r_1 + b)/(r_1 + r_2) \tag{5.7}$$

In this case: $2\alpha = 31$ degrees and $\cos \alpha = 0.964$, $b = (12.5/2) = 6.25$ mm or 0.00625 m and:

$$r_1 = (1.0/2) = 0.5 \, \text{m}.$$

Thus:

$$0.964 = (0.5 + 0.00625)/(0.5 + r_2).$$

and:

$$r_2 = 0.025 \, \text{m or } \underline{25 \, \text{mm}.}$$

The cross-sectional area for flow$=(0.0125 \times 0.4) = 0.005\,\text{m}^2$.

and the volumetric flowrate$=(2.0 \times 0.005) = 0.010\,\text{m}^3/\text{s}$.

Thus, the actual throughput$=(0.010 \times 12)/100 = 0.0012\,\text{m}^3/\text{s}$.

or: $(0.0012 \times 2500) = \underline{3.0\,\text{kg/s}}$.

The Symons disc crusher

The disc crusher shown in Fig. 5.17 employs two saucer-shaped discs mounted on horizontal shafts, one of which is rotated, and the other is mounted in an eccentric bearing so that the two crushing faces continuously approach and recede. The material is fed into the centre between the two discs, and the product is discharged by centrifugal action as soon as it is fine enough to escape through the opening between the faces.

5.3.3 Fine Crushers

The buhrstone mill

The buhrstone mill, shown in Fig. 5.18, is one of the oldest forms of fine crushing equipment, though it has been very largely superseded now by roller mills. Grinding takes place between two heavy horizontal wheels, one of which is stationary, and the other is driven. The surface of the stones is carefully dressed so that the material is continuously worked outwards from the centre of the circumference of the wheels. Size reduction takes place by a shearing action between the edges of the grooves on the two grinding stones. This equipment has been used for the grinding of grain, pigments for paints, pharmaceuticals, cosmetics, and printer's ink, although it is now used only where the quantity of material is very small.

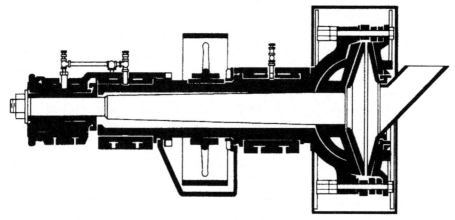

Fig. 5.17
Symons disc crusher.

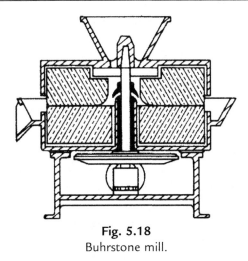

Fig. 5.18
Buhrstone mill.

Roller mills

The roller mill consists of a pair of rollers that rotate at different speeds, a 3:1 ratio for example, in opposite directions. As in crushing rolls, one of the rollers is held in a fixed bearing, whereas the other has an adjustable spring-loaded bearing. Because the rollers rotate at different speeds, size reduction is effected by a combination of compressive and shear forces. The roller mill is extensively used in the flour milling industry and for the manufacture of pigments for paints.

Centrifugal attrition mills

The Babcock mill. This mill, shown in Fig. 5.19, consists of a series of pushers which cause heavy cast iron balls to rotate against a bull ring like a ball race, with the pressure of the balls on the bull ring being produced by a loading applied from above. Material fed into the mill falls on the bull ring, and the product is continuously removed in an upward stream of air which carries the ground material between the blades of the classifier, which is shown toward the top of the photograph. The oversize material falls back and is reground. The air stream is produced by means of an external blower which may involve a considerable additional power consumption. The fineness of the product is controlled by the rate of feeding and the air velocity. This machine is used mainly for preparation of pulverised coal and sometimes for cement.

The Lopulco mill or ring-roll pulveriser. These mills are manufactured in large numbers for the production of industrial minerals such as limestone and gypsum. A slightly concave circular bull ring rotates at high speed, and the feed is thrown outwards by centrifugal action under the crushing rollers, which are shaped like truncated cones, as shown in Fig. 5.20. The rollers are spring-loaded, and the strength of the springs determines the grinding force available. There is a clearance between the rollers and the bull ring so that there is no wear

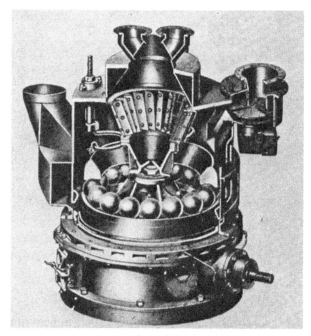

Fig. 5.19

Babcock mill. *Used with permission from Doosan Power Systems.*

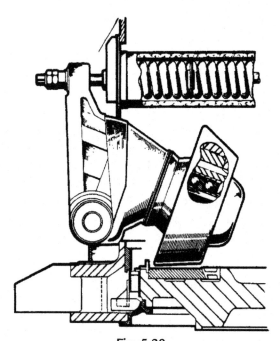

Fig. 5.20

Crushing roller in a Lopulco mill. *The author and editor thank NEI International Combustion Ltd, Derby, for this figure.*

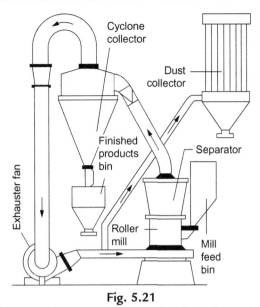

Fig. 5.21

Typical air classification for a Lopulco or pendulum mill. *The author and editor thank NEI International Combustion Ltd, Derby, for this figure.*

on the grinding heads if the plant is operated with a light load, and quiet operation is obtained. The product is continuously removed by means of a stream of air produced by an external fan and is carried into a separator fitted above the grinding mechanism. In this separator, the cross-sectional area for flow is gradually increased, and, as the air velocity fails, the oversize material falls back and is ground again. The product is separated in a cyclone separator from the air which is then recycled as shown in Fig. 5.21. Chemicals, dyes, cements, and phosphate rocks are also ground in the Lopulco mill. As the risk of sparking is reduced by the maintenance of a clearance between the grinding media, the mill may be used for grinding explosive materials.

Some typical figures for power consumption with the Lopulco mill are given in Table 5.3.

The NEI pendulum mill. The NEI pendulum mill, shown in Fig. 5.22, is slightly less economical in operation than the Lopulco mill, although it gives a rather finer and more uniform product. A central shaft driven by a bevel gear carries a yoke at the top and terminates in a foot-step bearing at the bottom. On the yoke are pivoted a number of heavy arms, shown in Figs 5.23 and 5.24, carrying the rollers which are thrown outwards by centrifugal action and bear on a circular bull ring. Both the rollers and the bull ring are readily replaceable. The material, which is introduced by means of an automatic feed device, is forced on to the bull ring by means of a plough which rotates on the central shaft. The ground material is removed by means of an air current, as in the case of the Lopulco mill, and the oversize material falls back and is again brought on to the bull ring by the plough.

As the mill operates at high speeds, it is not suitable for use with abrasive materials, nor will it handle materials that soften during milling. Although the power consumption and maintenance

Table 5.3 Typical power requirements for grinding

Material	Mill Size	Product Fineness	Output kg/s	tonne/h	Power Consumption MJ/tonne (Mill and Fan Installed)	HP h/tonne (Mill and Fan Installed)
Gypsum	LM12	99%–75 µm	4.5	1.3	134	50
	LM14	99%–150 µm	11.2	3.2	78	29
	LM16	80%–150 µm	27.5	7.8	48	18
Limestone	LM12	70%–75 µm	11.0	3.1	55	20.5
	LM14	80%–75 µm	11.5	3.2	75	28
	LM16	99%–75 µm	8.0	2.3	166	62
Phosphate	LM12	75%–150 µm	12	3.4	50	18.5
(Morocco)	LM14	90%–150 µm	13	3.7	67	25
	LM16	90%–150 µm	19	5.4	62	23
	LM16/3	90%–150 µm	35	9.9	54	20
Phosphate	LM16/3	97%–150 µm	27	7.6	70	26
(Nauru)						
Coal	LM12	96%–150 µm	9.4	2.7	56	21
	LM14	96%–150 µm	13.4	3.8	56	21
	LM16	96%–150 µm	20	5.6	56	21

100-mesh B.S.S. = 150 µm; 200-mesh B.S.S. = 75 µm.

costs are low, this machine does not compare favourably with the Lopulco mill under most conditions. It has another disadvantage in that wear will take place if the machine is run without any feed, because no clearance is maintained between the grinding heads and the bull ring. Originally used for the preparation of pulverised coal, the pendulum mill is now used extensively in the fine grinding of softer materials such as sulphur, bentonite, and ball clay as well as coal, barytes, limestone, and phosphate rock. In many industries, the sizing of the raw materials must be carried out within fine limits.

For example, the pottery industry might require a product whose size lies between 55 and 65 µm, and the pendulum mill is capable of achieving this. A comparison between the Lopulco and pendulum mills has shown that, whereas the Lopulco mill would give a product 98% of which was below 50 µm in size, the pendulum mill would give 100% below this size; in the latter case, however, the power consumption is considerably higher.

The crushing force in a pendulum mill may be obtained as follows.

Considering one arm rotating under uniform conditions, the sum of the moments of all the forces, acting on the roller and arm, about the point of support O shown in Fig. 5.25 will be zero.

Using the following notation:

M = the mass of the grinding head,

m = the mass per unit length of the arm,

ω = the angular speed of rotation,

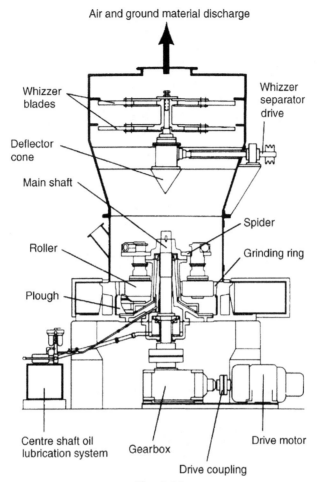

Air and ground material discharge

Whizzer blades

Whizzer separator drive

Deflector cone

Main shaft

Spider

Roller

Grinding ring

Plough

Centre shaft oil lubrication system

Gearbox

Drive motor

Drive coupling

Fig. 5.22

Sectional arrangement of an NEI pendulum 5-roller mill. *The author and editor thank NEI International Combustion Ltd, Derby, for this figure.*

$\theta =$ the angle between the arm and the vertical,

$l =$ the length of the arm,

$c =$ the radius of the yoke, and.

$R =$ the normal reaction of the bull ring on the grinding head.

For a length dy of the rod at a distance y from O:

Mass of the element $= m\, dy$

Centrifugal force acting on element $= m\, dy (c + y \sin\theta)\omega^2$

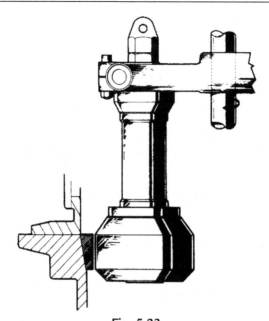

Fig. 5.23
Crushing roller in a pendulum mill. *The author and editor thank NEI International Combustion Ltd, Derby, for this figure.*

Fig. 5.24
Crushing heads of a pendulum mill. *The author and editor thank NEI International Combustion Ltd, Derby, for this figure.*

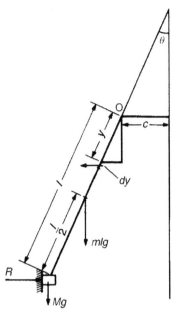

Fig. 5.25
Forces in the pendulum mill.

Moment of the force about O	$= -m\,dy(c + y\sin\theta)\,\omega^2\,y\cos\theta$
Total moment of the whole arm	$= -m\omega^2\cos\theta\left(\frac{1}{2}cl^2 + \frac{1}{3}l^3\sin\theta\right)$
Moment of the centrifugal force acting on the grinding head	$= -M\omega^2 l\cos\theta(c + l\sin\theta)$
Moment of the weight of the arm	$= \frac{1}{2}ml^2 g\sin\theta$
Moment of the weight of the grinding head	$= Mgl\sin\theta$
Moment of the normal reaction of the bull ring on the grinding head	$= Rl\cos\theta$
Total moment about O	$= \frac{1}{2}ml^2 g\sin\theta + Mgl\sin\theta + Rl\cos\theta$ $- ml^2\omega^2\cos\theta\left(\frac{1}{2}c + \frac{1}{3}l\sin\theta\right) - M\omega^2 l\cos\theta(c + l\sin\theta) = 0$

Thus:

$$R = M'\omega^2\left(\frac{1}{2}c + \frac{1}{3}l\sin\theta\right) + M\omega^2(c + l\sin\theta) - \frac{1}{2}M'g\tan\theta - Mg\tan\theta$$

where $M' = ml =$ the mass of arm.

Thus:

$$R = M' \left(\frac{1}{2}\omega^2 c + \frac{1}{3}l\omega^2 \sin\theta - \frac{1}{2}g\tan\theta \right) + M\left(\omega^2 c + \omega^2 l \sin\theta - g\tan\theta\right) \tag{5.8}$$

The Griffin mill

The Griffin mill is similar to the pendulum mill, other than it employs only one grinding head, and the separation of the product is effected using a screen mesh fitted around the grinding chamber. A product fineness from 8 to 240 mesh at an output from 0.15 to 1.5 kg/s may be obtained. These mills are widely used for dry fine grinding in many industries and are noted for their simplicity and reliability.

The Szego grinding mill

The Szego mill is a planetary ring-roller mill shown schematically in Fig. 5.26. It consists principally of a stationary, cylindrical grinding surface inside which a number of helically grooved rollers rotate. These radially mobile rollers are suspended from flanges connected to the central drive shaft; they are pushed outward by centrifugal force and roll on the grinding surface.

The material is fed by gravity, or pumped into a top-feed cylinder, and is discharged at the bottom of the mill. The particles, upon entering the grinding section, are repeatedly crushed between the rollers and the stationary grinding surface. The crushing force is created mainly by the radial acceleration of the rollers. Shearing action is induced by the high velocity gradients generated in the mill, and hence, the primary forces acting on the particles are the crushing and shearing force caused by rotational motion of the rollers. Attrition is also important, particularly in dry grinding, and impaction also occurs at higher rotational speeds.

An important feature is the ability of the roller grooves to aid the transport of material through the mill, thus, providing a means to control residence time and mill capacity. This transporting action is particularly important with materials which would not readily flow by gravity. Pastes and sticky materials fall into this category.

The mill has several design variables which may be utilised to meet specific product requirements. The important variables are the number of rollers, their mass, diameter and length, and the shape, size, and number of starts of the helical grooves on the rollers. As the number of rollers is increased, the product becomes finer. Heavier rollers and higher rotational speeds generate greater crushing forces and give higher reduction ratios. The ridge: groove size ratio may also be changed to increase or decrease the effective pressure acting

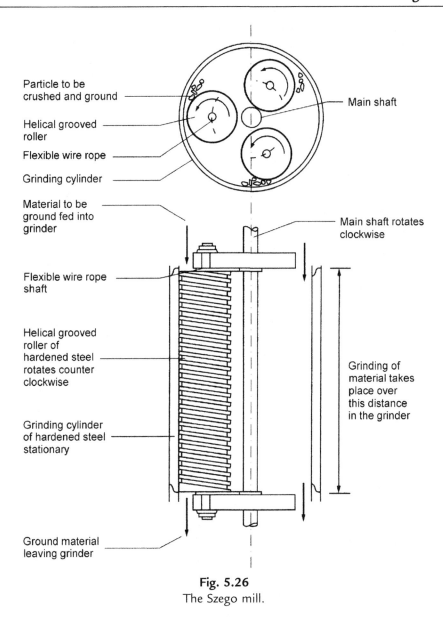

Fig. 5.26
The Szego mill.

on the particles. The common groove shapes are rectangular and tapered, and the latter will decrease the chance of particles getting stuck in the grooves.

The Szego mill has been used for dry as well as wet grinding of coal, in both water and oil, for the preparation of coal-slurry fuels. To grind coal in water to the standard boiler size (80% < 74 μm) takes about 7 MJ/Mg (~20 kWh/tonne). For "micronised" grinding, to a 15 μm median size, 2–3 passes through the mill are required, and the specific energy requirement increases to 20 MJ/Mg (~50 kWh/tonne). The mill has been used for simultaneous grinding of coal in oil and water where the carbonaceous matter agglomerates with oil as the bridging liquid, while the inorganic mineral matter stays suspended in the aqueous phase. Separating out the agglomerates allows good coal beneficiation.

The mill has also been used to grind industrial minerals and technical ceramics including limestone, lead zirconates, metal powders, fibrous materials (such as paper, wood chips, and peat), and chemicals and agricultural products (such as grains and oilseeds).

Because its grinding volume is some 30–40 times smaller than that of a ball mill for the same task, the Szego mill is compact for its capacity. It is characterised by relatively low specific energy consumption, typically 20%–30% lower than in a ball mill, and flexibility of operation. It can give a large reduction ratio, typically 10–20, and grinds material down to a 15–45 µm size range.

The mill has been developed, partly at the University of Toronto, Canada,[20, 21] and commercialised by General Comminution, Inc., in Toronto. Capacities range from 20 kg to 10 tonnes of dry material per hour. A small laboratory pilot mill has an inside diameter of 160 mm and fits on a bench. The large, 640-mm diameter unit has external dimensions of about 2 m × 2 m × 1 m in terms of height, length, and width.

The ball mill

In its simplest form, the ball mill consists of a rotating hollow cylinder, partially filled with balls, with its axis either horizontal or at a small angle to the horizontal. The material to be ground may be fed in through a hollow trunnion at one end, and the product leaves through a similar trunnion at the other end. The outlet is normally covered with a coarse screen to prevent the escape of the balls. Fig. 5.27 shows a section of an example of the Hardinge ball mill, which is also discussed later in this chapter.

The inner surface of the cylinder is usually lined with an abrasion-resistant material such as manganese steel, stoneware, or rubber. Less wear takes place in rubber-lined mills, and the coefficient of friction between the balls and the cylinder is greater than that with steel or stoneware linings. The balls are, therefore, carried further in contact with the cylinder and, thus, drop onto the feed from a greater height. In some cases, lifter bars are fitted to the inside of the cylinder. Another type of ball mill is used to an increasing extent, in which the mill is vibrated instead of being rotated, and the rate of passage of material is controlled by the slope of the mill.

The ball mill is used for the grinding of a wide range of materials, including coal, pigments, and felspar for pottery, and it copes with feed up to about 50 mm in size. The efficiency of grinding increases with the hold-up in the mill, until the voids between the balls are filled. Further increase in the quantity then lowers the efficiency.

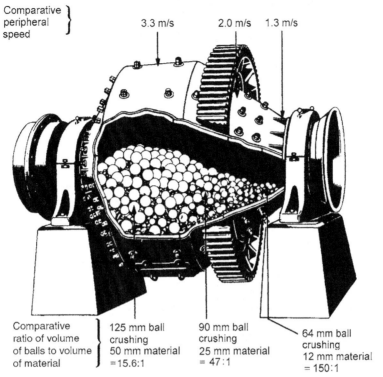

Comparative peripheral speed

3.3 m/s   2.0 m/s   1.3 m/s

Comparative ratio of volume of balls to volume of material

125 mm ball crushing 50 mm material =15.6:1

90 mm ball crushing 25 mm material = 47:1

64 mm ball crushing 12 mm material = 150:1

Fig. 5.27

Cut-away view of the Hardinge conical ball mill showing how energy is proportioned to the work required. *The author and editor thank NEI International Combustion Ltd, Derby, for this figure.*

The balls are usually made of flint or steel and occupy between 30% and 50% of the volume of the mill. The diameter of the balls used will vary between 12 and 125 mm, and the optimum diameter is approximately proportional to the square root of the size of the feed, with the proportionality constant being a function of the nature of the material.

During grinding, the balls wear and are constantly replaced by new ones so that the mill contains balls of various ages, and, hence, of various sizes. This is advantageous because the large balls deal effectively with the feed, and the small ones are responsible for giving a fine product. The maximum rate of wear of steel balls, using very abrasive materials, is about 0.3 kg/Mg of material for dry grinding, and 1–1.5 kg/Mg for wet grinding. The normal charge of balls is about 5 Mg/m^3. In small mills where very fine grinding is required, pebbles are often used in place of balls.

In the compound mill, the cylinder is divided into a number of compartments by vertical perforated plates. The material flows axially along the mill and can pass from one compartment to the next only when its size has been reduced to less than that of the perforations in the plate. Each compartment is supplied with balls of a different size. The large balls are at the entry end and, thus, operate on the feed material, while the small balls come into contact with the material immediately before it is discharged. This results in economical operation and the formation of a uniform product. It also gives an improved residence time distribution for the material, because a single stage ball mill closely approximates a completely mixed system.

In wet grinding, the power consumption is generally about 30% lower than that for dry grinding, and, additionally, the continuous removal of product as it is formed is facilitated. The rheological properties of the slurry are important, and the performance tends to improve as the apparent viscosity increases, reaching an optimum at about 0.2 Pa s. At very high volumetric concentrations (c.50 volume %), the fluid may exhibit shear-thickening behaviour or have a yield stress, and the behaviour may then be adversely affected.

Factors influencing the size of the product

(a) *The rate of feed.* With high rates of feed, less size reduction is effected because the material is in the mill for a shorter time.

(b) *The properties of the feed material.* The larger the feed, the larger is the product under given operating conditions. A smaller size reduction is obtained with a hard material.

(c) *Weight of balls.* A heavy charge of balls produces a fine product. The weight of the charge can be increased either by increasing the number of balls, or by using a material of higher density. Because optimum grinding conditions are usually obtained when the bulk volume of the balls is equal to 50% of the volume of the mill, variation of the weight of balls is normally effected by the use of materials of different densities. The effect on grinding performance of the loading of balls in a mill has been studied by Kano et al.,[22] who varied the proportion of the mill filled with balls from 0.2 to 0.8 of the volume of the mill. The grinding rates were found to be a maximum at loadings between 0.3 and 0.4. The effect of relative rotational speed, the ratio of actual speed to critical speed, was found to be complex. At loadings between 0.4 and 0.8, the grinding rate was a maximum at a relative rotation speed of about 0.8, although at lower loadings, the grinding rate increased to relative speeds of 1.1–1.6.

(d) *The diameter of the balls.* Small balls facilitate the production of fine material, although they do not deal so effectively with the larger particles in the feed. The limiting size reduction obtained with a given size of balls is known as the free grinding limit. For most economical operation, the smallest possible balls should be used.

(e) *The slope of the mill.* An increase in the slope of the mill increases the capacity of the plant because the retention time is reduced, although a coarser product is obtained.

(f) *Discharge freedom.* Increasing the freedom of discharge of the product has the same effect as increasing the slope. In some mills, the product is discharged through openings in the lining.

(g) *The speed of rotation of the mill.* At low speeds of rotation, the balls simply roll over one another, and little crushing action is obtained. At slightly higher speeds, the balls are projected short distances across the mill, and at still higher speeds, they are thrown greater distances, and considerable wear of the lining of the mill takes place. At very high speeds, the balls are carried right around in contact with the sides of the mill, and little relative movement or grinding takes place again. The minimum speed at which the balls are carried around in this manner is called the critical speed of the mill, and, under these conditions, there will be no resultant force acting on the ball when it is situated in contact with the lining of the mill in the uppermost position, that is, the centrifugal force will be exactly equal to the weight of the ball. If the mill is rotating at the critical angular velocity ω_c, then:

$$r\omega_c^2 = g$$

or:

$$\omega_c = \sqrt{\frac{g}{r}} \tag{5.9}$$

The corresponding critical rotational speed, N_c in revolutions per unit time, is given by:

$$N_c = \frac{\omega_c}{2\pi} = \frac{1}{2\pi}\sqrt{\frac{g}{r}} \tag{5.10}$$

In this equation, r is the radius of the mill less that of the particle. It is found that the optimum speed is between one-half and three-quarters of the critical speed. Fig. 5.28 illustrates conditions in a ball mill operating at the correct rate.

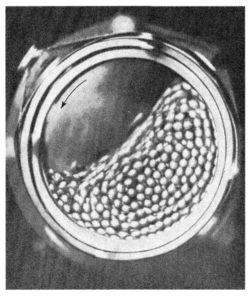

Fig. 5.28
A ball mill operating at the correct speed.

(h) *The level of material in the mill.* Power consumption is reduced by maintaining a low level of material in the mill, and this can be controlled most satisfactorily by fitting a suitable discharge opening for the product. If the level of material is raised, the cushioning action is increased, and power is wasted by the production of an excessive quantity of undersize material.

Example 5.3

A ball mill, 1.2 m in diameter, is operated at 0.80 Hz, and it is found that the mill is not working properly. Should any modification in the conditions of operation be suggested?

Solution

The critical angular velocity is given by:

$$\omega_c = \sqrt{(g/r)} \qquad (5.9)$$

In this equation, r=(radius of the mill − radius of the particle). For small particles, r=0.6 m, and, hence:

$$\omega_c = \sqrt{(9.81/0.6)} = 4.04\,\text{rad/s}$$

The actual speed$=(2\pi \times 0.80)=5.02\,\text{rad/s}$, and, hence, it may be concluded that the speed of rotation is too high, and that the balls are being carried around in contact with the sides of the mill with little relative movement or grinding taking place.

The optimum speed of rotation lies in the range $(0.5 - 0.75)\omega_c$, say $0.6\omega_c$ or:

$$(0.6 \times 4.04) = 2.42\,\text{rad/s}$$

This is equivalent to: $(2.42/2\pi)=0.39\,\text{Hz}$, or, in simple terms:

the speed of rotation should be halved.

Advantages of the ball mill

(i) The mill may be used wet or dry, although wet grinding facilitates the removal of the product.

(ii) The costs of installation and power are low.

(iii) The ball mill may be used with an inert atmosphere, and, therefore, can be used for the grinding of explosive materials.

(iv) The grinding medium is cheap.

(v) The mill is suitable for materials of all degrees of hardness.

(vi) It may be used for batch or continuous operation.

(vii) It may be used for open or closed circuit grinding. With open circuit grinding, a wide range of particle sizes is obtained in the product. With closed circuit grinding, the use of an external separator can be obviated by continuous removal of the product by means of a current of air or through a screen, as shown in Fig. 5.29.

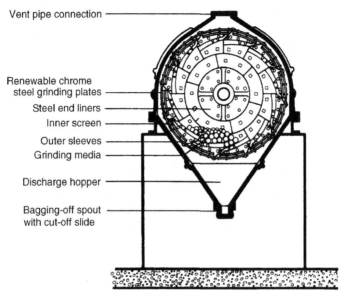

Fig. 5.29

End view of ball mill showing screens. *The author and editor thank Vaba Process Plant Ltd, Rotherham, Yorks for this figure.*

The tube mill

The tube mill is similar to the ball mill in construction and operation, although the ratio of length to the diameter is usually 3 or 4:1, as compared with 1 or 1.5:1 for the ball mill. The mill is filled with pebbles, rather smaller in size than the balls used in the ball mill, and the inside of the mill is so shaped that a layer of pebbles becomes trapped in it to form a self-renewing lining. The characteristics of the two mills are similar, although the material remains longer in the tube mill because of its greater length, and a finer product is therefore obtained.

The rod mill

In the rod mill, high carbon steel rods about 50 mm diameter and extending the whole length of the mill are used in place of balls. This mill gives a very uniform fine product and power consumption is low, although it is not suitable for very tough materials, and the feed should not exceed about 25 mm in size. It is particularly useful with sticky materials which would hold the balls together in aggregates, because the greater weight of the rods causes them to pull apart again. Worn rods must be removed from time to time and replaced by new ones, which are rather cheaper than balls.

The Hardinge mill

The Hardinge mill, shown in Fig. 5.30, is a ball mill in which the balls segregate themselves according to size. The main portion of the mill is cylindrical as in the ordinary ball mill, although the outlet end is conical and tapers toward the discharge point. The large balls collect

Fig. 5.30
The Hardinge mill. *The author and editor thank NEI International Combustion Ltd, Derby, for this figure.*

in the cylindrical portion while the smaller balls, in order of decreasing size, locate themselves in the conical portion as shown in Fig. 5.27. The material is, therefore, crushed by the action of successively smaller balls, and the mill is, thus, similar in characteristics to the compound ball mill. It is not known exactly how balls of different sizes segregate although it is suggested that, if the balls are initially mixed, the large ones will attain a slightly higher falling velocity and, therefore, strike the sloping surface of the mill before the smaller ones, and then run down toward the cylindrical section. The mill has an advantage over the compound ball mill in that the large balls are raised to the greatest height and, therefore, are able to exert the maximum force on the feed. As the size of the material is reduced, smaller forces are needed to cause fracture, and it is, therefore, unnecessary to raise the smaller balls as high. The capacity of the Hardinge mill is higher than that of a ball mill of similar size, and it gives a finer and more uniform product with a lower consumption of power. It is difficult to select an optimum speed, however, because of the variation in shell diameter. It is extensively used for the grinding of materials such as cement, fuels, carborundum, silica, talc, slate, and barytes.

The sand mill

The sand mill, or stirred ball mill, achieves fine grinding by continuously agitating the bed of grinding medium and charge by means of rotating bars which function as paddles. Because of its high density, zircon sand is frequently used as the grinding medium. A fluid medium, liquid or gas, is continuously passed through the bed to remove the fines, as shown in Fig. 5.31. A very fine product can be obtained at a relatively low energy input, and the mill is used for fine grades of ceramic oxides and china clay, and in the preparation of coal slurries.

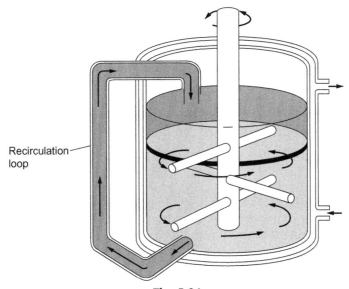

Fig. 5.31
Typical sand mill.

In cases where it is important that the product should not be contaminated with fine fragments of the grinding medium, *autogenous grinding* is used where coarse particles of the material are to be ground form the grinding medium.

The planetary mill

A serious limitation of the ball or tube mill is that it operates effectively only below its critical speed, as given by Eq. (5.10). In the planetary mill, described by Bradley et al.,[23] this constraint is obviated by rotating the mill simultaneously about its own axis and about an axis of gyration, as shown in Fig. 5.32. In practice, several cylinders are incorporated in the machine, all rotating about the same axis of gyration.

A ball of mass M in contact with the cylindrical wall is subject to the following two centrifugal forces acting simultaneously:

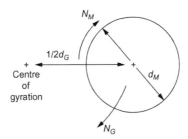

Fig. 5.32
Geometry of the planetary mill.

(a)   A centrifugal force attributable to rotation about the axis of gyration (radius $d_G/2$) $= (Md_G/2)(2\pi N_G)^2$, where $N_G =$ revolutions per unit time about the axis of gyration.

(b)   A centrifugal force attributable to rotation of the cylinder (radius $d_M/2$) $= (Md_M/2)[2\pi(N_M + N_G)]^2$, where $N_M =$ revolutions per unit time about the axis of the cylinder. In unit time, the total number of revolutions of the cylinder is $N_M + N_G$.

In addition, the gravitational force acts on the particle, although, under normal operating conditions, this is small compared with the centrifugal forces.

At the critical condition:

$$\left(M\frac{d_G}{2}\right)(2\pi N_G)^2 = \left(M\frac{d_M}{2}\right)[2\pi(N_M + N_G)]^2 \tag{5.11}$$

If $N_M/N_G = s$, say, which is determined by the gear ratio of the mill, then substituting into Eq. (5.11) and simplifying:

$$(s_c + 1)^2 = \frac{d_G}{d_M} \tag{5.12}$$

and:

$$s_c = \pm\sqrt{(d_G/d_M)} - 1 \tag{5.13}$$

where s_c is the value of s which gives rise to the critical condition for a given value of d_G/d_M.

The positive sign applies when $s_c > -1$ and the negative sign when $s_c < -1$.

It may be noted that it is necessary to take account of *Coriolis* forces in calculating the trajectory of a particle once it ceases to be in contact with the wall.

It is possible to achieve accelerations up to about $15g$ in practical operation. For further details of the operation of the planetary mill, reference may be made to Bradley et al.[23] and Kitschen et al.[24] The planetary mill is used for the preparation of stabilised slurries of coal in both oil and water, and it can also handle paste-like materials.

The vibration mill

Another way of increasing the value of the critical speed, and so improve the performance of a mill, is to increase the effective value of the acceleration due to gravity g. The rotation of the mill simultaneously about a vertical and a horizontal axis has been used to simulate the effect of an increased gravitational acceleration, although clearly, such techniques are applicable only to small machines.

By imparting a vibrating motion to a mill, either by the rotation of out-of-balance weights or by the use of electromechanical devices, accelerations many times the gravitational acceleration may be imparted to the machine. The body of the machine is generally supported on powerful springs and caused to vibrate in a vertical direction. Vibration frequencies of

6–60 Hz are common. In some machines, the grinding takes place in two stages, with the material falling from an upper to a lower chamber when its size has been reduced below a certain value.

The vibration mill has a very much higher capacity than a conventional mill of the same size, and consequently, either smaller equipment may be used, or a much greater throughput obtained. Vibration mills are well suited to incorporation in continuous grinding systems.

Colloid mills

Colloidal suspensions, emulsions, and solid dispersions are produced by means of colloid mills or dispersion mills. Droplets or particles of sizes <1 μm may be formed, and solids suspensions consisting of discrete solid particles are obtainable with feed material of approximately 100-mesh or 50 μm in size.

As shown in Fig. 5.33, the mill consists of a flat rotor and stator manufactured in a chemically inert synthetic abrasive material, and the mill can be set to operate at clearances from virtually zero to 1.25 mm, although in practice, the maximum clearance used is about 0.3 mm. When duty demands, steel working surfaces may be fitted, and in such cases, the minimum setting between rotor and stator must be 0.50–0.75 mm, otherwise "pick up" between the steel surfaces occurs.

The gap setting between rotor and stator is not necessarily in direct proportion to the droplet size or particle size of the end product. The thin film of material continually passing between the working surfaces is subjected to a high degree of shear, and consequently, the energy absorbed within this film is frequently sufficient to reduce the dispersed phase to a particle size far smaller than the gap setting used. The rotor speed varies with the physical size of the mill and the clearance necessary to achieve the desired result, although peripheral speeds

Fig. 5.33
Rotor and stator of a colloid mill. *Used with permission from NETZSCH Trockenmahltechnik GmbH.*

of the rotor of 18–36 m/s are usual. The required operating conditions and size of mill can only be found by experiment.

Some of the energy imparted to the film of material appears in the form of heat, and a jacketed shroud is frequently fitted around the periphery of the working surfaces so that some of the heat may be removed by coolant. This jacket may also be used for circulation of a heating medium to maintain a desired temperature of the material being processed.

In all colloid mills, the power consumption is very high, and the material should, therefore, be ground as finely as possible before it is fed to the mill.

Fluid energy mills

Another form of mill which does not give quite such a fine product is the jet pulveriser, in which the solid is pulverised in jets of high pressure superheated steam or compressed air, supplied at pressures up to 3.5 MN/m^2 (35 bar). The pulverising takes place in a shallow cylindrical chamber with a number of jets arranged tangentially at equal intervals around the circumference. The solid is thrown to the outside walls of the chamber, and the fine particles are formed by the shearing action resulting from the differential velocities within the fluid streams. The jet pulveriser will give a product with a particle size of 1–10 μm.

The microniser, probably the best known of this type of pulveriser, effects comminution by bombarding the particles of material against each other. Preground material, of about 500 μm in size, is fed into a shallow circular grinding chamber which may be horizontal or vertical, the periphery of which is fitted with a number of jets, equally spaced, and arranged tangentially to a common circle.

Gaseous fluid, often compressed air at approximately 800 kN/m^2 (8 bar) or superheated steam at pressures of 800–1600 kN/m^2 (8–16 bar) and temperatures ranging from 480 to 810 K, issues through these jets, thereby promoting high-speed reduction in the size of the contents of the grinding chamber, with turbulence and bombardment of the particles against each other. An intense centrifugal classifying action within the grinding chamber causes the coarser particles to concentrate toward the periphery of the chamber, while the finer particles leave the chamber, with the fluid, through the central opening.

The majority of applications for fluid energy mills are for producing powders in the sub-sieve range, on the order of 20 μm and less, and it may be noted that the power consumption per kilogram is proportionately higher than for conventional milling systems which grind to a top size of about 44 μm.

A section through a typical microniser is shown in Fig. 5.34.

Another pulveriser in this group is the Wheeler fluid energy mill, which is in the form of a vertical loop. The preground feed material is injected toward the bottom of the loop in which are

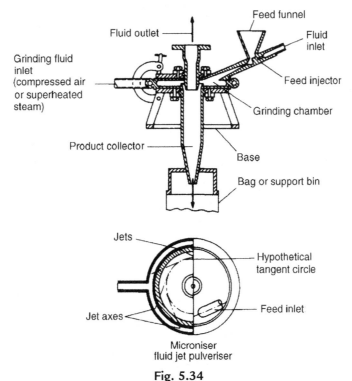

Fig. 5.34

Section through a microniser. *Used with permission from Baker Perkins Ltd.*

situated the nozzles for admitting the compressed air or superheated steam. Size reduction occurs as a result of bombardment of the particles against each other, and classification is effected by arranging for the fluid to leave the circulating gas stream through vanes which are situated just downstream of the top of the loop and on the inner face of the loop. Oversize particles continue their downward path with the circulating fluid and reenter the reduction chamber for further grinding.

5.3.4 Specialised Applications

Specialised techniques

Several techniques have been developed for specialised applications, as discussed by Prasher.[6] These include:

(i) *Electrohydraulic crushing* in which an underwater discharge is generated by the release of energy from a high-voltage capacitor. The spark length depends on the nature of the material to be crushed, though it is commonly 15–80 mm.

(ii) *Ultrasonic grinding* is a method in which the material to be ground is fed between a drive roll and a curved plate, both of which are ultrasonically activated. Experimental work has been carried out to produce coal particles smaller than 10 μm.

(iii) *Cryogenic grinding*. Size reduction is difficult to achieve by conventional means with many materials, such as plastics, rubber, waxes, and textile-based products, as they tend to distort rather than to fracture when subjected to compressive forces. However, it is frequently possible to effect a change in the structure of the material by subjecting it to very low temperatures. In *cryogenic grinding*, the material is cooled with liquid nitrogen at a temperature of about −196°C (77 K) to render it brittle before it enters the grinder. According to Butcher,[25] cooling causes the crystal lattice to shrink and to give rise to microscopic cracks which act as nuclei and then grow, thereby reducing the energy input required to cause fracturing to occur. As about 99% of the energy for size reduction finishes up as heat, economy in the use of liquid nitrogen is a critical consideration. It is, therefore, desirable, if practicable, to effect a preliminary reduction of particle size in order to reduce the time taken for the material to be cooled by the liquid nitrogen. Precooling of the feed material by using a conventional refrigeration plant is also advantageous. Both rotary cutting mills and pin mills show improved energy utilisation with nonbrittle materials as the temperature is reduced. The method is of new importance with the expanding market for frozen foods.

(iv) *Explosive shattering* is a method in which energy is transmitted to particles as shock waves set up on suddenly releasing steam from an explosion chamber containing the solid to be compressed. Equipment based on this technique is still at the development stage.

5.4 Size Enlargement of Particles

5.4.1 Agglomeration and Granulation

There are many situations in which fine particles are difficult to handle, mainly due to the fact that particles in bulk do not flow readily because of their tendency to adhere together as conglomerates as a result of the action of surface forces. In such cases, the finer the particles, the greater is their specific surface, and the gravitational forces acting on the particles may not be great enough to keep them apart during flow. The flowability of particulate systems can sometimes be improved by the use of "glidants," which are very fine powders which are capable of reducing interparticle friction by forming surface layers on the particles, thereby combating the effects of friction arising from surface roughness; they can also reduce the effects of electrostatic charges. However, the optimisation of particle size is by far the most important method of improving flow properties. As discussed in Chapter 9, fine particles are often difficult to fluidise in gases because channelling, rather than even dispersion of the particles, tends to occur.

Fine particles may be difficult to discharge from hoppers as particles may cling to the walls and form bridges at the point of discharge. Although such problems may be minimised, either by vibration or by mechanical stirring, it is very difficult to overcome them entirely, and the only satisfactory solution may be to increase the particle size by forming them into aggregates. In addition, very fine particles often give rise to serious environmental and health problems, particularly as they may form dust clouds when loaded into vehicles, and in windy conditions, may become dispersed over long distances. Although the particles involved may, in themselves, be inert, serious respiratory problems may result if these are inhaled. In such situations, particle size may be a critical factor because very fine particles may be exhaled, and very large particles may have a negligible effect on health. In this respect, it may be noted that the particular health hazard imposed by asbestos is largely associated with the size range and shape of the particles and their tendency to collect in the lungs (see chapter 14).

The size of particles may be increased from molecular dimensions by growing them by crystallisation from both solutions and melts as discussed in volume 2B. Here, dissolving and recrystallising may provide a mechanism for controlling both particle size and shape. It may be noted, as also discussed in Chapter 14, that fine particles may also be condensed out from both vapours and gases.

A desired particle size may also be achieved by building up from fine particles, and one example of this is the production of fertiliser granules by agglomeration, or by a repeated coating process. Another example is the formation of pellets or pills for medicinal purposes by the compression of a particulate mass, often with the inclusion of a binding agent that will impart the necessary strength to the pellet. von Smoluchowski[26] has characterised suspensions by a "half-time," t, defined as the time taken to halve the number of original particles in a mono-disperse system. Walton[27] arbitrarily defined an agglomerating system as one in which >10% of the original number of particles has agglomerated in <1000 s.

5.4.2 Growth Mechanisms

There are essentially two types of processes that can cause agglomeration of particles when they are suspended in a fluid:

(a) *Perikinetic processes* which are attributable to Brownian movement can, therefore, occur even in a static fluid. Double-layer repulsive forces and van der Waals attractive forces may both operate independently in disperse systems. Repulsion forces decrease exponentially with distance across the ionic double-layer, although attraction forces decrease, at larger distances from the particle surface, and are inversely proportional to the distance. Consequently, as Tadros[28] points out, attraction normally predominates both at very small and very large distances, and repulsion over intermediate distances. Fine particles may also be held together by electrostatic forces.

(b) *Orthokinetic processes* occur where the perikinetic process is supplemented by the action of eddy currents which may be set up in stirred vessels or in flowing systems. In these circumstances, the effects of the perikinetic mechanism are generally negligible.

According to Söhnel and Mullin,[29] the change in agglomerate size as a function of time may be represented by equations of the form:

For *perikinetic processes*:

$$d_t^3 = A_1 + B_1 t \tag{5.14}$$

and for *orthokinetic processes*:

$$\log \frac{d_t}{d_o} = A_2 + B_2 t \tag{5.15}$$

where d_t is the agglomerate size at time t. Thus, the shapes of the plots of d_t against t give an indication of the relative importance of the two processes. Eqs (5.14) and (5.15) will apply only in the initial stages of the enlargement process, because otherwise, they would indicate an indefinite increase of size d_t with time t. The dimensions of A_1 are $\mathbf{L}^3$, of B_1 are $\mathbf{L}^3\,\mathbf{T}^{-1}$ and of B_2 are $\mathbf{T}^{-1}$. A_2 is dimensionless. The limiting size is that at which the rates of aggregation and of breakdown of aggregates are in balance.

The stability of the aggregates may be increased by the effects of mechanical interlocking that may occur, especially between particles in the form of long fibres.

It is often desirable to add a liquid binder to fill the pore spaces between particles in order to increase the strength of the aggregate. The amount of binder is a function of the voidage of the particulate mass, a parameter that is strongly influenced by the size distribution and shape of the particles. Wide size distributions generally lead to close packing, requiring smaller amounts of binder, and, as a result, the formation of strong aggregates.

As discussed in volume 2B, the size distribution of particles in an agglomeration process is essentially determined by a *population balance* that depends on the kinetics of the various processes taking place simultaneously, some of which result in particle growth, and some in particle degradation. In a batch process, an equilibrium condition will eventually be established with the net rates of formation and destruction of particles of each size reaching an equilibrium condition. In a continuous process, there is the additional complication that the residence time distribution of particles of each size has an important influence.

In general, starting with a mixture of particles of uniform size, the following stages may be identified:

(a) *Nucleation* in which fresh particles are formed, generally by attrition.
(b) *Layering* or *coating* as material is deposited on the surfaces of the nuclei, thus increasing both the size and total mass of the particles.

(c) *Coalescence* of particles which results in an increase in particle size but not in the total mass of particles.

(d) *Attrition* results in degradation and the formation of small particles, thus generating nuclei that reenter the cycle again.

It is difficult to build these four stages into a mathematical model because the kinetics of each process is not generally known, and therefore, more empirical methods have to be adopted.

5.4.3 Size Enlargement Processes

Processes commonly used for size enlargement are listed in Table 5.4, taken from Perry.[30] For comprehensive overall reviews, reference may be made to Perry[30] and to the work of Browning.[31]

(a) *Spray drying* (as discussed in volume 2B).

In this case, particle size is largely determined by the size of the droplet of liquid or suspension, which may be controlled by a suitably designed spray nozzle. The aggregates of dried material

Table 5.4 Size-enlargement methods and applications

Method	Product Size (mm)	Granule Density	Scale of Operation	Additional Comments	Typical Application
Tumbling granulators drums discs	0.5–20	Moderate	0.5–800 ton/h	Very spherical granules	Fertilisers, iron ore, ferrous ore, agricultural chemicals
Mixer granulators Continuous high shear (e.g., Shugi mixer)	0.1–2	Low to high	Up to 50 ton/h	Handles very cohesive materials well, both batch and continuous	Detergents, clays, carbon black pharmaceuticals, ceramics
Batch high shear (e.g., paddle mixer)	0.1–2	High	Up to 500 kg batch		
Fluidised granulators Fluidised beds	0.1–2	Low (agglomerated)	100–900 kg batch	Flexible, relatively easy to scale, difficult for cohesive powders, good for coating applications	Continuous: fertilisers, inorganic salts, detergents
Spouted beds Wurster coaters		Moderate (layered)	50 ton/h continuous		Batch: pharmaceuticals, agricultural chemicals, nuclear wastes
Centrifugal granulators	0.3–3	Moderate to high	Up to 200 kg batch		

Table 5.4   Size-enlargement methods and applications—Cont'd

Method	Product Size (mm)	Granule Density	Scale of Operation	Additional Comments	Typical Application
Spray methods Spray drying	0.05–0.5	Low		Powder layering and coating applications Morphology of spray dried powders can vary widely	Pharmaceuticals, agricultural chemicals Instant foods, dyes, detergents, ceramics
Prilling	0.7–2	Moderate			Urea, ammonium nitrate
Pressure compaction Extrusion	>0.5	High to very high	Up to 5 ton/h	Very narrow size distributions, very sensitive to powder flow and mechanical properties	Pharmaceuticals, catalysts, inorganic chemicals, organic chemicals plastic performs, metal parts, ceramics, clay minerals, animal feeds
Roll press Tablet press	>1		Up to 50 ton/h		
Moulding press Pellet mill	10		Up to 1 ton/h		
Thermal processes sintering	2–50	High to very high	Up to 100 ton/h	Strongest bonding	Ferrous and nonferrous ores, cement clinker minerals, ceramics
Liquid systems immiscible wetting in mixers Sol–gel processes Pellet flocculation	<0.3	Low	Up to 10 ton/h	Wet processing based on flocculation properties of particulate feed	Coal fines, soot and oil removal from water Metal dicarbide, silica hydrogels Waste sludges and slurries

are held together as a result of the deposition of small amounts of solute on the surface of the particles. For a given nozzle, the drop sizes will be a function of both flow rate and liquid properties, particularly viscosity, and to a lesser extent, outlet temperature. In general, viscous liquids tend to form large drops, yielding large aggregates.

(b) *Prilling* is a method in which relatively coarse droplets are introduced into the top of a tall, narrow tower and allowed to fall against an upward flow of air. This results in somewhat larger particles than those formed in spray dryers.

(c) *Fluidized beds* (as discussed in Chapter 9). In this case, an atomised liquid or suspension is sprayed on to a bed of hot fluidised particles, and layers of solid build up to give enlarged particles, the size of which is largely dependent on their residence time, that is, the time over which successive layers of solids are deposited. *Spouted beds* (as discussed in Chapter 9). These are used particularly with large particles. In this case, the rapid circulation within the bed gives rise to a high level of interparticle impacts. These processes are discussed by Mortensen and Hovmand[32] and by Mathur and Epstein.[33]

(d) *Drum and pan agglomerators.*

In *drum agglomerators*, particles are "tumbled" in an open cylinder with roughened walls and subjected to a combination of gravitational and centrifugal forces. In order to aid agglomeration, liquid may be sprayed onto the surface of the bed or introduced through distribution pipes under the bed. Mean retention times in the equipment are in the range 60–120 s. A similar action is achieved in a *paddle mixer* where centrifugal forces dominate. In the *pan agglomerator*, a classifying action may be achieved which results in the fines having a preferentially longer retention time. Larger, denser, and stronger agglomerates are produced, as compared with those from the drum agglomerator.

(e) *Pug mills and extruders.*

Pug mills impart a complex kneading action that is a combination of ribbing, shearing, and mixing. Densification and extrusion are both achieved in a single operation. The feed, which generally has only a small water content, is subjected to a high energy input, which leads to a considerable rise in temperature. The action is similar to that occurring in an *extruder*. High degrees of compaction are achieved, leading to the production of pellets with low porosity, with the result that less binder is required.

(f) *Elevated temperatures.*

With many materials, agglomeration may be achieved by heating, as a result of which, softening occurs in the surface layers. For the formation of porous metal sheets and discs, high temperatures are required.

(g) *Pressure compaction.*

If a material is subjected to very high compaction forces, it may be formed into sheets, briquettes, or tablets. In the tableting machines used for producing pills of pharmaceuticals, the powder is compressed into dies, either with or without the addition of a binder.

Powder compaction may also be achieved in roll processes, including briquetting, in which compression takes place between two rollers rotating at the same speed—that is, without producing any shearing action. In pellet mills, a moist feed is forced through die holes where the resistance force is attributable to the friction between the powder and the walls of the dies.

A commercial pelleting process, used for powdery, lumpy, and pasty products, is illustrated in Fig. 5.35.

Fig. 5.35
The KAHL pelleting press. *Used with permission from Amandus–Kahl, Hamburg, Germany.*

5.5 Modelling of Breakage and Aggregation Using Population Balances[a]

5.5.1 Introduction

Quantitative models are frequently needed to predict the particle size distribution (PSD) after milling or agglomeration process. Such models are useful for evaluation of a new process or optimization and control of an existing process. The models usually predict the PSD as a function of various material and operational parameters such as rotational speed of the mill, type of mill, hardness of the material, etc. For particulate process, such parameters are large in number, and some of those are difficult to recognise at first glance. For example, cohesiveness

---

[a]A detail discussion on this topic is available at "Submicron Particles, Fundamental concepts and Models, ISBN:9781119296454".

of a powder may depend on the humidity of surrounding air (Landi et al.[34]). Because of the involvement of a large number of factors, conducting extensive experiments encompassing all such factors is very expensive and sometimes impossible. Hence, a good phenomenological model is extremely useful.

Unlike other chemical engineering systems, modelling of particulate systems requires a number balance equation along with mass, energy, and momentum balances. This number balance equation is known as Population Balance Equation (PBE). PBE describes the evolution of the particle size distribution (PSD) with time. In this section, we shall discuss a few key ideas and develop the population balance equation for breakage and aggregation processes. Although breakage and aggregation of solid particles will be in focus, the framework developed in this section is general and can be used for a variety of situations such as predicting the final molecular weight and its distribution from a polymerisation reactor, total available mass transfer surface in a mixer-settler, or the shape and size of a crystals from a crystallizer. A detailed discussion on the application of population balance modelling to wide variety of problems is provided by Ramkrishna and Singh.[35]

5.5.2 Representation of Particles

The first step for modelling of a system of particles is to represent a set of particles mathematically. Usually, a particle can be represented by its size. The simplest example is a spherical particle which can be described by its diameter. Because most particles are non-spherical, a representative size can be used. In Section 2.2, several such options have been discussed. Usually, a linear dimension such as the sphere equivalent diameter or particle volume is more suitable.

Particles are not stationary and always undergo motion in the mill. The location of the particle might also be important. For example, a ball mill has zones where vigorous impact occurs (the top layer of balls) as well as milder rolling action zones (between the mill and the balls). The size and size distribution of comminuted particles will be different in these two zones. The location of a particle in a mill with respect to a fixed coordinate system is called its *external coordinates*. The inherent characteristics of a particle, such as its diameter or mass, are called *internal coordinates*. The internal and external coordinates together comprise its *particle state vector*.

Although mills might have inhomogeneous regions, it is difficult to take such effects into account, especially when all three space dimensions are involved. Hence, for practical purposes, mills are often considered to be well mixed. In that case, only internal coordinates are needed. In this section, we shall assume mills to be well mixed and consider only internal coordinates. If spatial nonhomogeniety is to be considered, the multidimensional population balance equation (Kuvadia and Doherty[36]) must be used. The special distribution of particle velocity will also be needed for such cases, and mean field models such as PBE may not be entirely satisfactory. Other techniques such as Discrete Element Method (DEM) (Norouzi et al.[37]) may suit the purpose better.

Sometimes particles need to be characterised by more than one internal variable. For example, if we are interested in the aspect ratio of particles, we need both length and breadth of the particle. Such a need frequently arises for pharmaceutical powders. In the manufacturing of pharmaceutical powders, needle like shape ("habit") is to be strictly avoided because such forms create problems in downstream processing. Although multidimensional problems resulting from such descriptions can be formulated readily, their solution is computationally intensive. Hence, simplified assumptions should be used to reduce the dimensionality of the problem whenever possible. In this section, we shall assume that a one-dimensional description of a set of particles is adequate. The reader is referred to advanced texts and references (Chakraborty and Kumar[37a]) for treatment of multidimensional problems.

Example 5.4

(a) A concentrated slurry containing spherical particles is settling in a tall cylindrical vessel. The particles can form an agglomerate if in contact. Assign the appropriate internal and external coordinates to the particles.

(b) Flat, sheet-like particles (nanosheets Cu, shown in Fig. 5.36) are produced by a precipitation process in a well-mixed crystallizer. Determine the appropriate internal coordinates for these particles. How can you reduce the dimensionality of the problem?

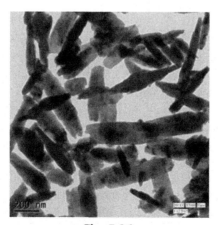

Fig. 5.36

Cu nanosheet powder: an example of two-dimensional particles. *Taken from Shaik AH, Chakraborty J. A simple room temperature fast reduction technique for preparation of a copper nanosheet powder.* RSC Adv 2016;**6**:14952–57.

Solution

(a) Because the slurry contains spherical particles, the particle diameter can be used as the internal coordinate. Settling of dense slurry produces crowding of particles toward the bottom of the vessel. Hence, external coordinates will be required. In a cylindrical vessel, the particle density does not vary appreciably in r or θ and hence only one external coordinate, z should suffice.

(b) Because the crystallizer is generally well mixed, the external coordinates are not required. Because the particles are nonspherical, more than one length dimension is needed to describe them. It appears that length, breadth, and thickness should be used to characterise the particles. This will require a 3-D population balance approach. It can be seen that the thickness of the particles does not vary appreciably, and hence, it can be taken to be constant. This keeps the dimensionality of the system to two. Further, if we can assume all particles are of the same aspect ratio, the dimensionality reduces to one. Such simplifications, if possible, are very useful for modelling of the breakage and aggregation processes.

Without the requirement of external coordinates and only one internal coordinate to describe the particle, powders can be represented by a size distribution curve (shown in Fig. 2.6 or Fig. 2.7). In this section, we shall take particle *volume* as an internal coordinate unless otherwise specified. We shall use the *size frequency curve* (shown in Fig. 2.6) to represent the size distribution, and the quantity dx/dd (shown in Fig. 2.6) will be referred to as the *number density function*. We shall denote it by $f_1(x,t)$. The explicit dependence on time implies that the particle size distribution (number density function) inside a mill evolves with time. We shall consider a batch process to develop the ideas, and the model should be modified appropriately for continuous grinding.

The number density function is very similar to probability density function. As probability can be obtained by integrating the probability density function within the required limits, the number of particles within a size range can be obtained by integrating the number density function within the limits. For example, the number of particles in between the size x_1 and x_2 is given by:

$$\int_{x_1}^{x_2} f_1(x,t)\,dx$$

It is also clear that the number of particles in a differential size range dx is $f_1(x,t)dx$. It may be noted that, in all cases, the number is the per unit volume of powder sample/slurry collected. Next, we focus on a mathematical description of the breakage and aggregation processes.

5.5.3 Breakage Functions and Aggregation Efficiency

For modelling breakage and aggregation, a set of mathematical functions is used. These functions are called breakage functions and aggregation efficiency for the respective processes.

As described by Ramkrishna,[38] three breakage functions are needed to complete the description of the breakage process: (1) breakage frequency, (2) the average number of daughter particles, and (3) the daughter distribution function.

(1) *Breakage frequency.* The first thing we need to know about breakage is how frequently breakage occurs. This is called breakage frequency. The breakage frequency is usually a function of particle size and is denoted by $\Gamma(x)$. The breakage frequency, $\Gamma(x)$, provides the fraction of particles that will break per unit time. Hence, for a differential size range dx, which contains $f_1(x,t)dx$ particles of size x, the number of breakage events per unit time will be $\Gamma(x) f_1(x,t)dx$.

This function quantitatively describes the propensity of particles of a certain size toward breakage. In Section 5.2.2, it was argued that, for solid particles, the propensity increases with particle size because of the availability of a larger number of weaker microcracks. Hence, larger particles break more frequently than smaller particles. Also, there should be a limiting size below which breakage becomes virtually impossible (grinding limit). This limiting size is different for different mill-material combinations. Hence, for a large class of systems, the following power law expression can be used for breakage frequency:

$$\Gamma(x) = a(x - x_l)^b \tag{5.16}$$

where x_l is the limiting size below which breakage is not possible. Clearly, the constants a, b, and x_l are to be determined for a particular mill-material combination. In addition to solid particles, many other systems, such as polymers, show similar behaviour, and a similar form of breakage frequency may be used.

(2) *Average number of daughter particles.* Once the frequency of breakage is obtained, the next quantity of interest is the number of daughter particles formed by breakage. This number may vary widely even for apparently identical particles under identical conditions. For example, it is impossible to specify a priori exactly how many daughter particle will form upon shattering of a piece of glass which is 1 in.^{2} and dropped from a height of 6 ft. But if we conduct many such experiments, we can find an average number of fragments around which observations fall. Such an average number of daughter particles is the second breakage function, which is denoted by $\nu(x)$. Like breakage frequency, the average number of daughter particles is dependent on particle size. However, there are cases where it remains constant. One such example is cell division, where $\nu(x)$ is identically 2.

(3) *Daughter distribution function.* The third function that completes the description of breakage process is daughter distribution function. This is the size distribution of daughter particles resulting from breakage of particles of a particular size. If we collect a large number of identical particles (of size x') and allow them to break, we will have a larger collection of daughter particles of varying size. If we obtain the size distribution of particles in this collection and convert it into a probability density function, we will obtain

the daughter distribution function for size x'. This is denoted by $p(x|x')$. The quantity $p(x|x')dx$ denotes the probability (fraction) of particles that will be generated in the differential size range dx upon breakage of one particle of size x'.

The daughter distribution function must obey the following constraints:

$$\int_0^{x'} p(x|x')dx = 1$$
$$p(x|x') = 0 \quad \text{for } x > x' \tag{5.17}$$
$$m(x') = \nu(x') \int_0^{x'} m(x)p(x|x')dx$$

In the above, $m(x')$ is the mass of a particle with internal coordinate x'. For binary breakage, an additional symmetry condition should also be satisfied:

$$p(x|x') = p(x' - x|x') \tag{5.18}$$

(4) *Aggregation efficiency.* Like breakage functions, aggregation is quantified by aggregation efficiency. For binary aggregation, this is the fraction of all possible pairs that are aggregating per unit time. Although aggregation of higher order cannot be ruled out, such aggregations are far less compared to binary aggregation for dilute dispersions. For concentrated dispersions, however, higher order aggregations are possible. Aggregation efficiency is a function of size of both the aggregating particles in the pair and is denoted by $a(x,x')$.

Example 5.5

Verify if the following daughter distribution function satisfies the required constraints for binary breakage: $P(x|x') = 1/x'$. In this case, x is the particle volume.

Solution

It can be readily seen that

$$\int_0^{x'} p(x|x')dx = \int_0^{x'} \frac{1}{x'} dx = \frac{x'}{x'} = 1$$

The second constraint

$$p(x|x') = 0 \quad \text{for } x > x'$$

has to be included in the definition of the kernel. The mass balance constraint is given as:

$$\int_0^{x'} x \, dx = \frac{2\rho}{x'} \int_0^{x'} x \, dx = \frac{2\rho x'^2}{x' \, 2} = \rho x'.$$

Hence, the given kernel satisfies the required constraints.

5.5.4 Number Balance for Breakage and Aggregation

In a grinding and/or aggregation process, the number of particles changes due to breakage, aggregation, or a combination of both. Other processes like in-out, nucleation, or growth may also be present, but we consider only breakage and aggregation in a batch mill in this chapter. First, we shall develop the number balance for breakage alone.

Number balance for breakage

To write the number balance for breakage, we consider a differential strip on the size axis, as shown in Fig. 5.37. As discussed in Section 5.5.2, the number of particles present in this strip is $f_1(x,t)\mathrm{d}x$, and this number changes because of breakage. There are only two ways in which the number of particles may change:

(a)   Particles in this size range break, become smaller, and are, therefore, removed from this size class.
(b)   Larger particles break, and some of them fall into the size ranging from x to $x+\mathrm{d}x$. The first process is called *breakage death*, and the second is called *breakage birth*. These processes are schematically shown in Fig. 5.37. Hence, we can write the number balance as:

$$\frac{\partial f_1(x, t)\mathrm{d}x}{\partial t} = -\text{breakage death} + \text{breakage birth} \qquad (5.19)$$

Now, we need to quantify these two processes by using the breakage functions. The number of particles in the differential strip of interest is given by $f_1(x,t)\mathrm{d}x$, and because $\Gamma(x)$ is the fraction of these particles breaking per unit time, the breakage death can be written as:

$$\text{Breakage death} = \Gamma(x)f_1(x, t) \qquad (5.20a)$$

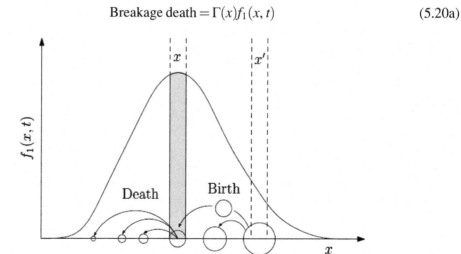

Fig. 5.37
Number balance for pure breakage. Schematic showing breakage birth and breakage death.

The breakage birth term is a little complicated. To elucidate this term, let us consider another differential size range, from x' to $x'+dx'$ where x' is larger than x. Now, we ask how many particles are produced in the size range from x to $x+dx$ by the breakage of particles in the size range from x' to $x'+dx'$. The number of breakages occurring per unit time at size x' to $x'+dx'$ is given by $\Gamma(x')f_1(x',t)dx'$. Now, because each breakage produces $\nu(x')$ particles on an average, a total of $\nu(x')\Gamma(x')f_1(x',t)dx'$ particles are produced. The fraction of these particles that fall in the size range dx is given by the daughter distribution function $p(x|x')dx$, and hence, the breakage birth term corresponding to the differential range dx' is given by:

$$\text{Breakage birth from } dx' = \nu(x)\Gamma(x')f_1(x',t)dx'p(x|x')dx \tag{5.20b}$$

Any such differential domain larger than x can undergo breakage to produce particles in the range of x to $x+dx$. Hence, to obtain the total contribution, all such domains have to be included by integrating the breakage birth term. Hence, the number balance equation becomes:

$$\frac{\partial f_1(x,t)}{\partial t} = -\Gamma(x)f_1(x,t) + \int_{x'=x}^{x'=\infty} \nu(x')\Gamma(x')f_1(x',t)p(x|x')dx' \tag{5.21}$$

In should be noted that the breakage number balance equation is a linear partial *integro* differential equation.

Number balance for aggregation

The number balance for the aggregation process can be written in a similar way. Like breakage birth and death, there will be aggregation birth and death terms. For particles of a specific size, aggregation death corresponds to a process in which particles in the said range aggregate with other particles and, therefore, grow in size and are removed from the size range of interest. Two smaller particles of size x' and x'' may also aggregate and provide particles of size x. This process leads to the aggregation birth term. These processes are shown in Fig. 5.38.

Expressions for aggregation birth and death terms can be written by recognising the binary aggregation as a second order process. Let us consider a pair of particles where one of the particles is from size x to $x+dx$ and another from size x' to $x'+dx'$. Hence, the number of pairs in this case will be $f_1(x,t)dxf_1(x',t)dx'$. A small fraction of these pairs will aggregate per unit time, as discussed before, and the actual number of aggregating particles can be found by multiplying the number of pairs by the aggregation efficiency. Hence, the number of particles aggregating per unit time for the said size ranges is given by: $a(x,x')f_1(x,t)dxf_1(x',t)dx'$. Because dx' was chosen arbitrarily, and any particle can form an aggregate with the particle of interest, the total aggregation death considering all possible pairs will be:

$$\text{Aggregation death} = \int_{x'=0}^{x'=\infty} a(x,x')f_1(x,t)dxf_1(x',t)dx' \tag{5.22}$$

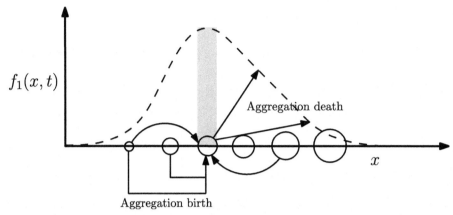

Fig. 5.38

Number balance for aggregation. Schematic showing aggregation birth and death.

In aggregation birth, two particles of size x' and x'' will aggregate and form a particle in x. Surely, all x' and x'' will not be able to produce a particle in x, and some constraint must exist which will identify the possible pairs. Such constraints are obtained by satisfying mass balance. For example, if x is particle volume and densities are constant, the statement of mass balance becomes:

$$x' + x'' = x \tag{5.23}$$

It is important to note that the form of the equation relating x, x' and x'' is dependent on the choice of internal coordinate. For example, if x is the particle diameter, the statement of mass balance translates into:

$$x'^3 + x''^3 = x^3 \tag{5.24}$$

The functional relation between x, x', and x'' implies that any two of them can be treated as independent variables. In general, this functional relation may be written as:

$$x'' = \phi(x, x') \tag{5.25}$$

With this understanding, the aggregation birth term may be written as:

$$a(x', x'')f_1(x', t)dx'f_1(x'', t)dx'' \tag{5.26}$$

Because x' and x'' are related to x, we can replace x'' and dx'' from above using Eq. (5.25):

$$a(x', \phi(x, x'))f_1(x', t)dx'f_1(\phi(x, x'), t)[\partial\phi/\partial x]dx \tag{5.27}$$

In this case, x' is one of the aggregating particles, and particles ranging from a very small size to size x can, in principle, produce particles of size x. Hence, the aggregation birth term should be integrated in the interval $[0, x]$. But in this case, we count each pair twice. For example, the pairs $[0, x]$ as well as $[x, 0]$ are counted toward making particles of size x.

But they are the same pair with reversal of order. Because order does not make any difference, we should multiply the terms by half while integrating from 0 to x. Hence, the aggregation birth term becomes:

$$\text{Aggregation birth} = \frac{1}{2} \int_{x'=0}^{x} a(x', \phi(x, x')) f_1(x', t) dx' f_1(\phi(x, x'), t) [\partial\phi/\partial x] dx \qquad (5.28)$$

Combining aggregation birth and death terms and cancelling dx from both sides gives the aggregation population balance equation:

$$\frac{\partial f_1(x, t)}{\partial t} = - \int_{x'=0}^{x'=\infty} a(x, x') f_1(x, t) f_1(x', t) dx' + \frac{1}{2} \int_{x'=0}^{x} a(x', \phi(x, x')) f_1(x', t) f_1(\phi(x, x'), t) [\partial\phi/\partial x] dx'$$

$$(5.29)$$

In both aggregation and breakage equations, the phenomenology of the process is tucked into the breakage and aggregation functions. These are also called breakage and aggregation *kernels*, respectively.

Example 5.6

Formulate the aggregation population balance equation where the internal coordinate is the particle radius.

Solution

All other parts of the aggregation equation remain the same except for the aggregation birth term. The functionality in Eq. (5.25) for this case will be:

$$\phi(x, x') \equiv \sqrt[3]{x^3 - x'^3} \qquad \therefore \frac{\partial\phi}{\partial x} = \left[\frac{x}{\sqrt[3]{x^3 - x'^3}} \right]^2$$

Therefore, the aggregation birth term becomes:

$$\frac{1}{2} \int_{x'=0}^{x} a(x', \phi(x, x')) f_1(x', t) f_1(\phi(x, x'), t) \left[\frac{x}{\sqrt[3]{x^3 - x'^3}} \right]^2 dx'$$

5.5.5 Specifying the Breakage and Aggregation Kernels

It can be seen that the number balance equations are useful if the kernels are known. There are various ways by which kernels can be obtained. The foremost one is through phenomenological models. For a few breakage and aggregation processes, such phenomenological functions can be obtained. For example, if particle breaks into two fragments distributed evenly around the middle size, a simple argument leads to analytical expressions for breakage functions. Such a situation arises for polymer degradation in high-shear extensional flow where a polymer chain has the highest probability to break at the midpoint. Because it is a binary breakage, $\nu(x) = 2.0$. Ziff and Mcgrady[39] modelled the other two breakage functions for this situation in the following manner:

The quantity $\Gamma(x')p(x|x')dx' \equiv F(x, x'-x)dx'$ may be interpreted as a rate constant for conversion of particles of size x' into x and $x'-x$. For extensional flow, $F(x, x'-x)$ was modelled as a symmetric function $F(x, x'-x) = x(x'-x)$ which peaks at $x'/2$. Because $\Gamma(x)$ may be interpreted as rate constant for breakage of particles of size x into all sizes upto x, it can be written as:

$$\Gamma(x) = \int_{\xi=0}^{x} F(\xi, x-\xi)d\xi = \frac{x^3}{6} \tag{5.30}$$

Note that the above expressions provide the "normalised" form of the kernels, and proportionality constants should be included for matching with the experimental data.

For most cases of solid particle breakage, the breakage process is more complicated, and kernels are mostly empirical. For breakage frequency, a power law is most frequently used:

$$\Gamma(x) = kx^a \tag{5.31}$$

It is generally observed that the rate of breakage is also dependent on the granular flow condition inside the mill (Tsoungui et al.[40]). For this reason, wet grinding is usually more efficient than dry grinding. However, consideration of such effects into the population balance model is difficult because the breakage rate becomes a function of particle size distribution: $\Gamma(x, f_1(x,t))$. This makes the breakage population balance a nonlinear integro differential equation (Bilgili and Scarlett[41]) whose solution becomes difficult.

For empirical breakage kernels, $\nu(x)$ is usually not specified separately, and a combined quantity $\nu(x) p(x|x')$ is specified. The notation is not unambiguous in the literature, and one has to exercise care while interpreting the quantity specified as the "daughter distribution."

For daughter size distribution, a cumulative probability distribution function is usually available. A widely used form (Bilgili et al.[42]) is:

$$\int_0^x p(\xi|x')d\xi \equiv P(x|x') = \phi\left(\frac{x}{x'}\right)^\alpha + (1-\phi)\left(\frac{x}{x'}\right)^\beta \tag{5.32}$$

The cumulative form is preferred because of its monotonic nature. Note that, in the above kernel, the cumulative probability is a function of the ratio of the mother and daughter particle size. The absolute size of the particles is immaterial for determining the size distribution.

Such properties are called similarity or scaling behaviour, and are frequently observed (see Kick's law in Section 5.2.3 for example) for comminution of solid particles. The parameters of the breakage functions (k, a, φ, α, β) are usually found by fitting the particle size distribution into data from a laboratory mill.

It is important to note that the form of the daughter distribution function should be determined by keeping the actual process in mind. Different breakage processes produce qualitatively different daughter particles, as shown in Fig. 5.39, and the function chosen as daughter distribution function should conform to the process.

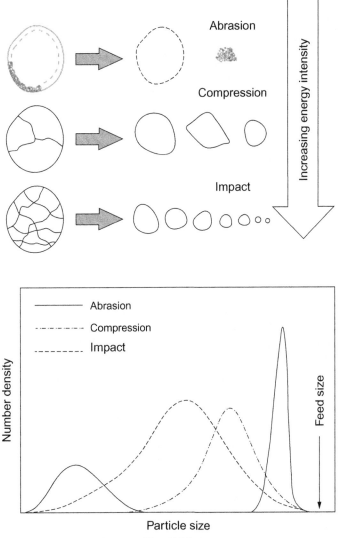

Fig. 5.39

Different mechanisms for breakage producing distinctly different daughter distribution. *Taken from Gao M, Forssberg E. Prediction of product size distributions for a stirred ball mill.* Powder Technol *1995;**84**:101–6.*

The aggregation kernel can also be obtained by writing a suitable phenomenological model. For perikinetic process, the random Brownian motion of particles can be described by a diffusion-like process, and it can be shown that the aggregation kernel is given by Ramkrishna[38]:

$$a(x, x') = \frac{2kT}{3\mu} \left[x^{-1/3} + x'^{-1/3} \right] \left[x^{1/3} + x'^{1/3} \right] \tag{5.33}$$

where x is the particle volume. For settling of a dense particulate system, the relative deterministic motion of particles can be modelled to obtain the aggregation kernel as (Ramkrishna[38]):

$$a(x, x') = \frac{2(\rho_p - \rho)g}{9\mu} \left[\frac{3\sqrt{\pi}}{4} \right]^{2/3} \left[x^{1/3} + x'^{1/3} \right] \left| x'^2 - x^2 \right| \tag{5.34}$$

Aggregation kernels for a variety of other situations are also available in standard handbooks (Lewis et al.[43]). In the absence of phenomenological models for aggregation kernel, empirical aggregation kernels may be used. Usually, three different degrees of particle size dependences are considered by these aggregation kernels.

(a) Constant kernel: The aggregation process is independent of size. $a(x,x') = a_0$. The Brownian kernel reduces to this form for equal size particles.
(b) Sum kernel: $a(x,x') = a_0(x+x')$. This should be used to empirically model mild dependence on particle size.
(c) Product kernel: This should be used for stronger size dependent aggregation: $a(x,x') = a_0 xx'$.

5.5.6 Solution of Population Balance Equation

It can be seen that the PBE is a nonlinear partial integro differential equation when breakage and aggregation are considered. Hence, its solution is not trivial. Analytical solution is very difficult except for a few simplified cases, and numerical solution is generally needed. Algorithms specifically designed for this equation should be used for its numerical solution. The most well-known algorithm for the solution of the population balance equation is the so-called Fixed Pivot Technique (FPT) (Kumar and Ramkrishna[44]).

In this algorithm, the population balance equation is integrated over a finite range of internal coordinates to yield a discretized equation. A geometric grid, as shown in Fig. 5.40, is almost always used for space discretization. The advantage of geometric grid is that it offers good

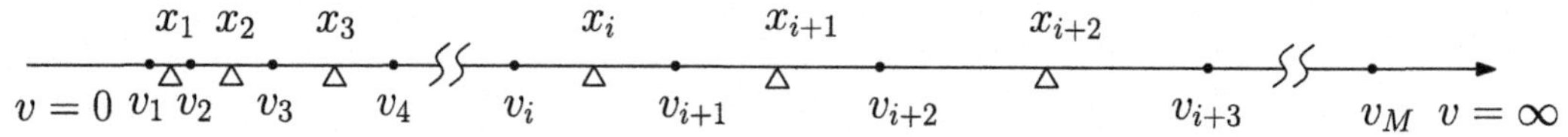

Fig. 5.40
Discretization of space using geometric grid in Fixed Pivot Technique.

resolution at small size range while covering several orders of magnitude using a manageable number of nodes. This is required, especially if aggregation and breakage are involved. Particle size changes over several orders of magnitude in these cases.

For a combined breakage and aggregation, the discretized equation becomes:

$$
\begin{aligned}
\frac{dN_i(t)}{dt} = {} & -\Gamma_i N_i(t) + \sum_{j=1}^{M} n_{ij}\Gamma_i N_i(t) \\
& - N_i(t) \sum_{j=1}^{M} a_{ij} N_j(t) + \sum_{\substack{j,k \\ x_{i-1} \le x_j + x_k \le x_{i+1}}}^{j \ge k} \left(1 - \frac{1}{2}\delta_{jk}\right)\eta a_{jk} N_j(t) N_k(t)
\end{aligned}
\tag{5.35}
$$

In this case, particle volume ($x \equiv v$) has been used as internal coordinate. The other terms are defined as:

$$
N_i(t) = \int_{v_i}^{v_{i+1}} f_1(x, t)dx
\tag{5.36}
$$

$$
x_i = {(v_i + v_{i+1})}\big/_2
\tag{5.37}
$$

$$
\Gamma_i = \Gamma(x_i)
\tag{5.38}
$$

$$
n_{ij} = \int_{x_i}^{x_{i+1}} \frac{x_{i+1} - v}{x_{i+1} - x_i} p(v| x_j)dv + \int_{x_{i-1}}^{x_i} \frac{v - x_{i-1}}{x_i - x_{i-1}} p(v| x_j)dv
\tag{5.39}
$$

$$
a_{ij} = a(x_i, x_j)
\tag{5.40}
$$

$$
\eta =
\begin{cases}
\dfrac{x_{i+1} - (x_j + x_k)}{x_{i+1} - x_i}, & x_i \le (x_j + x_k) \le x_{i+1} \\[2ex]
\dfrac{(x_j + x_k) - x_{i-1}}{x_i - x_{i-1}}, & x_{i-1} \le (x_j + x_k) \le x_i
\end{cases}
\tag{5.41}
$$

An ODE solver needs to be used to solve this equation. The users need to supply the initial condition and change the kernels according to the situation at hand and can obtain the evolution of numbers in each bin. The approximate number density function, if needed, can be obtained by dividing the number by bin width.

In some situations, evolution of the entire particle size distribution is not needed, and evolution of a few average properties such as the total number of particles and the number weighted average diameter, etc., need to be tracked as a function of time. These quantities are called moments, and the reader is advised to revisit Section 2.2.4 for a better understanding of these ideas. The *j*th moment of the number density function is defined as:

$$\mu_j \equiv \int_0^\infty x^j f_1(x, t)\mathrm{d}x \tag{5.42}$$

It can be readily seen that μ_0 is the total number of particles, and if x is the particle volume, μ_1 is the total volume of all the particles. Hence, the ratio μ_1/μ_0 is the average particle volume.

The evolution of jth moment can be obtained from the population balance if the kernels are known. Let us consider the case of a pure binary breakage where the breakage frequency is given by a simple power law of the form kx^a. The breakage produces uniformly distributed daughter particles (random breakage), and hence, the probability density of daughter particle is given by $1/x'$. The PBE for this case becomes:

$$\frac{\partial f_1(x, t)}{\partial t} = -kx^a f_1(x, t) + 2 \int_{x'=x}^{x'=\infty} kx'^a \frac{1}{x'} f_1(x', t)\mathrm{d}x'$$

$$= -kx^a f_1(x, t) + 2k \int_{x'=x}^{x'=\infty} x'^{a-1} f_1(x', t)\mathrm{d}x' \tag{5.43}$$

To obtain the equation for jth moment, we multiply this equation by $x^j \mathrm{d}x$ and integrate in the interval $[0, \infty]$:

$$\int_{x=0}^\infty x^j \mathrm{d}x \frac{\partial f_1(x, t)}{\partial t} = -\int_{x=0}^\infty x^j kx^a f_1(x, t)\mathrm{d}x + 2k \int_{x=0}^\infty x^j \mathrm{d}x \int_{x'=x}^{x'=\infty} x'^{a-1} f_1(x', t)\mathrm{d}x'$$

$$\therefore \frac{\mathrm{d}\mu_j}{\mathrm{d}t} = -k\mu_{j+a} + 2k \int_{x=0}^\infty \int_{x'=x}^{x'=\infty} x'^{a-1} x^j f_1(x', t)\mathrm{d}x'\mathrm{d}x \tag{5.44}$$

It can be noted that the second term has number density in x', but the integration in semi-infinite domain is with respect to x. Hence, we cannot translate this term into moment. However, a change in sequence in which the integrals are evaluated (as shown schematically in Fig. 5.41) resolves this problem:

After changing the sequence of evaluation of the integrals and changing limits appropriately:

$$\frac{\mathrm{d}\mu_j}{\mathrm{d}t} = -k\mu_{j+a} + 2k \int_{x'=0}^\infty \int_{x=0}^{x'} x'^{a-1} x^j \mathrm{d}x f_1(x', t)\mathrm{d}x'$$

$$= -k\mu_{j+1} + \frac{2k}{j+1} \int_{x'=0}^\infty x'^{a+j} f_1(x', t)\mathrm{d}x' \tag{5.45}$$

$$= -k\mu_{j+1} + \frac{2k}{j+1}\mu_{a+j}$$

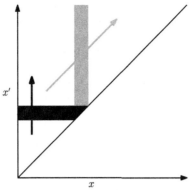

Fig. 5.41

Changing the limits of the double integration. The original sequence is shown in *grey*, while the new sequence is shown in *black*. Both cover the same portion of the domain.

It can be seen that the lower moment is dependent on the higher moment for this case, and hence, the system of moments does not form a closed set of equations. This occurs for most cases except a few. For example, if $a = 1$, the equation of moment becomes:

$$\frac{d\mu_j}{dt} = \frac{k(1-j)}{(1+j)}\mu_{j+1} \tag{5.46}$$

If we write the equations for the first two moments:

$$\frac{d\mu_0}{dt} = k\mu_1 \tag{5.47a}$$

$$\frac{d\mu_1}{dt} = 0 \tag{5.47b}$$

which forms a closed set. The second equation is simply the statement of mass balance. Approximate closure to a set of moment equations can also be obtained using Gaussian quadrature, and this method is known as the Quadrature Method of Moments (QMOM) (McGraw[45]).

Example 5.7

Obtain the equation for moments for aggregation equation for a constant kernel. Particle mass is the internal coordinate.

Solution

From Eq. (5.29), the aggregation equation for a constant kernel becomes:

$$\frac{\partial f_1(x, t)}{\partial t} = \frac{1}{2}\int_{x'=0}^{x'=x} a_0 f_1(x-x', t)f_1(x', t)dx' - \int_{x'=0}^{x'=\infty} a_0 f_1(x, t)f_1(x', t)dx'$$

Now, multiplying both sides of the equation with x^j and integrating from 0 to ∞, we get

$$\frac{\partial \mu_j(t)}{\partial t} = \frac{1}{2} \int_{x=0}^{x=\infty} \int_{x'=0}^{x'=x} a_0 x^j f_1(x-x',t) f_1(x',t) dx' dx - \int_{x=0}^{x=\infty} \int_{x'=0}^{x'=\infty} a_0 x^j f_1(x,t) f_1(x',t) dx' dx$$

$$= \frac{1}{2} \int_{x=0}^{x=\infty} \int_{x'=0}^{x'=x} a_0 x^j f_1(x-x',t) f_1(x',t) dx' dx - a_0 \mu_j \mu_0$$

Changing the limit of the first term:

$$\frac{\partial \mu_j(t)}{\partial t} = \frac{1}{2} \int_{x'=0}^{x'=\infty} \int_{x=x'}^{x=\infty} a_0 x^j f_1(x-x',t) f_1(x',t) dx dx' - a_0 \mu_j \mu_0$$

Denoting $\xi = x - x'$:

$$\frac{\partial \mu_j(t)}{\partial t} = \frac{1}{2} \int_{x'=0}^{x'=\infty} \int_{\xi=0}^{\xi=\infty} a_0 (x'+\xi)^j f_1(\xi,t) f_1(x',t) dx dx' - a_0 \mu_j \mu_0$$

Expanding the $(x'+\xi)^j = \sum_{k=0}^{j} \frac{j!}{k!(j-k)!} (x')^k (\xi)^{j-k}$ we get

$$\frac{\partial \mu_j(t)}{\partial t} = \frac{1}{2} \sum_{k=0}^{j} \int_{x'=0}^{x'=\infty} \int_{\xi=0}^{\xi=\infty} a_0 \frac{j!}{k!(j-k)!} (x')^k (\xi)^{j-k} f_1(\xi,t) f_1(x',t) dx dx' - a_0 \mu_j \mu_0$$

$$= \frac{a_0}{2} \sum_{k=0}^{j} \frac{j!}{k!(j-k)!} \mu_k \mu_{j-k} - a_0 \mu_j \mu_0$$

For $j = 0$ and 1:

$$\frac{\partial \mu_0(t)}{\partial t} = -\frac{a_0}{2} \mu_0(t)^2$$

$$\frac{\partial \mu_1(t)}{\partial t} = 0$$

As the total mass of the system is constant, the first moment remains constant. This serves as a check.

5.5.7 Discrete Breakage Equation

For breakage of solid particles in a mill, a discrete form of the breakage equation is more common than the continuous population balance equation (Austin[46]). In this case, the breakage equation is written as:

$$\frac{dw_i}{dt} = -s_i w_i + \sum_{j=1}^{i} b_{ij} s_j w_j \tag{5.48}$$

where w_i is the mass fraction of particles on ith sieve. Here, s_i is the breakage frequency for that size class, and b_{ij} is the discrete version of the daughter distribution function discussed above. It may be noted that $\nu(x)$ is included in b_{ij} here. As expected, s_i and b_{ij} are dependent on particle size but independent of time. The above set of coupled linear ODES can be written in a matrix form (Berthiaux et al.[47]) as:

$$\frac{\mathrm{d}W}{\mathrm{d}t} = (B - I)SW = AW \tag{5.49}$$

This equation can be readily solved by diagonalization (Kreyszig[48]). If we denote P as the matrix of all eigenvectors of A, it can be readily verified that $P^{-1}AP$ is a diagonal matrix with s_i as diagonal elements. Note that P is a lower triangular matrix because B is a lower triangular matrix.

Now, rearranging Eq. (5.49) as:

$$P^{-1}\frac{\mathrm{d}W}{\mathrm{d}t} = P^{-1}APP^{-1}W \tag{5.50}$$

and denoting $P^{-1}W$ as Z:

$$\frac{\mathrm{d}Z}{\mathrm{d}t} = -SZ \tag{5.51}$$

where $P^{-1}AP \equiv S$ is the diagonal matrix containing s_i, this equation is readily solved to obtain:

$$Z = Z_D E(t) \tag{5.52}$$

where Z_D is a diagonal matrix comprising the initial conditions, and $E(t)$ is a vector with elements as $\exp(-s_i t)$. Multiplying both sides of Eq. (5.52) by P and denoting $PZ_D \equiv T$:

$$W = TE(t) \tag{5.53}$$

Because P is a lower triangular matrix, T is also a lower triangular matrix, the solution can be written in a component form as:

$$\begin{aligned}
w_1(t) &= \tau_{11}\exp(-s_1 t) \\
w_2(t) &= \tau_{21}\exp(-s_1 t) \quad\quad +\tau_{22}\exp(-s_2 t) \\
&\;\;\vdots \\
w_n(t) &= \tau_{n1}\exp(-s_1 t) + \cdots \quad \cdots + \tau_{nn}\exp(-s_n t)
\end{aligned} \tag{5.54}$$

Denoting $R_i(t) = \sum_{j=1}^{i} w_j$ as the cumulative mass fraction *oversize*, the solution can be written as:

$$R_i(t) = \sum_{j=1}^{i} \left(\sum_{k=j}^{i} \tau_{kj} \right) \exp(-s_j t) \tag{5.55}$$

Similar to the continuous breakage equation, discrete breakage equation becomes useful if S and B are available. Empirical kernels similar to the continuous case can be used.

Usually, s_i can be taken as a power law. The selection function often shows similarity behaviour and can be approximated as $b_{ij} = b_{ij}(x_i/x_j)$. A frequently used form for the cumulative of b_{ij} is

$$B_{ij} = \phi \left(\frac{x_i}{x_j}\right)^\gamma + (1-\phi)\left(\frac{x_i}{x_j}\right)^\beta \quad i > j$$

$$b_{ij} = B_{i-1,j} - B_{i,j}$$

(5.56)

5.5.8 Estimation of Breakage Function From Laboratory Milling Data

Usually, a form of the breakage function is assumed, and the parameters are obtained by fitting the laboratory milling data to the model. For such fitting, an approximate solution is used instead of the exact solution as given in Eq. (5.55). There are several approximate solutions available including that by Kapur[49] and Berthiaux et al.[47] and in this text we shall use the approximate solution given by Berthiaux et al.[47] In this case, the approximate solution is applicable at early grinding time. It can be shown that under such conditions, the last term in the series of exponentials as given in Eq. (5.55) dominates. Hence, the approximate solution can be written as:

$$R_i(t) = R_i(0)\exp(-s_i t)$$

(5.57)

It can also be shown that b_{ij} is given by:

$$b_{ij} = \frac{s_{i-1} - s_i}{s_j}$$

(5.58)

With these simplifications, it is straightforward to obtain the size dependent parameters s_i and b_{ij}. A plot of $-\ln(R_i(t)/R_i(0))$ vs time will yield s_i as slope. Because breakage usually follows the power law relation, $(s_i = k(x_i)^a)$, a log–log plot of s_i vs x_i will yield the two parameters for breakage frequency. Similar exercises may be conducted to verify if the selection function fits to a simple power law function similar to that given in Eq. (5.56).

Example 5.8

Berthiaux et al.[47] presented the laboratory milling data for hydrargillite particles from a Dyno-mill. The cumulative weight fraction reatained (R_i) for four different screen sizes (x_i) at various times (min) was reported which is reproduced below:

Time (min)	R_1 ($x_1 = 39.1\ \mu m$)	R_2 ($x_2 = 18.2\ \mu m$)	R_3 ($x_3 = 8.48\ \mu m$)	R_4 ($x_4 = 3.95\ \mu m$)
0.0	0.35	0.62	0.80	0.92
0.5	0.30	0.57	0.77	0.91
1.0	0.12	0.44	0.69	0.85
2.0	0.11	0.40	0.64	0.84

3.0	0.09	0.35	0.61	0.82
4.0	0.08	0.31	0.58	0.79
5.0	0.08	0.30	0.58	0.78
6.0	0.07	0.26	0.54	0.75
7.0	0.05	0.24	0.51	0.73
9.0	0.03	0.19	0.43	0.69
11.0	0.01	0.15	0.39	0.65
15.0	0.01	0.14	0.35	0.63

Determine the selection and breakage matrices for this system. Assume a power law for breakage frequency ($s_i = kx_i^\alpha$) and obtain the values of the parameters k and α.

Solution

The plot of $Ln(R_i(t)/R_i(0))$ versus time is shown in Fig. 5.42.

It can be seen that straight lines can be fitted to each of the size classes. Using MATLAB curve fitting, the slopes of these lines are obtained, and s_i is evaluated using Eq. (5.56). This yields the selection matrix (S) as:

$$S = \begin{bmatrix} 0.232 & 0 & 0 & 0 & 0 \\ 0 & 0.101 & 0 & 0 & 0 \\ 0 & 0 & 0.055 & 0 & 0 \\ 0 & 0 & 0 & 0.026 & 0 \\ 0 & 0 & 0 & 0 & 0.000 \end{bmatrix}$$

We have added a size class corresponding to very small size for which the breakage rate is zero. This is required to prevent any mass loss through fragmentation. Using Eq. (5.57), the breakage matrix is obtained as:

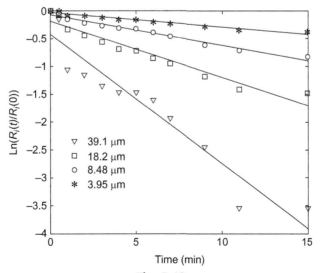

Fig. 5.42
Linear fit to milling data to obtain s_i.

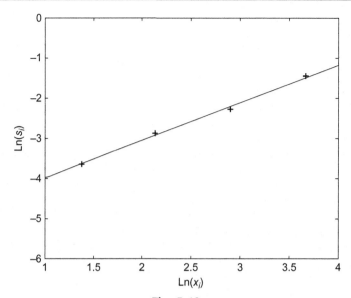

Fig. 5.43
Linear fit to the obtained values of s_i for extraction of k and α.

$$B = \begin{bmatrix} 0 & 0 & 0 & 0 & 0 \\ 0.564 & 0 & 0 & 0 & 0 \\ 0.198 & 0.454 & 0 & 0 & 0 \\ 0.125 & 0.288 & 0.527 & 0 & 0 \\ 0.112 & 0.258 & 0.473 & 1 & 1 \end{bmatrix}$$

A plot of $\ln(s_i)$ vs $\ln(x_i)$ will yield α as its slope. The value of k should be obtained by using the value of α, s_i, and x_i at all four points and finally averaging (Fig. 5.43). This exercise yields $k = 1.2 \times 10^{-4} \mathrm{s}^{-1} \mu\mathrm{m}^{-\alpha}$ and $\alpha = 0.94$.

5.6 Nomenclature

		Units in SI System	Dimension in **M, L, T**
a	Crack length	m	**L**
$a(x,x')$	Aggregation frequency	–	–
A_1	Parameter in Eq. (5.14)	m^3	$\mathbf{L}^3$
A_2	Parameter in Eq. (5.15)	–	–
b	Half distance between crushing rolls	m	**L**
bij	Discrete breakage parameter	–	–
B_1	Coefficient in Eq. (5.14)	m^3/s	$\mathbf{L}^3\,\mathbf{T}^{-1}$
B_2	Parameter in Eq. (5.15)	s^{-1}	$\mathbf{T}^{-1}$

Bij	Cumulative discrete breakage parameter	–	–
c	Radius of yoke of pendulum mill	m	$\mathbf{L}$
C	A coefficient	–	$\mathbf{L}^{1-p}\,\mathbf{T}^{-2}$
d_0	Initial size of agglomerate	m	$\mathbf{L}$
d_c	Twice radius of gyration of planetary mill	m	$\mathbf{L}$
d_M	Diameter of cylindrical mill unit	m	$\mathbf{L}$
d_t	Size of agglomerate at time t	m	$\mathbf{L}$
E	Energy per unit mass	J/kg	$\mathbf{L}^2\,\mathbf{T}^{-2}$
E_i	Work index	J/kg	$\mathbf{L}^2\,\mathbf{T}^{-2}$
F	Parameter in Eq. (5.1)	N/m	$\mathbf{MT}^{-2}$
$f_1(x,t)$	Number density function	–	–
f_c	Crushing strength of material	N/m^2	$\mathbf{ML}^{-1}\,\mathbf{T}^{-2}$
g	Acceleration due to gravity	m/s^2	$\mathbf{LT}^{-2}$
K_K	Kick's constant	m^3/kg	$\mathbf{M}^{-1}\,\mathbf{L}^3$
K_R	Rittinger's constant	m^4/kg	$\mathbf{M}^{-1}\,\mathbf{L}^4$
l	Length of arm of pendulum mill	m	$\mathbf{L}$
L	Characteristic linear dimension	m	$\mathbf{L}$
m	Mass per unit length of arm of pendulum mill	kg/m	$\mathbf{ML}^{-1}$
M	Mass of crushing head in pendulum mill, or of particle	kg	$\mathbf{M}$
M'	Mass of arm of pendulum mill	kg	$\mathbf{M}$
N_c	Critical speed of rotation of ball mill (rev/time)	s^{-1}	$\mathbf{T}^{-1}$
N_G	Speed of rotation of planetary mill about axis of gyration (rev/time)	s^{-1}	$\mathbf{T}^{-1}$
$N_i(t)$	Total number of particle in ith bin in a discretized particle size axis	–	–
N_M	Speed of rotation of cylindrical mill unit about own axis (rev/time)	s^{-1}	$\mathbf{T}^{-1}$
p	A constant used as an index in Eq. (5.2)	–	–
$p(x/x')dx$	Probability of formation of daughter particles in size range x to $x+dx$ due to breakage of a particle in the size rang x' to $x'+dx'$	–	–
q	Size reduction factor L_1/L_2	–	–
r	Radius of ball mill minus radius of particle	m	$\mathbf{L}$
R	Normal reaction	N	$\mathbf{MLT}^{-2}$
r_1	Radius of crushing rolls	m	$\mathbf{L}$
r_2	Radius of particle in feed	m	$\mathbf{L}$
$R_i(t)$	Cumulative weight fraction oversize for size x_i	–	–
s	Gear ratio in planetary mill (N_M/N_G)	–	–

s_c	Value of s at critical speed for given value of d_G/d_M	–	–
$S_i(t)$	Breakage frequency of ith size class	s^{-1}	$\mathbf{T}^{-1}$
t	Time	s	$\mathbf{T}$
v_i, v_{i+1}	Lower and upper boundary of ith bin in a discretized particle size axis	–	–
$w_i(t)$	Mass fraction of particles in ith bin	–	–
x	Internal coordinate (particle volume, diameter, mass, etc.) of particles	–	–
x_i	ith node in a discretized particle size axis	–	–
y	Distance along arm from point of support	m	$\mathbf{L}$
Y	Young's modulus	N/m^2	$\mathbf{ML}^{-1}\mathbf{T}^{-2}$
α	Half angle of nip	–	–
μ_j	jth moment of the particle size distribution	–	–
$\nu(x)$	Average number daughter particle formed due to breakage of a particle of size x	–	–
$\Gamma(x)$	Breakage frequency for particles of size x	s^{-1}	$\mathbf{T}^{-1}$
θ	Angle between axis and vertical	–	–
ρ_s	Density of solid material	kg/m^3	$\mathbf{ML}^{-3}$
ω	Angular velocity	s^{-1}	$\mathbf{T}^{-1}$
ω_c	Critical speed of rotation of ball mill	s^{-1}	$\mathbf{T}^{-1}$
τ	Stress	N	$\mathbf{ML}^{-1}\mathbf{T}^{-2}$

References

1. Heywood H. Some notes on grinding research. *J Imp Coll Chem Eng Soc* 1950;**6**:26.
2. Bemrose CR, Bridgwater J. A review of attrition and attrition test methods. *Powder Technol* 1987;**49**:97–126.
3. Hess W, Schönert K. In: *Plastic transition in small particles. Proc. 1981 Powtech Conf. on particle technology, Birmingham*; 1981. p. D2/I/1–9. EFCE Event No. 241.
4. Schönert K. Role of fracture physics in understanding comminution phenomena. *Trans Soc Min Eng AIME* 1972;**252**:21–6.
5. Piret EL. Fundamental aspects of crushing. *Chem Eng Prog* 1953;**49**:56.
6. Prasher CL. *Crushing and grinding process handbook.* Chichester: Wiley; 1987.
7. Kick F. *Das Gesetz der proportionalen Widerstande und seine Anwendungen.* Leipzig: Arthur Felix; 1885.
8. Von Rittinger PR. *Lehrbuch der Aufbereitungskunde in ihrer neuesten Entwicklung und Ausbildung systematisch dargestellt.* Berlin: Ernst und Korn; 1867.
9. Bond FC. Third theory of communtion. *Min Engng NY* 1952;**4**:484–94.
10. Bond FC. New grinding theory aids equipment selection. *Chem Eng, Albany* 1952;**59**:169.
11. Austin LG, Klimpel RR. The theory of grinding operations. *Ind Eng Chem* 1964;**56**:18–29.
12. Cutting GW. Grindability assessments using laboratory rod mill tests. *Chem Eng* 1977;**325**:702–4.
13. Owens JS. Notes on power used in crushing ore, with special reference to rolls and their behaviour. *Trans Inst Min Metall* 1933;**42**:407–25.
14. Gross J. Crushing and grinding. *US Bur Mines Bull* 1938;**402**.

15. Kwong JNS, Adams JT, Johnson JF, Piret EL. Energy–new surface relationship in crushing. I. Application of permeability methods to an investigation of the crushing of some brittle solids. *Chem Eng Prog* 1949;**45**:508.

16. Adams JT, Johnson JF, Piret EL. Energy—new surface relationship in the crushing of solids. II. Application of permeability methods to an investigation of the crushing of halite. *Chem Eng Prog* 1949;**45**:655.

17. Johnson JF, Axelson J, Piret EL. Energy–new surface relationship in crushing. III. Application of gas adsorption methods to an investigation of the crushing of quartz. *Chem Eng Prog* 1949;**45**:708.

18. Zeleny RA, Piret EL. Dissipation of energy in single particle crushing. *Ind Eng Chem Process Des Dev* 1962;**1**:37–41.

19. Work LT. Trends in particle size technology. *Ind Eng Chem* 1963;**55**:56–8.

19a. Brown GG, et al. *Unit Operations.* New York: John Wiley and Sons; 1950.

20. Gandolfi EAJ, Papachristodoulou G, Trass O. Preparation of coal slurry fuels with the Szego mill. *Powder Technol* 1984;**40**:269–82.

21. Koka VR, Trass O. Determination of breakage parameters and modelling of coal breakage in the Szego mill. *Powder Technol* 1987;**51**:201–4.

22. Kano J, Mio H, Saito F. Correlation of grinding rate of gibbsite with impact energy balls. *AIChE J* 2000;**46**:1694.

23. Bradley AA, Hinde AL, Lloyd PJD, Schymura K. *Proceedings of the European Symposium on Particle Technology, Amsterdam,* 1980.

24. Kitschen LP, Lloyd PJD, Hartmann R. The centrifugal mill: experience with a new grinding system and its application. In: *Proceedings of 14th International Mineral Processing Congress, Toronto*; 1982. p. I-9.

25. Butcher C. Cryogenic grinding: an independent voice. *Chem Eng* 2000;**713**.

26. von Smoluchowski M. Versuch einer mathematischen Theorie der Koagulationskinetik kolloider Losungen. *Z Physik Chem* 1917;**92**:129–68.

27. Walton JS. *The formation and properties of precipitates.* New York: Interscience; 1967.

28. Tadros TF. Rheology of concentrated suspensions. *Chem Ind (Lond)* 1985;**7**:210–8.

29. Söhnel O, Mullin JW. Agglomeration of batch precipitated suspensions. *AIChE Symp Ser* 1991;**87**:182–90.

30. Perry RH, Green DW, Maloney JO, editors. *Perry's chemical engineers' handbook.* 7th ed. New York: McGraw-Hill Book Company; 1997.

31. Browning JE. Agglomeration: growing larger in applications and technology. *Chem Eng, Albany* 1967;**74**:147–70.

32. Mortensen S, Hovmand S. In: Keairns DL, editor. *Fluidization technology.* Vol. II. Washington, DC: Hemisphere; 1976.

33. Mathur KB, Epstein N. *Spouted beds.* New York: Academic Press; 1974.

34. Landi G, Barletta D, Poletto M. Modelling and experiments on the effect of air humidity on the flow properties of glass powders. *Powder Technol* 2011;**207**:437–43.

35. Ramkrishna D, Singh MR. Population balance modeling: current status and future prospects. *Annu Rev Chem Biomol Eng* 2014;**5**:123–46.

36. Kuvadia ZB, Doherty MF. Reformulating multidimensional population balances for predicting crystal size and shape. *AIChE J* 2013;**59**:3468–74.

37. Norouzi HR, Zarghami R, Gharebagh RS, Mostoufi N. *Coupled CFD-DEM modeling formulation, implementation and application to multiphase flows.* Chichester: Wiley; 2016.

37a. Chakraborty J, Kumar S. A new framework for solution of multidimensional population balance equations. *Chemical engineering science* 2007;**62**:4112–25.

38. Ramkrishna D. *Population balances: theory and applications to particulate systems in engineering.* San Diego, CA: Academic Press; 2000.

39. Ziff RM, Mcgrady ED. Kinetics of polymer degradation. *Macromolecules* 1986;**19**:2513–9.

40. Tsoungui O, Vallet D, Charmet J. Numerical model of crushing of grains inside two-dimensional granular materials. *Powder Technol* 1999;**105**:190–8.

41. Bilgili E, Scarlett B. Population balance modeling of non-linear effects in milling processes. *Powder Technol* 2005;**153**:59–71.

42. Bilgili E, Yepes J, Scarlett B. Formulation of a non-linear framework for population balance modeling of batch grinding: beyond first order kinetics. *Chem Eng Sci* 2006;**61**:33–44.

43. Lewis A, Seckler MM, Kramer HJM, Rosmalen GV. *Industrial crystallization: fundamentals and applications.* London: Cambridge University Press; 2015.

44. Kumar S, Ramkrishna D. On the solution of population balance equations by discretization—I. A fixed pivot technique. *Chem Eng Sci* 1996;**51**:1311–32.

45. McGraw R. Description of aerosol dynamics by quadrature method of moments. *Aerosol Sci Technol* 1997;**27**:255–65.

46. Austin LG. A review: introduction to the mathematical description of grinding as a rate process. *Powder Technol* 1971;**5**:1–17.

47. Berthiaux H, Varinot C, Dodds J. Approximate calculation of breakage parameters from batch grinding tests. *Chem Eng Sci* 1996;**51**:4509–16.

48. Kreyszig E. *Advanced engineering mathematics.* India: Wiley; 2006.

49. Kapur PC. Self-preserving spectra of comminuted particles. *Chem Eng Sci* 1972;**27**:425–31.

Further Reading

Bond FC. Crushing and grinding calculations. *Br Chem Eng* 1961;**6**:378–85. 543–548.

Capes CE, Germain RJ, Coleman RD. Bonding requirements for agglomeration by tumbling. *Ind Eng Chem Proc Des Dev* 1977;**16**:517.

Capes CE. Particle size enlargement. In: Williams JC, Allen T, editors. *Handbook of power technology.* Vol. 1. Elsevier Scientific Publishing Company; 1980.

Kossen NWF, Heertjes AM. The determination of contact angle for systems with a powder. *Chem Eng Sci* 1965;**20**:593–9.

Kruis FE, Maisel SA, Fissan H. Direct simulation Monte Carlo method for particle coagulation and aggregation. *AIChE J* 2000;**46**:1735–42.

Lawn BR, Wilshaw TR. *Fracture of brittle solids.* Cambridge: Cambridge University Press; 1975.

Lowrison GC. *Crushing and grinding.* London: Butterworths; 1974.

Marshall VC, editor. *Comminution.* London: Institution of Chemical Engineers; 1974.

Newitt DM, Conway-Jones JM. A contribution to the theory and practice of granulation. *Trans IChemE* 1958;**36**:422–42.

Sherrington PJ. The granulation of sand as an aid to understanding fertilizer granulation. *Chem Eng* 1968;**220**: CE201–215.

Train D, Lewis CJ. Agglomeration of solids by compaction. *Trans IChemE* 1962;**40**:235–40.

Train D. Transmission of forces through a powder mass during the process of pelleting. *Trans IChemE* 1957;**35**:258–66.

Ramkrishna D. *Population balances: theory and applications to particulate systems in engineering.* Academic Press; 2000.

Randolph AD, Larson MA. *Theory of particulate processes: analysis and techniques of continuous crystallization.* Academic Press; 1988.

Vogel L, Peukert W. Breakage behaviour of different materials—construction of a mastercurve for the breakage probability. *Powder Technol* 2003;**129**:101–10.

Shaik AH, Chakraborty J. A simple room temperature fast reduction technique for preparation of a copper nanosheet powder. *RSC Adv* 2016;**6**:14952–7.

Gao M, Forssberg E. Prediction of product size distributions for a stirred ball mill. *Powder Technol* 1995;**84**:101–6.

CHAPTER 6

Motion of Particles in a Fluid

6.1 Introduction

Processes for the separation of particles of various sizes and shapes often depend on the variation in the behaviour of the particles when they are subjected to the action of a moving fluid. Further, many of the methods for the determination of the sizes of particles in the subsieve ranges involve relative motion between the particles and a fluid.

The flow problems considered in Volume 1A are unidirectional, with the fluid flowing along a pipe or channel, and the effect of an obstruction is discussed only in so far as it causes an alteration in the forward velocity of the fluid. In this chapter, the force exerted on a body as a result of the flow of fluid past it is considered, and, as the fluid is generally diverted all round it, the resulting three-dimensional flow is more complex. Conversely, it is frequently needed to estimate the force required to move an object (or a particle) in a fluid medium at a given velocity. The flow of fluid relative to an infinitely long cylinder, a spherical particle, and a nonspherical particle is considered, followed by a discussion of the motion of particles in both gravitational and centrifugal fields.

6.2 Flow Past a Cylinder and a Sphere

The crossflow of fluid past an infinitely long cylinder, in a direction perpendicular to its axis, is considered in the first instance because this involves only two-directional flow, with no flow parallel to the axis. For a nonviscous fluid flowing past a cylinder, as shown in Fig. 6.1, the velocity and direction of flow varies around the circumference. Thus, at points A and D, the fluid is brought to rest, and at B and C, the velocity is at a maximum. Because the fluid is nonviscous, there is no drag, and an infinite velocity gradient exists at the surface of the cylinder. If the fluid is incompressible, and the cylinder is small (so that the changes in the potential energy are negligible), the sum of the kinetic energy and the pressure energy is constant at all points on the surface. The kinetic energy is a maximum at B and C and zero at A and D, so that the pressure falls from A to B and from A to C and rises again from B to D and from C to D, the pressure at A and D being the same. No net force is, therefore, exerted by the fluid on the cylinder. It is found that, although the predicted pressure variation for a nonviscous fluid agrees well with the results obtained with a viscous fluid over the front face of the cylinder, very considerable differences occur at the rear face.

Coulson and Richardson's Chemical Engineering. https://doi.org/10.1016/B978-0-08-101098-3.00007-X

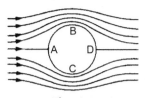

Fig. 6.1
Flow round a cylinder.

It is shown in Volume 1B, Chapter 3 that, when a viscous fluid flows over a surface, the fluid is retarded in the boundary layer which is formed near the surface, and that the boundary layer increases in thickness with increase in distance from the leading edge. If the pressure is falling in the direction of flow, the retardation of the fluid is less, and the boundary layer is thinner in consequence. If the pressure is rising, however, there will be a greater retardation, and the thickness of the boundary layer increases more rapidly. The force acting on the fluid at some point in the boundary layer may then be sufficient to bring it to rest or to cause flow in the reverse direction with the result that an eddy current is set up. A region of reverse flow then exists near the surface where the boundary layer has separated, as shown in Fig. 6.2. The velocity rises from zero at the surface to a maximum negative value and falls again to zero. It then increases in the positive direction until it reaches the main stream velocity at the edge of the boundary layer, as shown in Fig. 6.2. At PQ the velocity in the X-direction is zero and the direction of flow in the eddies must be in the Y-direction.

For the flow of a viscous fluid past the cylinder, the pressure decreases from A to B and from A to C so that the boundary layer is thin and the flow is similar to that obtained with a nonviscous fluid. From B to D and from C to D, the pressure is rising and therefore the boundary layer rapidly thickens with the result that it tends to separate from the surface. If separation occurs, eddies are formed in the wake of the cylinder and energy is thereby dissipated and an additional force, known

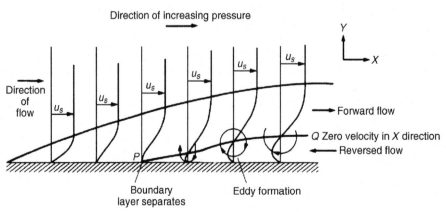

Fig. 6.2
Flow of fluid over a surface against a pressure gradient.

as form drag, is set up. In this way, on the forward surface of the cylinder, the pressure distribution is similar to that obtained with the ideal fluid of zero viscosity, although on the rear surface, the boundary layer is thickening rapidly and pressure variations are very different in the two cases.

All bodies immersed in a fluid are subject to a buoyancy force. In a flowing fluid, there is an additional force which is made up of two components: the skin friction (or viscous drag) and the form drag (due to the pressure distribution). At low rates of flow, no separation of the boundary layer takes place, although as the velocity is increased, separation occurs, and the skin friction forms a gradually decreasing proportion of the total drag. If the velocity of the fluid is very high, however, or if turbulence is artificially induced, the flow within the boundary layer will change from streamline to turbulent before separation takes place. Because the rate of transfer of momentum through a fluid in turbulent motion is much greater than that in a fluid flowing under streamline conditions, separation is less likely to occur, because the fast-moving fluid outside the boundary layer is able to keep the fluid within the boundary layer moving in the forward direction. If separation does occur, this takes place nearer to point D in Fig. 6.1, the resulting eddies are smaller, and the total drag will be reduced.

Turbulence may arise either from an increased fluid velocity or from artificial roughening of the forward face of the immersed body. Prandtl roughened the forward face of a sphere by fixing a hoop to it, with the result that the drag was considerably reduced. Further experiments have been carried out in which sand particles have been stuck to the front face, as shown in Fig. 6.3. The tendency for separation, and hence, the magnitude of the form drag, are also dependent on the shape of the body.

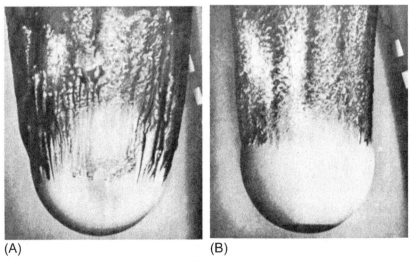

(A) (B)

Fig. 6.3

Effect of roughening front face of a sphere (A) 216 mm diameter ball entering water at 298 K (B) as above, except for 100 mm diameter patch of sand on nose.

Conditions of flow relative to a spherical particle are similar to those relative to a cylinder, except that the flow pattern is three-directional. The flow is characterised by the Reynolds number $Re'(=ud\rho/\mu)$, in which ρ is the density of the fluid, μ is the viscosity of the fluid, d is the diameter of the sphere, and u is the velocity of the fluid relative to the particle.

For the case of *creeping flow*, that is, flow at very low velocities relative to the sphere, the drag force F on the particle was obtained in 1851 by Stokes[1] who solved the hydrodynamic equations of motion, the Navier–Stokes equations, to give:

$$F = 3\pi\mu du \tag{6.1}$$

Eq. (6.1), which is known as Stokes' law, is applicable only at very low values of the particle Reynolds number, and deviations become progressively greater as Re' increases. Skin friction constitutes two-thirds of the total drag on the particle, as given by Eq. (6.1). Thus, the total force F is made up of two components:

$$\left.\begin{array}{l} \text{(i) skin \quad friction}: 2\pi\,\mu du \\ \text{(ii) form \quad drag}: \pi\,\mu du \end{array}\right\} \text{ total } 3\pi\mu du$$

As Re' increases, skin friction becomes proportionately less, and, at values greater than about 20, *flow separation* occurs with the formation of vortices in the wake of the sphere. At high Reynolds numbers, the size of the vortices progressively increases until, at values of between 100 and 200, instabilities in the flow give rise to *vortex shedding*. The effect of these changes in the nature of the flow on the force exerted on the particle is now considered.

6.3 The Drag Force on a Spherical Particle

6.3.1 Drag Coefficients

The most satisfactory way of representing the relation between drag force and velocity involves the use of two dimensionless groups, similar to those used for correlating information on the pressure drop for flow of fluids in pipes.

The first group is the particle Reynolds number $Re'(=ud\rho/\mu)$.

The second is the group $R'/\rho u^2$, in which R' is the force per unit projected area of particle in a plane perpendicular to the direction of motion. For a sphere, the projected area is that of a circle of the same diameter as the sphere.

$$\text{Thus}: \quad R' = \frac{F}{(\pi d^2/4)} \tag{6.2}$$

$$\text{and} \quad \frac{R'}{\rho u^2} = \frac{4F}{\pi d^2 \rho u^2} \tag{6.3}$$

$R'/\rho u^2$ is a form of *drag coefficient*, often denoted by the symbol C_D'. Frequently, a drag coefficient C_D is defined as the ratio of R' to $\frac{1}{2}\rho u^2$.

$$\text{Thus}: \quad C_D = 2C_D' = \frac{2R'}{\rho u^2} \tag{6.4}$$

It is seen that C_D' is analogous to the friction factor $\phi(=R/\rho u^2)$ for pipe flow, and C_D is analogous to the Fanning friction factor f.

When the force F is given by Stokes' law (Eq. 6.1), then:

$$\frac{R'}{\rho u^2} = 12\frac{\mu}{u d \rho} = 12 Re'^{-1} \tag{6.5}$$

Eqs (6.1) and (6.5) are applicable only at very low values of the Reynolds number Re'. Goldstein[2] has shown that, for values of Re' up to about 2, the relation between $R'/\rho u^2$ and Re' is given by an infinite series of which Eq. (6.5) is just the first term.

Thus:

$$\frac{R'}{\rho u^2} = \frac{12}{Re'}\left\{1 + \frac{3}{16}Re' - \frac{19}{1280}Re'^2 + \frac{71}{20,480}Re'^3 - \frac{30,179}{34,406,400}Re'^4 + \frac{122,519}{560,742,400}Re'^5 - \cdots\right\} \tag{6.6}$$

Oseen[3] employs just the first two terms of Eq. (6.6) to give:

$$\frac{R'}{\rho u^2} = 12 Re'^{-1}\left(1 + \frac{3}{16}Re'\right). \tag{6.7}$$

The correction factors for Stokes' law from both Eq. (6.6) and Eq. (6.7) are given in Table 6.1. It is seen that the correction becomes progressively greater as Re' increases.

Several researchers have used numerical methods for solving the equations of motion for flow at higher Reynolds numbers relative to spherical and cylindrical particles. These include, Jenson,[4] le Clair et al.,[5] Fornberg,[6,7] and Johnson and Patel.[8]

The relation between $R'/\rho u^2$ and Re' is conveniently given in graphical form by means of a logarithmic plot as shown in Fig. 6.4. The graph may be divided into four regions as shown. The four regions are now considered in turn.

Region (a) ($10^{-4} < Re' < 0.2$)

In this region, the relationship between $\dfrac{R'}{\rho u^2}$ and Re' is a straight line of slope -1 represented by Eq. (6.5):

$$\frac{R'}{\rho u^2} = 12 Re'^{-1} \quad (\text{Eq.6.5})$$

Table 6.1 Correction factors for Stokes' law

Re'	Goldstein (Eq. 6.6)	Oseen (Eq. 6.7)	Schiller and Naumann (Eq. 6.9)	Wadell (Eq. 6.12)	Khan and Richardson (Eq. 6.13)
0.01	1.002	1.002	1.007	0.983	1.038
0.03	1.006	1.006	1.013	1.00	1.009
0.1	1.019	1.019	1.03	1.042	1.006
0.2	1.037	1.037	1.05	1.067	1.021
0.3	1.055	1.056	1.07	1.115	1.038
0.6	1.108	1.113	1.11	1.346	1.085
1	1.18	1.19	1.15	1.675	1.137
2	1.40	1.38	1.24	1.917	1.240

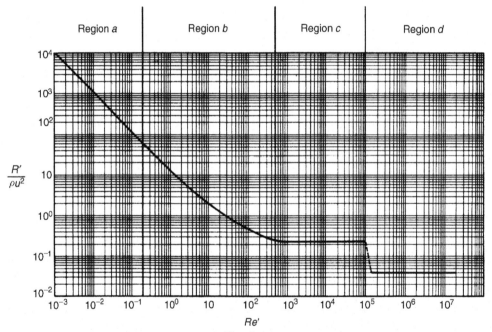

Fig. 6.4

$R'/\rho u^2$ versus Re' for spherical particles.

The limit of 10^{-4} is imposed because reliable experimental measurements have not been made at lower values of Re', although the equation could be applicable down to very low values of Re', provided that the dimensions of the particle are large compared with the mean free path of the fluid molecules so that the fluid behaves as a continuum.

The upper limit of $Re'=0.2$ corresponds to the condition where the error arising from the application of Stokes' law is about 4%. This limit should be reduced if a greater accuracy is required, and it may be raised if a lower level of accuracy is acceptable.

Region (b) (0.2 < Re' < 500–1000)

In this region, the slope of the curve changes progressively from -1 to 0 as Re' increases. Several researchers have suggested approximate equations for flow in this intermediate region. Dallavelle[9] proposed that $R'/\rho u^2$ may be regarded as being composed of two component parts, one due to Stokes' law and the other, a constant, due to additional nonviscous effects.

$$\text{Thus}: \quad \frac{R'}{\rho u^2} = 12Re'^{-1} + 0.22 \tag{6.8}$$

Schiller and Naumann[10] gave the following simple equation which gives a reasonable approximation for values of Re' up to about 1000:

$$\frac{R'}{\rho u^2} = 12Re'^{-1}\left(1 + 0.15Re'^{0.687}\right) \tag{6.9}$$

Region (c) (500–1000 < Re' < c.2 × 10⁵)

In this region, *Newton's law* is applicable, and the value of $R'/\rho u^2$ is approximately constant giving:

$$\frac{R'}{\rho u^2} = 0.22 \tag{6.10}$$

Region (d) (Re' > ~2 × 10⁵)

When Re' exceeds about 2×10^5, the flow in the boundary layer changes from streamline to turbulent and the separation takes place nearer to the rear of the sphere. The drag force is decreased considerably and:

$$\frac{R'}{\rho u^2} = 0.05 \tag{6.11}$$

A recent study[11] sheds some light on the underlying physics of the drag behaviour in this regime.

Values of $R'/\rho u^2$ using Eqs (6.5), (6.9), (6.10), and (6.11) are given in Table 6.2 and plotted in Fig. 6.4. The curve shown in Fig. 6.4 is really continuous, and its division into four regions is merely a convenient means by which a series of simple equations can be assigned to limited ranges of values of Re'.

Table 6.2 $R'/\rho u^2$, $(R'/\rho u^2)Re'^2$ and $(R'/\rho u^2)Re'^{-1}$ as a function of Re'

Re'	$R'/\rho u^2$	$(R'/\rho u^2)Re'^2$	$(R'/\rho u^2)Re'^{-1}$
10^{-3}	12,000		
2×10^{-3}	6000		
5×10^{-3}	2400		
10^{-2}	1200	1.20×10^{-1}	1.20×10^5
2×10^{-2}	600	2.40×10^{-1}	3.00×10^4
5×10^{-2}	240	6.00×10^{-1}	4.80×10^3
10^{-1}	124	1.24	1.24×10^3
2×10^{-1}	63	2.52	3.15×10^2
5×10^{-1}	26.3	6.4	5.26×10
10^0	13.8	1.38×10	1.38×10
2×10^0	7.45	2.98×10	3.73
5×10^0	3.49	8.73×10	7.00×10^{-1}
10	2.08	2.08×10^2	2.08×10^{-1}
2×10	1.30	5.20×10^2	6.50×10^{-2}
5×10	0.768	1.92×10^3	1.54×10^{-2}
10^2	0.547	5.47×10^3	5.47×10^{-3}
2×10^2	0.404	1.62×10^4	2.02×10^{-3}
5×10^2	0.283	7.08×10^4	5.70×10^{-4}
10^3	0.221	2.21×10^5	2.21×10^{-4}
2×10^3	0.22	8.8×10^5	1.1×10^{-4}
5×10^3	0.22	5.5×10^6	4.4×10^{-5}
10^4	0.22	2.2×10^7	2.2×10^{-5}
2×10^4	0.22		
5×10^4	0.22		
10^5	0.22		
2×10^5	0.05		
5×10^5	0.05		
10^6	0.05		
2×10^6	0.05		
5×10^6	0.05		
10^7	0.05		

A comprehensive review of the various equations proposed to relate drag coefficient to particle Reynolds number has been carried out by Clift et al.[12] and Michaelides.[13] One of the earliest equations applicable over a wide range of values of Re' is that due to Wadell,[14] which may be written as:

$$\frac{R'}{\rho u^2} = \left(0.445 + \frac{3.39}{\sqrt{Re'}}\right)^2 \tag{6.12}$$

Subsequently, Khan and Richardson[15] have examined the experimental data and suggest that a very good correlation between $R'/\rho u^2$ and Re', for values of Re' up to 10^5, is given by:

$$\frac{R'}{\rho u^2} = \left[1.84Re'^{-0.31} + 0.293Re'^{0.06}\right]^{3.45} \tag{6.13}$$

In Table 6.3, values of $R'/\rho u^2$, calculated from Eqs (6.12) and (6.13), together with values from the Schiller and Naumann Eq. (6.9), are given as a function of Re' over the range $10^{-2} < Re' < 10^5$. Values are plotted in Fig. 6.5, from which it will be noted that Eq. (6.13) gives a shallow minimum at Re' of about 10^4, with values rising to 0.21 at $Re' = 10^5$. This agrees with the limited experimental data which are available in this range.

Table 6.3 Values of drag coefficient $R'/\rho u^2$ as a function of Re'

Re'	Schiller and Naumann[10] (Eq. 6.9)	Wadell[14] (Eq. 6.12)	Khan and Richardson[15] (Eq. 6.13)
0.01	1208	1179	1246
0.1	124	125	121
1	13.8	14.7	13.7
10	2.07	2.3	2.09
100	0.55	0.62	0.52
500	0.281	0.356	0.283
1000	0.219	0.305	0.234
3000	0.151	0.257	0.200
10,000	0.10	0.229	0.187
30,000	–	0.216	0.191
100,000	–	0.208	0.210

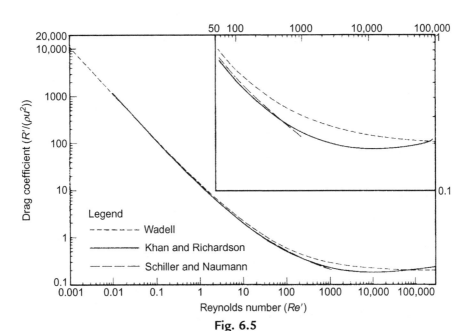

Fig. 6.5

$R'/\rho u^2$ versus Re' for spherical particles from Eqs (6.9) (Schiller and Naumann[10]), (6.12) (Wadell[14]) and (6.13) (Khan and Richardson[15]). The enlarged section covers the range of Re' from 50 to 10^5.

For values of $Re' < 2$, correction factors for Stokes' law have been calculated from Eqs (6.9), (6.12), and (6.13), and these are included in Table 6.1.

6.3.2 Total Force on a Particle

The force on a spherical particle may be expressed using Eqs (6.5), (6.9), (6.10), and (6.11) for each of the regions *a*, *b*, *c*, and *d* as follows.

$$\text{In region}(a): \quad R' = 12\rho u^2 \left(\frac{\mu}{ud\rho}\right) = \frac{12u\mu}{d} \tag{6.14}$$

The projected area of the particle is $\pi d^2/4$. Thus, the total force on the particle is given by:

$$F = \frac{12u\mu}{d}\frac{1}{4}\pi d^2 = 3\pi\mu du \tag{6.15}$$

This is the expression originally obtained by Stokes[1] already given as Eq. (6.1).

In region (*b*), from Eq. (6.9):

$$R' = \frac{12u\mu}{d}\left(1 + 0.15Re'^{0.687}\right) \tag{6.16}$$

$$\text{and therefore}: \quad F = 3\pi\mu du\left(1 + 0.15Re'^{0.687}\right) \tag{6.17}$$

$$\text{In region}(c): \quad R' = 0.22\rho u^2 \tag{6.18}$$

$$\text{and}: \quad F = 0.22\rho u^2 \frac{1}{4}\pi d^2 = 0.055\pi d^2\rho u^2 \tag{6.19}$$

This relation is often known as Newton's law.

In region (*d*):

$$R' = 0.05\rho u^2 \tag{6.20}$$

$$F = 0.0125\pi d^2\rho u^2 \tag{6.21}$$

Alternatively, using Eq. (6.13), which is applicable over the first three regions (*a*), (*b*), and (*c*) gives:

$$F = \frac{\pi}{4}d^2\rho u^2 \left(1.84Re'^{-0.31} + 0.293Re'^{0.06}\right)^{3.45} \tag{6.22}$$

6.3.3 Terminal Falling Velocities

If a spherical particle is allowed to settle in a fluid under gravity, its velocity will increase until the accelerating force is exactly balanced by the resistance force. Although this state is approached exponentially, the effective acceleration period is generally of short duration for very small particles. If this terminal falling velocity is such that the corresponding value of Re' is <0.2, the drag force on the particle is given by Eq. (6.15). If the corresponding value of Re' lies between 0.2 and 500, the drag force is given approximately by Schiller and Naumann in Eq. (6.17). It may be noted, however, that if the particle has started from rest, the drag force is given by Eq. (6.15) until Re' exceeds 0.2. Again, if the terminal falling velocity corresponds to a value of Re' greater than about 500, the drag on the particle is given by Eq. (6.19). Under terminal falling conditions, velocities are rarely high enough for Re' to approach 10^5, with the small particles generally used in industry.

The accelerating force due to gravity is given by:

$$= \left(\frac{1}{6}\pi d^3\right)(\rho_s - \rho)g \tag{6.23}$$

where ρ_s is the density of the solid.

The terminal falling velocity u_0 corresponding to region (a) is given by:

$$\left(\frac{1}{6}\pi d^3\right)(\rho_s - \rho)g = 3\pi\mu d u_0$$

$$\text{and}: \quad u_0 = \frac{d^2 g}{18\mu}(\rho_s - \rho) \tag{6.24}$$

The terminal falling velocity corresponding to region (c) is given by:

$$\left(\frac{1}{6}\pi d^3\right)(\rho_s - \rho)g = 0.055\pi d^2 \rho u_0^2$$

$$\text{or}: \quad u_0^2 = 3dg\frac{(\rho_s - \rho)}{\rho} \tag{6.25}$$

In the expressions given for the drag force and the terminal falling velocity, the following assumptions have been made:

(a)   That the settling is not affected by the presence of other particles in the fluid. This condition is known as 'free settling'. When the interference of other particles is appreciable, the process is known as 'hindered settling'.

(b)   That the walls of the containing vessel do not exert an appreciable retarding effect.

(c)   That the fluid can be considered as a continuous medium, that is the particle is large compared with the mean free path of the molecules of the fluid, otherwise the particles may occasionally 'slip' between the molecules and thus attain a velocity higher than that calculated using the aforementioned expressions.

These factors are considered further in Sections 6.3.4 and 6.3.5 and in Chapter 8.

From Eqs (6.24) and (6.25), it is seen that terminal falling velocity of a particle in a given fluid becomes greater as both particle size and density are increased. If for a particle of material **A** of diameter d_A and density ρ_A, Stokes' law is applicable, then the terminal falling velocity u_{0A} is given by Eq. (6.24) as:

$$u_{0A} = \frac{d_A^2 g}{18\mu}(\rho_A - \rho) \tag{6.26}$$

Similarly, for a particle of material **B**:

$$u_{0B} = \frac{d_B^2 g}{18\mu}(\rho_B - \rho) \tag{6.27}$$

The condition for the two terminal velocities to be equal is then:

$$\frac{d_B}{d_A} = \left(\frac{\rho_A - \rho}{\rho_B - \rho}\right)^{1/2} \tag{6.28}$$

If Newton's law is applicable, Eq. (6.25) holds, and:

$$u_{0A}^2 = \frac{3d_A g(\rho_A - \rho)}{\rho} \tag{6.29}$$

$$\text{and} \quad u_{0B}^2 = \frac{3d_B g(\rho_B - \rho)}{\rho} \tag{6.30}$$

For equal settling velocities:

$$\frac{d_B}{d_A} = \left(\frac{\rho_A - \rho}{\rho_B - \rho}\right) \tag{6.31}$$

In general, the relationship for equal settling velocities is:

$$\frac{d_B}{d_A} = \left(\frac{\rho_A - \rho}{\rho_B - \rho}\right)^S \tag{6.32}$$

where $S = \frac{1}{2}$ for the Stokes' law region, $S = 1$ for Newton's law, and, as an approximation, $\frac{1}{2} < S < 1$ for the intermediate region.

This method of calculating the terminal falling velocity is satisfactory provided that it is known a priori which equation should be used for the calculation of drag force or drag coefficient. It has already been seen that the equations give the drag coefficient in terms of the particle Reynolds number $Re_0'(=u_0 d\rho/\mu)$, which is itself a function of the terminal falling velocity u_0 which is to be determined. The problem is analogous to that discussed in Volume 1A, where the calculation of the velocity of flow in a pipe in terms of a known pressure difference presents difficulties, because the unknown velocity appears in both the friction factor and the Reynolds number.

The problem is most effectively solved by introducing a new dimensionless group which is independent of the particle velocity. The resistance force per unit projected area of the particle under terminal falling conditions R_0' is given by:

$$R_0' \frac{1}{4}\pi d^2 = \frac{1}{6}\pi d^3 (\rho_s - \rho)g$$

$$\text{or}: \quad R_0' = \frac{2}{3}d(\rho_s - \rho)g \tag{6.33}$$

$$\text{Thus}: \quad \frac{R_0'}{\rho u_0^2} = \frac{2dg}{3\rho u_0^2}(\rho_s - \rho) \tag{6.34}$$

The dimensionless group $(R_0'/\rho u_0^2)Re_0'^2$ does not involve u_0 because:

$$\frac{R_0'}{\rho u_0^2}\frac{u_0^2 d^2 \rho^2}{\mu^2} = \frac{2dg(\rho_s - \rho)}{3\rho u_0^2}\frac{u_0^2 d^2 \rho^2}{\mu^2}$$

$$= \frac{2d^3(\rho_s - \rho)\rho g}{3\mu^2} \tag{6.35}$$

The group $\dfrac{d^3\rho(\rho_s - \rho)g}{\mu^2}$ is known as the Galileo number Ga or sometimes the Archimedes number Ar.

$$\text{Thus}: \quad \frac{R_0'}{\rho u_0^2}Re_0'^2 = \frac{2}{3}Ga \tag{6.36}$$

Using Eqs (6.5), (6.9), and (6.10) to express $R'/\rho u^2$ in terms of Re' over the appropriate range of Re', then:

$$Ga = 18Re_0' \quad (Ga < 3.6) \tag{6.37}$$

$$Ga = 18Re_0' + 2.7Re_0'^{1.687} \left(3.6 < Ga <\sim 10^5\right) \tag{6.38}$$

$$Ga = \frac{1}{3}Re_0'^2 \quad (Ga >\sim 10^5) \tag{6.39}$$

$(R_0'/\rho u_0^2)Re_0'^2$ can be evaluated if the properties of the fluid and the particle are known.

Table 6.4 Values of log Re' as a function of $\log\{(R'/\rho u^2)Re'^2\}$ for spherical particles

$\log\{(R'/\rho u^2)Re'^2\}$	0.0	0.1	0.2	0.3	0.4	0.5	0.6	0.7	0.8	0.9
$\bar{2}$								$\bar{3}.620$	$\bar{3}.720$	$\bar{3}.819$
$\bar{1}$	$\bar{3}.919$	$\bar{2}.018$	$\bar{2}.117$	$\bar{2}.216$	$\bar{2}.315$	$\bar{2}.414$	$\bar{2}.513$	$\bar{2}.612$	$\bar{2}.711$	$\bar{2}.810$
0	$\bar{2}.908$	$\bar{1}.007$	$\bar{1}.105$	$\bar{1}.203$	$\bar{1}.301$	$\bar{1}.398$	$\bar{1}.495$	$\bar{1}.591$	$\bar{1}.686$	$\bar{1}.781$
1	$\bar{1}.874$	$\bar{1}.967$	0.008	0.148	0.236	0.324	0.410	0.495	0.577	0.659
2	0.738	0.817	0.895	0.972	1.048	1.124	1.199	1.273	1.346	1.419
3	1.491	1.562	1.632	1.702	1.771	1.839	1.907	1.974	2.040	2.106
4	2.171	2.236	2.300	2.363	2.425	2.487	2.548	2.608	2.667	2.725
5	2.783	2.841	2.899	2.956	3.013	3.070	3.127	3.183	3.239	3.295

In Table 6.4, values of log Re' are given as a function of $\log\{(R'/\rho u^2)Re'^2\}$ and the data taken from tables given by Heywood,[16] are represented in graphical form in Fig. 6.6. In order to determine the terminal falling velocity of a particle, $(R_0'/\rho u_0^2)Re_0'^2$ is evaluated, and the corresponding value of Re_0', and hence, of the terminal velocity, is found either from Table 6.4 or from Fig. 6.6.

Example 6.1

What is the terminal velocity of a spherical steel particle, 0.40 mm in diameter, settling in an oil of density 820 kg/m³ and viscosity 10 mN s/m²? The density of steel is 7870 kg/m³.

Solution

For a sphere:

$$\frac{R_0'}{\rho u_0^2}Re_0'^2 = \frac{2d^3(\rho_s - \rho)\rho g}{3\mu^2} \quad \text{(Eq.6.35)}$$

$$= \frac{2 \times 0.0004^3 \times 820(7870 - 820)9.81}{3(10 \times 10^{-3})^2}$$

$$= 24.2$$

$$\log_{10} 24.2 = 1.384$$

$$\text{From Table 6.4}: \quad \log_{10} Re_0' = 0.222$$

$$\text{Thus}: \quad Re_0' = 1.667$$

$$\text{and}: \quad u_0 = \frac{1.667 \times 10 \times 10^{-3}}{820 \times 0.004}$$

$$= 0.051 \,\text{m/s or } 51 \,\text{mm/s}$$

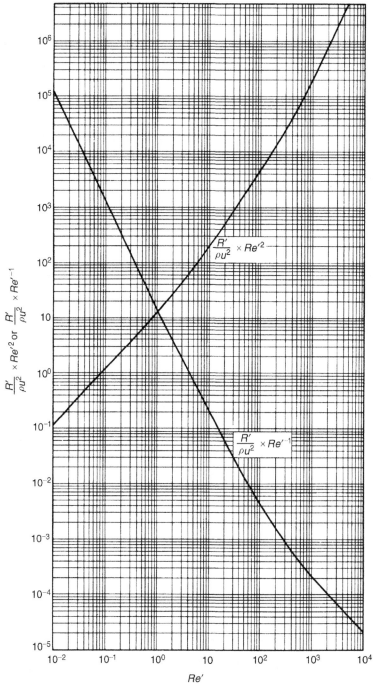

Fig. 6.6
$(R'/\rho u^2)Re'^2$ and $(R'/\rho u^2)Re'^{-1}$ versus Re' for spherical particles.

Example 6.2

A finely ground mixture of galena and limestone in the proportion of 1–4 by mass is subjected to elutriation by an upward-flowing stream of water flowing at a velocity of 5 mm/s. Assuming that the size distribution for each material is the same, and is as shown in the following table, estimate the percentage of galena in the material carried away and in the material left behind. The viscosity of water is 1 mN s/m² and Stokes' Eq. (6.1) may be used.

Diameter (μm)	20	30	40	50	60	70	80	100
Undersize (percent by mass)	15	28	48	54	64	72	78	88

The densities of galena and limestone are 7500 and 2700 kg/m³, respectively.

Solution

The first step is to determine the size of a particle which has a settling velocity equal to that of the upward flow of fluid, that is 5 mm/s.

Taking the largest particle, $d = (100 \times 10^{-6}) = 0.0001$ m

$$\text{and: } Re' = \left(5 \times 10^{-3} \times 0.0001 \times 1000\right)/\left(1 \times 10^{-3}\right) = 0.5$$

Thus, for the bulk of particles, the flow will be within region (a) in Fig. 6.4, and the settling velocity is given by Stokes' equation:

$$u_0 = \left(d^2 g/18\mu\right)(\rho_s - \rho) \quad \text{(Eq.6.24)}$$

For a particle of galena settling in water at 5 mm/s:

$$\left(5 \times 10^{-3}\right) = \left((d^2 \times 9.81)/(18 \times 10^{-3})\right)(7500 - 1000) = 3.54 \times 10^6 d^2$$
$$\text{and: } d = 3.76 \times 10^{-5} \text{ m or } 37.6 \text{ μm}$$

For a particle of limestone settling at 5 mm/s:

$$\left(5 \times 10^{-3}\right) = \left((d^2 \times 9.81)/(18 \times 10^{-3})\right)(2700 - 1000) = 9.27 \times 10^5 d^2$$
$$\text{and: } d = 7.35 \times 10^{-5} \text{ m or } 73.5 \text{ μm}$$

Thus, particles of galena of <37.6 μm and particles of limestone of <73.5 μm will be removed in the water stream.

Interpolation of the data given shows that 43% of the galena and 74% of the limestone will be removed in this way.

In 100 kg feed, there is 20 kg galena and 80 kg limestone.

Therefore, galena removed = (20 × 0.43) = 8.6 kg, leaving 11.4 kg, and limestone removed = (80 × 0.74) = 59.2 kg, leaving 20.8 kg.

Hence, in the *material removed*:

Concentration of galena = (8.6 × 100)/(8.6 + 59.2) = 12.7% by mass

and in the *material remaining*:

Concentration of galena = (11.4 × 100)/(11.4 + 20.8) = 35.4% by mass

As an alternative, the data used for the generation of Eq. (6.13) for the relation between drag coefficient and particle Reynolds number may be expressed as an explicit relation between Re_0' (the value of Re' at the terminal falling condition of the particle) and the Galileo number Ga. The equation takes the form[15]:

$$Re_0' = \left(2.33Ga^{0.018} - 1.53Ga^{-0.016}\right)^{13.3} \tag{6.40}$$

The Galileo number is readily calculated from the properties of the particle and the fluid, and the corresponding value of Re_0', from which u_0 can be found, is evaluated from Eq. (6.40).

A similar difficulty is encountered in calculating the size of a sphere having a given terminal falling velocity, because Re_0' and $R_0'/\rho u^2$ are both functions of the diameter d of the particle. This calculation is similarly facilitated by the use of another combination, $(R_0'/\rho u_0^2)Re'_0{}^{-1}$, which is independent of diameter. This is given by:

$$\frac{R_0'}{\rho u_0^2}Re'_0{}^{-1} = \frac{2\mu g}{3\rho^2 u_0^3}(\rho_s - \rho) \tag{6.41}$$

Log Re' is given as a function of $\log[(R'/\rho u^2)Re'^{-1}]$ in Table 6.5 and the functions are plotted in Fig. 6.6. The diameter of a sphere of known terminal falling velocity may be calculated by evaluating $(R_0'/\rho u_0^2)Re'_0{}^{-1}$, and then finding the corresponding value of Re_0', from which the diameter may be calculated.

As an alternative to this procedure, the data used for the generation of Eq. (6.13) may be expressed to give Re_0' as an explicit function of $\{(R_0'/\rho u_0^2)Re'_0{}^{-1}\}$, which from Eq. (6.40) is equal to $2/3[(\mu g/\rho^2 u_0^3)(\rho_s - \rho)]$. Then, writing $K_D = (\mu g/\rho^2 u_0^3)(\rho_s - \rho)]$, Re_0' may be obtained from:

$$Re_0' = \left(1.47K_D^{-0.14} + 0.11K_D^{0.4}\right)^{3.56} \tag{6.42}$$

d may then be evaluated because it is the only unknown quantity involved in the Reynolds number.

Table 6.5 Values of log Re' as a function of log{(R'/ρu²)Re'⁻¹} for spherical particles

$\log\{(R'/\rho u^2)Re'^{-1}\}$	0.0	0.1	0.2	0.3	0.4	0.5	0.6	0.7	0.8	0.9
$\overline{5}$										3.401
$\overline{4}$	3.316	3.231	3.148	3.065	2.984	2.903	2.824	2.745	2.668	2.591
$\overline{3}$	2.517	2.443	2.372	2.300	2.231	2.162	2.095	2.027	1.961	1.894
$\overline{2}$	1.829	1.763	1.699	1.634	1.571	1.508	1.496	1.383	1.322	1.260
$\overline{1}$	1.200	1.140	1.081	1.022	0.963	0.904	0.846	0.788	0.730	0.672
0	0.616	0.560	0.505	0.449	0.394	0.339	0.286	0.232	0.178	0.125
1	0.072	0.019	$\overline{1}$.969	$\overline{1}$.919	$\overline{1}$.865	$\overline{1}$.811	$\overline{1}$.760	$\overline{1}$.708	$\overline{1}$.656	$\overline{1}$.605
2	$\overline{1}$.554	$\overline{1}$.503	$\overline{1}$.452	$\overline{1}$.401	$\overline{1}$.350	$\overline{1}$.299	$\overline{1}$.249	$\overline{1}$.198	$\overline{1}$.148	$\overline{1}$.097
3	$\overline{1}$.047	$\overline{2}$.996	$\overline{2}$.946	$\overline{2}$.895	$\overline{2}$.845	$\overline{2}$.794	$\overline{2}$.744	$\overline{2}$.694	$\overline{2}$.644	$\overline{2}$.594
4	$\overline{2}$.544	$\overline{2}$.493	$\overline{2}$.443	$\overline{2}$.393	$\overline{2}$.343	$\overline{2}$.292				

6.3.4 Rising Velocities of Light Particles

Although there appears to be no problem in using the standard relations between drag coefficient and particle Reynolds number for the calculation of terminal falling velocities of particles denser than the liquid, Karamanev et al.[17] have shown experimentally that, for light particles rising in a denser liquid, an *overestimate* of the terminal rising velocity may result. This can occur in the Newton's law region and may be associated with an increase in the drag coefficient C_D' from the customary value of 0.22 for a spherical particle up to a value as high as 0.48. Vortex shedding behind the rising particle may cause it to take a longer spiral path, thus, reducing its vertical component of velocity. A similar effect is not observed with a falling dense particle because its inertia is too high for vortex-shedding to have a significant effect. Further experimental work by Dewsbury et al.[18] with shear-thinning power-law solutions of CMC (carboxymethylcellulose) has shown similar effects.

6.3.5 Effect of Boundaries

The discussion so far relates to the motion of a single spherical particle in an effectively infinite expanse of fluid. If other particles are present in the neighborhood of the sphere, the sedimentation velocity will be decreased, and the effect will become progressively more marked as the concentration is increased. There are three contributory factors. First, as the particles settle, they will displace an equal volume of fluid, and this gives rise to an upward flow of liquid. Second, the buoyancy force is influenced because the suspension has a higher density than the fluid. Finally, the flow pattern of the liquid relative to the particle is altered, and velocity gradients are affected. The settling of concentrated suspensions is discussed in detail in Chapter 8.

The boundaries of the vessel containing the fluid in which the particle is settling will also affect its settling velocity. If the ratio of diameter of the particle (d) to that of the tube (d_t) is significant, the motion of the particle is retarded. Two effects arise. First, as the particle moves downwards, it displaces an equal volume of liquid which must rise through the annular region between the particle and the wall. Second, the velocity profile in the fluid is affected by the presence of the tube boundary. There have been several studies[19–25] of the influence of the walls, most of them in connection with the use of the 'falling sphere' method of determining viscosity, in which the viscosity is calculated from the settling velocity of the sphere. The resulting correction factors have been tabulated by Clift et al.[12] and Chhabra.[25] The effect is difficult to quantify accurately because the particle will not normally follow a precisely uniform vertical path through the fluid. It is, therefore, useful also to take into account work on the sedimentation of suspensions of uniform spherical particles at various concentrations, and to extrapolate the results to zero concentration to obtain the free falling velocity for different values of the ratio d/d_t. The correction factor for the influence of the walls of the tube on the settling velocity of a particle situated at the axis of the tube was calculated by Ladenburg,[19] who has given the equation:

$$\frac{u_{0t}}{u_0} = \left(1 + 2.4\frac{d}{d_t}\right)^{-1} \quad (d/d_t < 0.1) \tag{6.43}$$

where u_{0t} is the settling velocity in the tube, and u_0 is the free falling velocity in an infinite expanse of fluid.

Eq. (6.43) was obtained for the Stokes' law regime. It overestimates the wall effect, however, at higher particle Reynolds number ($Re' > 0.2$).

Similar effects are obtained with noncylindrical vessels, although, in the absence of adequate data, it is best to use the correlations for cylinders, basing the vessel size on its hydraulic mean diameter, which is four times the ratio of the cross-sectional area to the wetted perimeter.

The particles also suffer a retardation as they approach the bottom of the containing vessel because the lower boundary then influences the flow pattern of the fluid relative to the particle. This problem has been studied by Ladenburg,[19] Tanner,[26] and Sutterby.[27] Ladenburg[19] gives the following equation:

$$\frac{u_{0t}}{u_0} = \left(1 + 1.65\frac{d}{L'}\right)^{-1} \tag{6.44}$$

where L' is the distance between the centre of the particle and the lower boundary, for the Stokes' law regime.

6.3.6 Behaviour of Very Fine Particles

Very fine particles, particularly in the submicron range ($d < 1\,\mu m$), are very readily affected by natural convection currents in the fluid, and great care must be taken in making measurements to ensure that temperature gradients are eliminated.

The behaviour is also affected by Brownian motion. The molecules of the fluid bombard each particle in a random manner. If the particle is large, the net resultant force acting at any instant may be large enough to cause a change in its direction of motion. This effect has been studied by Davies,[28] who has developed an expression for the combined effects of gravitation and Brownian motion on particles suspended in a fluid.

In the preceding treatment, it has been assumed that the fluid constitutes a continuum, and that the size of the particles is small compared with the mean free path λ of the molecules. Particles of diameter $d < 0.1\,\mu m$ in gases at atmospheric pressure (and for larger particles in gases at low pressures) can "slip" between the molecules, and, therefore, attain higher than predicted settling velocities. According to Cunningham,[29] the slip factor is given by:

$$1 + \beta \frac{\lambda}{d} \tag{6.45}$$

Davies[30] gives the following expression for β:

$$\beta = 1.764 + 0.562 e^{-0.785(d/\lambda)} \tag{6.46}$$

More details can be found in a book by Reist.[31]

6.3.7 Effect of Turbulence in the Fluid

If a particle is moving in a fluid which is in laminar flow, the drag coefficient is approximately equal to that in a still fluid, provided that the local relative velocity at the particular location of the particle is used in the calculation of the drag force. When the velocity gradient is sufficiently large to give a significant variation of velocity across the diameter of the particle; however, the estimated force may be somewhat in error.

When the fluid is in turbulent flow, or where turbulence is generated by some external agent such as an agitator, the drag coefficient may be substantially increased. Brucato et al.[32] have shown that the increase in drag coefficient may be expressed in terms of the Kolmogoroff scale of the eddies (λ_E) given by:

$$\lambda_E = \left[(\mu/\rho)^3 / \varepsilon \right]^{1/4} \tag{6.47}$$

where ε is the mechanical power generated per unit mass of fluid by an agitator, for example.

The increase in the drag coefficient C_D over that in the absence of turbulence C_{D0} is given by:

$$\eta = (C_D - C_{D0})/C_{D0} = 8.76 \times 10^{-4} (d/\lambda_E)^3 \tag{6.48}$$

Values of η of up to about 30, have been reported.

6.3.8 Effect of Motion of the Fluid

If the fluid is moving relative to some surface other than that of the particle, there will be a superimposed velocity distribution, and the drag on the particle may be altered. Thus, if the particle is situated at the axis of a vertical tube up which fluid is flowing in streamline motion, the velocity near the particle will be twice the mean velocity because of the parabolic velocity profile in the fluid. The drag force is then determined by the difference in the velocities of the fluid and the particle at the axis.

The effect of turbulence in the fluid stream has been studied by Richardson and Meikle,[33] who suspended a particle on a thread at the centre of a vertical pipe up which water was passed under conditions of turbulent flow. The upper end of the thread was attached to a lever fixed on a coil free to rotate in the field of an electromagnet. By passing a current through the coil, it was possible to bring the level back to a null position. After calibration, the current required could be related to the force acting on the sphere.

The results were expressed as the friction factor $(R'/\rho u^2)$, which was found to have a constant value of 0.40 for particle Reynolds numbers (Re') over the range from 3000 to 9000, and for tube Reynolds numbers (Re) from 12,000 to 26,000. Thus, the value of $R'/\rho u^2$ has been approximately doubled as a result of turbulence in the fluid.

By surrounding the particle with a fixed array of similar particles on a hexagonal spacing, the effect of neighboring particles was measured. The results are discussed in Chapter 8.

Rowe and Henwood[34] made similar studies by supporting a spherical particle 12.7 mm diameter, in water, at the end of a 100 mm length of fine nichrome wire. The force exerted by the water when flowing in a 150 mm square duct was calculated from the measured deflection of the wire. The experiments were carried out at low Reynolds numbers with respect to the duct ($<$1200), corresponding to between 32 and 96 relative to the particle. The experimental values of the drag force were about 10% higher than those calculated from the Schiller and Naumann equation, Eq. (6.9). The work was then extended to cover the measurement of the force on a particle surrounded by an assemblage of particles, as described in Chapter 8.

If Re' is of the order of 10^5, the drag on the sphere may be reduced if the fluid stream is turbulent. The flow in the boundary layer changes from streamline to turbulent and the size of the eddies in the wake of the particle is reduced. The higher the turbulence of the fluid, the lower is the value of Re' at which the transition from region (c) to region (d) occurs. The value of Re' at which $R'/\rho u^2$ is 0.15 is known as the *turbulence number* and is taken as an indication of the degree of turbulence in the fluid.

6.4 Nonspherical Particles

6.4.1 Effect of Particle Shape and Orientation on Drag

There are two difficulties which soon become apparent when attempting to assess the very large amount of experimental data which are available on drag coefficients and terminal falling velocities for nonspherical particles. The first is that an infinite number of nonspherical shapes exists, and the second is that each of these shapes is associated with an infinite number of orientations which the particle is free to take up in the fluid, and the orientation may oscillate during the course of settling.

In a recent comprehensive study, Chhabra et al. [35] have found that the most satisfactory characteristic linear dimension to use is the diameter of the sphere of equal volume, and that the most relevant characteristic of particle shape is the sphericity, (surface area of particle/surface area of sphere of equal volume). The limitation of this whole approach is that mean errors are often as high as about 16%, and maximum errors may be of the order of 100%. The extent of the errors may be reduced, however, by using separate shape factors in the Stokes' and Newton's law regions. Another problem is that, when settling, a nonspherical particle will not travel vertically in a fixed orientation unless it has a plane of symmetry which is horizontal. In general, the resistance force to movement in the gravitational field will not act vertically, and the particle will tend to spiral, to rotate, and to wobble.

A spherical particle is unique in that it presents the same area to the oncoming fluid whatever its orientation. For nonspherical particles, the orientation must be specified before the drag force can be calculated. The experimental data for the drag can be correlated in the same way as for the sphere, by plotting the dimensionless group $R'/\rho u^2$ against the Reynolds number, $Re' = ud'\rho/\mu$, using logarithmic coordinates, and a separate curve is obtained for each shape of particle and for each orientation. In these groups, R' is taken, as before, as the resistance force per unit area of particle, projected on to a plane perpendicular to the direction of flow. d' is defined as the diameter of the circle having the same area as the projected area of the particle and is, therefore, a function of the orientation, as well as the shape, of the particle.

The curve for $R'/\rho u^2$ against Re' may be divided into four regions, (a), (b), (c), and (d), as before. In region (a) the flow is entirely streamline, and, although no theoretical expressions have been developed for the drag on the particle, the practical data suggest that a law of the form:

$$\frac{R'}{\rho u^2} = K Re'^{-1} \tag{6.49}$$

is applicable. The constant K varies somewhat according to the shape and orientation of the particle, although it always has a value of about 12. In this region, a particle falling freely in the fluid under the action of gravity will normally move with its longest surface nearly parallel to the direction of motion.

At higher values of Re', the linear relation between $R'/\rho u^2$ and Re'^{-1} no longer holds and the slope of the curve gradually changes until $R'/\rho u^2$ becomes independent of Re' in region (c). Region (b) represents transition conditions and commences at a lower value of Re', and a correspondingly higher value of $R'/\rho u^2$, than in the case of the sphere. A freely falling particle will tend to change its orientation as the value of Re' changes, and some instability may be apparent. In region (c), the particle tends to fall so that it is presenting the maximum possible surface to the oncoming fluid. Typical values of $R'/\rho u^2$ for nonspherical particles in region (c) are given in Table 6.6.

It may be noted that all these values of $R'/\rho u^2$ are higher than the value of 0.22 for a sphere. Clift et al.[12] and Darby and Chhabra[36] have critically reviewed the information available on the drag of nonspherical particles.

Table 6.6 Drag coefficients for nonspherical particles

Configuration	Length/Breadth	$R'/\rho u^2$
Thin rectangular plates with their planes perpendicular to the direction of motion	1–5	0.6
	20	0.75
	∞	0.95
Cylinders with axes parallel to the direction of motion	1	0.45
Cylinders with axes perpendicular to the direction of motion	1	0.3
	5	0.35
	20	0.45
	∞	0.6

6.4.2 Terminal Falling Velocities

Heywood[16] has developed an approximate method for calculating the terminal falling velocity of a nonspherical particle, or for calculating its size from its terminal falling velocity. The method is an adaptation of his method for spheres.

A mean projected diameter of the particle d_p is defined as the diameter of a circle having the same area as the particle when viewed from above and lying in its most stable position. Heywood selected this particular dimension because it is easily measured by microscopic examination.

If d_p is the mean projected diameter, the mean projected area is $\pi d_p^2/4$ and the volume is $k'd_p^3$, where k' is a constant whose value depends on the shape of the particle. For a spherical particle, k' is equal to $\pi/6$. For rounded isometric particles, that is, particles in which the dimension in three mutually perpendicular directions is approximately the same, k' is about 0.5, and for angular particles, k' is about 0.4. For most minerals, the value of k' lies between 0.2 and 0.5.

The method of calculating the terminal falling velocity consists in evaluating $(R_0'/\rho u^2)Re_0'^2$, using d_p as the characteristic linear dimension of the particle and $\pi d_p^2/4$ as the projected area in a plane perpendicular to the direction of motion. The corresponding value of Re_0' is then found from Table 6.4 or from Fig. 6.6, which both refer to spherical particles, and a correction is then applied to the value of log Re_0' to account for the deviation from spherical shape. Values of this correction factor, which is a function both of k' and of $(R'/\rho u^2)Re'^2$, are given in Table 6.7. A similar procedure is adopted for calculating the size of a particle of given terminal velocity, using Tables 6.5 and 6.8.

The method is only approximate because it is assumed that k' completely defines the shape of the particle, whereas there are many different shapes of particle for which the value of k' is the same. Further, it assumes that the diameter d_p is the same as the mean projected diameter d'. This is very nearly so in regions (b) and (c), although in region (a), the particle tends to settle so

Table 6.7 Corrections to log Re' as a function of $\log\{(R'/\rho u^2)Re'^2\}$ for nonspherical particles

$\log\{(R'/\rho u^2)Re'^2\}$	$k' = 0.4$	$k' = 0.3$	$k' = 0.2$	$k' = 0.1$
$\bar{2}$	−0.022	−0.002	+0.032	+0.131
$\bar{1}$	−0.023	−0.003	+0.030	+0.131
0	−0.025	−0.005	+0.026	+0.129
1	−0.027	−0.010	+0.021	+0.122
2	−0.031	−0.016	+0.012	+0.111
2.5	−0.033	−0.020	0.000	+0.080
3	−0.038	−0.032	−0.022	+0.025
3.5	−0.051	−0.052	−0.056	−0.040
4	−0.068	−0.074	−0.089	−0.098
4.5	−0.083	−0.093	−0.114	−0.146
5	−0.097	−0.110	−0.135	−0.186
5.5	−0.109	−0.125	−0.154	−0.224
6	−0.120	−0.134	−0.172	−0.255

Table 6.8 Corrections to log Re' as a function of $\{\log(R'/\rho u^2)Re'^{-1}\}$ for nonspherical particles

$\log\{(R'/\rho u^2)Re'^{-1}\}$	$k' = 0.4$	$k' = 0.3$	$k' = 0.2$	$k' = 0.1$
$\bar{4}$	+0.185	+0.217	+0.289	
$\bar{4}.5$	+0.149	+0.175	+0.231	
$\bar{3}$	+0.114	+0.133	+0.173	+0.282
$\bar{3}.5$	+0.082	+0.095	+0.119	+0.170
$\bar{2}$	+0.056	+0.061	+0.072	+0.062
$\bar{2}.5$	+0.038	+0.034	+0.033	−0.018
$\bar{1}$	+0.028	+0.018	+0.007	−0.053
$\bar{1}.5$	+0.024	+0.013	−0.003	−0.061
0	+0.022	+0.011	−0.007	−0.062
1	+0.019	+0.009	−0.008	−0.063
2	+0.017	+0.007	−0.010	−0.064
3	+0.015	+0.005	−0.012	−0.065
4	+0.013	+0.003	−0.013	−0.066
5	+0.012	+0.002	−0.014	−0.066

that the longest face is parallel to the direction of motion and some error may, therefore, be introduced in the calculation, as indicated by Heiss and Coull.[37]

For a nonspherical particle settling under the influence of gravity:

$$\text{Total drag force, } F = R'_0 \frac{1}{4}\pi d_p^2 = (\rho_s - \rho)gk'd_p^3 \tag{6.50}$$

$$\text{Thus:} \quad \frac{R'_0}{\rho u_0^2} = \frac{4k'd_p g}{\pi \rho u_0^2}(\rho_s - \rho) \tag{6.51}$$

$$\frac{R'_0}{\rho u_0^2} Re'^2_0 = \frac{4k' \rho d_p^3 g}{\mu^2 \pi}(\rho_s - \rho) \qquad (6.52)$$

$$\text{and}: \quad \frac{R'_0}{\rho u_0^2} Re'^{-1}_0 = \frac{4k' \mu g}{\pi \rho^2 u_0^3}(\rho_s - \rho) \qquad (6.53)$$

Provided k' is known, the appropriate dimensionless group may be evaluated and the terminal falling velocity, or diameter, calculated.

Numerous other methods[38–40] have been proposed in the literature for estimating the terminal falling velocities of nonspherical particles, but none of these has proved to be entirely satisfactory. Depending upon the information available for the particle size and shape, one can use these methods to establish upper and lower bounds on the settling velocity of a particle in a given situation. Broadly, size is important in the viscous regime (low Reynolds numbers), and shape and orientation matter more in the inertial regime (high Reynolds numbers).

Most of the methods mentioned so far are applicable for regular-shaped particles for which it is possible to calculate/measure the volume and surface area. In many situations, one encounters particles of irregular shape with edges and corners, or agglomerates of fine particles such as flocs, composite particles, etc. It is, thus, not very practical to calculate or measure their surface area. Some ideas based on fractal dimensions, circularity, Corey shape factors, etc., are available in the literature for such situations.[41–45]

Example 6.3

What will be the terminal velocities of mica plates, 1 mm thick and ranging in area from 6 to 600 mm^2, settling in an oil of density 820 kg/m^3 and viscosity 10 mN s/m^2? The density of mica is 3000 kg/m^3.

Solution

	Smallest Particles	**Largest Particles**
A'	6×10^{-6} m^2	6×10^{-4} m^2
d_p	$\sqrt{(4 \times 6 \times 10^{-6}/\pi)} = 2.76 \times 10^{-3}$ m	$\sqrt{(4 \times 6 \times 10^{-4}/\pi)} = 2.76 \times 10^{-2}$ m
d_p^3	2.103×10^{-8} m^3	2.103×10^{-5} m^3
Volume	6×10^{-9} m^3	6×10^{-7} m^3
k'	0.285	0.0285

$$\left(\frac{R'_0}{\rho u^2}\right) Re'^2_0 = \frac{4k'}{\mu^2 \pi}(\rho_s - \rho)\rho d_p^3 g \qquad (Eq. 6.52).$$

$$= (4 \times 0.285/\pi \times 0.01^2)(3000 - 820)(820 \times 2.103 \times 10^{-8} \times 9.81)$$

$$= 1340 \text{ for the smallest particles and, similarly, } 134,000 \text{ for the largest particles.}$$

Thus:

	Smallest Particles	Largest Particles
$\log\left(\dfrac{R_0'}{\rho u_0^2}Re'^2_0\right)$	3.127	5.127
$\log Re_0'$	1.581	2.857 (from Table 6.4)
Correction from Table 6.6	−0.038	−0.300 (estimated)
Corrected Re_0'	1.543	2.557
Re_0'	34.9	361
u_0	0.154 m/s	0.159 m/s

Thus, it is seen that all the mica particles settle at approximately the same velocity.

6.5 Motion of Bubbles and Drops

The drag force acting on a gas bubble or a liquid droplet will not, in general, be the same as that acting on a rigid particle of the same shape and size because circulating currents are set up inside the bubble. The velocity gradient at the surface is thereby reduced, and the drag force is, therefore, less than that for the rigid particle. Hadamard[46] showed that, if the effects of surface tension are neglected, the terminal falling velocity of a drop, as calculated from Stokes' law, must be multiplied by a factor Q, to account for the internal circulation, where:

$$Q = \frac{3\mu + 3\mu_l}{2\mu + 3\mu_l} \tag{6.54}$$

In this equation, μ is the viscosity of the continuous fluid, and μ_l is the viscosity of the fluid forming the drop or bubble. This expression applies only in the range for which Stokes' law is valid.

If μ_l/μ is large, Q approaches unity. If μ_l/μ is small, Q approaches a value of 1.5. Thus, the effect of circulation is small when a liquid drop falls in a gas, and it is large when a gas bubble rises in a liquid. If the fluid within the drop is very viscous, the amount of energy which has to be transferred in order to induce circulation is large, and circulation effects are, therefore, small.

Hadamard's work was later substantiated by Bond[47] and by Bond and Newton,[48] who showed that Eq. (6.54) is valid provided that surface tension forces do not play a large role. With very small droplets, the surface tension forces tend to nullify the tendency for circulation, and the droplet falls at a velocity close to that of a solid sphere.

In addition, drops and bubbles are subject to deformation because of the differences in the pressures acting on various parts of the surface. Thus, when a drop is settling in a still fluid, both the hydrostatic and the impact pressures will be greater on the forward face than on the rear face and will tend to flatten the drop, whereas the viscous drag will tend to elongate it.

Deformation of the drop is opposed by the surface tension forces, so that very small drops retain their spherical shape, whereas large drops may be considerably deformed and the resistance to their motion thereby increased. For drops above a certain size, the deformation is so great that the drag force increases at the same rate as the volume, and the terminal falling velocity therefore becomes independent of size.

Garner and Skelland[49, 50] have shown the importance of circulation within a drop in determining the coefficient of mass transfer between the drop and the surrounding medium. The critical Reynolds number at which circulation commences has been shown[49] to increase at a rate proportional to the logarithm of viscosity of the liquid constituting the drop and to increase with interfacial tension. The circulation rate may be influenced by mass transfer because of the effect of concentration of diffusing material on both the interfacial tension and on the viscosity of the surface layers. As a result of circulation, the falling velocity may be up to 50% greater than for a rigid sphere, whereas oscillation of the drop between oblate and prolate forms will reduce the velocity of fall.[50] Terminal falling velocities of droplets have also been calculated by Hamielec and Johnson[51] and others[52] from approximate velocity profiles at the interface, and the values so obtained compare well with experimental values for droplet Reynolds numbers up to 80.

6.6 Drag Forces and Settling Velocities for Particles in Non-Newtonian Fluids

Only a very limited amount of data is available on the motion of particles in non-Newtonian fluids, and the following discussion is restricted to their behaviour in shear-thinning *power-law* fluids and in fluids exhibiting a *yield stress*, both of which are discussed in Volume 1A, Chapter 3.

6.6.1 Power-Law Fluids

Because most shear-thinning fluids, particularly polymer solutions and flocculated suspensions, have high apparent viscosities, even relatively coarse particles may have velocities in the *creeping-flow* or the Stokes' law regime. Chhabra[53, 54] has proposed that both theoretical and experimental results for the drag force F on an isolated spherical particle of diameter d moving at a velocity u may be expressed as a modified form of Stokes' law:

$$F = 3\pi\mu_c duY \tag{6.55}$$

where the apparent viscosity μ_c is evaluated at a characteristic shear rate u/d, and Y is a correction factor which is a function of the rheological properties of the fluid. The best available theoretical estimates of Y for power-law fluids are given in Table 6.9.

Table 6.9 Values of Y for power-law fluids[53]

n	1	0.9	0.8	0.7	0.6	0.5	0.4	0.3	0.2	0.1
Y	1	1.14	1.24	1.32	1.38	1.42	1.44	1.46	1.41	1.35

Several expressions of varying forms and complexity have been proposed[36, 53, 54] for the prediction of the drag on a sphere moving through a power-law fluid. These are based on a combination of numerical solutions of the equations of motion and extensive experimental results. In the absence of wall effects, dimensional analysis yields the following functional relationship between the variables for the interaction between a single isolated particle and a fluid:

$$2C_D' = C_D = f\left(Re_n', n\right) \tag{6.56}$$

where C_D and C_D' are drag coefficients defined by Eq. (6.4), n is the power-law index, and Re_n' is the particle Reynolds number given by:

$$Re_n' = \left(u^{2-n}d^n\rho\right)/k \tag{6.57}$$

where k is the consistency coefficient in the power-law relation. Combining Eqs (6.55) and (6.56):

$$C_D = \left(24Re_n'^{-1}\right)Y \tag{6.58}$$

$$C_D' = \left(12Re_n'^{-1}\right)Y \tag{6.59}$$

From Table 6.9 it is seen that, depending on the value of n, the drag on a sphere in a power-law fluid may be up to 46% higher than that in a Newtonian fluid at the same particle Reynolds number. Practical measurements lie in the range $1 < Y < 1.8$, with considerable divergences between the results of the various researchers.

In view of the general uncertainty concerning the value of Y, it may be noted that the unmodified Stokes' law expression gives a acceptable first approximation.

The terminal settling velocity u_0 of a particle in the gravitational field is then given by equating the buoyant weight of the particle to the drag force to give:

$$u_0 = \left\{\frac{gd^{n+1}(\rho_s - \rho)}{18kY}\right\}^{1/n} \tag{6.60}$$

where $(\rho_s - \rho)$ is the density difference between the particle and the fluid.

From Eq. (6.60), it is readily seen that in a shear-thinning fluid ($n < 1$), the terminal velocity is more strongly dependent on d, g and $\rho_s - \rho$ than in a Newtonian fluid, and a small change in any of these variables produces a larger change in u_0.

Outside the creeping flow regime, experimental results for drag on spheres in power-law fluids have been presented by Tripathi et al.,[55] Graham and Jones[56] and Song et al.[57] for values of Re'_n up to 100, and these are reasonably well correlated by the following expressions with an average error of about 10%:

$$C_D = \left[(35.2 Re'_n - 1.03)2^n\right] + n\left[1 - (20.9 Re'_n - 1.11)2^n\right]$$
$$(0.2 < (2^{-n} Re'_n) < 24) \tag{6.61a}$$

$$C_D = \left[(37 Re'_n - 1.1)2^n\right] + [0.36n + 0.25]$$
$$(24 < (2^{-n} Re'_n) < 100) \tag{6.61b}$$

$$C'_D = \left[\left(17.6 Re'^{-1.03}_n\right)2^n\right] + n\left[0.5 - \left(10.5 Re'^{-1.11}_n\right)2^n\right]$$
$$[0.2 < (2^{-n} Re'_n) < 24] \tag{6.62a}$$

$$C'_D = \left[\left(18.5 Re'^{-1.1}_n\right)2^n\right] + [0.18n + 0.125]$$
$$[24 < (2^{-n} Re'_n) < 100] \tag{6.62b}$$

It may be noted that these two equations do not reduce exactly to the relation for a Newtonian fluid ($n = 1$).

Extensive comparisons of predictions and experimental results for drag on spheres suggest that the influence of non-Newtonian characteristics progressively diminishes as the value of the Reynolds number increases, with inertial effects then becoming dominant, and the standard curve for Newtonian fluids may be used with little error. Experimentally determined values of the drag coefficient for power-law fluids ($1 < Re'_n < 1000; 0.4 < n < 1$) are within 30% of those given by the standard drag curve.[55, 56] Suffice it to add here that the other methods[36, 53, 58] available in the literature do not offer any significant improvement over Eqs (6.61) and (6.62).

While Eqs (6.62a) and (6.62b) are convenient for estimating the value of the drag coefficient, they need to be rearranged in order to enable the settling velocity u_0 of a sphere of given diameter and density to be calculated, because both $C_D(C'_D)$ and Re'_n are functions of the unknown settling velocity. By analogy with the procedure used for Newtonian fluids (Eq. 6.36), the dimensionless Galileo number Ga_n which is independent of u_0 may be defined by:

$$\frac{2}{3} Ga_n = C'_D Re'^{[2/(2-n)]} = g d^{[(n+2)/(2-n)]} (s_r - 1)(\rho/k)^{[2/(2-n)]} \tag{6.63}$$

where s_r is the ratio of the densities of the particle and of the fluid (ρ_s/ρ).

Eq. (6.56) may be written as:

$$Re'_n = f(Ga_n, n) \tag{6.64}$$

Experimental results comprising about 1000 data points from a large number of sources cover the following range of variables:

$$1 < d < 20 \text{(mm)}; 1190 < \rho_s < 16,600 \text{(kg/m}^3); \quad 990 < \rho < 1190 \text{(kg/m}^3);$$
$$0.4 < n < 1; \text{ and } 1 < Re'_n < 10^4.$$

These are satisfactorily correlated by:

$$Re'_n = r_1 \left(\frac{2}{3} Ga_n \right)^{r_2} \tag{6.65}$$

where $r_1 = 0.1\{\exp[(0.5/n) - 0.73n]$ and $r_2 = (0.954/n) - 0.16$.

Thus, in Eq. (6.65) only Re_n' includes the terminal falling velocity, which may then be calculated for a spherical particle in a power-law fluid.

6.6.2 Fluids With a Yield Stress

Much less is known about the settling of particles in fluids exhibiting a yield stress. Barnes[59] suggests that this is partly due to the fact that considerable confusion exists in the literature as to whether or not the fluids used in the experiments do have a true yield stress.[59] Irrespective of this uncertainty, which usually arises from the inappropriateness of the rheological techniques used for their characterisation, many industrially important materials, notably particulate suspensions, have rheological properties closely approximating to viscoplastic behavior.

By virtue of its yield stress, an unsheared viscoplastic material is capable of supporting the immersed weight of a particle for an indefinite period of time, provided that the immersed weight of the particle does not exceed the maximum upward force which can be exerted by virtue of the yield stress of the fluid. The conditions for the static equilibrium of a sphere are now discussed.

Static equilibrium

Many investigators[53] have reported experimental results on the necessary conditions for the static equilibrium of a sphere. The results of all such studies may be represented by a factor Z, which is proportional to the ratio of the forces due to the yield stress τ_Y and those due to gravity.

$$\text{Thus:} \quad Z = \frac{\tau_Y}{dg(\rho_s - \rho)} \tag{6.66}$$

The critical value of Z which indicates the point at which the particle starts to settle from rest appears to lie in the range $0.04 < Z < 0.2$.

Drag force

Under conditions where a spherical particle is not completely supported by the forces attributable to the yield stress, it will settle at a velocity such that the total force exerted by the fluid on the particle balances its weight.

For a fluid whose rheological properties may be represented by the Herschel-Bulkley model discussed in Volume 1A, Chapter 3, the shear stress τ is a function of the shear rate $\dot{\gamma}$ or:

$$\tau = \tau_Y + k'_{\text{HB}} \dot{\gamma}^{n_{\text{HB}}} \tag{6.67}$$

From dimensional considerations, the drag coefficient is a function of the Reynolds number for the flow relative to the particle, the exponent, n_{HB}, and the so-called Bingham number Bi, which is proportional to the ratio of the yield stress to the viscous stress attributable to the settling of the sphere. Thus:

$$C'_D = R'/\rho u^2 f(Re' m_{\text{HB}}, n_{\text{HB}}, Bi) \tag{6.68}$$

$$Bi = \tau_Y/k'_{\text{HB}}(u/d)^{n_{\text{HB}}} \tag{6.69}$$

Using the scant data in the literature and their own experimental results, Atapattu et al.[60] suggest the following expression for the drag on a sphere moving through a Herschel-Bulkley fluid in the creeping flow regime:

$$C'_D = 12Re'_{\text{HB}}(1 + Bi)^{-1} \tag{6.70}$$

It may be noted that an iterative solution to Eq. (6.70) is required for the calculation of the unknown settling velocity u_0, because this term appears in all three dimensionless groups, Re_{HB}', Bi, and C_D', for a given combination of properties of sphere and fluid. It is useful to mention here that the predictions of Eq. (6.70) are also in line with the recent numerical predictions[61] of drag on a sphere up to $Re_{\text{HB}}' = 100$.

The effect of particle shape on the forces acting when the particle is moving in a shear-thinning fluid has been investigated by Tripathi et al.,[55] and by Venumadhav and Chhabra.[62] In addition, some information is available on the effects of viscoelasticity of the fluid.[53]

In an extensive study dealing with the free settling of nonspherical particles in power-law fluids, Rajitha et al.[63] collated much of the literature data and modified a previous drag formula developed for spheres falling in power-law fluids.[64] Their approach hinges on evaluating three parameters to characterise the shape and orientation of the particle: equal volume sphere diameter (d_s), sphericity, ψ, defined as the ratio of the surface areas of the equal volume sphere and of the actual particle and the diameter, d_n, of a circle of area equal to the projected area of the falling particle normal to the direction of sedimentation. Furthermore, Rajitha et al.[63] introduced a composite shape parameter χ defined as:

$$\chi = \frac{4}{\psi}\left(\frac{d_s}{d_n}\right)^2 \tag{6.71}$$

Evidently, $\chi = 4$ for a sphere ($\psi = 1$; $d_s = d_n$) and χ can be greater or smaller than 4 depending upon the shape and orientation of the falling particle. The drag expression due to Rajitha et al.[63] is given as follows:

$$C_D = \frac{24}{Re'_n}Y + \chi C_{D\infty}\left(\frac{24Y}{Re'_n}\right)^{2\delta_o}\delta_1\left\{\frac{6Yb_o}{6Yb_o + \left(\frac{24Y}{Re'_n}\right)}\right\}^{\delta_o} + C_{D\infty}\left\{\frac{6Yb_o}{6Yb_o + 128\left(\frac{24Y}{Re'_n}\right)}\right\}^{\frac{11}{12}} \tag{6.72}$$

The three constants δ_o, δ_1 and b_o are evaluated using the numerical results of Tripathi et al.[55] as follows:

$$b_o = \exp\{3(\alpha - \ln 6)\} \tag{6.73a}$$

$$\delta_1 = \frac{\alpha_o - \alpha_1}{2\alpha_o\alpha_1}\exp\left\{3\left(\frac{\alpha_o - \alpha_1}{2\alpha_o\alpha_1}\right)\ln 3\right\} \tag{6.73b}$$

$$\delta_o = \frac{11}{48}\sqrt{6}\left\{1 - \exp\left[\left(\frac{\alpha_o - \alpha_1}{2(\alpha_o - 1)\alpha_1}\right)^2\ln\left(\frac{\sqrt{6} - 1}{\sqrt{6}}\right)\right]\right\} \tag{6.73c}$$

Here, $\alpha_o = 3$ and it corresponds to the value of α_1 for $n = 0$, see Eq. (6.73e).

The two functions Y and α_1 related to each other as follows:

$$Y = \left(\sqrt{6}\right)^{n-1}\left\{\frac{3}{n^2 + n + 1}\right\} \tag{6.73d}$$

This is based on the numerical results in the creeping flow summarised in Table 6.9. Finally, α_1 is given as:

$$\alpha_1 = \left\{6^{(1-n)/2}Y\right\}^{1/(n+1)} \tag{6.73e}$$

It is easily seen that for $n = 0$, $Y = \left(\frac{3}{\sqrt{6}}\right)$, which, in turn, yields the value of $\alpha_1 = \alpha_o = 3$.

Some comments about such a complex form of Eq. (6.72) are in order. First, in the limit of $n = 1$, Eq. (6.72) reproduces the standard drag curves for a sphere ($C_{D\infty} = 0.44$) and for a long cylinder ($C_{D\infty} = {\sim}1$) rather well.[65] Subsequently, it has been shown that in the limit of $\chi = 4$, Eq. (6.72) correlates much of the experimental data for spheres falling in power-law fluids with acceptable levels of accuracy.[64] Because with the increasing Reynolds number, the effects of fluid viscosity progressively diminish, Rajitha et al.[63] recommended the value of $C_{D\infty} = 0.44$ for spheres in power-law fluids also. Eq. (6.72) reproduced nearly 1500 data points relating to cylinders, cones, oblates, prolates, prisms, disks, etc., spanning the ranges of conditions as $0.33 \leq \psi \leq 0.98$; $0.3 \leq n \leq 1$ and $10^{-5} \leq Re_n' \leq 270$ with a mean error of 30%.

It is seen, therefore, that, in the absence of any entirely satisfactory theoretical approach or reliable experimental data, it is necessary to adopt a highly pragmatic approach to the estimation of the drag force on a particle in a non-Newtonian fluid.

6.7 Accelerating Motion of a Particle in the Gravitational Field

6.7.1 General Equations of Motion

The motion of a particle through a fluid may be traced because the value of the drag factor $R'/\rho u^2$ for a given value of the Reynolds number is fixed. The behaviour of a particle undergoing acceleration or retardation has been the subject of a very large number of investigations, which have been critically reviewed by Torobin and Gauvin[66] and others.[13] The results of different researchers are not consistent, although it is shown that the drag factor is often related, not only to the Reynolds number, but also to the number of particle diameters traversed by the particle since the initiation of the motion.

A relatively simple approach to the problem, which gives results closely in accord with practical measurements, is to consider the mass of fluid which is effectively given the same acceleration as the particle, as discussed by Mironer.[67] This is only an approximation because elements of fluid at different distances from the particle will not all be subject to the same acceleration. For a spherical particle, this added or *hydrodynamic* mass is equal to the mass of fluid whose volume is equal to one half of that of the sphere. This can give rise to a very significant effect in the case of movement through a liquid, and can result in accelerations substantially less than those predicted when the added mass is neglected. For movement through gases, the contribution of the added mass term is generally negligible. Added mass is most important for the motion of gas bubbles in a liquid because, in that case, the surrounding liquid has a much greater density than the bubble. The total mass of particle and associated fluid is sometimes referred to as the *virtual mass* (m').

Thus: Virtual mass = Mass of particle + Added mass.

For a sphere: $m' = \dfrac{\pi}{6}d^3 \rho_s + \dfrac{\pi}{12}d^3 \rho$

$$\text{or:} \quad m' = \frac{\pi}{6}d^3 \rho_s \left(1 + \frac{\rho}{2\rho_s}\right) = m\left(1 + \frac{\rho}{2\rho_s}\right) \tag{6.74}$$

Considering the motion of a particle of mass m in the earth's gravitational field, at some time t, the particle will be moving at an angle α to the horizontal with a velocity u as shown in Fig. 6.7. The velocity u may then be resolved into two components, $\dot{x}$ and $\dot{y}$, in the horizontal and vertical

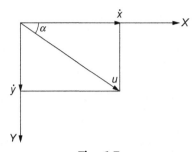

Fig. 6.7
Two-dimensional motion of a particle.

directions. $\dot{x}$ and $\ddot{x}$ will be taken to denote the first and second derivatives of the displacement x in the X-direction with respect to time, and $\dot{y}$ and $\ddot{y}$ the corresponding derivatives of y.

$$\text{Thus}: \quad \cos\alpha = \frac{\dot{x}}{u} \tag{6.75}$$

$$\sin\alpha = \frac{\dot{y}}{u} \tag{6.76}$$

$$\text{and}: \quad u = \sqrt{(\dot{x}^2 + \dot{y}^2)} \tag{6.77}$$

There are two forces acting on the body:

(a) In the vertical direction, the apparent weight of the particle,

$$W' = mg\left(1 - \frac{\rho}{\rho_s}\right) \tag{6.78}$$

(b) The drag force which is equal to $R'A'$ and acts in such a direction as to oppose the motion of the particle. Its direction, therefore, changes as α changes. Here, A' is the projected area of the particle on a plane at right angles to the direction of motion, and its value varies with the orientation of the particle in the fluid. The drag force can be expressed by:

$$F = \frac{R'}{\rho u^2}\rho u^2 A' \tag{6.79}$$

This has a component in the X-direction of:

$$\frac{R'}{\rho u^2}\rho u^2 A' \cos\alpha = \frac{R'}{\rho u^2}A'\rho\dot{x}\sqrt{(\dot{x}^2 + \dot{y}^2)}$$

and in the Y -direction of:

$$\frac{R'}{\rho u^2}\rho u^2 A' \sin\alpha = \frac{R'}{\rho u^2} A' \rho \dot{y} \sqrt{(\dot{x}^2 + \dot{y}^2)}$$

The equations of motion in the *X*- and *Y*-directions are, therefore:

$$m\ddot{x} = -\frac{R'}{\rho u^2}\rho A' \dot{x} \sqrt{(\dot{x}^2 + \dot{y}^2)} \tag{6.80}$$

$$\text{and}: \quad m\ddot{y} = -\frac{R'}{\rho u^2}\rho A' \dot{y} \sqrt{(\dot{x}^2 + \dot{y}^2)} + mg\left(1 - \frac{\rho}{\rho_s}\right) \tag{6.81}$$

If allowance is now made for the *added mass*, $m' = m[1 + (\rho/2\rho_s)]$ must be substituted for m. Eqs (6.80) and (6.81) refer to conditions where the particle is moving in the positive sense in the *X*-direction and in the positive (downward) sense in the *Y*-direction. If the particle is moving in the negative *X*-direction, the form of solution is unchanged, except that all increments of x will be negative. If, however, the particle is initially moving upwards, the sign of only the frictional term in Eq. (6.81) is changed, and the form of solution will, in general, be different from that for downward movement. Care must therefore be exercised in the application of the equation, particularly if a change of sense may occur during the motion of the particle.

It may be noted that the equations of motion for the two directions (*X* and *Y*) are coupled, with $\dot{x}$ and $\dot{y}$ appearing in each of the equations. General solutions are, therefore, not possible, except as will be seen later for motion in the Stokes' law region.

Putting $m = \frac{\pi}{6}d^3\rho_s$ and $A' = \frac{\pi}{4}d^2$ for a spherical particle, then:

$$\ddot{x} = -\frac{R'}{\rho u^2}\frac{3\rho}{2d\rho_s}\dot{x}\sqrt{(\dot{x}^2 + \dot{y}^2)} \tag{6.82}$$

If allowance is made for the added mass and m' is substituted for m in Eq. (6.81), then:

$$\ddot{x} = -\frac{R'}{\rho u^2}\frac{3}{d}\frac{\rho}{(2\rho_s + \rho)}\dot{x}\sqrt{(\dot{x}^2 + \dot{y}^2)} \tag{6.83}$$

$$\text{Similarly}: \quad \ddot{y} = \mp\frac{R'}{\rho u^2}\frac{3}{2d}\frac{\rho}{\rho_s}\dot{y}\sqrt{(\dot{x}^2 + \dot{y}^2)} + g\left(1 - \frac{\rho}{\rho_s}\right) \tag{6.84}$$

and allowing for the added mass:

$$\ddot{y} = \mp\frac{R'}{\rho u^2}\frac{3}{d}\frac{\rho}{(2\rho_s + \rho)}\dot{y}\sqrt{(\dot{x}^2 + \dot{y}^2)} + \frac{2g(\rho_s - \rho)}{(2\rho_s + \rho)} \tag{6.85}$$

where the minus sign in Eqs (6.84) and (6.85) is applicable for downward motion and the positive sign for upward motion (with downwards taken as the positive sense).

6.7.2 Motion of a Sphere in the Stokes' Law Region

Under these conditions, from Eq. (6.5):

$$\frac{R'}{\rho u^2} = 12Re'^{-1} = \frac{12\mu}{d\rho\sqrt{(\dot{x}^2 + \dot{y}^2)}} \tag{6.86}$$

Substituting in Eqs (6.82)–(6.86) gives:

$$\ddot{x} = -\frac{18\mu}{d^2\rho_s}\dot{x} = -a\dot{x} \tag{6.87}$$

$$\text{or:}\quad \ddot{x} = -\frac{36\mu}{d^2(2\rho_s + \rho)}\dot{x} = -a'\dot{x} \tag{6.88}$$

if allowance is made for the *added mass*.

$$\ddot{y} = -\frac{18\mu}{d^2\rho_s}\dot{y} + g\left(1 - \frac{\rho}{\rho_s}\right) = -a\dot{y} + b \tag{6.89}$$

$$\text{or:}\quad \ddot{y} = -\frac{36\mu}{d^2(2\rho_s + \rho)}\dot{y} + \frac{2g(\rho_s - \rho)}{(2\rho_s + \rho)} = -a'\dot{y} + b' \tag{6.90}$$

if allowance is made for the *added mass*.

In this particular case, the equations of motion for the X and Y directions are mutually independent and therefore can be integrated separately. Furthermore, because the frictional term is now a linear function of velocity, the sign will automatically adjust to take account of whether motion is downwards or upwards.

The equations are now integrated, ignoring the effects of *added mass* which can be accounted for by replacing a by a' and b by b'. For the Y-direction, integrating Eq. (6.89) with respect to t:

$$\dot{y} = -ay + bt + \text{constant.}$$

The axes are chosen so that the particle is at the origin at time $t=0$. If the initial component of the velocity of the particle in the Y-direction is v, then, when $t=0$, $y=0$ and $\dot{y}=v$, and the constant $=v$,

$$\text{or}\quad \dot{y} + ay = bt + v$$

$$\text{Thus:}\quad e^{at}\dot{y} + e^{at}ay = (bt + v)e^{at}$$

$$\text{Thus:}\quad e^{at}y = (bt + v)\frac{e^{at}}{a} - \int b\frac{e^{at}}{a}dt$$

$$= (bt + v)\frac{e^{at}}{a} - \frac{b}{a^2}e^{at} + \text{constant}$$

When $t=0$, $y=0$, and the constant $=\dfrac{b}{a^2}-\dfrac{v}{a}$.

Thus: $\quad y=\dfrac{b}{a}t+\dfrac{v}{a}-\dfrac{b}{a^2}+\left(\dfrac{b}{a^2}-\dfrac{v}{a}\right)e^{-at}$ $\hspace{2cm}$ (6.91)

where: $\quad a=18\dfrac{\mu}{d^2\rho_s}$ $\hspace{3cm}$ (6.92)

and: $\quad b=\left(1-\dfrac{\rho}{\rho_s}\right)g.$ $\hspace{3cm}$ (6.93)

It may be noted that $b/a=u_0$, the terminal falling velocity of the particle. This equation enables the displacement of the particle in the Y-direction to be calculated at any time t.

For the X-direction, Eq. (6.87) is of the same form as Eq. (6.89) with $b=0$. Substituting $b=0$ and writing w as the initial velocity in the X-direction, Eq. (6.91) becomes:

$$x=\dfrac{w}{a}(1-e^{-at})$$ $\hspace{3cm}$ (6.94)

Thus, the displacement in the X-direction may also be calculated for any time t.

By eliminating t between Eqs (6.92), (6.93) and (6.94), a relation between the displacements in the X- and Y-directions is obtained. Equations of this form are useful for calculating the trajectories of particles in size-separation equipment.

From Eq. (6.94):

$$e^{-at}=1-\dfrac{ax}{w}$$

and: $\quad t=-\dfrac{1}{a}\ln\left(1-\dfrac{ax}{w}\right)$ $\hspace{2cm}$ (6.95)

Substituting in Eq. (6.91) gives:

$$y=\dfrac{b}{a}\left\{-\dfrac{1}{a}\ln\left(1-\dfrac{ax}{w}\right)\right\}+\dfrac{v}{a}-\dfrac{b}{a^2}+\left(\dfrac{b}{a^2}-\dfrac{v}{a}\right)\left(1-\dfrac{ax}{w}\right)$$

$$=-\dfrac{b}{a^2}\ln\left(1-\dfrac{ax}{w}\right)-\dfrac{bx}{aw}+\dfrac{vx}{w}$$ $\hspace{2cm}$ (6.96)

The values of a and b can now be substituted and the final relation is:

$$y=-\dfrac{g\rho_s(\rho_s-\rho)d^4}{324\mu^2}\left\{\ln\left(1-\dfrac{18\mu x}{w\rho_s d^2}\right)+\dfrac{18\mu x}{w\rho_s d^2}\left(1-\dfrac{18v\mu}{d^2(\rho_s-\rho)g}\right)\right\}$$ $\hspace{1cm}$ (6.97)

If allowance is made for *added mass*, a' and b' are substituted for a and b, respectively.

Then:

$$y = -\frac{g(2\rho_s+\rho)(\rho_s-\rho)d^4}{648\mu^2}\left\{\ln\left(1-\frac{36\mu x}{w(2\rho_s+\rho)d^2}\right)+\frac{36\mu x}{w(2\rho_s+\rho)d^2}\left(1-\frac{18v\mu}{d^2(\rho_s-\rho)g}\right)\right\}$$

(6.98)

6.7.3 Vertical Motion (General Case)

For the *Stokes' law regime* Eqs (6.91)–(6.93) are applicable.

For the *Newton's law regime*, $R'/\rho u^2$ is a constant and equal to 0.22 for a spherical particle. Therefore, substituting in Eq. (6.84) and putting $\dot{x}=0$ for vertical motion, and using the *negative sign for downward motion* (and neglecting the effect of *added mass*):

$$\ddot{y} = -\frac{1}{3d}\frac{\rho}{\rho_s}\dot{y}^2 + g\left(1-\frac{\rho}{\rho_s}\right)$$

(6.99)

$$\ddot{y} = -c\dot{y}^2 + b$$

(6.100)

Thus:
$$\frac{d\dot{y}}{b-c\dot{y}^2} = dt$$

and:
$$\frac{d\dot{y}}{f^2-\dot{y}^2} = c\,dt$$

(6.101)

where:
$$c = \frac{1}{3d}\frac{\rho}{\rho_s}$$

(6.102)

and:
$$f = \sqrt{\left(\frac{b}{c}\right)} = \sqrt{\left(\frac{3d(\rho_s-\rho)g}{\rho}\right)}$$

(6.103)

Integrating Eq. (6.101) gives:

$$\frac{1}{2f}\ln\left(\frac{f+\dot{y}}{f-\dot{y}}\right) = ct + \text{constant}$$

When $t=0$, $\dot{y}=v$, say, and, therefore, the constant $= \frac{1}{2f}\ln\left(\frac{f+v}{f-v}\right)$.

Thus:
$$\frac{1}{2f}\ln\left(\frac{f+\dot{y}}{f-\dot{y}}\right)\left(\frac{f-v}{f+v}\right) = ct$$

Thus:
$$\left(\frac{f+\dot{y}}{f-\dot{y}}\right)\left(\frac{f-v}{f+v}\right) = e^{2fct}$$

$$\text{and:} \quad (f - \dot{y}) = \frac{2f}{1 + \left(\dfrac{f+v}{f-v}\right)e^{2fct}}$$

$$\text{Thus:} \quad \dot{y} = ft - 2f \int \frac{dt}{1 + \left(\dfrac{f+v}{f-v}\right)e^{2fct}}$$

$$= ft - 2f \int \frac{dt}{1 + je^{pt}} \quad \text{(say)}$$

$$\text{Putting:} \quad s = 1 + je^{pt}$$

$$\text{then:} \quad ds = pje^{pt}dt = p(s-1)dt$$

$$\text{Thus:} \quad y = ft - 2f \int \frac{ds}{ps(s-1)}$$

$$= ft - 2f\left\{\frac{1}{p}\ln\frac{s-1}{s} + \text{constant}\right\}$$

$$= ft - \frac{1}{c}\ln\frac{1}{1 + \left(\dfrac{f-v}{f+v}\right)e^{-2fct}} + \text{constant}$$

When $t = 0$, $y = 0$ and:

$$\text{constant} = \frac{1}{c}\ln\frac{1}{1 + \left(\dfrac{f-v}{f+v}\right)} = \frac{1}{c}\ln\frac{f+v}{2f}$$

$$\text{Thus:} \quad y = ft + \frac{1}{c}\ln\frac{f+v}{2f}\left\{1 + \frac{f-v}{f+v}e^{-2fct}\right\}$$

Thus, for downward motion:

$$y = ft + \frac{1}{c}\ln\frac{1}{2f}\left\{f + v + (f-v)e^{-2fct}\right\} \tag{6.104}$$

If the *added mass* is taken into account, f remains unchanged, but c must be replaced by c' in Eqs (6.100) and (6.104), where:

$$c' = \frac{2\rho}{3d(2\rho_s + \rho)}. \tag{6.105}$$

For *vertical upwards motion* in the *Newton's law* regime; the positive sign in Eq. (6.84) applies and, thus, by analogy with Eq. (6.101):

$$\frac{d\dot{y}}{f^2 + \dot{y}^2} = c\,dt \tag{6.106}$$

Integrating : $\dfrac{1}{f}\tan^{-1}\dfrac{\dot{y}}{f} = ct + \text{constant}$

When $t=0$, $\dot{y}=v$, say, and the constant $=(1/f)\tan^{-1}(v/f)$, v is a negative quantity. Thus:

$$\frac{1}{f}\tan^{-1}\frac{\dot{y}}{f} = ct + \frac{1}{f}\tan^{-1}\frac{v}{f}$$

$$\frac{\dot{y}}{f} = \tan\left(fct + \tan^{-1}\frac{v}{f}\right)$$

and : $y = \dfrac{f}{-fc}\ln\cos\left(fct + \tan^{-1}\dfrac{v}{f}\right) + \text{constant}$

When $t=0$, $y=0$, and:

$$\text{constant} = \frac{1}{c}\ln\cos\tan^{-1}\frac{v}{f},$$

then : $y = -\dfrac{1}{c}\ln\dfrac{\cos\left(fct + \tan^{-1}\dfrac{v}{f}\right)}{\cos\tan^{-1}\dfrac{v}{f}}$

$$= -\frac{1}{c}\ln\frac{\cos fct\cos\left[\tan^{-1}\dfrac{v}{f}\right] - \sin fct\sin\left[\tan^{-1}\dfrac{v}{f}\right]}{\cos\left[\tan^{-1}\dfrac{v}{f}\right]}$$

or : $y = -\dfrac{1}{c}\ln\left(\cos fct - \dfrac{v}{f}\sin fct\right) \tag{6.107}$

The relation between y and t may also be obtained graphically, though the process is more tedious than that of using the analytical solution appropriate to the particular case in question. When Re' lies between 0.2 and 500, there is no analytical solution to the problem, and a numerical or graphical method must be used.

When the spherical particle is moving downwards, that is when its velocity is positive:

$$\ddot{y} = -\frac{3}{2d}\frac{\rho}{\rho_s}\frac{R'}{\rho u^2}\dot{y}^2 + \left(1 - \frac{\rho}{\rho_s}\right)g \quad \text{(from Eq.6.84)}$$

$$\frac{\mu}{\rho d}\frac{dRe'}{dt} = -\frac{3}{2d}\frac{\rho}{\rho_s}\frac{R'}{\rho u^2}Re'^2\frac{\mu^2}{d^2\rho^2} + \left(1 - \frac{\rho}{\rho_s}\right)g$$

$$\text{and}: \quad t = \int_{Re_1'}^{Re_2'} \frac{dRe'}{\dfrac{d\rho(\rho_s - \rho)g}{\mu\rho_s} - \dfrac{3\mu}{2d^2\rho_s\rho u^2}\dfrac{R'}{}Re'^2} \tag{6.108}$$

If the particle is moving upwards, the corresponding expression for t is:

$$t = \int_{Re_1'}^{Re_2'} \frac{dRe'}{\dfrac{d\rho(\rho_s - \rho)g}{\mu\rho_s} + \dfrac{3\mu}{2d^2\rho_s\rho u^2}\dfrac{R'}{}Re'^2} \tag{6.109}$$

Eqs (6.108) and (6.109) do not allow for *added mass*. If this is taken into account, then:

$$t = \int_{Re_1'}^{Re_2'} \frac{dRe'}{\dfrac{2d(\rho_s - \rho)g}{\mu(2\rho_\rho + \rho)} \pm \dfrac{3\mu}{d^2(2\rho_s + \rho)\rho u^2}\dfrac{R'}{}Re'^2} \tag{6.110}$$

where the positive sign applies to upward motion and the negative sign to downward motion.

From these equations, Re' may be obtained as a function of t. The velocity $\dot{y}$ may then be calculated. By means of a second graphical integration, the displacement y may be found at any time t.

In using the various relations which have been obtained, it must be noted that the law of motion of the particle will change as the relative velocity between the particle and the fluid changes. If, for example, a particle is initially moving upwards with a velocity v, so that the corresponding value of Re' is greater than about 500, the relation between y and t will be given by Eq. (6.107). The velocity of the particle will progressively decrease, and, when Re' is <500, the motion is obtained by application of Eq. (6.109). The upward velocity will then fall still further until Re' falls below 0.2. While the particle is moving under these conditions, its velocity will fall to zero and will then gradually increase in the downward direction. The same Eq. (6.91) may be applied for the whole of the time the Reynolds group is <0.2, irrespective of sense. Then, for higher downward velocities, the particle motion is given by Eqs (6.108) and (6.104).

Unidimensional motion in the vertical direction, under the action of gravity, occurs frequently in elutriation and other size separation equipment, as described in Chapter 1.

Example 6.4

A material of density 2500 kg/m^3 is fed to a size separation plant where the separating fluid is water rising with a velocity of 1.2 m/s. The upward vertical component of the velocity of the particles is 6 m/s. How far will an approximately spherical particle, 6 mm diameter, rise relative to the walls of the plant before it comes to rest relative to the fluid?

Solution

Initial velocity of particle relative to fluid, $v = (6.0 - 1.2) = 4.8$ m/s.

$$\text{Thus}: \quad Re' = \left(6 \times 10^{-3} \times 4.8 \times 1000\right)/\left(1 \times 10^{-3}\right)$$
$$= 28,800$$

When the particle has been retarded to a velocity such that $Re' = 500$, the minimum value for which Eq. (6.107) is applicable:

$$\dot{y} = (4.8 \times 500)/28,800 = 0.083\,\text{m/s}$$

In this solution, the effect of *added mass* is not taken into account. Allowance may be made by adjustment of the values of the constants in the equations as indicated in Section 6.7.3.

When Re' is >500, the relation between the displacement of the particle y and the time t is:

$$y = -\frac{1}{c}\ln\left(\cos fct - \frac{v}{f}\sin fct\right) \quad \text{(Eq.6.107)}$$

$$\text{where}: \quad c = \frac{1}{3d}\frac{\rho}{\rho_s} = \left(0.33/6 \times 10^{-3}\right)(1000/2500) = 22.0 \quad \text{(Eq.6.102)}$$

$$f = \sqrt{\left(\frac{3d(\rho_s - \rho)g}{\rho}\right)} = \sqrt{\left[(6 \times 10^{-3} \times 1500 \times 9.81)/(0.33 \times 1000)\right]}$$
$$= 0.517 \quad \text{(Eq.6.103)}$$

$$\text{and}: \quad v = -4.8\,\text{m/s}$$

$$\text{Thus}: \quad y = -\frac{1}{22.0}\ln\left(\cos 0.517 \times 22t + \frac{4.8}{0.517}\sin 0.517 \times 22t\right)$$

$$= -0.0455\ln\left(\cos 11.37t + 9.28\sin 11.37t\right)$$

$$\dot{y} = -\frac{0.0455(-11.37\sin 11.37t + 9.28 \times 11.37\cos 11.37t)}{\cos 11.37t + 9.28\sin 11.37t}$$

$$= -\frac{0.517(9.28\cos 11.37t - \sin 11.37t)}{\cos 11.37t + 9.28\sin 11.37t}$$

The time taken for the velocity of the particle relative to the fluid to fall from 4.8 m/s to 0.083 m/s is given by:

$$-0.083 = -\frac{0.517(9.28\cos 11.37t - \sin 11.37t)}{\cos 11.37t + 9.28\sin 11.37t}$$

$$\text{or}: \quad \cos 11.37t + 9.28\sin 11.37t = -6.23\sin 11.37t + 57.8\cos 11.37t$$

$$\text{or}: \quad 56.8\cos 11.37 = 15.51\sin 11.37t$$

$$\therefore \quad \sin 11.37t = 3.66\cos 11.37t$$

$$\text{squaring}: \quad 1 - \cos^2 11.37t = 13.4\cos^2 11.37t$$

$$\therefore \quad \cos 11.37t = 0.264 \tag{i}$$

$$\text{and}: \quad \sin 11.37t = \sqrt{\left(1 - 0.264^2\right)} = 0.965$$

The distance moved by the particle relative to the fluid during this period is, therefore, given by:

$$y = -0.0455\ln\left(0.264 + 9.28 \times 0.965\right)$$
$$= -0.101\,\text{m}$$

If Eq. (6.107) is applied for a relative velocity down to zero, the time taken for the particle to come to rest is given by:

$$9.28 \cos 11.37t = \sin 11.37t$$

squaring: $1 - \cos^2 11.37t = 86.1 \cos^2 11.37t$

and: $\cos 11.37t = 0.107$

and: $\sin 11.37t = \sqrt{(1 - 0.107^2)} = 0.994$

The corresponding distance the particle moves relative to the fluid is then given by:

$$y = -0.0455 \ln(0.107 + 9.28 \times 0.994)$$
$$= -0.102\,\mathrm{m}$$

that is, the particle moves only a very small distance with a velocity of <0.083 m/s.

If form drag were neglected for all velocities <0.083 m/s, the distance moved by the particle would be given by:

$$y = \frac{b}{a}t + \frac{v}{a} - \frac{b}{a^2} + \left(\frac{b}{a^2} - \frac{v}{a^2}\right)e^{-at} \quad \text{(Eq.6.91)}$$

and: $\dot{y} = \frac{b}{a} - \left(\frac{b}{a} - v\right)e^{-at}$

where: $a = 18\dfrac{\mu}{d^2 \rho_s} = \left(\dfrac{18 \times 0.001}{0.006^2 \times 2500}\right) = 0.20 \quad \text{(Eq.6.92)}$

$b = \left(1 - \dfrac{\rho}{\rho_s}\right)g = [1 - (1000/2500)]9.81 = 5.89 \quad \text{(Eq.6.93)}$

Thus: $b/a = 29.43$

and: $v = -0.083\,\mathrm{m/s}$

Thus: $y = 29.43t - \left(\dfrac{0.083}{0.20} + \dfrac{29.43}{0.20}\right)(1 - e^{-0.20t})$

$$= 29.43t - \frac{29.51}{0.20}(1 - e^{-0.20t})$$

and: $\dot{y} = 29.43 - 29.51 e^{-0.20t}$

When the particle comes to rest in the fluid, $\dot{y} = 0$, and:

$$e^{-0.20t} = 29.43/29.51$$

and: $t = 0.0141\,\mathrm{s}$

The corresponding distance moved by the particle is given by:

$$y = 29.43 \times 0.0141 - (29.51/0.20)(1 - e^{-0.20 \times 0.0141})$$
$$= 0.41442 - 0.41550 = -0.00108\,\mathrm{m}$$

Thus, whether the resistance force is calculated by Eq. (6.15) or Eq. (6.19), the particle moves a negligible distance with a velocity relative to the fluid of <0.083 m/s. Further, the time is also negligible, and thus, the fluid also has moved through only a very small distance.

It may, therefore, be taken that the particle moves through 0.102 m before it comes to rest in the fluid. The time taken for the particle to move this distance, on the assumption that the drag force corresponds to that given by Eq. (6.19), is given by:

$$\cos 11.37t = 0.264 \quad \text{(from Eq.(i) above)}$$

$$\therefore \quad 11.37t = 1.304$$

$$\text{and}: \quad t = 0.115 \, s$$

The distance travelled by the fluid in this time $= (1.2 \times 0.115) = 0.138$ m.

Thus, the total distance moved by the particle relative to the walls of the plant

$$= 0.102 + 0.138 = 0.240 \, m \, \text{or} \, \underline{240 \, mm}$$

Example 6.5

Salt of density 2350 kg/m^3 is charged to the top of a reactor containing a 3 m depth of aqueous liquid of density 1100 kg/m^3 and viscosity 2 mN s/m^2, and the crystals must dissolve completely before reaching the bottom. If the rate of dissolution of the crystals is given by:

$$-dd/dt = \left(3 \times 10^{-6}\right) + \left(2 \times 10^{-4}u\right)$$

where d is the size of the crystal (m) at time t (s), and u its velocity in the fluid (m/s), calculate the maximum size of crystal which should be charged. The inertia of the particles may be neglected, and the resistance force may be taken as that given by Stokes' law ($3\pi\mu du$), where d is taken as the equivalent spherical diameter of the particle.

Solution

Assuming that the salt always travels at its terminal velocity, then for the Stokes' law region, this is given by Eq. (6.24). $u_0 = (d^2/g/18\mu)(\rho_8 - \rho)$ or, in this case, $u_0 = (d^2 \times 9.81)/(18 \times 2 \times 10^{-3})$ $(2350-1100) = 3.406 \times 10^5 d^2$ m/s

$$\text{The rate of dissolution}: -dd/dt = \left(3 \times 10^{-6}\right) + \left(2 \times 10^{-4}u\right) \text{m/s}$$

$$\text{and substituting}: dd/dt = \left(-3 \times 10^{-6}\right) - \left(2 \times 10^{-4} \times 3.406 \times 10^5 d^2\right)$$
$$= -3 \times 10^{-6} - 68.1d^2$$

The velocity at any point h from the top of the reactor is $u = dh/dt$ and:

$$\frac{dh}{dd} = \frac{dh}{dt}\frac{dt}{dd} = 3.406 \times 10^5 d^2 / \left(-3 \times 10^{-6} - 68.1d^2\right)$$

$$\text{Thus}: \quad \int_0^3 dh = -\int_d^0 \frac{\left(3.406 \times 10^5 d^2 \, dd\right)}{\left(3 \times 10^{-6} + 68.1d^2\right)}$$

$$\text{or:} \quad 3 = 3.406 \times 10^5 \left(\int_0^d \frac{dd}{C_2} - \frac{C_1}{C_2^2} \int_0^d \frac{dd}{(C_1/C_2) + d^2} \right)$$

where $C_1 = 3 \times 10^{-6}$ and $C_2 = 68.1$.

$$\text{Thus:} \quad 3 = (3.406 \times 10^5) \left\{ \left[\frac{d}{C_2} \right]_0^d - \left[\frac{C_1}{C_2^2} \frac{1}{(C_1/C_2)^{0.5}} \tan^{-1} \left(\frac{d}{(C_1/C_2)^{0.5}} \right) \right]_0^d \right\}$$

$$\text{and:} \quad 3 = (3.406 \times 10^5 / C_2) \left[d - (C_1/C_2)^{0.5} \tan^{-1} d (C_1/C_2)^{-0.5} \right]$$

Substituting for C_1 and C_2: $d = (6 \times 10^{-4}) + (2.1 \times 10^{-4}) \tan^{-1}(4.76 \times 10^3 d)$, and solving by trial and error: $d = 8.8 \times 10^{-4}$ m or 0.88 mm.

The integration may also be carried out numerically with the following results:

d (m)	d^2 (m²)	$\left(\dfrac{3.406 \times 10^5 d^2}{3 \times 10^{-6} + 68.1 d^2} \right)$	Interval of d (m)	Mean Value of Function in Interval	Integral Over Interval	Total Integral
0	0	0				
1×10^{-4}	1×10^{-8}	9.25×10^2	1×10^{-4}	4.63×10^2	0.0463	0.0463
2×10^{-4}	4×10^{-8}	2.38×10^3	1×10^{-4}	1.65×10^3	0.1653	0.2116
3×10^{-4}	9×10^{-8}	3.358×10^3	1×10^{-4}	2.86×10^3	0.2869	0.4985
4×10^{-4}	1.6×10^{-7}	3.922×10^3	1×10^{-4}	3.64×10^3	0.364	0.8625
5×10^{-4}	2.5×10^{-7}	4.25×10^3	1×10^{-4}	4.09×10^3	0.409	1.2715
6×10^{-4}	3.6×10^{-7}	4.46×10^3	1×10^{-4}	4.35×10^3	0.435	1.706
7×10^{-4}	4.9×10^{-7}	4.589×10^3	1×10^{-4}	4.52×10^3	0.452	2.158
8×10^{-4}	6.4×10^{-7}	4.679×10^3	1×10^{-4}	4.634×10^3	0.463	2.621
9×10^{-4}	8.1×10^{-7}	4.74×10^3	1×10^{-4}	4.709×10^3	0.471	3.09

From which $d = 0.9$ mm.

The acceleration of the particle to its terminal velocity has been neglected, and, in practice, the time taken to reach the bottom of the reactor will be slightly larger, allowing a somewhat larger crystal to dissolve completely.

6.8 Motion of Particles in a Centrifugal Field

In most practical cases where a particle is moving in a fluid under the action of a centrifugal field, gravitational effects are comparatively small and may be neglected. The equation of motion for the particles is similar to that for motion in the gravitational field, except that the gravitational acceleration g must be replaced by the centrifugal acceleration $r\omega^2$, where r is the radius of rotation, and ω is the angular velocity. It may be noted, however, that in this case, the acceleration is a function of the position r of the particle.

For a spherical particle in a fluid, the equation of motion for the Stokes' law region is:

$$\frac{\pi}{6}d^3(\rho_s-\rho)r\omega^2 - 3\pi\mu d\frac{dr}{dt} = \frac{\pi}{6}d^3\rho_s\frac{d^2r}{dt^2} \tag{6.111}$$

As the particle moves outwards, the accelerating force increases, and, therefore, it never acquires an equilibrium velocity in the fluid.

If the inertial terms on the right-hand side of Eq. (6.111) are neglected, then:

$$\frac{dr}{dt} = \frac{d^2(\rho_s-\rho)r\omega^2}{18\mu} \tag{6.112}$$

$$= \frac{d^2(\rho_s-\rho)g}{18\mu}\frac{r\omega^2}{g}$$

$$= u_0\left(\frac{r\omega^2}{g}\right) \tag{6.113}$$

Thus, the instantaneous velocity (dr/dt) is equal to the terminal velocity u_0 in the gravitational field, increased by a factor of $r\omega^2/g$.

Returning to the exact form (Eq. 6.111), this may be rearranged to give:

$$\frac{d^2r}{dt^2} + \frac{18\mu}{d^2\rho_s}\frac{dr}{dt} - \frac{\rho_s-\rho}{\rho_s}\omega^2 r = 0 \tag{6.114}$$

$$\text{or:}\quad \frac{d^2r}{dt^2} + a\frac{dr}{dt} - qr = 0 \tag{6.115}$$

The solution of Eq. (6.115) takes the form:

$$r = B_1 e^{-[a/2+\sqrt{(a^2/4+q)}]t} + B_2 e^{-[a/2-\sqrt{(a^2/4+q)}]t} \tag{6.116}$$

$$= e^{-at/2}\left\{B_1 e^{-kt} + B_2 e^{kt}\right\} \tag{6.117}$$

$$\text{where:}\quad a = \frac{18\mu}{d^2\rho_s}, \quad q = \left(1-\frac{\rho}{\rho_s}\right)\omega^2 \text{ and } k = \sqrt{[(a^2/4)+q]} \tag{6.118}$$

The effects of added mass, which have not been taken into account in these equations, require the replacement of a by a' and q by q', where:

$$a' = \frac{36\mu}{d^2(2\rho_s+\rho)} \quad \text{and} \quad q' = \frac{2(\rho_s-\rho)}{(2\rho_s+\rho)}\omega^2$$

Eq. (6.117) requires the specification of two boundary conditions so that the constants B_1 and B_2 may be evaluated.

If the particle starts ($t=0$) at a radius r_1 with zero velocity (dr/dt)=0, then from Eq. (6.117):

$$\frac{dr}{dt} = e^{-at/2}\left[-kB_1e^{-kt} + kB_2e^{kt}\right] - \frac{a}{2}e^{-at/2}\left[B_1e^{-kt} + B_2e^{kt}\right]$$

$$= e^{-at/2}\left\{\left(k - \frac{a}{2}\right)B_2e^{kt} - \left(k + \frac{a}{2}\right)B_1e^{-kt}\right\} \tag{6.119}$$

Substituting the boundary conditions into Eqs (6.117) and (6.119):

$$r_1 = B_1 + B_2$$

$$0 = B_2\left(k - \frac{a}{2}\right) - B_1\left(k + \frac{a}{2}\right)$$

$$\text{Hence}: \quad B_1 = \frac{k - a/2}{2k}r_1 \quad \text{and} \quad B_2 = \frac{k + a/2}{2k}r_1 \tag{6.120}$$

$$\text{Thus}: \quad r = e^{-at/2}\left\{\frac{k - a/2}{2k}r_1e^{-kt} + \frac{k + a/2}{2k}r_1e^{kt}\right\}$$

$$\text{and}: \quad \frac{r}{r_1} = e^{-at/2}\left\{\cosh kt + \frac{a}{2k}\sinh kt\right\} \tag{6.121}$$

Hence, r/r_1 may be directly calculated at any value of t, although a numerical solution is required to determine t for any particular value of r/r_1.

If the effects of particle acceleration may be neglected, Eq. (6.115) simplifies to:

$$a\frac{dr}{dt} - qr = 0$$

Direct integration gives:

$$\ln\frac{r}{r_1} = \frac{q}{a}t = \frac{d^2(\rho_s - \rho)\omega^2}{18\mu}t \tag{6.122}$$

Thus, the time taken for a particle to move to a radius r from an initial radius r_1 is given by:

$$t = \frac{18\mu}{d^2\omega^2(\rho_2 - \rho)}\ln\frac{r}{r_1} \tag{6.123}$$

For a suspension fed to a centrifuge, the time taken for a particle initially situated in the liquid surface ($r_1 = r_0$) to reach the wall of the bowl ($r = R$) is given by:

$$t = \frac{18\mu}{d^2\omega^2(\rho_s - \rho)}\ln\frac{R}{r_0} \tag{6.124}$$

If h is the thickness of the liquid layer at the walls, then:

$$h = R - r_0$$

$$\text{Then}: \ln\frac{R}{r_0} = \ln\frac{R}{R-h} = -\ln\left(1-\frac{h}{R}\right)$$

$$= \frac{h}{R} + \frac{1}{2}\left(\frac{h}{R}\right)^2 + \cdots$$

If h is small compared with R, then:

$$\ln\frac{R}{r_0} \approx \frac{h}{R}$$

Eq. (6.123) then becomes:

$$t = \frac{18\mu h}{d^2\omega^2(\rho_s-\rho)R} \tag{6.125}$$

For the Newton's law region, the equation of motion is:

$$\frac{\pi}{6}d^3(\rho_s-\rho)r\omega^2 - 0.22\frac{\pi}{4}d^2\rho\left(\frac{dr}{dt}\right)^2 = \frac{\pi}{6}d^3\rho_s\frac{d^2r}{dt^2}$$

This equation can only be solved numerically. If the acceleration term may be neglected, then:

$$\left(\frac{dr}{dt}\right)^2 = 3d\omega^2\left(\frac{\rho_s-\rho}{\rho}\right)r$$

$$\text{Thus}: \quad r^{-1/2}\frac{dr}{dt} = \left\{3d\omega^2\left(\frac{\rho_s-\rho}{\rho}\right)\right\}^{1/2} \tag{6.126}$$

Integration gives:

$$2\left(r^{1/2}-r_1^{1/2}\right) = \left\{3d\omega^2\frac{\rho_s-\rho}{\rho}\right\}^{1/2}t$$

$$\text{or}: \quad t = \left[\frac{\rho}{3d\omega^2(\rho_s-\rho)}\right]^{1/2}2\left(r^{1/2}-r_1^{1/2}\right) \tag{6.127}$$

6.9 Nomenclature

		Units in SI System	Dimensions in M, L, T
A'	Projected area of particle in plane perpendicular to direction of motion	m^2	L^2
a	$18\,\mu/d^2\rho_s$	s^{-1}	T^{-1}

a'	$36\,\mu/d^2(2\rho_s+\rho)$	s^{-1}	**T**$^{-1}$
B_1, B_2	Coefficients in Eq. (6.114)	m	**L**
b	$[1-(\rho/\rho_s)]g$	m/s^2	**LT**$^{-2}$
b'	$2\,g(\rho_s-\rho)/(2\rho_s+\rho)$	m/s^2	**LT**$^{-2}$
b_0	Constant, Eq. (6.72)	–	–
c	$\rho/3d\rho_s$	m^{-1}	**L**$^{-1}$
C_D	Drag coefficient $2R'/\rho u^2$	–	–
C_D'	Drag coefficient $R'/\rho u^2$	–	–
C_{D_0}	Drag coefficient in the absence of turbulence	–	–
$C_{D\infty}$	Limiting value of drag for a sphere	–	–
d	Diameter of sphere or characteristic dimension of particle	m	**L**
d_p	Mean projected diameter of particle	m	**L**
d_t	Diameter of tube or vessel	m	**L**
d'	Linear dimension of particle	m	**L**
d_n	Diameter of a circle of area equal to the projected area of a particle	m	L
d_s	Equal volume sphere diameter	m	L
F	Total force on particle	N	**MLT**$^{-2}$
f	$\sqrt{(b/c)}$	m/s	**LT**$^{-1}$
g	Acceleration due to gravity	m/s^2	**LT**$^{-2}$
h	Thickness of liquid layer	m	**L**
j	$(f+v)/(f-v)$	–	–
K	Constant for given shape and orientation of particle	–	–
K_D	$\frac{2}{3}(R'/\rho u^2)(Re_0'-1)$—see Eq. (6.42)	–	–
k	Consistency coefficient for power-law fluid	Ns^n/m^2	**ML**$^{-1}$**T**$^{n-2}$
k_{HB}	Consistency coefficient for Herschel–Bulkley fluid (Eq. 6.67)	Ns^n/m^2	**ML**$^{-1}$**T**$^{n-2}$
k'	Constant for calculating volume of particle	–	–
L'	Distance of particle from bottom of container	m	**L**
m	Mass of particle	kg	**M**
m'	Virtual mass (mass+added mass)	kg	**M**
n	Power-law index for non-Newtonian fluid	–	–
n_{HB}	Power-law index for Herschel–Bulkley fluid (Eq. 6.67)	–	–
q	$[1-(\rho/\rho_s)]\omega^2$ (Eqs 6.114 and 6.110)	s^{-2}	**T**$^{-2}$
q'	$2(\rho_s-\rho)\omega^2/(2\rho_s+\rho)$	s^{-2}	**T**$^{-2}$
p	$2fc$	s^{-1}	**T**$^{-1}$

Q	Correction factor for velocity of bubble	–	–
R	Radius of basket, *or*	m	**L**
	Shear stress at wall of pipe	N/m^2	**ML^{-1} T^{-2}**
R'	Resistance per unit projected area of particle	N/m^2	**ML^{-1} T^{-2}**
R'_0	Resistance per unit projected area of particle at free falling condition	N/m^2	**ML^{-1} T^{-2}**
r	Radius of rotation	m	**L**
r_0	Radius of inner surface of liquid	m	**L**
r_1	Coefficient in Eq. (6.65)	–	–
r_2	Exponent in Eq. (6.65)	–	–
S	Index in Eq. (6.32)	–	–
s	$1 + j\,e^{pt}$	–	–
s_r	Density ratio (ρ_s/ρ)	–	–
t	Time	s	**T**
u	Velocity of fluid relative to particle	m/s	**LT^{-1}**
u_0	Terminal falling velocity of particle	m/s	**LT^{-1}**
u_{0t}	Terminal falling velocity of particle in vessel	m/s	**LT^{-1}**
v	Initial component of velocity of particle in Y-direction	m/s	**LT^{-1}**
W'	Apparent (buoyant) weight of particle	N	**MLT^{-2}**
w	Initial component of velocity of particle in X-direction	m/s	**LT^{-1}**
x	Displacement of particle in X-direction at time t	m	**L**
$\dot{x}$	Velocity of particle in X-direction at time t	m/s	**LT^{-1}**
$\ddot{x}$	Acceleration of particle in X-direction at time t	m/s^2	**LT^{-2}**
Y	Correction factor in Stokes' law for power-law fluid	–	–
y	Displacement of particle in Y-direction at time t	m	**L**
$\dot{y}$	Velocity of particle in Y direction at time t	m/s	**LT^{-1}**
$\ddot{y}$	Acceleration of particle in Y-direction at time t	m/s^2	**LT^{-2}**
Z	Ratio of forces due to yield stress and to gravity (Eq. 6.66)	–	–
α	Angle between direction of motion of particle and horizontal	–	–
α_1	Constant, Eq. (6.73e)	–	–
β	Coefficient in Eqs (6.45) and (6.46)	–	–
$\dot{\gamma}$	Shear rate	s^{-1}	**T^{-1}**
δ_1	Constant, Eq. (6.72)	–	–
δ_0	Constant, Eq. (6.72)	–	–
η	$(C_D - C_{D_0})/C_{D_0}$	–	–

λ	Mean free path	m	**L**
λ_E	Kolmogoroff scale of turbulence (Eq. 6.46)	m	**L**
μ	Viscosity of fluid	N s/m^2	**ML^{-1} T^{-1}**
μ_1	Viscosity of fluid in drop or bubble	N s/m^2	**ML^{-1} T^{-1}**
μ_c	Viscosity at shear rate (u/d)	Ns/m^2	**ML^{-1} T^{-1}**
ϕ	Pipe friction factor $R/\rho u^2$	–	–
χ	Composite shape factor, Eq. (6.71)	–	–
τ_Y	Yield stress	N/m^2	**ML^{-1} T^{-2}**
ρ	Density of fluid	kg/m^3	**ML^{-3}**
ρ_s	Density of solid	kg/m^3	**ML^{-3}**
ψ	Sphericity	–	–
ω	Angular velocity	rad/s	**T^{-1}**
Bi	Bingham number (Eq. 6.69)	–	–
Ga	Galileo number $d^3(\rho_s - \rho)\rho g/\mu^2$	–	–
Ga_n	Galileo number for power-law fluid (Eq. 6.63)	–	–
Re'	Reynolds number $ud\rho/\mu$ or $ud'\rho/\mu$	–	–
Re'_0	Reynolds number $u_0 d\rho/\mu$	–	–
Re'_{HB}	Reynolds number for spherical particle in a Herschel–Bulkley fluid	–	–
Re'_n	Reynolds number for spherical particle in power-law fluid	–	–

Suffixes

A, B Particle **A, B**

References

1. Stokes GG. On the effect of the internal friction of fluids on the motion of pendulums. *Trans Cam Phil Soc* 1851;**9**:8.
2. Goldstein S. The steady flow of viscous fluid past a fixed spherical obstacle at small Reynolds numbers. *Proc Roy Soc* 1929;**A 123**:225.
3. Oseen CW. Über den Gültigkeitsbereich der Stokesschen Widerstandsformel. *Ark Mat Astr Fys* 1913;**9** (16):1–15.
4. Jenson VG. Viscous flow around a sphere at low Reynolds numbers (<40). *Proc Roy Soc* 1959;**249A**:346.
5. le Clair BP, Hamielec AE, Pruppacher HR. A numerical study of the drag on a sphere at low and intermediate Reynolds numbers. *J Atmos Sci* 1970;**27**:308.
6. Fornberg B. A numerical study of steady viscous flow past a circular cylinder. *J Fluid Mech* 1980;**98**:819–55.
7. Fornberg B. Steady viscous flow past a sphere at high Reynolds numbers. *J Fluid Mech* 1988;**190**:471–89.
8. Johnson TA, Patel VC. Flow past a sphere up to a Reynolds number of 300. *J Fluid Mech* 1999;**378**:19–70.
9. Dallavalle JM. *Micromeritics*. 2nd ed. Pitman; 1948.
10. Schiller L, Naumann A. Über die grundlegenden Berechnungen der Schwerkraftaufbereitung. *Z Ver deut Ing* 1933;**77**:318.
11. Deshpande R, Kanti V, Desai A, Mittal S. Intermittency of laminar separation bubble on a sphere during drag crisis. *J Fluid Mech* 2017;**812**:815–40.

12. Clift R, Grace JR, Weber ME. *Bubbles, drops and particles.* Academic Press; 1978.

13. Michaelides EE. *Particles, bubbles and drops—their motion, heat and mass transfer.* Singapore: World Scientific; 2006.

14. Wadell H. The coefficient of resistance as a function of Reynolds' number for solids of various shapes. *J Frankl Inst* 1934;**217**:459.

15. Khan AR, Richardson JF. The resistance to motion of a solid sphere in a fluid. *Chem Eng Commun* 1987;**62**:135.

16. Heywood H. Calculation of particle terminal velocities. *J Imp Coll Chem Eng Soc* 1948;**4**:17.

17. Karamanev DG, Chavarie C, Mayer RC. Free rise of a light solid sphere in liquid. *AIChE J* 1996;**42**:1789.

18. Dewsbury KH, Karamanev DG, Margaritis A. Dynamic behavior of freely rising buoyant solid sphere in non-Newtonian liquids. *AIChE J* 2000;**46**:46.

19. Ladenburg R. Über den Einfluss von Wänden auf die Bewegung einer Kugel in einer reibenden Flüssigkeit. *Ann Phys* 1907;**23**:447.

20. Francis AW. Wall effects in falling ball method for viscosity. *Physics* 1933;**4**:403.

21. Haberman WL, Sayre RM. *Motion of rigid and fluid spheres in stationary and moving liquids inside cylindrical tubes.* David Taylor Model Basin Report No. 1143 (1958).

22. Fidleris V, Whitmore RL. Experimental determination of the wall effect for spheres falling axially in cylindrical vessels. *Brit J Appl Phys* 1961;**12**:490.

23. Khan AR, Richardson JF. Fluid-particle interactions and flow characteristics of fluidized beds and settling suspensions of spherical particles. *Chem Eng Commun* 1989;**78**:111.

24. Wham RM, Basaran OA, Byers CH. Wall effects on flow past solid spheres at finite Reynolds numbers. *I&EC Res* 1995;**35**:864–74.

25. Chhabra RP. Wall effects on spheres falling axially in cylindrical tubes. In: Dekee D, Chhabra RP, editors. *Transport processes in bubbles, drops and particles.* 2nd ed. , (Chapter 13), New York: Taylor and Francis; 2002.

26. Tanner RI. End effects in falling ball viscometry. *J Fluid Mech* 1963;**17**:161.

27. Sutterby JL. Falling sphere viscometry. I. Wall and inertial corrections to Stokes' law in long tubes. *Trans Soc Rheol* 1973;**17**:559.

28. Davies CN. The sedimentation and diffusion of small particles. *Proc Roy Soc* 1949;**A200**:100.

29. Cunningham E. Velocity of steady fall of spherical particles. *Proc Roy Soc* 1910;**A83**:357.

30. Davies CN. Definitive equations for the fluid resistance of spheres. *Proc Phys Soc* 1945;**57**:259.

31. Reist PC. *Aerosol science and technology.* 2nd ed. Singapore: McGraw Hill; 1993.

32. Brucato A, Grisafi F, Montante G. Particle drag coefficients in turbulent fluids. *Chem Eng Sci* 1998;**53**:3295.

33. Richardson JF, Meikle RA. Sedimentation and fluidisation. Part IV. Drag force on individual particles in an assemblage. *Trans Inst Chem Eng* 1961;**39**:357.

34. Rowe PN, Henwood GN. Drag forces in a hydraulic model of a fluidised bed. Part 1. *Trans Inst Chem Eng* 1961;**39**:43.

35. Chhabra RP, Agarwal L, Sinha NK. Drag on non-spherical particles: an evaluation of available methods. *Powder Technol* 1999;**101**:288.

36. Darby R, Chhabra RP. *Chemical engineering fluid mechanics.* 3rd ed. New York: Taylor and Francis; 2017.

37. Heiss JF, Coull J. The effect of orientation and shape on the settling velocity of non-isometric particles in a viscous medium. *Chem Eng Prog* 1952;**48**:133.

38. Haider A, Levenspiel O. Drag coefficient and terminal velocity of spherical and non-spherical particles. *Powder Technol* 1989;**58**:63–70.

39. Ganser GH. A rational approach to drag prediction of spherical and non-spherical particles. *Powder Technol* 1993;**77**:143–52.

40. Hartman M, Trnka O, Svoboda K. Free settling of nonspherical particles. *Ind Eng Chem Res* 1994;**33**:1979–83.

41. Kasper G. Dynamics and measurement of smokes. I. Size characterization of nonspherical particles. *Aerosol Sci Technol* 1982;**1**:187–99.

42. Tran-Cong S, Gay M, Michaelides EE. Drag coefficient of irregularly shaped particles. *Powder Technol* 2004;**139**:21–32.

43. Nikora V, Aberle J, Green M. Sediment flocs: settling velocity, flocculation factor and optical backscatter. *J Hydraul Eng* 2004;**130**:1043–7.

44. Vahedi A. *Predicting the settling velocity of lime softening flocs using fractal geometry.* [PhD thesis]. Winnipeg, Canada: University of Manitoba; 2011.

45. Bagheri GH, Bonadonna C, Manzella I, Vonlanthen P. On the characterization of size and shape of irregular particles. *Powder Technol* 2015;**270**:141–53.

46. Hadamard J. Mouvement permanent lent d'une sphère liquide et visqueuse dans un liquide visqueux. *C R Acad Sci* 1911;**152**:1735.

47. Bond WN. Bubbles and drops and Stokes' law. *Phil Mag 7th Ser* 1927;**4**:889.

48. Bond WN, Newton DA. Bubbles, drops and Stokes' law (Paper 2). *Phil Mag 7th Ser* 1928;**5**:794.

49. Garner FH, Skelland AHP. Liquid–liquid mixing as affected by the internal circulation within drops. *Trans Inst Chem Eng* 1951;**29**:315.

50. Garner FH, Skelland AHP. Some factors affecting drop behaviour in liquid–liquid systems. *Chem Eng Sci* 1955;**4**:149.

51. Hamielec AE, Johnson AI. Viscous flow around fluid spheres at intermediate Reynolds numbers. *Can J Chem Eng* 1962;**40**:41.

52. Kelbaliyev GI. Drag coefficients of variously shaped solid particles, drops and bubbles. *Theor Found Chem Eng* 2011;**45**:248–66.

53. Chhabra RP. *Bubbles, drops and particles in non-Newtonian fluids.* 2nd ed. Boca Raton, FL: CRC Press; 2007.

54. Chhabra RP, Richardson JF. *Non-Newtonian flow and applied rheology.* 2nd ed. Oxford: Butterworth-Heinemann; 2008.

55. Tripathi A, Chhabra RP, Sundararajan T. Power law fluid flow over spherical particles. *Ind Eng Chem Res* 1994;**34**:403.

56. Graham DI, Jones TER. Settling and transport of spherical particles in power-law fluids at finite Reynolds number. *J Non-Newt Fluid Mech* 1994;**54**:465.

57. Song D, Gupta RK, Chhabra RP. Wall effects on a sphere falling in power-law fluids in cylindrical tubes. *I&EC Res* 2009;**48**:5845–56.

58. Shah SN, El-Fadili Y, Chhabra RP. New model for single spherical particle settling velocity in power-law (visco-inelastic) fluids. *Int J Multiphase Flow* 2007;**33**:51–66.

59. Barnes HA. The yield stress—a review. *J Non-Newt Fluid Mech* 1999;**81**:133.

60. Atapattu DD, Chhabra RP, Uhlherr PHT. Creeping sphere motion in Herschel-Bulkley fluids; flow field and drag. *J Non-Newt Fluid Mech* 1995;**59**:245.

61. Nirmalkar N, Chhabra RP, Poole RJ. Effect of shear-thinning behavior on heat transfer from a heated sphere in yield-stress fluids. *I&EC Res* 2013;**52**:13490–504.

62. Venumadhav G, Chhabra RP. Settling velocities of single nonspherical particles in non-Newtonian fluids. *Powder Technol* 1994;**78**:77.

63. Rajitha P, Chhabra RP, Sabiri NE, Comiti J. Drag on non-spherical particles in power-law non-Newtonian media. *Int J Miner Process* 2006;**78**:110–21.

64. Renaud M, Mauret E, Chhabra RP. Power-law fluid flow over a sphere; average shear rate and drag coefficient. *Can J Chem Eng* 2004;**82**:1066–70.

65. Mauret E, Renaud M. Transport phenomena in multiparticle system—II. Proposed new model based on flow around submerged objects for sphere and fiber beds-transition between the capillary and particulate representations. *Chem Eng Sci* 1997;**52**:1819–34.

66. Torobin LB, Gauvin WH. Fundamental aspects of solids-gas flow. Part I: introductory concepts and idealized sphere-motion in viscous regime. Part II: the sphere wake in steady laminar fluids. Part III: accelerated motion of a particle in a fluid. *Can J Chem Eng* 1959;**38**.129, 167, 224.

67. Mironer A. *Engineering fluid mechanics.* McGraw Hill; 1979.

Further Reading

Curle N, Davies HJ. *Modern fluid dynamics. Volume 1. Incompressible flow.* Van Nostrand; 1968.

Orr C. *Particulate technology.* Macmillan; 1966.

Ortega-Rivas E. *Unit operations of particulate solids.* Boca Raton, FL: CRC Press; 2012.

CHAPTER 7

Flow of Fluids Through Granular Beds and Packed Columns

7.1 Introduction

The flow of fluids through beds composed of stationary granular particles is a frequent occurrence in the chemical industry, and, therefore, expressions are needed to predict pressure drop across beds due to the resistance to fluid flow caused by the presence of the particles. For example, in fixed-bed catalytic reactors, such as SO_2–SO_3 converters, and drying columns containing silica gel or molecular sieves, gases are passed through a bed of particles. In the case of gas absorption into a liquid, the gas flows upward against a falling liquid stream, the fluids being contained in a vertical column packed with shaped particles. In the filtration of a suspension, liquid flows at a relatively low velocity through the spaces between the particles which have been retained by the filter medium, and, as a result of the continuous deposition of solids, the resistance to flow increases progressively throughout the operation. Furthermore, deep bed filtration is used on a very large scale in water treatment, for example, where the quantity of solids to be removed is small. In all these instances, it is necessary to estimate the size of the equipment required, and design expressions are required for the drop in pressure for a fluid flowing through a packing, either alone or as a two-phase system. The corresponding expressions for fluidised beds are discussed in Chapter 9. The drop in pressure for flow through a bed of small particles provides a convenient method for obtaining a measure of the external surface area of a powder, for example, cement or pigment.

The flow of either a single phase through a bed of particles or the more complex flow of two fluid phases is approached by using the concepts developed in Volume 1A for the flow of an incompressible fluid through regular pipes or ducts. It is found, however, that the problem is not, in practice, capable of complete analytical solution, and the use of experimental data obtained for a variety of different systems is essential. Later in the chapter, some aspects of the design of industrial packed columns involving countercurrent flow of liquids and gases are described.

Coulson and Richardson's Chemical Engineering. https://doi.org/10.1016/B978-0-08-101098-3.00008-1

7.2 Flow of a Single Fluid Through a Granular Bed

7.2.1 Darcy's Law and Permeability

The first experimental work on the subject was carried out by Darcy[1] in 1830 in Dijon, when he examined the rate of flow of water from the local fountains through beds of sand of various thicknesses. It was shown that the average velocity, as measured over the whole area of the bed, was directly proportional to the driving pressure and inversely proportional to the thickness of the bed. This relation, often termed Darcy's law, has subsequently been confirmed by a number of researchers and can be written as follows:

$$u_c = K\frac{(-\Delta P)}{l} \tag{7.1}$$

where $-\Delta P$ is the pressure drop across the bed, l is the thickness of the bed, u_c is the average velocity of flow of the fluid, defined as $(1/A)(dV/dt)$, A is the total cross sectional area of the bed, V is the volume of fluid flowing in time t, and K is a constant depending on the physical properties of the bed and fluid.

The linear relation between the rate of flow and the pressure difference leads one to suppose that the flow was streamline, as discussed in Volume 1A, Chapter 3. This would be expected because the Reynolds number for the flow through the pore spaces in a granular material is generally low, because both the velocity of the fluid and the width of the channels are normally small. The resistance to flow then arises mainly from viscous drag. Eq. (7.1) can then be expressed as:

$$u_c = \frac{K(-\Delta P)}{l} = B\frac{(-\Delta P)}{\mu l} \tag{7.2}$$

where μ is the viscosity of the fluid, and B is termed the permeability coefficient for the bed, and depends only on the properties of the bed.

The value of the permeability coefficient is frequently used to give an indication of the ease with which a fluid will flow through a bed of particles or a filter medium. Some values of B for various packings, taken from Eisenklam,[2] are shown in Table 7.1, and it can be seen that B can vary over a wide range of values. It should be noted that these values of B apply only to the laminar flow regime.

7.2.2 Specific Surface and Voidage

The general structure of a bed of particles can often be characterised by the specific surface area of the bed S_B and the fractional voidage of the bed e.

Table 7.1 Properties of beds of some regular-shaped materials[2]

No.	Description	Specific Surface Area S (m²/m³)	Fractional Voidage, e (−)	Permeability Coefficient B (m²)
	Spheres			
1	0.794 mm diam. ($\frac{1}{32}$ in.)	7600	0.393	6.2×10^{-10}
2	1.588 mm diam. ($\frac{1}{16}$ in.)	3759	0.405	2.8×10^{-9}
3	3.175 mm diam. ($\frac{1}{8}$ in.)	1895	0.393	9.4×10^{-9}
4	6.35 mm diam. ($\frac{1}{4}$ in.)	948	0.405	4.9×10^{-8}
5	7.94 mm diam. ($\frac{5}{16}$ in.)	756	0.416	9.4×10^{-8}
	Cubes			
6	3.175 mm ($\frac{1}{8}$ in.)	1860	0.190	4.6×10^{-10}
7	3.175 mm ($\frac{1}{8}$ in.)	1860	0.425	1.5×10^{-8}
8	6.35 mm ($\frac{1}{4}$ in.)	1078	0.318	1.4×10^{-8}
9	6.35 mm ($\frac{1}{4}$ in.)	1078	0.455	6.9×10^{-8}
	Hexagonal prisms			
10	4.76 mm × 4.76 mm thick ($\frac{3}{16}$ in. × $\frac{3}{16}$ in.)	1262	0.355	1.3×10^{-8}
11	4.76 mm × 4.76 mm thick ($\frac{3}{16}$ in. × $\frac{3}{16}$ in.)	1262	0.472	5.9×10^{-8}
	Triangular pyramids			
12	6.35 mm length × 2.87 mm ht. ($\frac{1}{4}$ in. × 0.113 in.)	2410	0.361	6.0×10^{-9}
13	6.35 mm length × 2.87 mm ht. ($\frac{1}{4}$ in. × 0.113 in.)	2410	0.518	1.9×10^{-8}
	Cylinders			
14	3.175 mm × 3.175 mm diam. ($\frac{1}{8}$ in. × $\frac{1}{8}$ in.)	1840	0.401	1.1×10^{-8}
15	3.175 mm × 6.35 mm diam. ($\frac{1}{8}$ in. × $\frac{1}{4}$ in.)	1585	0.397	1.2×10^{-8}
16	6.35 mm × 6.35 mm diam. ($\frac{1}{4}$ in. × $\frac{1}{4}$ in.)	945	0.410	4.6×10^{-8}

Continued

Table 7.1   Properties of beds of some regular-shaped materials—Cont'd

	Solid Constituents	Porous Mass		
No.	Description	Specific Surface Area $S(m^2/m^3)$	Fractional Voidage, $e\ (-)$	Permeability Coefficient $B\ (m^2)$
	Plates			
17	$6.35\,mm \times 6.35\,mm \times 0.794\,mm$ ($\frac{1}{4}$ in. $\times \frac{1}{4}$ in. $\times \frac{1}{32}$ in.)	3033	0.410	5.0×10^{-9}
18	$6.35\,mm \times 6.35\,mm \times 1.59\,mm$ ($\frac{1}{4}$ in. $\times \frac{1}{4}$ in. $\times \frac{1}{16}$ in.)	1984	0.409	1.1×10^{-8}
	Discs			
19	$3.175\,mm$ diam. $\times 1.59\,mm$ ($\frac{1}{8}$ in. $\times \frac{1}{16}$ in.)	2540	0.398	6.3×10^{-9}
	Porcelain Berl saddles			
20	6 mm (0.236 in.)	2450	0.685	9.8×10^{-8}
21	6 mm (0.236 in.)	2450	0.750	1.73×10^{-7}
22	6 mm (0.236 in.)	2450	0.790	2.94×10^{-7}
23	6 mm (0.236 in.)	2450	0.832	3.94×10^{-7}
24	Lessing rings (6 mm)	5950	0.870	1.71×10^{-7}
25	Lessing rings (6 mm)	5950	0.889	2.79×10^{-7}

S_B is the surface area presented to the fluid per unit volume of bed when the particles are packed in a bed. Its units are $(\text{length})^{-1}$.

e is the fraction of the volume of the bed not occupied by solid material and is termed the fractional voidage, voidage, or porosity. It is dimensionless. Thus, the fractional volume of the bed occupied by solid material is $(1 - e)$.

S is the specific surface area of the particles and is the surface area of a particle divided by its volume. Its units are again $(\text{length})^{-1}$. For a sphere, for example:

$$S = \frac{\pi d^2}{\pi (d^3/6)} = \frac{6}{d} \tag{7.3}$$

It can be seen that S and S_B are not equal due to the voidage which is present when the particles are packed into a bed. If point contact occurs between particles, so that only a very small fraction of surface area is lost by overlapping, then:

$$S_B = S(1 - e) \tag{7.4}$$

Some values of S and e for different beds of particles are listed in Table 7.1. Values of e much higher than those shown in Table 7.1, sometimes up to about 0.95, are possible in beds of fibres[3] and some ring packings. For a given shape of particle, S increases as the particle size is reduced, as shown in Table 7.1.

As e is increased, flow through the bed becomes easier, and so the permeability coefficient B increases; a relation between B, e, and S is developed in a later section of this chapter. If the particles are randomly packed, then e should be approximately constant throughout the bed, and the resistance to flow the same in all directions. Often, near containing walls, e is higher, and corrections for this should be made if the particle size is a significant fraction of the size of the containing vessel. This correction is discussed in more detail later.

7.2.3 General Expressions for Flow Through Beds in Terms of Carman–Kozeny Equations

Streamline flow—Carman–Kozeny equation

Many attempts have been made to obtain general expressions for pressure drop and mean velocity for flow through packings in terms of voidage and specific surface, as these quantities are often known or can be measured. Alternatively, measurements of the pressure drop, velocity, and voidage provide a convenient way of measuring the surface area of some particulate materials, as described later.

The analogy between streamline flow through a tube and streamline flow through the pores in a bed of particles is a useful starting point for deriving a general expression. From Volume 1A, Chapter 3, the equation for streamline flow through a circular tube is:

$$u = \frac{d_t^2}{32\mu} \frac{(-\Delta P)}{l_t} \tag{7.5}$$

where μ is the viscosity of the fluid, u is the mean velocity of the fluid, d_t is the diameter of the tube, and l_t is the length of the tube.

If the free space in the bed is assumed to consist of a series of tortuous channels arranged in parallel, as shown schematically in Fig. 7.1A and its idealisation in Fig. 7.1B, Eq. (7.5) may be rewritten for flow through a bed as:

$$u_1 = \frac{d'^2_m}{K'\mu} \frac{(-\Delta P)}{l'} \tag{7.6}$$

where d'_m is some equivalent diameter of the pore channels, K' is a dimensionless constant whose value depends on the structure of the bed, l' is the length of channel, and u_1 is the average velocity through the pore channels.

It should be noted that u_1 and l' in Eq. (7.6) now represent conditions in the pores and are not the same as u_c and l in Eqs. (7.1) and (7.2). However, it is a reasonable assumption that l' is directly proportional to l. Dupuit[4] related u_c and u_1 by the following argument.

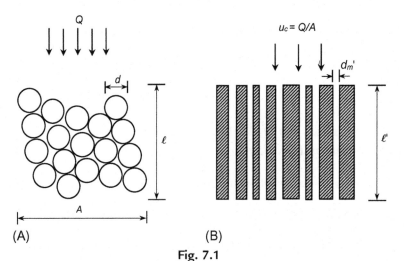

Fig. 7.1

Schematic representation of flow through a bed of uniform spheres (A) and the capillary model idealisation (B).

In a cube of side X, the volume of free space is eX^3, so that the mean cross-sectional area for flow is the free volume divided by the height, or eX^2. The volume flow rate through this cube is $u_c X^2$, so that the average linear velocity through the pores, u_1, is given by:

$$u_1 = \frac{u_c X^2}{eX^2} = \frac{u_c}{e} \tag{7.7}$$

Although Eq. (7.7) is reasonably true for random packings, it does not apply to all regular packings. Thus, with a bed of spheres arranged in cubic packing, $e = 0.476$, but the fractional free area varies continuously, from 0.215 in a plane across the diameters to 1.0 between successive layers.

For Eq. (7.6) to be generally useful, an expression is needed for d_m', the equivalent diameter of the pore space. Kozeny[5, 6] proposed that d_m' may be taken as:

$$d_m' = \frac{e}{S_B} = \frac{e}{S(1-e)}$$

$$\text{where} \quad : \frac{e}{S_B} = \frac{\text{volume of voids filled with fluid}}{\text{wetted surface area of the bed}} \tag{7.8}$$

$$= \frac{\text{cross} - \text{sectional area normal to flow}}{\text{wetted perimeter}}$$

The hydraulic mean diameter for such a flow passage has been shown in Volume 1A, Chapter 3 to be:

$$4\left(\frac{\text{cross} - \text{sectional area}}{\text{wetted perimeter}}\right)$$

It is then seen that:

$$\frac{e}{S_B} = \frac{1}{4}(\text{hydraulic mean diameter})$$

Then, taking $u_1 = u_c/e$ and $l' \propto l$, Eq. (7.6) becomes:

$$u_c = \frac{1}{K''} \frac{e^3}{S_B^2 \mu} \frac{1}{l} \frac{(-\Delta P)}{}$$

$$= \frac{1}{K''} \frac{e^3}{S^2(1-e)^2 \mu} \frac{1}{l} \frac{(-\Delta P)}{} \tag{7.9}$$

K'' is generally known as Kozeny's constant, and a commonly accepted value for K'' is 5. As will be shown later, however, K'' is dependent on porosity, particle shape, and other factors. Comparison with Eq. (7.2) shows that B, the permeability coefficient is given by:

$$B = \frac{1}{K''} \frac{e^3}{S^2(1-e)^2} \tag{7.10}$$

Inserting a value of 5 for K'' in Eq. (7.9):

$$u_c = \frac{1}{5} \frac{e^3}{(1-e)^2} \frac{-\Delta P}{S^2 \mu l} \qquad (7.11)$$

For spheres, $S = 6/d$ and thus,

$$u_c = \frac{1}{180} \frac{e^3}{(1-e)^2} \frac{-\Delta P d^2}{\mu l} \qquad (7.12)$$

$$= 0.0055 \frac{e^3}{(1-e)^2} \frac{-\Delta P d^2}{\mu l} \qquad (7.12a)$$

For nonspherical particles, the Sauter mean diameter d_s should be used in place of d. This is given in Chapter 1, Eq. (1.15).

Streamline and turbulent flow

Eq. (7.9) applies to streamline flow conditions, though Carman[7] and others have extended the analogy with pipe flow to cover both streamline and turbulent flow conditions through packed beds. In this treatment, a modified friction factor $R_1/\rho u_1^2$ is plotted against a modified Reynolds number Re_1. This is analogous to plotting $R/\rho u^2$ against Re for flow through a pipe, as discussed in Volume 1A, Chapter 3.

The modified Reynolds number Re_1 is obtained by taking the same velocity and characteristic linear dimension d_m' as were used in deriving Eq. (7.9). Thus:

$$Re_1 = \frac{u_c}{e} \frac{e}{S(1-e)} \frac{\rho}{\mu}$$
$$= \frac{u_c \rho}{S(1-e)\mu} \qquad (7.13)$$

The friction factor, which is plotted against the modified Reynolds number, is $R_1/\rho u_1^2$, where R_1 is the component of the drag force per unit area of particle surface in the direction of motion. R_1 can be related to the properties of the bed and pressure gradient as follows. Considering the forces acting on the fluid in a bed of unit cross-sectional area and thickness l, the volume of particles in the bed is $l(1-e)$, and, therefore, the total surface is $Sl(1-e)$. Thus, the hydrodynamic resistance force is $R_1 Sl(1-e)$. This force on the fluid must be equal to that produced by a pressure difference of ΔP across the bed. Then, because the free cross-section of fluid is equal to e:

$$(-\Delta P)e = R_1 Sl(1-e)$$

$$\text{and} \quad R_1 = \frac{e}{S(1-e)} \frac{(-\Delta P)}{l} \qquad (7.14)$$

$$\text{Thus} \quad \frac{R_1}{\rho u_1^2} = \frac{e^3}{S(1-e)} \frac{(-\Delta P)}{l} \frac{1}{\rho u_c^2} \tag{7.15}$$

Carman found that when $R_1/\rho u_1^2$ was plotted against Re_1 using logarithmic coordinates, his data for the flow through randomly packed beds of solid particles could be correlated approximately by a single curve (Curve A, Fig. 7.2), whose general equation is:

$$\frac{R_1}{\rho u_1^2} = 5Re_1^{-1} + 0.4Re_1^{-0.1} \tag{7.16}$$

The form of Eq. (7.16) is similar to that of Eq. (7.17) proposed by Forchheimer,[8] who suggested that the resistance to flow should be considered in two parts: that due to the viscous drag at the surface of the particles, and that due to loss in turbulent eddies and at the sudden changes in the cross-section of the channels. Thus:

$$(-\Delta P) = \alpha u_c + \alpha' u_c^{n'} \tag{7.17}$$

The first term in this equation will dominate at low rates of flow where the losses are mainly attributable to skin friction, and the second term will become significant at high flow

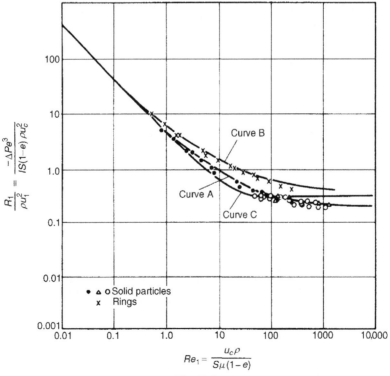

Fig. 7.2
Carman's graph of $R_1/\rho u_1^2$ against Re_1.

rates and in very thin beds where the enlargement and contraction losses become very important. At very high flow rates, the effects of viscous forces are negligible.

From Eq. (7.16), it can be seen that for values of Re_1 less than about 2, the second term is small, and, approximately:

$$\frac{R_1}{\rho u_1^2} = 5Re_1^{-1} \tag{7.18}$$

Eq. (7.18) can be obtained from Eq. (7.11) by substituting for $-\Delta P/l$ from Eq. (7.15). This gives:

$$u_c = \frac{1}{5}\left(\frac{1}{1-e}\right)\left(\frac{\rho u_c^2}{S\mu}\right)\left(\frac{R_1}{\rho u_1^2}\right)$$

$$\text{Thus}: \quad \frac{R_1}{\rho u_1^2} = 5\left(\frac{S(1-e)\mu}{u_c\rho}\right)$$

$$= 5Re_1^{-1} \quad \text{(from Eq.7.13)}$$

As the value of Re_1 increases from about 2 to 100, the second term in Eq. (7.16) becomes more significant, and the slope of the plot gradually changes from -1.0 to about $-\frac{1}{4}$. Above Re_1 of 100, the plot is approximately linear. The change from complete streamline flow to complete turbulent flow is very gradual because flow conditions are not the same in all the pores. Thus, the flow starts to become turbulent first in the larger pores, and subsequently, in successively smaller pores as the value of Re_1 increases. It is probable that the flow never becomes completely turbulent because some of the passages may be so small that streamline conditions prevail even at high flow rates.

Rings, which, as described later, are often used in industrial packed columns, tend to deviate from the generalised curve A on Fig. 7.2, particularly at high values of Re_1.

Sawistowski[9] compared the results obtained for flow of fluids through beds of hollow packings (discussed later) and has noted that Eq. (7.16) gives a consistently low result for these materials. He proposed:

$$\frac{R_1}{\rho u_1^2} = 5\,\text{Re}_1^{-1} + \text{Re}_1^{-0.1} \tag{7.19}$$

This equation is plotted as curve B in Fig. 7.2.

For flow through ring packings which as described later are often used in industrial packed columns, Ergun[10] obtained a good semi-empirical correlation for pressure drop as follows:

$$\frac{-\Delta P}{l} = 150\frac{(1-e)^2}{e^3}\frac{\mu u_c}{d^2} + 1.75\frac{(1-e)}{e^3}\frac{\rho u_c^2}{d} \tag{7.20}$$

Writing $d = 6/S$ (from Eq. 7.3):

$$\frac{-\Delta P}{Slpu_c^2}\frac{e^3}{1-e} = 4.17\frac{\mu S(1-e)}{\rho u_c} + 0.29$$

$$\text{or}: \quad \frac{R_1}{\rho u_1^2} = 4.17\,\text{Re}_1^{-1} + 0.29 \tag{7.21}$$

This equation is plotted as curve C in Fig. 7.2. The form of Eq. (7.21) is somewhat similar to that of Eqs (7.16) and (7.17), in that the first term represents viscous losses which are most significant at low velocities, and the second term represents kinetic energy losses which become more significant at high velocities. The equation is, thus, applicable over a wide range of velocities and was found by Ergun to correlate experimental data well for values of $Re_1/(1-e)$ from 1 to over 2000.

The form of the above equations suggests that the only properties of the bed on which the pressure gradient depends are its specific surface S (or particle size d) and its voidage e. However, the structure of the bed depends additionally on the particle size distribution, the particle shape, and the way in which the bed has been formed; in addition, both the walls of the container and the nature of the bed support can considerably affect the way the particles pack. It would be expected, therefore, that experimentally determined values of the pressure gradient would show a considerable scatter relative to the values predicted by the equations. The importance of some of these factors is discussed in the next section.

Furthermore, the rheology of the fluid is important in determining how it flows through a packed bed. Only Newtonian fluid behaviour has been considered hitherto. For non-Newtonian fluids, the effect of continual changes in the shape and cross-section of the flow passages may be considerable (particularly for viscoelastic fluids) and no simple relation may exist between the pressure gradient and flow rate. This problem has been the subject of extensive studies by several researchers including Kemblowski et al.[11]

In some applications, there may be simultaneous flow of two immiscible liquids, or of a liquid and a gas such as that encountered in oil wells and trickle bed reactors. In general, one of the liquids (or the liquid, in the case of liquid–gas systems) will preferentially wet the particles and flow as a continuous film over the surface of the particles, while the other phase flows through the remaining free space. The problem is complex, and the exact nature of the flow depends on the physical properties of the two phases, including their surface tensions. An analysis has been made by several researchers including Botset,[12] Glaser and Litt,[13] Dudukovic et al.,[14] and others.[15]

Dependence of K″ on bed structure

Tortuosity. Although it was implied in the derivation of Eq. (7.9) that a single value of the Kozeny constant K'' applied to all packed beds, in practice, this assumption does not hold.

Carman[7] has shown that:

$$K'' = \left(\frac{l'}{l}\right)^2 \times K_0 \tag{7.22}$$

where (l'/l) is the tortuosity and is a measure of the fluid path length through the bed compared with the actual depth of the bed, K_0 is a factor which depends on the shape of the cross-section of a channel through which fluid is passing.

For streamline fluid flow through a circular pipe where Poiseuille's equation applies (given in Volume 1A, Chapter 3), K_0 is equal to 2.0, and for streamline flow through a rectangle where the ratio of the lengths of the sides is 10: 1, $K_0 = 2.65$. Carman[16] has listed values of K_0 for other cross-sections. From Eq. (7.22), it can be seen that if, say, K_0 were constant, then K'' would increase with increase in tortuosity. The reason for K'' being near to 5.0 for many different beds is probably that changes in tortuosity from one bed to another have been compensated by changes in K_0 in the opposite direction.

Wall effect. In a packed bed, the particles will not pack as closely in the region near the wall as in the centre of the bed, so that the actual resistance to flow in a bed of small diameter is less than it would be in an infinite container for the same flow rate per unit area of bed cross-section. A correction factor f_w for this effect has been determined experimentally by Coulson.[17] This takes the form:

$$f_w = \left(1 + \frac{1}{2}\frac{S_c}{S}\right)^2 \tag{7.23}$$

where S_c is the surface of the container per unit volume of bed.

Eq. (7.9) then becomes:

$$u_c = \frac{1}{K''}\frac{e^3}{S^2(1-e)^2}\frac{1}{\mu}\frac{(-\Delta P)}{l}f_w \tag{7.24}$$

The values of K'' shown in Fig. 7.3 apply to Eq. (7.24).

Nonspherical particles. Coulson[17] and Wyllie and Gregory[18] have each determined values of K'' for particles of many different sizes and shapes, including prisms, cubes, and plates. Some of these values for K'' are shown in Fig. 7.3 where it is seen that they lie between 3 and 6, with the extreme values only occurring with thin plates. This variation of K'' with plates probably arises not only from the fact that area contact is obtained between the particles, but also because the plates tend to give greater tortuosities. For normal granular materials, Kihn[19] and Pirie[20] have found that K'' is reasonably constant and does not vary so widely as the K'' values for extreme shapes included in Fig. 7.3.

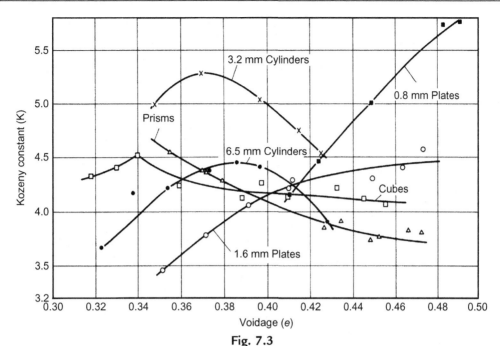

Fig. 7.3

Variation of Kozeny's constant K', with voidage for various shapes.

Spherical particles. Eq. (7.24) has been tested with spherical particles over a wide range of sizes and K'' has been found to be about 4.8 ± 0.3.[17, 21]

For beds composed of spheres of mixed sizes, the porosity of the packing can change very rapidly if the smaller spheres are able to fill the voids between the larger ones. Thus, Coulson[17] found that, with a mixture of spheres of size ratio 2:1, a bed behaves much in accordance with Eq. (7.19), but, if the size ratio is 5:1, and the smaller particles form <30% by volume of the larger ones, then K'' falls very rapidly, emphasising that only for uniform sized particles can bed behaviour be predicted with confidence.

Beds with high voidage. Spheres and particles which are approximately isometric do not pack to give beds with voidages in excess of about 0.6. With fibres and some ring packings, however, values of e near unity can be obtained, and for these high values, K'' rises rapidly. Some values are given in Table 7.2.

Deviations from the Carman–Kozeny Eq. (7.9) become more pronounced in these beds of fibres as the voidage increases, because the nature of the flow changes from one of channel flow to one in which the fibres behave as a series of obstacles in an otherwise unobstructed passage. The flow pattern is also different in expanded fluidised beds, and the Carman–Kozeny equation does not apply there either. As fine spherical particles move far apart in a fluidised bed, Stokes' law can be applied, whereas the Carman–Kozeny equation leads to no such limiting resistance. This problem is further discussed by Carman.[16]

Table 7.2 Experimental values of K'' for beds of high porosity

Voidage e	Experimental Value of K''		
	Brinkman[3]	Davies[22]	Silk Fibres[23]
0.5	5.5	–	–
0.6	4.3	–	–
0.8	5.4	6.7	5.35
0.9	8.8	9.7	6.8
0.95	15.2	15.3	9.2
0.98	32.8	27.6	15.3

Effect of bed support. The structure of the bed, and hence, K'', is markedly influenced by the nature of the support. For example, the initial condition in a filtration may affect the whole of a filter cake. Fig. 7.4 shows the difference in orientation of two beds of cubical particles, namely, on a plane surface (Fig. 7.4A) and above a bed of spheres (Fig. 7.4B). The importance of the packing support should not be overlooked in considering the drop in pressure through the column because the support may itself form an important resistance, and by orienting the particles as indicated, may also affect the total pressure drop.

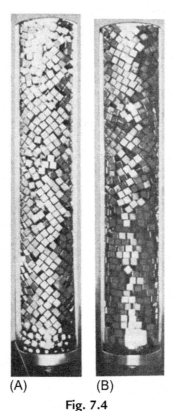

(A)   (B)

Fig. 7.4
Packing of cubes, stacked on (A) plane surface, and (B) on bed of spheres.

The application of Carman–Kozeny equations

Eqs (7.9) and (7.16), which involve e/S_B as a measure of the effective pore diameter, are developed from a relatively sound theoretical basis and are recommended for beds of small particles when they are nearly spherical in shape. The correction factor for wall effects, given by Eq. (7.23), should be included where appropriate. With larger particles which will frequently be far from spherical in shape, the correlations are not so reliable. As shown in Fig. 7.2, deviations can occur for rings at higher values of Re_1. Efforts to correct for nonsphericity, though frequently useful, are not universally effective, and in such cases, it will often be more rewarding to use correlations, such as Eq. (7.19), which are based on experimental data for large packings.

Use of Carman–Kozeny equation for measurement of particle surface area

The Carman–Kozeny equation relates the drop in pressure through a bed to the specific surface of the material and can, therefore, be used as a means of calculating S from measurements of the drop in pressure. This method is strictly only suitable for beds of uniformly packed particles, and it is not a suitable method for measuring the size distribution of particles in the subsieve range. A convenient form of apparatus developed by Lea and Nurse[24] is shown schematically in Fig. 7.5. In this apparatus, air or another suitable gas flows through the bed contained in a cell (25 mm diameter, 87 mm deep), and the pressure drop is obtained from h_1, and the gas flow rate from h_2.

If the diameters of the particles are below about 5 μm, then *slip* will occur, and this must be allowed for, as discussed by Carman and Malherbe.[25]

The method has been successfully developed for measurement of the surface area of cement and for such materials as pigments, fine metal powders, pulverised coal, and fine fibres.

7.2.4 Non-Newtonian Fluids

There is only a very limited amount of published work on the flow of non-Newtonian fluids through packed beds, and there are serious discrepancies between the results and conclusions of different researchers. The range of voidages studied is very narrow, in most cases, falling in the range $0.35 < e < 0.41$. For a detailed account of the current situation, reference should be made to work of Chhabra and coworkers.[26, 27]

Most published work relates to the flow of shear-thinning fluids whose rheological behaviour follows the two-parameter *power-law* model (discussed in Volume 1A, Chapter 3), in which the shear stress τ and shear rate $\dot{\gamma}$ are related by:

$$\tau = k\dot{\gamma}^n \tag{7.25}$$

where k is known as the consistency coefficient and n (<1 for a shear-thinning fluid) is the power-law index.

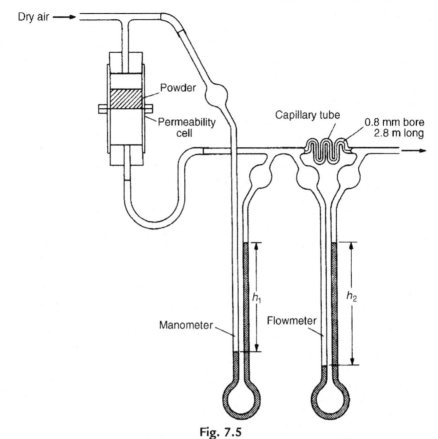

Fig. 7.5

The permeability apparatus of Lea and Nurse.[24]

The modelling of the flow of a non-Newtonian fluid through a packed bed follows a similar, though more complex, procedure to that adopted earlier in this chapter for the flow of a Newtonian fluid. It first involves a consideration of the flow through a cylindrical tube and then adapting this to the flow in the complex geometry existing in a packed bed. The procedure is described in detail elsewhere.[26, 27]

For the laminar flow of a power-law fluid through a cylindrical tube, the relation between mean velocity u and pressure drop $-\Delta P$ is given by:

$$u = \left(\frac{-\Delta P}{4kl}\right)^{1/n} \frac{n}{6n+2} d_t^{(n+1)/n} \tag{7.26}$$

and the so-called Metzner and Reed Reynolds number by:

$$\mathrm{Re}_{MR} = 8 \left(\frac{n}{6n+2}\right)^n \frac{\rho u^{2-n} d_t^n}{k} \tag{7.27}$$

(Corresponding to Volume 1A, Eqs 3.136 and 3.140)

For the laminar flow of a power-law fluid through a packed bed, Kemblowski et al.[11] have developed an analogous Reynolds number $(Re_1)_n$, which they have used as the basis for the calculation of the pressure drop for the flow of power-law fluids:

$$(\text{Re}_1)_n = \frac{\rho u_c^{2-n}}{kS^n(1-e)^n}\left(\frac{4n}{3n+1}\right)^n \left(\frac{b\sqrt{2}}{e^2}\right)^{1-n} \tag{7.28}$$

The last term in Eq. (7.28) is not a simple geometric characterisation of the flow passages, as it also depends on the rheology of the fluid (n). The constant b is a function of the shape of the particles constituting the bed, having a value of about 15 for particles of spherical, or near-spherical, shapes; there are insufficient reliable data available to permit values of b to be quoted for other shapes. Substitution of $n=1$ and of μ for k in Eq. (7.28) reduces it to Eq. (7.13), obtained earlier for Newtonian fluids.

Using this definition of the Reynolds number in place of Re_1, the value of the friction group $(R_1/\rho u_1^2)$ may be calculated from Eq. (7.18), developed previously for Newtonian fluids, and hence, the superficial velocity u_c for a power-law fluid may be calculated as a function of the pressure difference for values of the Reynolds number <2 to give:

$$u_c = \left(\frac{-\Delta P}{5kl}\right)^{1/n}\frac{1}{S^{(n+1)/n}}\frac{e^3}{(1-e)^2}\left(\frac{4n}{3n+1}\right)^{1/n}\left(\frac{b\sqrt{2}}{e^2}\right)^{(1-n)/n} \tag{7.29}$$

For Newtonian fluids $(n=1)$, Eq. (7.29) reduces to Eq. (7.9).

For polymer solutions, Eq. (7.29) applies only to flow through unconsolidated media because, otherwise, the pore dimensions may be of the same order of magnitude as those of the polymer molecules and additional complications, such as pore blocking and adsorption, may arise.

If the fluid has significant elastic properties, the flow may be appreciably affected because of the rapid changes in the magnitude and direction of flow as the fluid traverses the complex flow paths between the particles in the granular bed, as discussed by Chhabra and others.[26, 27]

7.2.5 Molecular Flow

In the relations given earlier, it is assumed that the fluid can be regarded as a continuum and that there is no slip between the wall of the capillary and the fluid layers in contact with it. However, when conditions are such that the mean free path of the molecules of a gas is a significant fraction of the capillary diameter, the flow rate at a given value of the pressure gradient becomes greater than the predicted value. If the mean free path exceeds the capillary diameter, the flow rate becomes independent of the viscosity, and the process is one of diffusion. Whereas these considerations apply only at very low pressures in normal tubes, in fine-pored materials, the pore diameter and the mean free path may be on the same order of magnitude, even at atmospheric pressure.

7.3 Dispersion

Dispersion is the general term which is used to describe the various types of self-induced mixing processes which can occur during the flow of a fluid through a pipe or vessel. The effects of dispersion are particularly important in packed beds, though they are also present under the simple flow conditions which exist in a straight tube or pipe. Dispersion can arise from the effects of molecular diffusion or as the result of the flow pattern existing within the fluid. An important consequence of dispersion is that the flow in a packed bed reactor deviates from plug flow, with an important effect on the characteristics of the reactor.

It is of interest to consider first what is happening in pipe flow. Random molecular movement gives rise to a mixing process, which can be described by Fick's law (given in Volume 1B, Chapter 2). If concentration differences exist, the rate of transfer of a component is proportional to the product of the molecular diffusivity and the concentration gradient. If the fluid is in laminar flow, a parabolic velocity profile is set up over the cross-section, and the fluid at the centre moves with twice the mean velocity in the pipe. This can give rise to dispersion because elements of fluid will take different times to traverse the length of the pipe, according to their radial positions. When the fluid leaves the pipe, elements that have been within the pipe for very different periods of time will be mixed together. Thus, if the concentration of a tracer material in the fluid is suddenly changed, the effect will first be seen in the outlet stream after an interval required for the fluid at the axis to traverse the length of the pipe. Then, as time increases, the effect will be evident in the fluid issuing at progressively greater distances from the centre. Because the fluid velocity approaches zero at the pipe wall, the fluid near the wall will reflect the change over only a very long period.

If the fluid in the pipe is in turbulent flow, the effects of molecular diffusion will be supplemented by the action of the turbulent eddies, and a much higher rate of transfer of material will occur within the fluid. Because the turbulent eddies also give rise to momentum transfer, the velocity profile is much flatter, and the dispersion due to the effects of the different velocities of the fluid elements will be correspondingly less.

In a packed bed, the effects of dispersion will generally be greater than in a straight tube. The fluid is flowing successively through constrictions in the flow channels and then through broader passages or cells. Radial mixing readily takes place in the cells because the fluid enters them with an excess of kinetic energy, much of which is converted into rotational motion within the cells. Furthermore, the velocity profile is continuously changing within the fluid as it proceeds through the bed. Wall effects can be important in a packed bed because the bed voidage will be higher near the wall, and flow will occur preferentially in that region.

At low rates of flow, the effects of molecular diffusion predominate and cell mixing contributes relatively little to the dispersion. At high rates, on the other hand, a realistic model is presented by considering the bed to consist of a series of mixing cells, the dimension of each of which is of the same order as the size of the particles forming the bed. Whatever the mechanism, however, the rate of dispersion can be conveniently described by means of a dispersion coefficient. The process is generally anisotropic, except at very low flow rates; that is, the dispersion rate is different in the longitudinal and radial directions, and, therefore, separate dispersion coefficients D_L and D_R are generally used to represent the behaviour in the two directions. The process is normally linear, with the rate of dispersion proportional to the product of the corresponding dispersion coefficient and concentration gradient. The principal factors governing dispersion in packed beds are discussed in a critical review by Gunn.[28]

The differential equation for dispersion in a cylindrical bed of voidage e may be obtained by taking a material balance over an annular element of height δl, inner radius r, and outer radius $r+\delta r$ (as shown in Fig. 7.6). On the basis of a dispersion model, it is seen that if C is the concentration of a reference material as a function of axial position l, radial position r, time t, and D_L and D_R are the axial and radial dispersion coefficients, then:

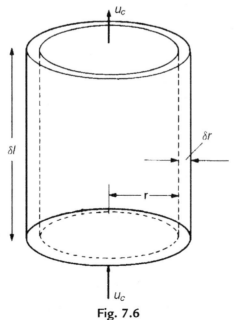

Fig. 7.6
Dispersion in packed beds.

Rate of entry of reference material due to flow in axial direction:

$$= u_c(2\pi r \delta r)C$$

Corresponding efflux rate:

$$= u_c(2\pi r \delta r)\left(C + \frac{\partial C}{\partial l}\delta l\right)$$

Net accumulation rate in element due to flow in the axial direction:

$$= -u_c(2\pi r \delta r)\frac{\partial C}{\partial l}\delta l \tag{7.30}$$

Rate of diffusion in axial direction across inlet boundary:

$$= -(2\pi r \delta r e)D_L\frac{\partial C}{\partial l}$$

Corresponding rate at outlet boundary:

$$= -(2\pi r \delta r e)D_L\left(\frac{\partial C}{\partial l} + \frac{\partial^2 C}{\partial l^2}\delta l\right)$$

Net accumulation rate due to diffusion from boundaries in axial direction:

$$= (2\pi r \delta r e)D_L\frac{\partial^2 C}{\partial l^2}\delta l \tag{7.31}$$

Diffusion in radial direction at radius r:

$$= (2\pi r \delta l e)D_R\frac{\partial C}{\partial r}$$

Corresponding rate at radius $r + \delta r$:

$$= [2\pi(r + \delta r)\delta l e]D_R\left[\frac{\partial C}{\partial r} + \frac{\partial^2 C}{\partial r^2}\delta r\right]$$

Net accumulation rate due to diffusion from boundaries in radial direction:

$$= -[2\pi r \delta l e]D_R\frac{\partial C}{\partial r} + [2\pi(r + \delta r)\delta l e]D_R\left(\frac{\partial C}{\partial r} + \frac{\partial^2 C}{\partial r^2}\delta r\right)$$

$$= 2\pi \delta l e D_R\left[\frac{\partial C}{\partial r}\delta r + r\delta r\frac{\partial^2 C}{\partial r^2} + (\delta r)^2\frac{\partial^2 C}{\partial r^2}\right]$$

$$= 2\pi \delta l e D_R\left[\delta r\frac{\partial}{\partial r}\left(r\frac{\partial C}{\partial r}\right)\right] \text{ (ignoring the last term)} \tag{7.32}$$

Now the total accumulation rate:

$$= (2\pi r\, \delta r\, \delta l)e\frac{\partial C}{\partial t} \tag{7.33}$$

Thus, from Eqs. (7.30)–(7.33):

$$(2\pi r\, \delta r\, \delta l)e\frac{\partial C}{\partial t} = -u_c(2\pi r\, \delta r)\frac{\partial C}{\partial l}\delta l + (2\pi r\, \delta r\, e)D_L\frac{\partial^2 C}{\partial l^2}\delta l + 2\pi\, \delta l e\, D_R\left[\delta r\frac{\partial}{\partial r}\left(r\frac{\partial C}{\partial r}\right)\right]$$

On dividing through by $(2\pi r \delta r \delta l)e$:

$$\frac{\partial C}{\partial t} + \frac{1}{e}u_c\frac{\partial C}{\partial l} = D_L\frac{\partial^2 C}{\partial l^2} + \frac{1}{r}D_R\frac{\partial}{\partial r}\left(r\frac{\partial C}{\partial r}\right) \tag{7.34}$$

Longitudinal dispersion coefficients can be readily obtained by injecting a pulse of tracer into the bed in such a way that radial concentration gradients are eliminated, and measuring the change in shape of the pulse as it passes through the bed. Because $\partial C/\partial r$ is then zero, Eq. (7.34) becomes:

$$\frac{\partial C}{\partial t} + \frac{u_c}{e}\frac{\partial C}{\partial l} = D_L\frac{\partial^2 C}{\partial l^2} \tag{7.35}$$

Values of D_L can be calculated from the change in shape of a pulse of tracer as it passes between two locations in the bed, and a typical procedure is described by Edwards and Richardson.[29] Gunn and Pryce,[30] on the other hand, imparted a sinusoidal variation to the concentration of tracer in the gas introduced into the bed. The results obtained by a number of researchers are shown in Fig. 7.7 as a Peclet number $Pe(=u_c\, d/eD_L)$ plotted against the particle Reynolds number $(Re'_c = u_c d\rho/\mu)$.

For gases, at low Reynolds numbers (<1), the Peclet number increases linearly with Reynolds number, giving:

$$\frac{u_c d}{eD_L} = K\frac{u_c d\rho}{\mu} = KSc^{-1}\frac{u_c d}{D} \tag{7.36}$$

$$\text{or}: \quad \frac{D_L}{D} = \text{constant}, \gamma \quad \text{which has a value of approximately } 0.7 \tag{7.37}$$

because Sc, the Schmidt number, is approximately constant for gases, and the voidage of a randomly packed bed is usually about 0.4. This is consistent with the hypothesis that, at low Reynolds numbers, molecular diffusion dominates. The factor 0.7 is a tortuosity factor which allows for the fact that the molecules must negotiate a tortuous path because of the presence of the particles.

At Reynolds numbers greater than about 10, the Peclet number becomes approximately constant, giving:

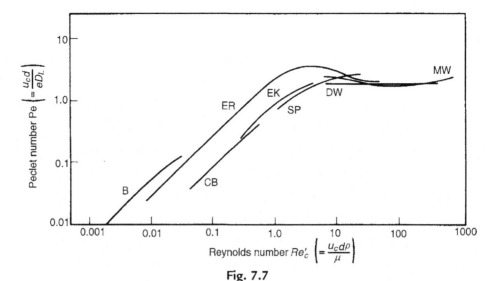

Fig. 7.7
Longitudinal dispersion in gases in packed beds. ER—Edwards and Richardson,[29] B—Blackwell et al.,[31] CB—Carberry and Bretton,[32] DW—de Maria and White,[33] MW—McHenry and Wilhelm,[34] SP—Sinclair and Potter,[35] EK—Evans and Kenney,[36] $N_2 + He$ in $N_2 + H_2$.

$$D_L \approx \frac{1}{2}\frac{u_c}{e}d \qquad (7.38)$$

This equation is predicted by the mixing cell model, and turbulence theories put forward by Aris and Amundson[37] and by Prausnitz.[38]

In the intermediate range of Reynolds numbers, the effects of molecular diffusivity and of macroscopic mixing are approximately additive, and the dispersion coefficient is given by an equation of the form:

$$D_L = \gamma D + \frac{1}{2}\frac{u_c d}{e} \qquad (7.39)$$

However, the two mechanisms interact, and molecular diffusion can reduce the effects of convective dispersion. This can be explained by the fact that with streamline flow in a tube, molecular diffusion will tend to smooth out the concentration profile arising from the velocity distribution over the cross-section. Similarly, radial dispersion can give rise to lower values of longitudinal dispersion than predicted by Eq. (7.39). As a result, the curves of Peclet number versus Reynolds number tend to pass through a maximum as shown in Fig. 7.7.

A comparison of the effects of axial and radial mixing is seen in Fig. 7.8, which shows results obtained by Gunn and Pryce[30] for dispersion of argon into air. The values of D_L were obtained as indicated earlier, and D_R was determined by injecting a steady stream of tracer at the axis and measuring the radial concentration gradient across the bed. It is seen that

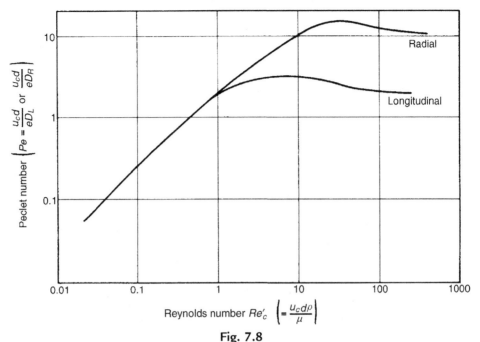

Fig. 7.8
Longitudinal and radial mixing coefficients for argon in air.[30]

molecular diffusion dominates at low Reynolds numbers, with both the axial and radial dispersion coefficients D_L and D_R equal to approximately 0.7 times the molecular diffusivity. At high Reynolds numbers, however, the ratio of the longitudinal dispersion coefficient to the radial dispersion coefficient approaches a value of about 5. That is:

$$\frac{D_L}{D_R} \approx 5 \qquad (7.40)$$

The experimental results for dispersion coefficients in gases show that they can be satisfactorily represented as Peclet number expressed as a function of particle Reynolds number, and that similar correlations are obtained, irrespective of the gases used. However, it might be expected that the Schmidt number would be an important variable, but it is not possible to test this hypothesis with gases as the values of Schmidt number are all approximately the same and equal to about unity.

With liquids, however, the Schmidt number is variable, and it is generally about three orders of magnitude greater than that for a gas. Results for longitudinal dispersion available in the literature, and plotted in Fig. 7.9, show that over the range of Reynolds numbers studied $(10^{-2} < Re'_c < 10^3)$, the Peclet number shows little variation and is of the order of unity. Comparison of these results with the corresponding ones for gases (shown in Fig. 7.7) shows that the effect of molecular diffusion in liquids is insignificant at Reynolds numbers up to unity.

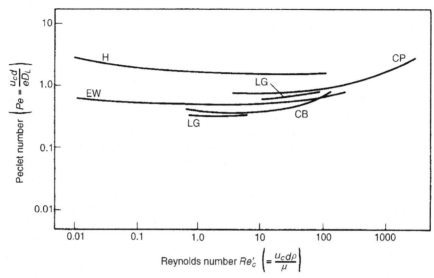

Fig. 7.9

Longitudinal dispersion in liquids in packed beds. CP—Cairns and Prausnitz,[39] CB—Carberry and Bretton,[32] EW—Ebach and White,[40] H—Hiby,[41] LG—Liles and Geankoplis.[42]

This difference can be attributed to the very different magnitudes of the Schmidt numbers for gases and liquids.

7.4 Heat Transfer in Packed Beds

For heat and mass transfer through a stationary or streamline fluid to a single spherical particle, it has been shown in Volume 1B, Chapter 1 that the heat and mass transfer coefficients reach limiting low values given by:

$$Nu' = Sh' = 2 \tag{7.41}$$

where $Nu'(= hd/k)$ and $Sh'(= h_D d/D)$ are the Nusselt and Sherwood numbers with respect to the fluid, respectively.

Kramers[43] has shown that, for conditions of forced convection, the heat transfer coefficient can be represented by:

$$Nu' = 2.0 + 1.3Pr^{0.15} + 0.66Pr^{0.31}Re_c'^{0.5} \tag{7.42}$$

where Re_c' is the particle Reynolds number $u_c d\rho/\mu$ based on the superficial velocity u_c of the fluid, and Pr is the Prandtl number $C_p\mu/k$.

This expression has been obtained on the basis of experimental results obtained with fluids of Prandtl numbers ranging from 0.7 to 380.

For natural convection, Ranz and Marshall[44] have given:

$$Nu' = 2.0 + 0.6Pr'^{1/3}Gr'^{1/4} \tag{7.43}$$

where Gr' is the Grashof number discussed in Volume 1B, Chapter 1.

Results for packed beds are much more difficult to obtain because the driving force cannot be measured very readily. Gupta and Thodos[45] suggest that the *j*-factor for heat transfer, j_h (Volume 1B, Chapter 1), forms the most satisfactory basis of correlation for experimental results and have proposed that:

$$ej_h = 2.06Re'^{-0.575}_c \tag{7.44}$$

where *e* is the voidage of the bed, $j_h = St' \, Pr^{2/3}$, and $St' = $ Stanton number $h/C_p \rho u_c$.

The *j*-factors for heat and mass transfer, j_h and j_d, are found to be equal, and, therefore, Eq. (7.44) can also be used for the calculation of mass transfer rates.

Reproducible correlations for the heat transfer coefficient between a fluid flowing through a packed bed and the cylindrical wall of the container are very difficult to obtain. The main difficulty is that a wide range of packing conditions can occur in the vicinity of the walls. However, the results quoted by Zenz and Othmer[46] suggest that:

$$Nu \propto Re'^{0.7-0.9}_c \tag{7.45}$$

It may be noted that in this expression the Nusselt number with respect to the tube wall Nu is related to the Reynolds number with respect to the particle Re'_c.

7.5 Packed Columns

Because packed columns consist of shaped particles contained within a column, their behaviour will in many ways be similar to that of packed beds which have already been considered. There are, however, several important differences which make the direct application of the equations for pressure gradient difficult. First, the size of the packing elements in the column will generally be very much larger, and the Reynolds number will, therefore, be such that the flow is turbulent. Secondly, the packing elements will normally be hollow, and, therefore, have a large amount of internal surface which will offer a higher flow resistance than their external surface. The shapes, too, are specially designed to produce good mass transfer characteristics with relatively small pressure gradients. Although some of the general principles already discussed can be used to predict pressure gradient as a function of flow rate, it is necessary to rely heavily on the literature issued by the manufacturers of the packings.

In general, packed towers are used for bringing two phases in contact with one another, and there will be strong interaction between the fluids. Normally, one of the fluids will preferentially wet the packing and will flow as a film over its surface; the second fluid then passes through the remaining volume of the column. With gas (or vapour)–liquid systems, the liquid will normally be the wetting fluid, and the gas or vapour will rise through the column, making close contact with the downflowing liquid and having little direct contact with the packing elements. An example of the liquid–gas system is an absorption process where a soluble gas is scrubbed from a mixture of gases by means of a liquid, as shown in Fig. 7.10. In a packed column used for distillation, the more volatile component of, say, a binary mixture is progressively transferred to the vapour phase, and the less volatile condenses out in the liquid. Packed columns have also been used extensively for liquid–liquid extraction processes where a solute is transferred from one solvent to another, as discussed in volume 2B. Some principles involved in the design and operation of packed columns will be illustrated by considering columns for gas absorption. In this chapter, an outline of the construction of the column, and the flow characteristics will be dealt with, whereas the magnitude of the mass transfer coefficients is discussed in Volume 2B. The full design process is outlined in Volume 6.

In order to obtain a good rate of transfer per unit volume of the tower, a packing is selected which will promote a high interfacial area between the two phases and a high degree of

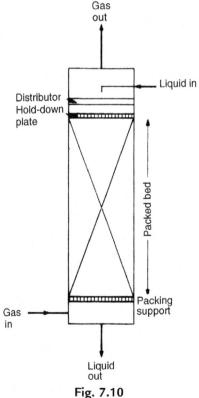

Fig. 7.10
Packed absorption column.

turbulence in the fluids. Usually, increased area and turbulence are achieved at the expense of increased capital cost and/or pressure drop, and a balance must be made between these factors when arriving at an economic design.

7.5.1 General Description

The construction of packed towers is relatively straightforward. The shell of the column may be constructed from metal, ceramics, glass, or plastics material, or from metal with a corrosion-resistant lining. The column should be mounted truly vertically to help uniform liquid distribution. Detailed information on the mechanical design and mounting of industrial scale column shells is given by Brownell and Young,[47] Molyneux,[48] and in BS 5500,[49] as well as in Volume 6.

The bed of packing rests on a support plate which should be designed to have at least 75% free area for the passage of the gas so as to offer as low a resistance as possible. The simplest support is a grid with relatively widely spaced bars on which a few layers of large Raschig or partition rings are stacked. One such arrangement is shown in Fig. 7.11. The gas injection plate described by Leva[50] shown in Fig. 7.12 is designed to provide separate passageways for gas and liquid so that they need not vie for passage through the same opening. This is achieved by providing the gas inlets to the bed at a point above the level at which liquid leaves the bed.

At the top of the packed bed, a liquid distributor of suitable design provides for the uniform irrigation of the packing which is necessary for satisfactory operation. Four examples of different distributors are shown in Fig. 7.13[51] and may be described as follows:

(a)   A simple orifice type, Fig. 7.13A, which gives very fine distribution, though it must be correctly sized for a particular duty and should not be used where there is any risk of the holes plugging,

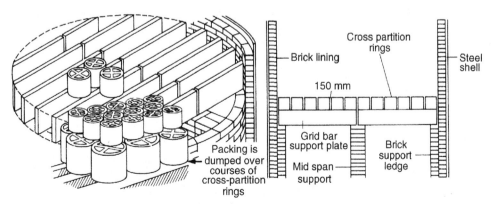

Fig. 7.11

Grid bar supports for packed towers.

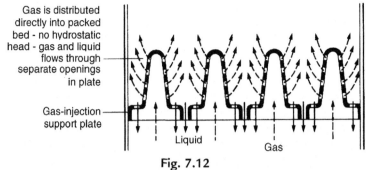

Gas is distributed directly into packed bed - no hydrostatic head - gas and liquid flows through separate openings in plate

Gas-injection support plate

Liquid

Gas

Fig. 7.12
The gas injection plate.[50]

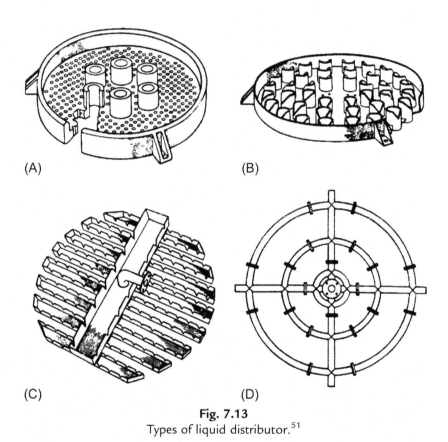

(A)

(B)

(C)

(D)

Fig. 7.13
Types of liquid distributor.[51]

(b) The notched chimney type of distributor, Fig. 7.13B, which has a good range of flexibility for the medium and upper flow rates, and is not prone to blockage,

(c) The notched trough distributor, Fig. 7.13C, which is specially suitable for the larger sizes of tower, and, because of its large free area, it is also suitable for the higher gas rates, and

(d)   The perforated ring type of distributor, Fig. 7.13D, for use with absorption columns where high gas rates and relatively small liquid rates are encountered. This type is especially suitable where pressure loss must be minimised. For the larger size of tower, where installation through manholes is necessary, it may be made up in flanged sections.

Uniform liquid flow is essential if the best use is to be made of the packing, and, if the tower is high, redistributing plates are necessary. These plates are needed at intervals of about 2–3 times the column diameters for Raschig rings and about 5–10 column diameters for Pall rings, but are usually not more than 6 m apart.[52] A "hold-down" plate is often placed at the top of a packed column to minimise movement and breakage of the packing caused by surges in flow rates. The gas inlet should also be designed for uniform flow over the cross-section, and the gas exit should be separate from the liquid inlet. Further details on internal fittings are given by Leva[50] and by Mackowiak and Hall.[53]

Columns for both absorption and distillation vary in diameter from about 25 mm for small laboratory purposes to over 4.5 m for large industrial operations; these industrial columns may be 30 m or more in height. Columns may operate at pressures ranging from high vacuum to high pressure, the optimum pressure depending on both the chemical and the physical properties of the system.

7.5.2 Packings

Packings can be divided into four main classes—broken solids, shaped packings, grids, and structured packings. Broken solids are the cheapest form and are used in sizes from about 10 to 100 mm according to the size of the column. Although they frequently form a good corrosion-resistant material, they are not as satisfactory as shaped packings either in regard to liquid flow or to effective surface offered for mass transfer. The packing should be of as uniform size as possible so as to produce a bed of uniform characteristics with a desired voidage.

The most commonly used packings are Raschig rings, Pall rings, Lessing rings, and Berl saddles. Newer packings include Nutter rings, Intalox, Intalox metal saddles, Hy-Pak, and mini rings, and, because of their high performance characteristics and low pressure drop, these packings now account for a large share of the market. Commonly used packing elements are illustrated in Fig. 7.14. Most of these packings are available in a wide range of materials such as ceramics, metals, glass, plastics, carbon, and sometimes rubber. Ceramic packings are resistant to corrosion and comparatively cheap, but are heavy and may require a stronger packing support and foundations. The smaller metal rings made from wire mesh are also available, and these give much-improved transfer characteristics in small columns.

A nonporous solid should be used if there is any risk of crystal formation in the pores when the packing dries, as this can give rise to serious damage to the packing elements. However, some

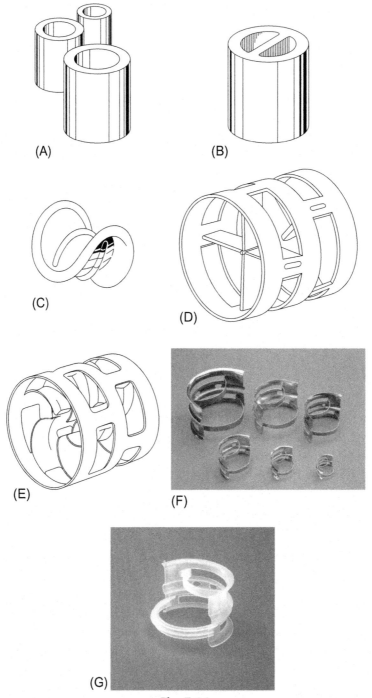

Fig. 7.14
(A) Ceramic Raschig rings; (B) Ceramic Lessing ring; (C) Ceramic Berl saddle; (D) Pall ring (plastic); (E) Pall ring (metal); (F) Metal Nutter rings; (G) Plastic Nutter ring. *Image Courtesy: Sulzer (UK) Ltd, Farnborough, Hants.*

plastics are not very good because they are not wetted by many liquids. Channelling, that is, nonuniform distribution of liquid across the column cross-section, is much less marked with shaped packings, and their resistance to flow is also much less. Shaped packings also give a more effective surface per unit volume because surface contacts are reduced to a minimum, and the film flow is much improved compared with that with broken solids. On the other hand, the shaped packings are more expensive, particularly when small sizes are used. The voidage obtainable with these packings varies from about 0.45 to 0.95. Ring packings are either dumped into a tower, dropped in small quantities, or may be individually stacked if 75 mm or larger in size. To obtain high and uniform voidage and to prevent breakage, it is often found better to dump the packings into a tower full of liquid. Stacked packings, as shown in Fig. 7.11, have the advantage that the flow channels are vertical, and there is much less tendency for the liquid to flow to the walls than with random packings. The properties of some commonly used industrial packings are shown in Table 7.3.

The size of packing used influences the height and diameter of a column, the pressure drop and cost of packing. Generally, as the packing size is increased, the cost per unit volume of packing and the pressure drop per unit height of packing are reduced, and the mass transfer efficiency is reduced. Reduced mass transfer efficiency results in a taller column being needed, so that the overall column cost is not always reduced by increasing the packing size. Normally, in a column in which the packing is randomly arranged, the packing size should not exceed one-eighth of the column diameter. Above this size, liquid distribution, and hence, the mass transfer efficiency, deteriorates rapidly. Because cost per unit volume of packing does not fall much for sizes above 50 mm, whereas efficiency continues to fall, there is seldom any advantage in using packings much larger than 50 mm in a randomly packed column.

For laboratory purposes, a number of special packings have been developed which are, in general, too expensive for large diameter towers. Dixon packings, which are Lessing rings made from wire mesh, and KnitMesh, a fine wire mesh packing, are typical examples. These packings give very high interfacial areas, and, if they are flooded with liquid before operation, all of the surface is active, so that the transfer characteristics are very good even at low liquid rates. The volume of liquid held up in such a packing is low, and the pressure drop is also low. Some of these high-efficiency woven wire packings have been used in columns up to 500 mm diameter.

Grid packings, which are relatively easy to fabricate, are usually used in columns of square section, and frequently, in cooling towers, which are described in Volume 1B, Chapter 5. They may be made from wood, plastics, carbon, or ceramic materials, and, because of the relatively large spaces between the individual grids, they give low pressure drops. Further advantages lie in their ease of assembly, their ability to accept fluids with suspended solids, and their ease of wetting even at very low liquid rates. The main problem is that of obtaining good liquid distribution because, at high liquid rates, the liquid tends to cascade

Table 7.3 Design data for various packings

	Size (in.)	Size (mm)	Wall Thickness (in.)	Wall Thickness (mm)	Number Density (/ft³)	Number Density (/m³)	Bed Density (lb/ft³)	Bed Density (kg/m³)	Contact Surface S_B (ft²/ft³)	Contact Surface S_B (m²/m³)	Free Space (%) (100 e)	Packing Factor F (ft²/ft³)	Packing Factor F (m²/m³)
Ceramic Raschig Rings	0.25	6	0.03	0.8	85,600	3,020,000	60	960	242	794	62	1600	5250
	0.38	9	0.05	1.3	24,700	872,000	61	970	157	575	67	1000	3280
	0.50	12	0.07	1.8	10,700	377,000	55	880	112	368	64	640	2100
	0.75	19	0.09	2.3	3090	109,000	50	800	73	240	72	255	840
	1.0	25	0.14	3.6	1350	47,600	42	670	58	190	71	160	525
	1.25	31			670	23,600	46	730			71	125	410
	1.5	38			387	13,600	43	680			73	95	310
	2.0	50	0.25	6.4	164	5790	41	650	29	95	74	65	210
	3.0	76			50	1765	35	560			78	36	120
Metal Raschig Rings	0.25	6	0.03	0.8	88,000	3,100,000	133	2130			72	700	2300
	0.38	9	0.03	0.8	27,000	953,000	94	1500			81	390	1280
	0.50	12	0.03	0.8	11,400	402,000	75	1200	127	417	85	300	980
	0.75	19	0.03	0.8	3340	117,000	52	830	84	276	89	185	605
(Bed densities are for mild steel;	0.75	19	0.06	1.6	3140	110,000	94	1500			80	230	750
multiply by 1.105,	1.0	25	0.03	0.8	1430	50,000	39	620	63	207	92	115	375
1.12, 1.37, 1.115	1.0	25	0.06	1.6	1310	46,200	71	1130			86	137	450
for stainless steel,	1.25	31	0.06	1.6	725	25,600	62	990			87	110	360
copper,	1.5	38	0.06	1.6	400	14,100	49	780			90	83	270
aluminium, and	2.0	50	0.06	1.6	168	5930	37	590	31	102	92	57	190
monel	3.0	76	0.06	1.6	51	1800	25	400	22	72	95	32	105
respectively)													
Carbon Raschig Rings	0.25	6	0.06	1.6	85,000	3,000,000	46	730	212	696	55	1600	5250
	0.50	12	0.06	1.6	10,600	374,000	27	430	114	374	74	410	1350
	0.75	19	0.12	3.2	3140	110,000	34	540	75	246	67	280	920
	1.0	25	0.12	3.2	1325	46,000	27	430	57	187	74	160	525
	1.25	31			678	23,000	31	490			69	125	410
	1.5	38			392	13,800	34	540			67	130	425
	2.0	50	0.25	6.4	166	5860	27	430	29	95	74	65	210
	3.0	76	0.31	8.0	49	1730	23	370	19	62	78	36	120
Metal Pall Rings	0.62	15	0.02	0.5	5950	210,000	37	590	104	341	93	70	230
	1.0	25	0.025	0.6	1400	49,000	30	480	64	210	94	48	160
(Bed densities are	1.25	31	0.03	0.8	375	13,000	24	380	39	128	95	28	92
for mild steel)	2.0	50	0.035	0.9	170	6000	22	350	31	102	96	20	66
	3.5	76	0.05	1.2	33	1160	17	270	20	65	97	16	52

Packing	Size										ε (%)		F
Plastic Pall Rings (Bed densities are for polypropylene)	0.62	16	0.03	0.8	6050	213,000	7.0	112	104	341	87	97	320
	1.0	25	0.04	1.0	1440	50,800	5.5	88	63	207	90	52	170
	1.5	38	0.04	1.0	390	14,000	4.75	76	39	128	91	40	130
	2.0	50	0.06	1.5	180	6350	4.25	68	31	102	92	25	82
	3.5	88	0.06	1.5	33	1160	4.0	64	26	85	92	16	52
Ceramic Intalox Saddles	0.25	6			117,500	4,150,000	54	860			65	725	2400
	0.38	9			49,800	1,750,000	50	800			67	330	1080
	0.50	12			18,300	646,000	46	730			71	200	660
	0.75	19			5640	199,000	44	700			73	145	475
	1.0	25			2150	76,000	42	670			73	92	300
	1.5	38			675	23,800	39	620	59	194	76	52	170
	2.0	50			250	8820	38	600			76	40	130
	3.0	76			52	1830	36	570			79	22	72
Plastic Super Intalox	No. 1				1620	57,200	6.0	96	63	207	90	33	108
	No. 2				190	6710	3.75	60	33	108	93	21	210
	No. 3				42	1480	3.25	52	27	88	94	16	52
Intalox Metal	25			4770		168,000					96.7	41	135
	40			1420		50,100					97.3	25	82
	50			416		14,600					97.8	16	52
	70			131		4600					98.1	13	43
Hy-Pak (Bed densities are for mild steel)	No. 1				850	30,000	19	304			96	43	140
	No. 2				107	3770	14	224			97	18	59
	No. 3				31	1090	13	208			97	15	49
Plastic Cascade Mini Rings	No. 1												25	82
	No. 2											15	49
	No. 3											12	39
Metal Cascade Mini Rings	No. 0											55	180
	No. 1											34	110
	No. 2											22	72
	No. 3											14	46
	No. 4											10	33
Ceramic Cascade Mini Rings	No. 2											38	125
	No. 3											24	79
	No. 5											18	59

The packing factor F replaces the term S_B/e^3. Use of the given value of F in Fig. 7.19 permits more predictable performance of designs incorporating packed beds because the values quoted are derived from the operating characteristics of the packings rather than from their physical dimensions.

from one grid to the next without being broken up into fine droplets which are desirable for a high interfacial surface. An example of a cooling tower packing, "Coolflo 3,"[54] is shown in Fig. 7.15. This is similar to the structured packings described later, and consists of vacuum-formed PVC sheets clamped together within a metal and plastics frame to form a module which can be 0.6 or 1.2 m in depth. Structured packings may be broadly classified into either the knitted or the nonknitted type, and both types may be assembled in a segmented way or in a spiral form. In the latter, corrugated strips or ribbons coil about a centre axis to form a flat cake of the requisite tower diameter, which is usually <1 m. These elements are then stacked one upon the other to provide the necessary bed depth. In the rigid type of structured packing, these corrugated sheets of metal or plastic are assembled to form intersecting open channels. The sheets may, in addition, be perforated, and they provide uniform liquid flow over both sides while vapour flows upward and provides intimate contact with the liquid. One such type of packing, Mellapak[55] is shown in Fig. 7.16, and others such as Gempak[56] are also available. Low pressure drops of typically $50 \, N/m^2$ per theoretical stage are possible with HETPs, ranging from 0.2 to 0.6 m, voidages in excess of 95%, and high specific surface areas. The resulting higher capacity and efficiency with structured packings is, however, achieved at higher initial capital cost than with the other packings discussed in this section.[57]

Fig. 7.15

Visco Coolflo 3 extended surface, cooling tower packing. *Image Courtesy: Sulzer (UK) Ltd, Farnborough, Hants.*

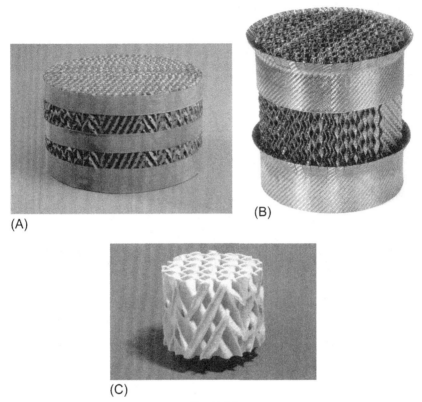

Fig. 7.16

Structured packings (A) metal gauze (B) carbon (C) corrosion-resistant plastic. *Image Courtesy: Sulzer (UK) Ltd, Farnborough, Hants.*

7.5.3 Fluid Flow in Packed Columns

Pressure drop

It is important to be able to predict the drop in pressure for the flow of the two fluid streams through a packed column. Earlier in this chapter, the drop in pressure arising from the flow of a single phase through granular beds is considered, and the same general form of approach is usefully adopted for the flow of two fluids through packed columns. It was noted that the expressions for flow through ring-type packings are less reliable than those for flow through beds of solid particles. For the typical absorption column, there is no very accurate expression, but there are several correlations that are useful for design purposes. In the majority of cases, the gas flow is turbulent, and the general form of the relation between the drop in pressure $-\Delta P$ and the volumetric gas flow rate per unit area of column u_G is shown on curve A of Fig. 7.17. $-\Delta P$ is then proportional to $u_G^{1.8}$ approximately, in agreement with the curve A of Fig. 7.17 at high Reynolds numbers. If, in addition to the gas flow, liquid flows down the tower, the passage of the gas is not significantly affected at low liquid rates, and the pressure drop line is similar to

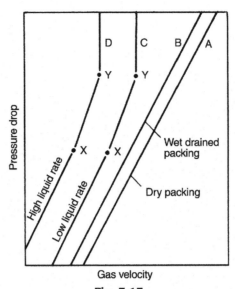

Fig. 7.17
Pressure drops in wet packings (logarithmic axes).

line A, although for a given value of u_G the value of $-\Delta P$ is somewhat increased. When the gas rate reaches a certain value, the pressure drop then rises very much more quickly and is proportional to $u_G^{2.5}$, as shown by the section XY on curve C. Over this section, the liquid flow is interfering with the gas flow, and the hold-up of liquid is progressively increasing. The free space in the packings is, therefore, being continuously taken up by the liquid, and, thus, the resistance to gas flow rises quickly. At gas flows beyond Y, $-\Delta P$ rises very steeply, and the liquid is held up in the column. The point X is known as the loading point, and point Y as the flooding point for the given liquid flow. If the flow rate of liquid is increased, a similar plot D is obtained in which the loading point is achieved at a lower gas rate though at a similar value of $-\Delta P$. While it is advantageous to have a reasonable liquid hold-up in the column, as this promotes interphase contact, it is not practicable to operate under flooding conditions, and columns are best operated over the section XY. Because this is a section with a relatively short range in gas flow, the safe practice is to design for operation at the loading point X. It is of interest to note that, if a column is flooded and then allowed to drain, the value of $-\Delta P$ for a given gas flow is increased over that for an entirely dry packing as shown by curve B. Rose and Young[58] correlated their experimental pressure drop data for Raschig rings by the following equation:

$$-\Delta P_w = -\Delta P_d \left(1 + \frac{3.30}{d_n} \right) \qquad (7.46)$$

where $-\Delta P_w$ is the pressure drop across the wet drained column, $-\Delta P_d$ is the pressure drop across the dry column, and d_n is the nominal size of the Raschig rings in mm.

This effect will thus be most significant for small packings.

There are several ways of calculating the pressure drop across a packed column when gas and liquid are flowing simultaneously, and the column is operating below the loading point.

One approach is to calculate the pressure drop for gas flow only and then multiply this pressure drop by a factor which accounts for the effect of the liquid flow. Eq. (7.19) may be used for predicting the pressure drop for the gas only, and then, the pressure drop with gas and liquid flowing is obtained by using the correction factors for the liquid flow rate given by Sherwood and Pigford.[59]

Another approach is that of Morris and Jackson,[60] who arranged experimental data for a wide range of ring and grid packings in a graphical form convenient for the calculation of the number of velocity heads N lost per unit height of packing. N is substituted in the equation:

$$-\Delta P = \frac{1}{2} N \rho_G u_G^2 l \tag{7.47}$$

where: $-\Delta P$ = pressure drop, ρ_G = gas density, u_G = gas velocity, based on the empty column cross-sectional area, and l = height of packing.

Eq. (7.47) is in consistent units. For example, with ρ_G in kg/m^3, u_G in m/s, l in m, and N in m^{-1}, $-\Delta P$ is then in N/m^2.

Empirical correlations of experimental data for pressure drop have also been presented by Leva,[61] and by Eckert et al.[62] for Pall rings. Where the data are available, the most accurate method of obtaining the pressure drop for flow through a bed of packing is from the manufacturer's own literature. This is usually presented as a logarithmic plot of gas rate against pressure drop, with a parameter of liquid flow rate on the graphs, as shown in Fig. 7.17, although it should be stressed that all of these methods apply only to conditions at or below the loading point X on Fig. 7.17. If applied to conditions above the loading point, the calculated pressure drop would be too low. It is, therefore, necessary first to check whether the column is operating at or below the loading point, and some of the methods of predicting loading points are now considered.

Loading and flooding points

Although the loading and flooding points have been shown on Fig. 7.17, there is no completely generalised expression for calculating the onset of loading, although one of the following semi-empirical correlations is often found to be adequate. Morris and Jackson[60] gave their results in the form of plots of $\psi(u_G/u_L)$ at the loading rate for various wetting rates L_w (m^3/s m). u_G and u_L are the average gas and liquid velocities based on the empty column, and $\psi = (\sqrt{(\rho_G/\rho_A)})$ is a gas density correction factor, where ρ_A is the density of air at 293 K.

A useful graphical correlation for flooding rates was first presented by Sherwood et al.[63] and later developed by Lobo et al.[64] for random-dumped packings, as shown in Fig. 7.18 in which:

$$\left(\frac{u_G^2 S_B}{ge^3}\right)\left(\frac{\rho_G}{\rho_L}\right)\left(\frac{\mu_L}{\mu_w}\right)^{0.2} \text{ is plotted against } \frac{L'}{G'}\sqrt{\left(\frac{\rho_G}{\rho_L}\right)}$$

where u_G is the superficial velocity of the gas, S_B is the surface area of the packing per unit volume of bed, g is the acceleration due to gravity, L' is the mass rate of flow per unit area of the liquid, G' is the mass rate of flow per unit area of the gas, and μ_w is the viscosity of water at 293 K (approximately 1 mN s/m^2), and suffix G refers to the gas and suffix L to the liquid.

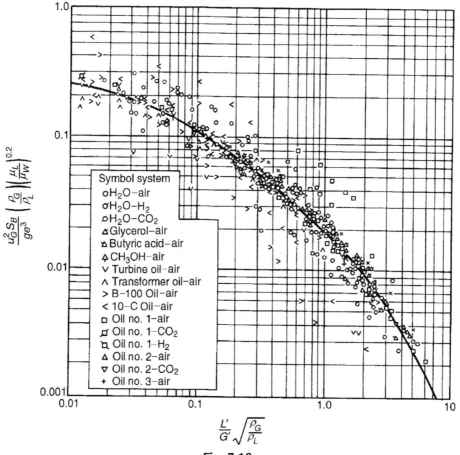

Fig. 7.18

Generalised correlation for flooding rates in packed towers.[64]

The area inside the curve represents possible conditions of operation. In these expressions, the ratios ρ_G/ρ_L and μ_L/μ_w have been introduced so that the relationship can be applied for a wide range of liquids and gases. It may be noted that, if the effective value of g is increased by using a rotating bed, then higher flow rates can be achieved before the onset of flooding.

The generalised pressure drop correlation

The generalised pressure drop correlation by Eckert[65] has been developed as a practical aid to packed tower design and incorporates flow rates, physical properties of the fluid, a wide range of packings and pressure drop on one chart presented in dimensionally consistent form, as shown Fig. 7.19. The broken line representing flooding lies above the top curve in Fig. 7.19, and hence, the correlation may be used with safety in design procedures. Most of the data on which it is based are obtained for cases where the liquid is water and the correction factor $[(\mu_L/\mu_w)/(\rho_w/\rho_L)]^{0.1}$, in which μ_w and ρ_w refer to water at 293 K, is introduced to enable it to be used for other liquids. The packing factor F which is employed in the correlation is a modification of the specific surface of the packing that is used in Fig. 7.18. Values of F are included in Table 7.3. In practice, a pressure drop is selected for a given duty, and use is

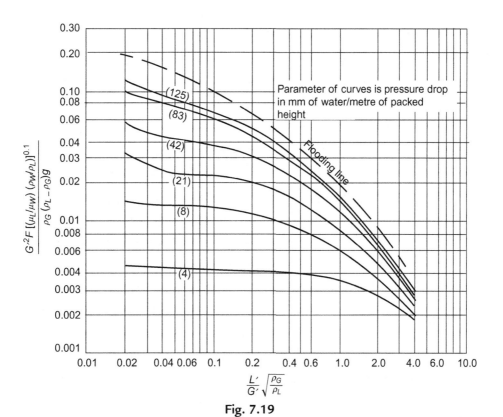

Fig. 7.19
Generalised pressure drop correlation. *Adapted from a figure by the Norton Co., with permission.*

made of the correlation to determine the gas flow rate per unit area G', from which the tower diameter may be calculated for the required flows. The method is discussed in detail in Volume 6.

Leva[66] has extended the isobars in Fig. 7.19 to a value of 0.01 for the abscissa and, additionally, tabulates the limiting asymptotic values of the ordinate which are reached as $L'/G' \to 0$, corresponding to dry packings. He also makes a comparison of the different methods of predicting pressure drops.

Liquid distribution

Provision of a packing with a high surface area per unit volume may not result in good contacting of gas and liquid unless the liquid is distributed uniformly over the surface of the packing. The need for liquid distribution and redistribution and correct packing size has been noted previously. The effective wetted area decreases as the liquid rate is decreased, and, for a given packing, there is a minimum liquid rate for effective use of the surface area of the packing. A useful measure of the effectiveness of wetting of the available area is the wetting rate L_w defined as:

$$\frac{\text{Volumetric liquid rate per unit cross} - \text{sectional area of column}}{\text{Packing surface area per unit volume of column}}$$

$$\text{or}: \; L_W = \frac{L}{A\rho_L S_B} = \frac{u_L}{S_B} \qquad (7.48)$$

Thus, the wetting rate is analogous to the volumetric liquid rate per unit length of circumference in a wetted-wall column in which the liquid flows down the surface of a cylinder. If the liquid rate were too low, a continuous liquid film would not be formed around the circumference of the cylinder, and some of the area would be ineffective.

Similar effects occur in a packed column, although the flow patterns and arrangement of the surfaces are then obviously much more complex. Morris and Jackson[60] have recommended minimum wetting rates of 2×10^{-5} m^3/s m for rings 25–75 mm in diameter and grids of pitch <50 mm, and 3.3×10^{-5} m^3/s m for larger packings.

The distribution of liquid over packings has been studied experimentally by many researchers; for instance, Tour and Lerman[67, 68] showed that for a single point feed, the distribution is given by:

$$Q_x = c \exp\left(-a^2 x^2\right) \qquad (7.49)$$

where Q_x is the fraction of the liquid collected at a distance x from the centre, and c and a are constants depending on the packing arrangement. Norman,[69] Manning and Cannon,[70] and others have shown that this maldistribution is one cause of falling transfer coefficients with tall towers.

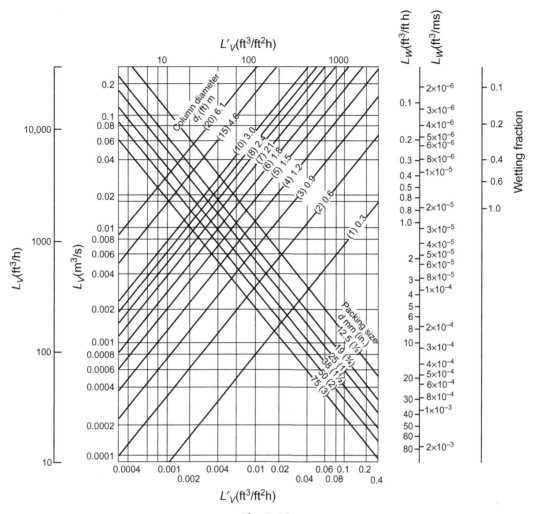

Fig. 7.20

Nomograph for the estimation of the degree of wetting in a packed column.[51]

A nomograph which relates liquid rate, tower diameter, and packing size is given in Fig. 7.20.[51] The wetting rate L_W may be obtained as an absolute value from the inner right-hand axis or as *wetting fraction* from the outer scale. A value of wetting fraction exceeding unity on that scale indicates that the packing is satisfactorily wet. It should be noted that many organic liquids have favourable wetting properties, and wetting may be effective at much lower rates, though materials such as plastics and polished stainless steel are difficult to wet. Fig. 7.20 does, however, represent the best available data on the subject of wetting. In the example shown in Fig. 7.20, the arrowed line corresponds to the case of a liquid flow of 0.018 m³/s in a column of 1.6 m diameter and a packing size of 25 mm, which gives an approximate wetting rate of 5×10^{-5} m³/m s, corresponding to a total wetting of >1 on the outside right-hand scale; this is satisfactory.

7.6 Hold-Up

In many industrial applications of packed columns, it is desirable to know the volumetric hold-up of the liquid phase in the column. This information might be needed, for example, if the liquid were involved in a chemical reaction, or if a control system for the column were being designed. For gas–liquid systems, the hold-up of liquid H_w for conditions below the loading point has been found[50] to vary approximately as the 0.6 power of the liquid rate, and for rings and saddles, this is given approximately by:

$$H_w = 0.143 \left(\frac{L'}{d}\right)^{0.6} \tag{7.50}$$

where L' is the liquid flow rate (kg/m^{2}s), d is the equivalent diameter of the packing (mm), and H_w is the hold-up (m^3 of liquid/m^3 of column).

Thus, with 25 mm Raschig rings, L' of 1.0 kg/m^2 s, and $d = 20$ mm, H_w has a value of 0.021 m^3/m^3.

For further information on the design of packed columns, reference should be made to Volume 6. In addition, manufacturers' data are often available on hold-up for specific packings, and these should be consulted whenever possible for design purposes.

7.6.1 Economic Design of Packed Columns

In designing industrial scale packed columns, a balance must be made between the capital cost of the column and ancillary equipment on the one side, and the running costs on the other. Generally, reducing the diameter of the column will reduce the capital cost, though increase the cost of pumping the gas through the column due to the increased pressure drop.

For columns operating at atmospheric and subatmospheric pressure, and where the mass transfer rate is controlled by transfer through the gas film, Morris and Jackson[60] calculated ranges of economic gas velocities for various packings under specified conditions. In selecting a gas velocity, and hence, the column cross-sectional area, it is necessary to check that the liquid rate is above the minimum wetting rate, as discussed in the previous section. Selection of the appropriate packing will help in achieving the minimum wetting rate. The loading condition should also be calculated to ensure that the column would not be operating above this condition.

For columns operating at high pressures, the capital cost of the column shell becomes much more significant, and it may be more economic to operate at gas velocities above the loading condition. Morris and Jackson[60] suggest a gas velocity about 75%–80% of the flooding velocity for normal systems, and <40% of the flooding rate, if foaming is likely to occur. The height of the column would have to be taken into consideration in making an economically optimum design. The height is usually determined by the mass transfer duty of the columns and mass transfer rates per unit height of packing.

7.6.2 Vacuum Columns

Sawistowski[9] has shown that the curve in Fig. 7.18 may be converted to a straight line by plotting:

$$\ln\left\{\ln\left[\frac{u_G^2 S_B}{ge^3}\left(\frac{\rho_G}{\rho_L}\right)\left(\frac{\mu_L}{\mu_w}\right)^{0.2}\right]\right\} \text{ against } \ln\left\{\frac{L'}{G'}\sqrt{\frac{\rho_G}{\rho_L}}\right\}$$

The equation of the curve is then found to be:

$$\ln\left\{\frac{u_G^2 S_B}{ge^3}\left(\frac{\rho_G}{\rho_L}\right)\left(\frac{\mu_L}{\mu_w}\right)^{0.2}\right\} = -4\left(\frac{L'}{G'}\right)^{1/4}\left(\frac{\rho_G}{\rho_L}\right)^{1/8} \tag{7.51}$$

When a column is operating under reduced pressure, and the pressure drop is of the same order of magnitude as the absolute pressure, it is not immediately obvious whether the onset of flooding will be determined by conditions at the top or the bottom of the column. If G_F is the gas flow rate under flooding conditions in the column, then:

$$G_F = u_G\rho_G A \tag{7.52}$$

Substituting in Eq. (7.51) gives:

$$\ln\left\{\frac{G_F^2 S_B}{A^2 ge^3}\frac{1}{\rho_G\rho_L}\left(\frac{\mu_L}{\mu_w}\right)^{0.2}\right\} = -4\left(\frac{L'}{G'}\right)^{1/4}\left(\frac{\rho_G}{\rho_L}\right)^{1/8} \tag{7.53}$$

For a column operating at a given reflux ratio, L'/G' is constant, and the only variables over the length of the column are now the minimum flooding rate G_F and the gas density ρ_G. In order to find the condition for a minimum or maximum value of G_F, $d(G_F^2)/d\rho_G$ is obtained from Eq. (7.53) and equated to zero. Thus:

$$\left(\frac{G_F^2}{\rho_G}\right)^{-1}\left\{G_F^2(-\rho_G)^{-2} + \rho_G^{-1}\frac{dG_F^2}{d\rho_G}\right\} = -4\left(\frac{L'}{G'}\right)^{1/4}\rho_L^{-1/8}\left(\frac{1}{8}\rho_G^{-7/8}\right) \tag{7.54}$$

This gives $(L'/G')\sqrt{(\rho_G/\rho_L)}=16$, when $dG_F^2/d\rho_G=0$. As the second differential coefficient is negative at this point, G_F^2 is a maximum.

A value of $(L'/G')\sqrt{(\rho_G/\rho_L)}=16$ is well in excess of the normal operating range of the column (especially of a distillation column operating at reduced pressure), as seen in Fig. 7.18. Thus, over the whole operating range of a column, the value of G' which just gives rise to flooding increases with gas density, and hence, with the absolute pressure. The tendency for a column to flood will always be greater, therefore, at the low pressure end, that is, at the top.

Calculation of the pressure drop and flooding rate is particularly important for vacuum columns, in which the pressure may increase severalfold from the top to the bottom of the column. When a heat-sensitive liquid is distilled, the maximum temperature, and hence, the pressure, at the bottom of the column is limited, and hence, the vapour rate must not exceed a certain value. In a vacuum column, the throughput is very low because of the high specific volume of the vapour, and the liquid reflux rate is generally so low that the liquid flow has little effect on the pressure drop. The pressure drop can be calculated by applying Eq. (7.15) over a differential height and integrating. Thus:

$$-\frac{dP}{dl} = \left(\frac{R_1}{\rho u_1^2}\right) S\rho_G u^2 \left(\frac{1-e}{e^3}\right) \tag{7.55}$$

Writing $G' = \rho_G u$ and $P/\rho_G = P_0/\rho_0$ for isothermal operation, where ρ_0 is the vapour density at some arbitrary pressure P_0:

$$-\frac{dP}{dl} = \left(\frac{R_1}{\rho u_1^2}\right) S\frac{(1-e)}{e^3}\frac{G'^2}{P}\frac{P_0}{\rho_0} \tag{7.56}$$

The Reynolds number, and hence, $R_1/\rho u_1^2$, will remain approximately constant over the column.

Integrating Eq. (7.56):

$$P_1^2 - P_2^2 = 2\left(\frac{R_1}{\rho u_1^2}\right)\frac{(1-e)}{e^3}\frac{SG'^2 2P_0}{\rho_0}l \tag{7.57}$$

It may be noted that, when the pressure at the top of the column is small compared with that at the bottom, the pressure drop is directly proportional to the vapour rate.

Example 7.1

Two-heat sensitive organic liquids, of average molecular weight of 155 kg/kmol, are to be separated by vacuum distillation in a 100 mm diameter column packed with 6 mm stoneware Raschig rings. The number of theoretical plates required is 16, and it has been found that the HETP is 150 mm. If the product rate is 0.5 g/s at a reflux ratio of 8, calculate the pressure in the condenser so that the temperature in the still does not exceed 395 K, which is equivalent to a pressure of 8 kN/m². It may be assumed that $a = 800$ m²/m³, $\mu = 0.02$ mN s/m², $e = 0.72$, and that temperature changes and the correction for liquid flow may be neglected.

Solution

The modified Reynolds number Re_1 is defined by:

$$Re_1 = \frac{u_c\rho}{S(1-e)\mu} = \frac{G'}{S(1-e)\mu} \quad \text{(Eq.7.13)}$$

The Ergun equation may be rewritten as:

$$\frac{R_1}{\rho u_1^2} = \frac{4.17}{Re_1} + 0.29 \quad \text{(Eq.7.21)}$$

$$\text{Hence:} \quad \frac{R}{\rho u_1^2} = \frac{4.17S(1-e)\mu}{G'} + 0.29$$

Eq. (7.15), written in its differential form, becomes:

$$\frac{R_1}{\rho u_1^2} = \frac{e^3}{S(1-e)}\left(-\frac{dP}{dl}\right)\frac{1}{\rho u_c^2}$$

$$= \frac{e^3}{S(1-e)}\left(-\frac{dP}{dl}\right)\frac{\rho}{G'^2}$$

$$-\rho\frac{dP}{dl} = \frac{R_1}{\rho u_1^2}\frac{S(1-e)}{e^3}G'^2$$

$$\text{Thus:} \quad -\int_{P_C}^{P_S} \rho dP = \frac{R_1}{\rho u_1^2}\frac{S(1-e)}{e^3}G'^2\int_0^l dl$$

(where suffix C refers to the condenser and S to the still)

$$= \frac{R_1}{\rho u_1^2}\frac{S(1-e)}{e^3}G'^2 l$$

In this example: $a = 800\,\text{m}^2/\text{m}^3 = S(1-e)$
Product rate $= 0.5\,\text{g/s}$ and, if the reflux ratio $= 8$, then:

$$\text{Vapour rate} = 4.5\,\text{g/s}$$

$$\text{and:} \quad G' = 4.5 \times 10^{-3}/(\pi/4)(0.1)^2 = 0.573\,\text{kg/m}^2\text{s},$$

$$\mu = 0.02 \times 10^{-3}\,\text{Ns/m}^2 \text{ and } e = 0.72$$

$$\text{Hence:} \quad Re_1 = 0.573/800 \times 0.28 \times 0.02 \times 10^{-3} = 128$$

$$\frac{R_1}{\rho u_1^2} = (4.17/128) + 0.29 = 0.32$$

$$\text{Because:} \quad l = (16 \times 0.15) = 2.4\,\text{m}$$

$$-\int_{P_C}^{P_S} \rho dP = (0.32 \times 800 \times 0.28) \times (0.573)^2 \times 2.4/(0.72)^3$$

$$\text{giving:} \quad -\int_{P_C}^{P_S} \rho dP = 151.3\,\text{N/m}^2$$

$$\rho = \rho_s \times P/P_s$$

The vapour density in the still is given by:

$$\rho_s = \left(\frac{155}{22.4}\right)\left(\frac{273}{395}\right)\left(\frac{P_S}{101.3 \times 10^3}\right) = 4.73 \times 10^{-5}P_S\,\text{kg/m}^3$$

$$\therefore \quad \rho = 4.73 \times 10^{-5}P\,\text{kg/m}^3$$

$$\therefore \quad -\int_{P_C}^{P_S} \rho dP = -\int_{P_C}^{P_S} 4.73 \times 10^{-5}PdP = 4.73 \times 10^{-5}\left(P_S^2 - P_C^2\right)$$

$$\text{Thus:} \quad 151.3 = \left(4.73 \times 10^{-5}\right)\left(P_S^2 - P_C^2\right)$$

$$\text{Because} \quad P_S = 8000 \text{ N/m}^2$$

$$\text{then:} \quad 151.3 = \left(4.73 \times 10^{-5}\right)\left(8 \times 10^3\right)^2 - \left(4.73 \times 10^{-5}\right)P_C^2$$

$$\text{and:} \quad P_C = 7790 \text{ N/m}^2 \text{ or } 7.79 \text{ kN/m}^2$$

7.7 Pressure Drop in Fibrous Porous Systems

Undoubtedly, most porous media and packed columns are made up of granular particles; there are, however, similar systems composed of long particles, fibres, etc., and are characterised by high values of voidage, as large as $e = 0.98\text{–}0.99$. The flow of fluids in such media is encountered in a broad spectrum of applications, including aerosol filtration, production and processing of fibre-reinforced polymer and metal composites, drying of fibrous foodstuffs, cotton batting, textile fibres, biomedical settings, etc. In such cases, additional advantages accrue on account of their high porosity (and hence, large specific surface area), relatively low resistance to flow, high rates of mass and heat transfer, etc. Consequently, over the years, a reasonable body of knowledge has accrued which has been reviewed by Jackson and James,[71] Chen et al.,[72] Johnson and Deen,[73] Mattern and Deen,[74] etc.; the latter two references focus on aerosol gels, ligaments and tendons, etc.

In the streamline flow regime, Jackson and James[71] have collated much of the literature data and represented it in the form of dimensionless permeability, B/r^2, defined by Eq. (7.1), r being the radius of the fibre. They correlated (B/r^2) with the mean voidage of the system. Notwithstanding considerable scatter of data, especially at small values of e, they put forward the following empirical correlation for $e > \sim 0.6$:

$$\left(B/r^2\right) = -10.68 + 39.33e - 58.75e^2 + 32.05e^3 \tag{7.58}$$

The available experimental results suggest the Kozney constant K'' to lie in the range of 4 to 5 for these systems also. Detailed discussion on this topic in the context of the flow of both the Newtonian and non-Newtonian fluids in such porous media is available in Ref. 27. Suffice it to add here that the flow resistance is influenced by the wettability of the fibres, their stiffness and cross-section, orientation, etc., in addition to the overall voidage and the flow conditions.

7.8 Nomenclature

		Units in SI System	Dimensions in $\mathbf{M, L, T,}\ \theta$
A	Total cross-sectional area of bed or column	m^2	$\mathbf{L}^2$
a	Coefficient in Eq. (7.49)	m^{-1}	$\mathbf{L}^{-1}$
B	Permeability coefficient (Eq. (7.2))	m^2	$\mathbf{L}^2$
b	Constant in Eq. (7.28) (15 for spherical particles)	–	–
C	Concentration	kg/m^3	$\mathbf{ML}^{-3}$
C_p	Specific heat at constant pressure	J/kg K	$\mathbf{L}^2\mathbf{T}^{-2}\theta^{-1}$
c	Coefficient in Eq. (7.49)	–	–
D	Molecular diffusivity	m^2/s	$\mathbf{L}^2\mathbf{T}^{-1}$
D_L	Axial dispersion coefficient	m^2/s	$\mathbf{L}^2\mathbf{T}^{-1}$
D_R	Radial dispersion coefficient	m^2/s	$\mathbf{L}^2\mathbf{T}^{-1}$
d	Diameter of particle	m	$\mathbf{L}$
d'_m	Equivalent diameter of pore space $= e/S_B$ as used by Kozeny	m	$\mathbf{L}$
d_n	Nominal packing size (e.g., diameter for a Raschig ring)	m	$\mathbf{L}$
d_t	Tube or column diameter	–	$\mathbf{L}$
e	Fractional voidage of bed of particles or packing	–	–
F	Packing factor	m^2/m^3	$\mathbf{L}^{-1}$
f_w	Wall correction factor (Eq. (7.23))	–	–
G	Gas mass flow rate	kg/s	$\mathbf{MT}^{-1}$
G_F	Gas mass flow rate under flooding conditions	kg/s	$\mathbf{MT}^{-1}$
G'	Gas mass velocity	kg/s m^2	$\mathbf{ML}^{-2}\mathbf{T}^{-1}$
g	Acceleration due to gravity	m/s^2	$\mathbf{LT}^{-2}$
H_w	Liquid hold-up in bed, volume of liquid per unit volume of bed	–	–
h	Heat transfer coefficient	W/m^{2}K	$\mathbf{MT}^{-3}\theta^{-1}$
h_D	Mass transfer coefficient	m/s	$\mathbf{LT}^{-1}$
j_d	j-factor for mass transfer	–	–
j_h	j-factor for heat transfer	–	–
K	Constant in flow Eqs (7.1) and (7.2)	m^{3}s/kg	$\mathbf{M}^{-1}\mathbf{L}^3\mathbf{T}$
K'	Dimensionless constant in Eq. (7.6)	–	–
K''	Kozeny constant in Eq. (7.9)	–	–
K_0	Shape factor in Eq. (7.22)	–	–

k	Thermal conductivity	W/m K	$\mathbf{MLT}^{-3}\theta^{-1}$
k	Consistency coefficient for power-law fluid	Ns^n/m^2	$\mathbf{ML}^{-1}\mathbf{T}^{n-2}$
L	Liquid mass flow rate	kg/s	$\mathbf{MT}^{-1}$
L'	Liquid mass velocity	kg/s m^2	$\mathbf{ML}^{-2}\mathbf{T}^{-1}$
L_v	Volumetric liquid rate	m^3/s	$\mathbf{L}^3\mathbf{T}^{-1}$
L_v'	Volumetric liquid rate per unit area	m/s	$\mathbf{LT}^{-1}$
L_w	Wetting rate (u_L/S_B)	m^2/s	$\mathbf{L}^2\mathbf{T}^{-1}$
l	Length of bed or height of column packing	m	$\mathbf{L}$
l'	Length of flow passage through bed	m	$\mathbf{L}$
l_t	Length of circular tube	m	$\mathbf{L}$
N	Number of velocity heads lost through unit height of bed (Eq. 7.47)	m^{-1}	$\mathbf{L}^{-1}$
n	Flow behaviour index for power-law-fluid	–	–
n'	Exponent in Eq. (7.17)	–	–
P	Pressure	N/m^2	$\mathbf{ML}^{-1}\mathbf{T}^{-2}$
$-\Delta P$	Pressure drop across bed or column	N/m^2	$\mathbf{ML}^{-1}\mathbf{T}^{-2}$
$-\Delta P_d$	Pressure drop across bed of dry packing	N/m^2	$\mathbf{ML}^{-1}\mathbf{T}^{-2}$
$-\Delta P_w$	Pressure drop across bed of wet packing	N/m^2	$\mathbf{ML}^{-1}\mathbf{T}^{-2}$
Q_x	Fraction of liquid collected at distance x from centre line of packing in Eq. (7.49)	–	–
R	Drag force per unit area of tube wall	N/m^2	$\mathbf{ML}^{-1}\mathbf{T}^{-2}$
R_1	Drag force per unit surface area of particles	N/m^2	$\mathbf{ML}^{-1}\mathbf{T}^{-2}$
r	Radius	m	$\mathbf{L}$
S	Surface area per unit volume of particle or packing	m^2/m^3	$\mathbf{L}^{-1}$
S_B	Surface area per unit volume of bed (specific surface)	m^2/m^3	$\mathbf{L}^{-1}$
S_c	Surface area of container per unit volume of bed time	m^2/m^3	$\mathbf{L}^{-1}$
t	Time	s	$\mathbf{T}$
u_c	Average fluid velocity based on cross-sectional area A of empty column	m/s	$\mathbf{LT}^{-1}$
u_t	Mean velocity of fluid in tube	m/s	$\mathbf{LT}^{-1}$
u_G	Volumetric flow rate of gas per unit area of cross-section	m/s	$\mathbf{LT}^{-1}$
u_L	Volumetric flow rate of liquid per unit area of cross-section	m/s	$\mathbf{LT}^{-1}$
u_1	Mean velocity in pore channel	m/s	$\mathbf{LT}^{-1}$
V	Volume of fluid flowing in time t	m^3	$\mathbf{L}^3$
X	Side of cube	m	$\mathbf{L}$

x	Distance from centre	m	**L**
α	Coefficient in Eq. (7.17)	(N/m^2)/(m/s)	$\mathbf{ML^{-2}T^{-1}}$
α'	Coefficient in Eq. (7.17)	(N/m^2)/(m/s)$^{n'}$	$\mathbf{ML^{-(n'+1)}T^{(n'-2)}}$
γ	Coefficient of D in Eqs. (7.37), (7.39)	–	–
$\dot{\gamma}$	Shear rate	s^{-1}	$\mathbf{T^{-1}}$
μ	Fluid viscosity	Ns/m^2	$\mathbf{ML^{-1}T^{-1}}$
ρ	Density of fluid	kg/m^3	$\mathbf{ML^{-3}}$
τ	Shear stress	N/m^2	$\mathbf{ML^{-1}T^{-2}}$
ψ	Density correction factor $\sqrt{(\rho_G/\rho_A)}$	–	–
Gr'	Grashof number (particle) (Volume 1B, Chapter 1)	–	–
Nu	Nusselt number for tube wall (hd_t/k)	–	–
Nu'	Nusselt number (particle) (hd/k)	–	–
Pe	Peclet number ($u_c d/eD_L$) or ($u_c d/eD_R$)	–	–
Pr	Prandtl number ($C_p\mu/k$)	–	–
Re	Reynolds number for flow through tube ($ud_t\rho/\mu$)	–	–
Re_1	Modified Reynolds number based on pore size as used by Carman (Eq. (7.13))	–	–
Re'_c	Modified Reynolds number based on particle size ($u_c d\rho/\mu$)	–	–
Re_{MR}	Metzner and Reed Reynolds number (Eq. 7.27)	–	–
$(Re_1)_n$	Reynolds number for power-law fluid in a granular bed (Eq. 7.28)	–	–
Sc	Schmidt number ($\mu/\rho D$)	–	–
Sh'	Sherwood number (particle) ($h_D d/D$)	–	–
St'	Stanton number (particle) ($h/C_p\rho u_c$)	–	–

Subscripts

A	refers to air at 293 K	–	–
G	refers to gas		
L	refers to liquid		
w	refers to water at 293 K		
0	refers to standard conditions		

References

1. Darcy HPG. *Les Fontaines publiques de la ville de Dijon. Exposition et application à suivre et des formules à employer dans les questions de distribution d'eau.* Victor Dalamont; 1856.
2. Eisenklam P. Porous masses. In: Cremer HW, Davies T, editors. *Chemical engineering practice.* Vol. 2. Butterworths; 1956 (Chapter 9).

3. Brinkman HC. On the permeability of media consisting of closely packed porous particles. *Appl Sci Res* 1948;**1A**:81–6.

4. Dupuit AJEJ. *Etudes théoriques et pratiques sur le mouvement des eaux.* 1863.

5. Kozeny J. Über kapillare Leitung des Wassers im Boden (Aufstieg, Versicherung, und Anwendung auf die Bewässerung). *Sitzb Akad Wiss, Wien, Math-naturw Kl* 1927;**136**(Abt, IIa):271–306.

6. Kozeny J. Über Bodendurchlässigkeit. *Z Pfl-Ernähr Düng Bodenk* 1933;**28A**:54–6.

7. Carman PC. Fluid flow through granular beds. *Trans Inst Chem Eng* 1937;**15**:150–66.

8. Forchheimer P. *Hydraulik.* Teubner; 1930.

9. Sawistowski H. Flooding velocities in packed columns operating at reduced pressures. *Chem Eng Sci* 1957;**6**:138.

10. Ergun S. Fluid flow through packed columns. *Chem Eng Prog* 1952;**48**:89–94.

11. Kemblowski Z, Dziubinski M, Mertl J. Flow of non-Newtonian fluids through granular media. *Adv Transport Proc* 1987;**5**:117.

12. Botset HG. Flow of gas–liquid mixtures through consolidated sand. *Trans Am Inst Mining Met Eng* 1940;**136**:41.

13. Glaser MB, Litt M. A physical model for mixed phase flow through beds of porous particles. *AIChE J* 1963;**9**:103.

14. Dudukovic MP, Larachi F, Mills P. Multiphase catalytic reactors: a perspective on current knowledge and future trends. *Catal Rev Sci Eng* 2002;**44**:123–246.

15. Fanchi JR, Seidle JP. Multiphase flow in porous media. In: Michaelides EE, Crowe CT, Schwarzkopf JD, editors. *Handbook of multiphase systems.* 2nd ed. Boca Raton, FL: CRC Press; 2016 [Chapter 11].

16. Carman PC. *Flow of gases through porous media.* Butterworths; 1956.

17. Coulson JM. The flow of fluids through granular beds; effect of particle shape and voids in streamline flow. *Trans Inst Chem Eng* 1949;**27**:237–57.

18. Wyllie MRJ, Gregory KR. Fluid flow through unconsolidated porous aggregates—effect of porosity and particle shape on Kozeny–Carman constants. *Ind Eng Chem* 1955;**47**:1379–88.

19. Kihn E. *Streamline flow of fluids through beds of granular materials.* University of London; 1939 (Ph.D. Thesis).

20. Pirie JM. In discussion of Coulson (ref. 17 above).

21. Muskat M, Botset HG. Flow of gas through porous material. *Physics* 1931;**1**:27–47.

22. Davies CN. in discussion of Hutchison, H. P., Nixon, I. S., and Denbigh, K. G.: The thermosis of liquids through porous materials. *Discuss Faraday Soc* 1948;**3**:86–129.

23. Lord E. Air flow through plugs of textile fibres. Part I. General flow relations. *J Text Inst* 1951;**46**:T191.

24. Lea FM, Nurse RW. Permeability methods of fineness measurement. *Trans Inst Chem Eng* 1947;**25**:47–63. Supplement: Symposium on Particle Size Analysis.

25. Carman PC, Malherbe PlR. Routine measurement of surface of paint pigments and other fine powders, I. *J Soc Chem Ind Trans* 1950;**69**:134T–43T.

26. Chhabra RP, Comiti JC, Machac I. Flow of non-Newtonian fluids in fixed and fluidised beds. *Chem Eng Sci* 2001;**50**:1.

27. Chhabra RP. *Bubbles, drops and particles in non-Newtonian fluids.* 2nd ed. Roca Raton, FL: CRC Press; 2007.

28. Gunn DJ. Mixing in packed and fluidised beds. *Chem Eng* 1968;**219**:CE153.

29. Edwards MF, Richardson JF. Gas dispersion in packed beds. *Chem Eng Sci* 1968;**22**:109.

30. Gunn DJ, Pryce C. Dispersion in packed beds. *Trans Inst Chem Eng* 1969;**47**:T341.

31. Blackwell RJ, Rayne JR, Terry MW. Factors influencing the efficiency of miscible displacement. *J Pet Technol* 1959;**11**:1–8.

32. Carberry JJ, Bretton RH. Axial dispersion of mass in flow through fixed beds. *AIChE J* 1958;**4**:367.

33. de Maria F, White RR. Transient response study of gas flowing through irrigated packing. *AIChE J* 1960;**6**:473.

34. McHenry KW, Wilhelm RH. Axial mixing of binary gas mixtures flowing in a random bed of spheres. *AIChE J* 1957;**3**:83.

35. Sinclair RJ, Potter OE. The dispersion of gas in flow through a bed of packed solids. *Trans Inst Chem Eng* 1965;**43**:T3.

36. Evans EV, Kenney CN. Gaseous dispersion in packed beds at low Reynolds numbers. *Trans Inst Chem Eng* 1966;**44**:T189.
37. Aris R, Amundson NR. Some remarks on longitudinal mixing or diffusion in fixed beds. *AIChE J* 1957;**3**:280.
38. Prausnitz JM. Longitudinal dispersion in a packed bed. *AIChE J* 1958;**4**:14M.
39. Cairns EJ, Prausnitz JM. Longitudinal mixing in packed beds. *Chem Eng Sci* 1960;**12**:20.
40. Ebach EE, White RR. Mixing of fluids flowing through beds of packed solids. *AIChE J* 1958;**4**:161.
41. Hiby JW. In: *Longitudinal and transverse mixing during single-phase flow through granular beds*. Proceedings of the symposium on interaction between fluids and particles. London: I. Chem. E.; 1962. p. 312
42. Liles AW, Geankoplis CJ. Axial diffusion of liquids in packed beds and end effects. *AIChE J* 1960;**6**:591.
43. Kramers H. Heat transfer from spheres to flowing media. *Physica* 1946;**12**:61.
44. Ranz WE, Marshall WR. Evaporation from drops. *Chem Eng Prog* 1952;**48**:141. 173.
45. Gupta AS, Thodos G. Direct analogy between mass and heat transfer to beds of spheres. *AIChE J* 1963;**9**:751.
46. Zenz FA, Othmer DF. *Fluidization and fluid-particle systems*. Reinhold; 1960.
47. Brownell LE, Young EH. *Process equipment design, vessel design*. Chapman & Hall; 1959.
48. Molyneux F. *Chemical plant design*. Vol. 1. Butterworths; 1963.
49. BS 5500. *Fusion welded pressure vessels*. London: British Standards Institution; 1978.
50. Leva M. *Tower packings and packed tower design*. U.S. Stoneware Co.; 1953
51. Norton Chemical Process Products Div, Box 350, Akron, Ohio; Hydronyl Ltd., King St., Fenton, Stokeon-Trent, U.K.
52. Eckert JS. Design techniques for designing packed towers. *Chem Eng Prog* 1961;**57**(9):54.
53. Mackowiak J, Hall C. *Fluid dynamics of packed columns: principles of the fluid dynamic design of columns for gas/liquid and liquid/liquid systems*. Berlin: Springer; 2010.
54. Coolflo 3 is a product of Visco Ltd., Croydon Surrey.
55. Mellapak is a registered trademark of Sulzer (UK) Ltd., Farnborough, Hants.
56. Gempak is a registered trademark of Glitsch (UK) Ltd., Kirkby Stephen, Cumbria.
57. Cheng GK. Packed column internals. *Chem Eng Albany* March 5 1984;**91**(5):40.
58. Rose HE, Young PH. Hydraulic characteristics of packed towers operating under countercurrent flow conditions. *Proc Inst Mech Eng* 1952;**1B**:114.
59. Sherwood TK, Pigford RL. *Absorption and extraction*. New York: McGraw-Hill; 1952.
60. Morris GA, Jackson J. *Absorption towers*. Butterworths; 1953.
61. Leva M. Flow through irrigated dumped packing pressure drop, loading, flooding. *Chem Eng Prog Symp Ser* 1954;**50**(10):51–9. Also see, Chem. Eng. Prog., 88(1992) 65–72.
62. Eckert JS, Foote EH, Huntington RL. Pall rings—new type of tower packing. *Chem Eng Prog* 1958;**54** (1):70–5.
63. Sherwood TK, Shipley GH, Holloway FAL. Flooding velocities in packed columns. *Ind Eng Chem* 1938;**30**:765–9.
64. Lobo WE, Friend L, Hashmall F, Zenz F. Limiting capacity of dumped tower packings. *Trans Am Inst Chem Eng* 1945;**41**:693–710.
65. Eckert JS. Tower packings—comparative performance. *Chem Eng Prog* 1963;**59**(5):76.
66. Leva M. Reconsider packed-tower pressure-drop correlations. *Chem Eng Prog* 1992;**88**(1):65.
67. Tour RS, Lerman F. An improved device to demonstrate the laws of frequency distribution. With special reference to liquid flow in packed towers. *Trans Am Inst Chem Eng* 1939;**35**:709–18.
68. Tour RS, Lerman F. The unconfined distribution of liquid in tower packing. *Trans Am Inst Chem Eng* 1939;**35**:719–42.
69. Norman WS. The performance of grip packed towers. *Trans Inst Chem Eng* 1951;**29**:226–39.
70. Manning RE, Cannon MR. Distillation improvement by control of phase channelling in packed columns. *Ind Eng Chem* 1957;**49**:347–9.
71. Jackson GW, James DF. The permeability of fibrous media. *Can J Chem Eng* 1986;**64**:364–74.
72. Chen C-T, Malkus DS, Vanderby Jr R. A fiber matrix model for interstitial fluid flow and permeability in ligaments and tendons. *Biorheology* 1998;**35**:103–18.

73. Johnson EM, Deen WM. Hydraulic permeability of agarose gels. *AIChE J* 1996;**42**:1220–4.
74. Mattern KJ, Deen WM. Mixing rules for estimating the hydraulic permeability of fiber mixtures. *AIChE J* 2008;**54**:32–41.

Further Reading

Chhabra RP, Richardson JF. *Non-Newtonian flow and applied rheology,* 2nd ed. Butterworth-Heinemann; 2008.

Dullien FAL. *Porous media: fluid transport and pore structure.* 2nd ed. New York: Academic Press; 1992.

Leva M. *Tower packings and packed tower design.* 2nd ed. U.S. Stoneware Co; 1953.

Morris GA, Jackson J. *Absorption towers.* Butterworths; 1953.

Norman WS. *Absorption, distillation and cooling towers.* Longmans; 1961.

Strigle RF. *Random packings and packed towers. Design and applications.* Gulf Publishing Company; 1987.

CHAPTER 8

Sedimentation

8.1 Introduction

In Chapter 6, consideration is given to the forces acting on an isolated particle moving relative to a fluid, and it is seen that the hydrodynamic drag may be expressed in terms of a friction factor (i.e., drag coefficient), which is, in turn, a function of the particle Reynolds number. If the particle is settling in the gravitational field, it rapidly reaches its terminal falling velocity when the fluid resistance force has become equal to the net gravitational force. In a centrifugal field, the particle may reach a very much higher velocity because the centrifugal force may be many thousands of times greater than the gravitational force.

In practice, the concentrations of suspensions used in industry will usually be high enough for there to be significant interaction between the particles, and the frictional force exerted at a given velocity of the particles relative to the fluid may be greatly increased as a result of modifications of the flow pattern, so that *hindered settling* takes place. As a corollary, the sedimentation rate of a particle in a concentrated suspension may be considerably less than its terminal falling velocity under *free settling* conditions when the effects of mutual interference between the particles are negligible. In this chapter, the behaviour of concentrated suspensions in a gravitational field is discussed, and the equipment used industrially for concentrating or *thickening* such suspensions will be described. Sedimentation in a centrifugal field is considered in Chapter 11.

It is important to note that suspensions of fine particles tend to behave rather differently from coarse suspensions in that a high degree of flocculation may occur as a result of the very high specific surface of the particles. For this reason, fine and coarse suspensions are considered separately, and the factors giving rise to flocculation are discussed in Section 8.2.2.

Although the sedimentation velocity of particles tends to decrease steadily as the concentration of the suspension is increased, it has been shown by Kaye and Boardman[1] that particles in very dilute suspensions may settle at velocities up to 1.5 times the normal terminal falling velocities, due to the formation of clusters of particles which settle in well-defined streams. This effect is important when particle size is determined by a method involving the measurement of the settling velocity of particles in dilute concentration, though is not significant with concentrated suspensions.

Coulson and Richardson's Chemical Engineering. https://doi.org/10.1016/B978-0-08-101098-3.00009-3

8.2 Sedimentation of Fine Particles

8.2.1 Experimental Studies

The sedimentation of metallurgical slimes has been studied by Coe and Clevenger,[2] who concluded that a concentrated suspension may settle in one of two different ways. In the first, after an initial brief acceleration period, the interface between the clear liquid and the suspension moves downward at a constant rate, and a layer of sediment builds up at the bottom of the container. When this interface approaches the layer of sediment, its rate of fall decreases until the 'critical settling point' is reached, when a direct interface is formed between the sediment and the clear liquid. Further sedimentation then results solely from a consolidation of the sediment, with liquid being forced upward around the solids which are then forming a loose bed with the particles in contact with one another. Because the flow area is gradually being reduced, the rate progressively diminishes. In Fig. 8.1A, a stage in the sedimentation process is illustrated. A is clear liquid, B is suspension of the original concentration, C is a layer through which the concentration gradually increases, and D is sediment. The sedimentation rate remains constant until the upper interface corresponds with the top of zone C, and it then falls until the critical settling point is reached, when both zones B and C will have disappeared. A second and rather less common mode of sedimentation, as shown in Fig. 8.1B, is obtained when the range of particle size in the suspension is very great. The sedimentation rate progressively decreases throughout the whole operation because there is no zone of constant composition, and zone C extends from the top interface to the layer of sediment.

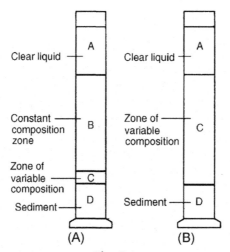

Fig. 8.1
Sedimentation of concentrated suspensions (A) Type 1 settling (B) Type 2 settling.

The main reasons for the modification of the settling rate of particles in a concentrated suspension are as follows:

(i)   If a significant size range of particles is present in the suspension, the large particles are settling relative to a suspension of smaller ones so that the effective density and viscosity of the fluid are increased.

(ii) The upward velocity of the fluid displaced during settling is appreciable in a concentrated suspension, and the apparent settling velocity is less than the actual velocity relative to the fluid.

(iii) The velocity gradients in the fluid close to the particles are increased as a result of the change in the area and shape of the flow passages.

(iv) The smaller particles tend to be dragged downward by the motion of the large particles and are, therefore, accelerated.

(v)   Because the particles are closer together in a concentrated suspension, flocculation is more marked in an ionised solvent, and the effective size of the small particles is increased.

If the range of particle size is not more than about 6:1, a concentrated suspension settles with a sharp interface, and all the particles fall at the same velocity. This is in contrast to the behaviour of a dilute suspension, for which the rates of settling of the particles can be calculated by the methods given in Chapter 6, and where the settling velocity is greater for the large particles. The two types of settling are often referred to as *sludge line settling* and *selective settling*, respectively. The overall result is that in a concentrated suspension, the large particles are retarded, and the small ones are accelerated.

Several attempts have been made to predict the apparent settling velocity of a concentrated suspension. In 1926, Robinson[3] suggested a modification of Stokes' law and used the density (ρ_c) and viscosity (μ_c) of the suspension in place of the properties of the fluid to give:

$$u_c = \frac{K'' d^2 (\rho_s - \rho_c) g}{\mu_c} \tag{8.1}$$

where K'' is a constant.

The effective buoyancy force is readily calculated because:

$$(\rho_s - \rho_c) = \rho_s - \{\rho_s(1 - e) + \rho e\} = e(\rho_s - \rho) \tag{8.2}$$

where e is the voidage of the suspension.

Robinson determined the viscosity of the suspension μ_c experimentally, although it may be obtained approximately from the following formula of Einstein[4]:

$$\mu_c = \mu(1 + k''C) \tag{8.3}$$

where k'' is a constant for a given shape of particle (2.5 for spheres), C is the volumetric (fractional) concentration of particles, and μ is the viscosity of the fluid. Evidently, the voidage e and concentration C are related as $C + e = 1$.

This equation, Eq. (8.3), holds for values of C up to 0.02. For more concentrated suspensions, Vand[5] gives the following equation:

$$\mu_c = \mu e^{k''C/(1-a'C)} \tag{8.4}$$

in which a' is a second constant, equal to $(39/64) = 0.609$ for spheres.

Steinour,[6] who studied the sedimentation of small uniform particles, adopted a similar approach, using the viscosity of the fluid, the density of the suspension, and a function of the voidage of the suspension to take account of the character of the flow spaces, and obtained the following expression for the velocity of the particle relative to the fluid u_p:

$$u_p = \frac{d^2(\rho_s - \rho_c)g}{18\mu} f(e) \tag{8.5}$$

Because the fraction of the area available for flow of the displaced fluid is e, its upward velocity is $u_c(1-e)/e$ so that:

$$u_p = u_c + u_c \frac{1-e}{e} = \frac{u_c}{e} \tag{8.6}$$

From his experiments on the sedimentation of tapioca in oil, Steinour found:

$$f(e) = 10^{-1.82(1-e)} \tag{8.7}$$

Substituting in Eq. (8.5), from Eqs. (8.2), (8.6), and (8.7):

$$u_c = \frac{e^2 d^2(\rho_s - \rho)g}{18\mu} 10^{-1.82(1-e)} \tag{8.8}$$

Hawksley[7] also used a similar method and gave:

$$u_p = \frac{u_c}{e} = \frac{d^2(\rho_s - \rho_c)g}{18\mu_c} \tag{8.9}$$

In each of these cases, it is correctly assumed that the upthrust acting on the particles is determined by the density of the suspension rather than that of the fluid. The use of an effective viscosity, however, is valid only for a large particle settling in a fine suspension. For the sedimentation of uniform particles, the increased drag is attributable to the steepening of the velocity gradients rather than to a change in the viscosity of the base fluid.

The rate of sedimentation of a suspension of fine particles is difficult to predict because of the large number of factors involved. Thus, for instance, the presence of an ionised solute in the liquid and the nature of the surface of the particles will affect the degree of flocculation and hence, the mean size and density of the flocs. The flocculation of a suspension is usually completed quite rapidly so that it is not possible to detect an increase in the sedimentation rate in the early stages after the formation of the suspension. Most fine suspensions flocculate readily in tap water, and it is generally necessary to add a deflocculating agent to maintain the particles individually dispersed. The factors involved in flocculation are discussed later in this chapter. A further factor influencing the sedimentation rate is the degree of agitation of the suspension. Gentle stirring may produce accelerated settling if the suspension behaves as a non-Newtonian fluid, in which the apparent viscosity is a function of the rate of shear. The change in apparent viscosity can probably be attributed to the re-orientation and/or rearrangement of the particles. The effect of stirring is, however, most marked on the consolidation of the final sediment, in which 'bridge formation' by the particles can be prevented by gentle stirring. During these final stages of consolidation of the sediment, liquid is being squeezed out through a bed of particles which are gradually becoming more tightly packed.

A number of empirical equations have been obtained for the rate of sedimentation of suspensions, as a result of tests carried out in vertical tubes. For a given solid and liquid, the main factors which affect the process are the height of the suspension, the diameter of the containing vessel, and the volumetric concentration. An attempt at coordinating the results obtained under a variety of conditions has been made by Wallis.[8]

Height of suspension

The height of suspension does not generally affect either the rate of sedimentation or the consistency of the sediment ultimately obtained. If, however, the position of the sludge line is plotted as a function of time for two different initial heights of slurry, curves of the form shown in Fig. 8.2 are obtained in which the ratio OA' : OA'' is everywhere constant. Thus, if the curve is obtained for any one initial height, the curves can be drawn for any other height.

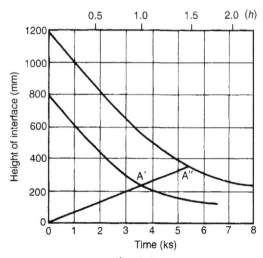

Fig. 8.2

Effect of height on sedimentation of a 3% (by volume) suspension of calcium carbonate.

Diameter of vessel

If the ratio of the diameter of the vessel to the diameter of the particle is greater than about 100, the walls of the container appear to have no effect on the rate of sedimentation. For smaller values, the sedimentation rate may be reduced because of the retarding influence of the walls, similar to the wall effects on a freely setting particle.

Concentration of suspension

As already indicated, the higher the concentration, the lower is the rate of fall of the sludge line, because the greater is the upward velocity of the displaced fluid, and the steeper are the velocity gradients in the fluid. Typical curves for the sedimentation of a suspension of precipitated calcium carbonate in water are shown in Fig. 8.3, and in Fig. 8.4, the mass rate of sedimentation (kg/m²s) is plotted against the concentration. This curve has a maximum value, corresponding to a volumetric concentration of about 2%. Egolf and McCabe,[9] Work and Kohler,[10] and others have given empirical expressions for the rate of sedimentation at the various stages, although these are generally applicable over a narrow range of conditions and involve constants which need to be determined experimentally for each suspension.

The final consolidation of the sediment is the slowest part of the process because the displaced fluid has to flow through the small spaces between the particles, similar to the flow in a packed bed. As consolidation occurs, the rate falls off because the resistance to the flow of liquid progressively increases. The porosity of the sediment is smallest at the bottom because the compressive force due to the weight of particles is greatest, and because the lower portion was formed at an earlier stage in the sedimentation process. The rate of sedimentation during this period is given approximately by:

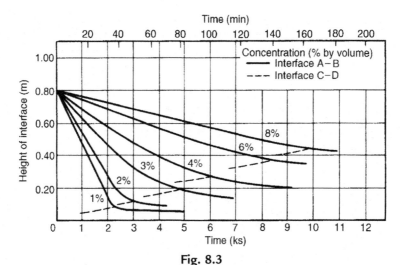

Fig. 8.3

Effect of concentration on the sedimentation of calcium carbonate suspensions.

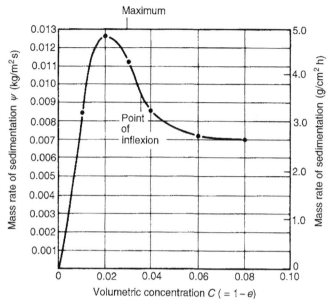

Fig. 8.4

Effect of concentration on mass rate of sedimentation of calcium carbonate.

$$-\frac{dH}{dt} = b(H - H_\infty) \tag{8.10}$$

where H is the height of the sludge line at time t, H_∞ is the final height of the sediment, and b is a constant for a given suspension.

The time taken for the sludge line to fall from a height H_c, corresponding to the critical settling point, to a height H is given by:

$$-bt = \ln(H - H_\infty) - \ln(H_c - H_\infty) \tag{8.11}$$

Thus, if $\ln(H - H_\infty)$ is plotted against t, a straight line of slope $-b$ is obtained.

The values of H_∞ are determined largely by the nature of the surface film of the liquid adhering to the particles.

Shape of vessel

Provided that the walls of the vessel are vertical, and that the cross-sectional area does not vary with depth, the shape of the vessel has little effect on the sedimentation rate. However, if parts of the walls of the vessel face downward, as in an inclined tube, or if part of the cross-section is obstructed for a portion of the height, the effect on the sedimentation process may be considerable.

Pearce[11] studied the effect of a downward-facing surface by considering an inclined tube as shown in Fig. 8.5. Starting with a suspension reaching a level AA, if the sludge line falls to a

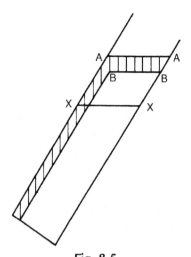

Fig. 8.5
Sedimentation in an inclined tube.

new level BB, then material will tend to settle out from the whole of the shaded area. This configuration is not stable, and the system tends to adjust itself so that the sludge line takes up a new level XX, the volume corresponding to the area AAXX being equal to that corresponding to the shaded area. By applying this principle, it is seen that it is possible to obtain an accelerated rate of settling in an inclined tank by inserting a series of inclined plates. The phenomenon has been studied further by several researchers including Schaflinger.[12]

The effect of a nonuniform cross-section was considered by Robins,[13] who studied the effect of reducing the area in part of the vessel by immersing a solid body, as shown in Fig. 8.6. If the cross-sectional area, sedimentation velocity, and fractional volumetric concentration are A, u_c, and C below the obstruction, and A', u'_c, and C' at the horizontal level of the obstruction, and ψ and ψ' are the corresponding rates of deposition of solids per unit area, then:

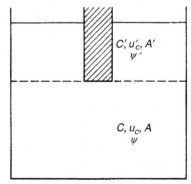

Fig. 8.6
Sedimentation in partially obstructed vessel.

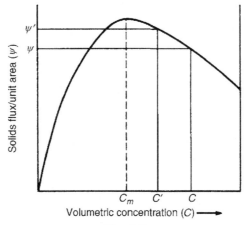

Fig. 8.7

Solids flux per unit area as a function of volumetric concentration.

$$A\psi = ACu_c \tag{8.12}$$

$$\text{and}: \quad A'\psi' = A'C'u'_c \tag{8.13}$$

For continuity at the bottom of the obstruction:

$$\psi' = \frac{A}{A'}\psi \tag{8.14}$$

A plot of ψ versus C will have the same general form as Fig. 8.4, and a typical curve is given in Fig. 8.7. If the concentration C is appreciably greater than the value C_m at which ψ is a maximum, C' will be less than C, and the system will be stable. On the other hand, if C is less than C_m, C' will be greater than C, and mixing will take place because of the greater density of the upper portion of the suspension. The range of values of C for which Eq. (8.14) is valid in practice, and for which mixing currents are absent, may be very small.

8.2.2 Flocculation

Introduction

The behaviour of suspensions of fine particles is very considerably influenced by whether the particles flocculate. The overall effect of flocculation is to create large conglomerations of elementary particles with occluded liquid. The flocs, which easily become distorted, are effectively enlarged particles of a density intermediate between that of the constituent particles and the liquid.

The tendency of the particulate phase of colloidal dispersions to aggregate is an important physical property which finds practical application in solid–liquid separation processes, such as sedimentation and filtration. The aggregation of colloids is known as coagulation, or flocculation. Particles dispersed in liquid media collide due to their relative motion, and the stability (that is stability against aggregation) of the dispersion is determined by the interaction between particles during these collisions. Attractive and repulsive forces can be operative between the particles; these forces may react in different ways depending on the environment conditions, such as salt concentration and pH. The commonly occurring forces between colloidal particles are van der Waals forces, electrostatic forces, and forces due to adsorbed macromolecules. In the absence of macromolecules, aggregation is largely due to van der Waals attractive forces, whereas stability is due to repulsive interaction between similarly charged electrical double layers.

The electrical double layer

Most particles acquire a surface electric charge when in contact with a polar medium. Ions of opposite charge (counter-ions) in the medium are attracted toward the surface, and ions of like charge (co-ions) are repelled. This process, together with the mixing tendency due to thermal motion, results in the creation of an electrical double layer which comprises the charged surface and a neutralising excess of counter-ions over co-ions distributed in a diffuse manner in the polar medium. The quantitative theory of the electrical double layer, which deals with the distribution of ions and the magnitude of electric potentials, is beyond the scope of this text, although an understanding of it is essential in an analysis of colloid stability.[14–17]

For present purposes, the electrical double layer is represented in terms of Stern's model (Fig. 8.8), wherein the double layer is divided into two parts separated by a plane (Stern plane)

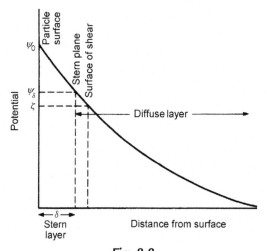

Fig. 8.8
Schematic representation of the Stern's model.

located at a distance of about one hydrated-ion radius from the surface. The potential changes from ψ_0 (surface) to ψ_δ (Stern potential) in the Stern layer and decays to zero in the diffuse double layer; quantitative treatment of the diffuse double layer follows the Gouy–Chapman theory.[18, 19]

ψ_δ can be estimated from electrokinetic measurements, such as electrophoresis and streaming potential. In such measurements, surface and liquid move tangentially with respect to each other. For example, in electrophoresis, the liquid is stationary, and the particles move under the influence of an applied electric field. A thin layer of liquid, a few molecules thick, moves together with the particle so that the actual hydrodynamic boundary between the moving unit and the stationary liquid is a *slipping plane* inside the solution. The potential at the slipping plane is termed the *zeta potential*, ζ, as shown in Fig. 8.8.

Lyklema[20] considers that the slipping plane may be identified with the Stern plane so that $\psi_\delta \simeq \zeta$. Thus, because the surface potential ψ_0 is inaccessible, zeta potentials find practical application in the calculation of V_R from Eq. (8.16). In practice, electrokinetic measurements must be carried out with considerable care if reliable estimates of ζ are to be made.[21]

Interactions between particles

The interplay of forces between particles in lyophobic sols may be interpreted in terms of the theory of Derjaguin and Landau[22] and Verwey and Overbeek.[14] Their theory (the DLVO theory) considers that the potential energy of interaction between a pair of particles consists of two components:

(a) a repulsive component V_R arising from the overlap of the electrical double layers, and
(b) a component V_A due to van der Waals attraction arising from electromagnetic effects.

These are considered to be additive so that the total potential energy of interaction V_T is given by:

$$V_T = V_R + V_A \tag{8.15}$$

In general, as pointed out by Gregory,[23] the calculation of V_R is complex, but a useful approximation for identical spheres of radius a is given by[14]:

$$V_R = \frac{64\pi a n_i K T \gamma^2 e^{-\kappa H_s}}{\kappa^2} \tag{8.16}$$

$$\text{where:} \quad \gamma = \frac{\exp\left(Z e_c \psi_\delta / 2KT\right) - 1}{\exp\left(Z e_c \psi_\delta / 2KT\right) + 1} \tag{8.17}$$

$$\text{and:} \quad \kappa = \left(\frac{2e_c^2 n_i Z^2}{\varepsilon K T}\right)^{1/2} \tag{8.18}$$

For identical spheres with $H_s \leq 10\text{--}20\,\text{nm}$ $(100\text{--}200\,\text{Å})$, and when $H_s \ll a$, the energy of attraction V_A is given by the approximate expression[24]:

$$V_A = -\frac{\mathcal{A}a}{12H_s} \tag{8.19}$$

where $\mathcal{A}$ is the Hamaker[25] constant whose value depends on the nature of the material of the particles. The presence of liquid between particles reduces V_A, and an effective Hamaker constant is calculated from:

$$\mathcal{A} = \left(\mathcal{A}_2^{1/2} - \mathcal{A}_1^{1/2}\right)^2 \tag{8.20}$$

where subscripts 1 and 2 refer to the dispersion medium and particles, respectively. Eq. (8.19) is based on the assumption of complete additivity of intermolecular interactions; this assumption is avoided in the theoretical treatment of Lifshitz,[26] which is based on macroscopic properties of materials.[27] Tables of $\mathcal{A}$ are available in the literature[28]; values are generally found to lie in the range 0.1×10^{-20} to $10 \times 10^{-20}\,\text{J}$.

The general form of V_T versus distance of separation between particle surfaces H_s is shown schematically in Fig. 8.9. At very small distances of separation, repulsion due to overlapping electron clouds (Born repulsion)[14] dominates, and consequently, a deep minimum (primary minimum) occurs in the potential energy curve. For smooth surfaces, this limits the distance of closest approach (H_{smin}) to $\sim 0.4\,\text{nm}$ $(4\,\text{Å})$. Aggregation of particles occurring in this primary minimum, for example, aggregation of lyophobic sols in the presence of NaCl, is termed *coagulation*.[27]

At high surface potentials, low ionic strengths, and intermediate distance, the electrical repulsion term is dominant, and so a maximum (primary maximum) occurs in the potential

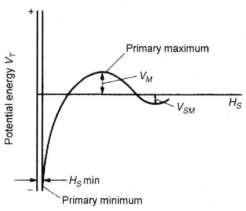

Fig. 8.9

Potential energy as a function of separation.

energy curve. At larger distances of separation, V_R decays more rapidly than V_A, and a secondary minimum appears. If the potential energy maximum is large compared with the thermal energy KT of the particles ($\sim 4.2 \times 10^{-20}$ J), the system should be stable; otherwise, the particles would coagulate. The height of this energy barrier to coagulation depends upon the magnitude of ψ_δ (and ζ) and upon the range of the repulsive forces (that is, upon $1/\kappa$). If the depth of the secondary minimum is large compared with KT, it should produce a loose, easily reversible form of aggregation, which is termed flocculation. This term also describes aggregation of particles in the presence of polymers,[27] as discussed later in this chapter.

It is of interest to note that both V_A and V_R increase as particle radius a becomes larger, and thus, V_m in Fig. 8.9 would be expected to increase with the sol becoming more stable; also, if a increases, then V_{SM} increases and may become large enough to produce "secondary minimum" flocculation.

Coagulation concentrations

Coagulation concentrations are the electrolyte concentrations required just to coagulate a sol. Clearly, V_M in Fig. 8.9 must be reduced, preferably to zero, to allow coagulation to occur. This can be achieved by increasing the ionic strength of the solution, thus, increasing κ and thereby reducing V_R in Eq. (8.16). The addition of salts with multivalent ions (such as Al^{3+}, Ca^{2+}, Fe^{3+}) is most effective because of the effect of charge number Z and κ (Eq. 8.18). Taking as a criterion that $V_T = 0$ and $dV_T/dH_s = 0$ for the same value of H_s, it may be shown[14] that the coagulation concentration c_c' is given by:

$$c_c' = \frac{9.75 B^2 \varepsilon^3 K^5 T^5 \gamma^4}{e_c^2 \mathrm{N} \mathcal{A}^2 Z^6} \tag{8.21}$$

where $B = 3.917 \times 10^{39}$ coulomb^{-2}. At high values of surface potentials, $\gamma \approx 1$, and Eq. (8.21) predicts that the coagulation concentration should be inversely proportional to the sixth power of the valency Z. Thus coagulation concentrations of those electrolytes whose counter-ions have charge numbers 1, 2, and 3 should be in the ratio 100: 1.6: 0.13. It may be noted that, if an ion is specifically adsorbed on the particles, ψ_δ can be drastically reduced and coagulation effected without any great increase in ionic strength. For example, minute traces of certain hydrolysed metal ions can cause coagulation of negatively charged particles.[29] In such cases, charge reversal often occurs, and the particles can be restabilised if excess coagulant is added.

Kinetics of coagulation

The rate of coagulation of particles in a liquid depends on the frequency of collisions between particles due to their relative motion. When this motion is due to Brownian movement, coagulation is termed *perikinetic*; when the relative motion is caused by velocity gradients, coagulation is termed *orthokinetic*.

Modern analyses of perikinesis and orthokinesis take account of hydrodynamic forces as well as interparticle forces. In particular, the frequency of binary collisions between spherical particles has received considerable attention.[24, 30–32]

The frequency of binary encounters during perikinesis is determined by considering the process as that of diffusion of spheres (radius a_2 and number concentration n_2) toward a central reference sphere of radius a_1, whence the frequency of collision I is given by[24]:

$$I = \frac{4\pi D_{12}^{(\infty)} n_2(a_1 + a_2)}{1 + \frac{a_2}{a_1} \int_{1+(a_2/a_1)}^{\infty} \left(D_{12}^{(\infty)}/D_{12}\right) \exp\left(V_T/KT\right)\frac{ds}{s^2}} \tag{8.22}$$

where $s = a_r/a_1$ and the coordinate a_r has its origin at the centre of sphere 1.

Details of D_{12}, the relative diffusivity between unequal particles, are given by Spielman,[24] who illustrates the dependence of D_{12} on the relative separation a_r between particle centres. At infinite separation, where hydrodynamic effects vanish:

$$D_{12} = D_{12}^{(\infty)} = D_1 + D_2 \tag{8.23}$$

where D_1 and D_2 are absolute diffusion coefficients given by the Stokes–Einstein equation.

$$\left.\begin{array}{l} D_1 = KT/(6\pi\mu a_1) \\ D_2 = KT/(6\pi\mu a_2) \end{array}\right\} \tag{8.24}$$

When long-range particle interactions and hydrodynamics effects are ignored, Eq. (8.22) becomes equivalent to the solution of von Smoluchowski,[33] who obtained the collision frequency I_s as:

$$I_s = 4\pi D_{12}^{(\infty)} n_2(a_1 + a_2) \tag{8.25}$$

and who assumed an attractive potential only given by:

$$V_A = -\infty \quad s \leq +\frac{a_2}{a_1}$$

$$V_A = 0 \quad s > 1 + \frac{a_2}{a_1} \tag{8.26}$$

$$\text{Thus:} \quad I = \alpha_p I_s \tag{8.27}$$

where the ratio α_p is the reciprocal of the denominator in Eq. (8.22); values of α_p are tabulated by Spielman.[24]

Assuming an attractive potential only, given by Eq. (8.26), Smoluchowski[33] showed that the frequency of collisions per unit volume between particles of radii a_1 and a_2 in the presence of a laminar shear gradient $\dot{\gamma}$ is given by:

$$J_s = \frac{4}{3} n_1 n_2 (a_1 + a_2)^3 \dot{\gamma} \tag{8.28}$$

Analyses of the orthokinetic encounters between equisized spheres[32] have shown that, as with perikinetic encounters, Eq. (8.28) can be modified to include a ratio α_0 to give the collision frequency J as:

$$J = \alpha_0 J_s \tag{8.29}$$

where α_0, which is a function of $\dot{\gamma}$, corrects the Smoluchowski relation for hydrodynamic interactions and interparticle forces. Zeichner and Schowalter[32] present α_0^{-1} graphically as a function of a dimensionless parameter $N_F (= 6\pi\mu a^3 \dot{\gamma}/\mathcal{A})$ for the condition $V_R = 0$, whence it is possible to show that, for values of $N_F > 10$, J is proportional to $\dot{\gamma}$ raised to the 0.77 power instead of the first power as given by Eq. (8.28).

Perikinetic coagulation is normally too slow for economic practical use in such processes as wastewater treatment, and orthokinetic coagulation is often used to produce rapid growth of aggregate of floc size. In such situations, floc–floc collisions occur under nonuniform turbulent flow conditions. A rigorous analysis of the kinetics of coagulation under these conditions is not available at present. A widely used method of evaluating a mean shear gradient in such practical situations is given by Camp and Stein,[34] who propose that:

$$\dot{\gamma} = [\mathbf{P}/\mu]^{1/2} \tag{8.30}$$

where $\mathbf{P} =$ power input/unit volume of fluid.

Effect of polymers on stability

The stability of colloidal dispersions is strongly influenced by the presence of adsorbed polymers. Sols can be stabilised or destabilised depending on a number of factors including the relative amounts of polymer and sol, the mechanism of adsorption of polymer, and the method of mixing polymer and dispersion.[35] Adsorption of polymer onto colloidal particles may increase their stability by decreasing V_A,[36, 37] increasing V_R,[38] or by introducing a *steric* component of repulsion V_S.[39, 40]

Flocculation is readily produced by linear homopolymers of high molecular weight. Although they may be nonionic, they are commonly polyelectrolytes; polyacrylamides and their derivatives are widely used in practical situations.[41] Flocculation by certain high molecular weight polymers can be interpreted in terms of a *bridging* mechanism; polymer molecules may be long and flexible enough to adsorb onto several particles. The precise nature of the attachment between polymer and particle surface depends on the nature of the surfaces of particle and polymer, and on the chemical properties of the solution. Various types of interaction between polymer segments and particle surfaces may be envisaged. In the case of polyelectrolytes, the strongest of these interactions would be ionic association between a charged site on the surface and an oppositely charged polymer segment, such as polyacrylic acid and positively charged silver iodide particles.[42–45]

Polymers may show an optimum flocculation concentration which depends on molecular weight and concentration of solids in suspension. Overdosing with flocculant may lead to restabilisation,[46] as a consequence of particle surfaces becoming saturated with polymer. Optimum flocculant concentrations may be determined by a range of techniques including sedimentation rate, sedimentation volume, filtration rate, and clarity of supernatant liquid.

Effect of flocculation on sedimentation

In a flocculated, or coagulated, suspension, the aggregates of fine particles (or flocs) are the basic structural units, and in a low shear rate process, such as gravity sedimentation, their settling rates and sediment volumes depend largely on volumetric concentration of flocs and on interparticle forces. The type of settling behaviour exhibited by flocculated suspensions depends largely on the initial solids concentration and chemical environment. Two kinds of batch settling curves are frequently seen. At low initial solids concentration, the flocs may be regarded as discrete units consisting of particles and immobilised fluid. The flocs settle initially at a constant settling rate, though as they accumulate on the bottom of the vessel, they deform under the weight of the overlying flocs. The curves shown earlier in Fig. 8.3 for calcium carbonate suspensions relate to this type of sedimentation. When the solids concentration is very high, the maximum settling rate is not immediately reached and thus, may increase with the increasing initial height of suspension.[47] Such behaviour appears to be characteristic of structural flocculation associated with a continuous network of flocs extending to the walls of the vessel. In particular, the first type of behaviour, giving rise to a constant settling velocity of the flocs, has been interpreted quantitatively by assuming that the flocs consist of aggregates of particles and occluded liquid. The flocs are considerably larger than the primary particles and of density intermediate between that of the water and the particles themselves. Michaels and Bolger[47] found good agreement between their experimental results and predicted sedimentation rates. The latter were calculated from the free settling velocity of an individual floc, corrected for the volumetric concentration of the flocs using Eq. (8.71) (Section 8.3.2), which has been developed for the sedimentation of systems of fully-dispersed monosize particles.

8.2.3 The Kynch Theory of Sedimentation

The behaviour of concentrated suspensions during sedimentation has been analysed by Kynch,[48] largely using considerations of continuity. The basic assumptions made here are as follows:

(a)   Particle concentration is uniform across any horizontal layer,
(b)   Wall effects can be ignored,

(c) There is no differential settling of particles as a result of differences in shape, size, or composition,

(d) The velocity of fall of particles depends only on the local concentration of particles,

(e) The initial concentration is either uniform or increases toward the bottom of the suspension, and

(f) The sedimentation velocity tends to zero as the concentration approaches a limiting value corresponding to that of the sediment layer deposited at the bottom of the vessel.

If at some horizontal level where the volumetric concentration of particles is C and the sedimentation velocity is u_c, the volumetric rate of sedimentation per unit area, or flux, is given by:

$$\psi = Cu_c \tag{8.31}$$

Then a material balance taken between a height H above the bottom, at which the concentration is C, and the mass flux is ψ, and at a height $H+dH$, where the concentration is $C+(\partial C/\partial H)dH$, and the mass flux is $\psi+(\partial \psi/\partial H)dH$, gives:

$$\left\{ \left(\psi + \frac{\partial \psi}{\partial H}dH \right) - \psi \right\} dt = \left(dH\frac{\partial C}{\partial t} \right) dt$$

$$\text{That is:} \quad \frac{\partial \psi}{\partial H} = \frac{\partial C}{\partial t} \tag{8.32}$$

$$\text{Hence:} \quad \frac{\partial \psi}{\partial H} = \frac{\partial \psi}{\partial C} \cdot \frac{\partial C}{\partial H} = \frac{d\psi}{dC} \cdot \frac{\partial C}{\partial H} \text{(since } \psi \text{ depends only on } C) \tag{8.33}$$

$$\text{Thus:} \quad \frac{\partial C}{\partial t} - \frac{d\psi}{dC} \cdot \frac{\partial C}{\partial H} = 0 \tag{8.34}$$

In general, the concentration of particles will be a function of position and time, and thus:

$$C = f(H, t)$$

$$\text{and:} \quad dC = \frac{\partial C}{\partial H}dH + \frac{\partial C}{\partial t}dt$$

Conditions of constant concentration are, therefore, defined by setting $dC = 0$:

$$\frac{\partial C}{\partial H}dH + \frac{\partial C}{\partial t}dt = 0$$

$$\text{Thus:} \quad \frac{\partial C}{\partial H} = -\frac{\partial C}{\partial t}\bigg/\frac{dH}{dt} \tag{8.35}$$

Substituting in Eq. (8.34) gives the following relation for constant concentration:

$$\frac{\partial C}{\partial t} - \frac{d\psi}{dC}\left(-\frac{\partial C}{\partial t}\Big/\frac{dH}{dt}\right) = 0$$

$$\text{or}: \quad -\frac{d\psi}{dC} = \frac{dH}{dt} = u_w \tag{8.36}$$

Because Eq. (8.36) refers to a constant concentration, $d\psi/dC$ is constant, and $u_w(=dH/dt)$ is, therefore, also constant for any given concentration and is the velocity of propagation of a zone of constant concentration C. Thus, lines of constant slope, on a plot of H versus t, will refer to zones of constant composition, each of which will be propagated at a constant rate, dependent only on the concentration. Then, because $u_w = -(d\psi/dC)$ (Eq. (8.36)) when $d\psi/dC$ is negative (as it is at volumetric concentrations >0.02 in Fig. 8.4), u_w is positive, and the wave will propagate upward. At lower concentrations, $d\psi/dC$ is positive, and the wave will propagate downward. Thus, waves originating at the base of the sedimentation column will propagate upward to the suspension interface if $d\psi/dC$ is negative, but will be prevented from propagating if $d\psi/dC$ is positive because of the presence of the base. Although Kynch's arguments may be applied to any suspension in which the initial concentration increases continuously from top to bottom, consideration will be confined to suspensions initially of uniform concentration.

In an initially uniform suspension of concentration C_0, the interface between the suspension and the supernatant liquid will fall at a constant rate until a zone of composition, greater than C_0, has propagated from the bottom to the free surface. The sedimentation rate will then fall off progressively as zones of successively greater concentrations reach the surface, until eventually, sedimentation will cease when the C_{max} zone reaches the surface. This assumes that the propagation velocity decreases progressively with increase of concentration.

However, if zones of higher concentration propagate at velocities greater than those of lower concentrations, they will automatically overtake them, giving rise to a sudden discontinuity in concentration. In particular, if the propagation velocity of the suspension of maximum possible concentration C_{max} exceeds that of all of the intermediate concentrations between C_0 and C_{max}, sedimentation will take place at a constant rate, corresponding to the initial uniform concentration C_0, and will then cease abruptly as the concentration at the interface changes from C_0 to C_{max}.

Because the propagation velocity u_w is equal to $-(d\psi/dC)$, the sedimentation behaviour will be affected by the shape of the plot of ψ versus C. If this is consistently concave to the time-axis, $d\psi/dC$ will become increasingly negative as C increases, u_w will increase monotonically, and consequently, there will be a discontinuity because the rate of propagation of a zone of concentration C_{max} exceeds that for all lower concentrations; this is the condition referred to in the previous paragraph. On the other hand, if there is a point of inflexion in the curve (as at $C = 0.033$ in Fig. 8.4), the propagation rate will increase progressively up to the condition given by this point of inflexion (concentration C_i) and will then decrease as the concentration is further increased. There will again be a discontinuity, although this time when the wave corresponding to concentration C_i reaches the interface, the sedimentation rate will then fall off gradually as zones of successively higher concentration reach the interface, and sedimentation will finally cease when the concentration at the interface reaches C_{max}.

It is possible to apply this analysis to obtain the relationship between the flux of solids and concentration over the range where $-(d\psi/dC)$ is decreasing with increase of concentration, using the results of a single sedimentation test. By taking a suspension of initial concentration $C_0(\geq C_i)$, it is possible to obtain the $\psi - C$ curve over the concentration range C_0 to C_{max} from a single experiment. Fig. 8.10 shows a typical sedimentation curve for such a suspension. The H-axis represents the initial condition ($t=0$), and lines such as KP and OB representing constant concentrations have slopes of $u_w\ (=dH/dt)$. Lines from all points between A and O corresponding to the top and bottom of the suspension, respectively, will be parallel because the concentration is constant, and their location and slope are determined by the initial concentration of the suspension. As solids become deposited at the bottom, the concentration there will rapidly rise to the maximum possible value C_{max} (ignoring the effects of possible sediment consolidation), and the line OC represents the line of constant concentration C_{max}. Other lines, such as OD of greater slope, all originate at the base of the suspension and correspond to intermediate concentrations.

Considering a line, such as KP, which refers to the propagation of a wave corresponding to the initial uniform composition from an initial position K in the suspension, this line terminates on the curve ABDC at P, which is the position of the top interface of the suspension

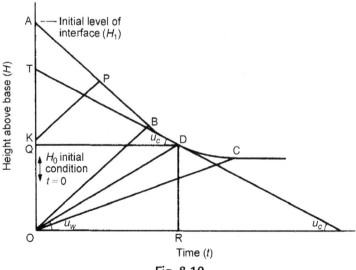

Fig. 8.10
Construction for Kynch theory.

at time t. The location of P is determined by the fact that KP represents the upward propagation of a zone of constant composition at a velocity u_w through which particles are falling at a sedimentation velocity u_c. Thus, the total volume of particles passing per unit area through the plane in time t is given by:

$$V = C_0(u_c + u_w)t \tag{8.37}$$

Because P corresponds to the surface of the suspension, V must be equal to the total volume of particles which was originally above the level indicated by K.

$$\text{Thus}: \quad C_0(u_c + u_w)t = C_0(H_t - H_0)$$

$$\text{or}: \quad (u_c + u_w)t = H_t - H_0 \tag{8.38}$$

Because the concentration of particles is initially uniform, and the sedimentation rate is a function solely of the particle concentration, the line APB will be straight, having a slope of $(-dH/dt)$ equal to u_c.

After point B, the sedimentation curve has a decreasing negative slope, reflecting the increasing concentration of solids at the interface. Line OD represents the locus of points of some concentration C, where $C_0 < C < C_{max}$. It corresponds to the propagation of a wave at a velocity u_w, from the bottom of the suspension. Thus, when the wave reaches the interface, point D, all the particles in the suspension must have passed through the plane of the wave. Thus, considering unit area:

$$C(u_c + u_w)t = C_0 H_t \tag{8.39}$$

In Fig. 8.10:

$$H_t = \text{OA}$$

By drawing a tangent to the curve ABDC at D, the point T is located.

Then:

$$u_c t = \text{QT}(\text{since} - u_c \text{ is the slope of the curve at D})$$

$$u_w t = \text{RD} = \text{OQ}(\text{since } u_w \text{ is the slope of line OD})$$

$$\text{and}: \quad (u_c + u_w)t = \text{OT}$$

Thus, the concentration C corresponding to the line OD is given by:

$$C = C_0 \frac{\text{OA}}{\text{OT}} \tag{8.40}$$

and the corresponding solids flux is given by:

$$\psi = Cu_c = C_0 \frac{\text{OA}}{\text{OT}} u_c \tag{8.41}$$

Thus, by drawing the tangent at a series of points on the curve BDC and measuring the corresponding slope $-u_c$ and intercept OT, it is possible to establish the solids flux ψ for any concentration $C(C_i < C < C_{max})$.

It is shown in Section 8.3.3 that, for coarse particles, the point of inflexion does not occur at a concentration which would be obtained in practice in a suspension, and, therefore, the particles will settle throughout at a constant rate until an interface forms between the clear liquid and the sediment, when sedimentation will abruptly cease. With a highly flocculated suspension, the point of inflexion may occur at a very low volumetric concentration. In these circumstances, there will be a wide range of concentrations for which the constant rate sedimentation is followed by a period of falling rate.

8.2.4 The Thickener

The thickener is the industrial unit in which the concentration of a suspension is increased by sedimentation, with the formation of a clear liquid. In most cases, the concentration of the suspension is high, and hindered settling takes place. Thickeners may operate as batch or continuous units, and consist of tanks from which the clear liquid is taken off at the top, and the thickened liquor, at the bottom.

In order to obtain the largest possible throughput from a thickener of given size, the rate of sedimentation should be as high as possible. In many cases, the rate may be artificially increased by the addition of small quantities of an electrolyte, which causes precipitation of colloidal particles and the formation of flocs. The suspension is also frequently heated because this lowers the viscosity of the liquid and encourages the larger particles in the suspension to grow in size at the expense of the more soluble small particles. Further, the thickener frequently incorporates a slow stirrer, which causes a reduction in the apparent viscosity of the suspension and also aids in the consolidation of the sediment.

The batch thickener usually consists of a cylindrical tank with a slightly conical bottom. After sedimentation has proceeded for an adequate time, the thickened liquor is withdrawn from the bottom, and the clear liquid is taken off through an adjustable offtake pipe from the upper part of the tank. The conditions prevailing in the batch thickener are similar to those in the ordinary laboratory sedimentation tube, and during the initial stages, there will generally be a zone in which the concentration of the suspension is the same as that in the feed.

The continuous thickener consists of a cylindrical tank with a flat bottom. The suspension is fed in at the centre, at a depth of from 0.3 to 1 m below the surface of the liquid, with as little disturbance as possible. The thickened liquor is continuously removed through an outlet at the bottom, and any solids which are deposited on the floor of the tank may be directed toward the outlet by means of a slowly rotating rake mechanism incorporating scrapers. The rakes are often hinged so that the arms fold up automatically if the torque

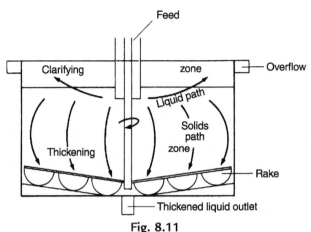

Fig. 8.11
Flow in continuous thickener.

exceeds a certain value; this prevents it from being damaged if it is overloaded. The raking action can increase the degree of thickening achieved in a thickener of given size. The clarified liquid is continuously removed from an overflow which runs round the whole of the upper edge of the tank. The solids are, therefore, moving continuously downward, and then inward toward the thickened liquor outlet; the liquid is moving upward and radially outward, as shown in Fig. 8.11. In general, there will be no region of constant composition in a continuous thickener.

Thickeners may vary from a few metres to several hundred metres in diameter. Small ones are made of wood or metal, and the rakes rotate at about 0.02 Hz (1 rpm). Very large thickeners generally consist of large concrete tanks, and the stirrers and rakes are driven by means of traction motors which drive on a rail running round the whole circumference; the speed of rotation may be as low as 0.002 Hz (0.1 rpm).

The thickener has a twofold function. First, it must produce a clarified liquid, and, therefore, the upward velocity of the liquid must, at all times, be less than the settling velocity of the particles. Thus, for a given throughput, the clarifying capacity is determined by the diameter of the tank. Secondly, the thickener is required to produce a given degree of thickening of the suspension. This is controlled by the time of residence of the particles in the tank, and hence, by the depth below the feed inlet.

There are, therefore, two distinct requirements in the design—first, the provision of an adequate diameter to obtain satisfactory clarification, and, second, sufficient depth to achieve the required degree of thickening of the underflow. Frequently, the high diameter: height ratios which are employed result in only the first requirement being adequately met. The Dorr thickener is an example of a relatively shallow equipment employing a rake mechanism.

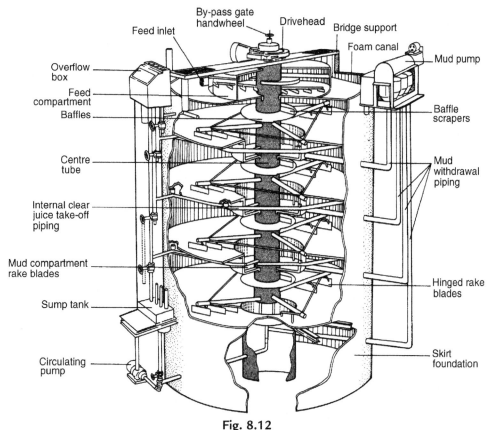

Fig. 8.12
Four-tray Dorr thickener.

In order to save ground space, a number of trays may be mounted above one another and a common drive shaft employed. A four-tray thickener is illustrated in Fig. 8.12; this type of equipment generally gives a better performance for clarification than for thickening.

The satisfactory operation of the thickener as a clarifier depends upon the existence of a zone of negligible solids content toward the top. In this zone, conditions approach those under which free settling takes place, and the rate of sedimentation of any particles which have been carried to this height is, therefore, sufficient for them to settle against the upward current of liquid. If this upper zone is too shallow, some of the smaller particles may escape in the liquid overflow. The volumetric rate of flow of liquid upward through the clarification zone is equal to the difference between the rate of feed of liquid in the slurry and the rate of removal in the underflow. Thus the required concentration of solids in the underflow, as well as the throughput, determines the conditions in the clarification zone.

Thickening zone

In a continuous thickener, the area required for thickening must be such that the total solids flux (volumetric flowrate per unit area) at any level does not exceed the rate at which the solids can be transmitted downward. If this condition is not met, solids will build up, and steady-state operation will not be possible. If no solids escape in the overflow, this flux must be constant at all depths below the feed point. In the design of a thickener, it is, therefore, necessary to establish the concentration at which the total flux is a *minimum* in order to calculate the required area.

The total flux ψ_T may be expressed as the product of the volumetric rate per unit area at which thickened suspension is withdrawn (u_u) and its volumetric concentration is C_u.

$$\text{Thus:} \quad \psi_T = u_u C_u \tag{8.42}$$

This flux must also be equal to the volumetric rate per unit area at which solids are fed to the thickener.

$$\text{Thus:} \quad \psi_T = \frac{Q_0}{A} C_0 \tag{8.43}$$

where Q_0 is the volumetric feed rate of suspension, A is the area of the thickener, and C_0 is the volumetric concentration of solids in the feed.

At *any horizontal plane* in a continuous thickener operating under steady-state conditions, the total flux of solids ψ_T is made up of two components:

(a)   That attributable to the sedimentation of the solids in the liquid—as measured in a batch sedimentation experiment.

This is given by:

$$\psi = u_c C \quad (\text{Eq.8.31})$$

where u_c is the sedimentation velocity of solids at a concentration C. ψ corresponds to the flux in a batch thickener at that concentration.

(b)   That arising from the bulk downward flow of the suspension which is drawn off as underflow from the base of the thickener which is given by:

$$\psi_u = u_u C \tag{8.44}$$

Thus, the total flux:

$$\psi_T = \psi + \psi_u = \psi + u_c C \tag{8.45}$$

Fig. 8.13 shows a typical plot of sedimentation flux ψ against volumetric concentration C; this relationship needs to be based on experimental measurements of u_c as a function of C. The curve must always pass through a maximum, and usually exhibits a point of inflexion at high

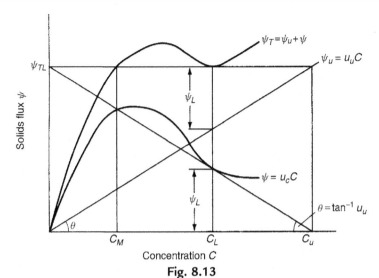

Fig. 8.13

Solids fluxes as functions of concentration and Yoshioka construction.[49]

concentrations. At a given withdrawal rate per unit area (u_u), the bulk flux (ψ_u) is given by a straight line, of slope u_u, passing through the origin as given by Eq. (8.44). The total solids flux ψ_T, obtained as the summation of the two curves, passes through a maximum, followed by a minimum (ψ_{TL}) at a higher concentration (C_L). For all concentrations exceeding C_M (shown in Fig. 8.13), ψ_{TL} is the parameter which determines the capacity of the thickener when operating at the fixed withdrawal rate u_u.

It may be noted that, because no further sedimentation occurs below the level of the exit of the thickener, there will at that position be a discontinuity, and the solids concentration will undergo a step change from its value in the suspension to that in the underflow (C_u).

It is now necessary to determine the limiting total flux ψ_{TL} for a specified concentration C_u of overflow. The required area of the thickener is then obtained by substituting this value into Eq. (8.43) to give:

$$A = \frac{Q_0 C_0}{\psi_{TL}} \qquad (8.46)$$

A simple construction, proposed by Yoshioka et al.[49] has been described by Hassett,[50] among others. By differentiation of Eq. (8.45) at a fixed value of u_u:

$$\frac{\partial \psi_T}{\partial C} = \frac{\partial \psi}{\partial C} + u_u \qquad (8.47)$$

The minimum value of $\psi_T(=\psi_{TL})$ occurs when $\dfrac{\partial \psi_T}{\partial C} = 0$;

$$\text{that is when:} \quad \frac{\partial \psi}{\partial C} = -u_u \qquad (8.48)$$

If a tangent is drawn from the point on the abscissa corresponding to the required underflow concentration C_u, it will meet the ψ curve at a concentration value C_L, at which ψ_T has the minimum value ψ_{TL} and will intersect the ordinate axis at a value equal to ψ_{TL}. The construction is dependent on the fact that the slopes of the tangent and of the ψ_u line are equal and opposite ($\mp u_u$). Thus, in order to determine both C_L and ψ_{TL}, it is not necessary to plot the total flux curve (ψ_T), but only to draw the tangent to the batch sedimentation (ψ) curve. The value of ψ_{TL} determined in this way is then inserted in Eq. (8.46) to obtain the required area of the thickener A.

From the geometry of Fig. 8.13:

$$\frac{\psi_{TL}}{C_u} = \frac{\psi_L}{C_u - C_L} \tag{8.49}$$

where ψ_L is the value of ψ at the concentration C_L.

Because $\psi_L = u_{cL}C_L$ (from Eq. 8.31), again from Eq. (8.46):

$$A = \frac{Q_0 C_0}{\psi_{TL}}$$

$$= Q_0 C_0 \left[\frac{\dfrac{1}{C_L} - \dfrac{1}{C_u}}{u_{cL}}\right] \tag{8.50}$$

where u_{cL} is the value of u_c at the concentration C_L. Thus, the minimum necessary area of the thickener may be obtained from the maximum value of

$$\left[\frac{\dfrac{1}{C} - \dfrac{1}{C_u}}{u_c}\right] \text{ which is designated } \left[\frac{\dfrac{1}{C} - \dfrac{1}{C_u}}{u_c}\right]_{max.}$$

Concentrations may also be expressed as mass per unit volume (using c in place of C) to give:

$$A = Q_0 c_0 \left[\frac{[(1/c) - (1/c_u)]}{u_c}\right]_{max} \tag{8.51}$$

Overflow

The liquid flowrate in the overflow (Q') is the difference between that in the feed and in the underflow.

$$\text{Thus:} \quad Q' = Q_0(1 - C_0) - (Q_0 - Q')(1 - C_u)$$

$$\text{or:} \quad \frac{Q'}{Q_0} = 1 - \frac{C_0}{C_u} \tag{8.52}$$

At any depth below the feed point, the upward liquid velocity must not exceed the settling velocity of the particles (u_c). Where the concentration is C, the required area is, therefore, given by:

$$A = Q_0 \frac{1}{u_c} \left[1 - \frac{C}{C_u} \right] \qquad (8.53)$$

It is, therefore, necessary to calculate the *maximum* value of A for all the values of C which may be encountered in a given application.

Eq. (8.53) can usefully be rearranged in terms of the mass ratio of liquid to solid in the feed (Y) and the corresponding value (U) in the underflow to give:

$$Y = \frac{1-C}{C} \left(\frac{\rho}{\rho_s} \right) \text{ and } U = \frac{1-C_u}{C_u} \left(\frac{\rho}{\rho_s} \right)$$

$$\text{Then: } C = \frac{1}{1 + Y(\rho_s/\rho)} \text{ and } C_u = \frac{1}{1 + U(\rho_s/\rho)}$$

$$\text{and: } A = \frac{Q_0}{u_c} \left\{ 1 - \frac{1 + U(\rho_s/\rho)}{1 + Y(\rho_s/\rho)} \right\}$$

$$= \frac{Q_0(Y - U)C\rho_s}{u_c \rho} \qquad (8.54)$$

The values of A should be calculated for the whole range of concentrations present in the thickener, and the design should then be based on the maximum value so obtained.

The above procedure for the determination of the required cross-sectional area of a continuous thickener is illustrated in Example 8.2.

Great care should be used, however, in applying the results of batch sedimentation tests carried out in the laboratory to the design of large-scale continuous industrial thickeners as the conditions in the two cases are different. In a batch experiment, the suspension is initially well mixed, and the motion of both fluid and particles takes place in the vertical direction only. In a continuous thickener, the feed is normally introduced near the centre at some depth (usually between 0.3 and 1 m) below the free surface, and there is a significant radial velocity component in the upper regions of the suspension. The liquid flows predominantly outward and upward toward the overflow; this is generally located around the whole of the periphery of the tank. In addition, the precise design of the feed device will exert some influence on the flow pattern in its immediate vicinity.

Underflow

In many operations, the prime requirement of the thickener is to produce a thickened product of as high a concentration as possible. This necessitates the provision of sufficient depth to allow time for the required degree of consolidation to take place; the critical dimension is the vertical distance between the feed point and the outlet at the base of the tank. In most cases, the time of compression of the sediment will be large compared with the time taken for the critical settling conditions to be reached.

The time required to concentrate the sediment after it has reached the critical condition can be determined approximately by allowing a sample of the slurry at its critical composition to settle in a vertical glass tube, and measuring the time taken for the interface between the sediment and the clear liquid to fall to such a level that the concentration is that required in the underflow from the thickener. The use of data so obtained assumes that the average concentration in the sediment in the laboratory test is the same as that which would be obtained in the thickener after the same time. This is not quite so because, in the thickener, the various parts of the sediment have been under compression for different times. Further, it assumes that the time taken for the sediment to increase in concentration by a given amount is independent of its depth.

The top surface of the suspension should always be sufficiently far below the level of the overflow weir to provide a clear zone deep enough to allow any entrained particles to settle out. As they will be present only in very low concentrations, they will settle approximately at their terminal falling velocities. Provided that this requirement is met, the depth of the thickener does not have any appreciable effect on its clarifying capacity.

However, it is the depth of the thickener below the clarifying zone that determines the residence time of the particles and the degree of thickening which is achieved at any given throughput. Comings[51] has carried out an experimental study of the effect of underflow rate on the depth of the thickening, (compression) zone and has concluded that it is the retention time of the particles within the thickening zone, rather than its depth, per se, which determines the underflow concentration.

An approximate estimate of the mean residence time of particles in the thickening zone may be made from the results of a batch settling experiment on a suspension which is of the *critical concentration* (the concentration at which the settling rate starts to fall off with time). Following a comprehensive experimental program of work, Roberts[52] found that the rate of sedimentation dH/dt decreases linearly with the difference between its height at any time t and its ultimate height H_∞ which would be achieved after an infinite time of settling:

$$\text{Thus:} \quad -\frac{dH}{dt} = k(H - H_\infty) \tag{8.55}$$

where k is a constant.

$$\text{On integration:} \quad \ln\frac{H - H_\infty}{H_c - H_\infty} = -kt \tag{8.56}$$

$$\text{or:} \quad \frac{H - H_\infty}{H_c - H_\infty} = e^{-kt} \tag{8.57}$$

where t is time from the start of the experiment, H is the height of the interface at time t, and H_c is the initial height corresponding to the critical height at which the constant settling rate gives way to a diminishing rate.

H_∞, which can be seen from Eq. (8.57) to be approached exponentially, cannot be estimated with any precision, and a trial-and-error method must be used to determine at what particular value of H_∞, the plot of the left hand side of Eq. (8.56) against t yields a straight line. The slope of this line is equal to $-k$, which is constant for any particular suspension.

If the required fractional volumetric concentration in the underflow is C_u, and C_c is the value of the critical concentration, a simple material balance gives the value of H_u the height of the interface when the concentration is C_u.

$$\text{Thus:}\quad H_u = H_c \frac{C_c}{C_u} \tag{8.58}$$

The corresponding value of the residence time t_R to reach this condition is obtained by substituting from Eq. (8.58) into Eq. (8.56), giving:

$$
\begin{aligned}
t_R &= \frac{1}{k} \ln \frac{H_c - H_\infty}{H_u - H_\infty} \\
&= \frac{1}{k} \ln \frac{H_c - H_\infty}{H_c(C_c/C_u) - H_\infty}
\end{aligned}
\tag{8.59}
$$

Thus, Eq. (8.59) may be used to calculate the time required for the concentration to be increased to such a value that the height of the suspension is reduced from H_0 to a desired value H_u, at which height the concentration corresponds with the value C_u required in the underflow from the thickener.

In a batch sedimentation experiment, the sediment builds up gradually, and the solids which are deposited in the early stages are those which are subjected to the compressive forces for the longest period of time. In the continuous thickener, on the other hand, all of the particles are retained for the same length of time, with fresh particles continuously being deposited at the top of the sediment, and others being removed at the same rate in the underflow, with the inventory thus remaining constant. Residence time distributions are, therefore, not the same in batch and continuous systems. Therefore, the value of t_R calculated from Eq. (8.59) will be subject to some inaccuracy because of the mismatch between the models for batch and continuous operation.

An approximate value for the depth of the thickening zone is then found by adding the volume of the liquid in the sediment to the corresponding volume of solid, and dividing by the area which has already been calculated, in order to determine the clarifying capacity of the thickener. The required depth of the thickening region is thus:

$$\left\{\frac{Wt_R}{A\rho_s} + W\frac{t_R}{A\rho}X\right\} = \frac{Wt_R}{A\rho_s}\left(1+\frac{\rho_s}{\rho}X\right) \tag{8.60}$$

where t_R is the required time of retention of the solids, as determined experimentally, W is the mass rate of feed of solids to the thickener, X is the average value of the mass ratio of liquid to solids in the thickening portion, and ρ and ρ_s are the densities of the liquid and solid, respectively.

This method of design is only approximate, and, therefore, in a large tank, about 1 m should be added to the calculated depth as a safety margin and to allow for the depth required for the suspension to reach the critical concentration. In addition, the bottom of the tanks may be slightly pitched to assist the flow of material toward the thickened liquor outlet.

The use of slowly rotating rakes is beneficial in large thickeners to direct the underflow to the central outlet at the bottom of the tank. At the same time, the slow motion of the rakes tends to give a gentle agitation to the sediment which facilitates its consolidation and water removal. The height requirement of the rakes must be added to that needed for the thickening zone to achieve the desired underflow concentration.

Additional height (up to 1 m) should also be allowed for the depth of submergence of the feed pipe and to accommodate fluctuations in the feed rate to the thickener (approximately 0.5 m).

The limiting operating conditions for continuous thickeners have been studied by a number of researchers including Tiller and Chen,[53] and the height of the compression zone has been the subject of a paper by Font.[54]

A paper by Fitch[55] gives an analysis of existing theories and identifies the domains in which the various models which have been proposed give reasonable approximations to observed behaviour.

The importance of using deep thickeners for producing thickened suspensions of high concentrations has been emphasised by Dell and Keleghan,[56] who used a tall tank in the form of an inverted cone. They found that consolidation was greatly facilitated by the use of stirring, which created channels through the sediment for the escape of water in an upward direction, and eliminated frictional support of the solids by the walls of the vessel. The conical shape is clearly uneconomic for large equipment, however, because of the costs of both fabrication and supports, and because of the large area at the top.

Chandler[57] has reported on the use of deep cylindrical tanks of height to diameter ratios of 1.5 and up to 19 m in height, fitted with steep conical bases of half angles of about 30 degrees. Using this method for the separation of 'red mud' from caustic liquors in the aluminium industry, it was found to be possible to dispense with the conical section because the slurry tended to form its own cone at the bottom of the tank as a result of the buildup of stagnant solids. Rakes were not used because of the very severe loading which would have been imposed at the high concentrations achieved in the sediments. With flocculated slurries, the water liberated as the result of compaction was able to pass upward through channels in the sediment—an essential feature of the operation because the resistance to flow through the sediment itself would have been much too large to permit such a high degree of thickening. The system was found to be highly effective, and much more economic than using large diameter tanks, where it has been found that the majority of the solids tend to move downward through the central zone, leaving the greater part of the cross-section ineffective.

More details on thickener design and operation can be found in a recent book.[58]

Example 8.1

A slurry containing 5 kg of water/kg of solids is to be thickened to a sludge containing 1.5 kg of water/kg of solids in a continuous operation. Laboratory tests using five different concentrations of the slurry yielded the following data:

Concentration (kg water/kg solid)	5.0	4.2	3.7	3.1	2.5
Rate of sedimentation (mm/s)	0.20	0.12	0.094	0.070	0.050

Calculate the minimum area of a thickener required to effect the separation of a flow of 1.33 kg/s of solids.

Solution

Basis: 1 kg solids

$$\text{Mass rate of feed of solids} = 1.33 \, \text{kg/s}$$

1.5 kg water is carried away in the underflow, with the balance in the overflow. Thus, $U = 1.5$ kg water/kg solids

Concentration (Y) (kg Water/kg Solids)	Water to Overview ($Y - U$) (kg Water/kg Solids)	Sedimentation Rate u_c (m/s)	$\dfrac{(Y-U)}{u_c}$ (s/m)
5.0	3.5	2.00×10^{-4}	1.75×10^4
4.2	2.7	1.20×10^{-4}	2.25×10^4
3.7	2.2	0.94×10^{-4}	2.34×10^4
3.1	1.6	0.70×10^{-4}	2.29×10^4
2.5	1.0	0.50×10^{-4}	2.00×10^4

Maximum value of $\dfrac{(Y-U)}{u_c} = 2.34 \times 10^4 \, \text{s/m}$.

From Eq. (8.54):

$$A = \left(\frac{Y-U}{u_c}\right)\left(\frac{Q_0 C \rho_s}{\rho}\right)$$

$Q_0 C \rho_s = 1.33 \text{ kg/s}$, and taking ρ as 1000 kg/m^3, then:

$$A = 2.34 \times 10^4 \times \left(\frac{1.33}{1000} \right)$$

$$= \underline{\underline{31.2 . \text{m}^2}}$$

Example 8.2

A batch test on the sedimentation of a slurry containing 200 kg solids/m³ gave the results shown in Fig. 8.14 for the position of the interface between slurry and clear liquid as a function of time.

Using the Kynch theory, tabulate the sedimentation velocity and solids flux due to sedimentation as a function of concentration. What area of tank will be required to give an underflow concentration of 1200 kg/m³ for a feed rate of 2 m³/min of slurry?

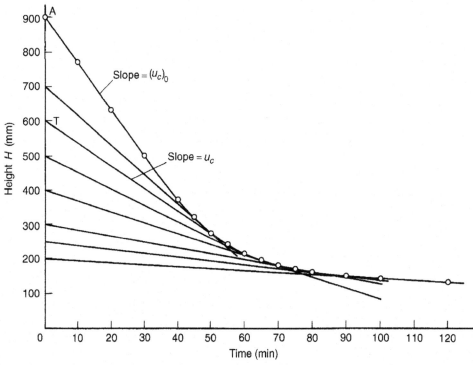

Fig. 8.14
Graphical data for Example 8.2.

Solution

On the diagram, the height H above the base is plotted against time. The initial height is given by OA (900 mm), and the initial constant slope of the curve gives the sedimentation velocity $(u_c)_0$ for a concentration of 200 kg/m^3 (c_0).

For some other height, such as OT (600 mm), the slope of the tangent gives the sedimentation velocity u_c for a concentration of:

$$c = c_0 \frac{OA}{OT} = 200 \times \frac{OA}{OT} \, kg/m^3 \quad \text{(from Eq.8.40)}$$

Thus, for each height, the corresponding concentration may be calculated and the slope of the tangent measured to give the sedimentation velocity. The solids flux in kg/m^{2}s is then:

$$c\left(kg/m^3\right) \times u_c(mm/min) \times \frac{1}{1000 \times 60}$$

These quantities are tabulated as follows:

H (mm)	c (kg/m^3)	u_c (mm/min)	Sedimentation Flux (kg/m^{2}s)	$1000\left(\frac{1}{c}-\frac{1}{c_u}\right)=x$ (m^3/kg $\times 10^3$)	$\frac{u_c}{(1/c)-(1/c_u)}$ (kg/m^{2}s)	$\frac{(1/c)-(1/c_u)}{u_c}$ (m^{2}s/kg)
900	200	13.4	0.0447	4.167	0.0536	18.7
800	225	10.76	0.0403	3.611	0.0497	20.1
700	257	8.6	0.0368	3.058	0.0468	21.4
600	300	6.6	0.0330	2.500	0.0440	22.7
500	360	4.9	0.0294	1.944	0.0420	23.8
400	450	3.2	0.0240	1.389	0.0383	26.1
300	600	1.8	0.0180	0.833	0.0360	27.8
260	692	1.21	0.0140	0.612	0.0330	30.3
250	720	1.11	0.0133	0.556	0.0333	30.0
220	818	0.80	0.0109	0.389	0.0342	29.2
200	900	0.60	0.0090	0.278	0.0358	27.9
180	1000	0.40	0.0067	0.167	0.0398	25.1

Now $c_u = 1200 \, kg/m^3$ and the maximum value of $\dfrac{(1/c) - (1/c_u)}{u_c} \approx 30.3 \, m^2 s/kg$.

From Eq. (8.51), the area A required is given by:

$$A = Q_0 c_0 \left[\frac{(1/c - (1/c_u))}{u_c}\right]_{max}$$

$$= \frac{2}{60} \times 200 \times 30.3$$

$$= \underline{\underline{202 \, m^2}}$$

8.3 Sedimentation of Coarse Particles

8.3.1 Introduction

Coarse particles have a much lower specific surface, and consequently, surface forces and electrical interactions between particles are of very much less significance than in the fine particle systems considered in the previous sections. Flocculation will be absent, and, generally, the particles will not influence the rheology of the liquid. The dividing point between coarse and fine particles is somewhat arbitrary, although it is on the order of 0.1 mm (100 μm).

There has been considerable discussion in the literature as to whether the buoyancy force acting on a particle in a suspension is determined by the density of the liquid or by that of the suspension. For the sedimentation of a coarse particle in a suspension of much finer particles, the suspension behaves effectively as a continuum. As the large particle settles, it displaces an equal volume of suspension in which there will be comparatively little relative movement between the fine particles and the liquid. The effective buoyancy may, in these circumstances, be attributed to a fluid of the same density as the suspension of fines. On the other hand, the pressure distribution around a small particle settling in a suspension of much coarser particles will hardly be affected by the presence of the large particles, which will, therefore, contribute little to the buoyancy force. If the surrounding particles are all of a comparable size, an intermediate situation will exist.

In a sedimenting suspension, however, the gravitational force on any individual particle is balanced by a combination of buoyancy and fluid friction forces, both of which are influenced by the flow field and pressure distribution in the vicinity of the particle. It is a somewhat academic exercise to apportion the total force between its two constituent parts, both of which are influenced by the presence of neighbouring particles, the concentration of which will determine the magnitude of their effect.

8.3.2 Suspensions of Uniform Particles

Several experimental and analytical studies have been made of the sedimentation of uniform particles. The following approach is a slight modification of that originally used by Richardson and Zaki.[59]

For a single spherical particle settling at its terminal velocity u_0, a force balance gives:

$$R_0' \frac{\pi}{4} d^2 = \frac{\pi}{6} d^3 (\rho_s - \rho) g \tag{8.61}$$

$$\text{or:} \quad R_0' = \frac{2}{3} d (\rho_s - \rho) g \tag{8.62}$$

(see Eq. 6.33)

where R_0' is the drag force per unit projected area of particle, d is the particle diameter, ρ_s is the density of the particle, ρ is the density of the liquid, and g is the acceleration due to gravity.

R' has been shown in Chapter 6 to be determined by the diameter of the particle, the viscosity and density of the liquid, and the velocity of the particle relative to the liquid. Thus, at the terminal falling condition:

$$R_0' = f(\rho, \mu, u_0, d)$$

In a tube of diameter d_t, the wall effects may be significant and:

$$R_0' = f_1(\rho, \mu, u_0, d, d_t) \tag{8.63}$$

In a concentrated suspension, the drag force on a particle will be a function of its velocity u_p relative to the liquid and will be influenced by the concentration of particles; that is, it will be a function of the voidage e of the suspension.

$$\text{Thus:} \quad R_0' = f_2(\rho, \mu, u_p, d, d_t, e) \tag{8.64}$$

The argument holds equally well if the buoyancy force is attributable to the density of the suspension as opposed to that of the liquid. The force balance, Eq. (8.61), then becomes:

$$R_{0s}' \frac{\pi}{4} d^2 = \frac{\pi}{6} d^3 (\rho_s - \rho_c) g \quad \text{(from Eq.6.2)}$$

$$= \frac{\pi}{6} d^3 (\rho_s - \rho) e g \tag{8.65}$$

where ρ_c is the density of the suspension.

Comparing Eqs (8.65) and (8.61):

$$R_{0s}' = e R_0' \tag{8.66}$$

Thus, no additional variable would be introduced into Eq. (8.64).

Rearranging Eqs (8.63) and (8.64) to give u_0 and u_p explicitly:

$$u_0 = f_3(R_0', \rho, \mu, d, d_t) \tag{8.67}$$

$$u_p = f_4(R_0', \rho, \mu, d, d_t, e) \tag{8.68}$$

Dividing Eq. (8.68) by Eq. (8.67) gives:

$$\frac{u_p}{u_0} = f_5(R_0', \rho, \mu, d, d_t, e) \tag{8.69}$$

because all the variables in function f_3 also appear in function f_4.

Because u_p/u_0 a dimensionless ratio, f_5 must also be dimensionless, and Eq. (8.64) becomes:

$$\frac{u_p}{u_0} = f_5\left(\frac{R_0'd^2\rho}{\mu^2}, \frac{d}{d_t}, e\right)$$

because $R_0'd^2\rho/\mu^2$ is the only possible dimensionless derivative of R_0'.

The sedimentation velocity u_c of the particles relative to the walls of the containing vessel will be less than the velocity u_p of the particles relative to the fluid.

$$\text{From Eq.} (8.6): \quad u_c = e u_p$$

$$\text{Thus}: \quad \frac{u_c}{u_0} = f_6\left(\frac{R_0'd^2\rho}{\mu^2}, \frac{d}{d_t}, e\right)$$

From Eq. (8.62):

$$\frac{R_0'd^2\rho}{\mu^2} = \frac{2}{3}\frac{d^3\rho(\rho_s - \rho)g}{\mu^2}$$

$$= \frac{2}{3}Ga \quad (\text{see Eq.}6.36)$$

$$\text{where}: \quad Ga = \frac{d^3\rho(\rho_s - \rho)g}{\mu^2} \text{ is the Galileo number}$$

$$\text{Thus}: \quad \frac{u_c}{u_0} = f_7\left(Ga, \frac{d}{d_t}, e\right) \tag{8.70}$$

$$\text{However}: \quad \frac{R_0'd^2\rho}{\mu^2} = \frac{R_0'}{\rho u_0^2}\left(\frac{u_0 d\rho}{\mu}\right)^2$$

Because there is a unique relation between $R_0'/\rho u_0^2$ and $u_0\,d\rho/\mu$ for spherical particles:

$$Ga = f_8\left(\frac{u_0 d\rho}{\mu}\right) = f_8(Re_0') \tag{8.71}$$

$$\text{and}: \quad \frac{u_c}{u_0} = f_9\left(Re_0', \frac{d}{d_t}, e\right) \tag{8.72}$$

Eq. (8.72) is just an alternative way of expressing Eq. (8.70).

Two special cases of Eqs (8.70) and (8.72) are now considered.

(a) *At low values of $Re_0'(<\sim 0.2)$.* This corresponds to the Stokes' law region in which drag force, and hence, R_0', is independent of the density of the liquid. Eq. (8.69) then becomes:

$$\frac{u_p}{u_0}=f_{10}\left(R'_0,\mu,d,d_t,e\right) \tag{8.73}$$

Now R'_0, μ and d cannot be formed into a dimensionless group, and, therefore, in this region, u_p/u_0 must be independent of R'_0 and μ.

(b) Similarly, *at high values of* $Re'_0(>500)$, the Newton's law region, drag force is independent of viscosity, and for similar reasons to (a) above, u_p/u_0 is independent of R'_0 and ρ.

Thus $(Re'_0<0.2$ and $Re'_0>500)$ both Eqs (8.70) and (8.72) become:

$$\frac{u_c}{u_0}=f_{11}\left(\frac{d}{d_t},e\right) \tag{8.74}$$

The form of the functions $(f_7, f_9, \text{and } f_{11})$ has been obtained from experimental results for both the sedimentation and the fluidisation (Chapter 9) of suspensions of uniform spheres. For a given suspension, a plot of log u_c versus log e does, in most cases, gives a good straight line, as shown in Fig. 8.15.

$$\text{Thus}: \quad \log u_c = n\log e + \log u_i \tag{8.75}$$

where n is the slope and log u_i is the intercept at $e=1$, that is, at infinite dilution.

$$\text{Hence}: \quad \frac{u_c}{u_i}=e^n \tag{8.76}$$

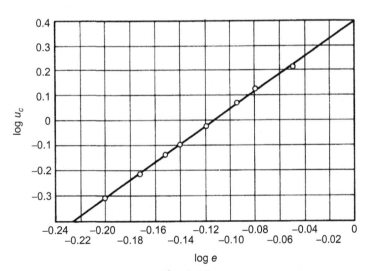

Fig. 8.15
Sedimentation velocity as a function of voidage.

$$\text{Now}: \quad \frac{u_i}{u_0} = f\left(\frac{d}{d_t}\right) \text{only} \tag{8.77}$$

because u_0 and u_i are the free falling velocities in an infinite fluid and in a tube of diameter d_t, respectively.

Substituting from Eqs (8.76) and (8.77) in Eq. (8.70) gives:

$$e^n = f_{12}\left(Ga, \frac{d}{d_t}, e\right)$$

Because n is found experimentally to be independent of e (see Fig. 8.15):

$$n = f_{13}\left(Ga, \frac{d}{d_t}\right) \tag{8.78}$$

Eqs (8.72) and (8.74) similarly become, respectively:

$$n = f_{14}\left(Re'_0, \frac{d}{d_t}\right) \tag{8.79}$$

$$\text{and}: \quad n = f_{15}\left(\frac{d}{d_t}\right) \tag{8.80}$$

Eq. (8.80) is applicable for:

$$\begin{cases} Re'_0 < 0.2 \\ Ga < 3.6 \end{cases} \text{and} \begin{cases} Re'_0 > 500 \\ Ga > 8.3 \times 10^4 \end{cases}$$

Experimental results generally confirm the validity of Eq. (8.80) over these ranges, with $n \approx 4.8$ at low Reynolds numbers and $n \approx 2.4$ at high values. Eq. (8.78) is to be preferred to Eq. (8.79) as the Galileo number can be calculated directly from the properties of the particles and of the fluid, whereas Eq. (8.79) necessitates the calculation of the terminal falling velocity u_0.

Values of the exponent n were determined by Richardson and Zaki[59] by plotting n against d/d_t with Re'_0 as parameter, as shown in Fig. 8.16. Equations for the evaluation of n are given in Table 8.1, over the stated ranges of Ga and Re'_0.

As an alternative to the use of Table 8.1, it is convenient to be able to use a single equation for the calculation of n over the whole range of values of Ga or Re'_0 of interest. Thus, Garside and Al-Dibouni[60] have proposed:

$$\frac{5.1 - n}{n - 2.7} = 0.1 Re'^{0.9}_0 \tag{8.81}$$

They suggest that particle interaction effects are underestimated by Eq. (8.81), particularly at high voidages ($e > 0.9$) and for the turbulent region, and that, for a given value of voidage (e),

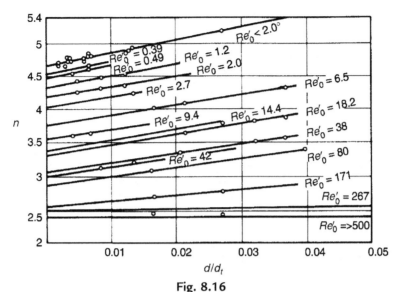

Fig. 8.16

Dependence of exponent n on the ratio of particle-to-container diameter.[59]

Table 8.1 n as a function of Ga or Re'_0 and d/d_t

Range of Ga	Range of Re'_0	n as Function of Ga, d/d_t	n as Function of Re'_0, d/d_t
0–3.6	0–0.2	$4.6 + 20\ d/d_t$	$4.6 + 20\ d/d_t$
3.6–21	0.2–1	$(4.8 + 20\ d/d_t)Ga^{-0.03}$	$(4.4 + 18d/d_t)Re'^{-0.03}_0$
21–2.4×10^4	1–200	$(5.5 + 23\ d/d_t)Ga^{-0.075}$	$(4.4 + 18d/d_t)Re'^{-0.1}_0$
2.4×10^4–8.3×10^4	200–500	$5.5Ga^{-0.075}$	$4.4Re'^{-0.1}_0$
$>8.3 \times 10^4$	>500	2.4	2.4

the equation predicts too high a value for the sedimentation (or fluidisation) velocity (u_c). They prefer the following equation, which they claim overall gives a higher degree of accuracy:

$$\frac{u_c/u_0 - e^{5.14}}{e^{2.68} - u_c/u_0} = 0.06Re'_0 \tag{8.82}$$

Eq. (8.82) is less easy to use than Eq. (8.81) and it is doubtful whether it gives any significantly better accuracy, there being evidence[61] that with viscous oils Eq. (8.76) is more satisfactory.

Rowe[62] has subsequently given:

$$\frac{4.7 - n}{n - 2.35} = 0.175Re'^{0.75}_0 \tag{8.83}$$

Subsequently Khan and Richardson[63] have examined the published experimental results for both sedimentation and fluidisation of uniform spherical particles and recommend the following equation from which n may be calculated in terms of both the Galileo number Ga and the particle to vessel diameter ratio d/d_t:

$$\frac{4.8-n}{n-2.4} = 0.043Ga^{0.57}\left[1-1.24\left(\frac{d}{d_t}\right)^{0.27}\right] \qquad (8.84)$$

As already indicated, it is preferable to employ an equation involving Ga which can be directly calculated from the properties of the solid and of the liquid.

Values of n from Eq. (8.84) are plotted against Galileo number for $d/d_t=0$ and $d/d_t=0.1$ in Fig. 8.17. It may be noted that the value of n is critically dependent on d/d_t in the intermediate range of Galileo number, but is relatively insensitive at the two extremities of the curves where n attains a constant value of about 4.8 at low values of Ga and about 2.4 at high values. It is seen that n becomes independent of Ga in these regions, as predicted by dimensional analysis and by Eq. (8.80).

In Chapter 6, Eq. (6.40) was proposed for the calculation of the free falling velocity of a particle in an infinite medium.[64] This equation which was shown to apply over the whole range of values of Ga of interest takes the form:

$$Re_0' = \left[2.33Ga^{0.018} - 1.53Ga^{-0.016}\right]^{13.3} \qquad (8.85)$$

(see Eq. 6.40)

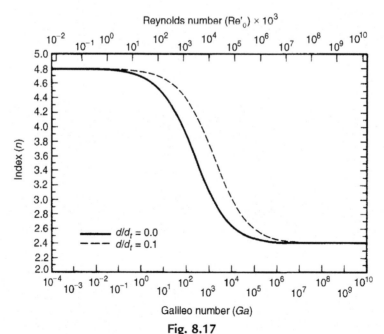

Fig. 8.17

Sedimentation index n as function of Galileo number Ga (from Eq. 8.84).

The correction factor for the effect of the tube wall is given by substituting u_i for u_{0t} in Eq. (6.43) to give:

$$\frac{u_i}{u_0} = \left[1 + 1.24\frac{d}{d_t}\right]^{-1} \tag{8.86}$$

Thus, combining Eqs. (8.84), (8.85), and (8.86):

$$u_c = \frac{\mu}{\rho d}\left[2.33Ga^{0.018} - 1.53Ga^{-0.016}\right]^{13.3}\left[1 + 1.24\frac{d}{d_t}\right]^{-1} e^n \tag{8.87}$$

where n is given by Eq. (8.84).

Eq. (8.87) permits the calculation of the sedimentation velocity of uniform spherical particles in a liquid at any voidage e in a vessel of diameter d_t.

For a more detailed account of the various approaches to the problem of sedimentation of concentrated suspensions, reference may be made to the work Khan and Richardson.[65]

8.3.3 Solids Flux in Batch Sedimentation

In a sedimenting suspension, the sedimentation velocity u_c is a function of fractional volumetric concentration C, and the volumetric rate of sedimentation per unit area or flux ψ is equal to the product $u_c C$.

$$\text{Thus:} \quad \psi = u_c C = u_c(1-e) \tag{8.88}$$

Then, if the relation between settling velocity and concentration can be expressed in terms of a terminal falling velocity (u_0) for the particles, substituting for u_0 using Eq. (8.71) gives:

$$\psi = u_0 e^n(1-e) \tag{8.89}$$

From the form of the function, it is seen that ψ should have a maximum at some value of e lying between 0 and 1.

Differentiating;

$$\frac{d\psi}{de} = u_0\left\{ne^{n-1}(1-e) + e^n(-1)\right\} = u_0\left\{ne^{n-1} - (n+1)e^n\right\} \tag{8.90}$$

When $d\psi/de$ is zero, then:

$$ne^{n-1}(1-e) - e^n = 0$$

$$\text{or:} \quad n(1-e) = e$$

$$\text{and:} \quad e = \frac{n}{1+n} \tag{8.91}$$

If n ranges from 2.4 to 4.8 as for suspensions of uniform spheres, the maximum flux should occur at a voidage between 0.71 and 0.83 (volumetric concentration 0.29 to 0.17). Furthermore, there will be a point of inflexion if $d^2\psi/de^2$ is zero for real values of e. Differentiating Eq. (8.90):

$$\frac{d^2\psi}{de^2} = u_0\{n(n-1)e^{n-2} - (n+1)ne^{n-1}\} \tag{8.92}$$

When $d^2\psi/de^2 = 0$: $n - 1 - (n+1)e = 0$

$$\text{or:} \quad e = \frac{n-1}{n+1} \tag{8.93}$$

If n ranges from 2.4 to 4.8, there should be a point of inflexion in the curve occurring at values of e between 0.41 and 0.65, corresponding to very high concentrations ($C = 0.59$ to 0.35). It may be noted that the point of inflexion occurs at a value of voidage e below that at which the mass rate of sedimentation ψ is a maximum. For coarse particles ($n \approx 2.4$), the point of inflexion is not of practical interest because the concentration at which it occurs ($C \approx 0.59$) usually corresponds to a packed bed rather than a suspension.

The form of variation of flux (ψ) with voidage (e) and volumetric concentration (C) is shown in Fig. 8.18 for a value of $n = 4.8$. This corresponds to the sedimentation of uniform spheres for which the free-falling velocity is given by Stokes' law. It may be compared with Fig. 8.4 obtained for a flocculated suspension of calcium carbonate.

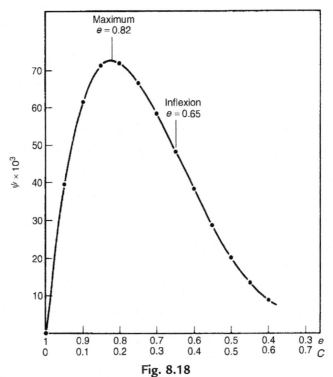

Fig. 8.18

Flux (ψ)-concentration (C) curve for a suspension for which $n = 4.8$.

Example 8.3

For the sedimentation of a suspension of uniform particles in a liquid, the relation between observed sedimentation velocity u_c and fractional volumetric concentration C is given by:

$$\frac{u_c}{u_0} = (1 - C)^{4.8}$$

where u_0 is the free falling velocity of an individual particle.

Calculate the concentration at which the rate of deposition of particles per unit area will be a maximum, and determine this maximum flux for 0.1 mm spheres of glass (density 2600 kg/m^3) settling in water (density 1000 kg/m^3, viscosity 1 mNs/m^2).

It may be assumed that the resistance force F on an isolated sphere is given by Stokes' law:

$$F = 3\pi\mu du$$

where μ = fluid viscosity, d = particle diameter, u = velocity of particle relative to fluid.

Solution

The mass rate of sedimentation per unit area $= \psi = u_c C$

$$= u_0 (1 - C)^{4.8} C.$$

The maximum flux occurs when $\dfrac{d\psi}{dC} = 0$.

$$\text{that is, when:} \quad -\left[4.8(1 - C)^{3.8} C\right] + (1 - C)^{4.8} = 0$$

$$\text{or:} \quad -4.8C + (1 - C) = 0$$

$$\text{Then:} \quad C = \frac{1}{5.8} = 0.172$$

The maximum flux $\psi_{max} = u_0 (1 - 0.172)^{4.8} \times 0.172$

$$= 0.0695 u_0$$

The terminal falling velocity u_0 is given by a force balance, as explained in Chapter 6:

$$\text{Gravitational force} = \text{resistance force}$$

$$\frac{\pi}{6} d^3 (\rho_s - \rho)g = 3\pi\mu du_0$$

$$u_0 = \frac{d^2 g}{18\mu}(\rho_s - \rho) \quad \text{(Eq.6.24)}$$

$$= \frac{(10^{-4})^2 9.81}{(18 \times 10^{-3})}(2600 - 10,00)$$

$$= 0.00872 \text{ m/s, i.e., } 8.72 \text{ mm/s}$$

The maximum flux $\psi_{max} = (0.0695 \times 0.00872)$

$$= 6.06 \times 10^{-4} \text{m}^3/\text{m}^2\text{s}$$

8.3.4 Comparison of Sedimentation With Flow Through Fixed Beds

Richardson and Meikle[66] have shown that, at high concentrations, the results of sedimentation and fluidisation experiments can be represented in a manner similar to that used by Carman[67] for fixed beds, discussed in Chapter 7. Using the interstitial velocity and a linear dimension given by the reciprocal of the surface of particles per unit volume of fluid, the Reynolds number is defined as:

$$Re_1 = \frac{(u_c/e)e/[(1-e)S]\rho}{\mu} = \frac{u_c\rho}{S\mu(1-e)} \tag{8.94}$$

The friction group ϕ'' is defined in terms of the resistance force per unit particle surface (R_1) and the interstitial velocity.

$$\text{Thus:} \quad \phi'' = \frac{R_1}{\rho(u_c/e)^2} \tag{8.95}$$

A force balance on an element of system containing unit volume of particles gives:

$$R_1 S = (\rho_s - \rho_c)g = e(\rho_s - \rho)g \tag{8.96}$$

$$\phi'' = \frac{e^3(\rho_s - \rho)g}{S\rho u_c^2} \tag{8.97}$$

In Fig. 8.19, ϕ'' is plotted against Re_1 using results obtained in experiments on sedimentation and fluidisation. On a logarithmic scale, a linear relation is obtained, for values of $Re_1 < 1$, with the following equation:

$$\phi'' = 3.36 Re_1^{-1} \tag{8.98}$$

Results for flow through a fixed bed are also shown. These may be represented by:

$$\phi'' = 5 Re_1^{-1} \quad (Eq.7.18)$$

Because Eq. (7.18) is applicable to low Reynolds numbers at which the flow is streamline, it appears that the flow of fluid at high concentrations of particles in a sedimenting or fluidised system is also streamline. The resistance to flow in the latter case appears to be about 30% lower, presumably because the particles are free to move relative to one another.

Eq. (8.98) can be arranged to give:

$$u_c = \frac{e^3}{3.36(1-e)} \frac{(\rho_s - \rho)g}{S^2\mu} \tag{8.99}$$

In the Stokes' law region, from Eq. (6.24):

$$u_0 = \frac{d^2 g}{18\mu}(\rho_s - \rho) = \frac{2g}{S^2\mu}(\rho_s - \rho) \tag{8.100}$$

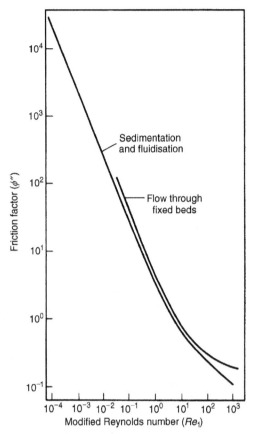

Fig. 8.19
Correlation of data on sedimentation and fluidisation and comparison with results for flow through fixed granular beds.

$$\text{Dividing}: \quad \frac{u_c}{u_0} = \frac{e^3}{6.7(1-e)} \tag{8.101}$$

From Eqs (8.76) and (8.84) and neglecting the tube wall effect:

$$\frac{u_c}{u_0} = e^{4.8} \quad (\text{for } Re_1 < 0.2) \tag{8.102}$$

The functions $e^3/[6.7(1-e)]$ and $e^{4.8}$ are both plotted as a function of e in Fig. 8.20, from which it is seen that they correspond closely for voidages <0.75, i.e., $C>0.25$.

The application of the relations obtained for monodisperse systems to fine suspensions containing particles of a wide range of sizes was examined by Richardson and Shabi,[68] who studied the behaviour of aqueous suspensions of zirconia of particle size ranging from $<1\,\mu\text{m}$ to $20\,\mu\text{m}$. A portion of the solid forming the suspension was irradiated, and the change in concentration with time at various depths below the surface was followed by means of a

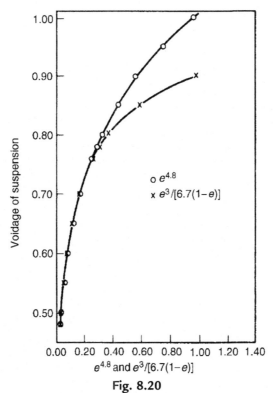

Fig. 8.20

Comparison of two functions of voidage used as correction factors for sedimentation velocities.

Geiger-Müller counter. It was found that selective settling took place initially, and that the effect of concentration on the falling rate of a particle was the same as in a monodisperse system, provided that the total concentration of particles of all sizes present was used in the calculation of the correction factor. Agglomeration was found to occur rapidly at high concentrations, and the conclusions were, therefore, applicable only to the initial stages of settling.

8.3.5 Model Experiments

The effect of surrounding a particle with an array of similar particles has been studied by Richardson and Meikle[69] and by Rowe and Henwood.[70] In both investigations, the effect of the proximity and arrangement of fixed neighbouring particles on the drag was examined.

Richardson and Meikle[69] surrounded the test particle by a uniplanar hexagonal arrangement of identical spheres which were held in position by means of rods passing through glands in the tube walls, as shown in Fig. 8.21. This arrangement permitted the positioning of the spheres to be altered without interfering with the flow pattern in the hexagonal space. For each arrangement of the spheres, the drag force was found to be proportional to the square of the liquid velocity. The effective voidage e of the system was calculated by considering a

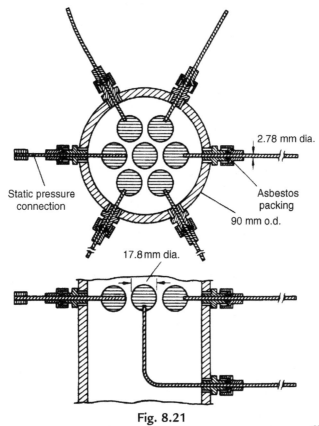

Fig. 8.21
Arrangement of hexagonal spacing of particles in tube.[69]

hexagonal cell located symmetrically about the test sphere, of depth equal to the diameter of the particle, as shown in Fig. 8.22. The drag force F_c on a particle of volume v settling in a suspension of voidage e is given by:

$$F_c = v(\rho_s - \rho_c)g = ev(\rho_s - \rho)g \tag{8.103}$$

Thus, for any voidage, the drag force on a sedimenting particle can be calculated, and the corresponding velocity required to produce this force on a particle at the same voidage in the model is obtained from the experimental results. All the experiments were carried out at particle Reynolds number >500, and under these conditions, the observed sedimentation velocity is given by Eqs (8.76) and (8.84) as:

$$\frac{u_c}{u_0} = e^{2.4} \tag{8.104}$$

Writing $u_c = eu_p$, where u_p is the velocity of the particle relative to the fluid:

$$\frac{u_p}{u_0} = e^{1.4} \tag{8.105}$$

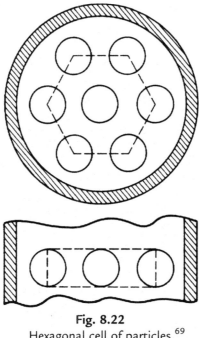

Fig. 8.22
Hexagonal cell of particles.[69]

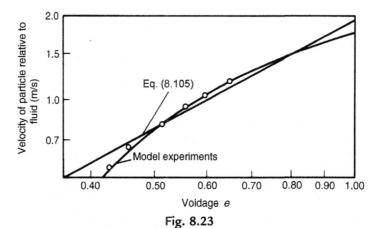

Fig. 8.23
Comparison of velocity in model required to produce a drag force equal to buoyant weight of particle, with velocity determined experimentally.

In Fig. 8.23, the velocity of the fluid relative to the particle, as calculated for the model experiments and from Eq. (8.100), is plotted against voidage. It may be seen that reasonable agreement is obtained at voidages between 0.45 and 0.90, indicating that the model does fairly closely represent the conditions in a suspension.

Rowe and Henwood[70] made assemblies of particles from cast blocks of polythene and found that the force on a single particle was increased by an order of magnitude as a result of surrounding it with a close assembly. They also found that adjacent spheres tend to repel one another, and that surfaces facing downstream tend to expel particles, whereas those facing upstream attract particles. This was offered as an explanation for the stable upper surface and rather diffuse lower surface of a bubble in a fluidised bed.

Over the years, numerous numerical studies have been reported in which the effect of neighbouring particles on the drag of a sphere has been estimated for simple cubic, body-centred, face-centred cubic arrays of spheres.[71–73] At low Reynolds numbers, the drag force is predicted to be some what insensitive to the actual arrangement and is determined mainly by the voidage, e, of the system, which is also in line with the experimental results presented here. However, it is not so at high Reynolds numbers.

8.3.6 Sedimentation of Two-Component Mixtures

Particles of different size but same density

Several researchers[74–76] have studied the sedimentation of suspensions consisting of particles of two different sizes, but of the same densities. The large particles have higher settling velocities than the small ones, and, therefore, four zones form, in order, from the top downward:

(a) Clear liquid.
(b) Suspension of the fine particles.
(c) Suspension of mixed sizes.
(d) Sediment layer.

For relatively coarse particles, the rates of fall of the interface between (a) and (b) and between (b) and (c) can be calculated approximately if the relation between sedimentation velocity and voidage, or concentration, is given by Eq. (8.76). Richardson and Shabi[68] have shown that, in a suspension of particles of mixed sizes, it is the total concentration which controls the sedimentation rate of each species.

A suspension of a mixture of large particles of terminal falling velocity u_{0L} and of small particles of terminal falling velocity u_{0S} may be considered, in which the fractional volumetric concentrations are C_L and C_s, respectively, if the value of n in Eq. (8.76) is the same for each particle. For each of the spheres settling on its own:

$$\frac{u_{cL}}{u_{0L}} = e^n \qquad (8.106)$$

and:

$$\frac{u_{cS}}{u_{0S}} = e^n \qquad (8.107)$$

Then, considering the velocities of the particles relative to the fluid, u_{cL}/e and u_{cS}/e, respectively:

$$\frac{u_{cL}}{e} = u_{0L}e^{n-1} \qquad (8.108)$$

and:

$$\frac{u_{cS}}{e} = u_{0S}e^{n-1} \qquad (8.109)$$

When the particles of two sizes are settling together, the upflow of displaced fluid is caused by the combined effects of the sedimentation of the large and small particles. If this upward velocity is u_F, the sedimentation rates u_{ML} and u_{MS} will be obtained by deducting u_F from the velocities relative to the fluid.

$$\text{Thus}: \quad u_{ML} = u_{0L}e^{n-1} - u_F \qquad (8.110)$$

$$\text{and}: \quad u_{MS} = u_{0S}e^{n-1} - u_F \qquad (8.111)$$

Then, because the volumetric flow of displaced fluid upward must be equal to the total volumetric flowrate of particles downward, then:

$$u_F e = (u_{0L}e^{n-1} - u_F)C_L + (u_{0S}e^{n-1} - u_F)C_S$$
$$\therefore u_F = e^{n-1}(u_{0L}C_L + u_{0S}C_S) \qquad (8.112)$$
$$(\text{since } e + C_L + C_S = 1)$$

Then substituting from Eq. (8.112) into Eqs (8.110) and (8.111):

$$u_{ML} = e^{n-1}[u_{0L}(1 - C_L) - u_{0S}C_S] \qquad (8.113)$$

$$\text{and}: \quad u_{MS} = e^{n-1}[u_{0S}(1 - C_S) - u_{0L}C_L] \qquad (8.114)$$

Eq. (8.113) gives the rate of fall of the interface between zones (b) and (c); that is, the apparent rate of settling of the zone of mixed particles.

The velocity of fall u_f of the interface between zones (a) and (b) is the sedimentation rate of the suspension composed only of fine particles, and will, therefore, depend on the free-falling velocity u_{0S} and the concentration C_f of this zone.

$$\text{Now}: \quad C_f = \frac{\text{Volumetric rate at which solids are entering zone}}{\text{Total volumetric growth rate of zone}}$$
$$= \frac{(u_{ML} - u_{MS})C_S}{u_{ML} - u_f} \qquad (8.115)$$

$$\text{Thus}: \quad u_f = u_{0S}(1 - C_f)^n \tag{8.116}$$

u_f and C_f are determined by solving Eqs (8.115) and (8.116) simultaneously. Further study has been carried out of the behaviour of particles of mixed sizes by several groups of researchers including Selim et al.[77]

Particles of equal terminal falling velocities

By studying suspensions containing two different solid components, it is possible to obtain a fuller understanding of the process of sedimentation of a complex mixture. Richardson and Meikle[66] investigated the sedimentation characteristics of suspensions of glass ballotini and polystyrene particles in a 22% by mass ethanol–water mixture. The free-falling velocity, and the effect of concentration on sedimentation rate, were identical for each of the two solids alone in the liquid.

The properties of the components are given in Table 8.2.

The sedimentation of mixtures containing equal volumes of the two solids was then studied, and it was found that segregation of the two components tended to take place, the degree of segregation increasing with concentration. This arises because the sedimentation velocity of an individual particle in the suspension is different from its free-falling velocity, first because the buoyancy force is greater, and, second, because the flow pattern in the proximity of particle is different. For a monodisperse suspension of either constituent of the mixture, the effect of flow pattern as determined by concentration is the same, but the buoyant weights of the two species of particles are altered in different proportions. The settling velocity of a particle of polystyrene or ballotini in the mixture (u_{PM} or u_{BM}) can be written in terms of its free-falling velocity (u_{p0} or u_{B0}) as:

$$u_{PM} = u_{P0}\frac{\rho_P - \rho_c}{\rho_P - \rho}f(e) \tag{8.117}$$

$$\text{and}: \quad u_{BM} = u_{B0}\frac{\rho_B - \rho_c}{\rho_B - \rho}f(e) \tag{8.118}$$

Table 8.2 **Properties of solids and liquids in the sedimentation of two-component mixtures**

	Glass Ballotini	Polystyrene	22% Ethanol in Water
Density (kg/m^3)	$\rho_B = 1921$	$\rho_P = 1045$	$\rho = 969$
Particle size (μm)	71.1	387	–
Viscosity (mN s/m^2)	–	–	1.741
Free-falling velocity (u_0) (mm/s)	3.24	3.24	–

In these equations, $f(e)$ represents the effects of concentration, other than those associated with an alteration of buoyancy arising from the fact that the suspension has a higher density than the liquid. In a uniform suspension, the density of the suspension is given by:

$$\rho_c = \rho e + \frac{1-e}{2}(\rho_B + \rho_P) \tag{8.119}$$

Substitution of the numerical values of the densities in Eqs (8.117), (8.118), and (8.119) gives:

$$u_{PM} = u_{P0}(13.35e - 12.33)f(e) \tag{8.120}$$

$$\text{and}: \quad u_{BM} = u_{B0}(0.481 + 0.520)f(e) \tag{8.121}$$

Noting that u_{P0} and u_{B0} are equal, it is seen that the rate of fall of the polystyrene becomes progressively less than that of the ballotini as the concentration is increased. When $e = 0.924$, the polystyrene particles should remain suspended in the mixed suspension, and settling should occur only when the ballotini have separated out from that region. At lower values of e, the polystyrene particles should rise in the mixture. At a voidage of 0.854, the polystyrene should rise at the rate at which the ballotini are settling, so that there will then be no net displacement of liquid. When the ballotini are moving at the higher rate, the net displacement of liquid will be upward, and the zone of polystyrene suspension will become more dilute. When the polystyrene is moving at the higher rate, the converse is true.

Because there is no net displacement of liquid at a voidage of 0.854, corresponding to a total volumetric concentration of particles of 14.6%, the concentration of each zone of suspension, after separation, should be equal. This has been confirmed within the limits of experimental accuracy. It has also been confirmed experimentally that upward movement of the polystyrene particles does not take place at concentrations <8% ($e > 0.92$).

On the above basis, the differences between the velocities of the two types of particles should increase with concentration and, consequently, the concentration of the two separated zones should become increasingly different. This is found to be so at volumetric concentrations up to 20%. At higher concentrations, the effect is not observed, probably because the normal corrections for settling velocities cannot be applied when particles are moving in opposite directions in suspensions of high concentration.

The behaviour of a suspension of high concentration ($e < 0.85$) is shown in Fig. 8.24, where it is seen that the glass ballotini settles out completely before any deposition of polystyrene particles occurs.

These experiments clearly show that the tendency for segregation to occur in a two-component mixture becomes progressively greater as the concentration is increased. This behaviour is in distinct contrast to that observed in the sedimentation of a suspension of multisized particles of a given material, when segregation becomes less as the concentration is increased.

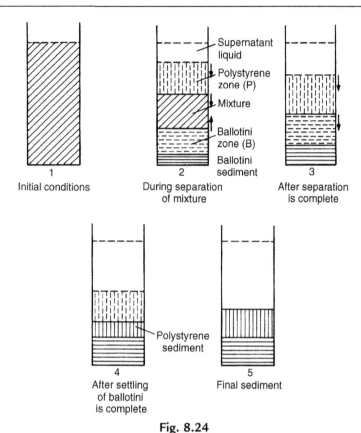

Fig. 8.24

Settling behaviour of a mixture consisting of equal volumes of polystyrene and ballotini at a volumetric concentrations exceeding 15%.

8.4 Sedimentation of Suspensions of Rod-Like Particles or Fibrous Systems

The discussion thus far has been restricted to the suspension of granular particles. There are situations where the suspensions comprising rod-like particles or fibres are encountered, with paper pulp being one example that immediately comes to mind. Other examples are found in food, pharmaceutical, textile, and tobacco processing applications. In spite of their pragmatic significance, little is known about their setting behaviour. Due to their large aspect ratios, these systems often exhibit inhomogeneities and anisotropy during their sedimentation. Furthermore, due to their large surface areas, these systems are generally flocculated. Even when flocculation is avoided by the use of suitable dispersing agents, the available scant experimental studies[78–81] suggest that their sedimentation behaviour differs significantly from that of a suspension of granular particles. Thus, for instance, Kumar and Ramarao[78] working with 10 μm diameter glass fibres (length to diameter ratio of 5 and 25) suspension in 50% aqueous glycerol solution reported that a 0.5% suspension settled much faster than a single fibre thereby suggesting "negative" "hindrance effect." This was attributed to the formation of "packets of fibers, similar to flocs." However, as the concentration of fibres was gradually increased, the settling rate dropped due to the hydrodynamic interactions between the packets. Also, they reported up to 30% higher settling velocity of the suspensions of longer fibres than that of the shorter fibres at the same volumetric concentration. In order to develop a rational framework to classify the settling behaviour of such systems, Herzhaft and Guazzelli[81] introduced three regimes depending the number density of fibres. As the disturbance introduced by a fibre persists over a distance of the order of fibre length (l), it is convenient to define an effective particle concentration (volume fraction) of $n_f (l/2)^3$ where n_f is the number of fibres per unit volume. For $n_f (l/2)^3 << 1$, the fibre-fibre interactions will be rather weak, and a fibre is free to rotate. This is the so-called dilute regime, and, in this case, the suspension reaches a steady state where the packets are formed, and most of the fibres are aligned with the direction of gravity, thereby settling at a faster rate than a single fibre, as revealed by the scant experimental studies.[78, 81] On the other hand, the suspension is called concentrated when the factor $n_f (l/2)^2 (d/2)$ is of order unity, which, in fact, translates into a rather low volume fraction, C. For the intermediate concentrations, that is, $n_f (l/2)^3 \gg 1 \gg n_f (l/2)^2 (d/2)$, the semidilute regime prevails. Using this criterion, the experiments of Kumar and Ramarao[78] belong to the dilute regime. On the other hand, the results of Turney et al.[79] relate to the semidilute regime. They used a NMR imaging technique to study the sedimentation of suspension of rayon fibres (of aspect ratio 17) and reported hindrance effect which was greater than that for a suspension of spheres at the same concentration. This suggests much larger values of the index, n in Eq. (8.76). Indeed, the experimental results of Herzhaft and Guazzelli[81] indicate the value of n of the order of 9 ± 2, which is almost twice the commonly used value of 4.8 in the low Reynolds number regime for spherical particles. Following the initial observations of Whitmore[82] and subsequently studied systematically by Weiland and colleagues,[83, 84] the sedimentation rates of fibrous suspension can be enhanced five- to sevenfold by the addition of neutrally buoyant or lighter particles in small amounts,[78] and this is attributed to the formation of streams of upward-moving light particles and downward-moving heavy fibres in these systems.

8.5 Nomenclature

		Units in SI System	Dimensions in **M, N, L, T,** θ, **A**
A	Cross-sectional area of vessel or tube	m^2	L^2
A'	Cross-sectional area at level of obstruction	m^2	L^2
$\mathcal{A}$	Hamaker constant	J	ML^{2}T^{-2}
a	Particle radius	m	L
a_r	Separation distance between particle centres	m	L

a	Constant in Vand's equation (8.4)	–	–
B	Constant in Eq. (8.21) ($=3.917 \times 10^{39} \text{C}^{-2}$)	C^{-2}	$\mathbf{T}^{-2}\mathbf{A}^{-2}$
b	Constant in Eq. (8.10)	s^{-1}	$\mathbf{T}^{-1}$
C	Fractional volumetric concentration $(1-e)$	–	–
C_0	Initial uniform concentration	–	–
C_c	Critical concentration	–	–
C_f	Concentration of fines in upper zone	–	–
C_i	Concentration corresponding to point of inflexion on $\psi - C$ curve	–	–
C_L	Limit value of C	–	–
C_M	Concentration on Fig. 8.14	–	–
C_m	Value of C at which ψ is a maximum	–	–
C_{max}	Concentration corresponding to sediment layer	–	–
C_u	Concentration of underflow in continuous thickener	–	–
C'	Value of C at constriction	–	–
c	Concentration of solids (mass/volume)	kg/m^3	$\mathbf{ML}^{-3}$
c_u	Concentration of solids in underflow (mass/volume)	kg/m^3	$\mathbf{ML}^{-3}$
c'_c	Critical coagulation concentration	kmol/m^3	$\mathbf{NL}^{-3}$
D_1, D_2	Absolute diffusivities	m^2/s	$\mathbf{L}^2\,\mathbf{T}^{-1}$
D_{12}	Relative diffusivity	m^2/s	$\mathbf{L}^2\,\mathbf{T}^{-1}$
d	Diameter of sphere or equivalent spherical diameter or fibre	m	$\mathbf{L}$
d_t	Diameter of tube or vessel	m	$\mathbf{L}$
e	Voidage of suspension $(=1-C)$	–	–
e_c	Elementary charge ($=1.60 \times 10^{-19}\text{C}$)	C	$\mathbf{TA}$
F_c	Drag force on particle of volume v	N	$\mathbf{MLT}^{-2}$
g	Acceleration due to gravity	m/s^2	$\mathbf{LT}^{-2}$
H	Height in suspension above base	m	$\mathbf{L}$
H_0	Value of H at $t=0$	m	$\mathbf{L}$
H_c	Height corresponding to critical settling point	m	$\mathbf{L}$
H_s	Separation between particle surfaces	m	$\mathbf{L}$
H_t	Initial total depth of suspension	m	$\mathbf{L}$
H_u	Height of interface when underflow concentration is reached	m	$\mathbf{L}$
H_∞	Final height of sediment	m	$\mathbf{L}$
I	Collision frequency calculated from Eq. (8.22)	s^{-1}	$\mathbf{T}^{-1}$

I_s	Collision frequency calculated from Eq. (8.25)	s^{-1}	$\mathbf{T}^{-1}$	
J	Collision frequency calculated from Eq. (8.29)	$m^{-3}\,s^{-1}$	$\mathbf{L}^{-3}\,\mathbf{T}^{-1}$	
J_s	Collision frequency calculated from Eq. (8.28)	$m^{-3}\,s^{-1}$	$\mathbf{L}^{-3}\,\mathbf{T}^{-1}$	
K	Boltzmann's constant ($= 1.38 \times 10^{-23}$ J/K)	J/K	$\mathbf{ML}^2\mathbf{T}^{-2}\boldsymbol{\theta}^{-1}$	
K''	Constant in Robinson's equation (8.1)	–	–	
k	Constant in Eq. (8.55)	s^{-1}	$\mathbf{T}^{-1}$	
k''	Constant in Einstein's equation (8.3)	–	–	
l	length of fibre	m	L	
$\mathbf{N}$	Avogadro number (6.023×10^{26} molecules per kmol or 6.023×10^{23} molecules per mol)	$kmol^{-1}$	$\mathbf{N}^{-1}$	
N_F	Dimensionless parameter ($6\pi\mu a^3\dot{\gamma}/\mathcal{A}$)	–	–	
n	Index in Eqs (8.70) and (8.71)	–	–	
n_f	number density of fibres per unit volume	–	–	
n_i	Number of ions per unit volume	m^{-3}	$\mathbf{L}^{-3}$	
n_1, n_2	Number of particles per unit volume	m^{-3}	$\mathbf{L}^{-3}$	
$\mathbf{P}$	Power input per unit volume of fluid	W/m^3	$\mathbf{ML}^{-1}\,\mathbf{T}^{-3}$	
Q_0	Volumetric feed rate of suspension to continuous thickener	m^3/s	$\mathbf{L}^3\,\mathbf{T}^{-1}$	
Q'	Volumetric flowrate of overflow from continuous thickener	m^3/s	$\mathbf{L}^3\,\mathbf{T}^{-1}$	
R'_0	Drag force per unit projected area of isolated spherical particle	N/m^2	$\mathbf{ML}^{-1}\,\mathbf{T}^{-2}$	
R'_c	Drag force per unit projected area of spherical particle in suspension	N/m^2	$\mathbf{ML}^{-1}\,\mathbf{T}^{-2}$	
R_1	Drag force per unit surface of particle	N/m^2	$\mathbf{ML}^{-1}\,\mathbf{T}^{-2}$	
S	Specific surface	m^2/m^3	$\mathbf{L}^{-1}$	
s	Dimensionless separation distance ($=a_r/a_1$)	–	–	
T	Absolute temperature	K	$\boldsymbol{\theta}$	
t	Time	s	$\mathbf{T}$	
t_R	Residence time	s	$\mathbf{T}$	
U	Mass ratio of liquid to solid in underflow from continuous thickener	–	–	
u	Velocity	m/s	$\mathbf{LT}^{-1}$	
u_0	Free-falling velocity of particle	m/s	$\mathbf{LT}^{-1}$	
u_c	Sedimentation velocity of particle in suspension	m/s	$\mathbf{LT}^{-1}$	
u_{cL}	Limit value of u_c	m/s	$\mathbf{LT}^{-1}$	
u'_c	Sedimentation velocity at level of obstruction	m/s	$\mathbf{LT}^{-1}$	
u_F	Velocity of displaced fluid	m/s	$\mathbf{LT}^{-1}$	

u_f	Velocity of sedimentation of small particles in upper layer	m/s	$\mathbf{LT^{-1}}$
u_i	Sedimentation velocity at infinite dilution	m/s	$\mathbf{LT^{-1}}$
u_p	Velocity of particle relative to fluid	m/s	$\mathbf{LT^{-1}}$
u_u	Sedimentation velocity at concentration C_u of underflow	m/s	$\mathbf{LT^{-1}}$
u_w	Velocity of propagation of concentration wave	m/s	$\mathbf{LT^{-1}}$
V	Volume of particles passing plane per unit area	m/s	$\mathbf{LT^{-1}}$
V_A	van der Waals attraction energy	J	$\mathbf{ML^2T^{-2}}$
V_R	Electrical repulsion energy	J	$\mathbf{ML^2T^{-2}}$
V_T	Total potential energy of interaction	J	$\mathbf{ML^2T^{-2}}$
v	Volume of particle	m³	$\mathbf{L^3}$
X	Average value of mass ratio of liquid to solids	–	–
Y	Mass ratio of liquid to solids in feed to continuous thickener	–	–
Z	Valence of ion	–	–
α_0	Ratio defined by Eq. (8.29)	–	–
α_p	Ratio defined by Eq. (8.27)	–	–
γ	Constant defined by Eq. (8.17)	–	–
$\dot{\gamma}$	Shear rate	s^{-1}	$\mathbf{T^{-1}}$
δ	Stern layer thickness	m	$\mathbf{L}$
ε	Permittivity	s⁴A²/ kg m³	$\mathbf{M^{-1}L^{-3}T^4A^2}$
κ	Debye–Hückel parameter (Eq. 8.18)	m^{-1}	$\mathbf{L^{-1}}$
ζ	Electrokinetic or zeta potential	V	$\mathbf{ML^2T^{-3}A^{-1}}$
μ	Viscosity of fluid	Ns/m²	$\mathbf{ML^{-1}T^{-1}}$
μ_c	Viscosity of suspension	Ns/m²	$\mathbf{ML^{-1}T^{-1}}$
ρ	Density of fluid	kg/m³	$\mathbf{ML^{-3}}$
ρ_c	Density of suspension	kg/m³	$\mathbf{ML^{-3}}$
ρ_s	Density of solid	kg/m³	$\mathbf{ML^{-3}}$
Φ''	Friction factor defined by Eq. (8.90)	–	–
ψ	Volumetric rate of *sedimentation* per unit area	m/s	$\mathbf{LT^{-1}}$
ψ_L	Limit value of ψ	m/s	$\mathbf{LT^{-1}}$
ψ_T	*Total* volumetric rate of solids movement per unit area	m/s	$\mathbf{LT^{-1}}$
ψ_{TL}	Limit value of ψ_T	m/s	$\mathbf{LT^{-1}}$
ψ_u	Solids volumetric flowrate per unit area due to underflow	m/s	$\mathbf{LT^{-1}}$
ψ'	Mass rate of sedimentation per unit area at level of obstruction	kg/m²s	$\mathbf{ML^{-2}T^{-1}}$

ψ_s	Stern plane potential	V	$ML^2T^{-3}A^{-1}$
ψ_0	Surface potential	V	$ML^2T^{-3}A^{-1}$
Ga	Galileo number ($d^3g(\rho_s-\rho)\rho/\mu^2$)	–	–
Re'	Reynolds number ($ud\rho/\mu$)	–	–
Re'_0	Reynolds number for particle under terminal falling conditions ($u_0d\rho/\mu$)	–	–
Re_1	Bed Reynolds number ($u_c\rho/S\mu(1-e)$)	–	–

Subscripts

B	Glass ballotini particles
L	Large particles
M	Mixture
P	Polystyrene particles
S	Small particles
0	Value under free-falling conditions

References

1. Kaye BH, Boardman RP. In: *Cluster formation in dilute suspensions. Third congress of the European federation of chemical Engineering*, 1962. Symposium on the Interaction between Fluids and Particles 17.
2. Coe HS, Clevenger GH. Methods for determining the capacities of slime-settling tanks. *Trans Am Inst Min Met Eng* 1916;**55**:356.
3. Robinson CS. Some factors influencing sedimentation. *Ind Eng Chem* 1926;**18**:869.
4. Einstein A. Eine neue Bestimmung der Molekuldimensionen. *Ann Phys* 1906;**19**:289–306.
5. Vand V. Viscosity of solutions and suspensions. *J Phys Coll Chem* 1948;**52**:277.
6. Steinour HH. Rate of sedimentation. *Ind Eng Chem* 1944;**36**: 618, 840, and 901.
7. Hawksley PGW. The effect of concentration on the settling of suspensions and flow through porous media. *Inst Phys Symp* 1950;114.
8. Wallis GB. In: *A simplified one-dimensional representation of two-component vertical flow and its application to batch sedimentation. Third congress of the European federation of chemical engineering*, 1962. Symposium on the interaction between fluids and particles, 9.
9. Egolf CB, McCabe WL. Rate of sedimentation of flocculated particles. *Trans Am Inst Chem Eng* 1937;**33**:620.
10. Work LT, Kohler AS. The sedimentation of suspensions. *Trans Am Inst Chem Eng* 1940;**36**:701.
11. Pearce KW. In: *Settling in the presence of downward-facing surfaces. Third congress of the European federation of chemical engineering*, 1962. Symposium on the interaction between fluids and particles, 30.
12. Schaflinger U. Influence of non-uniform particle size on settling beneath downward-facing walls. *Int J Multiphase Flow* 1985;**11**:783.
13. Robins WHM. In: *The effect of immersed bodies on the sedimentation of suspensions. Third congress of the European federation of chemical engineering*, 1962. Symposium on the interaction between fluids and particles, 26.
14. Verwey EJW, Overbeek JTHG. *Theory of the stability of lyophobic colloids.* Amsterdam: Elsevier; 1948.
15. Sparnaay MJ. *The electrical double layer.* Oxford: Pergamon Press; 1972.
16. Hunter RJ. *Foundations of colloid science.* 2nd ed. Oxford: Oxford University Press; 2001.
17. Berg JC. *An introduction to interfaces and colloids.* New York: World Scientific; 2009.
18. Shaw DJ. *Introduction to colloid and surface chemistry.* London: Butterworths; 1970.
19. Smith AL. Electrical phenomena associated with the solid–liquid interface. In: Parfitt GD, editor. *Dispersion of powders in liquids.* London: Applied Science Publishers; 1973.

20. Lyklema J. Water at interfaces: a colloid chemical approach. *J Colloid Interface Sci* 1977;**58**:242.

21. Williams DJA, Williams KP. Electrophoresis and zeta potential of kaolinite. *J Colloid Interface Sci* 1978;**65**:79.

22. Derjaguin BV, Landau L. Theory of the stability of strongly charged lyophobic solutions and of the adhesion of strongly charged particles in solution of electrolytes. *Acta Physicochim* 1941;**14**:663.

23. Gregory J. Interfacial phenomena. In: Ives KJ, editor. *The scientific basis of filtration.* Leyden: Noordhoff; 1975.

24. Spielman LA. Viscous interactions in Brownian coagulation. *J Colloid Interface Sci* 1970;**33**:562.

25. Hamaker HC. The London-van der Waals attraction between spherical particles. *Physica* 1937;**4**:1058.

26. Lifshitz EM. Theory of molecular attractive forces between solid. *Soviet Phys JETP* 1956;**2**:73.

27. Ottewill RH. Particulate dispersions. In: *Colloid science.* Vol. 2. London: The Chemical Society; 1973. p. 173.

28. Visser J. On Hamaker constants: a comparison between Hamaker constants and Lifshitz–van der Waals constants. *Adv Colloid Interface Sci* 1972;**3**:331.

29. Matijevic E, Janauer GE, Kerker M. Reversal of charge of hydrophobic colloids by hydrolyzed metal ions, I. Aluminium nitrate. *J Colloid Sci* 1964;**19**:333.

30. Honig EP, Roebersen GJ, Wiersema PH. Effect of hydrodynamic interaction on the coagulation rate of hydrophobic colloids. *J Colloid Interface Sci* 1971;**36**:97.

31. van de Ven TGM, Mason SG. The microrheology of colloidal dispersions. IV. Pairs of interacting spheres in shear flow. *J Colloid Interface Sci* 1976;**57**:505.

32. Zeichner GR, Schowalter WR. Use of trajectory analysis to study stability of colloidal dispersions in flow fields. *AIChE J* 1977;**23**:243.

33. von Smoluchowski M. Versuch einer mathematischen Theorie der Koagulationskinetik Kolloider Lösungen. *Z Physik Chem* 1917;**92**:129.

34. Camp TR, Stein PC. Velocity gradients and internal work in fluid motion. *J Boston Soc Civ Eng* 1943;**30**:219.

35. La Mer VK, Healy TW. Adsorption–flocculation reactions of macromolecules at the solid–liquid interface. *Rev Pure Appl Chem (Australia)* 1963;**13**:112.

36. Vold MJ. The effect of adsorption on the van der Waals interaction of spherical particles. *J Colloid Sci* 1961;**16**:1.

37. Osmond DWJ, Vincent B, Waite FA. The van der Waals attraction between colloid particles having adsorbed layers, I. A re-appraisal of the "Vold" effect. *J Colloid Interface Sci* 1973;**42**:262.

38. Vincent B. The effect of adsorbed polymers on dispersion stability. *Adv Colloid Interface Sci* 1974;**4**:193.

39. Napper DH. Steric stabilisation. *J Colloid Interface Sci* 1977;**58**:390.

40. Dobbie JW, Evans RE, Gibson DV, Smitham JB, Napper DH. Enhanced steric stabilisation. *J Colloid Interface Sci* 1973;**45**:557.

41. O'Gorman JV, Kitchener JA. The flocculation and dewatering of Kimberlite clay slimes. *Int J Miner Process* 1974;**1**:33.

42. Williams DJA, Ottewill RH. The stability of silver iodide solutions in the presence of polyacrylic acids of various molecular weights. *Kolloid Z Z Polym* 1971;**243**:141.

43. Gregory J. Flocculation of polystyrene particles with cationic polyelectrolytes. *Trans Faraday Soc* 1969;**65**:2260.

44. Gregory J. The effect of cationic polymers on the colloid stability of latex particles. *J Colloid Interface Sci* 1976;**55**:35.

45. Griot O, Kitchener JA. Role of surface silanol groups in the flocculation of silica by polycrylamide. *Trans Faraday Soc* 1965;**61**:1026.

46. la Mer VK. Filtration of colloidal dispersions flocculated by anionic and cationic polyelectrolytes. *Discuss Faraday Soc* 1966;**42**:248.

47. Michaels AS, Bolger JC. Settling rates and sediment volumes of flocculated kaolinite suspensions. *Ind Eng Chem Fundam* 1962;**1**:24.

48. Kynch GJ. A theory of sedimentation. *Trans Faraday Soc* 1952;**48**:166.

49. Yoshioka N, Hotta Y, Tanaka S, Naito S, Tsugami S. Continuous thickening of homogeneous flocculated slurries. *Kagaku Kogaku (J Soc Chem Eng, Japan 2)* 1957;**21**:66.

50. Hassett NJ. Theories of the operation of continuous thickeners. *Industr Chem* 1961;**37**:25.

51. Comings EW. Thickening calcium carbonate slurries. *Ind Eng Chem* 1940;**32**:1663.

52. Roberts EJ. Thickening–art or science? *Trans Amer Inst Min, Met Eng* 1949;**184**:61 [*Trans Min Eng* **1** (1949) 61].

53. Tiller FM, Chen W. Limiting operating conditions for continuous thickeners. *Chem Eng Sci* 1988;**43**:1695.

54. Font R. Calculation of the compression zone height in continuous thickeners. *AIChE J* 1990;**36**:3.

55. Fitch B. Thickening theories—an analysis. *AIChE J* 1993;**39**:27.

56. Dell CC, Keleghan WTH. The dewatering of polyclay suspensions. *Powder Technol* 1973;**7**:189.

57. Chandler JL. Dewatering by deep thickeners without rakes. World Filtration Congress III. (Downingtown, Pennsylvania). *Proceedings* 1982;**1**:372.

58. Fernando Concha A. *Solid-liquid separation in the mining industry.* New York: Springer; 2014.

59. Richardson JF, Zaki WN. Sedimentation and fluidisation: Part I. *Trans Inst Chem Eng* 1954;**32**:35.

60. Garside J, Al-Dibouni MR. Velocity–voidage relationships for fluidization and sedimentation in solid–liquid systems. *Ind Eng Chem Proc Des Dev* 1977;**16**:206.

61. Khan AR. *Heat transfer from immersed surfaces to liquid-fluidised beds.* University of Wales; 1978 [Ph.D. thesis].

62. Rowe PN. A convenient empirical equation for estimation of the Richardson-Zaki exponent. *Chem Eng Sci* 1987;**42**:2795.

63. Khan AR, Richardson JF. Fluid-particle interactions and flow characteristics of fluidized beds and settling suspensions of spherical particles. *Chem Eng Commun* 1989;**78**:111.

64. Khan AR, Richardson JF. The resistance to motion of a solid sphere in a fluid. *Chem Eng Commun* 1987;**62**:135.

65. Khan AR, Richardson JF. Pressure gradient and friction factor for sedimentation and fluidisation of uniform spheres in liquids. *Chem Eng Sci* 1990;**45**:255.

66. Richardson JF, Meikle RA. Sedimentation and fluidisation. Part III. The sedimentation of uniform fine particles and of two-component mixtures of solids. *Trans Inst Chem Eng* 1961;**39**:348.

67. Carman PC. Fluid flow through granular beds. *Trans Inst Chem Eng* 1937;**15**:150.

68. Richardson JF, Shabi FA. The determination of concentration distribution in a sedimenting suspension using radioactive solids. *Trans Inst Chem Eng* 1960;**38**:33.

69. Richardson JF, Meikle RA. Sedimentation and fluidisation. Part IV. Drag force on individual particles in an assemblage. *Trans Inst Chem Eng* 1961;**39**:357.

70. Rowe PN, Henwood GN. Drag forces in a hydraulic model of a fluidised bed. Part I. *Trans Inst Chem Eng* 1961;**39**:43.

71. Sangani AS, Acrivos A. Slow flow through a periodic array of spheres. *Int J Multiphase Flow* 1982;**8**:343–60.

72. Hill RJ, Koch DL, Ladd AJC. The first effect of fluid inertia on flows in ordered and random arrays of spheres. *J Fluid Mech* 2001;**448**:213–41; Also, see ibid pp. 243–278.

73. Hill RJ, Koch DL. The transition from steady to weakly turbulent flow in a close-packed ordered arrays of spheres. *J Fluid Mech* 2002;**465**:59–97.

74. Smith TN. The differential sedimentation of particles of various species. *Trans Inst Chem Eng* 1967;**45**: T311.

75. Lockett MJ, Al-Habbooby HM. Differential settling by size of two particle species in a liquid. *Trans Inst Chem Eng* 1973;**51**:281.

76. Richardson JF, Mirza S. Sedimentation of suspensions of particles of two or more sizes. *Chem Eng Sci* 1979;**34**:447.

77. Selim MS, Kothari AC, Turian RM. Sedimentation of multisized particles in concentrated suspensions. *AIChE J* 1983;**29**:1029.

78. Kumar P, Ramarao BV. Enhancement of sedimentation rates of fibrous suspensions. *Chem Eng Commun* 1991;**108**:381–401.

79. Turney MA, Cheung MK, Powell RL, McCarthy MJ. Hindered settling of rod-like particles measured with magnetic resonance imaging. *AIChE J* 1995;**41**:251–7.

80. Herzhaft B, Guazzelli E, Mackaplow MB, Shaqfeh ESG. Experimental investigation of the sedimentation of a dilute fiber suspension. *Phys Rev Lett* 1996;**77**:290–3.

81. Herzhaft B, Guazzelli E. Experimental study of the sedimentation of dilute and semi-dilute suspensions of fibers. *J Fluid Mech* 1999;**384**:133–58.

82. Whitmore RL. The sedimentation of suspensions of spheres. *Brit J Appl Phys* 1955;**6**:239–45.
83. Weiland RH, McPherson RR. Accelerated settling by addition of buoyant particles. *Ind Eng Chem Fundam* 1979;**18**:45–9.
84. Venkataraman S, Weiland RH. Buoyant-particle-promoted settling of industrial suspensions. *Ind Eng Chem Process Des Dev* 1985;**24**:966–70.

Further Reading

Clift R, Grace JR, Weber ME. *Bubbles, drops and particles.* Academic Press; 1978.
Dallavalle JM. *Micromeritics.* 2nd ed. London: Pitman; 1948.
Orr C. *Particulate technology.* New York: Macmillan; 1966.
Ottewill RH. Particulate dispersions. In: *Colloid Science.* Vol. 2. London: The Chemical Society; 1973. p. 173–219.
Rhodes MJ. *Introduction to particle technology.* 2nd ed. London: Wiley; 2008.
Smith AL, editor. *Particle growth in suspensions.* New York: Academic Press; 1973.
Svarovsky L, editor. *Solid–liquid separation.* 4th ed. Oxford: Butterworth-Heinemann; 2000.

CHAPTER 9

Fluidisation

9.1 Characteristics of Fluidised Systems

9.1.1 General Behaviour of Gas Solids and Liquid Solids Systems

When a fluid is passed downwards through a bed of solids, no relative movement between the particles takes place, unless the initial orientation of the particles is unstable, and where the flow is streamline, the pressure drop across the bed is directly proportional to the rate of flow, although at higher rates the pressure drop increases more rapidly. The pressure drop under these conditions may be obtained using the equations presented in Chapter 7.

When a fluid is passed upwards through a bed, the pressure drop is the same as that for downward flow at relatively low rates (Fig. 9.1A). When, however, the frictional drag on the particles becomes equal to their apparent weight—that is, the actual weight less the buoyancy force—the particles become rearranged, thus offering less resistance to the flow of fluid, and the bed starts to expand with a corresponding increase in voidage. This process continues with increase in velocity, with the total frictional force remaining equal to the weight of the particles, until the bed has assumed its loosest stable form of packing. If the velocity is then increased further, the individual particles separate from one another and become freely supported in the fluid. At this stage, the bed is described as *fluidised*, as shown schematically in Fig. 9.1B. Further increase in the velocity causes the particles to separate still further from one another, although the pressure difference remains approximately equal to the weight per unit area of the bed. In practice, the transition from the fixed to the fluidised bed condition is not uniform, mainly due to irregularities in the packing and, over a range of velocities, fixed and fluidised bed regions may co-exist. In addition, with gases, surface-related forces give rise to the formation of conglomerates of particles through which there is a little flow and, as a result, much of the gas may pass through the bed in channels. This is an unstable condition, since the channels that offer a relatively low resistance to flow tend to open up as the gas flowrate is increased, and regions of the bed may remain in an unfluidised state even though the overall superficial velocity may be much higher than the minimum fluidising velocity.

Up to this point, the system behaves in a similar way with both liquids and gases, although at high fluid velocities, there is usually a fairly sharp distinction between the behaviour of the two systems. With a liquid, the bed continues to expand as the velocity is increased and it maintains its uniform character, with the degree of agitation of the particles increasing progressively.

Coulson and Richardson's Chemical Engineering. https://doi.org/10.1016/B978-0-08-101098-3.00010-X

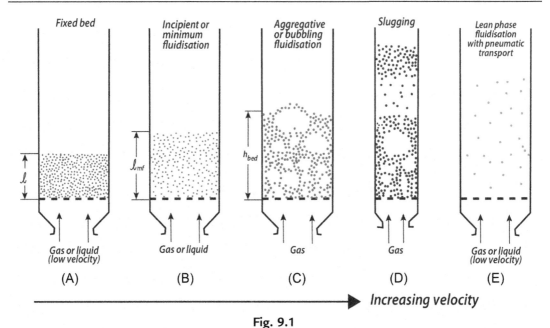

Fig. 9.1

Schematics of the various flow regimes in fluid–solid systems. Fluid velocity is increasing from left to right. *Adapted from Kunii D, Levenspiel O. Fluidization engineering, 2nd ed., Butterworth-Heinmann, Boston, 1991.*

This type of fluidisation is known as *particulate fluidisation*. With a gas, however, uniform fluidisation is frequently obtained only at low velocities. At higher velocities two separate *phases* may form—a continuous phase, often referred to as the *dense* or *emulsion* phase, and a discontinuous phase known as the *lean* or *bubble* phase. The fluidisation is then said to be *aggregative* or bubbling type (Fig. 9.1C). At much higher velocities, the bubbles tend to break down—a feature that leads to a much more chaotic structure. When gas bubbles pass through a relatively high-density fluidised bed, the system closely resembles a boiling liquid, with the lean phase corresponding to the vapour and the dense or continuous phase corresponding to the liquid. The bed is then often referred to as a *boiling bed*, as opposed to the *quiescent bed* usually formed at low flowrates. As the gas flowrate is further increased, the velocity relative to the particles in the dense phase does not change appreciably, and streamline flow may persist even at very high overall rates of flow because a high proportion of the total flow is then in the form of bubbles. At high flowrates in deep beds, coalescence of the bubbles takes place, and in narrow vessels, slugs of gas occupying the whole cross-section may be produced. These slugs of gas alternate with slugs of fluidised solids that are carried upwards and subsequently collapse, releasing the solids, which fall back, as shown schematically in Fig. 9.1D.

Finally, when a bed of fine particles is fluidised at sufficiently high velocities (well above the minimum fluidisation), the terminal velocity of the solids is exceeded, the bed surface becomes diffused and essentially disappears. The entrainment of particles is significant, and there is turbulent motion of solid clusters and gas voids of various shapes and sizes leading to the so-called turbulent fluidised bed. This type of bed has high heat and mass transfer rates. Further increase in the fluid velocity results in nearly complete entrainment of particles, requiring recycling or continuous feed of the particles into the bed. This is a dilute or lean-phase fluidised bed state with pneumatic transport of solids, shown schematically in Fig. 9.1E.

In an early attempt to differentiate between the conditions leading to particulate or aggregative fluidisation, Wilhelm and Kwauk[1] suggested using the value of the Froude number (u_{mf}^2/gd) as a criterion, where:

u_{mf} is the minimum superficial velocity of fluid at which fluidisation occurs;

d is the diameter of the particles; and.

g is the acceleration due to gravity.

At values of a Froude group of less than unity, particulate fluidisation normally occurs and, at higher values, aggregative fluidisation takes place. Much lower values of the Froude number are encountered with liquids because the minimum velocity required to produce fluidisation is less. A theoretical justification for using the Froude number as a means of distinguishing between particulate and aggregative fluidisation has been provided by Jackson[2] and Murray.[3]

Although the possibility of forming fluidised beds had been known for many years, the subject remained of academic interest only, until the adoption of fluidised catalysts by the petroleum industry for the cracking of heavy hydrocarbons and for the synthesis of fuels from natural gas, or from carbon monoxide and hydrogen. In many ways, the fluidised bed behaves as a single fluid of a density equal to that of the mixture of solids and fluid. Such a bed will flow, it is capable of transmitting hydrostatic forces, and solid objects with densities less than that of the bed will float at the surface. Intimate mixing occurs within the bed and heat and mass transfer rates are very high, with the result that uniform temperatures are quickly attained throughout the system. The easy control of temperature is the feature that has led to the use of fluidised systems for highly exothermic processes, where uniformity of temperature is important.

In order to understand the properties of a fluidised system, it is necessary to study the flow patterns of both the solids and the fluid. The mode of formation and behaviour of fluid bubbles is of particular importance because these usually account for the flow of a high proportion of the fluid in a gas–solids system.

In any study of the properties of a fluidised system, it is necessary to select conditions that are reproducible, and the lack of agreement between the results of different researchers, particularly those relating to heat transfer, is largely attributable to the existence of widely different uncontrolled conditions within the bed. The fluidisation should be of good quality, that is to say, that the bed should be free from irregularities and channelling. Many solids, particularly those of appreciably non-isometric shape and those that have a tendency to form agglomerates, seldom fluidise readily in a gas. Furthermore, the fluid must be evenly distributed at the bottom of the bed and it is usually necessary to provide a distributor across which the pressure drop is equal to at least that across the bed. This condition is much more readily achieved in a small laboratory apparatus than in large-scale industrial equipment.

As already indicated, when a liquid is the fluidising agent, substantially uniform conditions pervade in the bed, although with a gas, bubble formation tends to occur except at very low fluidising velocities. In an attempt to improve the reproducibility of conditions within a bed, much of the earlier research work with gas-fluidised systems was carried out at gas velocities sufficiently low for bubble formation to be absent. Subsequently, however, it has been recognised that bubbles normally tend to form in such systems, that they exert an important influence on the flow pattern of both gas and solids, and that the behaviour of individual bubbles can often be predicted with reasonable accuracy.

9.1.2 Effect of Fluid Velocity on Pressure Gradient and Pressure Drop

When a fluid flows slowly upwards through a bed of very fine particles, the flow is streamline and a linear relation exists between the pressure gradient and flowrate as discussed in Chapter 7, Section 7.2.3. If the pressure gradient $(-\Delta P/l)$ is plotted against the superficial velocity (u_c) using logarithmic co-ordinates, a straight line of unit slope is obtained, as shown in Fig. 9.2. As the superficial velocity approaches the *minimum* fluidising velocity (u_{mf}), the bed starts to expand and when the particles are no longer in physical contact with one another, the bed *is fluidised*. The pressure *gradient* then becomes lower because of the increased voidage and, consequently, the weight of particles per unit height of bed is smaller. This fall continues until the velocity is high enough for transport of the material to take place, and the pressure gradient then starts to increase again because the frictional drag of the fluid at the walls of the column starts to become significant. When the bed is composed of large particles, the flow will be laminar only at very low velocities, and the slope s of the lower part of the curve will be greater $(1 < s < 2)$ and may not be constant, particularly if there is a progressive change in flow regime as the velocity increases.

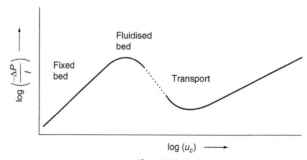

Fig. 9.2

Pressure gradient across a bed as a function of fluid velocity.

If the pressure across the whole bed, instead of the pressure gradient, is plotted against the fluid velocity, also using logarithmic coordinates as shown in Fig. 9.3, a linear relation is again obtained, up to the point where expansion of the bed starts to take place (point A), although the slope of the curve then gradually diminishes as the bed expands and its porosity increases. As the velocity is further increased, the pressure drop passes through a maximum value (point B) and then falls slightly and attains an approximately constant value that is independent of the fluid velocity (CD). If the fluid velocity is reduced again, the bed contracts until it reaches the condition where the particles are just resting on one another (E). The porosity then has the maximum stable value that can occur for a fixed bed of the particles. If the velocity is further decreased, the structure of the bed then remains unaffected, provided that the bed is not subjected to vibration. The pressure drop (EF) across this reformed fixed bed at any fluid velocity is then less than that before fluidisation. If the velocity is now increased again, it might be expected that the curve (FE) would be retraced and that the slope would suddenly change from 1 (over FE) to 0 (corresponding to ED) at the fluidising point. This condition is difficult to reproduce, however, because the bed tends to become consolidated again unless it is completely free from channelling. In the absence of channelling, it is the shape and size of the particles that determine both the maximum porosity and the pressure drop across a given height of a fluidised bed of a given depth. In an ideal fluidised bed, the pressure drop corresponding to ECD is equal to the buoyant weight of particles per unit area. In practice, it may deviate appreciably from this value as a result of channelling and the effect of particle-wall friction. Point B lies above CD because the frictional forces between the particles have to be overcome before the bed rearrangement can take place.

The minimum fluidising velocity, u_{mf}, may be determined experimentally by measuring the pressure drop across the bed for both increasing and decreasing velocities and plotting the results, as shown in Fig. 9.3. The two 'best' straight lines are then drawn through the experimental points and the velocity at their point of intersection is taken as the minimum fluidising velocity. Linear rather than logarithmic plots are generally used to minimise error in

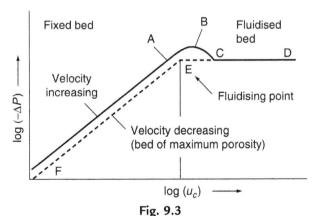

Fig. 9.3
Pressure drop over fixed and fluidised beds.

the determination of the minimum fluidizing velocity, although it is necessary to use logarithmic plots if the plot of pressure gradient against velocity in the fixed bed is not linear.

The theoretical value of the minimum fluidising velocity may be calculated from the equations given in Chapter 7 for the relation between pressure drop and velocity in a fixed packed bed, with the pressure drop through the bed made equal to the apparent weight of particles per unit area, and the porosity set at the maximum value that can be attained in the fixed bed.

In a fluidised bed, the total frictional force on the particles must equal the effective weight of the bed. Thus, in a bed of unit cross-sectional area, depth l, and porosity e, the additional pressure drop across the bed attributable to the layout weight of the particles is given by:

$$-\Delta P = (1 - e)(\rho_s - \rho)lg \tag{9.1}$$

where g is the acceleration due to gravity and ρ_s and ρ are the densities of the particles and the fluid, respectively.

This relation applies from the initial expansion of the bed until transport of solids takes place. There may be some discrepancy between the calculated and measured minimum velocities for fluidisation. This may be attributable to:

- channelling, as a result of which the drag force acting on the bed is reduced;
- the action of electrostatic forces in the case of gaseous fluidisation, which is particularly important in the case of sands;
- agglomeration, which is often considerable with small particles; or
- friction between the fluid and the walls of the containing vessel, i.e., wall effects.

This last factor is of greatest importance with beds of small diameters. Leva et al.[4] introduced a term, $(G_F - G_E)/G_F$, which is a fluidisation efficiency, in which G_F is the minimum flowrate required to produce fluidisation and G_E is the rate required to produce the initial expansion of the bed.

If flow conditions within the bed are streamline, the relation between fluid velocity u_c, pressure drop $(-\Delta P)$, and voidage e is given, for a fixed bed of spherical particles of diameter d, by the Carman–Kozeny Eq. (7.12a) which takes the form:

$$u_c = 0.0055 \left(\frac{e^3}{(1 - e)^2} \right) \left(\frac{-\Delta P d^2}{\mu l} \right) \tag{9.2}$$

For a fluidised bed, the buoyant weight of the particles is counterbalanced by the frictional drag. Substituting for $-\Delta P$ from Eq. (9.1) into Eq. (9.2) gives:

$$u_c = 0.0055 \left(\frac{e^3}{1 - e} \right) \left(\frac{d^2 (\rho_s - \rho) g}{\mu} \right) \tag{9.3}$$

There is evidence in the work reported in Chapter 8 on sedimentation[5] to suggest that where the particles are free to adjust their orientations with respect to one another and to the fluid, as in sedimentation and fluidisation, the equations for pressure drop in fixed beds overestimate the values where the particles can 'choose' their orientation. A value of 3.36 rather than 5 for the Carman–Kozeny constant is in closer accord with experimental data. The coefficient in Eq. (9.3) then takes on the higher value of 0.0089. The experimental evidence is limited to a few measurements, however, and Eq. (9.3), with its possible inaccuracies, is used here.

9.1.3 Minimum Fluidising Velocity

As the upward velocity of flow of fluid through a packed bed of uniform spheres is increased, the point of *incipient fluidisation* is reached when the particles are just supported in the fluid. The corresponding value of the *minimum fluidising velocity* (u_{mf}) is then obtained by substituting e_{mf} into Eq. (9.3) to give:

$$u_{mf} = 0.0055 \left(\frac{e_{mf}^3}{1 - e_{mf}} \right) \frac{d^2 (\rho_s - \rho) g}{\mu} \tag{9.4}$$

Since Eq. (9.4) is based on the Carman–Kozeny equation, it applies only to conditions of laminar flow, and hence to low values of the Reynolds number for flow in the bed. In practice, this restricts its application to fine particles.

The value of e_{mf} will be a function of the shape, size distribution, and surface properties of the particles. Substituting a typical value of 0.4 for e_{mf} in Eq. (9.4) gives:

$$\left(u_{mf} \right)_{e_{mf}=0.4} = 0.00059 \left(\frac{d^2 (\rho_s - \rho) g}{\mu} \right) \tag{9.5}$$

When the flow regime at the point of incipient fluidisation is outside the range over which the Carman–Kozeny equation is applicable, it is necessary to use one of the more general equations for the pressure gradient in the bed, such as the Ergun equation given in Eq. (7.20) as:

$$\frac{-\Delta P}{l} = 150 \left(\frac{(1-e)^2}{e^3} \right) \left(\frac{\mu u_c}{d^2} \right) + 1.75 \left(\frac{(1-e)}{e^3} \right) \left(\frac{\rho u_c^2}{d} \right) \tag{9.6}$$

where d is the diameter of the sphere with the same volume to surface area ratio as the particles.

Substituting $e = e_{mf}$ at the incipient fluidisation point and for $-\Delta P$ from Eq. (9.1), Eq. (9.6) is then applicable at the minimum fluidisation velocity u_{mf}, and gives:

$$(1 - e_{mf})(\rho_s - \rho) g = 150 \left(\frac{(1 - e_{mf})^2}{e_{mf}^3} \right) \left(\frac{\mu u_{mf}}{d^2} \right) + 1.75 \left(\frac{(1 - e_{mf})}{e_{mf}^3} \right) \left(\frac{\rho u_{mf}^2}{d} \right) \tag{9.7}$$

Multiplying both sides by $\frac{\rho d^3}{\mu^2\left(1-e_{mf}\right)}$ gives:

$$\frac{\rho(\rho_s-\rho)gd^3}{\mu^2}=150\left(\frac{1-e_{mf}}{e_{mf}^3}\right)\left(\frac{u_{mf}d\rho}{\mu}\right)+\left(\frac{1.75}{e_{mf}^3}\right)\left(\frac{u_{mf}d\rho}{\mu}\right)^2 \tag{9.8}$$

In Eq. (9.8):

$$\frac{d^3\rho(\rho_s-\rho)g}{\mu^2}=Ga \tag{9.9}$$

where Ga is the 'Galileo number' (also known as the Archimedes number), and:

$$\frac{u_{mf}d\rho}{\mu}=Re'_{mf}. \tag{9.10}$$

where Re_{mf} is the Reynolds number at the minimum fluidising velocity. Eq. (9.8) then becomes:

$$Ga=150\left(\frac{1-e_{mf}}{e_{mf}^3}\right)Re'_{mf}+\left(\frac{1.75}{e_{mf}^3}\right)Re_{mf}^2 \tag{9.11}$$

For a typical value of $e_{mf}=0.4$:

$$Ga=1406Re'_{mf}+27.3Re'^2_{mf} \tag{9.12}$$

Thus:

$$Re'^2_{mf}+51.4Re'_{mf}-0.0366Ga=0 \tag{9.13}$$

and:

$$\left(Re'_{mf}\right)_{e_{mf}=0.4}=25.7\left\{\sqrt{(1+5.53\times10^{-5}Ga)}-1\right\} \tag{9.14}$$

and similarly for $e_{mf}=0.45$:

$$\left(Re'_{mf}\right)_{e_{mf}=0.45}=23.6\left\{\sqrt{(1+9.39\times10^{-5}Ga)}-1\right\} \tag{9.15}$$

By rearranging Eq. (9.10):

$$u_{mf}=\frac{\mu}{d\rho}Re'_{mf}$$

It is possible that the Ergun equation, like the Carman–Kozeny equation, also overpredicts pressure drop for fluidised systems, although no experimental evidence is available on the basis of which the values of the coefficients may be amended.

Wen and Yu[6] have examined the relationship between voidage at the minimum fluidising velocity, e_{mf}, and particle shape, φ_s, which is defined as the ratio of the diameter of a sphere of the same volume to surface area ratio as the particle d, as used in the Ergun equation, to the diameter of a sphere with the same volume as the particle d_p.

Thus:

$$\varphi_s = d/d_p \qquad (9.16)$$

where:

$$d = 6V_p/A_p \text{ and } d_p = \left(6V_p/\pi\right)^{1/3}.$$

In practice, the particle size d can be determined only by measuring both the volumes V_p and the areas A_p of the particles. Since this operation involves a somewhat tedious experimental technique, it is more convenient to measure the particle volume only and then work in terms of d_p and the shape factor.

The minimum fluidising velocity is a function of both e_{mf} and ϕ_s, neither of which is easily measured or estimated, and Wen and Yu[6] have shown that these two quantities are, in practice, inter-related. These authors have published experimental data of e_{mf} and ϕ_s for a wide range of well-characterised particles, and it has been shown that the relation between these two quantities is essentially independent of particle size over a wide range. It has also been established that the following two expressions give reasonably good correlations between e_{mf} and ϕ_s, as shown in Fig. 9.4:

$$\left(\frac{1-e_{mf}}{e_{mf}^3}\right)\frac{1}{\varphi_s^2} = 11 \qquad (9.17)$$

$$\left(\frac{1}{e_{mf}^3 \varphi_s}\right) = 14 \qquad (9.18)$$

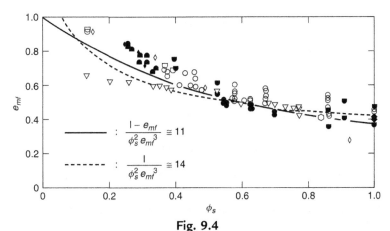

Fig. 9.4
Relation between e_{mf} and ϕ_s.

Niven[7] discusses the significance of the two dimensionless groups in Eqs (9.17) and (9.18), and also suggests that d and u_{mf} in Eqs (9.8), (9.9), and (9.10) are more appropriately replaced by a mean linear dimension of the pores and the mean pore velocity at the point of incipient fluidisation.

Using Eq. (9.16) to substitute $\varphi_s d_p$ for d in Eq. (9.6) gives:

$$(1 - e_{mf})(\rho_s - \rho)g = 150 \left(\frac{(1 - e_{mf})^2}{e_{mf}^3}\right) \left(\frac{\mu u_{mf}}{\varphi_s^2 d_p^2}\right) + 1.75 \left(\frac{1 - e_{mf}}{e_{mf}^3}\right) \frac{\rho u_{mf}^2}{\varphi_s d_p}$$

Thus:

$$\frac{(\rho_s - \rho)\rho g d_p^3}{\mu^2} = 150 \left(\frac{1 - e_{mf}}{e_{mf}^3}\right) \frac{1}{\varphi_s^2} \left(\frac{\rho d_p u_{mf}}{\mu}\right) + 1.75 \left(\frac{1}{e_{mf}^3 \varphi_s}\right) \left(\frac{\rho^2 d_p^2 u_{mf}^2}{\mu^2}\right)$$

Substituting from Eqs (9.17) and (9.18):

$$Ga_p = (150 \times 11)Re'_{mfp} + (1.75 \times 14)Re'^2_{mfp}$$

where Ga_p and Re_{mfp} are the Galileo number and the particle Reynolds number at the point of incipient fluidisation, respectively, in both cases with the linear dimension of the particles expressed as d_p. Thus:

$$Re'^2_{mfp} + 67.3Re'_{mfp} - 0.0408Ga_p = 0$$

giving:

$$Re'_{mfp} = 33.65 \left[\sqrt{(1 + 3.6 \times 10^{-5} Ga_p)} - 1\right] \qquad (9.19)$$

Before leaving this section, it is appropriate to add here that one can generalise Eq. (9.19) as:

$$Re'_{mfp} = \frac{\rho d_p u_{mf}}{\mu} = \sqrt{C_1^2 + C_2 Ga_p} - C_1 \qquad (9.20)$$

Naturally, the values of the constants C_1 and C_2 vary from one system to another and much of the literature on this has been collated by Kunii and Levenspiel.[8] For fine particles, Wen and Yu[9] recommend $C_1 = 33.65$ and $C_2 = 0.0408$, consistent with Eq. (9.19). However, Xu and Zhu[10] have argued that, in the case of fine cohesive powders, Eq. (9.20) is not very satisfactory and one must take into account the inter-particle cohesive forces vis-à-vis the gravitational forces. For such systems, they were able to correlate the minimum fluidisation velocity directly with the particle size.

Example 9.1

A bed consists of uniform spherical particles of diameter 3 mm and density 4200 kg/m³. What will be the minimum fluidising velocity in a liquid of viscosity 3 mNs/m² and density 1100 kg/m³?

Solution

By definition:

Galileo number, $Ga = d^3 \rho (\rho_s - \rho) g / \mu^2$

$$= \left((3 \times 10^{-3})^3 \times 1100 \times (4200 - 1100) \times 9.81 \right) / (3 \times 10^{-3})^2$$

$$= 1.003 \times 10^5$$

Assuming a value of 0.4 for e_{mf}, Eq. (9.14) gives:

$$Re'_{mf} = 25.7 \left\{ \sqrt{\left(1 + (5.53 \times 10^{-5})(1.003 \times 10^5)\right)} - 1 \right\} = 40$$

and:

$$u_{mf} = (40 \times 3 \times 10^{-3})/(3 \times 10^{-3} \times 1100) = 0.0364 \,\text{m/s or } 36.4 \,\text{mm/s}$$

Example 9.2

Oil, of density 900 kg/m³ and viscosity 3 mNs/m² is passed vertically upwards through a bed of catalyst consisting of approximately spherical particles of diameter 0.1 mm and density 2600 kg/m³. At approximately what mass rate of flow per unit area of bed will (a) fluidisation and (b) transport of particles occur?

Solution

(a) Eqs (7.9) and (9.1) may be used to determine the fluidising velocity, u_{mf}.

$$u = (1/K'')(e^3/(S^2(1-e)^2(1/\mu)(-\Delta P/l) \quad (equation\ 7.9)$$

$$-\Delta P = (1-e)(\rho_s - \rho) l g \quad (equation\ 9.1)$$

where S = surface area/volume, which, for a sphere, $= \pi d^2/(\pi d^3/6) = 6/d$.

Substituting $K'' = 5$, $S = 6/d$ and $-\Delta P/l$ from Eq. (9.1) into Eq. (7.9) gives:

$$u_{mf} = 0.0055 \left(e^3/(1-e) \right) \left(d^2 (\rho_s - \rho) g \right) / \mu$$

Hence:

$$G'_{mf} = \rho u = \left(0.0055 e^3/(1-e) \right) \left(d^2 (\rho_s - \rho) g \right) / \mu$$

In this problem, $\rho_s = 2600$ kg/m³, $\rho = 900$ kg/m³, $\mu = 3.0 \times 10^{-3}$ Ns/m² and $d = 0.1$ mm $= 1.0 \times 10^{-4}$ m.

As no value of the voidage is available, e will be estimated by considering eight closely packed spheres of diameter d in a cube of side $2d$. Thus:

$$\text{volume of spheres} = 8(\pi/6)d^3.$$

$$\text{volume of the enclosure} = (2d)^3 = 8d^3.$$

and hence:

$$\text{voidage, } e = [8d^3 - 8(\pi/6)d^3]/8d^3 = 0.478, \text{ say, } 0.48.$$

Thus:

$$G'_{mf} = 0.0055(0.48)^3 (10^{-4})^2 ((900 \times 1700) \times 9.81)/((1-0.48) \times 3 \times 10^{-3})$$
$$= 0.059 \, \text{kg/m}^2\text{s}$$

(b) Transport of the particles will occur when the fluid velocity is equal to the terminal falling velocity of the particle.

Using Stokes' law:

$$u_0 = d^2 g(\rho_s - \rho)/18\mu \quad (equation\ 6.24)$$
$$= \left((10^{-4})^2 \times 9.81 \times 1700\right)/(18 \times 3 \times 10^{-3})$$
$$= 0.0031 \, \text{m/s}$$

The Reynolds number $= ((10^{-4} \times 0.0031 \times 900)/(3 \times 10^{-3}) = 0.093$ and hence Stokes' law applies.

The required mass flow $= (0.0031 \times 900) = 2.78 \, \text{kg/m}^2\text{s}$.

An alternative approach is to make use of Fig. 6.6 and Eq. (6.35):

$$(R/\rho u^2)Re^2 = 2d^3 \rho g(\rho_s - \rho)/3\mu^2$$
$$= \left(2 \times (10^{-4})^3 \times (900 \times 9.81) \times 1700\right)/\left(3(3 \times 10^{-3})^2\right) = 1.11$$

From Fig. 6.6, $Re = 0.09$. Hence:

$$u_0 = Re(\mu/\rho d) = \left(0.09 \times 3 \times 10^{-3}\right)/\left(900 \times 10^{-4}\right) = 0.003 \, \text{m/s}$$

and:

$$G' = (0.003 \times 900) = 2.7 \, \text{kg/m}^2\text{s}$$

Example 9.3

Calculate the minimum fluidisation velocity u_{mf} for beds of FCC particles used by Lettieri et al.[11] and compare with the experimental values reported by them.

Solution

Given data:

Particles:	FCC catalyst characterised by wide size distribution
$d_p = 71, 57, 49 \, \mu\text{m}$,	$\rho_s = 1420 \, \text{kg/m}^3$
Fluidising gas:	Nitrogen at 20°C
$\rho = 1.16 \, \text{kg/m}^3$,	$\mu = 1.76 \times 10^{-5} \, \text{Pa.s}$

Table 9.1 Comparison of experimental and calculated values of V_{mf} for FCC catalysts

Values of Constants Used in Eq. (9.20)	Catalyst Particle Size (μm)	Experimental,[11] u_{mf} (mm/s)	Calculated, u_{mf} (mm/s)	% Relative Error $\dfrac{(calc. - exptl.)}{exptl.} \times 100$
$C_1 = 28.7$	71	4.1	3.43	−16.3
$C_2 = 0.0494^{12}$	57	2.9	2.2	−24.13
$C_1 = 33.7$	71	4.1	2.41	−41.2
$C_2 = 0.0408^{9}$	57	2.9	1.55	−46.55

Eq. (9.20) can be rearranged as below to calculate minimum fluidisation velocity.

$$u_{mfp} = \left\{ \left[C_1^2 + C_2 Ga_p \right]^{0.5} - C_1 \right\} \left(\frac{\mu}{d_p \rho} \right)$$

Two sets of values for the constants recommended in the literature were selected for comparison. The calculated values are compared with experimental values in Table 9.1. The predicted values are lower by about 20%, with the values of constants given by Chitester et al.[12] The table also shows that relative errors increase to about 40% when the constants proposed by Wen and Yu[9] are used. Therefore, there is a need to make a proper selection of constants to be used in Eq. (9.20) and apply a proper safety factor. Methods for more accurate prediction of the minimum fluidisation velocity in gas–solid fluidised beds covering a wide range of conditions have been reported in more recent studies.[13, 14]

9.1.4 Minimum Fluidising Velocity in Terms of Terminal Falling Velocity

The minimum fluidising velocity, u_{mf}, may be expressed in terms of the free-falling velocity u_0 of the particles in the fluid. The Ergun equation (Eq. 9.11) relates the Galileo number Ga to the Reynolds number Re'_{mf} in terms of the voidage e_{mf} at the incipient fluidisation point.

In Chapter 6, relations are given that permit the calculation of $Re'_0(u_0 d\rho/\mu)$, the particle Reynolds number for a sphere at its terminal falling velocity u_0, also as a function of the Galileo number. Thus, it is possible to express Re'_{mf} in terms of Re'_0 and u_{mf} in terms of u_0.

For a spherical particle the Reynolds number Re'_0 is expressed in terms of the Galileo number Ga by Eq. (6.40) which covers the whole range of values of Re' of interest. This takes the form:

$$Re'_0 = \left(2.33 Ga^{0.018} - 1.53 Ga^{-0.016} \right)^{13.3} \tag{9.21}$$

Eq. (9.21) applies when the particle motion is not significantly affected by the walls of the container; that is, when d/d_t tends to zero.

Thus, for any value of Ga, Re'_0 may be calculated from Eq. (9.21) and Re'_{mf} from Eq. (9.11) for a given value of e_{mf}. The ratio $Re'_0/Re'_{mf}(=u_0/u_{mf})$ may then be plotted against Ga with emf as the parameter. Such a plot is given in Fig. 9.5 which includes some experimental data from a number of sources including Refs. [14–18]. Some scatter is evident, largely attributable to the fact that the diameter of the vessel (d_t) was not always sufficiently large compared with that of the particle. Nevertheless, it is seen that the experimental results straddle the curves covering a range of values of e_{mf} from about 0.38 to 0.42. The agreement between the experimental and calculated values is quite good, especially in view of the uncertainty of the actual values of e_{mf} in the experimental work, and the fact that the Ergun equation does not necessarily give an accurate prediction of pressure drop in a fixed bed, especially near the incipient fluidisation points.

It is seen in Chapter 6 that it is also possible to express Re'_0 in terms of Ga by means of three simple equations, each covering a limited range of values of Ga (Eqs 6.37, 6.38, and 6.39) as follows:

$$Ga = 18Re'_0 \quad (Ga < 3.6) \tag{9.22}$$

$$Ga = 18Re'_0 + 2.7Re'^{1.687}_0 \quad (3.6 < Ga < 10^5) \tag{9.23}$$

$$Ga = \frac{1}{3}Re'^2_0 \quad (Ga > c.10^5) \tag{9.24}$$

It is convenient to use Eqs (9.22) and (9.24) as these enable very simple relations for Re'_0/Re'_{mf} to be obtained at both low and high values of Ga.

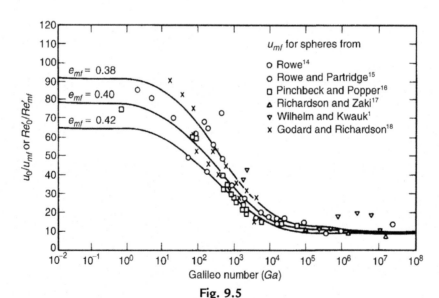

Fig. 9.5

Ratio of terminal falling velocity to minimum fluidising velocity, as a function of Galileo number.

Taking a typical value of e_{mf} of 0.4, the relation between Re'_{mf} and Ga is given by Eq. (9.13). For low values of Re'_{mf} (<0.003) and of Ga (<3.6), the first term may be neglected and:

$$Re'_{mf} = 0.000712 Ga \qquad (9.25)$$

Eq. (9.22) gives:

$$Re'_0 = 0.0556 Ga \qquad (9.26)$$

Combining Eqs (9.25) and (9.26):

$$\frac{Re'_0}{Re'_{mf}} = \frac{u_0}{u_{mf}} = 78 \qquad (9.27)$$

Again, for high values of Re'_{mf} ($>\sim 200$) and Ga ($>10^5$), Eq. (9.13) gives:

$$Re'_{mf} = 0.191 Ga^{1/2} \qquad (9.28)$$

Eq. (9.24) gives:

$$Re'_0 = 1.732 Ga^{1/2} \qquad (9.29)$$

Thus:

$$\frac{Re'_0}{Re'_{mf}} = \frac{u_0}{u_{mf}} = 9.1 \qquad (9.30)$$

This shows that u_0/u_{mf} is much larger for low values of Ga,—generally obtained with small particles—than with high values. For particulate fluidisation with liquids, the theoretical range of fluidising velocities is from a minimum of u_{mf} to a maximum of u_0. It is thus seen that there is a far greater range of velocities possible in the streamline flow region. In practice, it is possible to achieve flow velocities greatly in excess of u_0 for gases, because a high proportion of the gas can pass through the bed as bubbles and effectively bypass the particles.

9.2 Liquid–Solids Systems

9.2.1 Bed Expansion

Liquid-fluidised systems are generally characterised by the regular expansion of the bed that takes place as the velocity increases from the minimum fluidisation velocity to the terminal falling velocity of the particles. The general relation between the velocity and volumetric concentration, or voidage, is found to be similar to that between sedimentation velocity and concentration for particles in a suspension. The two systems are hydrodynamically similar in that in the fluidised bed the particles undergo no net movement and are maintained in suspension by the upward flow of liquid, whereas in a sedimenting suspension the particles move downwards and the only flow of liquid is the upward flow of that liquid, which is displaced by the settling particles. Richardson and Zaki[17] showed that, for sedimentation or fluidisation of uniform particles:

$$\frac{u_c}{u_i} = e^n = (1 - C)^n \tag{9.31}$$

where u_c is the observed sedimentation velocity or the empty tube fluidisation velocity; u_i is the corresponding velocity at infinite dilution $(C \rightarrow 0)$; e is the voidage of the system; C is the volumetric fractional concentration of solids; and n is an index.

The existence of a relationship of the form of Eq. (9.31) had been established way back in 1948 by Wilhelm and Kwauk,[1] who fluidised particles of glass, sand, and lead shot with water. On plotting the particle Reynolds number against the bed voidage using logarithmic scales, good straight lines were obtained over the range of conditions for which the bed was fluidised. A similar equation had previously been given by Lewis and Bowerman.[19]

Eq. (9.31) is similar to Eq. (8.76) for a sedimenting suspension. Values of the index n range from 2.4 to 4.8 and are the same for sedimentation and for fluidisation at a given value of the Galileo number Ga. These may be calculated using Eq. (9.32), which is identical to Eq. (8.84) in Chapter 8:

$$\frac{(4.8 - n)}{(n - 2.4)} = 0.043 Ga^{0.57} \left[1 - 1.24 \left(\frac{d}{d_t} \right)^{0.27} \right] \tag{9.32}$$

Richardson and Zaki[17] found that u_i corresponded closely to u_0, the free settling velocity of a particle in an infinite medium, for work on sedimentation as discussed in Chapter 8, although u_i was somewhat less than u_0 in fluidisation. The following equation for fluidisation was presented:

$$\log_{10} u_0 = \log_{10} u_i + \frac{d}{d_t} \tag{9.33}$$

The difference is likely to be attributed to the fact that d/d_t was very small in the sedimentation experiments. Subsequently, Khan and Richardson[20] have proposed the following relation to account for the effect of the walls of the vessel in fluidisation:

$$\frac{u_i}{u_0} = 1 - 1.15 \left(\frac{d}{d_t} \right)^{0.6} \tag{9.34}$$

If logarithmic co-ordinates are used to plot the voidage e of the bed against the superficial velocity u_c (Fig. 9.6), the resulting curve can be represented approximately by two straight lines joined by a short transitional curve. At low velocities, the voidage remains constant, corresponding

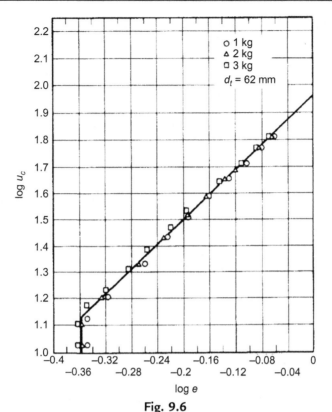

Fig. 9.6

Relation between fluid velocity (u_c) and voidage (e) for the fluidisation of 6.4 mm steel spheres in water.

to that of the fixed bed, and for the fluidised state there is a linear relation between $\log u_c$ and $\log e$. The curve shown in Fig. 9.6 refers to the fluidisation of steel spheres in water. It should be noted that, whereas in the absence of channelling the pressure drop across a bed of a given expansion is directly proportional to its depth, the fluidising velocity is independent of depth.

An alternative way of calculating the value of the index n in Eq. (9.31) for the expansion of particulately fluidised systems is now considered. Neglecting effects due to the container walls then:

$$\frac{u_c}{u_0} = \frac{Re'_c}{Re'_0} = e^n \tag{9.35}$$

where Re'_c is the Reynolds number $u_c d\rho/\mu$.

Taking logarithms:

$$n = \frac{\log(u_c/u_0)}{\log e} = \frac{-\log(Re'_0/Re'_c)}{\log e} \tag{9.36}$$

On the assumption that Eq. (9.31) may be applied at the point of incipient fluidisation:

$$n = \frac{\log\left(u_{mf}/u_0\right)}{\log e_{mf}} = \frac{-\log\left(Re'_0/Re'_{mf}\right)}{\log e_{mf}}$$ (9.37)

For a typical value of e_{mf} of 0.4, Re'_{mf} is given by Eq. (9.14). Furthermore, Re'_0 is given by Eq. (9.21). Substitution into Eq. (9.37) then gives:

$$n = 2.51 \log\left\{\frac{\left(1.83Ga^{0.018} - 1.2Ga^{-0.016}\right)^{13.3}}{\sqrt{\left(1 + 5.53 \times 10^{-5}Ga\right)} - 1}\right\}$$ (9.38)

Eq. (9.38), which applies to small values of d/d_t, is plotted in Fig. 9.7, together with experimental points from the literature, annotated according to the d/d_t range, which is applicable.[20] The scatter, and the low experimental values of n, are attributable partly to the wider range of d/d_t values covered and also inaccuracies in the experimental measurements, which are obtained from the results of a number of workers. For $e_{mf} = 0.43$, the calculated values of n are virtually unchanged over the range $10 < Ga < 10^5$.

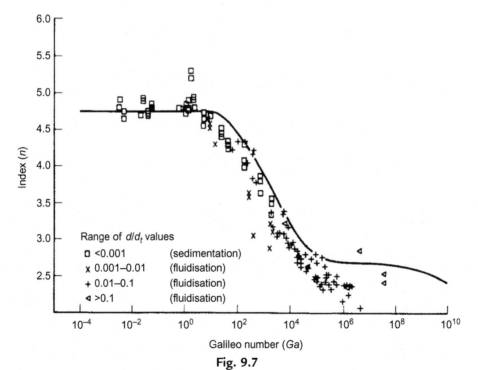

Fig. 9.7

Comparison of values of the index n calculated from Eq. (9.37) with experimental data.

An alternative method of calculating the value of Re'_{mf} (and hence u_{mf}) is to substitute for Re'_0 from Eq. (9.21) into Eq. (9.35), and to set the voidage e equal to its value e_{mf} at the minimum fluidising velocity.

In this way:

$$Re'_{mf} = \left(2.33Ga^{0.018} - 1.53Ga^{0.016}\right)^{13.3} e_{mf}^n \tag{9.39}$$

where n is given by Eq. (9.32).

The same procedure may be adopted for calculating the minimum fluidising for a shear-thinning non-Newtonian fluid which exhibits *power-law* behaviour, although it is necessary to use the modified Reynolds number $(Re_1)_n$ given in Chapter 7, Eq. (7.28).

For inelastic fluids exhibiting power-law behaviour, the bed expansion that occurs as the velocity is increased above the minimum fluidising velocity follows a similar pattern to that obtained with a Newtonian liquid, with the exponent in Eq. (9.31) differing by no more than about 10%. There is some evidence, however, that with viscoelastic polymer solutions, the exponent may be considerably higher. Reference may be made to the work by Srinivas and Chhabra[21] for further details.

Example 9.4

Glass particles of 4 mm diameter are fluidised by water at a velocity of 0.25 m/s. What will be the voidage of the bed?

The density of glass $= 2500\,\text{kg/m}^3$, the density of water $= 1000\,\text{kg/m}^3$, and the viscosity of water $= 1\,\text{mNs/m}^2$.

Solution

Galileo number for particles in water:

$$Ga = \frac{d^3 \rho (\rho_s - \rho)g}{\mu^2} \quad (equation\ 9.9)$$

$$= \frac{(4 \times 10^{-3})^3 \times 1000 \times 1500 \times 9.81}{(1 \times 10^{-3})^2} = 9.42 \times 10^5$$

Reynolds number Re'_0 at terminal falling velocity is given by Eq. (9.21):

$$Re'_0 = \left(2.33Ga^{0.018} - 1.53Ga^{-0.016}\right)^{13.3}$$

Thus:

$$u_0 = 1800 \left(\frac{1 \times 10^{-3}}{4 \times 10^{-3} \times 1000}\right) = 0.45\,\text{m/s}$$

The value of n in Eq. (9.31) is given by Eq. (9.32) for small values of d/d_t as:

$$\frac{(4.8-n)}{(n-2.4)} = 0.043Ga^{0.57} = 109.5$$

$$n = 2.42$$

The voidage e at a velocity of 0.25 m/s is then given by Eq. (9.31) as:

$$\frac{0.25}{0.45} = e^{2.42}$$

and:

$$e = \underline{\underline{0.784}}$$

9.2.2 Non-uniform Fluidisation

Regular and even expansion of the bed does not always occur when particles are fluidised by a liquid. This is particularly so for solids of high densities, and non-uniformities are most marked with deep beds of small particles. In such cases, there are significant deviations from the relation between bed voidage and velocity predicted by Eq. (9.31).

Stewart (referred to in Stewart and Davidson[22]) has shown that well-defined bubbles of liquid and slugs are formed when tungsten beads (density 19,300 kg/m³, and particle sizes 776 and 930 μm) are fluidised with water. Simpson and Rodger,[23] Harrison et al.,[24] Lawther and Berglin,[25] and Richardson and Smith,[26] amongst others, have observed that lead shot fluidised by water gives rise to non-uniform fluidised beds. Anderson and Jackson[27] have shown that this system would be expected to be transitional in behaviour. Hassett[28] and Lawson and Hassett[29] have also noted instabilities and non-uniformities in liquid–solids systems, particularly in beds of narrow diameter. Similar observations have also been made by Cairns and Prausnitz,[30] Kramers et al.,[31] and Reuter,[32] who have published photographs of bubbles in liquid–solids systems. Gibilaro et al.[33] have made experimental measurements of one-dimensional waves in liquid–solids fluidised beds. Bailey[34] has studied the fluidisation of lead shot with water and has reported the occurrence of non-uniformities, though not of well-defined bubbles. He has also shown that the logarithmic plots of voidage against velocity are no longer linear and that the deviations from the line given by Eq. (9.31) increase with:

(a) increase in bed weight per unit area; and
(b) decrease in particle size.

The deviation passes through a maximum as the velocity is increased, as shown in Fig. 9.8.

The importance of particle density in determining the nature of fluidised systems is well established, and an increase in density generally results in a less uniform fluidised system. It is, however, surprising that a reduction in particle size should also cause increased deviations from

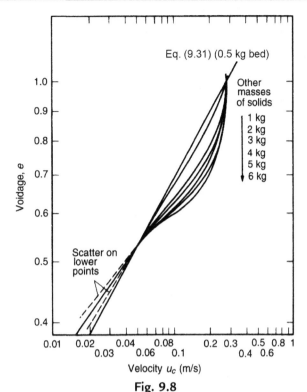

Fig. 9.8

Bed expansion for the fluidisation of 0.5–0.6 mm lead shot in water in a 100 mm tube.

the ideal behaviour. It may be noted from Fig. 9.8 that, over a wide range of liquid velocities, the mean voidage of the bed is less than that predicted by Eq. (9.31). This may be explained in terms of part of the fluid taking a low resistance path through the bed, remaining there for less than the average residence time, and not therefore contributing fully to the expansion of the bed. The effect of partial channelling will certainly be more marked with fine solids than with coarse, since the ratio of the resistance of the bed to that of the channel will be much greater, and a comparatively small channel will accommodate the flow of a proportionately larger amount of fluid.

A simple model may be developed to represent what is happening under these circumstances. The bed may be considered to be divided into two portions, one with uniformly dispersed particles and the other consisting of fluid channels. It is assumed that the voidage of the region of uniformly dispersed particles is determined, according to Eq. (9.31), by the flowrate through that part of the bed. If, then, a fraction f of the fluid introduced to the bottom of the bed flows through the channels at a velocity u_f, it can be readily shown that the relation between the mean voidage of the bed e and the mean superficial velocity of the liquid u_c is given by:

$$e = f\frac{u_c}{u_f} + \left(\frac{u_c}{u_i}(1-f)\right)^{1/n}\left(1 - f\frac{u_c}{u_f}\right)^{1-(1/n)} \tag{9.40}$$

Eq. (9.40) gives the relation between all possible corresponding values of u_f and f. For a typical experiment on the fluidisation of 5 kg of lead shot ($d = 0.55$ mm) in a 100 mm diameter tube with water flowing at a mean superficial velocity of 0.158 m/s, the measured voidage of the bed was 0.676. This would give a value of $f = 0.53$ for a channel velocity $u_f = 1.58$ m/s, or 0.68 for a channel velocity $u_f = 0.80$ m/s.

Local variations in voidage in a liquid–solids fluidised bed have also been observed by Volpicelli et al.[35] who fluidised steel, aluminium and plastic beads with water and glycerol in a column only 3.55 mm thick, using particles ranging from 2.86 to 3.18 mm diameter, which effectively gave a bed one particle thick (i.e., significant wall effects). This system facilitated observation of the flow patterns within the bed. It was found that the velocity–voidage relationship was of the same form as Eq. (9.31), but that it was necessary to use the actual measured falling velocity of the particle in the apparatus to represent u_i. Non-uniformities within the bed were not apparent at voidages near unity or near u_{mf}, but rose to a maximum value somewhere in between; this is generally in line with Bailey's work (Fig. 9.8). The local variations of bed voidage were found to be highly dependent on the arrangement of the liquid distributor used at the bottom of the column.

Subsequent work by Foscolo et al.[36] has shown that instabilities can also arise in the fluidisation of particles where densities are only slightly greater than that of the fluidising liquid.

9.2.3 Segregation in Beds of Particles of Mixed Sizes

When a bed consists of particles with a significant size range, stratification occurs, with the largest particles forming a bed of low voidage near the bottom and the smallest particles forming a bed of high voidage near the top. If the particles are in the form of sharp-cut size fractions, segregation will be virtually complete with what is, in effect, a number of fluidised beds of different voidages, stacked one above the other. If the size range is small, there will be a continuous variation in both composition and concentration of particles throughout the depth of the bed.

It has been shown experimentally[17] that a mixture of equal masses of 1 mm and 0.5 mm particles when fluidised by water will segregate almost completely over the whole range of velocities, for which particles of each size will, on their own, form a fluidised bed. Wen and Yu[6] have shown that this behaviour is confined to mixtures in which the ratio of the minimum fluidising velocities of the components exceeds about 2. The tendency for classification has been examined experimentally and theoretically by several workers, including Jottrand,[37] Pruden and Epstein,[38] Kennedy and Bretton,[39] Al-Dibouni and Garside,[40] Juma and Richardson,[41] Gibilaro et al.,[42] Moritomi et al.,[43] and others.[44, 45]

In a mixture of large and small particles fluidised by a liquid, one of three situations may exist:

(a) Complete (or virtually complete) segregation, with a high voidage bed of small particles above a bed of lower voidage containing the large particles.
(b) Beds of small and of large particles as described in (a), but separated by a transition region in which the proportion of small particles and the voidage both increase from bottom to top.
(c) A bed in which there are no fully segregated regions, but with the transition region described in (b) extending over the whole extent of the bed. If the range of particle sizes in the bed is small, this transition region may be of nearly constant composition, with little segregation occurring.

Indeed, Escudie et al.[45] introduced a range of terms including homogeneous mixing (for nearly perfect mixing) and heterogeneous mixing, in line with the preceding description, to describe the complex behaviour of fluidised systems.

At any level in the transition region, there will be a balance between the mixing effects attributable to (a) axial dispersion and to (b) the segregating effect, which will depend on the difference between the interstitial velocity of the liquid and that interstitial velocity which would be required to produce a bed of the same voidage for particles of that size on their own. On this basis, a model may be set up to give the vertical concentration profile of each component in terms of the axial mixing coefficients for the large and the small particles.

Experimental measurements[41] of concentration profiles within the bed have been made using a pressure transducer attached to a probe, whose vertical position in the bed could be varied. The voidage e of the bed at a given height may then be calculated from the local value of the pressure gradient using Eq. (9.1), from which:

$$-\frac{dP}{dl} = (1-e)(\rho_s - \rho)g \tag{9.41}$$

It has been established that the tendency for segregation increases, not only with the ratio of the sizes of particles (largest:smallest), but also as the liquid velocity is raised. Thus, for a given mixture, there is more segregation in a highly expanded bed than in a bed of low voidage.

Binary mixtures—particles differing in both size and density

The behaviour of a fluidised bed consisting of particles differing in both size and density can be extremely complex. This situation may be illustrated by considering the simplest case—the fluidisation of a binary mixture of spherical particles. If the heavy particles are also the larger, they will always constitute the denser bottom layer. On the other hand, if the mixture consists of small, high-density particles H, and larger particles of lower density L, the relative densities of the two layers is a function of the fluidising velocity. In either case, for both species of solids to be fluidised simultaneously, the superficial velocity u_c of the liquid must lie between the minimum fluidising velocity u_{mf} and the terminal falling velocity u_0 for each of the solids of $u_{mf} < u_c < u_0$. In general, segregation tends to occur, resulting in the formation of two fluidised beds of different densities, possibly separated by a transition zone, with the bed of higher density forming the bottom layer. In all cases, the interface between the two beds may be diffuse, as a result of the effect of dispersion.

The densities of the two beds are then given by:

$$\rho_{bH} = (1 - e_H)\rho_{sH} + e_H\rho = \rho_{sH} - e_H(\rho_{sH} - \rho) \tag{9.42a}$$

$$\rho_{bL} = (1 - e_L)\rho_{sL} + e_L\rho = \rho_{sL} - e_L(\rho_{sL} - \rho) \tag{9.42b}$$

where the suffix L refers to the light particles and the suffix H to the heavy particles.

Applying Eq. (9.31) to each fluidised bed gives:

$$\frac{u_c}{u_{0H}} = e_H^{n_H} \quad \text{and} \quad \frac{u_c}{u_{0L}} = e_L^{n_L}$$

Noting that the superficial velocity u_c is the same in each case, and assuming that $n_H \approx n_L \approx n$, then:

$$\frac{e_L}{e_H} = \left(\frac{u_{0H}}{u_{0L}}\right)^{1/n} \tag{9.43}$$

As the fluidising velocity is progressively increased, the voidage of both beds increases, although not generally at the same rate. Two cases are considered here:

(a) $u_{0H} > u_{0L}$

From Eq. (9.43), $e_H < e_L$ and therefore, from Eqs (9.42a) and (9.42b), $\rho_{bH} > \rho_{bL}$ at all fluidising velocities u_c, and the heavy particles will always form the bottom layer.

(b) $u_{0H} < u_{0L}$

If, with increase in velocity, the density of the upper zone decreases more rapidly than that of the bottom zone, the two beds will maintain the same relative orientation. If the reverse situation applies, there may be a velocity u_{INV} where the densities of the two layers become equal, with virtually complete mixing of the two species taking place. Any further increase in velocity above u_{INV} then causes the beds to invert, as shown diagrammatically in Fig. 9.9A.

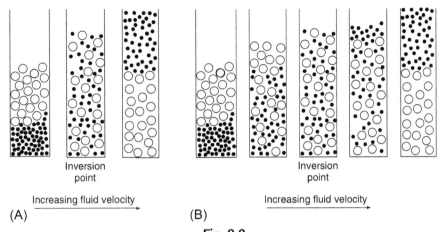

Inversion point

Inversion point

Increasing fluid velocity

Increasing fluid velocity

(A)

(B)

Fig. 9.9

Bed inversion (A) complete segregation (B) complete and partial segregation.[43]

The relative rates at which the bed densities change as the fluidising velocity is increased may be obtained by differentiating Eqs (9.42a) and (9.42b) with respect to u_c, and dividing to give:

$$r = -\frac{d\rho_{bH}}{du_c} \bigg/ -\frac{d\rho_{bL}}{du_c} = \frac{d\rho_{bH}}{d\rho_{bL}} = \frac{(\rho_{sH} - \rho)}{\rho_{sL} - \rho} \left(\frac{u_{0L}}{u_{0H}}\right)^{1/n} = \frac{(\rho_{sH} - \rho) e_H}{(\rho_{sL} - \rho) e_L} \qquad (9.44)$$

As $e_H > e_L$ and $p_{sH} > \rho_{sL}$, then from Eq. (9.44), r, which is independent of the fluidising velocity, must be greater than unity. It is thus the bed of heavy particles that expands more rapidly as the velocity is increased, and which must, therefore, form the bottom layer at low velocities if inversion is possible. That is, it is the small, heavy particles that move from the lower to the upper layer, and vice versa, as the velocity is increased beyond the inversion velocity u_{Inv}. Richardson and Afiatin[46] have analysed the range of conditions over which segregation of spherical particles can occur, and these are shown diagrammatically in Fig. 9.10 for the Stokes' law region (A) and for the Newton's law region (B).

It has been observed by several workers, including by Moritomi et al.[43] and Epstein and Pruden,[47] that a sharp transition between two mono-component layers does not always occur and that, on each side of the transition point, there may be a condition where the lower zone consists of a mixture of both species of particles, the proportion of heavy particles becoming progressively smaller as the velocity is increased. This situation, depicted in Fig. 9.9B, can arise when, at a given fluidising velocity, there is a stable two-component bed that has a higher density than a bed composed of either of the two species on its own. Fig. 9.11, taken from the work of Epstein and Pruden,[47] shows how the bed densities for the two mono-component layers change as the liquid velocity is increased, with point C then defining the inversion point when complete segregation can take place. Between points A and D

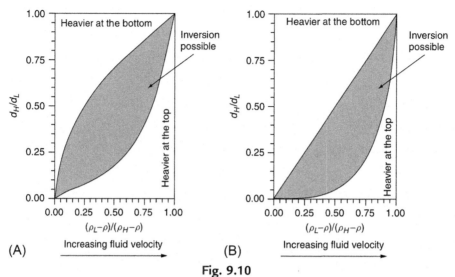

Fig. 9.10
The possibility of inversion (A) Stokes' law region (B) Newton's law region.[46]

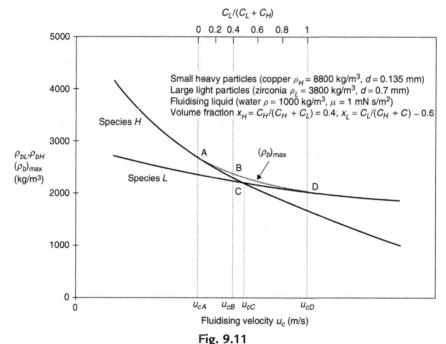

Fig. 9.11
Bed densities as a function of fluidising velocity, showing the mixed particle region.[47]

(corresponding to velocities u_{cA} and u_{cB}), however, a two-component hybrid bed (represented by curve ABD) may be formed, which has a density greater than that of either mono-component bed over this velocity range. In moving along this curve from A to D, the proportion of light particles in the lower layer decreases progressively from unity to zero, as shown on the top scale of the diagram. This proportion is equal to that in the total mix of solids at point B, where the whole bed is of uniform composition, and the velocity u_{cB} therefore represents the effective inversion velocity.

If the flow of the fluidising liquid to a completely segregated bed is suddenly stopped, the particles will all then start to settle at a velocity equal to that at which they have been fluidised, because Eq. (9.31) is equally applicable to sedimentation and fluidisation.

Thus, since the voidages of the two beds will both be greater at higher fluidisation velocities, the subsequent sedimentation velocity will then also be greater. Particles in both beds will settle at the same velocity and segregation will be maintained. Eventually, two packed beds will be formed, one above the other. Thus, if the fluidising velocity is less than the transition (inversion) velocity, a packed bed of large, light particles will form above a bed of small, dense particles, and conversely, if the fluidising velocity is greater than the inversion velocity. Thus, fluidisation followed by sedimentation can provide a means of forming two completely segregated mono-component beds, the relative configuration of which depends solely on the liquid velocity at which the particles would have been fluidised.

9.2.4 Liquid and Solids Mixing

Kramers et al.[31] have studied longitudinal dispersion in the liquid in a fluidised bed composed of glass spheres of 0.5 mm and 1 mm diameter. A step change was introduced by feeding a normal solution of potassium chloride into the system. The concentration at the top of the bed was measured as a function of time by means of a small conductivity cell. On the assumption that the flow pattern could be regarded as longitudinal diffusion superimposed on piston flow, an eddy longitudinal diffusivity was calculated. This was found to range from 10^{-4} to 10^{-3} m^2/s, increasing with both voidage and particle size.

The movement of individual particles in a liquid–solid fluidised bed has been measured by Handley et al.,[48] Carlos,[49, 50] and Latif.[51] In all cases, the method involved fluidising transparent particles in a liquid of the same refractive index so that the whole system became transparent. The movement of coloured tracer particles, whose other physical properties were identical to those of the bed particles, could then be followed photographically.

Handley et al.[48] fluidised soda glass particles using methyl benzoate, and obtained data on the flow pattern of the solids and the distribution of vertical velocity components of the particles. It was found that a bulk circulation of solids was superimposed on their random movement. Particles normally tended to move upwards in the centre of the bed and downwards at the walls, following a circulation pattern, which was less marked in regions remote from the distributor.

Carlos[49, 50] and Latif[51] both fluidised glass particles in dimethyl phthalate. Data on the movement of the tracer particles, in the form of spatial coordinates as a function of time, were used as direct input to calculate the vertical, tangential, and radial velocities of the particles as a function of the location. When plotted as a histogram, the total velocity distribution was found to be of the same form as that predicted by the kinetic theory for the molecules in a gas. A typical result is shown in Fig. 9.12.[49] Effective diffusion or mixing coefficients for the particles were then calculated from the product of the mean velocity and mean free path of the particles, using the simple kinetic theory.

Solids mixing was also studied by Carlos[50] in the same apparatus, starting with a bed composed of transparent particles and a layer of tracer particles at the base of the bed. The concentration of particles in a control zone was then determined at various intervals of time after the commencement of fluidisation. The mixing process was described by a diffusion-type equation. This was then used to calculate the mixing coefficient. A comparison of the values of mixing coefficient obtained by the two methods then enabled the *persistence of velocity* factor to be calculated. A typical value of the mixing coefficient was 1.5×10^{-3} m^2/s for 9 mm glass ballotini fluidised at a velocity of twice the minimum fluidising velocity.

Latif[51] represented the circulation currents of the particles in a fluidised bed by plotting stream functions for the particles on the assumption that the particles could be regarded as behaving as a continuum. A typical result for the fluidisation of 6-mm glass particles by dimethyl phthalate is shown in Fig. 9.13; in this case, the velocity has been adjusted to give a bed voidage of 0.65. Because the bed is symmetrical about its axis, only the pattern over a radial slice is shown. It may be noted that the circulation patterns are concentrated mainly in the lower portion of the bed, with particles moving upwards in the centre and downwards at the walls. As the bed voidage is decreased, the circulation patterns tend to occupy progressively smaller portions of the bed, but there is a tendency for a small reverse circulation pattern to develop in the upper regions of the bed.

Further work on axial dispersion of particles has been carried out by Dorgelo et al.,[52] who used a random-walk approach.

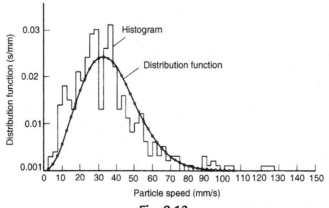

Fig. 9.12

Distribution of particle velocities in a fluidised bed.[49]

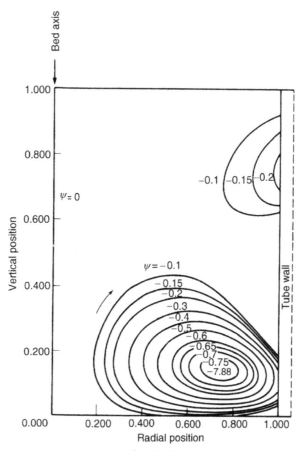

Fig. 9.13

Particle stream functions, $\psi(e=0.65)$ (Radial position is expressed as fraction of radial distance from centre line, and axial position as fraction of bed height measured from the bottom).[51]

9.3 Gas–Solids Systems

9.3.1 General Behaviour

In general, the behaviour of gas-fluidised systems is considerably more complex than that of liquid-fluidised systems, which exhibit a gradual transition from fixed bed to fluidised bed followed by particle transport, without a series of transition regions, and with bed expansion and pressure drop conforming reasonably closely to values calculated for ideal systems.

Part of the complication with gas–solid systems arises from the fact that the purely hydrodynamic forces acting on the particles are relatively small compared with the frictional forces between particles, electrostatic forces, and surface forces, which play a much more dominant role when the particles are very fine. As the gas velocity in a fluidised bed is increased, the system tends to go through various stages:

(a) *Fixed bed*, in which the particles remain in contact with one another (Fig. 9.1A) and the structure of the bed remains stable until the velocity is increased to the point where the pressure drop is equal to the weight per unit area of the particles (Fig. 9.1B).
(b) *Particulate* and regular predictable expansion over a limited range of gas velocities.
(c) A *bubbling* region characterised by a high proportion of the gas passing through the bed as bubbles, which cause rapid mixing in the dense particulate phase (Fig. 9.1C).
(d) A *turbulent* chaotic region in which the gas bubbles tend to coalesce and lose their identity.
(e) A region where the dominant pattern is one of *vertically upward transport of particles*—essentially, gas–solids transport or pneumatic conveying. This condition, sometimes referred to as *fast fluidisation*, lies outside the range of true fluidization (Fig. 9.1E).

9.3.2 Particulate Fluidisation

Although fine particles generally form fluidised beds more readily than coarse particles, surface-related forces tend to dominate over the hydrodynamic ones with very fine particles. It is very difficult to fluidise some very fine particles as they tend to form large stable conglomerates that are almost entirely bypassed by the gas. In some extreme cases, particularly with small diameter beds, the whole of the particulate mass may be lifted as a solid 'piston'. The uniformity of the fluidised bed is often critically influenced by the characteristics of the gas distributor or bed support. Fine mesh distributors are generally to be preferred to a series of nozzles at the base of the bed, although the former are generally more difficult to install in larger beds because they are less robust from a mechanical standpoint.

Good distribution of gas over the whole cross-section of the bed may often be difficult to achieve, although this is enhanced by ensuring that the pressure drop across the distributor is large compared with that across the bed of particles. In general, the quality of gas distribution improves with increased flowrate, because the pressure drop across the bed when it is fluidised is, theoretically, independent of the flowrate. The pressure drop across the distributor will increase, however, approximately in proportion to the square of the flowrate, and therefore the fraction of the total pressure drop that occurs across the distributor increases rapidly as the flowrate increases.

Apart from the non-uniformities that characterise many gas–solid fluidised beds, it is in the low fluidising-velocity region that the behaviour of the gas–solid and liquid–solid beds are most similar. At low gas flowrates, the bed may exhibit a regular expansion as the flowrate increases, with the relation between fluidising velocity and voidage following the form of Eq. (9.31), although, in general, the values of the exponent n are higher than those for liquid–solids systems, partly because particles have a tendency to form small agglomerates, thereby increasing the effective particle size. The range of velocities over which particulate expansion occurs is, however, quite narrow in most cases.

9.3.3 Bubbling Fluidisation

The region of particulate fluidisation usually comes to an abrupt end as the gas velocity is increased, with the formation of gas bubbles. These bubbles are usually responsible for the flow of almost all of the gas in excess of that flowing at the minimum fluidising velocity. If bed expansion has occurred before bubbling commences, the excess gas will be transferred to the bubbles, whilst the continuous phase reverts to its voidage at the minimum fluidising velocity and, in this way, it contracts. Thus, the expanded bed appears to be in a meta-stable condition, which is analogous to that of a supersaturated solution reverting to its saturated concentration when fed with small seed crystals, with the excess solute being deposited on to the seed crystals, which then increase in size as a result.

The upper limit of gas velocity for particulate expansion is termed the *minimum bubbling velocity*, u_{mb}. Determining this can present difficulties, as its value may depend on the nature of the distributor, on the presence of even tiny obstructions in the bed, and even on the immediate pre-history of the bed. The ratio u_{mb}/u_{mf}, which gives a measure of the degree of expansion which may be effected, usually has a high value for fine light particles and a low value for large dense particles.

For a cracker catalyst ($d = 55$ μm, density $= 950$ kg/m^3) fluidised by air, values of u_{mb}/u_{mf} of up to 2.8 have been found by Davies and Richardson[53]. During the course of this work it was found that there is a minimum size of bubble that is stable. Small bubbles injected into a non-bubbling bed tend to become assimilated in the dense phase, whilst, on the other hand, larger bubbles tend to grow at the expense of the gas flow in the dense phase. If a bubble larger than the critical size is injected into an expanded bed, the bed will initially expand by an amount equal to the volume of the injected bubble. When, however, the bubble breaks the surface, the bed will fall back below the level existing before injection and will therefore have acquired a reduced voidage.

Thus, the bubbling region, which is an important feature of beds operating at gas velocities in excess of the minimum fluidising velocity, is usually characterised by two phases: a continuous emulsion phase with a voidage approximately equal to that of a bed at its minimum fluidising velocity, and a discontinuous or bubble phase that accounts for most of the excess flow of gas. This is sometimes referred to as the *two-phase theory of fluidisation*.

The bubbles exert a very strong influence on the flow pattern in the bed and provide the mechanism for the high degree of mixing of solids that occurs in such systems. The properties and behaviour of the bubbles are described later in this section.

When the gas flowrate is increased to a level at which the bubbles become very large and unstable, the bubbles tend to lose their identity and the flow pattern changes to a chaotic form without well-defined regions of high and low concentrations of particles. This is commonly described as the *turbulent* region, which has, until fairly recently, been the subject of relatively few studies.[54–56] The transition from the bubbling to the turbulent fluidising regime is characterised by large pressure fluctuations, which, after reaching peak values, fall and eventually level off in the fully turbulent region. This transition is strongly influenced by the particle size and density.[55] Generally, the corresponding transition velocity increases with both particle size and density (i.e., particle inertia).[54] As noted above, clusters and strands of particles, as well as voids of elongated and irregular shapes, characterise the structure of the bed. Bed voidage in such a bed is high and, from the indistinct bed surface, clusters and strands of particles are continually ejected into the free board. Smaller particles in the clusters are conveyed out of the bed. Therefore, cyclones and diplegs are installed to prevent the loss of solids. A detailed review of gas and solid mixing in such systems has been provided by Bi et al.[56]

Categorisation of solids

The ease with which a powder can be fluidised by a gas is highly dependent on the properties of the particles. Whilst it is not possible to forecast just how a given powder will fluidise without carrying out tests on a sample, it is possible to indicate some trends. In general, fine, low-density particles fluidise more evenly than large, dense ones, provided that they are not so small that the London–van der Waals attractive forces are great enough for the particles to adhere together strongly. For very fine particles, these attractive forces can be three or more orders of magnitude greater than their weight. Generally, the more nearly spherical the particles, the better they will fluidise. In this respect, long needle-shaped particles or fibres are the most difficult to fluidise. Particles of mixed sizes will usually fluidise more evenly than those of a uniform size. Furthermore, the presence of a small proportion of fines will frequently aid the fluidisation of coarse particles by coating them with a 'lubricating' layer.

In classifying particles into four groups, Geldart[57] has used the following criteria:

(a) Whether or not, as the gas flowrate is increased, the fluidised bed will expand significantly before bubbling takes place. This property may be quantified by the ratio u_{mb}/u_{mf}, where u_{mb} is the minimum velocity at which bubbling occurs. This assessment can only be qualitative as the value of u_{mb} is very critically dependent on the conditions under which it is measured.

(b) Whether the rising velocity of the majority of the bubbles, is greater or less than the interstitial gas velocity. The significance of this factor is discussed in Section 9.3.5.

(c) Whether the adhesive forces between particles are so great that the bed tends to channel rather than to fluidise. Channelling depends on a number of factors, including the degree to which the bed has consolidated and the condition of the surface of the particles at the time.

With powders that channel badly, it is sometimes possible to initiate fluidisation by mechanical stirring, as discussed in Section 9.3.4.

The classes into which powders are grouped are given in Table 9.2, which is taken from the work of Geldart,[57] and shown in Fig. 9.14. In this figure, different types of particles are located approximately on a particle density–particle size chart.

The effect of pressure

The effect of pressure on the behaviour of a fluidised bed is important because many industrial processes, including fluidised bed combustion, which is discussed in Section 9.8.4., are carried out at elevated pressures. Several workers have reported measurements of bed expansion as a function of gas rate for elevated pressures when very much higher values of the ratio u_{mb}/u_{mf} may be obtained.[23, 58–60]

Because minimum fluidising velocity is not very sensitive to the pressure in the bed, much greater mass flowrates of gas may be obtained by increasing the operating pressure.

The influence of pressure over the range 100–1600 kPa, on the fluidisation of three grades of sand in the particle size range 0.3 to 1 mm has been studied by Olowson and Almstedt[61] and

Table 9.2 Categorisation of powders in relation to fluidisation characteristics[57]

	Typical Particle Size (μm)	Fluidisation/Powder Characteristics	Examples of Materials
Group A	30–100	Particulate expansion of bed will take place over significant velocity range. Small particle size and low density ($\rho_s < 1400\,kg/m^3$).	Cracker catalyst
Group B	100–800	Bubbling occurs at velocity $>u_{mf}$. Most bubbles have velocities greater than interstitial gas velocity. No evidence of maximum bubble size.	Sand
Group C	20	Fine cohesive powders, difficult to fluidise and readily form channels.	Flour, fine silica
Group D	1000	All but largest bubbles rise at velocities less than interstitial gas velocity. Can be made to form spouted beds. Particles large and dense.	Wheat, metal shot

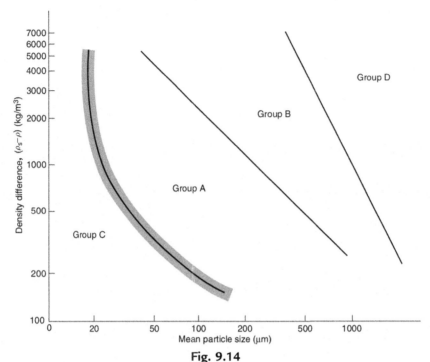

Fig. 9.14

Powder classification diagram for fluidisation by air at ambient conditions.[57]

it was shown that the minimum fluidising velocity became less as the pressure was increased. The effect, most marked with the coarse solids, was in agreement with that predicted by standard relations such as Eq. (9.14). For fine particles, the minimum fluidising velocity is independent of gas density (Eq. (9.5) with $\rho_s \gg \rho$), and hence of pressure.

Tapered beds

Where there is a wide range of particle sizes present in the powder, fluidisation will be more even in a bed that is tapered so as to provide the minimum cross-section at the bottom. If the pressure gradient is low and the gas does not, therefore, expand significantly, the velocity will decrease in the direction of flow. Coarse particles, which will then tend to become fluidised near the bottom of the bed, assist in the dispersion of the fluidising gas. At the same time, the carry-over of fines from the top will be reduced because of the lower velocity at the exit.

When deep beds of solids are fluidised by a gas, the use of a tapered bed can counterbalance the effects of gas expansion. For example, the pressure drop over a 5 m deep bed of solids of density $4000 \, \text{kg/m}^3$ is about $10^5 \, \text{N/m}^2$. Thus, with atmospheric pressure at the outlet, the volumetric flowrate will double from the bottom to the top of an isothermal cylindrical bed. If the area at the outlet is twice that at the base, the velocity will be maintained approximately constant throughout.

The effect of magnetic and electrical fields

Magnetic particles may form much more stable beds when subjected to a magnetic field. Saxena and Shrivastava[62] have examined the complex behaviour of spherical steel particles of a range of sizes when subjected to fields of different strengths, considering in particular the bed pressure drop, the quality of fluidisation and the structure of the surface of the bed.

Dielectric particles show a reduced tendency for bubbling and a larger range of velocities over which particulate expansion occurs when an alternating electrical field is applied.

The effect of baffles

It is possible substantially to eliminate the fluctuations that are characteristic of beds of coarse solids by incorporating baffles into the bed. The nature and arrangement of the baffles is critical, and it is generally desirable to avoid downward-facing horizontal surfaces because these can give rise to regimes of defluidisation by blocking the upward flow of gas. For this reason, horizontal tubes immersed in a fluidised bed tend to exhibit low values of heat transfer coefficients because of the partial defluidisation that occurs.

9.3.4 The Effect of Stirring

Stirring can be effective in improving the quality of fluidisation. In particular, if the agitator blades lift the particles as they rotate, fluidisation can be effected at somewhat lower gas velocities. In addition, the fluidisation of very fine particles that tend to aggregate can be substantially improved by a slow stirrer fitted with blades that provide only a small clearance at the gas distributor. Godard and Richardson[63] found that it was possible to fluidise fine silica particles ($d = 0.05$ μm) only if the stirrer was situated <10 mm from the support. In the absence of stirring, the fluidising gas passed through channels in the bed and the solids were completely unfluidised. Once fluidisation had been established, the stirrer could be stopped and uniform fluidisation would then be maintained indefinitely. With such fine solids, a high degree of bed expansion could be achieved with u_{mb}/u_{mf} ratios up to 18. Over this range, Eq. (9.31) is followed with a value of n ranging from 6.7 to 7.5, as compared with a value of 3.7 calculated from Eq. (9.32) for a liquid at the same value of the Galileo number.

The rotation of a paddle in a fluidised bed provides a means of measuring an effective viscosity of the bed in terms of the torque required to rotate the paddle at a controlled speed.[64] A good discussion concerning the viscosity of fluidised systems is available in the literature.[65]

9.3.5 Properties of Bubbles in the Bed

The formation of bubbles at orifices in a fluidised bed, including measurement of their size, the conditions under which they will coalesce with one another, and their rate of rise in the bed has been investigated. Davidson et al.[66] injected air from an orifice into a fluidised bed composed of particles of sand (0.3–0.5 mm) and glass ballotini (0.15 mm) fluidised by air at a velocity just above the minimum required for fluidisation. By varying the depth of the injection point from the free surface, it was shown that the injected bubble rises through the bed with a constant velocity, which is dependent only on the volume of the bubble. In addition, this velocity of rise corresponds with that of a spherical cap bubble in an inviscid liquid of zero surface tension, as determined from the equation of Davies and Taylor[67]:

$$u_b = 0.792 V_B^{1/6} g^{1/2} \tag{9.45}$$

The velocity of rise is independent of the velocity of the fluidising air and of the properties of the particles making up the bed. Eq. (9.45) is applicable, provided that the density of the gas in the bubbles may be neglected in comparison with the density of the solids. In other cases, the expression must be multiplied by a factor of $(1 - \rho/\rho_c)^{1/2}$.

In bubbling beds, Davidson and Harrison[68] used the two phase concept and suggested a correction to Eq. (9.45) by adding the excess velocity above the minimum fluidisation velocity to the value obtained from Eq. (9.45). In case the bubble volume is not known, some of the available methods[69–72] can be used to estimate the bubble diameter and hence the volume. Furthermore, if the estimated bubble diameter is larger than 40% of the column diameter, the column is said to operate in the slugging regime (Fig. 9.10), and one must use the slug velocity—approximately 50% of the value predicted by Eq. (9.45)—instead of the bubble velocity.[22]

Harrison et al.[24] applied these results to the problem of explaining why gas and liquid fluidised systems behave differently. Photographs of bubbles in beds of lead shot fluidised with air and with water have shown that an injected bubble is stable in the former case, though it tends to collapse in the latter. The water–lead system tends to give rise to inhomogeneities, and it is therefore interesting to note that bubbles, as such, are apparently not stable. As the bubble rises in a bed, internal circulation currents are set up because of the shear stresses existing at the boundary of the bubble. These circulation velocities are of the same order of magnitude as the rising velocity of the bubble. If the circulation velocity is appreciably greater than the falling velocity of the particles, the bubble will tend to draw in particles at the wake and will, therefore, be destroyed. On the other hand, if the rising velocity is lower, particles will not be drawn in at the wake and the bubble will be stable.

As a first approximation, a bubble is assumed to be stable if its rising velocity is less than the free-falling velocity of the particles, and therefore, for any system, the limiting size of stable bubble may be calculated using Eq. (9.45). If this is of the same order of size as the particle diameter, the bubble will not readily be detected. On the other hand, if it is more than about 10 times the particle diameter, it will be visible and the system will be seen to contain stable bubbles. On the basis of this argument, in gases, large bubbles are generally stable, whereas in liquids, the largest size of stable bubble is comparable with the diameter of the particles. It should be possible to achieve fluidisation free of bubbles with very light particles by using a gas of high density. Leung[59] succeeded in reaching this condition by fluidising hollow phenolic microballoons at pressures of about 4500 kN/m^2. It was found possible to form stable bubbles with glycerine-water mixtures and lead shot of 0.77 mm particle size. This transitional region has also been studied by Simpson and Rodger.[23]

Harrison and Leung[73] have shown that the frequency of formation of bubbles at an orifice (size range 1.2–25 mm) is independent of the bed depth, the flowrate of gas, and the properties of the particles constituting the continuous phase, although the frequency of formation depends on the injection rate of gas, tending to a frequency of 18–21 s^{-1} at high flowrates.

Subsequently,[74] it has been shown that a wake extends for about 1.1 bubble diameters behind each rising bubble. If a second bubble follows in this wake, its velocity is increased by an amount equal to the velocity of the leading bubble, and in this way, coalescence takes place.

The differences between liquid and gas fluidised systems have also been studied theoretically by Jackson,[2] who showed that small discontinuities tend to grow in a fluidised bed, although the rate of growth is greater in a gas–solids system.

Rowe and Wace,[75] and Wace and Burnett[76] have examined the influence of gas bubbles on the flow of gas in their vicinity. By constructing a thin bed 300 mm wide, 375 mm deep, and only 25 mm across, it was possible to take photographs through the transparent Perspex wall, showing the behaviour of a thin filament of nitrogen dioxide gas injected into the bed. In a bed consisting of 0.20 mm ballotini, it was found that the filament tended to be drawn towards a rising bubble, as shown in Fig. 9.15, and through it, if sufficiently close. This establishes that there is a flow of gas from the continuous phase into the bubble and out through the roof of the bubble, as shown in the figure, and that the gas tends to flow in definite streamlines. As a result, the gas is accelerated towards the bubble and is given a horizontal velocity component, and the gas velocity in the continuous phase close to the bubble is reduced.

In a bubbling bed, there is a tendency for the bubbles to form 'chains' and for successive bubbles to follow a similar track, thus creating a relatively low-resistance path for the gas flow. The bubbles are usually considerably larger at the top than at the bottom of the bed, mainly as a result of coalescence, with larger bubbles catching up and absorbing smaller ones, particularly in the bubble-chains. In addition, the gas expands as the hydrostatic pressure falls, although this is usually a much smaller effect, except in very deep beds.

The pressure distribution round a stationary bubble has been measured by inserting a gauze sphere 50 mm in diameter in the bed and measuring the pressure throughout the bed using the pressure inside the sphere as a datum, as shown in Fig. 9.16. It has been found that, near the bottom of the bubble, the pressure was less than that remote from it at the same horizontal level, and that the situation was reversed, though to a smaller degree, towards the top of the bubble. Although the pressure distribution is somewhat modified in a moving bubble, the model serves qualitatively to explain the observed flow patterns of the tracer gas.

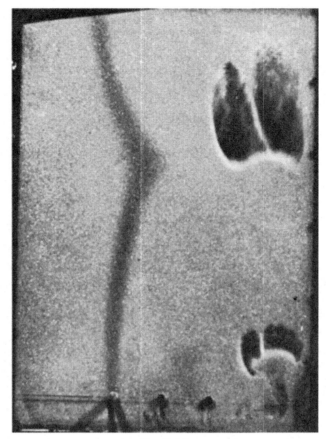

Fig. 9.15
Photograph of tracer and bubble.[75]

When the rate of rise of the bubble exceeds the velocity of the gas in the continuous phase, the gas leaving the top of the bubble is recycled and it re-enters the base. As a result, the gas in the bubble comes into contact with only those solid particles that immediately surround the bubble. Davidson[77] has analyzed this problem and shown that if the inertia of the gas is neglected, the diameter d_c of the cloud of recycling gas surrounding a bubble of diameter d_b is given by:

$$\frac{d_c}{d_b} = \left(\frac{\alpha+2}{\alpha-1}\right)^{1/3}$$

(9.46)

where α is the ratio of the linear velocity of the gas in the bubble to that in the emulsion phase, that is:

$$\alpha = e\left(\frac{u_b}{u_c}\right)$$

(9.47)

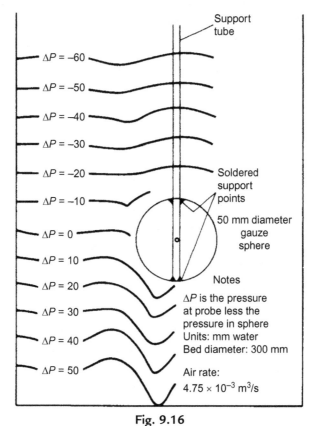

Fig. 9.16
Isobars round a gauze sphere in a bed of mixed sand.[75]

The corresponding expression for a thin, essentially two-dimensional, bed is:

$$\frac{d_c}{d_b} = \left(\frac{\alpha+2}{\alpha-1}\right)^{1/2} \tag{9.48}$$

Rowe et al.[78] have studied the behaviour of the cloud surrounding a bubble by using nitrogen dioxide gas as a tracer, and found that Davidson's theory consistently overestimated the size of the cloud. A more detailed theory, developed by Murray,[3] predicted the observed size of the cloud much more closely over the forward face, although neither approach satisfactorily represented conditions in the wake of the bubble. The gas cloud was found to break off at intervals, and the gas so detached became dispersed in the emulsion phase. In addition, it was shown by X-ray photographs that the bubbles would, from time to time, become split by fingers of particles falling through the roof of the bubble.

In most practical gas–solid systems, the particles are fine and the rising velocities of the individual bubbles are considerably greater than the velocity of the gas in the continuous phase.

Thus, the value of α in Eq. (9.47) is considerably greater than unity. Furthermore, the bubbles will tend to flow in well-defined paths at velocities considerably in excess of the rising velocity of individual bubbles. The whole pattern is, in practice, very complex, because of the large number of bubbles present simultaneously, and also because of the size range of the bubbles.

The work of Rowe and Henwood[79] on the drag force exerted by a fluid on a particle, discussed in Chapter 7, showed why a bubble tended to be stable during its rise. Surfaces containing particles that face downstream, corresponding to the bottom of a bubble, tend to expel particles and are therefore diffuse. Surfaces facing upstream tend to attract particles and thus, the top surface of a bubble will be sharp. Particles in a close-packed array were found to be subjected to a force 68.5 times greater than that on an isolated particle for the same relative velocity. It should, therefore, be possible to evaluate the minimum fluidising velocity, as the fluid velocity at which the drag force acting on a single isolated particle would be equal to 1/68.5 of its buoyant weight. Values calculated on this basis agree well with experimental determinations. A similar method of calculating minimum fluidising velocities has also been proposed by Davies and Richardson,[53] who give a mean value of 71.3 instead of 68.5 for the drag force factor.

From Fig. 9.5 it is seen that for $e_{mf}=0.40$, u_0/u_{mf} is about 78 at low values of the Galileo number and about 9 for high values. In the first case, the drag on the particle is directly proportional to velocity and in the latter case proportional to the square of the velocity. Thus the force on a particle in a fluidised bed of voidage 0.4 is about 80 times that on an isolated particle for the same velocity.

9.3.6 Turbulent Fluidisation

The upper limit of velocity for a bubbling fluidised bed is reached when the bubbles become so large that they are unstable, and break down and merge with the continuous phase to give a highly turbulent, chaotic type of flow. Pressure fluctuations become less violent and the bed is no longer influenced by the effect of large bubbles bursting at the surface, thereby causing a sudden reduction of the inventory of gas within the bed. The transition point occurs at a flowrate at which the standard deviation of the pressure fluctuations reaches a maximum. This occurs where gas bubbles occupy roughly half of the total volume of the bed, a value which corresponds to very much more than half of the overall volumetric flow because, for beds composed of fine particles, the velocity of rise of the bubbles is much greater than the interstitial velocity of the gas in the continuous phase. The breakdown of the bubbles tends to occur when they reach a maximum sustainable size, which is often reached as a result of coalescence. This may frequently occur when a following bubble is drawn into the wake of a bubble that is already close to its maximum stable size.

The turbulent region has been the subject of comparatively few studies until recent years. A comprehensive critical review of the present state of knowledge has been presented by Bi et al.,[56] who emphasise the apparent inconsistencies between the experimental results and correlations of different workers. Bi et al.[56] define the regime of turbulent fluidisation as 'that in which there is no clear continuous phase, but, instead, either, via intermittency or by interspersing voids and dense regions, a competition between dense and dilute phases takes place in which neither gains the ascendancy'.

The onset of *turbulent* fluidisation appears to be almost independent of bed height, or height at the minimum fluidisation velocity, if this condition is sufficiently well defined. It is, however, strongly influenced by the bed diameter, which clearly imposes a maximum on the size of the bubble that can form. The critical fluidising velocity tends to become smaller as the column diameter and gas density, and hence pressure, increase. Particle size distribution appears to assert a strong influence on the transition velocity. With particles of wide size distributions, pressure fluctuations in the bed are smaller and the transition velocity tends to be lower.

As already noted, the onset of the turbulent region seems to occur at the superficial gas velocity at which the standard deviation of the fluctuations of local pressure in the bed is at maximum. It may also be identified by an abrupt change in the degree of bed expansion as the fluidising velocity is increased. Both of these methods give results that are in reasonable accord with visual observations. Measurements based on differential pressures in the bed, rather than on point values, tend to give higher values for the transition velocity. Thus, Bi and Grace[80] give the following correlations for the Reynolds number at the commencement of turbulent fluidisation, Re_c, in terms of the Galileo (or Archimedes) number, Ga, of the gas–solids system. The Reynolds number is based on a 'mean' particle diameter and the superficial velocity of the fluidising gas. The correlations (based, respectively, on differential pressures and on point values) take the form:

$$Re_c = 1.24 Ga^{0.447} \tag{9.49}$$

and:

$$Re_c = 0.56 Ga^{0.461} \tag{9.50}$$

Eqs (9.49) and (9.50) should be used with caution in view of the considerable scatter of the experimental data.

The upper boundary of the turbulent region, based on velocity, is set by the condition at which particle entrainment increases so rapidly as to give rise to what is essentially a gas–solids transport system, sometimes referred to as *fast-fluidisation*. There is no sharp transition because entrainment is always occurring once the gas velocity exceeds the terminal falling velocity of the smallest particles though, in the bubbling region in particular, a high proportion of the particles may still be retained in the bed at high superficial velocities because these do not necessarily 'see' the high proportion of the gas which rises in the bubbles.

Because of the complex flow pattern, it is not generally possible to determine visually the velocity at which the system ceases, in effect, to be a fluidised bed. From measurements of pressure fluctuations, the following approximate relation may be used, although with considerable care:

$$Re_c \approx 2Ga^{0.45} \tag{9.51}$$

It may be seen by comparison with Eqs (9.49) and (9.50) that the turbulent fluidised bed region covers an approximately 2–4-fold range of fluidising velocities.

As the gas velocity increases in a turbulent fluidised column, the bed of solids continues to expand and an increasing fraction of the particles is carried out from the top of the column. In order to maintain a stable operation under these conditions, it is necessary to continuously add solids into the bottom of the riser, or to capture the exiting particles using a cyclone at the top and return them to the bed resulting in the so-called a circulating fluidised bed. In this regime, solid distribution is characterised by a lower dense-phase region and an upper dilute-phase region along the bed height. Furthermore, the upper region has a lean core and a dense annulus region across the bed cross-section. In such circulating fluidised beds, the typical gas velocity is in the range 1–5 m/s, resulting in fast fluidisation. The concentration of solids in the lower, dense region is of the order to 0.16 to 0.22, as compared to the bubbling bed solid fractions of 0.4 to 0.55. A major advantage of circulating fluidised fluid-bed boilers is in their ability to process a range of feedstocks with varying composition and moisture content.

9.3.7 Gas and Solids Mixing

Flow pattern of solids

The movement of the solid particles in a gas-fluidised bed has been studied by a number of workers using a variety of techniques, although very few satisfactory quantitative studies have been made. In general, it has been found that a high degree of mixing is usually present and that, for most practical purposes, the solids in a bed may be regarded as completely mixed. One of the earliest quantitative investigations was that of Bart,[81] who added some particles impregnated with salt to a bed 32 mm in diameter, composed of spheres of cracker catalyst. The tracer particles were fed continuously to the middle of the bed and a sample was removed at the top. Massimilla and Bracale[82] fluidised glass beads in a bed of diameter 100 mm and showed that, in the absence of bubbles, solids mixing could be represented by a diffusional type of equation. Most other workers report almost complete mixing of solids, and fail to account for any deviations by assuming a diffusional model.

May[83] studied the flow patterns of solids in beds up to 1.5 m in diameter by introducing radioactive particles near the top of the bed, and monitoring the radioactivity at different levels by means of a series of scintillation counters. With small beds, it was found that the solids mixing could be represented as a diffusional type of process, although in large-diameter beds there appeared to be an overall circulation superimposed. Values of longitudinal diffusivity ranged from 0.0045 m^2/s, in a bed of 0.3 m diameter, to about 0.015 m^2/s, in a bed 1.5 m in diameter. May[83] also investigated the residence time distribution of gas in a fluidised bed by causing a step-change in the concentration of helium in the fluidising air and following the change in concentration in the outlet air using a mass spectrometer.

Sutherland[84] used nickel particles as tracers in a bed of copper shot. The nickel particles were virtually identical to the copper shot in all the relevant physical properties although, being magnetic, they could be readily separated again from the bed. About 1% of nickel particles was added to the top of a fluidised bed of copper spheres (140 mm diameter and up to 2 m deep). Mixing was found to be rapid, although the results were difficult to interpret because the flow patterns were complex. It was established, however, that vertical mixing was less in a bed that was tapered towards the base. In this case, the formation of bubbles was reduced, because the gas velocity was maintained at a nearly constant value by increasing the flow area in the direction in which the pressure was falling. Gas bubbles appear to play a very large part in promoting circulation of the solids, although the type of flow obtained depends very much on the geometry of the system.

Rowe and Partridge[85] made a photographic study of the movement of solids produced by the rise of a single bubble. As the bubble forms and rises, it gathers a wake of particles, and then draws up a spout of solids behind it, as shown in Fig. 9.17. The wake grows by the addition of particles and then becomes so large that it sheds a fragment of itself, usually in the form of a complete ring. This process is repeated at intervals as the bubble rises, as shown in Fig. 9.18. In a relatively short period ~0.1–0.2 s or so a bubble pases through various stages from its formation (Fig. 9.18A) to its rise to the free surface (Fig. 9.18F) including the entrainment of particles in its wake (Fig. 9.18B, C) and splitting (Fig. 9.18D). A single bubble causes roughly half its own volume of particles to be raised in the spout through a distance of half its diameter, and one-third of its volume in the wake through a bubble diameter. The overall effect is to move a quantity of particles equal to a bubble volume through one and a half diameters. In a fluidised bed, there will normally be many bubbles present simultaneously and their interaction results in a very complex process, which gives rise to a high degree of mixing.

Flow pattern of gas

The existence and the movement of gas bubbles in a fluidised bed has an appreciable influence on the flow pattern within the bed. Several studies have been made of gas–flow patterns, although, in most cases, these have suffered from the disadvantage of having been carried out in small equipment where the results tend to be specific to the equipment used.

Some measurements have been made in industrial equipment and, for example, Askins et al.[86] measured the gas composition in a catalytic regenerator, and concluded that much of the gas passed through the bed in the form of bubbles. Danckwerts et al.[87] used a tracer injection technique, also in a catalytic regenerator, to study the residence time distribution of gas, and

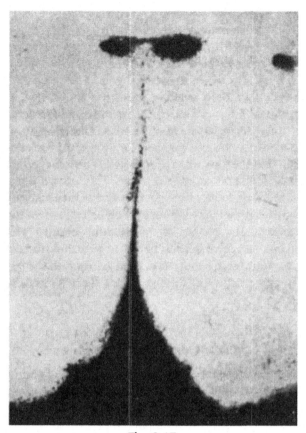

Fig. 9.17
Photograph of the solids displacement caused by a single bubble.[85]

concluded that the flow was approximately piston-type and that the amount of back-mixing was small. Gilliland et al.[88–90] carried out a series of laboratory-scale investigations, although the results were not conclusive. In general, information on the flow patterns within the bed is limited, since these are very much dependent on the scale and geometry of the equipment and on the quality of fluidisation obtained. The gas flow pattern may be regarded, however, as a combination of piston-type flow, of back-mixing, and of bypassing. Back-mixing appears to be associated primarily with the movement of the solids and to be relatively unimportant at high gas rates. This has been confirmed by Lanneau.[91]

9.3.8 Transfer Between Continuous and Bubble Phases

There is no well-defined boundary between the bubble and continuous phase, although the bubble exhibits many properties in common with a gas bubble in a liquid. If the gas in the bubble did not interchange with the gas in the continuous phase, there would be the possibility

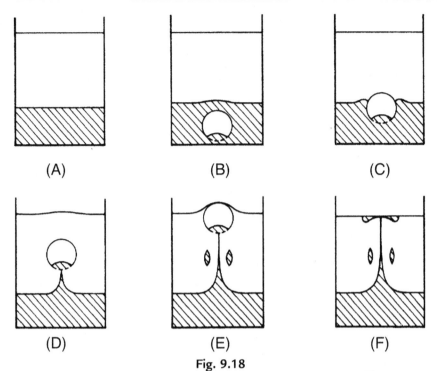

Fig. 9.18

Stages of particle movement caused by a rising bubble.[85]

of a large proportion of the gas in the bed effectively bypassing the continuous phase and of not coming into contact with the solids. This would obviously have a serious effect where a chemical reaction involving the solids was being carried out. The rate of interchange between the bubble and continuous phases has been studied by Szekely[92] and by Davies and Richardson.[53]

In a preliminary investigation, Szekely[92] injected bubbles of air containing a known concentration of carbon tetrachloride vapour into a 86 mm diameter bed of silica-alumina microspheres fluidised by air. By assuming that the vapour at the outside of the bubble was instantaneously adsorbed on the particles, and by operating with very low concentrations of adsorbed material, it was possible to determine the transfer rate between the two phases from the vapour concentrations in the exit gas. By studying beds of differing depths, it was established that most of the mass transfer occurs during the formation of the bubble rather than during its rise. Mass transfer coefficients were calculated on the assumption that the bubbles were spherical during formation and then rose as spherical cap bubbles. Values of mass transfer coefficients of the order of 20 or 30 mm/s were obtained. The variation of the average value of the coefficient with depth is shown in Fig. 9.19. At low bed depths, the coefficient is low because the gas is able to bypass the bed during the formation of the bubble.

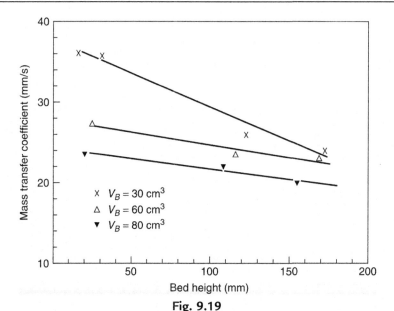

Fig. 9.19

Mass transfer coefficient between the bubble and continuous phases as function of the bed height for various bubble volumes.[92]

Davies and Richardson[53] found that, with a range of fine materials, a bubble injected into a fluidised bed tends to grow as a result of the net transfer of gas to it from the continuous phase. The effect becomes progressively less marked as the minimum fluidising velocity of the system increases. It is found that the growth in bubble volume from V_{B1} to V_{B2} in a height of bed Δz for a system with a minimum fluidising velocity u_{mf}, is given by:

$$\ln\left(\frac{V_{B1}}{V_{B2}}\right) = \frac{1}{K}\Delta z \qquad (9.52)$$

where K represents the distance the bubble must travel for its volume to increase by a factor of e (exponential). Typically, K has a value of 900 mm for a cracker catalyst of mean particle size 55 μm fluidised at a velocity of $2.3u_{mf}$.

As a result of the gas flow into a bubble, the mean residence time of the gas in such systems is reduced because the bubble rises more rapidly than the gas in the continuous phase. Thus, the injection of a single bubble into the bed will initially cause the bed to expand by an amount equal to the volume of the bubble. When this bubble breaks at the surface of the bed, however, the bed volume decreases to a value less than its initial value, the gas content of the bed being reduced. If the value of u_c is only slightly in excess of u_{mf}, the gas in a small injected bubble may become dispersed throughout the continuous phase, so that no bubble appears at the surface of the bed.

Eq. (9.52) enables the net increase in the volume of a bubble to be calculated. The bubble grows because more gas enters through the base than leaves through the cap. By injecting bubbles of carbon dioxide, which is not adsorbed by the particles, and analyzing the concentration in the continuous phase at various heights in the bed, it has been possible to determine the actual transfer from the bubble when rising in a bed of cracker catalyst fluidised at an air velocity of u_{mf}. For bubbles in the size range of 80–250 cm^3, the transfer velocity through the roof of the bubble is constant at about 20 mm/s, which compares very closely with the results of Szekely,[92] who used a quite different technique and made different assumptions.

From a more general perspective, the gas carrying its solute (the diffusing component) passes through the bed in two phases: emulsion phase and bubble phase (Fig. 9.20). Bubble phase includes both the bubbles and the associated cloud (due to boundary layer and wake phenomena). The overall mass transfer coefficient between bubble and emulsion (K_{be}) consists of contributions from the mass transfer coefficient between bubble and cloud (K_{bc}) and that between the cloud and emulsion phase (K_{ce}).

The following rate equations describe the removal of solute from a bubble to emulsion for a differential height.

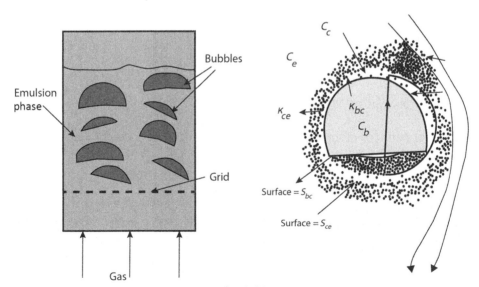

Fig. 9.20

Schematics of the bubble and emulsion phases and the resulting mass transfer processes in a gas–solid fluidised bed. *Redrawn from Kunnii D, Levenspiel O. Fluidization engineering, 2nd ed., Butterworth-Heinmann, Boston, 1991.*

$$-u_b \frac{dC_b}{dz} = K_{be}(C_b - C_e) \tag{9.53a}$$

$$= K_{bc}(C_b - C_e) \tag{9.53b}$$

$$= K_{ce}(C_b - C_e) \tag{9.53c}$$

The K values in the above equations have the dimensions of s^{-1} and C_b, C_c, and C_e are the mean concentrations of the solute in the bubble, in the gas cloud, and in the emulsion phase, respectively. From the above equations, the following relationship between the individual mass transfer coefficients can be derived:

$$\frac{1}{K_{be}} = \frac{1}{K_{bc}} + \frac{1}{K_{ce}} \tag{9.54}$$

For the estimation of the coefficient between the bubble and the cloud, the following equation is proposed by Davidson and Harrison[68]:

$$K_{bc} = 4.5\left(\frac{u_{mf}}{d_b}\right) + 5.85\left(\frac{D_e^{0.5} g^{0.25}}{d_b^{1.25}}\right) \tag{9.55}$$

It consists of two terms. The first term represents the contribution of the convection flux due to the gas flow in and out of bubble and cloud, and the second term accounts for diffusional transfer. For the coefficient between the cloud and the emulsion phase, the following equation based on the diffusional flux is available in literature (see Kunii and Levenspiel[8]):

$$K_{ce} = 6.8\left(\frac{D_e e_{mf} u_b}{d_b^3}\right) \tag{9.56}$$

9.3.9 Beds of Particles of Mixed Sizes

The behaviour of particles of mixed sizes and of mixed densities is highly complex and it has been the subject of many investigations. The works of Chen and Keairns[93], Hoffman et al.,[94] and Wu and Baeyens[95] indicate the state of knowledge in this area.

The most uniform fluidisation might be expected to occur when all the particles are of approximately the same size, so that there is no great difference in their terminal falling velocities. The presence of a very small quantity of fines is often found to improve the quality of fluidisation of gas–solids systems however, although, if fines are present in the bed, bubble formation may occur at a lower velocity. If the sizes differ appreciably, elutriation occurs, and the smaller particles are continuously removed from the system. If the particles forming the bed are initially of the same size, fines will often be produced as a result of mechanical attrition or as a result of breakage due to high thermal stresses. If the particles themselves take part in a chemical reaction, their sizes may alter as a result of the elimination of part of the material, such as during carbonisation or combustion, for example. Any fines that are produced should be recovered using a cyclone separator. Final traces of fine material may then be eliminated with an electrostatic precipitator, as discussed in Chapter 4.

Leva[96] measured the rate of elutriation from a bed composed of particles of two different sizes fluidised in air and found that, if the height of the containing vessel above the top of the bed was small, the rate of elutriation was high; if the height was greater than a certain value, the rate was not affected. This is due to the fact that the small particles were expelled from the bed with a velocity higher than the equilibrium value in the unobstructed area above the bed, because the linear velocity of the fluid in the bed is much higher than that in the empty tube. The tests showed that the concentration of the fine particles in the bed varied with the time of elutriation, according to a law of the form:

$$C = C_0 e^{-Mt} \tag{9.57}$$

where C is the concentration of particles at time t; C_0 is the initial concentration of particles; and M is a constant.

Further work on elutriation has been carried out by Pemberton and Davidson.[97]

A bed containing particles of different sizes behaves in a similar manner to a mixture of liquids of different volatilities. Thus the finer particles, when associated with the fluidising medium, correspond to the lower boiling point liquid and are more readily elutriated, and the rate of their removal from the system and the degree of separation are affected by the height of the reflux column. A law analogous to the Henry's law for the solubility of gases in liquids is obeyed, with the concentration of solids of a given size in the bed bearing, at equilibrium, a constant relation to the concentration of solids in the gas that is passing upwards through the bed. Thus, if clean gas is passed upwards through a bed containing fine particles, these particles are continuously removed. On the other hand, if a dust-laden gas is passed through the bed, particles will be deposited until the equilibrium condition is reached. Fluidised beds are therefore sometimes used for removing suspended dusts and mist droplets from gases, having the advantage that the resistance to flow is, in many cases, less than in equipment employing a fixed filter medium. In this way, clay particles and small glass beads may be used for the removal of sulfuric acid mists.

In fluidisation with a liquid, a bed of particles of mixed sizes will become sharply stratified, with the small particles on top and the large ones at the bottom. The pressure drop is that which would be expected for each of the layers in series. If the size range is small, however, no appreciable segregation will occur.

9.3.10 The Centrifugal Fluidised Bed

One serious limitation of the fluidised bed is that gas throughput must be kept below the level at which a high level of entrainment of particles takes place. In addition, at high gas rates there can be significant bypassing because of the rapid rate of rise of large bubbles, even though there is interchange of gas with the continuous phase. One way of increasing the gas throughput and of reducing the tendency for the formation of large bubbles is to operate at high pressures. Another method is to operate in a centrifugal field.[98, 99]

In the centrifugal fluidised bed, the solids are rotated in a basket and the gravitational field is replaced by a centrifugal field that is usually sufficiently strong for gravitational effects to be neglected. Gas is fed in at the periphery and travels inwards through the bed. At a radius in the bed, the centrifugal acceleration is $r\omega^2$ for an angular speed of rotation of ω. Thus $r\omega^2/g = E$ is a measure of the enhancement factor for the accelerating force on the particles. From Eq. (9.13), it may be seen that, since the effective value of the Galileo number is increased by the factor E, the minimum fluidising velocity is also increased. If the bed is shallow, it behaves essentially as a fluidised bed, although it has a very much higher minimum fluidising velocity, a wider range of operating flowrates, and less tendency for bubbling. From Eq. (9.1) it is seen that the pressure drop over the bed is increased by the factor E.

In a deep bed, the situation is more complex, and the value of E may be considerably greater at the bed inlet than that at its inner surface. In addition, the gas velocity increases towards the centre because of the progressively reducing area of flow. These factors combine so that the value of the minimum fluidising velocity increases with radius in the bed. Thus, as flowrate is increased, fluidisation will first occur at the inner surface of the bed and it will then take place progressively further outwards, until eventually the whole bed will become fluidised. In effect, there are, therefore, two minimum fluidising velocities of interest: that at which fluidisation first occurs at the inner surface, and that at which the whole bed becomes fluidised.

The equipment required for a centrifugal fluidised bed is much more complex than that used for a conventional fluidised bed, and, therefore, it has found use in highly specialised situations only. One of these is in zero-gravity situations, such as on a spacecraft, where carbon dioxide needs to be absorbed from the atmosphere.

9.3.11 The Spouted Bed

The spouted bed represents the ultimate in channelling in a fluidised bed. Gas enters at the conical base, as shown in Fig. 9.21, with much of it flowing rapidly upwards through a centre core, and the remainder percolating through the surrounding solids, which may not be in a fluidised state. The solids, therefore, form a wall for the central spout of gas, gradually become entrained, and then disengage from the gas stream in the space above the bed, where the gas velocity falls as the flow area increases from the spout area to the vessel cross-sectional area. As a result of the erosion of the walls of the spout, the surrounding solids gradually settle to give a solids circulation pattern consisting of rapid upward movement in the spout and slow downward movement in the surrounding bed.

The pressure drop over a spouted bed is normally lower than that for a fluidised bed, because part of the weight of the solids is supported by the frictional force between the solids and the wall of the vessel. Good contact is achieved between the solids and the gas, and high rates of

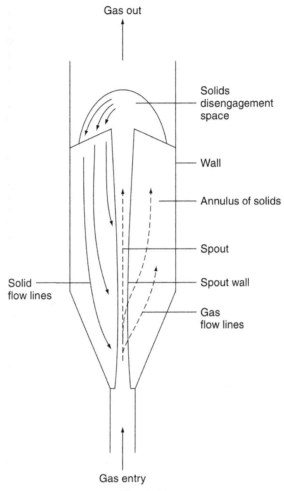

Gas out

Solids
disengagement
space

Wall

Annulus of solids

Spout

Solid
flow lines

Spout wall

Gas
flow lines

Gas entry

Fig. 9.21
Schematic layout of a spouted bed.

heat and mass transfer occur because the relative velocity between the gas and the particles at the wall of the spout is considerably higher than that in a fluidised bed. Furthermore, spouted beds can be operated with sticky materials in which the adhesion between particles is too strong for the possibility of fluidisation taking place. In this way, spouted beds are frequently used for the drying of relatively coarse or sticky solids.

A critical minimum volumetric flowrate of gas per unit area is needed to maintain the spouted condition, and its value tends to increase as the bed becomes deeper. At elevated pressures, lower gas velocities are required although, because of the increased gas density, mass flowrates are higher. In general, a higher gas flowrate is required for the start-up of the spouted bed compared with that needed for its steady state operation.

The effect of operating parameters cannot be predicted reliably, although a substantial amount of work has been published on the operation of spouted beds.[100–102]

9.4 Gas–Liquid–Solids Fluidised Beds

If a gas is passed through a liquid–solids fluidised bed, it is possible to disperse the gas in the form of small bubbles and thereby obtain good contacting between the gas, the liquid, and the solid. Such systems are often referred to as *three-phase fluidised beds*. An important application is in a biological fluidised bed reactor in which oxygen transfer to the biomass takes place, first by its dissolution from the air, which is bubbled through the bed, and then its subsequent transfer from the solution to the biomass particles. Good overall mass transfer characteristics are obtained in this way. Three-phase fluidised beds are also used for carrying out gas–liquid reactions that are catalysed by solids, such as hydrogenation of vegetable oils and organic liquids.

The hydrodynamics of three-phase fluidised systems is complex and reference should be made to specialised publications for a detailed treatment.[103–106] Only a brief summary of their main characteristics is given here.

If a gas is introduced at the bottom of a bed of solids fluidised by a liquid, the expansion of the bed may either decrease or increase, depending on the nature of the solids, and particularly their inertia. It is generally found that the minimum fluidising velocity of the liquid is reduced by the presence of the gas stream. Accurate measurements are difficult to make, however, because of the fluctuating flow pattern that develops.

At *low gas flow rates*, it is found that, with relatively large particles of high inertia such as, for example, 6 mm glass beads fluidised by water, any large gas bubbles are split up by the solids to give a dispersion of fine gas bubbles ($\sim$ 2 mm in diameter), which, therefore, present a high interfacial area for mass transfer. Even smaller bubbles are produced if the surface tension of the liquid is reduced. On the other hand, small particles appear to be unable to overcome the surface tension forces and do not penetrate into the large bubbles. As discussed in Section 9.3.3, large gas bubbles in a fluidised bed have extensive wakes. The same effect is observed in three-phase fluidised systems and, in this case, the bubbles draw liquid rapidly upwards in their wake, thus reducing the liquid flow on the remainder of the bed and causing the bed to contract.

At very *high gas rates*, even large particles do not break down the large gas bubbles and the amount of liquid dragged up in their wake can be so great that some regions of the bed may become completely defluidised. This effect occurs when the ratio of the volumetric flowrate of liquid to that of gas is less than about 0.4.

In gas–liquid–solid fluidised beds, particles are in suspension due to the combined liquid and gas flow. Gas and liquid can enter the column concurrently or counter-currently. Moreover, solids can be introduced batch-wise or continuously. These combinations of gas, liquid, and solids flows lead to several different modes of operations, as discussed in Fan.[106] A good understanding of the resulting hydrodynamics is necessary for successful design and operation of these reactors. The important hydrodynamic parameters include: 1) minimum fluidising velocity; 2) flow regimes; and 3) phase holdups.

9.4.1 Minimum Liquid Fluidisation Velocity-With Gas Flow

In the presence of a gas flowing concurrently with the liquid, fluidisation usually occurs at a lower liquid velocity (u_{lmf}) due to the extra upward lift provided by the gas flow. Increasing gas velocity decreases the minimum liquid velocity in three-phase systems. Following correlation by Song et al.,[107] relating the minimum fluidisation velocity at zero gas flow (u_{lmfo}) to u_{lmf} over a wide range of operating conditions:

$$\frac{u_{lmf}}{u_{lmfo}} = 1 - 376 u_g^{0.327} \mu_l^{0.227} d^{0.213} (\rho_s - \rho_l)^{-0.423} \tag{9.58}$$

Eq. (9.58) is based on the use of S.I. units of variables appearing in it. A typical operating regime for a three-phase fluidised bed based on the particle terminal settling velocity (u_t) is shown in Fig. 9.22 based on the results from Refs. [108, 109]. For a given gas velocity, increasing the liquid velocity takes the bed from the fixed bed regime, through the expanded bed regime, to the transport regime. Increasing the gas velocity leads to a reduction in both the liquid minimum fluidisation velocity and the particle terminal velocity. It is also observed from Fig. 9.22 that the minimum fluidisation velocity drops quickly with the decreasing terminal velocity. Therefore, fine particles can be fluidised or entrained at low liquid velocities. Generally, a distinct bed surface is difficult to maintain for these fine particles and the suspension behaves more like a slurry—hence the name 'slurry bubble column'. The unique feature of the slurry bubble column system is that the particles are so fine (usually <100 μm) that they can be suspended by gas flow alone and usually follow the flow of the liquid in the bed.

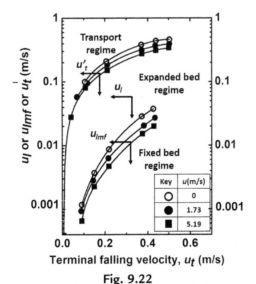

Fig. 9.22

Different regimes of operation for upward flow of gas and liquid in a bed of particles. *Based on Fan LS, Jean RH, Kitano K. On the operating regimes of cocurrent upward gas-liquid-solid systems with liquid as the continuous phase. Chem Eng Sci* **1987***;42:1853.*

9.4.2 Hydrodynamics of Gas–Liquid–Solid Fluidised Beds

Three phase sparged reactors show hydrodynamic behaviour similar to gas–liquid bubble columns. Three-phase fluidised beds and slurry bubble columns can operate in three main flow regimes, depending on the operating conditions:

1. dispersed bubble regime (homogeneous)
2. coalesced bubble regime (heterogeneous)
3. slugging regime.

In the dispersed bubble regime, small bubbles are well dispersed in the bed. There is little bubble-to-bubble collision, thus reducing the possibility of bubble coalescence. This regime is encountered at low superficial gas velocities (< 0.05 m/s), large liquid velocities, and with large particles (3–5 mm). In the coalesced bubble regime, there is a continuous bubble coalescence and break up along the column height, with the dynamic mean bubble size remaining nearly constant. This regime is likely to occur at high superficial gas velocities (> 0.1 m/s), low superficial liquid velocities, and with small particles (< 2 mm). The slugging regime occurs mainly in small experimental columns (< 0.05 m) and is seldom encountered in industrial scale reactors. A few literature studies have presented flow regime diagrams for concurrent gas–liquid–solid fluidised beds.[110, 111] Often, these flow regimes (and their boundaries) are determined visually, requiring caution with their use.

9.4.3 Phase Holdups

The phase holdup represents a fraction of the total volume occupied by each individual phase in the system. In three-phase gas–liquid–solid fluidisation, the individual holdups satisfy the following expression:

$$\varepsilon_g + \varepsilon_l + \varepsilon_s = 1 \tag{9.59}$$

In a fluidised bed with low solids entrainment rates, the solids holdup can be expressed in terms of the total mass of the solids (W_s), the solid density (ρ_s), the cross-sectional area of the column (A_c), and the expanded bed height (H_t) as:

$$\varepsilon_s = 1 - \varepsilon = \frac{W_s}{\rho_s A_c H_t} \tag{9.60}$$

A third equation is needed to obtain the phase holdup fractions for all three phases in the system. This equation can be obtained from the static pressure gradient along the bed as:

$$\frac{\Delta P}{\Delta z} = \left(\varepsilon_s \rho_s + \varepsilon_l \rho_l + \varepsilon_g \rho_g \right) g \tag{9.61}$$

The product $\rho_g \varepsilon_g$ is usually negligible compared to the other two terms in the above equation, which can be used to calculate ε_l together with Eq. (9.60). Then the gas phase holdup can be obtained as:

$$\varepsilon_g = 1 - \varepsilon_s - \varepsilon_l \tag{9.62}$$

In a slurry bubble column, there is no distinct bed of solids, since the fine particles are dispersed over the entire height of the dispersion. The average solid volume fraction in the dispersion can be obtained from Eq. (9.63), if the bed height (H_t) is replaced by the dispersion height (H_d). The liquid holdup can be obtained from the pressure gradient measured along the column height—similarl to that in a three-phase fluidised bed. The average gas holdup in the column can also be calculated by measuring the static and expanded bed heights prior to and during the experimental runs:

$$\varepsilon_g = \frac{(H_d - H_s)}{H_d} \tag{9.63}$$

Phase holdup correlations

Numerous empirical correlations for the estimation of gas holdup in three-phase fluidised beds are available in the literature.[106] Following correlation by Hu et al.,[112] based on about a thousand data points from the literature studies, provides a conservative estimate of gas holdup in a coalesced flow regime:

$$\varepsilon_g = 0.0645 u_g^{0.842} u_l^{0.096} d_p^{0.168} d_t^{-0.419} \tag{9.64}$$

Another correlation often recommended in the literature is by Begovich and Watson[113]:

$$\varepsilon_g = 1.61 u_g^{0.72} d_p^{0.168} d_t^{-0.125} \tag{9.65}$$

Both these equations employ S.I. units.

Bed porosity correlations

Bed porosity in gas–liquid–solid fluidised beds refers to the bed volume fraction not occupied by solid particles, i.e., combined gas and liquid phase holdup. Three-phase fluidised beds using small particles (e.g., 1–2 mm glass beads) exhibit unique behaviour, i.e., upon initial introduction of the gas into the liquid–solid fluidised bed, contraction instead of expansion of the bed has been observed.[114, 115] Further increase in the gas flowrate can lead to more bed contraction, up to a critical gas flowrate, beyond which the bed expands. This behaviour has been attributed to the formation of large coalesced bubbles, which accelerate the liquid out of the bed in their wake. A distinct bed contraction occurs at a combination of high liquid velocities and a gas velocity corresponding to the flow regime change from the dispersed bubble flow to the coalesced bubble flow. The following bed porosity correlation of Begovich and Watson[113] is based on a large set of data points combined from 10 experimental studies:

$$\varepsilon = 3.93 u_l^{0.2711} u_g^{0.041} (\rho_s - \rho_l) d_p^{-0.268} \mu_l^{0.055} d_t^{-0.033} \tag{9.66}$$

This correlation (which is based on the use of S.I. units) is not applicable at zero gas velocity and does not account for the initial bed contraction effect. Bed porosity correlations that account for the initial bed contraction effects are also available in the literature.[116]

Example 9.6

A three phase fluidised bed bioreactor contains biofilm-coated sand particles. The diameter of sand particles is 0.6 mm and the thickness of biofilm is limited to 100 μm. Density of sand and biofilm are 2.4 and 1.2 g/cc, respectively. Liquid and gas phase are water and air at ambient conditions. Following additional information is given:

Bed diameter:	1.0 m
Bed height:	1.25 m
Reactor height:	3.0 m
Weight of solids in bed:	700 kg

Pressure gradient $(-\Delta P/\Delta z)$ measured along bed height $= 11{,}500$ Pa/m.

Calculate the volume fraction of each phase in the bed.

$$\text{Diameter of sand core} = 0.6 \text{ mm.}$$

$$\text{Diameter of particle with biofilm} = 0.6 + 0.2 = 0.8 \text{ mm.}$$

Weight of particle (w_p):

$$= (\pi/6) \times (0.6)^3 \times 10^{-9} \times 2400 + (\pi/6) \times (0.8^3 - 0.6^3) \times 10^{-9} \times 1200.$$
$$= 2.71 \times 10^{-7} + 1.86 \times 10^{-7} = 4.57 \times 10^{-7} \text{ kg.}$$

Volume of particle (v_p):

$$= (\pi/6) \times (0.6)^3 \times 10^{-9} \text{ m}^3.$$
$$= 0.268 \times 10^{-9} \text{ m}^3.$$

Particle density $(\rho_p) = (4.5 \times 10^{-7}/2.68 \times 10^{-10}) = 1700$ kg/m^3.

Solids holdup:

$$\varepsilon_s = 1 - \varepsilon = W_s / \left(\rho_p A_c H_{bed} \right).$$
$$= (700 \times 4)/\left(1700 \times 1.25 \times \pi \times 1^2\right).$$
$$= 0.42.$$

Axial pressure gradient:

$$-(\Delta P/\Delta z) = g\left(\varepsilon_s \rho_s + \varepsilon_l \rho_l + \varepsilon_g \rho_g \right).$$

The last term in the above equation (i.e., $\varepsilon_g \rho_g$) is usually negligible compared to the other terms, substituting for known values:

$$11{,}500/9.8 = 0.42 \times 1700 + 997\varepsilon_l$$

or

$$\varepsilon_l = 0.46 \varepsilon_g = 1 - \varepsilon_l - \varepsilon_s$$
$$= 1 - 0.42 - 0.46 = 0.12$$

9.5 Heat Transfer to a Boundary Surface

9.5.1 Mechanisms Involved

The good heat transfer properties of fluidised systems have led to their adoption in circumstances where close control of temperature is required. The presence of the particles in a fluidised system results in an increase of up to one-hundredfold in the heat transfer coefficient, as compared with the value obtained with a gas alone at the same velocity. In a liquid-fluidised system, the increase is, however, not so marked.

Many investigations of heat transfer between a gas-fluidised system and a heat transfer surface have been carried out, although the agreement between the correlations proposed by different workers is very poor, with differences of one or even two orders of magnitude occurring at times. The reasons for these large discrepancies appear to be associated with the critical dependence of heat transfer coefficients on the geometry of the system, on the quality of fluidisation, and consequently, on the flow patterns obtained. Much of the work was carried out before any real understanding existed of the nature of the flow patterns within the bed. There is, however, almost universal agreement that the one property which has virtually no influence on the process is the thermal conductivity of the solids.

Three main mechanisms have been suggested for the improvement in the heat transfer coefficients brought about by the presence of the solids. First, the particles—having a heat capacity per unit volume many times greater than that of the gas—act as heat-transferring agents. As a result of their rapid movement within the bed, the particles pass from the bulk of the bed to the layers of gas in close contact with the heat transfer surface, exchanging heat at this point and then returning to the body of the bed. Because of their short residence time and their high heat capacity, they change little in temperature, and this fact, coupled with the extremely short physical contact time of the particle with the surface, accounts for the unimportance of their thermal conductivity. The second mechanism is the erosion of the laminar sub-layer at the heat transfer surface by the particles, and the consequent reduction in its effective thickness. The third mechanism, suggested by Mickley and Fairbanks,[117] is that 'packets' of particles move to the heat transfer surface, where an unsteady state heat transfer process takes place.

9.5.2 Liquid–Solids Systems

The heat transfer characteristics of liquid–solid fluidised systems, in which the heat capacity per unit volume of the solids is of the same order as that of the fluid, are of considerable interest. The first investigation into such a system was carried out by Lemlich and Caldas,[118] although most of their results were obtained in the transitional region between streamline and turbulent flow and are therefore difficult to assess. Richardson and Mitson[119] and Richardson and Smith[26] measured heat transfer coefficients for systems in which a number of different solids were fluidised by water in a 50 mm diameter brass tube, fitted with an annular heating jacket. The apparatus was fitted with thermocouples to measure the wall and liquid temperatures at various points along the heating section. Heat transfer coefficients at the tube wall were calculated and plotted against volumetric concentration, as shown in Fig. 9.23 for gravel. The coefficients were found to increase with concentration and to pass through a maximum at a volumetric concentration of 25%–30%. Since in a liquid–solid fluidised system there is a unique relation between concentration and velocity, as the concentration is increased the velocity necessarily falls, and the heat transfer coefficient for liquid alone at the corresponding velocity shows a continuous decrease as the concentration is increased. The difference in the two values— that is, the increase in coefficient attributable to the presence of the particles, $(h - h_l)$—is plotted against concentration, in Fig. 9.24, for ballotini. These curves also pass through a maximum.

Experimental results in the region of turbulent flow may be conveniently correlated in terms of the specific heat C_s of the solid (kJ/kg K) by the equation:

$$(h - h_t) = 24.4 \left(1 + 1.71 C_s^{2.12}\right) (1 - e)^m \left(\frac{u_c}{e}\right)^{1.15} \qquad (9.67)$$

Here, the film coefficients are in kW/m^2 K and the fluidising velocity u_c is in m/s. As Eq. (9.67) is not in dimensionally consistent units (because some of the relevant properties were not

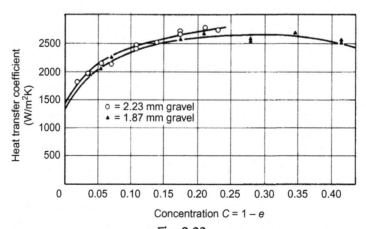

Fig. 9.23

Heat transfer coefficients for gravel particles fluidised in water.[26]

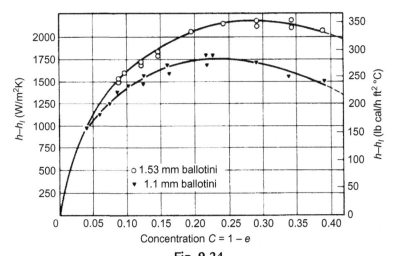

Fig. 9.24

Increase in heat transfer coefficient caused by glass ballotini fluidised in water.[26]

varied), the coefficient, 24.4, is valid only for the units stated. The value of the index m is given by:

$$m = 0.079 \left(\frac{u_0 d \rho}{\mu} \right)^{0.36} \tag{9.68}$$

where u_0 is the terminal falling velocity of the particle.

The maximum value of the ratio of the coefficient for the fluidised system to that for liquid alone at the same velocity is about 3.

In a modified system, in which a suspension of solids is conveyed through the heat transfer section, the heat transfer coefficient is greater than that obtained with liquid alone, though lower than that obtained at the same concentration in a fluidised system. Similar conclusions have been reached by Jepson et al.,[120] who measured the heat transfer to a suspension of solids in gas.

Kang et al.[121] measured coefficients for heat transfer from a cone-shaped heater to beds of glass particles fluidised by water. They also found that the heat transfer coefficient passed through a maximum as the liquid velocity was increased. The heat transfer rate was strongly influenced by the axial dispersion coefficient for the particles, indicating the importance of convective heat transfer by the particles. The region adjacent to the surface of the heater was found to contribute the greater part of the resistance to heat transfer.

Average values of heat-transfer coefficients to liquid–solids systems[122–124] have been measured using small electrically heated surfaces immersed in the bed. The temperature of the element is obtained from its electrical resistance, provided that the temperature coefficient of its electrical resistance is known. The heat supplied is obtained from the measured applied voltage (V) and resistance (R), and is equal to $V^2 = R$.

The energy given up by the element to the bed in which it is immersed, Q, may be expressed as the product of the heat transfer coefficient h, the area A, and the temperature difference between the element and the bed $(T_E - T_B)$.

Thus:

$$Q = hA(T_E - T_B) \tag{9.69}$$

At equilibrium, the energy supplied must be equal to that given up to the bed and:

$$\frac{V^2}{R} = hA(T_E - T_B) \tag{9.70}$$

Thus, the heat-transfer coefficient may be obtained from the slope of the straight line relating V^2 and T_E. This method has been successfully used for measuring the heat transfer coefficient from a small heated surface, 25 mm square, to a wide range of fluidised systems. Measurements have been made of the coefficient for heat transfer between the 25 mm square surface and fluidised beds formed in a tube of 100 mm diameter. Uniform particles of sizes between 3 mm and 9 mm were fluidised by liquids consisting of mixtures of kerosene and lubricating oil, whose viscosities ranged from 1.55 to 940 mN s/m^2. The heat transfer coefficient increases as the voidage and hence the velocity, is increased and passes through a maximum. This effect has been noted by most of the workers in the area. The voidage e_{max} at which the maximum heat transfer coefficient occurs becomes progressively greater as the viscosity of the liquid is increased.

The experimental results obtained for a wide range of systems[122–124] are correlated by Eq. (9.71), in terms of the Nusselt number $(Nu = hd/k)$ for the particle, expressed as a function of the Reynolds number $(Re'_c = u_c d\rho/\mu)$ for the particle, the Prandtl number Pr for the liquid, and the voidage of the bed. This takes the form:

$$Nu' = \left(0.0325 Re'_c + 1.19 Re'^{0.43}_c\right) Pr^{0.37}(1-e)^{0.725} \tag{9.71}$$

A plot showing agreement between this equation and the experimental results is given in Fig. 9.25.

Eq. (9.71) covers the following range of variables:

$$10^{-1} < Re'_c < 10^3$$
$$22 < Pr < 14,000$$
$$0.4 < e < 0.9$$

This equation predicts that the heat transfer coefficient should pass through a maximum as the velocity of the liquid increases. It may be noted that for liquid–solids fluidised beds, $(1-e)$ falls as Re increases, and a maximum value of the heat transfer coefficient is usually obtained at a voidage e of about 0.6 to 0.8.

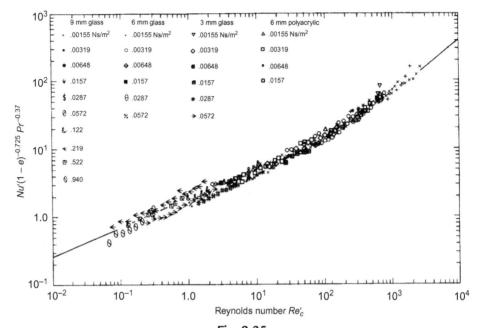

Fig. 9.25

Data for heat transfer to fluidised systems.

Although the parameters in Eq. (9.71) were varied over a wide range, the heat capacities per unit volume and the thermal conductivities of the liquids were almost constant, as they are for most organic liquids, and the dimensions of the surface and of the tube were not varied. Nevertheless, for the purposes of comparison with other results, it is useful to work in terms of dimensionless groups.

The maximum values of the heat transfer coefficients, and of the corresponding Nusselt numbers, may be predicted satisfactorily from Eq. (9.71) by differentiating with respect to voidage and making the derivative equal to zero.

9.5.3 Gas–Solids Systems

With gas–solids systems, the heat-transfer coefficient to a surface is very much dependent on the geometrical arrangement and the quality of fluidisation; furthermore, in many cases, the temperature measurements are suspect. Leva[125] plotted the heat transfer coefficient for a bed, composed of silica sand particles of diameter 0.15 mm fluidised in air, as a function of gas rate, using the correlations put forward as a result of ten different investigations, as shown in Fig. 9.26.

For details of the experimental conditions relating to these studies, reference should be made to Leva.[125] For a gas flow of 0.3 kg/m^2 s, the values of the heat transfer coefficient ranged

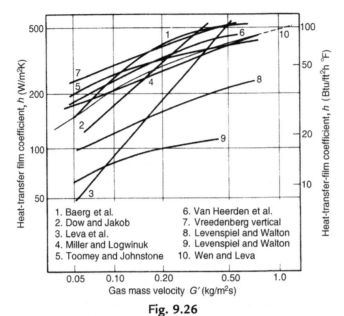

Fig. 9.26

Comparison of heat transfer correlations. Silica sand (0.15 mm) fluidised in air.[125]

from about 75 W/m² K when calculated by the formula of Levenspiel and Walton[126] to about 340 W/m² K when Vreedenberg's[127] expression was used. The equation of Dow and Jakob,[128] gives a value of about 200 W/m² K for this case. The equation takes the form:

$$\frac{hd_t}{k} = 0.55 \left(\frac{d_t}{l}\right)^{0.65} \left(\frac{d_t}{d}\right)^{0.17} \left\{\frac{(1-e)\rho_s C_s}{e\rho C_\rho}\right\}^{0.25} \left(\frac{u_c d_t \rho}{\mu}\right)^{0.80} \tag{9.72}$$

where h is the heat transfer coefficient; k is the thermal conductivity of the gas; d is the diameter of the particle; d_t is the diameter of the tube; l is the depth of the bed; e is the voidage of the bed; ρ_s is the density of the solid; ρ is the density of the gas; C_s is the specific heat of the solid; C_p is the specific heat of the gas at constant pressure; μ is the viscosity of the gas; and u_c is the superficial velocity based on the empty tube.

The Nusselt number with respect to the tube $Nu(=hd_t/k)$ is expressed as a function of four dimensionless groups: the ratio of tube diameter to length, the ratio of tube to particle diameter, the ratio of the heat capacity per unit volume of the solid to that of the fluid, and the tube Reynolds number, $Re_c = (u_c d_t \rho/\mu)$. However, Eq. (9.72) and other equations quoted in the literature should be used with extreme caution, as the value of the heat transfer coefficient will be highly dependent on the flow patterns of gas and solid and the precise geometry of the system.

Mechanism of heat transfer

The mechanism of heat transfer to a surface has been studied in a fixed and a fluidised bed by Botterill and co-workers.[129–132] An apparatus was constructed in which particle replacement at a heat transfer surface was obtained by means of a rotating stirrer with blades close to the surface. A steady-state system was employed using an annular bed, with heat supplied at the inner surface and removed at the outer wall by means of a jacket. The average residence time of the particle at the surface in a fixed bed was calculated, assuming the stirrer to be perfectly efficient, and in a fluidised bed it was shown that the mixing effects of the gas and the stirrer were additive. The heat transfer process was considered to be one of unsteady-state thermal conduction during the period of residence of the particle and its surrounding layer of fluid at the surface. Virtually all the heat transferred between the particle and the surface passes through the intervening fluid, as there is only point of contact between the particle and the surface. It is shown that the heat transfer rate falls off in an exponential manner with time and that, with gases, the heat transfer is confined to the region surrounding the point of contact. With liquids, however, the thermal conductivities and diffusivities of the two phases are comparable, and appreciable heat flow occurs over a more extended region. Thus, the heat transfer process is capable of being broken down into a series of unsteady-state stages, at the completion of each of which the particle with its attendant fluid layer becomes mixed again with the bulk of the bed. This process is very similar to that assumed in the penetration theory of mass transfer proposed by Higbie[133] and by Danckwerts,[134] and described in Chapter 2 of Volume 1B. The theory shows how the heat-carrying effect of the particles, and their disruption of the laminar sub-layer at the surface, must be considered as component parts of a single mechanism. Reasonable agreement is obtained by Botterill et al.[129, 131] between the practical measurements of heat transfer coefficient and the calculated values.

Mickley et al.[135] made measurements of heat transfer between an air–fluidised bed of 104 mm diameter and a concentric heater 6.4 mm diameter and 600 mm long, divided into six 100 mm lengths, each independently controllable. The mean value of the heat transfer coefficient was found to decrease with increase in height, and this was probably attributable to an increased tendency for slugging. The instantaneous value of the heat transfer coefficient was found by replacing a small segment of one of the heaters, of arc 120° and depth 7.9 mm, by a piece of platinum foil of low heat capacity with a thermocouple mounted behind it. The coefficient was found to fluctuate from <50 to about 600 W/m^2 K. A typical curve is shown in Fig. 9.27, from which it is seen that the coefficient rises rapidly to a peak, and then falls off slowly before there is a sudden drop in value as a bubble passes the foil. It is suggested that heat transfer takes place by unsteady-state conduction to a packet of particles, which are then displaced by a gas slug. Heat transfer coefficients were calculated on the assumption that heat flowed for a mean exposure time into a mixture whose physical and thermal properties corresponded with those of an element of the continuous phase, and that the heat flow into the gas slug could be neglected. The fraction of the time during which the surface was in contact with a gas slug ranged from zero to 40% with an easily fluidised solid, but appeared to continue increasing with gas rate to much higher values for a solid, which gave uneven fluidisation. It is easy on this basis to see why very variable values of heat transfer coefficients have been obtained by different workers. For good heat transfer, rapid replacement of the solids is required, without an appreciable coverage of the surface by gas alone.

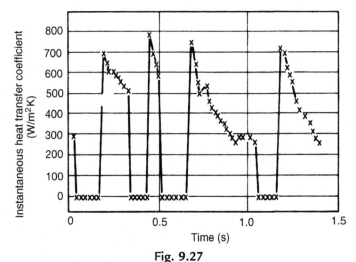

Fig. 9.27
Instantaneous heat transfer coefficients for glass beads fluidised in air.[135]

Bock and Molerus[136] also concluded that the heat transfer coefficient decreases with increase in contact time between elements of bed and the heat transfer surface. In order to observe the effects of long contact times, tests were also carried out with non-fluidised solids. Vertical single tubes and a vertical tube bundle were used. It was established that it was necessary to allow for the existence of a 'gas-gap' between the fluidised bed and the surface to account for the observed values of transfer coefficients. The importance of having precise information on the hydrodynamics of the bed before a reasonable prediction can be made of the heat transfer coefficient was emphasised.

Heat transfer coefficients for a number of gas–solids systems were measured by Richardson and Shakiri,[137] using an electrically heated element 25 mm square, over a range of pressures from sub-atmospheric ($0.03\ \text{MN/m}^2$) to elevated pressures (up to $1.5\ \text{MN/m}^2$). The measuring technique was essentially similar to that employed earlier for liquid–solids systems, as described in Section 9.5.2.

In all cases, there was a similar pattern as the fluidising velocity was increased, as typified by the results shown in Fig. 9.28 for the fluidisation of 78 μm glass spheres by air at atmospheric pressure. Region AB corresponds to a fixed bed. As the velocity is increased beyond the minimum fluidising velocity (B), the frequency of formation of small bubbles increases rapidly and the rate of replacement of solids in the vicinity of the surface rises. This is reflected in a rapid increase in the heat transfer coefficient (BC). A point is then reached where the rate of bubble formation is so high that an appreciable proportion of the heat transfer surface is blanketed by gas. The coefficient thus rises to a maximum (C), then falls as the gas rate is further increased (CD). This fall occurs over only a comparatively small velocity range, following which the coefficient passes through a minimum (D), and then starts to rise again

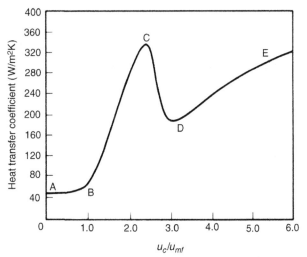

Fig. 9.28

Heat transfer coefficient from a surface to 78 μm glass spheres fluidised by air. AB = fixed bed; B = minimum fluidising velocity; C = maximum coefficient; CD = falling coefficient; D = minimum coefficient; DE = final region of increasing coefficient.

(DE) because of the effect of the increased rate of circulation of the solids in the bed. Values of the heat transfer coefficient show considerable fluctuations at all velocities, as a result of the somewhat random behaviour of the gas bubbles, which are responsible for the continual replacement of the solids in the vicinity of the heat transfer surface.

The nature of the curves depicted in Figs. 9.27 and 9.28 shows just how difficult it is to compare heat transfer coefficients obtained under even marginally different operating conditions. Any correlation should therefore be treated with extreme caution. Some attempts have been made to correlate values of the maximum heat transfer coefficient (Point C in Fig. 9.28), although it is difficult then to relate these to the mean coefficient experienced under practical circumstances.

Effect of pressure

The operation of fluidised beds at elevated pressures not only permits greater mass flowrates through a bed of a given diameter and gives rise to more uniform fluidisation, but also results in improved heat transfer coefficients. Xavier et al.[138] fluidised a variety of solid particles in the size range 0.06–0.7 mm in a bed of 100 mm diameter. The heat transfer surface was provided by a small vertical flat element arranged near the top of the axis of the bed, as shown in the illustration on the right of Fig. 9.29. Experimental results for the fluidisation of glass beads 0.475 mm diameter by nitrogen at pressures of 100–2000 kN/m² are shown. It is seen that the plots of heat transfer coefficient as a function of fluidising velocity all retain the general shape of those seen in Fig. 9.28, and show an improvement in heat transfer at all

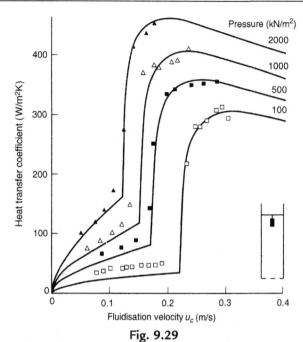

Fig. 9.29

Effect of pressure on heat transfer between a 40 mm vertical flat surface and a fluidised bed.

velocities as the pressure is increased. The discrepancy between the lines and the points arises from the fact that the former were calculated from the model proposed by Xavier et al.[138]

The preceding discussion clearly shows that heat transfer to immersed surfaces in gas–solid fluidised beds shows a complex dependence on numerous factors. Broadly, the overall heat transfer can be broken down into three components, namely, particle convective, gas convective, and radiative. The radiative mechanism only makes contributions when the bed temperature is in excess of 600 °C or so.[55] Similarly, the gas convective contribution is considered to be relevant to large group B and D (Geldart classification) materials. For group A and small (<100 μm) group B particles, this contribution is negligible under normal operating conditions. In addition to the above representative studies, the literature abounds with numerous predictive correlations. However, most of these have limited applicability and the experimental results of various studies often deviate from each other appreciably. Also, much of the literature relates to the maximum values of the heat transfer coefficient that occur at the axis of the column.[139, 140] Limited results are, however, available that demonstrate the variation of the heat transfer coefficient in the axial and radial directions. Thus, for instance, Pisters and Prakash[140] reported that, at a fixed axial location, the value of the heat transfer coefficient decreases (Fig. 9.30) roughly proportional to the third power of the radial position, up to about $(r/R) \leq 0.75$ where r is measured from the axis of the tube. The experimental errors increased significantly in the wall region.

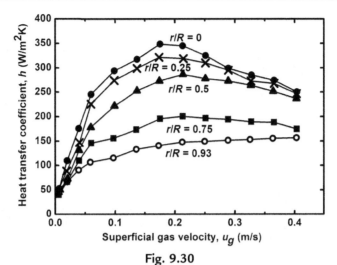

Fig. 9.30

Variation of heat transfer coefficient with gas velocity at different radial locations. *Replotted from Pisters K, Prakash AL. Investigations of axial and radial variations of heat transfer coefficient in bubbling fluidized bed with fast response probe. Powder Tech. 2011;207:224.*

9.6 Mass and Heat Transfer Between Fluid and Particles

9.6.1 Introduction

The calculation of coefficients for the transfer of heat or mass between the particles and the fluid stream requires a knowledge of the heat or mass flow, the interfacial area, and the driving force expressed either as a temperature or a concentration difference. Many early investigations are unsatisfactory, in that one or more of these variables was inaccurately determined. This applies particularly to the driving force, which was frequently based on completely erroneous assumptions about the nature of the flow in the bed.

One difficulty in making measurements of transfer coefficients is that equilibrium is rapidly attained between the particles and the fluidising medium. This has in some cases been obviated by the use of very shallow beds. In addition, in measurements of mass transfer, the methods of analysis have been inaccurate, and the particles used have frequently been of such a nature that it has not been possible to obtain fluidisation of good quality.

9.6.2 Mass Transfer Between Fluid and Particles

Richardson and Bakhtiar[141] adsorbed toluene and isooctane vapours from a vapour-laden air stream on to the surface of synthetic alumina microspheres and followed the change of concentration of the outlet gas with time, using a sonic gas analyser. It was found that equilibrium was attained between outlet gas and solids in all cases, and therefore transfer coefficients could not be calculated. The progress of the adsorption process was still followed, however.

Richardson and Szekely[142] modified the system so that equilibrium was not achieved at the outlet. Thin beds and low concentrations of vapour were used, so that the slope of the adsorption isotherm was greater. Particles of charcoal of different pore structures, and of silica gel, were fluidised by means of air or hydrogen containing a known concentration of carbon tetrachloride or water vapour. A small glass apparatus was used so that it could be readily dismantled, and the adsorption process was followed by weighing the bed at intervals. The inlet concentration was known and the outlet concentration was determined as a function of time from a material balance, using the information obtained from the periodic weighing of the bed. The driving force was then obtained at the inlet and the outlet of the bed, on the assumption that the solids were completely mixed and that the partial pressure of vapour at their surface was given by the adsorption isotherm.

At any height z above the bottom of the bed, the mass transfer rate per unit time, on the assumption of *piston flow* of gas, is given by:

$$dN_A = h_D \Delta C a' dz \tag{9.73a}$$

where a' is the transfer area per unit height of bed.

Integrating over the whole depth of the bed gives:

$$N_A = h_D a' \int_0^z \Delta C dz \tag{9.73b}$$

The integration may be carried out only if the variation of driving force throughout the depth of the bed can be estimated. It was not possible to make measurements of the concentration profiles within the bed, although as the value of ΔC did not vary greatly from the inlet to the outlet, no serious error was introduced by using the logarithmic mean value ΔC_{lm}.

Thus:

$$N_A \approx h_D a' Z \Delta C_{lm} \tag{9.74}$$

Values of mass transfer coefficients were calculated using Eq. (9.74), and it was found that the coefficient progressively became less as each experiment proceeded and as the solids became saturated. This effect was attributed to the gradual buildup of the resistance to transfer in the solids. In all cases, the transfer coefficient was plotted against the relative saturation of the bed, and the values were extrapolated back to zero relative saturation, corresponding to the commencement of the test. These maximum extrapolated values were then correlated by plotting the corresponding value of the Sherwood number ($Sh' = h_D d/D$) against the particle Reynolds number ($Re'_c = u_c d\rho/\mu$) to give two lines, as shown in Fig. 9.31, which could be represented by the following equations:

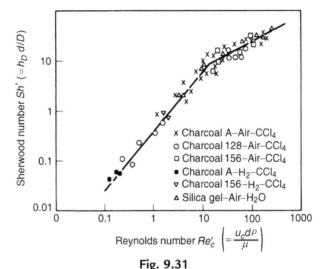

Fig. 9.31
Sherwood number as a function of Reynolds number for adsorption experiments.[142]

$$(0.1 < Re'_c < 15) \quad \frac{h_D d}{D} = Sh' = 0.37 Re'^{1.2}_c \tag{9.75}$$

$$(15 < Re'_c < 250) \quad \frac{h_D d}{D} = Sh' = 2.01 Re'^{0.5}_c \tag{9.76}$$

These correlations are applicable to all the systems employed, provided that the initial maximum values of the transfer coefficients are used. This suggests that the extrapolation gives the true gas-film coefficient. This is borne out by the fact that the coefficient remained unchanged for a considerable period when the pores were large, though it fell off extremely rapidly with solids with a fine pore structure. It was not possible, to relate the behaviour of the system quantitatively to the pore size distribution, however.

The values of the Sherwood number fall below the theoretical minimum value of 2 for mass transfer to a spherical particle, and this indicates that the assumption of piston flow of gases is not valid at low values of the Reynolds number. In order to obtain realistic values in this region, information on the axial dispersion coefficient is required.

A study of mass transfer between a liquid and a particle forming part of an assemblage of particles was made by Mullin and Treleaven,[143] who subjected a sphere of benzoic acid to the action of a stream of water. For a fixed sphere, or a sphere free to circulate in the liquid, the mass transfer coefficient was given, for $50 < Re'_c < 700$, by:

$$Sh' = 0.94 Re'^{1/2}_c Sc^{1/3} \tag{9.77}$$

The presence of adjacent spheres caused an increase in the coefficient because the turbulence was thereby increased. The effect became progressively greater as the concentration increased, although the results were not influenced by whether or not the surrounding particles were free to move. This suggests that the transfer coefficient was the same in a fixed or a fluidised bed.

The results of earlier work by Chu et al.[144] suggested that transfer coefficients were similar in fixed and fluidised beds. Apparent differences at low Reynolds numbers were probably due to the fact that there could be appreciable back-mixing of fluid in the fluidised bed.

Example 9.7

In a fluidised bed, isooctane vapour is adsorbed from an air stream on to the surface of alumina microspheres. The mole fraction of isooctane in the inlet gas is 1.442×10^{-2} and the mole fraction in the outlet gas is found to vary with time as follows:

Time from start (s)	Mole fraction in outlet gas ($\times 10^2$)
250	0.223
500	0.601
750	0.857
1000	1.062
1250	1.207
1500	1.287
1750	1.338
2000	1.373

Show that the results may be interpreted on the assumptions that the solids are completely mixed, that the gas leaves in equilibrium with the solids, and that the adsorption isotherm is linear over the range considered. If the gas flowrate is 0.679×10^{-6} kmol/s and the mass of solids in the bed is 4.66 g, calculate the slope of the adsorption isotherm.

What evidence do the results provide concerning the flow pattern of the gas?

Solution

A mass balance over a bed of particles at any time t after the start of the experiment, gives;

$$G_m(y_0 - y) = \frac{d(WF)}{dt}$$

where G_m is the molar flowrate of gas; W is the mass of solids in the bed; F is the number of moles of vapour adsorbed on unit mass of solid; and y_0, y are the mole fraction of vapour in the inlet and outlet stream, respectively.

If the adsorption isotherm is linear, and if equilibrium is reached between the outlet gas and the solids and if none of the gas bypasses the bed, then F is given by:

$$F = f + by$$

where f and b are the intercept and slope of the isotherm, respectively.

Combining these equations and integrating gives:

$$\ln(1 - y/y_0) = -(G_m/Wb)t$$

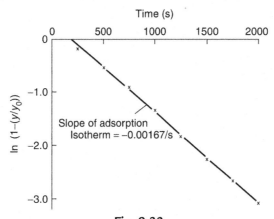

Fig. 9.32
Adsorption isotherm for Example 9.7.

If the assumptions outlined previously are valid, a plot of $\ln(1 - y/y_0)$ against t should yield a straight line of slope $-G_m/Wb$. As $y_0 = 0.01442$, the following table may be produced:

Time (s)	y	y/y_0	$1 - (y/y_0)$	$\ln(1 - (y/y_0))$
250	0.00223	0.155	0.845	−0.168
500	0.00601	0.417	0.583	−0.539
750	0.00857	0.594	0.406	−0.902
1000	0.0106	0.736	0.263	−1.33
1250	0.0121	0.837	0.163	−1.81
1500	0.0129	0.893	0.107	−2.23
1750	0.0134	0.928	0.072	−2.63
2000	0.0137	0.952	0.048	−3.04

These data are plotted in Fig. 9.32 and a straight line is obtained, with a slope of $-0.00167/s$ If $G_m = 0.679 \times 10^{-6}$ kmol/s and $W = 4.66$ g, then:

$-0.00167 = (-0.679 \times 10^{-6})/4.66b$.
from which:

$b = 87.3 \times 10^{-6}$ kmol/g or 0.0873 kmol/kg.

9.6.3 Heat Transfer Between Fluid and Particles

In measuring heat transfer coefficients, many workers failed to measure any temperature difference between gas and solid in a fluidised bed. Frequently, an incorrect area for transfer was assumed, since it was not appreciated that thermal equilibrium existed everywhere in a fluidised bed, except within a thin layer immediately above the gas distributor. Kettenring et al.[145] and Heertjes and McKibbins[146] measured heat transfer coefficients for the evaporation of water from particles of alumina or silica gel fluidised by heated air. In the former investigation, it is probable that considerable errors arose from the conduction of heat along the leads of the thermocouples used for measuring the gas temperature. Heertjes and McKibbins[146] found that any temperature gradient was confined to the bottom part of the bed. A suction thermocouple was used for measuring gas temperatures, although this probably caused some disturbance to the flow pattern in the bed. Frantz[147] has reviewed many of the investigations in this field.

Richardson and Ayers[148] used a steady-state system in which spherical particles were fluidised in a rectangular bed by means of hot air. A continuous flow of solids was maintained across the bed, and the particles on leaving the system were cooled and then returned to the bed. Temperature gradients within the bed were measured using a fine thermocouple assembly, with a junction formed by welding together 40-gauge wires of copper and constantan. The thermo-junction leads were held in an approximately isothermal plane to minimise the effect of heat conduction. After steady-state conditions had been reached, it was found that the temperature gradient was confined to a shallow zone, not >2.5 mm deep at the bottom of the bed. Elsewhere, the temperature was uniform and equilibrium existed between the gas and the solids. A typical temperature profile is shown in Fig. 9.33.

At any height z above the bottom of the bed, the heat transfer rate between the particles and the fluid, on the assumption of complete mixing of the solids and *piston flow* of the gas, is given by:

$$dQ = h\Delta T a' dz \tag{9.78}$$

Integrating gives:

$$Q = ha' \int_0^z \Delta T dz \tag{9.79}$$

In Eq. (9.79), Q may be obtained from the change in temperature of the gas stream, and a', the area for transfer per unit height of bed, from measurements of the surface of solids in the bed. In this case, the integration may be carried out graphically, because the relation

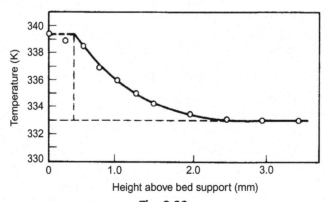

Fig. 9.33
Vertical temperature gradient in a fluidised bed.[148]

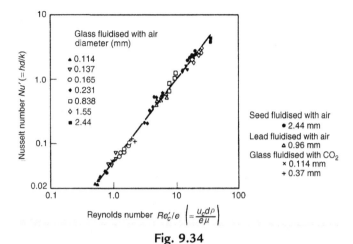

Fig. 9.34

Correlation of experimental results for heat transfer to particles in a fluidised bed.[148]

between ΔT and z was obtained from the readings of the thermocouple. The temperature recorded by the thermocouple was assumed to be the gas temperature, and, if the solids were completely mixed, their temperature would be the same as that of the gas in the upper portion of the bed.

The results for the heat transfer coefficient were satisfactorily correlated by Eq. (9.80) as shown in Fig. 9.34. Since the resistance to heat transfer in the solid could be neglected compared with that in the gas, the coefficients that were calculated were gas-film coefficients, correlated by:

$$Nu' = \frac{hd}{k} = 0.054 \left(\frac{u_c d\rho}{e\mu}\right)^{1.28} = 0.054 \left(\frac{Re'_c}{e}\right)^{1.28} \tag{9.80}$$

Taking an average value of 0.57 for the voidage of the bed, this equation may be rewritten as:

$$Nu' = 0.11 Re'^{1.28}_c \tag{9.81}$$

This equation was found to be applicable for values of Re'_e from 0.25 to 18.

As in the case of mass transfer, the assumption of piston flow is not valid, certainly not at low values of the Reynolds number (<0) at which the Nusselt number is less than the theoretical minimum value of 2. This question is discussed further at the end of this section.

Example 9.8

Cold particles of glass ballotini are fluidised with heated air in a bed in which a constant flow of particles is maintained in a horizontal direction. When steady conditions have been reached, the temperatures recorded by a bare thermocouple immersed in the bed are as follows:

Distance above bed support (mm)	Temperature (K)
0	339.5
0.64	337.7
1.27	335.0
1.91	333.6
2.54	333.3
3.81	333.2

Calculate the coefficient for heat transfer between the gas and the particles, and the corresponding values of the particle Reynolds and Nusselt numbers. Comment on the results and on any assumptions made.

The gas flowrate is 0.2 kg/m^2 s, the specific heat of air is 0.88 kJ/kg K, the viscosity of air is 0.015 mNs/m^2, the particle diameter is 0.25 mm and the thermal conductivity of air is 0.03 W/m K.

Solution

For the system described in this problem, the rate of heat transfer between the particles and the fluid is given by:

$$dQ = ha'\Delta T dz \quad (Eq.9.78)$$

and on integration:

$$Q = ha' \int_0^z \Delta T dz \quad (Eq.9.79)$$

where Q is the heat transferred; h is the heat transfer coefficient; a' is the area for transfer/unit height of bed; and ΔT is the temperature difference at height z.

From the data given, ΔT may be plotted against z as shown in Fig. 9.35 where the area under the curve gives the value of the integral as 8.82 mm K.

Thus:
$$\text{Heat transferred} = 0.2 \times 0.88(339.5 - 332.2)$$
$$= 1.11 \, \text{kW/m}^2 \text{of bed cross} - \text{section.}$$

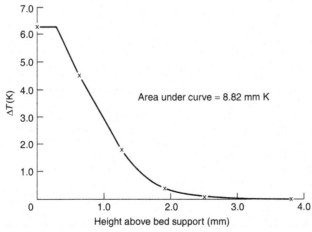

Fig. 9.35

Temperature rise as a function of bed height for Example 9.8.

If the bed voidage $= 0.57$, and a bed 1 m^2 × 1 m high, is considered with a volume $= 1$ m^3, then:

$$\text{Volume of particles} = (1 - 0.57) \times 1 = 0.43 \, \text{m}^3.$$

$$\text{Volume of 1 particle} = (\pi/6)(0.25 \times 10^{-3})^3 = 8.18 \times 10^{-12} \, \text{m}^3.$$

Thus:

$$\text{Number of particles} = 0.43/(8.18 \times 10^{-12}) = 5.26 \times 10^{10} \, \text{per m}^3.$$

$$\text{Area of particles} \, a' = 5.26 \times 10^{10} \times (\pi/4)(0.25 \times 10^{-3})^2 = 1.032 \times 10^4 \, \text{m}^2/\text{m}^3.$$

Substituting in Eq. (9.79) gives:

$$1100 = h \times (1.03 \times 10^4 \times 8.82 \times 10^{-3})$$

and:

$$h = \underline{12.2 \text{W}/\text{m}^2 \, \text{K}}$$

From Eq. (9.81):

$$Nu = 0.11 \, Re^{1.28}.$$

$$Re = G'd/\mu = (0.2 \times 0.25 \times 10^{-3})/(0.015 \times 10^{-3}) = \underline{3.33}$$

Thus:

$$Nu = 0.11 \times (3.33)^{1.28} = \underline{0.513}$$

and:

$$h = (0.513 \times 0.03)/(0.25 \times 10^{-3}) = \underline{61.6 \text{W}/\text{m}^2 \, \text{K}}$$

Example 9.9

Ballotini particles, 0.25 mm in diameter, are fluidised by hot air flowing at the rate of 0.2 kg/m^2 s to give a bed of voidage 0.5 and a cross-flow of particles is maintained to remove the heat. Under steady state conditions, a small bare thermocouple immersed in the bed gives the following temperature data:

Distance above bed support	Temperature	
(mm)	(K)	(°C)
0	339.5	66.3
0.625	337.7	64.5
1.25	335.0	61.8
1.875	333.6	60.4
2.5	333.3	60.1
3.75	333.2	60.0

Assuming plug flow of the gas and complete mixing of the solids, calculate the coefficient for heat transfer between the particles and the gas. This specific heat capacity of air is 0.85 kJ/kg K.

A fluidised bed of total volume 0.1 m^3 containing the same particles is maintained at an approximately uniform temperature of 423 K (150 °C) by external heating, and a dilute aqueous solution at 373 K (100 °C) is fed to the bed at the rate of 0.1 kg/s so that the water is completely evaporated at atmospheric pressure. If the heat transfer coefficient is the same as that previously determined, what volumetric fraction of the bed is effectively carrying out the evaporation?

The latent heat of vaporisation of water is 2.6 MJ/kg.

Solution

Considering unit area of the bed with voidage e and a mass flowrate G', then a heat balance over an increment of height dz gives:

$$G'C_p dT = hS(1-e)dz(T_p - T)$$

where C_p = specific heat, (J/kg K); T_p = Particle temperature, (K); $S(1-e)$ = surface area/volume of bed (m^2/m^3); and h = heat transfer coefficient (W/m^2 K).

Thus:

$$\int_0^z \frac{dT}{T_p - T} = \frac{hS(1-e)z}{G'C_p}$$

or:

$$G'C_p\Delta T = hS(1-e)\int_0^z (T_p - T)dz$$

$\int(T_p - T)dz$ may be found from a plot of the experimental data shown in Fig. 9.36 as 6.31 K mm.

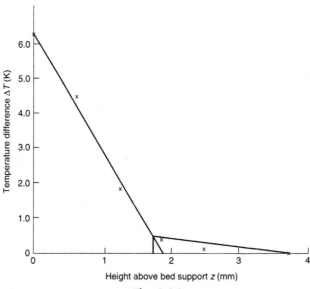

Fig. 9.36

ΔT as a function of bed height z in Example 9.9.

$$S(1-e) = \frac{6}{d}(1-e) = 6(1-0.5)/(0.25 \times 10^{-3}) = 1.2 \times 10^4 /m$$
$$G' = 0.2 \, kg/m^2 s, \, C_p = 850J/kgK$$

Hence:

$$(0.2 \times 850 \times 6.3) = h \times (1.2 \times 10^4 \times 6.31 \times 10^{-3})$$

and:

$$h = 14.1 W/m^2 K$$

If the evaporation rate is 0.1 kg/s at a temperature difference, $\Delta T = 50$ deg. K, the heat flow $=$ $(0.1 \times 2.6 \times 10^6) = 2.6 \times 10^5$ W.

If the effective area of the bed is A, then:

$$(14.1 \times A \times 50) = 2.6 \times 10^5 \, and \, A = 369 m^2$$

The surface area of the bed $= (1.2 \times 10^4 \times 0.1) = 1200$ m^2.

Hence, the fraction of bed which is used $= (369/1200) = 0.31$ or 31%.

9.6.4 Analysis of Results for Heat and Mass Transfer to Particles

A comparison of Eqs (9.75) and (9.81) shows that similar forms of equations describe the processes of heat and mass transfer. The values of the coefficients are, however, different in the two cases, largely due to the fact that the average value for the Prandtl number, Pr, in the heat transfer work was lower than the value of the Schmidt number, Sc, in the mass transfer tests.

It is convenient to express results for tests on heat transfer and mass transfer to particles in the form of j-factors. If the concentration of the diffusing component is small, then the j-factor for mass transfer may be defined by:

$$j_d' = \frac{h_D}{u_c} Sc^{0.67} \tag{9.82}$$

where h_D is the mass transfer coefficient; u_c is the fluidising velocity; Sc is the Schmidt number ($\mu/\rho D$); μ is the fluid viscosity; ρ is the fluid density; and D is the diffusivity of the transferred component in the fluid.

The corresponding relation for heat transfer is:

$$j_h' = \frac{h}{C_p \rho u_c} Pr^{0.67} \tag{9.83}$$

where h is the heat transfer coefficient; C_p is the specific heat of the fluid at constant pressure; Pr is the Prandtl number ($C_p\mu/k$); and k is the thermal conductivity of the fluid.

The significance of j-factors is discussed in detail in Volume 1B, Chapters 1 and 2.

Rearranging Eqs (9.76), (9.77), and (9.81) in the form of (9.82), and (9.83) and substituting mean values of 2.0 and 0.7, respectively, for Sc and Pr, gives:

$$(0.1 < Re'_c < 15)$$

$$j'_d = \frac{Sh'}{ScRe'_c}Sc^{0.67} = 0.37Re'^{0.2}_c Sc^{-0.33} = 0.29Re'^{0.2}_c \tag{9.84}$$

$$(15 < Re'_c < 250)$$

$$j'_d = \frac{Sh'}{ScRe'_c}Sc^{0.67} = 2.01Re'^{-05}_c Sc^{-0.33} = 1.59Re'^{-0.5}_c \tag{9.85}$$

$$(0.25 < Re'_c < 18)$$

$$j'_h = \frac{Nu'}{PrRe'_c}Pr^{0.67} = 0.11Re'^{0.28}_c Pr^{-0.33} = 0.13Re'^{0.28}_c \tag{9.86}$$

These relations are plotted in Fig. 9.37 as lines A, B, and C, respectively.

Resnick and White[149] fluidised naphthalene crystals of five different size ranges (between 1000 and 250 μm) in air, hydrogen, and carbon dioxide at a temperature of 298 K. The gas was passed through a sintered disc, which served as the bed support, at rates of between 0.01 and 1.5 kg/m^2 s. Because of the nature of the surface and of the shape of the particles, uneven

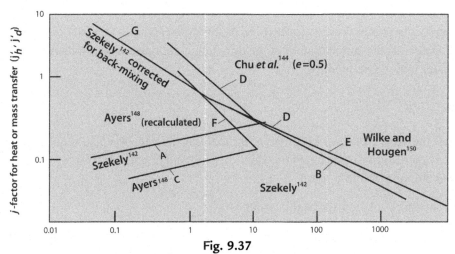

Fig. 9.37
Heat and mass transfer results expressed as j-factors.

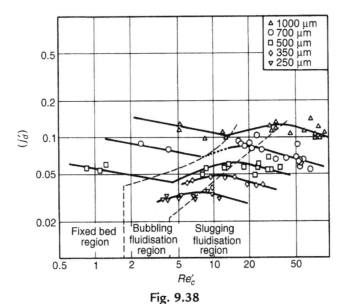

Fig. 9.38

j'_d, for the transfer of naphthalene vapour to air in fixed and fluidised beds.[149]

fluidisation would have been obtained. The rate of vaporisation was determined by a gravimetric analysis of the outlet gas, and mass transfer coefficients were calculated. These were expressed as j-factors and plotted against Reynolds number $Re'_c (= u_c d\rho/\mu)$ in Fig. 9.38. It may be seen that, whilst separate curves were obtained for each size fraction of particles, each curve was of the same general shape, showing a maximum in the fluidisation region, roughly at the transition between the bubbling and slugging conditions.

Chu et al.[144] obtained an improved quality of fluidisation by coating spherical particles with naphthalene, although it is probable that some attrition occurred. Tests were carried out withparticles ranging in size from 0.75 to 12.5 mm and voidages from 0.25 to 0.97. Fixed beds were also used. Again, it was found that particle size was an important parameter in the relation between j-factor and Reynolds number. When these and the other literature results[152, 153] are plotted as shown in Fig. 9.39 against a modified Reynolds number $Re_1^*[= (u_c d\rho/(1 - e)\mu)]$, however, a single correlation was obtained. In addition, it was possible to represent with a single curve the results of a number of workers, obtained in fixed and fluidised beds with both liquids and gases as the fluidising media. A range of 0.6–1400 for the Schmidt number was covered. It may be noted that the results for fluidised systems are confined to values obtained at relatively high values of the Reynolds number. The curve may be represented approximately by the equations:

$$\left(1 < Re_1^* < 30\right) \quad j'_d = 5.7 Re_1^{*-0.78} \tag{9.87}$$

$$\left(30 < Re_1^* < 5000\right) \quad j'_d = 1.77 Re_1^{*-0.44} \tag{9.88}$$

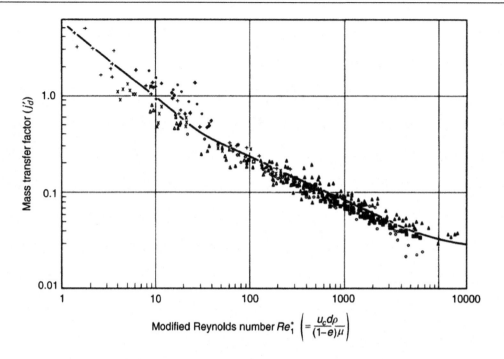

Symbol	System	Schmidt no.	Type of particle	State of bed	Ref.
⊗	Naphthalene-air	2.57	Spheres, cylinders	Fixed, fluidised	144
■	Water-air	0.60	Spheres, cylinders	Fixed	150
○	2-naphthol-water	1400	Modified spheres	Fixed, fluidised	151
●	Isobutyl alcohol-water	866	Spheres	Fixed	152
Ø	Methyl ethyl ketone-water	776	Spheres	Fixed	152
+	Salicylic acid-benzene	368	Modified spheres	Fixed	153
×	Succinic acid-n-butyl alcohol	690	Modified spheres	Fixed	153
△	Succinic acid-acetone	164	Modified spheres	Fixed	153

Fig. 9.39

j-factor, j'_d, for fixed and fluidised beds.[144]

These two relations are also shown as curve D in Fig. 9.37 for a voidage of 0.5.

A number of other workers have measured mass transfer rates. McCune and Wilhelm[151] studied transfer between naphthol particles and water in fixed and fluidised beds. Hsu and Molstad[154] absorbed carbon tetrachloride vapour on activated carbon particles in very shallow beds, which were sometimes less than one particle diameter deep. Wilke and Hougen[150] dried Celite particles (size range approximately 3–19 mm) in a fixed bed by means of a stream of air, and found that their results were represented by:

$$\left(50 < Re'_c < 250\right) \quad J'_d = 1.82 Re'^{-0.51}_c \tag{9.89}$$

$$(Re'_c > 350) \quad J'_d = 1.99 Re'^{-0.41}_c \tag{9.90}$$

These relations are plotted as curve E in Fig. 9.37.

It may be seen that the general trend of the results of different workers is similar, but that the agreement is not good. In most cases a direct comparison of results is not possible because the experimental data are not available in the required form.

The importance of the flow pattern on the experimental data is clearly apparent, and the reasons for discrepancies between the results of different workers are largely attributable to the rather different characters of the fluidised systems. It is of particular interest to note that, at high values of the Reynolds number when the effects of back-mixing are unimportant, similar results are obtained in fixed and fluidised beds. This conclusion was also reached by Mullin and Treleaven[143] in their tests with models.

There is apparently an inherent anomaly in the heat and mass transfer results in that, at low Reynolds numbers, the Nusselt and Sherwood numbers (Figs. 9.34 and 9.31) are very low, and substantially below the theoretical minimum value of 2 for transfer by thermal conduction or molecular diffusion to a spherical particle when the driving force is spread over an infinite distance (Volume 1B, Chapter 1). The most probable explanation is that, at low Reynolds numbers, there is appreciable back-mixing of gas associated with the circulation of the solids. If this is represented as a diffusional type of process with a longitudinal diffusivity of D_L, the basic equation for the heat transfer process is:

$$D_L c_p \rho \frac{d^2 T}{dz^2} - u_c c_p \rho \frac{dT}{dz} - ha(T - T_s) = 0 \tag{9.91}$$

Eq. (9.91) may be obtained in a similar manner to Eq. (7.34), but with the addition of the last term which represents the transfer of sensible heat from the gas to the solids. The time derivative is zero because a steady-state process is considered.

On integration, Eq. (9.91) gives a relation between h and D_L, and h may only be evaluated if D_L is known. If it is assumed that, at low Reynolds numbers, the value of the Nusselt or Sherwood numbers approaches the theoretical minimum value of 2, it is possible to estimate the values of D_L at low Reynolds numbers, and then to extrapolate these values over the whole range of Reynolds numbers used. This provides a means of recalculating all the results using Eq. (9.91). It is then found that the results for low Reynolds numbers are substantially modified and the anomaly is eliminated, whereas the effect at high Reynolds numbers is small. Recalculated values[142] of the Nusselt number for heat transfer tests[148] are shown in Fig. 9.40. This confirms the view already expressed, and borne out by the work of Lanneau,[91] that back-mixing is of importance only at low flowrates.

Recalculated values of j'_h and j'_d obtained from the results of Ayers[148] and Szekely[142] are shown in Fig. 9.37 as curves F and G. It will be seen that the curves B, D, E, F, and G follow the same trend.

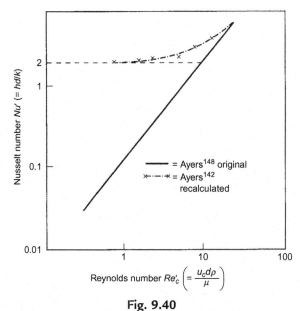

Fig. 9.40

Recalculated values of Nusselt number, taking into account the effects of back-mixing.[142]

Cornish[155] considered the minimum possible value of the Nusselt number in a multiple particle system. By regarding an individual particle as a source and the remote fluid as a sink, it was shown that values of Nusselt number <2 may then be obtained. In a fluidised system, however, the inter-particle fluid is usually regarded as the sink and, under these circumstances, the theoretical lower limit of 2 for the Nusselt number applies. Zabrodsky[156] has also discussed the fallacy of Cornish's argument.

9.6.5 Heat and Mass Transfer in Gas–Liquid–Solid Fluidised Beds

- A good understanding of heat and mass transfer characteristics is required for proper design and optimisation of a multiphase reactor system with respect to its operating conditions and the reactor geometry. Following is a summary of the current knowledge of heat and mass transfer in three-phase fluidised beds and bubble column reactors. Three-phase fluidised bed data refers to concurrent upflow of liquid and gas.

Heat transfer

Heat transfer coefficient in three-phase fluidised beds[157] exhibits a maximum value as a function of the increasing liquid velocity (Fig. 9.41). These results correspond to cylindrical particles 52 mm long and ~1 mm in diameter and of 2000 kg/m^3 density. The liquid velocity at which the heat transfer coefficient, h, exhibits a maximum value is mainly dependent on the properties of the liquid and solid particles; it increases as the size and density of particles increase, but falls as the liquid viscosity is increased. The initial increase in heat transfer

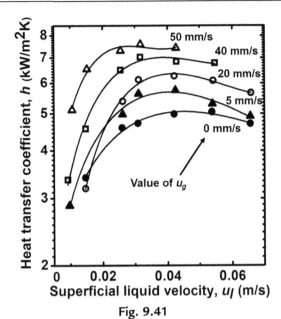

Fig. 9.41

Dependence of the heat transfer coefficient on the liquid velocity in a three-phase fluidised bed. *Replotted from Magiliotou M, Chen Y-M, Fan L-S, Bed-immersed object heat transfer in a three-phase fluidized bed. AIChE J* **1988**;34:1043.

	○	●	□	■	△	▲
d_p (mm)	1.7	1.7	6.0	6.0	6.0	6.0
u_l (mm/s)	60	60	60	60	120	120
μ_l (Pa s)	0.012	0.039	0.012	0.039	0.012	0.039

coefficient h with u_l has been generally attributed to the increase of turbulence, and the oscillatory motion of the fluidised solid particles. Similar observations on heat transfer rate with the increasing liquid velocity have been made in liquid–solid fluidised beds as well, i.e., initially there is an increase, it passes through a maximum value (depending on particle size and liquid viscosity), and then decreases with further increase in the liquid velocity, finally approaching the value for a single liquid flow. These observations have been reported by several researchers in the literature.[123, 158, 159]

The introduction of a gas into a liquid–solid fluidised bed increases heat transfer coefficient h between the wall and the bed, and between an immersed heating surface and the bed. In general, the increase of heat transfer coefficient is initially rapid, then slows down, and tends finally towards an asymptotic value at higher gas velocities (Fig. 9.42). The increase of 'h' can be attributed to the turbulence generated with the injection of the gas in the bed.[158, 160, 161] At low gas flowrates u_g (< 0.05 m/s), bubble size is small and with less turbulence intensity, but the bubbles are more uniformly distributed in the bed (bubbly flow regime).

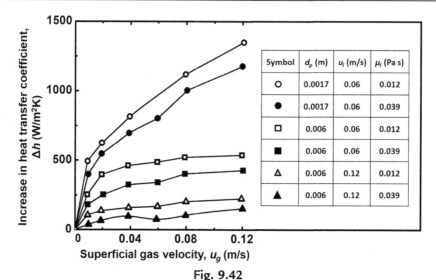

Fig. 9.42

Increase in heat transfer coefficient with gas velocity. *Replotted from Kang Y, Suh IS, Kim SD. Heat transfer characteristics of three phase fluidized beds. Chem Eng Commun 1985;34:1.*

At higher gas velocities u_g, the degree of turbulence distribution becomes poorer, owing to the irregular distribution of bubbles in the bed (coalesced bubble regime). Higher gas flowrates lead to less isotropic turbulence, which dampens the effect of turbulence intensity on the heat transfer. Therefore, the heat transfer coefficient initially increases rapidly and then approaches an asymptotic value with the increasing gas velocity. It is also observed from Fig. 9.42 that the increase in heat transfer is faster in beds of smaller particles. Heat transfer coefficient decreased with the increasing liquid viscosity (as expected) in beds of different solids.[158] This decrease may be attributed to the increase of boundary sub-layer thickness of laminar flow around the heater due to the reduced turbulence and increase of viscous friction loss between phases.

Numerous correlations of varying forms and complexity are available in the literature that can be employed to estimate the transfer coefficients for heat transfer and mass transfer in these systems. Most of these have been reviewed in references.[162–167]

Example 9.10

An ebullating bed reactor is being evaluated for heavy oil upgrading by hydrocracking. The heavy oil feed will be 150,000 barrels/day. The rector will operate at 400 °C and 55 bar pressure. The following data is available from the lab and pilot scale studies:

Space velocity $(h^{-1}) = 1$ $[(m^3 \text{ feed/h})]/[m^3 \text{ catalyst}]$.

Catalyst particle diameter:	1.5 mm
Particle density:	1400 kg/m^3
Liquid density:	920 kg/m^3
Liquid viscosity:	3.5×10^{-3} Pa.s

The superficial liquid velocity is 50% above the minimum fluidisation velocity and the gas superficial velocity is 0.08 m/s.

Calculate:

(a) reactor diameter;
(b) number of reactors and reactor diameter if the permissible maximum diameter is limited to 6 m; and
(c) fluidised bed height.

Solution

(a) Reactor diameter

$$D_R = \sqrt{\frac{4Q_l}{\pi u_l}}$$

$(\pi D^2{}_R/4)\, u_l = Q_l$ (volumetric liquid feed, m^3/s)Q_l
$$= 150{,}000\ (\text{bbl/d}) \times 42\ (\text{gal/bbl}) \times 3.785\ (\text{L/gal}) \times 10^{-3}\ (\text{m}^3/\text{L})(d/24\,\text{h})$$
$$\times (h/3600\,\text{s})$$
$$= 0.276\,\text{m}^3/\text{s or } 994\,\text{m}^3/\text{h}$$

Minimum liquid fluidisation velocity[9]:

$$Re_{mfo} = \left[C_1^2 + C_2 Ga\right]^{0.5} - C_1$$

where:

$$Re_{mfo} = \frac{d_p u_{lmfo} \rho_l}{\mu_l}$$

$$Ga = \frac{d_p^3 \rho_l \left(\rho_p - \rho_l\right)}{\mu_l^2}$$

Recommended values for the empirical constants, C_1 and C_2:

$$C_1 = 33.7 \quad C_2 = 0.0408$$

$$Ga = \frac{0.0015^3 \times 920 \times (1400 - 920) \times 9.8}{(3.5 \times 10^{-3})^2}$$

$$= 1.19 \times 10^4$$

$$Re_{mfo} = \left[(33.7)^2 + 0.0408 \times 1.19 \times 10^4\right]^{0.5} - 33.7$$

$$= 0.714$$

Minimum liquid fluidisation velocity:

$$u_{mfo} = \frac{0.714 \times 3.5 \times 10^{-3}}{0.0015 \times 920}$$

$$= 1.8 \times 10^{-3}\,\text{m/s.}$$

Actual liquid velocity:

$$u_l = 1.8 \times 10^{-3} \times 1.5 = 2.7 \times 10^{-3} \,\text{m/s}.$$

Substituting for u_l and Q_l:

$$D_R = 11.4 \,\text{m}$$

(b) Number of reactors and reactor diameter, if maximum reactor diameter is 6 m.
   Let $D_R = 6$ m.

$$\left[(\pi D^2{}_R/4)\, u_l \right] \times N\,(\text{number of reactors}) = 0.276.$$

$$N = 3.6$$

Since number of reactors is an integer, let $N = 4$.

Then:

$$D_R = 5.7 \,\text{m}$$

(c) Fluidised bed height:

$$\varepsilon_s = 1 - \varepsilon_g - \varepsilon_l = W_s / \left(\rho_p A_c H_{bed} \right)$$

$H_{bed} = (W_s/\rho_p)/A_c/(1 - \varepsilon_g - \varepsilon_l).$

where $(W_s/\rho_p) =$ Catalyst volume $(\text{m}^3) = SV\,(1/h) \times$ Feed rate $(\text{m}^3/h) = 994\,\text{m}^3/h$; catalyst volume per reactor $= 994/4 = 248.5\,\text{m}^3/h$; and reactor cross-sectional area $= (\pi D_R^2/4) = 25.52\,\text{m}^2$.

Calculate bed voidage by Equation of Begovich and Watson.[113]

Assumption: no bed contraction due to gas injection.

$\varepsilon_b = \varepsilon_g + \varepsilon_l$

$$= 3.93 u_l^{0.2711} u_g^{0.041} (\rho_s - \rho_l)^{-0.316} d_p^{-0.268} \mu_l^{0.055} d_t^{-0.033}$$

$$= 3.93 \times (2.7 \times 10^{-3})^{0.2711} \times (0.08)^{0.041} \times (480)^{-0.316} \times (0.0015)^{-0.268}$$

$$\times (0.0035)^{0.055} \times (5.7)^{-0.033}.$$

$$= 0.443.$$

$$H_{bed} = 248.5/25.52/(1 - 0.443).$$

$$= 17.5 \,\text{m}.$$

9.7 Summary of the Properties of Fluidised Beds

Over the past 50–60 years, much data on the operation, performance, and new emerging applications of fluidised beds have accrued in the literature, albeit there are still gaps in our knowledge particularly with regard to the scale-up and design of such systems.

Fluidised beds may be divided into two classes. In the first, there is a uniform dispersion of the particles within the fluid and the bed expands in a regular manner as the fluid velocity is increased. This behaviour, termed *particulate fluidisation*, is exhibited by most liquid–solids systems, the only important exceptions being those composed of fine particles of high density. This behaviour is also exhibited by certain gas–solids systems over a very small range of velocities just in excess of the minimum fluidising velocity—particularly where the particles are approximately spherical and have very low free-falling velocities. In particulate fluidisation, the rate of movement of the particles is comparatively low, and the fluid is predominantly in piston-type flow with some back-mixing, particularly at low flowrates. Overall turbulence normally exists in the system.

In the other form of fluidisation—*aggregative fluidisation*—two phases are present in the bed: a continuous or emulsion phase, and a discontinuous or bubble phase. This is the pattern normally encountered with gas–solids systems. Bubbles tend to form at gas rates above the minimum fluidising rate and grow as they rise through the bed. The bubbles grow because the hydrostatic pressure is falling, as a result of coalescence with other bubbles, and by flow of gas from the continuous to the bubble phase. The rate of rise of the bubble is approximately proportional to the one-sixth power of its volume. If the rising velocity of the bubble exceeds the free-falling velocity of the particles, it will tend to draw in particles at its wake and to destroy itself. There is, therefore, a maximum stable bubble size in a given system. If this exceeds about 10 particle diameters, the bubble will be obvious and aggregative fluidisation will exist; this is the usual condition with a gas–solids system. Otherwise, the bubble will not be observable and particulate fluidisation will occur. In aggregative fluidisation, the flow of the fluid in the continuous phase is predominantly streamline.

In a gas–solids system, the gas distributes itself between the bubble phase and the continuous phase, which generally has a voidage a little greater than at the point of incipient fluidisation. If the rising velocity of the bubbles is less than that of the gas in the continuous phase, it behaves as a rising void through which the gas will tend to flow preferentially. If the rising velocity exceeds the velocity in the continuous phase—and this is usually the case—the gas in the bubble is continuously recycled through a cloud surrounding the bubble. Partial bypassing, therefore, occurs and the gas comes into contact with only a limited quantity of solids. The gas cloud surrounding the bubble detaches itself from time to time, however.

The bubbles appear to be responsible for a large amount of mixing of the solids. A rising bubble draws up a spout of particles behind it and carries a wake of particles equal to about one-third of the volume of the bubble and wake together. This wake detaches itself at intervals. The pattern in a bed containing a large number of bubbles is, of course, very much more complex.

One of the most important properties of the fluidised bed is its good heat transfer characteristics. For a liquid–solids system, the presence of the particles may increase the coefficient by a factor of two or three. In a gas–solids system, the factor may be about two orders of magnitude, with the coefficient being raised by the presence of the particles, from a value for the gas to one normally associated with a liquid. The improved heat transfer is associated with the movement of the particles between the main body of the bed and the heat transfer surface. The particles act as heat-transferring elements and bring material at the bulk temperature in close proximity to the heat transfer surface. A rapid circulation, therefore, gives a high heat transfer coefficient. In a gas–solids system, the amount of bubbling within the bed should be sufficient to give adequate mixing, and at the same time should not be sufficient to cause an appreciable blanketing of the heat transfer surface by gas.

9.8 Applications of the Fluidised Solids Technique

9.8.1 General

The use of the fluidised solids technique was developed mainly by the petroleum and chemical industries, for processes where the very high heat transfer coefficients and the high degree of uniformity of temperature within the bed enabled the development of processes that would otherwise be impracticable. Fluidised solids are now used quite extensively in many industries where it is desirable to bring about intimate contact between small solid particles and a gas stream. In many cases, it is possible to produce the same degree of contact between the two phases with a very much lower pressure drop over the system. Drying of finely divided solids is carried out in a fluidised system, and some carbonisation and gasification processes are in operation. Fluidised beds are employed in gas purification work, in the removal of suspended dusts and mists from gases, in lime burning, and in the manufacture of phthalic anhydride.

9.8.2 Fluidised Bed Catalytic Cracking

The existence of a large surplus of high boiling material after the distillation of crude oil led to the introduction of a cracking process to convert these materials into compounds of lower molecular weight and lower boiling point—in particular, into petroleum spirit. The cracking was initially carried out using a fixed catalyst, although local variations in temperature in the bed led to a relatively inefficient process, and the deposition of carbon on the surface of the catalyst particles necessitated taking the catalyst bed out of service periodically so that the carbon could be burned off. Many of these difficulties are obviated by the use of a fluidised catalyst, since it is possible to remove catalyst continuously from the reaction vessel and to supply regenerated catalyst to the plant. The high heat transfer coefficients for fluidised systems account for the very uniform temperatures within the reactors and make it possible to control conditions very closely. The fluidised system has one serious drawback, however, in that some longitudinal mixing occurs, which gives rise to a number of side reactions.

A diagram of the plant used for the catalytic cracking process is given in Fig. 9.43.[168] Hot oil vapour, containing the required amount of regenerated catalyst, is introduced into the reactor,

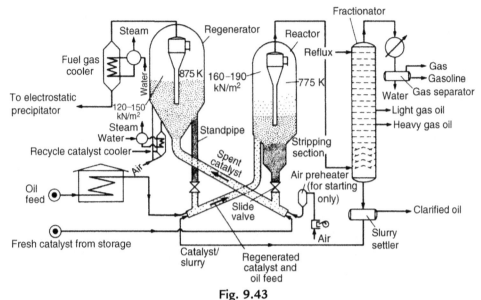

Fig. 9.43
Fluidised-bed catalytic cracking plant.[168]

which is at a uniform temperature of about 775 K. As the velocity of the vapour falls in the reactor, because of the greater cross-sectional area for flow, a fluidised bed is formed and thesolid particles are maintained in suspension. The vapours escape from the top of the unit and the flowrate is such that the vapour remains in the reactor for about 20 s. It is necessary to provide a cyclone separator in the gas outlet to remove entrained catalyst particles and droplets of heavy oil. The vapour from the cracked material is then passed to the fractionating unit, whilst the catalyst particles and the heavy residue are returned to the bed. Some of the catalyst is continuously removed from the bottom of the reactor and, together with any fresh catalyst required, is conveyed in a stream of hot air into the regenerator where the carbon deposit and any adhering film of heavy oil are burned off at about 875 K.

In the regenerator, the particles are again suspended as a fluidised bed. The hot gases leave the regenerator through a cyclone separator—from which the solids return to the bed—and then flow to waste through an electrostatic precipitator, which removes any very fine particles still in suspension. The temperature in the regenerator remains constant to within about 3 °K, even where the dimensions of the fluidised bed are as high as 6 m deep and 15 m in diameter. Catalyst is continuously returned from the regenerator to the reactor by introducing it into the supply of hot vapour. The complete time cycle for the catalyst material is about 600 s. By this process, a product consisting of between about 50% and 75% of petroleum spirit of high octane number is obtained. The quality of the product may be controlled by the proportion of catalyst used and the exact temperature in the reactor.

Subsequent tests have shown that much of the cracking takes place in the transfer line in which the regenerated catalyst is conveyed into the reactor in the stream of oil vapour. The chemical reaction involved is very fast, and the performance of the reactor is not sensitive to the hydrodynamic conditions.

From the diagram of the catalytic cracking plant in Fig. 9.43, it may be noted that there is a complete absence of moving parts in the reactor and the regenerator. The relative positions of components are such that the catalyst is returned to the reactor under the action of gravity.

9.8.3 Applications in the Chemical and Process Industries

Fluidised catalysts are also used in the synthesis of high-grade fuels from mixtures of carbon monoxide and hydrogen, obtained either by coal carbonisation or by partial oxidation of methane. An important application in the chemical industry is the oxidation of naphthalene to phthalic anhydride, as discussed by Riley.[169] The kinetics of this reaction are much slower than those of catalytic cracking, and considerable difficulties have been experienced in correctly designing the system.

Purely physical operations are also frequently carried out in fluidised beds. Thus, fluidised bed dryers, discussed in volume 2B, are successfully used, frequently for heat-sensitive materials that must not be subjected to elevated temperatures for prolonged periods.

The design of a fluidised bed for the carrying out of an exothermic reaction involving a long reaction time has been considered by Rowe[170] and, as an example, the reaction between gaseous hydrogen fluoride and solid uranium dioxide to give solid uranium tetrafluoride and water vapour is considered. This is a complex reaction which, as an approximation, may be considered as first order with respect to hydrogen fluoride. The problem here is to obtain the required time of contact between the two phases by the most economical method. The amount of gas in the bubbles must be sufficient to give an adequate heat transfer coefficient, although the gas in the bubbles does have a shorter contact time with the solids because of its greater velocity of rise. In order to increase the contact time of the gas, the bed may be deep, although this results in a large pressure drop. Bubble size may be reduced by the incorporation of baffles and, as described by Volk et al.,[171] this is frequently an effective manner both of increasing the contact time and of permitting a more reliable scale-up from small scale experiments. The most effective control is obtained by careful selection of the particle size of the solids. If the particle size is increased, there will be a higher gas flow in the emulsion phase and less in the bubble phase. Thus, the ratio of bubble to emulsion phase gas velocity will be reduced, and the size of the gas cloud will increase. If the particle size is increased too much, however, there will be insufficient bubble phase to give good mixing. In practice, an overall gas flowrate of about twice that required for incipient fluidisation is frequently suitable.

9.8.4 Fluidised Bed Combustion

An important application of fluidisation that has attracted considerable interest over the years is fluidised bed combustion. The combustible material is held in a fluidised bed of inert material and the air for combustion is the fluidising medium. The system has been developed for steam rising on a very large scale for electricity generation and for incineration of domestic refuse.

The particular features of fluidised combustion of coal that have given rise to the great interest are, firstly, its suitability for use with very low-grade coals, including those with very high ash contents and sulphur concentrations, and secondly, the very low concentrations of sulphur dioxide that are attained in the stack gases. This situation arises from the very much lower bed temperatures ($\sim$1200 K) than those existing in conventional grate-type furnaces, and the possibility of reacting the sulphur in the coal with limestone or dolomite to enable its discharge as part of the ash.

Much of the basic research, development studies, and design features of large-scale fluidised bed combustors is discussed in the Proceedings of the Symposium on Fluidised Bed Combustion, organised by the Institute of Fuel as long ago as 1975,[172] and subsequently in the literature.[173–177] Pilot scale furnaces with ratings up to 0.5 MW have been operated and large-scale furnaces have outputs of up to 30 MW.

The bed material normally consists initially of an inert material, such as sand or ash, of particle size between 500 and 1500 μm. This gradually becomes replaced by ash from the coal and additives used for sulphur removal. Ash is continuously removed from the bottom of the bed and, in addition, there is a considerable carry-over by elutriation and this fly ash must be collected in cyclone separators. Bed depths are usually kept below about 0.6 m in order to limit power requirements.

Coal has a lower density than the bed material and, therefore, tends to float, although in a vigorously bubbling bed, the coal can become well mixed with the remainder of the material and the degree of mixing determines the number of feed points that are required. In general, the combustible material does not exceed about 5% of the total solids content of the bed. The maximum size of coal that has been successfully used is 25 mm, which is about two orders of magnitude greater than the particle size in pulverised fuel. The ability to use coal directly from the colliery eliminates the need for pulverising equipment, with its high capital and operating costs.

The coal, when it first enters the bed, gives up its volatiles, and the combustion process thus involves both vapour and char. The mixing of volatiles with air in the bed is not usually very good, with the result that there is considerable flame burning above the surface of the bed. Because the air rates are chosen to give vigorously bubbling beds, much of the oxygen for combustion must pass from the bubble phase to the dense phase before it can react with the char.

Then, in the dense phase, there will be a significant diffusional resistance to transfer of oxygen to the surface of the particles. The combustion process is, as a result, virtually entirely diffusion-controlled at temperatures above about 1120 K. The reaction is one of oxidation to carbon monoxide near the surface of the particles and subsequent reaction to carbon dioxide. Despite the diffusional limitations, up to 90% utilisation of the inlet oxygen may be achieved. The residence time of the gas in the bed is of the order of 1 s, whereas the coal particles may remain in the bed for many minutes. There is generally a significant amount of carry-over char in the fly ash, which is then usually recycled to a burner operating at a rather higher temperature than the main bed. An additional advantage of using a large size of feed coal is that the proportion carried over is correspondingly small.

Whilst fluidised bed furnaces can be operated in the range 1075–1225 K, most operate close to 1175 K. Some of the tubes are immersed in the bed and others are above the free surface. Heat transfer to the immersed tubes is good. Tube areas are usually 6–10 m^2/m^3 of furnace, and transfer coefficients usually range from 300 to 500 W/m^2 K. The radiation component of heat transfer is highly important and heat releases in large furnaces are about 10^6 W/m^3 of furnace.

One of the major advantages of fluidised bed combustion of coal is that it is possible to absorb the sulphur dioxide formed. Generally, limestone or dolomite is added, which breaks up in the bed to yield calcium oxide or magnesium and calcium oxide, which then react with the sulphur dioxide as follows:

$$CaO + SO_2 + \frac{1}{2}O_2 \rightarrow CaSO_4$$

It is possible to regenerate the solid in a separate reactor using a reducing gas consisting of hydrogen and carbon monoxide. There is some evidence that the reactivity of the limestone or dolomite is improved by the addition of chloride, although its use is not generally favoured because of corrosion problems. Fluidised combustion also gives less pollution because less oxides of nitrogen are formed at the lower temperatures existing in the beds.

Corrosion and erosion of the tubes immersed in the bed are at a low level, although there is evidence that the addition of limestone or dolomite causes some sulphide penetration. The chief operating problem is corrosion by chlorine.

Experimental and pilot scale work has been carried out on pressurised operation, and plants have been operated up to 600 kN/m^2, and in at least one case up to 1 MN/m^2 pressure. High pressure operation permits the use of smaller beds. The fluidising velocity required to produce a given condition in the bed is largely independent of pressure, and thus the mass rate of feed of oxygen to the bed is approximately linearly related to the pressure. It is practicable to use deeper beds for pressure operation. Because of the low temperature of operation of fluidised beds, the ash is friable and relatively non-erosive, so that the combustion products can be passed directly through a gas turbine. This inclusion of a gas turbine is an essential feature of the economic operation of pressurised combustors. In general, it is better to use dolomite in place of limestone as an absorbent for sulphur dioxide, because the higher pressures of carbon dioxide lead to inhibition of the breakdown of calcium carbonate to oxide.

It appears likely that fluidised bed combustion of coal may, in the near future, be one of the most important applications of fluidised systems and it may well be that many new coal-fired generating stations will incorporate fluidised bed combustors.

Fluidised bed combustors are now commonplace for large-scale operations and are extensively used in large electricity generating stations. More recently, smaller scale units have been developed for use by individual industrial concerns and, in operating as combined heat and power units, can give overall thermal efficiencies of up to 80%.

Pressurised fluidised combustion units operating at 1–2 MN/m^2 (10–20 bar) are now gaining favour because they give improved combustion and the hot gases can be expanded through a gas turbine, which may be used for generating up to 20% of the total electrical power. Thus, there are marked improvements in efficiency. Because of the reduced specific volume of the high pressure air, it is necessary to use tall narrow combustion units in order to obtain the required gas velocity through the bed, with the minimum fluidising velocity being approximately proportional to the square root of the density (or pressure) of the fluidising air. The flowsheet of a typical pressurised fluidised bed combustor is given in Fig. 9.44,[176] where it is seen that the exhaust gas from the gas turbine is used to preheat the boiler feed water. Sometimes gas velocities are sufficient to entrain part of the fuel from the bed which is then recycled from the outlet of the cyclone to the combustion unit. Fuel may constitute only about 1% of the total volume of the solids in the bed and the particle size needs to be reduced to about 10 mm, compared with 100 µm in pulverised coal burners.

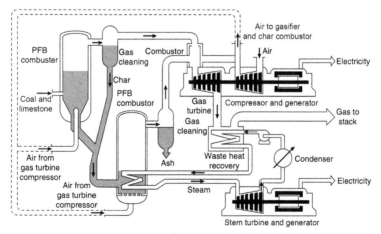

Fig. 9.44
Pressurised fluidised bed combustor with waste heat recovery.[176]

9.8.5 Production of Acrylonitrile by the Sohio Process

Production of acrylonitrile by the Sohio process is an important example of successful application of fluidised beds to carry out exothermic synthesis reactions. Acrylonitrile is produced by a highly exothermic (heat of reaction ~515 kJ/mol) catalytic oxidation of propylene and ammonia.

$$H_2C = CH - CH_3 + NH_3 + (3/2)O_2 \rightarrow CH_2 = CH.CN + 3H_2O$$

A schematic of the fluidised bed reactor in Fig. 9.46 shows air being fed through a bottom distributor, while propylene and ammonia enter the bed above the bottom distributor. The lower section of the bed, being oxygen rich, also serves for carbon burn-off and catalyst regeneration. The mean particle size is in the 50–80 μm range and the reaction is carried out in the 400–500°C temperature range. The heat of reaction is removed by boiling off water passing through the in-bed cooling tubes. The high pressure steam generated is used to drive the air compressor and to supply heat to the downstream process units. Additional details of the process are available in literature sources given in Kunii and Levenspiel.[8]

9.8.6 Polymerisation of Olefins

Another important large-scale application of fluidised bed reactors is in the production of polyolefins, such as polyethylene and polypropylene, using highly active and selective catalysts. The polymerisation reaction shown below for polyethylene production is also highly exothermic (heat of reaction is 92 kJ/mol).

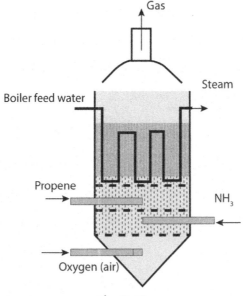

Fig. 9.45
Schematics of a fluidised bed reactor system for the production of Acetonitrile.

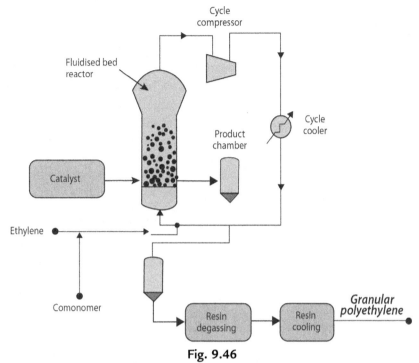

Fig. 9.46
Schematics of process flow sheet used in the production of polyethylene.

$$nC_2H_4 \rightarrow [CH_2 - CH_2]_n$$
$$\underset{\text{Ethylene}}{} \qquad \underset{\text{Polyethylene}}{}$$

A schematic of the process in Fig. 9.46 shows the reactant gas (ethylene and comonomer) joining the recycle gas stream before entering the reactor at its bottom. The combined flow of unreacted recycle gas and fresh feed fluidises the bed of polymer particles (250–1000 μm). Polymerisation occurs on the surface of catalyst particles also fed continuously into the bed. The reaction heat is removed by the cold recycle gas.

9.8.7 Industrial Applications of Three-Phase Fluidised Bed Reactors

Three-phase gas–liquid–solid reactors with suspended particles find applications in a number of processes in chemical, biochemical, and petrochemical industries, and environmental pollution control.[55, 106, 116, 178, 179] In fluidised-bed bioreactors, the bioparticles are either polymeric gel particles with cells entrapped within their porous structure, or particles with biofilm attached to their surfaces. These bioparticles have densities ranging from 1020 to 1300 kg/m^3. Some of these applications require improved circulation and contact between various phases relative to the conventional designs. This can be achieved either by means of a draft tube located concentrically inside the bubble column (internal tube) or by an external tube connected to the top and bottom sections of the bubble column (external loop).

The hydroprocessing of heavy crude oils is one of the most important applications of three-phase fluidised beds. In a hydrotreating process, hydrogen gas, liquid (heavy feed stock), and a solid catalyst are contacted in the reactor and many complex reactions take place, including thermal cracking, hydrodesulfurisation, hydrodeoxygenation, hydrodenitrogenation, hydrodemetallisation, and carbonaceous solid formation. Industrial hydrotreating units employ the fluidised bed or the ebullated bed (H-Oil and LC-Fining) reactor technology. Three-phase fluidised beds offer the following major advantages for this process:

- entrained solids can be tolerated in the heavy oil feed stock;
- on-line catalyst addition/removal maintains constant catalyst activity; and
- isothermal conditions of operation can be maintained in the bed.

Typically, hydrotreating of heavy crude is conducted at temperatures between 300 and 425°C and at pressures between 5.5 and 21 MPa. Fig. 9.47 shows the basic design of an ebullated bed for hydrotreating of petroleum residue.[180] Cylindrical catalyst particles with an average

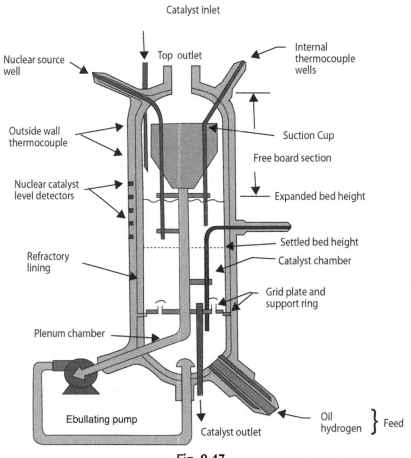

Fig. 9.47
Ebullated bed reactor for heavy oil upgrading.[180]

diameter of 0.8 mm and length 3 mm are fed at the top of the bed and are withdrawn from the bottom of the bed (continuously or periodically). Fresh feed from the preheater is introduced into the reactor plenum chamber along with the internally recycled products. The extent of bed expansion ranges from near the minimum fluidisation condition to a moderate expansion. The bed height is monitored by nuclear catalyst level detectors. Radiation sources are positioned within the vessel to monitor the bed density.

9.9 Nomenclature

		units in SI System	Dimensions in $\mathbf{M, N, L, T, \theta, A}$
A	Area of heat transfer surface	m^2	$\mathbf{L}^2$
A_P	Area of particle	m^2	$\mathbf{L}^2$
a	Area for transfer per unit volume of bed	m^2/m^3	$\mathbf{L}^{-1}$
a'	Area for transfer per unit height of bed	m^2/m	$\mathbf{L}$
C	Fractional volumetric concentration of solids	–	–
C_0	Value of C at $t=0$	–	–
ΔC	Driving force expressed as a molar concentration difference	kmol/m^3	$\mathbf{N L}^{-3}$
ΔC_{lm}	Logarithmic mean value of C	kmol/m^3	$\mathbf{N L}^{-3}$
C_p	Specific heat of gas at constant pressure	J/kg K	$\mathbf{L}^2 \mathbf{T}^{-2} \theta^{-1}$
C_s	Specific heat of solid particle	J/kg K	$\mathbf{L}^2 \mathbf{T}^{-1} \theta^{-1}$
D	Gas phase diffusivity	m^2/s	$\mathbf{L}^2 \mathbf{T}^{-1}$
D_L	Longitudinal diffusivity	m^2/s	$\mathbf{L}^2 \mathbf{T}^{-1}$
d	Particle diameter or diameter of sphere with same surface as particle	m	$\mathbf{L}$
d_b	Bubble diameter	m	$\mathbf{L}$
d_c	Cloud diameter	m	$\mathbf{L}$
d_p	Diameter of sphere of same volume as particle	m	$\mathbf{L}$
d_t	Tube diameter	m	$\mathbf{L}$
E	Factor $r\omega^2/g$ for centrifugal fluidised bed	–	–
e	Voidage	–	–
e_{mf}	Voidage corresponding to minimum fluidising velocity	–	–
f	Fraction of fluid passing through channels in bed	–	–
G	Mass flowrate of fluid	kg/s	$\mathbf{MT}^{-1}$
G_E	Mass flowrate of fluid to cause initial expansion of bed	kg/s	$\mathbf{MT}^{-1}$
G_F	Mass flowrate of fluid to initiate fluidisation	kg/s	$\mathbf{MT}^{-1}$

G'	Mass flowrate of fluid per unit area	kg/m^{2}s	$\mathbf{ML^{-2}T^{-1}}$
g	Acceleration due to gravity	m/s^2	$\mathbf{LT^{-2}}$
h	Heat transfer coefficient	W/m^{2}K	$\mathbf{MT^{-3}\theta^{-1}}$
h_D	Mass transfer coefficient	m/s	$\mathbf{LT^{-1}}$
h_L	Heat transfer coefficient for liquid alone at same rate as in bed	W/m^{2}K	$\mathbf{MT^{-3}\theta^{-1}}$
j'_d	j-factor for mass transfer to particles	–	–
j'_h	j-factor for heat transfer to particles	–	–
K	Distance travelled by bubble to increase its volume by a factor e	m	$\mathbf{L}$
k	Thermal conductivity of fluid	W/m K	$\mathbf{MLT^{-3}\theta^{-1}}$
l	Depth of fluidised bed	m	$\mathbf{L}$
M	Constant in Eq. (9.57)	1/s	$\mathbf{T^{-1}}$
m	Index of $(1-e)$ in Eq. (9.67)	–	–
N_A	Molar rate of transfer of diffusing component	kmol/s	$\mathbf{NT^{-1}}$
n	Index of e in Eq. (9.31)	–	–
P	Pressure	N/m^2	$\mathbf{ML^{-1}T^{-2}}$
$-\Delta P$	Pressure drop across bed due to the presence of solids	N/m^2	$\mathbf{ML^{-1}T^{-2}}$
Q	Rate of transfer of heat	W	$\mathbf{ML^2T^{-3}}$
R	Electrical resistance	W/A^2	$\mathbf{ML^2T^{-3}A^{-2}}$
r	Ratio of rates of change of bed density with velocity for two species (Eq. 9.44)		
S	Specific surface of particles	m^2/m^3	$\mathbf{L^{-1}}$
T	Temperature of gas	K	$\boldsymbol{\theta}$
T_B	Bed temperature	K	$\boldsymbol{\theta}$
T_E	Element temperature	K	$\boldsymbol{\theta}$
T_s	Temperature of solid	K	$\boldsymbol{\theta}$
ΔT	Temperature driving force	K	$\boldsymbol{\theta}$
t	Time	s	$\mathbf{T}$
u_b	Velocity of rise of bubble	m/s	$\mathbf{LT^{-1}}$
u_c	Superficial velocity of fluid (empty tube)	m/s	$\mathbf{LT^{-1}}$
u_i	Value of u_c at infinite dilution	m/s	$\mathbf{LT^{-1}}$
u_{mb}	Minimum value of u_c at which bubbling occurs	m/s	$\mathbf{LT^{-1}}$
u_{mf}	Minimum value of u_c at which fluidisation occurs	m/s	$\mathbf{LT^{-1}}$
u_0	Free-falling velocity of particle in infinite fluid	m/s	$\mathbf{LT^{-1}}$
V	Voltage applied to element	W/A	$\mathbf{ML^2T^{-3}A^{-1}}$
V_B	Volume of bubble	m^3	$\mathbf{L^3}$

V_p	Volume of particle	m^3	**L**^{3}
W	Mass of solids in bed	kg	**M**
x	Volume fraction of spheres	–	–
y	Mole fraction of vapour in gas stream	–	–
y_0	Mole fraction of vapour in inlet gas	–	–
y^*	Mole fraction of vapour in equilibrium with solids	–	–
Z	Total height of bed	m	**L**
z	Height above bottom of bed	m	**L**
α	Ratio of gas velocities in bubble and emulsion phases	–	–
μ	Viscosity of fluid	Ns/m^2	**ML**$^{-1}$ **T**$^{-1}$
ρ	Density of fluid	kg/m^3	**ML**$^{-3}$
ρ_b	Bed density	kg/m^3	**ML**$^{-3}$
ρ_c	Density of suspension	kg/m^3	**ML**$^{-3}$
ρ_s	Density of particle	kg/m^3	**ML**$^{-3}$
ψ	Particle stream function	–	–
ϕ_s	Ratio of diameter of the same specific surface as particles to that of same volume (Eq. 9.16)		–
ω	Angular speed of rotation	s^{-1}	**T**$^{-1}$
Ga	Galileo number $[d^3\rho(\rho_s - \rho)g/\mu^2]$ (Eq. 9.9)	–	–
Ga_p	Galileo number at minimum fluidising velocity	–	–
Nu	Nusselt number (hd_t/k)	–	–
Nu'	Nusselt number (hd/k)	–	–
Pr	Prandtl number ($C_p\mu/k$)	–	–
Re_c	Tube Reynolds number ($u_c d_t\rho/\mu$)	–	–
Re'_c	Particle Reynolds number ($u_c d\rho/\mu$)	–	–
Re'_{mf}	Particle Reynolds number at minimum fluidising velocity ($u_{mf}\rho d/\mu$)	–	–
Re'_{mfh}	Particle Reynolds number Re'_h at minimum fluidising velocity	–	–
Re_1	Bed Reynolds number ($u_c\rho/S\mu(1-e)$)	–	–
Re'_p	Particle Reynolds number with d_p as characteristic diameter	–	–
Re_{1n}	Re_1 for power-law fluid	–	–
Re'_0	Particle Reynolds number ($u_0 d\rho/\mu$)	–	–
Re_1^*	Particle Reynolds number ($u_c d\rho/(1-e)\mu$)	–	–
Sc	Schmidt number ($\mu/\rho D$)	–	–
Sh'	Sherwood number ($h_D d/D$)	–	–

References

1. Wilhelm RH, Kwauk M. Fluidization of solid particles. *Chem Eng Prog* 1948;**44**:201.
2. Jackson R. The mechanisms of fluidised beds. Part 1. The stability of the state of uniform fluidisation. Part 2. The motion of fully developed bubbles. *Trans Inst Chem Eng* 1963;**41**:13, 22.
3. Murray JD. On the mathematics of fluidization. Part 1. Fundamental equations and wave propagation. Part 2. Steady motion of fully developed bubbles. *J Fluid Mech* 1965;**21**:465. 22 (1965) 57.
4. Leva M, Grummer M, Weintraub M, Pollchik M. Fluidization of non-vesicular particles. *Chem Eng Prog* 1948;**44**:619.
5. Richardson JF, Meikle RA. Sedimentation and fluidisation. Part III. The sedimentation of uniform fine particles and of two-component mixtures of solids. *Trans Inst Chem Eng* 1961;**39**:348.
6. Wen CY, Yu YH. Mechanics of fluidization. *Chem Eng Prog Symp Ser* 1960;**62**:100.
7. Niven RK. Physical insight into the Ergun and Wen & Yu equations for fluid flow in packed or fluidised beds. *Chem Eng Sci* 2002;**57**:527.
8. Kunnii D, Levenspiel O. *Fluidization engineering.* 2nd ed. Boston: Butterworth-Heinmann; 1991.
9. Wen CY, Yu YH. A generalized method for predicting the minimum fluidization velocity. *AIChE J* 1966;**12**:610.
10. Xu CC, Zhu J. Prediction of the minimum fluidization velocity for fine particles of various degrees of cohesiveness. *Chem Eng Commun* 2009;**196**:499.
11. Lettieri P, Newton D, Yates JG. Homogeneous bed expansion of FCC catalysts, influence of temperature on the parameters of the Richardson–Zaki equation. *Powder Technol* 2002;**123**:221.
12. Chitester DC, Kornosky RM, Fan LS, Danko JP. Characteristics of fluidization at high pressure. *Chem Eng Sci* 1984;**39**:253.
13. Coltters R, Rivas AL. Minimum fluidation velocity correlations in particulate systems. *Powder Technol* 2004;**147**:34.
14. Rowe PN. Drag forces in a hydraulic model of a fluidised bed—Part II. *Trans Inst Chem Eng* 1961;**39**:175.
15. Rowe PN, Partridge BA. An X-ray study of bubbles in fluidised beds. *Trans Inst Chem Eng* 1965;**43**:T157.
16. Pinchbeck PH, Popper F. Critical and terminal velocities in fluidization. *Chem Eng Sci* 1956;**6**:57.
17. Richardson JF, Zaki WN. Sedimentation and fluidisation. Part 1. *Trans Inst Chem Eng* 1954;**32**:35.
18. Godard KE, Richardson JF. Correlation of data for minimum fluidising velocity and bed expansion in particulately fluidised beds. *Chem Eng Sci* 1969;**24**:363.
19. Lewis EW, Bowerman EW. Fluidization of solid particles in liquids. *Chem Eng Prog* 1952;**48**:603.
20. Khan AR, Richardson JF. Fluid–particle interactions and flow characteristics of fluidized beds and settling suspensions of spherical particles. *Chem Eng Commun* 1989;**78**:111.
21. Srinivas BK, Chhabra RP. An experimental study of non-Newtonian fluid flow in fluidised beds: minimum fluidisation velocity and bed expansion. *Chem Eng Process* 1991;**29**:121–31.
22. Stewart PSB, Davidson JF. Slug flow in fluidised beds. *Powder Technol* 1967;**1**:61.
23. Simpson HC, Rodger BW. The fluidization of light solids by gases under pressure and heavy solids by water. *Chem Eng Sci* 1961;**16**:153.
24. Harrison D, Davidson JF, de Kock JW. On the nature of aggregative and particulate fluidisation. *Trans Inst Chem Eng* 1961;**39**:202.
25. Lawther KP, Berglin CLW. *Fluidisation of lead shot with water.* United Kingdom Atomic Energy Authority Report, A.E.R.E., CE/R 2360. 1957.
26. Richardson JF, Smith JW. Heat transfer to liquid fluidised systems and to suspensions of coarse particles in vertical transport. *Trans Inst Chem Eng* 1962;**40**:13.
27. Anderson TB, Jackson R. The nature of aggregative and particulate fluidization. *Chem Eng Sci* 1964;**19**:509.
28. Hassett NJ. The mechanism of fluidization. *Br Chem Eng* 1961;**6**:777.
29. Lawson A, Hassett NJ. Discontinuities and flow patterns in liquid–fluidized beds, In: Proc. Intl. Symp. on Fluidization, Eindhoven: Netherlands Univ. Press; 1967. p. 113.
30. Cairns EJ, Prausnitz JM. Longitudinal mixing in fluidization. *AIChE J* 1960;**6**:400.

31. Kramers H, Westermann MD, de Groot JH, Dupont FAA. The longitudinal dispersion of liquid in a fluidised bed. In: *Third congress of the european federation of chemical engineering*; 1962. The Interaction between Fluids and Particles 114.

32. Reuter H. On the nature of bubbles in gas and liquid fluidized beds. *Chem Eng Prog Symp Ser* 1966;**62**:92.

33. Gibilaro LG, di Felice R, Hossain I, Foscolo PU. The experimental determination of one-dimensional wave velocities in liquid fluidized beds. *Chem Eng Sci* 1989;**44**:101.

34. Bailey, C. Private Communication.

35. Volpicelli G, Massimilla L, Zenz FA. Non-homogeneities in solid–liquid fluidization. *Chem Eng Prog Symp Ser* 1966;**62**:63.

36. Foscolo PU, di Felice R, Gibilaro LG. The pressure field in an unsteady state fluidized bed. *AIChE J* 1989;**35**:1921.

37. Jottrand R. Etude de quelques aspects de la fluidisation dans les liquides. *Chem Eng Sci* 1954;**3**:12.

38. Pruden BB, Epstein N. Stratification by size in particulate fluidization and in hindered settling. *Chem Eng Sci* 1964;**19**:696.

39. Kennedy SC, Bretton RH. Axial dispersion of spheres fluidized with liquids. *AIChE J* 1966;**12**:24.

40. Al-Dibouni MR, Garside J. Particle mixing and classification in liquid fluidised beds. *Trans Inst Chem Eng* 1979;**57**:94.

41. Juma AKA, Richardson JF. Particle segregation in liquid-solid fluidised beds. *Chem Eng Sci* 1979;**34**:137.

42. Gibilaro LG, Hossain I, Waldram SP. On the Kennedy and Bretton model for mixing and segregation in liquid fluidized beds. *Chem Eng Sci* 1985;**40**:2333.

43. Moritomi H, Yamagishi T, Chiba T. Prediction of complete mixing of liquid-fluidized binary solid particles. *Chem Eng Sci* 1986;**41**:297.

44. Barghi S, Briens CL, Bergougnou MA. Mixing and segregation of binary mixtures of particles in liquid–solid fluidized beds. *Powder Technol* 2003;**131**:223.

45. Escudie R, Epstein N, Grace JR, Bi HT. Effect of particle shape on liquid-fluidized beds of binary (and ternary) solids mixtures: segregation vs. mixing. *Chem Eng Sci* 2006;**61**:1528.

46. Richardson, J. F. and Afiatin, E. Fluidisation and sedimentation of mixtures of particles. 9th T & S. International Conference on Solid Particles, 2–5 September, 1997, Crakow, Poland, 486.

47. Epstein N, Pruden BB. Liquid fluidisation of binary particle mixtures—III. Stratification by size and related topics. *Chem Eng Sci* 1999;**54**:401.

48. Handley D, Doraisamy A, Butcher KL, Franklin NL. A study of the fluid and particle mechanics in liquid-fluidised beds. *Trans Inst Chem Eng* 1966;**44**:T260.

49. Carlos CR, Richardson JF. Particle speed distribution in a fluidised system. *Chem Eng Sci* 1967;**22**:705.

50. Carlos CR. *Solids mixing in fluidised beds*. University of Wales; 1967. Ph.D. thesis.

51. Latif BAJ, Richardson JF. Circulation patterns and velocity distributions for particles in a liquid fluidised bed. *Chem Eng Sci* 1972;**27**:1933.

52. Dorgelo EAH, van der Meer AP, Wesselingh JA. Measurement of the axial dispersion of particles in a liquid fluidized bed applying a random walk method. *Chem Eng Sci* 1985;**40**:2105.

53. Davies L, Richardson JF. Gas interchange between bubbles and the continuous phase in a fluidised bed. *Trans Inst Chem Eng* 1966;**44**:T293.

54. Yerushalmi J, Cankurt NT. Further studies of the regimes of fluidization. *Powder Technol* 1979;**24**:187.

55. Yang WC, editor. *Handbook of fluidization and fluid-particle systems*. Marcel Dekker Inc.; 2003.

56. Bi HT, Ellis N, Abba IA, Grace JR. A state-of-the-art review of gas-solid turbulent fluidization. *Chem Eng Sci* 2000;**55**:4789.

57. Geldart D. Types of fluidization. *Powder Technol* 1973;**7**:285.

58. Godard KE, Richardson JF. The behaviour of bubble-free fluidised beds. *I Chem E Symp Ser* 1968;**30**:126.

59. Leung LS. *Bubbles in fluidised beds*. University of Cambridge; 1961. Ph.D. thesis.

60. Jacob KV, Weiner AW. High pressure particulate expansion and minimum bubbling of fine carbon powders. *AIChE J* 1987;**33**:1698.

61. Olowson PA, Almstedt AE. Influence of pressure on the minimum fluidization velocity. *Chem Eng Sci* 1991;**46**:637.
62. Saxena SC, Shrivastava S. The influence of an external magnetic field on an air-fluidized bed of ferromagnetic particles. *Chem Eng Sci* 1990;**45**:1125.
63. Godard K, Richardson JF. The use of slow speed stirring to initiate particulate fluidisation. *Chem Eng Sci* 1969;**24**:194.
64. Matheson GL, Herbst WA, Holt PH. Characteristics of fluid–solid systems. *Ind Eng Chem* 1949;**41**: 1099–104.
65. Leltieri P, Macri D. Effect of process conditions on fluidization. *KONA Powder Part J* 2016;**33**:86.
66. Davidson JF, Paul RC, Smith MJS, Duxbury HA. The rise of bubbles in a fluidised bed. *Trans Inst Chem Eng* 1959;**37**:323.
67. Davies, R. M. and Taylor, G. I.: The mechanics of large bubbles rising through extended liquids and liquids in tubes. Proc Roy Soc A200 (1950) 375.
68. Davidson JF, Harrison D. *Fluidised particles*. Cambridge: Cambridge University Press; 1963.
69. Darton RC, LaNauze RD, Davidson JF, Harrison D. Bubble growth due to coalescence in fluidized beds. *Trans Inst Chem Eng* 1977;**55**:274.
70. Horio M, Nonaka A. A generalized bubble diameter correlation for gas-solid fluidized beds. *AIChE J* 1987;**33**:1865.
71. Mori S, Wen CY. Estimation of bubble diameter in gaseous fluidized beds. *AIChE J* 1975;**21**:109.
72. Rowe PN. Prediction of bubble size in a gas fluidised bed. *Chem Eng Sci* 1976;**31**:285.
73. Harrison D, Leung LS. Bubble formation at an orifice in a fluidised bed. *Trans Inst Chem Eng* 1961;**39**:409.
74. Harrison D, Leung LS. The coalescence of bubbles in fluidised beds. In: *Third congress of the european federation of chemical engineering*; 1962. The Interaction between Fluids and Particles 127.
75. Rowe PN, Wace PF. Gas-flow patterns in fluidised beds. *Nature* 1960;**188**:737.
76. Wace PF, Burnett ST. Flow patterns in gas-fluidised beds. *Trans Inst Chem Eng* 1961;**39**:168.
77. Davidson JF. In discussion of symposium on fluidisation. *Trans Inst Chem Eng* 1961;**39**:230.
78. Rowe PN, Partridge BA, Lyall E. Cloud formation around bubbles in gas fluidized beds. *Chem Eng Sci* 1964;**19**:973. 20 (1965) 1151.
79. Rowe PN, Henwood GN. Drag forces in a hydraulic model of a fluidised bed. Part 1. *Trans Inst Chem Eng* 1961;**39**:43.
80. Bi HT, Grace JR. Effects of measurement method on velocities used to demarcate the transition to turbulent fluidization. *Chem Eng J* 1995;**57**:261.
81. Bart R. *Mixing of fluidized solids in small diameter columns*. Massachusetts Institute of Technology; 1950. Sc.D. thesis.
82. Massimilla L, Bracale S. Il mesolamento della fase solida nei sistemi: Solido–gas fluidizzati, liberi e frenati. *La Ricerca Scientifica* 1957;**27**:1509.
83. May WG. Fluidized bed reactor studies. *Chem Eng Prog* 1959;**55**:49.
84. Sutherland KS. Solids mixing studies in gas fluidised beds. Part 1. A preliminary comparison of tapered and non-tapered beds. *Trans Inst Chem Eng* 1961;**39**:188.
85. Rowe PN, Partridge BA. Particle movement caused by bubbles in a fluidised bed. In: *Third congress of the european federation of chemical engineering*; 1962. The Interaction between Fluids and Particles 135.
86. Askins JW, Hinds GP, Kunreuther F. Fluid catalyst–gas mixing in commercial equipment. *Chem Eng Prog* 1951;**47**:401.
87. Danckwerts PV, Jenkins JW, Place G. The distribution of residence times in an industrial fluidised reactor. *Chem Eng Sci* 1954;**3**:26.
88. Gilliland ER, Mason EA. Gas and solid mixing in fluidized beds. *Ind Eng Chem* 1949;**41**:1191.
89. Gilliland ER, Mason EA. Gas mixing in beds of fluidized solids. *Ind Eng Chem* 1952;**44**:218.
90. Gilliland ER, Mason EA, Oliver RC. Gas-flow patterns in beds of fluidized solids. *Ind Eng Chem* 1953;**45**:1177.
91. Lanneau KP. Gas–solids contacting in fluidized beds. *Trans Inst Chem Eng* 1960;**38**:125.

92. Szekely J. Mass transfer between the dense phase and lean phase in a gas–solid fluidised system. In: *Third congress of the european federation of chemical engineering*; 1962. The Interaction between Fluids and Particles 197.

93. Chen JL-P, Keairns DL. Particle segregation in a fluidized bed. *Can J Chem Eng* 1975;**53**:395.

94. Hoffman AC, Janssen LPBM, Prins J. Particle segregation in fluidized binary mixtures. *Chem Eng Sci* 1993;**48**:1583.

95. Wu SY, Baeyens J. Segregation by size difference in gas fluidized beds. *Powder Technol* 1998;**98**:139.

96. Leva M. Elutriation of fines from fluidized systems. *Chem Eng Prog* 1951;**47**:39.

97. Pemberton ST, Davidson JF. Elutriation from fluidized beds. I Particle ejection from the dense phase into the freeboard. II. Disengagement of particles from gas in the freeboard. *Chem Eng Sci* 1986;**41**:243. 253.

98. Chen Y-M. Fundamentals of a centrifugal fluidized bed. *AIChE J* 1987;**33**:722.

99. Fan LT, Chang CC, Yu YS. Incipient fluidization condition for a centrifugal fluidized bed. *AIChE J* 1985;**31**:999.

100. Bridgwater J. Spouted beds. In: Davidson JF, Clift R, Harrison D, editors. *Fluidization*. 2nd ed. Academic Press; 1985. p. 201.

101. Mathur KB, Epstein N. *Spouted beds*. Academic Press; 1974.

102. Mathur KB, Epstein N. Dynamics of spouting beds. In: Drew TB, Cokelet GR, Hoopes JW, Vermeulen T, editors. *Advances in chemical engineering*. **9**. Academic Press; 1974. p. 111.

103. Darton RC. The physical behaviour of three-phase fluidized beds. In: Davidson JF, Clift R, Harrison D, editors. *Fluidization*. 2nd ed. Academic Press; 1985. p. 495.

104. Epstein N. Three phase fluidization: some knowledge gaps. *Can J Chem Eng* 1981;**59**:649.

105. Lee JC, Buckley PS. Fluid mechanics and aeration characteristics of fluidised beds. In: Cooper PF, Atkinson B, editors. *Biological fluidised bed treatment of water and wastewater*. Chichester: Ellis Horwood; 1981. p. 62.

106. Fan L-S. *Gas–liquid–solid fluidization engineering*. Butterworth; 1989.

107. Song GH, Bavarian F, Fan LS, Buttke RD, Peck LB. *Paper presented at* A.I.Ch.E. *Annual Meeting, New York, November 15–20, 1987*

108. Fan LS, Jean RH, Kitano K. On the operating regimes of cocurrent upward gas-liquid-solid systems with liquid as the continuous phase. *Chem Eng Sci* 1987;**42**:1853.

109. Fan LS, Yamashita T, Jean RH. Solids mixing and segregation in a gas-liquid-solid fluidized bed. *Chem Eng Sci* 1987;**42**:17.

110. Muroyama K, Fan L-S. Fundamentals of gas–liquid–solid fluidization. *AIChE J* 1985;**31**:1.

111. Zhang JP, Grace JR, Epstein N, Lim KS. Flow regime identification in gas-liquid flow and three-phase fluidized beds. *Chem Eng Sci* 1997;**42**:3979.

112. Hu T, Yu B, Wang Y. Holdups and models of three-phase fluidized beds. In: Ostergaard K, Sorensen A, editors. *Fluidization V*. Engineering Foundation Press; 1986. p. 353.

113. Begovich JM, Watson JS. Hydrodynamic characteristics of three-phase fluidized beds. In: Davidson JF, Keairns DL, editors. *Fluidization*. Cambridge Univ. Press; 1978. p. 190.

114. Østergaard K. On bed porosity in gas—liquid fluidization. *Chem Eng Sci* 1965;**20**:165.

115. Turner R. *Fluidisation*. Society of Chemical Industry; 1964, p. 47.

116. Nigam KDP, Schumpe A. *Three-phase sparged reactors*. Netherlands: Gordon and Breach; 1996.

117. Mickley HS, Fairbanks DF. Mechanism of heat transfer to fluidized beds. *AIChE J* 1955;**1**:374.

118. Lemlich R, Caldas I. Heat transfer to a liquid fluidized bed. *AIChE J* 1958;**4**:376.

119. Richardson JF, Mitson AE. Sedimentation and fluidisation. Part II. Heat transfer from a tube wall to a liquid-fluidised system. *Trans Inst Chem Eng* 1958;**36**:270.

120. Jepson G, Poll A, Smith W. Heat transfer from gas to wall in a gas–solids transport line. *Trans Inst Chem Eng* 1963;**41**:207.

121. Kang Y, Fan LT, Kim SD. Immersed heater-to-bed heat transfer in liquid–solid fluidized beds. *AIChE J* 1991;**37**:1101.

122. Romani MN, Richardson JF. Heat transfer from immersed surfaces to liquid-fluidized beds. *Lett Heat Mass Transf* 1974;**1**:55.

123. Richardson JF, Romani MN, Shakiri KJ. Heat transfer from immersed surfaces in liquid fluidised beds. *Chem Eng Sci* 1976;**31**:619.

124. Khan AR, Juma AKA, Richardson JF. Heat transfer from a plane surface to liquids and to liquid-fluidised beds. *Chem Eng Sci* 1983;**38**:2053.

125. Leva M. *Fluidization.* McGraw-Hill; 1959.

126. Levenspiel O, Walton JS. *Heat transfer coefficients in beds of moving solids.* Berkeley, CA: Proc. Heat Transf. Fluid Mech. Inst.; 1949, pp. 139–46

127. Vreedenberg HA. Heat transfer between fluidised beds and vertically inserted tubes. *J Appl Chem* 1952;**2**:S26.

128. Dow WM, Jakob M. Heat transfer between a vertical tube and a fluidized air–solid mixture. *Chem Eng Prog* 1951;**47**:637.

129. Botterill JSM, Redish KA, Ross DK, Williams JR. The mechanism of heat transfer to fluidised beds. In: *Third congress of the european federation of chemical engineering*; 1962. The Interaction between Fluids and Particles 183.

130. Botterill JSM, Williams JR. The mechanism of heat transfer to gas-fluidised beds. *Trans Inst Chem Eng* 1963;**41**:217.

131. Botterill JSM, Brundrett GW, Cain GL, Elliott DE. Heat transfer to gas-fluidized beds. *Chem Eng Prog Symp Ser* 1966;**62**:1.

132. Williams JR. *The mechanism of heat transfer to fluidised beds.* University of Birmingham; 1962. Ph.D. thesis.

133. Higbie R. The rate of absorption of a pure gas into a still liquid during periods of exposure. *Trans Am Inst Chem Eng* 1935;**31**:365.

134. Danckwerts PV. Significance of liquid-film coefficients in gas absorption. *Ind Eng Chem* 1951;**43**:1460.

135. Mickley, H. S., Fairbanks, D. F., and Hawthorn, R. D.: The relation between the transfer coefficient and thermal fluctuations in fluidized-bed heat transfer. Chem Eng Prog Symp Ser 32, 57 (1961) 51.

136. Bock, H.-J. and Molerus, O.: Influence of hydrodynamics on heat transfer in fluidized beds. In Fluidization by Grace, J. R. and Matsen, J. M. (eds.), Plenum Press, New York (1980), 217.

137. Richardson JF, Shakiri KJ. Heat transfer between a gas–solid fluidised bed and a small immersed surface. *Chem Eng Sci* 1979;**34**:1019.

138. Xavier AM, King DF, Davidson JF, Harrison D. Surface–bed heat transfer in a fluidised bed at high pressure. In: Grace JR, Matsen JM, editors. *Fluidization.* New York: Plenum Press; 1980. p. 209.

139. Stefanova A, Bi HT, Lim CJ, Grace JR. Heat transfer from immersed vertical tube in a fluidised bed of group A particles near the transition to the turbulent fluidization flow regime. *Int J Heat Mass Transf* 2008;**51**:2020.

140. Pisters K, Prakash A. Investigations of axial and radial variations of heat transfer coefficient in bubbling fluidized bed with fast response probe. *Powder Technol* 2011;**207**:224.

141. Richardson JF, Bakhtiar AG. Mass transfer between fluidised particles and gas. *Trans Inst Chem Eng* 1958;**36**:283.

142. Richardson JF, Szekely J. Mass transfer in a fluidised bed. *Trans Inst Chem Eng* 1961;**39**:212.

143. Mullin JW, Treleaven CR. Solids–liquid mass transfer in multi-particulate systems. In: *Third congress of the european federation of chemical engineering*; 1962. The Interaction between Fluids and Particles 203.

144. Chu JC, Kalil J, Wetteroth WA. Mass transfer in a fluidized bed. *Chem Eng Prog* 1953;**49**:141.

145. Kettenring, K. N., Manderfield, E. L., and Smith, J. M.: Heat and mass transfer in fluidized systems. Chem Eng Prog 46 (1950) 139.

146. Heertjes PM, McKibbins SW. The partial coefficient of heat transfer in a drying fluidized bed. *Chem Eng Sci* 1956;**5**:161.

147. Frantz JF. Fluid-to-particle heat transfer in fluidized beds. *Chem Eng Prog* 1961;**57**:35.

148. Richardson JF, Ayers P. Heat transfer between particles and a gas in a fluidised bed. *Trans Inst Chem Eng* 1959;**37**:314.

149. Resnick W, White RR. Mass transfer in systems of gas and fluidized solids. *Chem Eng Prog* 1949;**45**:377.

150. Wilke CR, Hougen OA. Mass transfer in the flow of gases through granular solids extended to low modified Reynolds Numbers. *Trans Am Inst Chem Eng* 1945;**41**:445.

151. McCune LK, Wilhelm RH. Mass and momentum transfer in solid–liquid system. Fixed and fluidized beds. *Ind Eng Chem* 1949;**41**:1124.

152. Hobson M, Thodos G. Mass transfer in flow of liquids through granular solids. *Chem Eng Prog* 1949;**45**:517.

153. Gaffney BJ, Drew TB. Mass transfer from packing to organic solvents in single phase flow through a column. *Ind Eng Chem* 1950;**42**:1120.

154. Hsu CT, Molstad MC. Rate of mass transfer from gas stream to porous solid in fluidized beds. *Ind Eng Chem* 1955;**47**:1550.

155. Cornish ARH. Note on minimum possible rate of heat transfer from a sphere when other spheres are adjacent to it. *Trans Inst Chem Eng* 1965;**43**:T332.

156. Zabrodsky SS. On solid-to-fluid heat transfer in fluidized systems. *Int J Heat Mass Transfer* 1967;**10**:1793.

157. Magiliotou M, Chen Y-M, Fan L-S. Bed-immersed object heat transfer in a three-phase fluidized bed. *AIChE J* 1988;**34**:1043.

158. Kang Y, Suh IS, Kim SD. Heat transfer characteristics of three phase fluidized beds. *Chem Eng Commun* 1985;**34**:1.

159. Kato Y, Uchida K, Kago T, Morooka S. Liquid holdup and heat transfer coefficient between bed and wall in liquid-solid and gas-liquid-solid fluidized beds. *Powder Technol* 1981;**28**:173.

160. Baker CGJ, Armstrong ER, Bergougnou MA. Heat transfer in three-phase fluidized beds. *Powder Technol* 1978;**21**:195.

161. Chiu TM, Ziegler EN. Heat transfer in three-phase fluidized beds. *AIChE J* 1983;**29**:677.

162. Kim SD, Kang Y. Heat and mass transfer in three-phase fluidized bed reactors- an overview. *Chem Eng Sci* 1997;**52**:3639.

163. Deckwer WD, Louisi Y, Zaidi A, Ralek M. Hydrodynamic properties of the fischer-tropsch slurry process. *Ind Eng Chem Process Des Dev* 1980;**19**:699.

164. Kim SD, Kang Y, Kwon HK. Heat transfer characteristics in two and three-phase slurry fluidized beds. *AIChE J* 1986;**32**:1397.

165. Suh IS, Jin GT, Kim SD. Heat transfer coefficients in three phase fluidized beds. *Int J Multiphase Flow* 1985;**11**:255.

166. Lee DH, Kim JO, Kim SD. Mass transfer and hold-up characteristics in three phase fluidized beds. *Chem Eng Commun* 1993;**119**:179.

167. Fukuma M, Sato M, Muroyama K, Yasunishi A. Particle to liquid mass transfer in gas-liquid-solid fluidization. *J Chem Eng Jap* 1988;**21**:231.

168. Windebank CS. The fluid catalyst technique in modern petroleum refining. *J Imp Coll Chem Eng Soc* 1948;**4**:31.

169. Riley HL. Design of fluidised reactors for naphthalene oxidation: a review of patent literature. *Trans Inst Chem Eng* 1959;**37**:305.

170. Rowe, P. N.: A theoretical study of a batch reaction in a gas-fluidised bed. *Soc Chem Ind, Fluidisation* (1964) 15.

171. Volk, W., Johnson, C. A., Stotler, H. H. Effect of reactor internals on quality of fluidization. *Chem Eng Prog Symp Ser* 38, 58 (1962) 38.

172. Institute of Fuel. *Symposium series no. 1: Fluidised combustion 1.* 1975.

173. Turnbull E, Davidson JF. Fluidized combustion of char and volatiles from coal. *AIChE J* 1984;**30**:881.

174. Wu RL, Grace JR, Lim CJ, Brereton MH. Surface-to-surface heat transfer in a circulating-fluidized-bed combustor. *AIChE J* 1989;**35**:1685.

175. Agarwal PK, la Nauze RD. Transfer processes local to the coal particle. A review of drying, devolatilization and mass transfer in fluidized bed combustion. *Chem Eng Res Des* 1989;**67**:457.

176. Redman J. Fluidised bed combustion, SOx and NOx. *Chem Eng* 1989;**467**:32.

177. Valk M, editor. *Atmospheric fluidized bed coal combustion.* Amsterdam: Elsevier; 1994.

178. Nicolella C, Van Loosdrecht MCM, Heijnen JJ. Wastewater treatment with particulate biofilm reactors. *J Biotechnol* 2000;**80**:1.

179. Wild G, Saberian M, Schwartz JL, Charpentier JC. Gas-liquid-solid fluidized bed reactors. State of the art and industrial possibilities. *Int Chem Eng* 1984;**24**:639.

180. Li A, Liu D. *Proceedings of the 2nd world congress of chemical engineering, IV.* **170**.

Further Reading

Botterill JSM. *Fluid bed heat transfer*. Academic Press; 1975.

Davidson JF, Clift R, Harrison D, editors. *Fluidization*. 2nd ed. Academic Press; 1985.

Kwauk M. *Fluidization. Idealized and bubbleless, with applications*. Science Press and Ellis Horwood; 1992.

Molerus O. *Principles of flow in disperse systems*. Berlin: Chapman and Hall; 1993.

Molerus O, Wirth KE. *Heat transfer in fluidized beds*. Chapman and Hall; 1997.

Zenz FA, Othmer DF. *Fluidization and fluid–particle systems*. Reinhold; 1960.

CHAPTER 10

Liquid Filtration

10.1 Introduction

The separation of solids from a suspension in a liquid by means of a porous medium or screen which retains the solids and allows the liquid to pass is termed filtration.

In general, the pores of the medium are larger than the particles which are to be removed, and the filter works efficiently only after an initial deposit has been trapped in the medium. In the laboratory, filtration is often carried out using a form of Buchner funnel, and the liquid is sucked through the thin layer of particles using a source of vacuum. In even simpler cases, the suspension is poured into a conical funnel fitted with a filter paper. In the industrial equivalent, difficulties are encountered in the mechanical handling of much larger quantities of suspensions and solids. A thicker layer of solids has to form, and, in order to achieve a high rate of passage of liquid through the solids, higher pressures are needed, and a far greater area has to be provided. A typical filtration operation is illustrated in Fig. 10.1, which shows the filter medium, in this case a cloth, its support and the layer of solids, or filter cake, which has already formed.

Volumes of the suspensions to be handled vary from the extremely large quantities involved in water purification and ore handling in the mining industry to relatively small quantities, as in the fine and specialty chemical industry, where the variety of solids is considerable. In most industrial applications, it is the solids that are required, and their physical size and properties are of paramount importance. Thus, the main factors to be considered when selecting equipment and operating conditions are:

- The properties of the fluid, particularly its viscosity, density, and corrosive properties.
- The nature of the solid—its particle size and shape, size distribution, and packing characteristics.
- The concentration of solids in suspension.
- The quantity of material to be handled, and its value.
- Whether the valuable product is the solid, the fluid, or both.
- Whether it is necessary to wash the filtered solids.

Coulson and Richardson's Chemical Engineering. https://doi.org/10.1016/B978-0-08-101098-3.00011-1

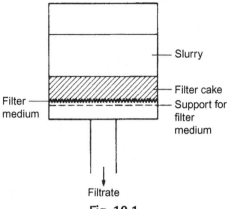

Fig. 10.1
Principle of filtration.

- Whether very slight contamination caused by contact of the suspension or filtrate with the various components of the equipment is detrimental to the product.
- Whether the feed liquor may be heated.
- Whether any form of pretreatment might be helpful.

Filtration is essentially a mechanical operation and is less demanding in energy than evaporation or drying, where the high latent heat of the liquid, which is usually water, has to be provided. In the typical operation shown in Fig. 10.1, the cake gradually builds up on the medium, and the resistance to flow progressively increases. During the initial period of flow, particles are deposited in the surface layers of the cloth to form the true filtering medium. This initial deposit may be formed from a special initial flow of precoat material, which is discussed later. The most important factors on which the rate of filtration then depends will be:

(a) The drop in pressure from the feed to the far side of the filter medium.
(b) The area of the filtering surface.
(c) The viscosity of the filtrate.
(d) The resistance of the filter cake.
(e) The resistance of the filter medium and initial layers of cake.

Two basic types of filtration processes may be identified, although there are cases where the two types appear to merge. In the first, frequently referred to as *cake filtration*, the particles from the suspension (which usually has a high proportion of solids) are deposited on the surface of a porous septum, which should ideally offer only a small resistance to flow. As the solids build up on the septum, the initial layers form the effective filter medium, preventing the particles from embedding themselves in the filter cloth, and ensuring that a particle-free filtrate is obtained.

In the second type of filtration, *depth* or *deep-bed filtration*, the particles penetrate into the pores of the filter medium, where impacts between the particles and the surface of the medium are largely responsible for their removal and retention. This configuration is commonly used for the removal of fine particles from very dilute suspensions, where the recovery of the particles is not of primary importance. Typical examples here include air and water filtration. The filter bed gradually becomes clogged with particles, and its resistance to flow eventually reaches an unacceptably high level. For continued operation, it is, therefore, necessary to remove the accumulated solids, and it is important that this can be readily achieved. For this reason, the filter commonly consists of a bed of particulate solids, such as sand, which can be cleaned by back-flushing, often accompanied by fluidisation. In this chapter, the emphasis is on cake filtration, although deep-bed filtration, which has been discussed in detail by Ives,[1, 2] is considered in the section on bed filters.

There are two principal modes under which deep-bed filtration may be carried out. In the first, *dead-end filtration*, which is illustrated in Fig. 10.1, the slurry is filtered in such a way that it is fed perpendicularly to the filter medium, and there is little flow parallel to the surface of the medium. In the second, termed *cross-flow filtration*, which is discussed in Section 10.3.5 and which is used particularly for very dilute suspensions, the slurry is continuously recirculated so that it flows essentially across the surface of the filter medium at a rate considerably in excess of the flow rate through the filter cake.

10.2 Filtration Theory

10.2.1 Introduction

Equations are given in Chapter 7 for the calculation of the rate of flow of a fluid through a bed of granular material, and these are now applied to the flow of filtrate through a filter cake. Some differences in general behaviour may be expected, however, because the cases so far considered relate to uniform fixed beds, whereas in filtration, the bed is steadily growing in thickness. Thus, if the filtration pressure is constant, the rate of flow progressively diminishes, whereas, if the flow rate is to be maintained as constant, the pressure must be gradually increased.

The mechanical details of the equipment, particularly of the flow channel and the support for the medium, influence the way the cake is built up and the ease with which it may be removed. A uniform structure is very desirable for good washing, and cakes formed from particles of very mixed sizes and shapes present special problems. Although filter cakes are complex in their structure and cannot truly be regarded as being composed of rigid non-deformable particles, the method of relating the flow parameters developed in Chapter 7 is useful in describing the flow within the filter cake. The general theory of filtration and its importance in design has been considered by Suttle.[3] It may be noted that there are two quite different methods of operating a batch filter. If the pressure is kept constant, then the rate of flow progressively diminishes, whereas if the flow rate is kept constant, then the pressure must be gradually increased. Because the particles forming the cake are small and the flow through the bed is slow, streamline conditions are almost invariably obtained, and, at any instant, the flow rate of the filtrate may be represented by the following form of Eq. (7.9):

$$u_c = \frac{1}{A}\frac{dV}{dt} = \frac{1}{5}\frac{e^3}{(1-e)^2}\frac{-\Delta P}{S^2 \mu l} \tag{10.1}$$

where V is the volume of filtrate which has passed in time t, A is the total cross-sectional area of the filter cake, u_c is the superficial velocity of the filtrate, l is the cake thickness, S is the specific surface of the particles, e is the voidage, μ is the viscosity of the filtrate, and ΔP is the applied pressure difference.

In deriving this equation, it is assumed that the cake is uniform, and that the voidage is constant throughout. In the deposition of a filter cake, this is unlikely to be the case, and the voidage, e, will depend on the nature of the support, including its geometry and surface structure, and on the rate of deposition. The initial stages in the formation of the cake are, therefore, of special importance for the following reasons:

(a)   For any filtration pressure, the rate of flow is greatest at the beginning of the process because the resistance is then at a minimum.

(b)   High initial rates of filtration may result in plugging of the pores of the filter cloth and cause a very high resistance to flow.

(c)   The orientation of the particle in the initial layers may appreciably influence the structure of the whole filter cake.

Filter cakes may be divided into two classes—incompressible cakes and compressible cakes. For an incompressible cake, the resistance to flow of a given volume of cake is not appreciably affected either by the pressure difference across the cake or by the rate of deposition of material. On the other hand, with a compressible cake, increase of the pressure difference or of the rate of flow causes the formation of a denser cake with a higher resistance. For incompressible cakes, e in Eq. (10.1) may be taken as constant, and the quantity $e^3/[5(1-e)^2 S^2]$ is then a property of the particles forming the cake and should be constant for a given material.

Thus:

$$\frac{1}{A}\frac{dV}{dt} = \frac{-\Delta P}{\mathbf{r}\mu l} \tag{10.2}$$

where

$$\mathbf{r} = \frac{5(1-e)^2 S^2}{e^3} \tag{10.3}$$

It may be noted that, when there is a hydrostatic pressure component such as with a horizontal filter surface, this should be included in the calculation of $-\Delta P$.

Eq. (10.2) is the basic filtration equation, and $\mathbf{r}$ is termed the specific resistance, which is seen to depend on e and S. For incompressible cakes, $\mathbf{r}$ is taken as constant, although it depends on the rate of deposition, the nature of the particles, and on the forces between the particles. $\mathbf{r}$ has the dimensions of $\mathbf{L}^{-2}$ and the units m^{-2} in the SI system.

10.2.2 Relation Between Thickness of Cake and Volume of Filtrate

In Eq. (10.2), the variables l and V are connected, and the relation between them may be obtained by making a material balance between the solids in both the slurry and the cake as follows.

Mass of solids in filter cake $= (1-e)Al\rho_s$, where ρ_s is the density of the solids.

Mass of liquid retained in the filter cake $= eAl\rho$, where ρ is the density of the filtrate.

If J is the mass fraction of solids in the original suspension then:

$$(1-e)lA\rho_s = \frac{(V+eAl)\rho J}{1-J}$$

or:

$$(1-J)(1-e)Al\rho_s = JV\rho + AeJl\rho$$

so that:

$$l = \frac{JV\rho}{A\{(1-J)(1-e)\rho_s - Je\rho\}} \tag{10.4}$$

and:

$$V = \frac{\{\rho_s(1-e)(1-J) - e\rho J\}Al}{\rho J} \tag{10.5}$$

If v is the volume of cake deposited by unit volume of filtrate, then:

$$v = \frac{lA}{V} \quad \text{or} \quad l = \frac{vV}{A} \tag{10.6}$$

and from Eq. (10.5):

$$v = \frac{J\rho}{(1-J)(1-e)\rho_s - Je\rho} \tag{10.7}$$

Substituting for l in Eq. (10.2):

$$\frac{1}{A}\frac{dV}{dt} = \frac{(-\Delta P)}{\mathbf{r}\mu}\frac{A}{vV}$$

or :

$$\frac{dV}{dt} = \frac{A^2(-\Delta P)}{\mathbf{r}\mu vV} \qquad (10.8)$$

Eq. (10.8) may be regarded as the basic relation between $-\Delta P$, V, and t. Two important types of operation are: (i) where the pressure difference is maintained constant, and (ii) where the rate of filtration is maintained constant.

For a filtration at constant rate

so that :

$$\frac{dV}{dt} = \frac{V}{t} = \text{constant}$$
$$\frac{V}{t} = \frac{A^2(-\Delta P)}{\mathbf{r}\mu Vv} \qquad (10.9)$$

or :

$$\frac{t}{V} = \frac{\mathbf{r}\mu v}{A^2(-\Delta P)}V \qquad (10.10)$$

and $-\Delta P$ is directly proportional to V.

For a filtration at constant pressure difference, Eq. (10.8) may be integrated:

$$\frac{V^2}{2} = \frac{A^2(-\Delta P)t}{\mathbf{r}\mu v} \qquad (10.11)$$

or :

$$\frac{t}{V} = \frac{\mathbf{r}\mu v}{2A^2(-\Delta P)}V \qquad (10.12)$$

Thus, for a constant pressure filtration, there is a linear relation between V^2 and t or between t/V and V.

Filtration at a constant pressure is more frequently adopted in practice, although the pressure difference is normally gradually built up to its ultimate value.

If this takes a time t_1 during which a volume V_1 of filtrate passes, then integration of Eq. (10.8) gives:

$$\frac{1}{2}\left(V^2 - V_1^2\right) = \frac{A^2(-\Delta P)}{\mathbf{r}\mu v}(t - t_1) \qquad (10.13)$$

or :

$$\frac{t - t_1}{V - V_1} = \frac{\mathbf{r}\mu v}{2A^2(-\Delta P)}(V - V_1) + \frac{\mathbf{r}\mu vV_1}{A^2(-\Delta P)} \qquad (10.14)$$

Thus, there is a linear relation between V^2 and t and between $(t - t_1)/(V - V_1)$ and $(V - V_1)$, where $(t - t_1)$ represents the time of the constant pressure filtration and $(V - V_1)$ the corresponding volume of filtrate obtained.

Ruth et al.[4–7] have made measurements on the flow in a filter cake and have concluded that the resistance is somewhat greater than that indicated by Eq. (10.1). It was assumed that part of the pore space is rendered ineffective for the flow of filtrate because of the adsorption of ions on the surface of the particles. This is not borne out by the results of Grace[8] or Hoffing and Lockhart,[9] who determined the relation between flow rate and pressure difference, both by means of permeability tests on a fixed bed and by filtration tests using suspensions of quartz and diatomaceous earth.

Typical values of the specific resistance **r** of filter cakes, taken from the work of Carman,[10] are given in Table 10.1. In the absence of details of the physical properties of the particles and of the conditions under which they had been formed, these values are approximate, although they do provide an indication of the orders of magnitude.

10.2.3 Flow of Liquid Through the Cloth

Experimental work on the flow of the liquid under streamlined conditions[10] has shown that the flow rate is directly proportional to the pressure difference. It is the resistance of the cloth plus initial layers of deposited particles that is important because the latter not only form the true medium, but also tend to block the pores of the cloth, thus increasing its resistance. Cloths may have to be discarded because of high resistance well before they are mechanically worn. No true

Table 10.1 Typical values of specific resistance[10]

Material	Upstream Filtration Pressure (kN/m^2)	r(m^{-2})
High-grade kieselguhr	–	2×10^{12}
Ordinary kieselguhr	270	1.6×10^{14}
	780	2.0×10^{14}
Carboraffin charcoal	110	4×10^{13}
	170	8×10^{13}
Calcium carbonate	270	3.5×10^{14}
(precipitated)	780	4.0×10^{14}
Ferric oxide (pigment)	270	2.5×10^{15}
	780	4.2×10^{15}
Mica clay	270	7.5×10^{14}
	780	13×10^{14}
Colloidal clay	270	8×10^{15}
	780	10×10^{15}
Gelatinous	270	5×10^{15}
magnesium hydroxide	780	11×10^{15}
Gelatinous aluminium	270	3.5×10^{16}
hydroxide	780	6.0×10^{16}
Gelatinous ferric	270	3.0×10^{16}
hydroxide	780	9.0×10^{16}
Thixotropic mud	650	2.3×10^{17}

(Based on the information provided by Stockdale Filtration systems Ltd., Macclesfield, Cheshire).

analysis of the buildup of resistance is possible because the resistance will depend on the way in which the pressure is developed, and small variations in support geometry can have an important influence. It is, therefore, usual to combine the resistance of the cloth with that of the first few layers of particles and suppose that this corresponds to a thickness L of cake as deposited at a later stage. The resistance to flow through the cake and cloth combined is now considered.

10.2.4 Flow of Filtrate Through the Cloth and Cake Combined

If the filter cloth and the initial layers of cake are together equivalent to a thickness L of cake, as deposited at a later stage in the process, and if $-\Delta P$ is the pressure drop across the cake and cloth combined, then:

$$\frac{1}{A}\frac{dV}{dt} = \frac{(-\Delta P)}{\mathbf{r}\mu(l+L)}$$

(10.15)

which may be compared with Eq. (10.2).

Thus:

$$\frac{dV}{dt} = \frac{A(-\Delta P)}{\mathbf{r}\mu\left(\dfrac{Vv}{A}+L\right)} = \frac{A^2(-\Delta P)}{\mathbf{r}\mu v\left(V+\dfrac{LA}{v}\right)}$$

(10.16)

This equation may be integrated between the limits $t=0$, $V=0$ and $t=t_1$, $V=V_1$ for constant rate filtration, and $t=t_1$, $V=V_1$ and $t=t$, $V=V$ for a subsequent constant pressure filtration.

For the period of *constant rate filtration*:

$$\frac{V_1}{t_1} = \frac{A^2(-\Delta P)}{\mathbf{r}\mu v\left(V_1+\dfrac{LA}{v}\right)}$$

or:

$$\frac{t_1}{V_1} = \frac{\mathbf{r}\mu v}{A^2(-\Delta P)}V_1 + \frac{\mathbf{r}\mu L}{A(-\Delta P)}$$

or:

$$V_1^2 + \frac{LA}{v}V_1 = \frac{A^2(-\Delta P)}{\mathbf{r}\mu v}t_1$$

(10.17)

For a subsequent *constant pressure filtration*:

$$\frac{1}{2}\left(V^2 - V_1^2\right) + \frac{LA}{v}(V-V_1) = \frac{A^2(-\Delta P)}{\mathbf{r}\mu v}(t-t_1)$$

(10.18)

or :
$$(V - V_1 + 2V_1)(V - V_1) + \frac{2LA}{v}(V - V_1) = \frac{2A^2(-\Delta P)}{r\mu v}(t - t_1)$$

or :
$$\frac{t - t_1}{V - V_1} = \frac{r\mu v}{2A^2(-\Delta P)}(V - V_1) + \frac{r\mu v V_1}{A^2(-\Delta P)} + \frac{r\mu L}{A(-\Delta P)} \qquad (10.19)$$

Thus, there is a linear relation between $(t - t_1)/(V - V_1)$ and $V - V_1$, as shown in Fig. 10.2, and the slope is proportional to the specific resistance, as in the case of the flow of the filtrate through the filter cake alone given by Eq. (10.14), although the line does not now go through the origin.

The intercept on the $(t - t_1)/(V - V_1)$ axis should enable L, the equivalent thickness of the cloth, to be calculated, although reproducible results are not obtained because this resistance is critically dependent on the exact manner in which the operation is commenced. The time at which measurement of V and t is commenced does not affect the slope of the curve, only the intercept. It may be noted that a linear relation between t and V^2 is no longer obtained when the cloth resistance is appreciable.

10.2.5 Compressible Filter Cakes

Nearly all filter cakes are compressible to at least some extent, although in many cases, the degree of compressibility is so small that the cake may, for practical purposes, be regarded as incompressible. The evidence for compressibility is that the specific resistance is a function of

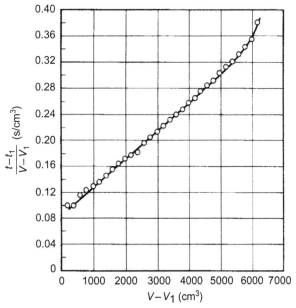

Fig. 10.2
A typical filtration curve.

the pressure difference across the cake. Compressibility may be a reversible or an irreversible process. Most filter cakes are inelastic, and the greater resistance offered to flow at high pressure differences is caused by the more compact packing of the particles forming the filter cake. Thus, the specific resistance of the cake corresponds to that for the highest pressure difference to which the cake is subjected, even though this maximum pressure difference may be maintained for only a short time. It is, therefore, important that the filtration pressure should not be allowed to exceed the normal operating pressure at any stage. In elastic filter cakes, the elasticity is attributable to compression of the particles themselves. This is less usual, although some forms of carbon can give rise to elastic cakes.

As the filtrate flows through the filter cake, it exerts a drag force on the particles, and this force is transmitted through successive layers of particles right up to the filter cloth. The magnitude of this force increases progressively from the surface of the filter cake to the filter cloth because, at any point, it is equal to the summation of the forces on all the particles up to that point. If the cake is compressible, then its voidage will decrease progressively in the direction of flow of the filtrate, giving rise to a corresponding increase in the local value of the specific resistance, r_z, of the filter cake. The structure of the cake is, however, complex and may change during the course of the filtration process. If the feed suspension is flocculated, the flocs may become deformed within the cake, and this may give rise to a change in the effective value of the specific surface, S. In addition, the particles themselves may show a degree of compressibility. Whenever possible, experimental measurements should be made to determine how the specific resistance varies over the range of conditions which will be employed in practice.

It is usually possible to express the voidage e_z at a depth z as a function of the difference between the pressure at the free surface of the cake P_1 and the pressure P_z at that depth, that is, e_z as a function of $(P_1 - P_z)$. The nomenclature is as defined in Fig. 10.3.

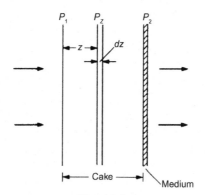

Fig. 10.3
Flow through a compressible filter cake.

For a compressible cake, Eq. (10.1) may be written as:

$$\frac{1}{A}\frac{dV}{dt} = \frac{e_z^3}{5(1-e_z)^2 S^2 \mu}\frac{1}{}\left(-\frac{dP_z}{dz}\right) \tag{10.20}$$

where e_z is now a function of depth z from the surface of the cake.

In a compressible cake, the volume v of cake deposited per unit area as a result of the flow of unit volume of filtrate will not be constant, but will vary during the filtration cycle. If the particles themselves are not compressible, however, the volume of *particles* (v') will be *almost* independent of the conditions under which the cake is formed, assuming a dilute feed suspension. Any small variations in v' arise because the volume of filtrate retained in the cake is a function of its voidage, although the effect will be very small, except possibly for the filtration of very highly concentrated suspensions. The increase in cake thickness, dz resulting from the flow of a volume of filtrate dV is given by:

$$dz = dV\frac{v'}{(1-e_z)A} \tag{10.21}$$

By comparison with Eq. (10.6), it may be seen that:

$$\frac{v'}{v} = 1 - e_z \tag{10.22}$$

Substituting from Eq. (10.21) into Eq. (10.20) gives:

Thus:
$$\frac{1}{A}\frac{dV}{dt} = \frac{e_z^3}{5(1-e_z)^2 S^2}\frac{(1-e_z)A}{v'}\frac{1}{\mu}\left(-\frac{dP_z}{dV}\right) \tag{10.23}$$

$$\frac{dV}{dt} = \frac{e_z^3}{5(1-e_z)S^2 \mu v'}A^2\left(-\frac{dP_z}{dV}\right)$$

$$= \frac{A^2}{\mu v' \mathbf{r}_z}\left(-\frac{dP_z}{dV}\right) \tag{10.24}$$

where:
$$\mathbf{r}_z = \frac{5(1-e_z)S^2}{e_z^3} \tag{10.25}$$

Comparing Eqs (10.8) and (10.24) shows that, for an incompressible cake:

$$v'\mathbf{r}_z = v\mathbf{r}$$
or:
$$\mathbf{r}_z = \mathbf{r}\frac{v}{v'}$$

At any instant in a constant pressure filtration, integration of Eq. (10.24) through the whole depth of the cake gives:

$$\int_0^v \frac{dV}{dt}dV = \frac{A^2}{\mu v'}\int_{P_1}^{P_2}\frac{(-dP_z)}{\mathbf{r}_z} \tag{10.26}$$

At any time t, dV/dt is approximately constant throughout the cake, unless the rate of change of holdup of liquid within the cake is comparable with the filtration rate dV/dt, such as is the case with very highly compressible cakes and concentrated slurries, and therefore:

$$V\frac{dV}{dt} = \frac{A^2}{\mu v'V}\int_{P_1}^{P_2}\frac{(-dP_z)}{\mathbf{r}_z}$$

(10.27)

$\mathbf{r}_z$ has been shown to be a function of the pressure difference $(P_1 - P_z)$, although it is independent of the absolute value of the pressure. Experimental studies frequently show that the relation between $\mathbf{r}_z$ and $(P_1 - P_z)$ is of the form:

$$\mathbf{r}_z = \mathbf{r}'(P_1 - P_z)^{n'}$$

(10.28)

where $\mathbf{r}'$ is independent of P_z and $0 < n' < 1$.

Thus:

$$\begin{aligned}\int_{P_1}^{P_2}\frac{(-dP_z)}{\mathbf{r}_z} &= \frac{1}{\mathbf{r}'}\int_{P_2}^{P_1}\frac{dP}{(P_1 - P_z)^{n'}}\\ &= \frac{1}{\mathbf{r}'}\frac{(P_1 - P_2)^{1-n'}}{1 - n'}\\ &= \frac{1}{\mathbf{r}'}\frac{(-\Delta P)^{1-n'}}{1 - n'}\end{aligned}$$

(10.29)

Thus:

$$\begin{aligned}\frac{dV}{dt} &= \frac{A^2}{V\mu v'\mathbf{r}'}\frac{(-\Delta P)}{(1 - n')(-\Delta P)^{n'}}\\ &= \frac{A^2(-\Delta P)}{V\mu v'\mathbf{r}''(-\Delta P)^{n'}}\end{aligned}$$

(10.30)

where $\mathbf{r}'' = (1 - n')\mathbf{r}'$

and:

$$\frac{dV}{dt} = \frac{A^2(-\Delta P)}{V\mu v'\bar{\mathbf{r}}}$$

(10.31)

where $\bar{\mathbf{r}}$ is the mean resistance defined by:

$$\bar{\mathbf{r}} = \bar{\mathbf{r}}''(-\Delta P)^{n'}$$

(10.32)

Heertjes[11] has studied the effect of pressure on the porosity of a filter cake and suggested that, as the pressure is increased above atmospheric, the porosity decreases in proportion to some power of the excess pressure.

Grace[8] has related the anticipated resistance to the physical properties of the feed slurry. Valleroy and Maloney[12] have examined the resistance of an incompressible bed of spherical particles when measured in a permeability cell, a vacuum filter, and a centrifuge, and emphasised the need for caution in applying laboratory data to units of different geometry.

Tiller and Huang[13] give further details of the problem of developing a usable design relationship for filter equipment. Studies by Tiller and Shirato,[14] Tiller and Yeh,[15] and Rushton and Hameed[16] show the difficulty in presenting practical conditions in a way which can be used analytically. It is very important to note that tests on slurries must be made with equipment that is geometrically similar to that proposed. This means that specific resistance is very difficult to define in practice, because it is determined by the nature of the filtering unit and the way in which the cake is initially formed and then built up.

10.3 Filtration Practice

10.3.1 The Filter Medium

The function of the filter medium is generally to act as a support for the filter cake, and the initial layers of cake provide the true filter. The filter medium should be mechanically strong and resistant to the corrosive action of the fluid, and should offer as little resistance as possible to the flow of filtrate. Woven materials are commonly used (shown schematically in Fig. 10.4A–G), though granular materials and porous solids are useful for filtration of corrosive liquids in batch units. An important feature in the selection of a woven material is the ease of cake removal, because this is a key factor in the operation of modern automatic units. Ehlers[17] has discussed the selection of woven synthetic materials and Wrotnowski[18] that of nonwoven materials. Further details of some more materials are given in the literature,[19–22] and a useful summary is presented in Volume 6.

10.3.2 Blocking Filtration

In the previous discussion, it is assumed that there is a well-defined boundary between the filter cake and the filter cloth. The initial stages in the buildup of the filter cake are important, however, because these may have a large effect on the flow resistance and may seriously affect the useful life of the cloth.

The blocking of the pores of the filter medium by particles is a complex phenomenon, partly because of the complicated nature of the surface structure of the usual types of filter media, and partly because the lines of movement of the particles are not well defined. At the start of filtration, the manner in which the cake forms will lie between two extremes—the penetration of the pores by particles and the shielding of the entry to the pores by the particles forming bridges. Heertjes[11] considered a number of idealised cases in which suspensions of specified pore size distributions were filtered on a cloth with a regular pore distribution. First, it was assumed that an individual particle was capable on its own of blocking a single pore, then, as filtration proceeded, successive pores would be blocked, so that the apparent value of the specific resistance of the filter cake would depend on the amount of solids deposited.

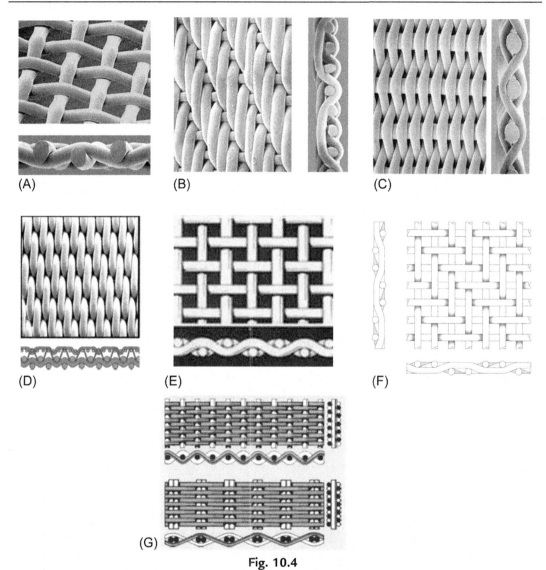

Fig. 10.4

Typical filter media: (A) plain, square weave, (B) twill weave, (C) plain, reverse Dutch, (D) double layer weave, (E) plain weave in metal media, (F) twilled weave in metal media, (G) twilled weave in metal media.

The pore and particle size distributions might, however, be such that more than one particle could enter a particular pore. In this case, the resistance of the pore increases in stages as successive particles are trapped, until the pore is completely blocked. In practice, however, it is much more likely that many of the pores will never become completely blocked, and a cake of relatively low resistance will form over the entry to the partially blocked pore.

One of the most important variables affecting the tendency for blocking is the concentration of particles. The greater the concentration, the smaller will be the average distance between the particles, and the smaller will be the tendency for the particle to be drawn into the streamlines directed toward the open pores. Instead, the particles in the concentrated suspension tend to distribute themselves fairly evenly over the filter surface and form bridges. As a result, suspensions of high concentration generally give rise to cakes of lower resistance than those formed from dilute suspensions.

10.3.3 Effect of Particle Sedimentation on Filtration

There are two important effects due to particle sedimentation which may affect the rate of filtration. First, if the sediment particles are all settling at approximately the same rate, as is frequently the case in a concentrated suspension in which the particle size distribution is not very wide, a more rapid buildup of particles will occur on an upward-facing surface, and a correspondingly reduced rate of buildup will take place if the filter surface is facing downwards. Thus, there will be a tendency for accelerated filtration with downward-facing filter surfaces and reduced filtration rates for upward-facing surfaces. On the other hand, if the suspension is relatively dilute, so that the large particles are settling at a higher rate than the small ones, there will be a preferential deposition of large particles on an upward-facing surface during the initial stages of filtration, giving rise to a low resistance cake. Conversely, for a downward-facing surface, fine particles will initially be deposited preferentially, and the cake resistance will be correspondingly increased. It is, thus, seen that there can be complex interactions where sedimentation is occurring at an appreciable rate, and that the orientation of the filter surface is an important factor.

10.3.4 Delayed Cake Filtration

In the filtration of a slurry, the resistance of the filter cake progressively increases and consequently, in a constant pressure operation, the rate of filtration falls. If the buildup of solids can be reduced, the effective cake thickness will be less and the rate of flow of filtrate will be increased.

In practice, it is sometimes possible to incorporate moving blades in the filter equipment so that the thickness of the cake is limited to the clearance between the filter medium and the blades. Filtrate then flows through the cake at an approximately constant rate, and the solids are retained in suspension. Thus, the solids concentration in the feed vessel increases until the particles are in permanent physical contact with one another. At this stage, the boundary between the slurry and the cake becomes ill-defined, and a significant resistance to the flow of liquid develops within the slurry itself with a consequent reduction in the flow rate of filtrate.

By the use of this technique, a much higher rate of filtration can be achieved than is possible in a filter operated in a conventional manner. In addition, the resulting cake usually has a lower porosity because the blades effectively break down the bridges or arches, which give rise to a structure in the filter cake, and the final cake is significantly drier as a result.

If the scrapers are in the form of rotating blades, the outcome differs according to whether they are moving at low or at high speed. At low speeds, the cake thickness is reduced to the clearance depth each time the scraper blade passes, although cake then builds up again until the next passage of the scraper. If the blade is operated at high speed, there is little time for solids to build up between successive passages of the blade, and the cake reaches an approximately constant thickness. Because particles tend to be swept across the surface of the cake by the moving slurry, they will be trapped in the cake only if the drag force which the filtrate exerts on them is great enough. As the thickness of the cake increases, the pressure gradient becomes less, and there is a smaller force retaining particles in the cake surface. Thus, the thickness of the cake tends to reach an equilibrium value, which can be considerably less than the clearance between the medium and the blades.

Experimental results for the effect of stirrer speed on the rate of filtration of a 10% by mass suspension of clay are shown in Fig. 10.5, taken from the work of Tiller and Cheng,[23] in which the filtrate volume collected per unit cross-section of filter is plotted against time, for several stirrer speeds.

The concentration of solids in the slurry in the feed vessel to the filter at any time can be calculated by noting that the volumetric rate of feed of slurry must be equal to the rate at which filtrate leaves the vessel. For a rate of flow of filtrate of dV/dt out of the filter, the rate of flow of slurry into the vessel must also be dV/dt, and the corresponding influx of solids is $(1 - e_0)\, dV/dt$, where $(1 - e_0)$ is the volume fraction of solids in the feed slurry. At any time t, the volume of solids in the vessel is $\mathbf{V}(1 - e_V)$, where $\mathbf{V}$ is its volume, and $(1 - e_V)$ is the volume fraction of solids at that time. Thus, a material balance on the solids gives:

$$(1 - e_0)\frac{dV}{dt} = \frac{d}{dt}[\mathbf{V}(1 - e_V)] \tag{10.33}$$

$$\frac{d(1 - e_V)}{dt} = \frac{1}{\mathbf{V}}(1 - e_0)\frac{dV}{dt} \tag{10.34}$$

For a constant filtration rate dV/dt, the fractional solids hold-up $(1 - e_V)$ increases linearly with time, until it reaches a limiting value when the resistance to flow of liquid within the slurry becomes significant. The filtration rate then drops rapidly to a near-zero value.

10.3.5 Crossflow Filtration

An alternative method of reducing the resistance to filtration is to recirculate the slurry and thereby maintain a high velocity of flow parallel to the surface of the filter medium. Typical recirculation rates may be 10–20 times the filtration rate. By this means the cake is prevented

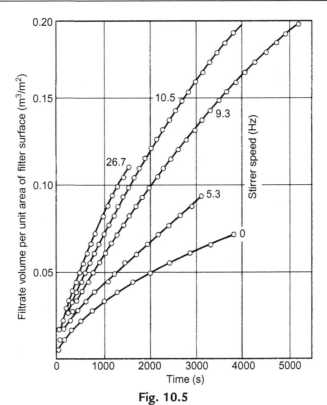

Fig. 10.5

Volume as function of time for delayed cake and constant pressure filtration as a function of stirrer speed.[23]

from forming during the early stages of filtration. This can be particularly beneficial when the slurry is flocculated and exhibits shear-thinning, non-Newtonian properties. This method of operation is discussed by Mackley and Sherman[24] and by Holdich et al.[25]

In cases where a dilute solution containing small quantities of solids which tend to blind the filter cloth is to be filtered, crossflow filtration is extensively used. This is the normal mode of operation for ultrafiltration using membranes, a topic which is discussed in Chapter 11.

10.3.6 Preliminary Treatment of Slurries Before Filtration

If a slurry is dilute, and the solid particles settle readily in the fluid, it may be desirable to effect a preliminary concentration in a thickener, as discussed in Chapter 8. The thickened suspension is then fed from the thickener to the filter, and the quantity of material to be handled is thereby reduced.

Theoretical treatment has shown that the nature of the filter cake has a very pronounced effect on the rate of flow of filtrate, and that it is, in general, desirable that the particles forming the filter cake should have as large a size as possible. More rapid filtration is, therefore, obtained if a suitable agent is added to the slurry to cause coagulation. If the solid material is formed in a chemical reaction by precipitation, the particle size can generally be controlled to a certain extent by the actual conditions of formation. For example, the particle size of the resultant precipitate may be controlled by varying the temperature and concentration, and sometimes, the pH of the reacting solutions. As indicated by Grace,[8] a flocculated suspension gives rise to a more porous cake, although the compressibility is greater. In many cases, crystal shape may be altered by adding traces of material which is selectively adsorbed onto particular faces.

Filter aids are extensively used where the filter cake is relatively impermeable to the flow of filtrate. These are materials which pack to form beds of very high voidages, and therefore, they are capable of increasing the porosity of the filter cake if added to the slurry before filtration. Apart from economic considerations, there is an optimum quantity of filter aid which should be added in any given case. Whereas the presence of the filter aid reduces the specific resistance of the filter cake, it also results in the formation of a thicker cake. The actual quantity used will, therefore, depend on the nature of the material. The use of filter aids is normally restricted to operations in which the filtrate is valuable, and the cake is a waste product. In some circumstances, however, the filter aid must be readily separable from the rest of the filter cake by physical or chemical means. Filter cakes incorporating filter aid are usually very compressible, and care should, therefore, be taken to ensure that the good effect of the filter aid is not destroyed by employing too high a filtration pressure. Kieselguhr (type of diatomaceous earth), which is a commonly used filter aid, has a voidage of about 0.85. Addition of relatively small quantities increases the voidage of most filter cakes, and the resulting porosity normally lies between that of the filter aid and that of the filter solids. Sometimes the filter medium is "precoated" with filter aid, and a thin layer of the filter aid is removed with the cake at the end of each cycle.

In some cases, the filtration time can be reduced by diluting the suspension in order to reduce the viscosity of the filtrate. This does, of course, increase the bulk to be filtered, and is applicable only when the value of the filtrate is not affected by dilution. Raising the temperature may be advantageous in that the viscosity of the filtrate is reduced.

10.3.7 Washing of the Filter Cake

When the wash liquid is miscible with the filtrate and has similar physical properties, the rate of washing at the same pressure difference will be about the same as the final rate of filtration. If the viscosity of the wash liquid is less, a somewhat greater rate will be obtained. Channelling sometimes occurs, however, with the result that much of the cake is incompletely washed, and the fluid passes preferentially through the channels, which are gradually enlarged by its continued passage. This does not occur during filtration because channels are self-sealing by virtue of deposition of solids from the slurry. Channelling is most marked with compressible filter cakes and can be minimised by using a smaller pressure difference for washing than for filtration.

Washing may be regarded as taking place in two stages. First, filtrate is displaced from the filter cake by wash liquid during the period of *displacement washing*, and in this way, up to 90% of the filtrate may be removed. During the second stage, *diffusional washing*, solvent diffuses into the wash liquid from the less accessible voids, and the following relation applies:

$$\left(\frac{\text{volume of wash liquid passed}}{\text{cake thickness}}\right) = (\text{constant}) \times \log\left(\frac{\text{initial concentration of solute}}{\text{concentration at particular time}}\right)$$

$$(10.35)$$

Although an immiscible liquid is seldom used for washing, air is often used to effect partial drying of the filter cake. The rate of flow of air must normally be determined experimentally.

10.4 Filtration Equipment

10.4.1 Filter Selection

The most suitable filter for any given operation is the one which will fulfil the requirements at minimum overall cost. Because the cost of the equipment is closely related to the filtering area, it is normally desirable to obtain a high overall rate of filtration. This involves the use of relatively high pressures, although the maximum pressures are often limited by mechanical design considerations. Although a higher throughput from a given filtering surface is obtained from a continuous filter than from a batch operated filter, it may sometimes be necessary to use a batch filter, particularly if the filter cake has a high resistance, because most continuous filters operate under reduced pressure, and the maximum filtration pressure is, therefore, limited. Other features which are desirable in a filter include ease of discharge of the filter cake in a convenient physical form, and a method of observing the quality of the filtrate obtained from each section of the plant. These factors are important in considering the types of equipment available. The most common types are filter presses, leaf filters, and continuous rotary filters. In addition, there are filters for special purposes, such as bag filters, and the disc type of filter, which is used for the removal of small quantities of solids from a fluid.

The most important factors in filter selection are the specific resistance of the filter cake, the quantity to be filtered, and the solids concentration. For free-filtering materials, a rotary vacuum filter is generally the most satisfactory because it has a very high capacity for its size and does not require any significant manual attention. If the cake has to be washed, the rotary drum is to be preferred to the rotary leaf. If a high degree of washing is required, however, it is usually desirable to repulp the filter cake and to filter a second time.

For large-scale filtration, there are three principal cases in which a rotary vacuum filter will not be used. First, if the specific resistance is high, a positive pressure filter will be required, and a filter press may well be suitable, particularly if the solid content is not so high that frequent dismantling of the press is necessary. Second, when efficient washing is required, a leaf filter is effective, because very thin cakes can be prepared, and the risk of channelling during washing is reduced to a minimum. Finally, where only very small quantities of solids are present in the liquid, an edge filter may be employed.

While it may be possible to predict qualitatively the effect of the physical properties of the fluid and the solid on the filtration characteristics of a suspension, it is necessary in all cases to carry out a test on a sample before the large-scale plant can be designed. A simple vacuum filter with a filter area of 0.0065 m^2 is used to obtain laboratory data, as illustrated in Fig. 10.6. The information on filtration rates and specific resistance obtained in this way can be directly applied to industrial filters provided due account is taken of the compressibility of the filter cake. It cannot be stressed too strongly that data from any laboratory test cell must not be used without practical experience in the design of industrial units where the geometry of the flow channel is very different. The laying down of the cake influences the structure to a very marked extent.

A "compressibility–permeability" test cell has been developed by Ruth[7] and Grace[8] for testing the behavior of slurries under various conditions of filtration. A useful guide to the selection of a filter type based on slurry characteristics is given in Volume 6.

10.4.2 Bed Filters

Bed filters provide an example of the application of the principles of *deep-bed filtration*, in which the particles penetrate into the interstices of the filter bed where they are trapped following impingement on the surfaces of the material of the bed.

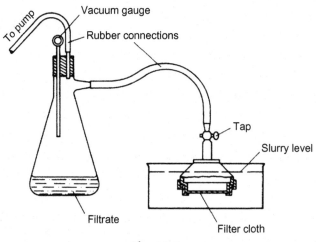

Fig. 10.6
Laboratory test filter.

For the purification of water supplies and for waste water treatment where the solid content is about 10 g/m^3 or less, as noted by Cleasby,[26] granular bed filters have largely replaced the formerly used very slow sand filters. The beds are formed from granular material of grain size 0.6–1.2 mm in beds 0.6–1.8 m deep. The very fine particles of solids are removed by mechanical action, although the particles finally adhere as a result of surface electric forces or adsorption, as Ives[27] points out. This operation has been analysed by Iwasaki,[28] who proposed the following equation:

$$-\frac{\partial C}{\partial l} = \lambda C \tag{10.36}$$

On integration:

$$C/C_0 = e^{-\lambda l} \tag{10.37}$$

where C is the volume concentration of solids in suspension in the filter, C_0 is the value of C at the surface of the filter, l is the depth of the filter, and λ is the filter coefficient.

If u_c is the superficial velocity of the slurry, then the rate of flow of solids through the filter at depth l is $u_c C$ per unit area. Thus, the rate of accumulation of solids in a distance $dl = -u_c(\partial C/\partial l) \, dl$. If σ is the volume of solids deposited per unit volume of filter at a depth l, the rate of accumulation may also be expressed as $(\partial \sigma/\partial t) \, dl$.

Thus:

$$-\frac{\partial C}{\partial l} = u_c \frac{\partial \sigma}{\partial t} \tag{10.38}$$

The problem is discussed further by Ives[27] and by Spielman and Friedlander.[29] The backwashing of these beds has presented problems and several techniques have been adopted. These include a backflow of air followed by water, the flow rate of which may be high enough to give rise to fluidisation, with the maximum hydrodynamic shear occurring at a voidage of about 0.7.

10.4.3 Bag Filters

Bag filters have now been almost entirely superseded for liquid filtration by other types of filters, although one of the few remaining types is the Taylor bag filter, which has been widely used in the sugar industry. A number of long, thin bags are attached to a horizontal feed tray, and the liquid flows under the action of gravity so that the rate of filtration per unit area is very low. It is possible, however, to arrange a large filtering area in the plant of up to about 700 m^2. The filter is usually arranged in two sections, so that each may be inspected separately without interrupting the operation.

Bag filters are still extensively used for the removal of dust particles from gases and can be operated either as pressure filters or as suction filters. Their use is discussed in Chapter 4.

10.4.4 The Filter Press

The filter press is one of the two main types, the *plate and frame press* and the *recessed plate* or *chamber press.*

The plate and frame press

This type of filter press consists of plates and frames arranged alternately and supported on a pair of rails, as shown in Fig. 10.7. The plates have a ribbed surface, and the edges stand slightly proud and are carefully machined. The hollow frame is separated from the plate by the filter cloth, and the press is closed either by means of a hand screw or hydraulically, using the minimum pressure in order to reduce wear on the cloths. A chamber is, therefore, formed between each pair of successive plates, as shown in Fig. 10.8. The slurry is introduced through a port in each frame, and the filtrate passes through the cloth on each side so that two cakes are formed simultaneously in each chamber, and these join when the frame is full. The frames are usually square and may be 100 mm–2.5 m across and 10–75 mm thick.

The slurry may be fed to the press through the continuous channel formed by the holes in the corners of the plates and frames, in which case, it is necessary to cut corresponding holes in the cloths, which themselves act as gaskets. Cutting of the cloth can be avoided by feeding through a channel at the side, although rubber bushes must then be fitted so that a leak-tight joint is formed.

Fig. 10.7
A large filter press with 2 m by 1.5 m plates. *Image Courtesy: Johnson-Progress Ltd., Stoke-on-Trent, Staffs.*

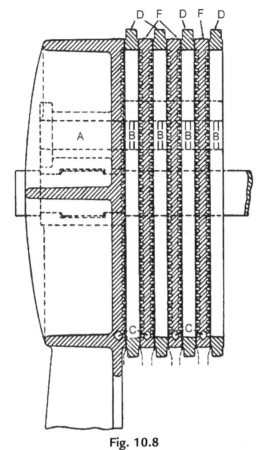

Fig. 10.8

Plate and frame press. A—inlet passage. B—feed ports. C—filtrate outlet. D—frames. F—plates.

The filtrate runs down the ribbed surface of the plates and is then discharged through a cock into an open launder, so that the filtrate from each plate may be inspected, and any plate can be isolated if it is not giving a clear filtrate. In some cases, the filtrate is removed through a closed channel, although it is not then possible to observe the discharge from each plate separately.

In many filter presses, provision is made for steam heating, so that the viscosity of the filtrate is reduced, and a higher rate of filtration obtained. Materials, such as waxes, that solidify at normal temperatures may also be filtered in steam-heated presses. Steam heating also facilitates the production of a dry cake.

Optimum time cycle. The optimum thickness of cake to be formed in a filter press depends on the resistance offered by the filter cake and on the time taken to dismantle and refit the press. Although the production of a thin filter cake results in a high average rate of filtration, it is necessary to dismantle the press more often, and a greater time is, therefore, spent on this operation. For a filtration carried out entirely at constant pressure, a rearrangement of Eq. (10.19) gives:

$$\frac{t}{V} = \frac{r\mu v}{2A^2(-\Delta P)}V + \frac{r\mu L}{A(-\Delta P)} \tag{10.39}$$

$$= B_1 V + B_2 \tag{10.40}$$

where B_1 and B_2 are constants.

Thus, the time of filtration t is given by:

$$t = B_1 V^2 + B_2 V \tag{10.41}$$

The time of dismantling and assembling the press, say t', is substantially independent of the thickness of cake produced. The total time of a cycle in which a volume V of filtrate is collected is then $(t+t')$, and the overall rate of filtration is given by:

$$W = \frac{V}{B_1 V^2 + B_2 V + t'}$$

W is a maximum when $dW/dV = 0$.

Differentiating W with respect to V and equating to zero:

$$B_1 V^2 + B_2 V + t' - V(2B_1 V + B_2) = 0$$

or:
$$t' = B_1 V^2 \tag{10.42}$$

or:
$$V = \sqrt{\left(\frac{t'}{B_1}\right)} \tag{10.43}$$

If the resistance of the filter medium is neglected, $t = B_1 V^2$, and the time during which filtration is carried out is exactly equal to the time the press is out of service. In practice, in order to obtain the maximum overall rate of filtration, the filtration time must always be somewhat greater in order to allow for the resistance of the cloth, represented by the term $B_2 V$. In general, the lower the specific resistance of the cake, the greater will be the economic thickness of the frame.

The application of these equations is illustrated later in Example 10.5, which is based on the work of Harker.[30]

It is shown in Example 10.5, which appears later in the chapter, that, provided the cloth resistance is very low, adopting a filtration time equal to the downtime will give the maximum throughput. Where the cloth resistance is appreciable, then the term $B_2(t'/B_1)^{0.5}$ becomes significant, and a longer filtration time is desirable. It may be seen in Fig. 10.9, which is based on data from Example 10.5, that neither of these values represents the minimum cost condition however, except for the unique situation where $t' = $ (cost of shutdown)/(cost during filtering),

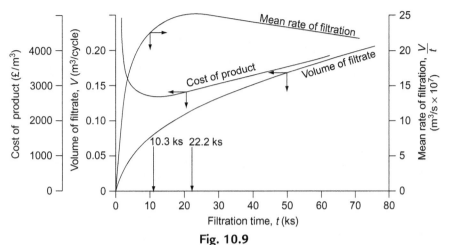

Fig. 10.9

Optimisation of plate and frame press (data from Example 10.5).[30]

and a decision has to be made as to whether cost or throughput is the overriding consideration. In practice, operating schedules are probably the dominating feature, although significant savings may be made by operating at the minimum cost condition.

Washing

Two methods of washing may be employed, "simple" washing and "through" or "thorough" washing. With simple washing, the wash liquid is fed in through the same channel as the slurry, although, as its velocity near the point of entry is high, erosion of the cake takes place. The channels which are thus formed gradually enlarge, and uneven washing is usually obtained. Simple washing may be used only when the frame is not completely full.

In *thorough* washing, the wash liquid is introduced through a separate channel behind the filter cloth on alternate plates, known as washing plates, shown in Fig. 10.10, and flows through the whole thickness of the cake, first in the opposite direction, and then in the same direction as the filtrate. The area during washing is one-half of that during filtration, and, in addition, the wash liquid has to flow through twice the thickness, so that the rate of washing should, therefore, be about one-quarter of the final rate of filtration. The wash liquid is usually discharged through the same channel as the filtrate, though sometimes, a separate outlet is provided. Even with thorough washing, some channelling occurs, and several inlets are often provided, so that the liquid is well distributed. If the cake is appreciably compressible, the minimum pressure should be used during washing, and in no case should the final filtration pressure be exceeded. After washing, the cake may be made easier to handle by removing excess liquid with compressed air.

For ease in identification, small buttons are embossed on the sides of the plates and frames, one on the nonwashing plates, two on the frames, and three on the washing plates, as shown in Fig. 10.11.

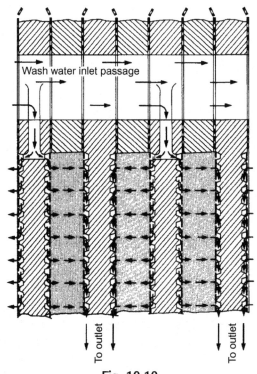

Fig. 10.10
Schematics of thorough washing.

Example 10.1

A slurry is filtered in a plate and frame press containing 12 frames, each 0.3 m square and 25 mm thick. During the first 180 s, the pressure difference for filtration is slowly raised to the final value of 400 kN/m², and, during this period, the rate of filtration is maintained as constant. After the initial period, filtration is carried out at constant pressure, and the cakes are completely formed in a further 900 s. The cakes are then washed with a pressure difference of 275 kN/m² for 600 s using *thorough washing* (See the plate and frame press in Section 10.4.4). What is the volume of filtrate collected per cycle, and how much wash water is used?

A sample of the slurry had previously been tested with a leaf filter of 0.05 m² filtering surface using a vacuum giving a pressure difference of 71.3 kN/m². The volume of filtrate collected in the first 300 s was 250 cm³, and, after a further 300 s, an additional 150 cm³ was collected. It may be assumed that the cake is incompressible, and that the cloth resistance is the same in the leaf filter as in the filter press.

Solution

In the leaf filter, filtration occurs at constant pressure from the start.

Thus:
$$V^2 + 2\frac{AL}{v}V = 2\frac{(-\Delta P)A^2}{\mathbf{r}\mu v}t \qquad \text{(from Eq.10.18)}$$

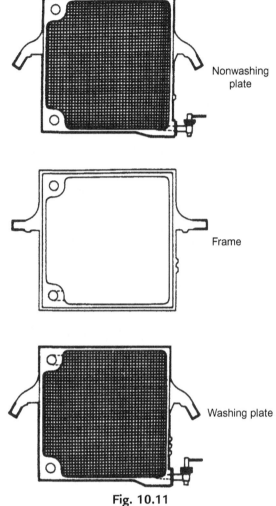

Fig. 10.11
Schematics of plates and frames.

In the filter press, a volume V_1 of filtrate is obtained under constant rate conditions in time t_1, and filtration is then carried out at a constant pressure.

Thus:
$$V_1^2 + \frac{AL}{v}V_1 = \frac{(-\Delta P)A^2}{r\mu v}t_1 \qquad \text{(from Eq.10.17)}$$

and:
$$\left(V^2 - V_1^2\right) + 2\frac{AL}{v}(V - V_1) = 2\frac{(-\Delta P)A^2}{r\mu v}(t - t_1) \qquad \text{(from Eq.10.18)}$$

For the leaf filter

When $t = 300$ s, $V = 250$ cm^3 $= 2.5 \times 10^{-4}$ m^3 and when $t = 600$ s, $V = 400$ cm^3 $= 4 \times 10^{-4}$ m^3, $A = 0.05$ m^2 and $-\Delta P = 71.3$ kN/m^2 or 7.13×10^4 N/m^2.

Thus:
$$(2.5 \times 10^{-4})^2 + 2(0.05L/v)2.5 \times 10^{-4} = 2(7.13 \times 10^4 \times 0.05^2/r\mu v)300$$

and:
$$(4 \times 10^{-4})^2 + 2(0.05L/v)4 \times 10^{-4} = 2(7.13 \times 10^4 \times 0.05^2/r\mu v)600$$

That is:
$$6.25 \times 10^{-8} + 2.5 \times 10^{-5}\frac{L}{v} = \frac{1.07 \times 10^5}{r\mu v}$$

and:
$$16 \times 10^{-8} + 4 \times 10^{-5}\frac{L}{v} = \frac{2.14 \times 10^5}{r\mu v}$$

Hence:
$$L/v = 3.5 \times 10^{-3} \quad \text{and} \quad r\mu v = 7.13 \times 10^{11}$$

For the filter press

$A = (12 \times 2 \times 0.3^2) = 2.16$ m^2, $-\Delta P = 400$ kN/m^2 $= 4 \times 10^5$ N/m^2, $t = 180$ s. The volume of filtrate V_1 collected during the constant rate period on the filter press is given by:

$$V_1^2 + (2.16 \times 3.5 \times 10^{-3}V_1) = [(4 \times 10^5 \times 2.16^2)/(7.13 \times 10^{11})]180$$
$$V_1^2 + (7.56 \times 10^{-3}V_1) - (4.711 \times 10^{-4}) = 0$$

or:
$$V_1 = -(3.78 \times 10^{-3}) + \sqrt{(1.429 \times 10^{-5} + 4.711 \times 10^{-4})} = 1.825 \times 10^{-2} \text{m}^3$$

For the constant pressure period:

$$(t - t_1) = 900 \text{ s}$$

The total volume of filtrate collected is, therefore, given by:

or:
$$(V^2 - 3.33 \times 10^{-4}) + (1.512 \times 10^{-2})(V - 1.825 \times 10^{-2}) = 5.235 \times 10^{-6} \times 900$$
$$V^2 + (1.512 \times 10^{-2}V) - (4.712 \times 10^{-3}) = 0$$

Thus:
$$V = -0.756 \times 10^{-2} + \sqrt{(0.572 \times 10^{-4} + 4.712 \times 10^{-3})}$$
$$= 6.15 \times 10^{-2} \text{ or } \underline{0.062 \text{ m}^3}$$

The final rate of filtration is given by:

$$\frac{-\Delta P A^2}{r\mu v(V + AL/v)} = \frac{4 \times 10^5 \times 2.16^2}{7.13 \times 10^{11}(6.15 \times 10^{-2} + 2.16 \times 3.5 \times 10^{-3})}$$
$$= 3.79 \times 10^{-5} \text{m}^3/\text{s} \quad \text{(from Eq.10.16)}$$

If the viscosity of the filtrate is the same as that of the wash water, then:

$$\text{Rate of washing at 400 kN/m}^2 = \frac{1}{4} \times 3.79 \times 10^{-5} = 9.5 \times 10^{-6} \text{m}^3/\text{s}$$

$$\text{Rate of washing at 275 kN/m}^2 = 9.5 \times 10^{-6} \times (275/400) = 6.5 \times 10^{-6} \text{m}^3/\text{s}$$

Thus the amount of wash water passing in 600 s $= (600 \times 6.5 \times 10^{-6})$
$$= 3.9 \times 10^{-3} \text{m}^3 \text{ or } \underline{0.004 \text{m}^3}$$

The recessed plate filter press

The recessed type of press is similar to the plate and frame type, except that the use of frames is obviated by recessing the ribbed surface of the plates so that the individual filter chambers are formed between successive plates. In this type of press, therefore, the thickness of the cake cannot be varied, and it is equal to twice the depth of the recess on individual plates.

The feed channel shown in Fig. 10.12 usually differs from that employed on the plate and frame press. All the chambers are connected by means of a comparatively large hole in the centre of each of the plates, and the cloths are secured in position by means of screwed unions. Slurries containing relatively large solid particles may readily be handled in this type of press without fear of blocking the feed channels. As described by Cherry,[31] developments in filter presses have been toward the fabrication of larger units, made possible by mechanisation and the use of newer, lighter materials of construction. The plates of wood used in earlier times were limited in size because of limitations of pressures, and large cast-iron plates presented difficulty in handling. Large plates are now frequently made of rubber mouldings or of polypropylene, although distortion may be a problem, particularly if the temperature is high.

The second area of advance is in mechanisation, which enables the opening and closing to be done automatically by a ram driven hydraulically or by an electric motor. Plate transportation is effected by fitting triggers to two endless chains operating the plates, and labour costs have consequently been reduced considerably. Improved designs have given better drainage, which has led to improved washing. Much shorter time cycles are now obtained, and the cakes are

Fig. 10.12
A recessed chamber plate, $2\,m^2$. *Image Courtesy: Filtration Systems Ltd., Mirfield, W.Yorks.*

thinner, more uniform, and drier. These advantages have been rather more readily obtained with recessed plates where the cloth is subjected to less wear.

Advantages of the filter press

(a) Because of its basic simplicity, the filter press is versatile and may be used for a wide range of materials under varying operating conditions of cake thickness and pressure.
(b) Maintenance cost is low.
(c) It provides a large filtering area on a small floor space, and few additional associated units are needed.
(d) Most joints are external and leakage is easily detected and managed.
(e) High pressure operation is usually possible.
(f) It is equally suitable whether the cake or the liquid is the main product.

Disadvantages of the filter press

(a) It is intermittent in operation, and continual dismantling is apt to cause high wear on the cloths.
(b) Despite the improvements mentioned previously, it is fairly heavy on labour.

Example 10.2

A slurry containing 100 kg of whiting, with a density of 3000 kg/m^3, per m^3 of water, is filtered in a plate and frame press, which takes 900 s to dismantle, clean, and re-assemble. If the cake is incompressible and has a voidage of 0.4, what is the optimum thickness of cake for a filtration pressure $(-\Delta P)$ of 1000 kN/m^2? If the cake is washed at 500 kN/m^2, and the total volume of wash water employed is 25% of that of the filtrate, how is the optimum thickness of the cake affected? The resistance of the filter medium may be neglected, and the viscosity of water is 1 mNs/m^2. In an experiment, a pressure difference of 165 kN/m^2 produced a flow of water of 0.02 cm^3/s through a centimetre cube of filter cake.

Solution

The basic filtration equation may be written as:

$$\frac{1}{A}\frac{dV}{dt}=\frac{(-\Delta P)}{r\mu l}$$ (Eq.10.2)

where **r** is defined as the specific resistance of the cake.

The slurry contains 100 kg whiting/m^3 of water.

Volume of 100 kg whiting $=(100/3000)=0.0333$ m^3.

Volume of cake $=0.0333/(1-0.4)=0.0556$ m^3.

Volume of liquid in cake $=(0.0556 \times 0.4)=0.0222$ m^3.

Volume of filtrate $=(1-0.0222)=0.978$ m^3.

Thus: volume of cake/volume of filtrate $v = 0.0569$

In the experiment:

$$A = 10^{-4} \text{m}^2, (-\Delta P) = 1.65 \times 10^5 \text{N/m}^2, l = 0.01 \text{m},$$
$$\frac{dV}{dt} = 2 \times 10^{-8} \text{m}^3/\text{s}, \quad \mu = 10^{-3} \text{Ns/m}^2$$

Inserting these values in Eq. (10.2) gives:

$$\left(\frac{1}{10^{-4}}\right)(2 \times 10^{-8}) = \frac{1}{r} \frac{(1.65 \times 10^5)}{(10^{-3})(10^{-2})}, \quad \text{and} \quad r = 8.25 \times 10^{13} \text{m}^{-2}$$

From Eq. (10.2):

$$V^2 = \frac{2A^2(-\Delta P)t}{r\mu v} \qquad \text{(Eq.10.11)}$$

But: $L = $ half frame thickness $= Vv/A$ (Eq.10.6)

$$L^2 = \frac{2A(-\Delta P)vt}{r\mu}$$

Thus:

$$= \frac{2 \times (1 \times 10^6) \times 0.0569}{(8.25 \times 10^{13})(1 \times 10^{-3})} t_f$$

$$= 1.380 \times 10^{-6} t_f \text{ (where } t_f \text{ is the filtration time)}$$

$$L = 1.161 \times 10^{-3} t_f^{1/2}$$

It is shown in Section 10.4.4 that, if the resistance of the filter medium is neglected, the optimum cake thickness occurs when the filtration time is equal to the downtime,

Thus: $t = 900$ s, $t^{1/2} = 30$

∴ $L_{opt} = 34.8 \times 10^{-3}$ m $= 34.8$ or 35 mm

and: optimum frame thickness $= \underline{70 \text{mm}}$

For the washing process, if the filtration pressure is halved, the rate of washing is halved. The wash water has twice the thickness to penetrate and half the area for flow that is available to the filtrate, so that, considering these factors, the washing rate is one-eighth of the final filtration rate.

The final filtration rate $\dfrac{dV}{dt} = \dfrac{A^2(-\Delta P)}{r\mu vV}$

and:

$$= \frac{1 \times 10^6 A^2}{(8.25 \times 10^{13}) \times 10^{-3} \times 0.0569V} = \frac{(2.13 \times 10^{-4} A^2)}{V}$$

The washing rate $=$ (final rate of filtration/8) $= 2.66 \times 10^{-5} A^2/V$

The volume of wash water $= V/4$.

Hence: washing time $t_w = (V/4)/(2.66 \times 10^{-5}A^2/V)$

That is: $t_w = 940V^2/A^2$

$$V^2 = L^2 A^2/v^2$$

Therefore: $$t_w = \left(\frac{L^2 A^2}{(0.0569)^2}\right)\left(\frac{940}{A^2}\right) = 2.90 \times 10^5 L^2$$

The filtration time t_f was shown earlier to be: $t_f = L^2/1.380 \times 10^{-6} = 7.25 \times 10^5 L^2$

Thus: total cycle time $= L^2(2.90 \times 10^5 + 7.25 \times 10^5) + 900$
$$= 1.015 \times 10^6 L^2 + 900$$

The rate of cake production is then:

$$= \frac{L}{1.025 \times 10^6 L^2 + 900} = R$$

For $dR/dL = 0$, then: $1.025 \times 10^6 L^2 + 900 - 2.050 \times 10^6 L^2 = 0$

Thus: $$L^2 = \frac{900}{1.025 \times 10^6} \quad \text{and} \quad L = 29.6 \times 10^{-3}\,\text{m} = 29.6\,\text{mm}$$

Frame thickness $= 59.2 \approx \underline{60\,\text{mm}}$

10.4.5 Pressure Leaf Filters

Pressure leaf filters are designed for final discharge of solids in either a dry or wet state, under totally enclosed conditions, with fully automatic operation.

Each type of pressure leaf filter features a pressure vessel in which are located one or more filter elements or leaves of circular or rectangular construction. The filter media may be in the form of a synthetic fibre or other fabrics, or metallic mesh. Supports and intermediate drainage members are in coarse mesh with all components held together by edge binding. Leaf outlets are connected individually to an outlet manifold, which passes through the wall of the pressure vessel.

The material to be filtered is fed into the vessel under pressure, and separation takes place, with the solids being deposited on the leaf surface, and the liquid passing through the drainage system and out of the filter. Cycle times are determined by pressure, cake capacity, or batch quantity. Where particularly fine solids must be removed, a layer of precoat material may be deposited on the leaves prior to filtration, using diatomaceous earth, Perlite, or other suitable precoat materials.

Cake washing for recovery of mother liquor or for removal of solubles may be carried out before discharge of the solids as a slurry or a dry cake.

Pressure leaf filters are supplied in a wide range of size and materials of construction. One typical design is the "Verti-Jet" unit with a vertical tank and vertical leaf filter, as shown in Fig. 10.13, with rectangular leaves mounted individually but connected to a common outlet manifold. For sluice cleaning, either a stationary or oscillating jet system using high efficiency spray nozzles is fitted so as to give complete cake removal. For recovery of dry solids, vibration of the leaves allows automatic discharge of the solids through a bottom discharge port provided with a quick opening door.

In the "Auto-Jet" design, circular leaves are mounted on a horizontal shaft which serves as the filtrate outlet manifold. The leaves are rotated during the cleaning cycle, although, in addition, extra low-speed continuous rotation during operation ensures uniform cake buildup in difficult applications. The leaves are of metallic or plastic construction covered with fabric or wire cloth for direct or precoat operation, and rotation of the leaves during cleaning promotes fast, efficient sluice discharge with minimum power consumption. As an alternative, the leaves may be rotated over knife blades which remove the cake in a dry state. Units of this type are used for handling foodstuffs and also for the processing of minerals and effluents.

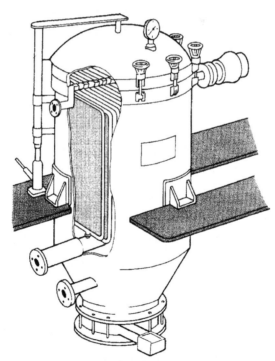

Fig. 10.13
"Verti-jet" pressure leaf filter. *Image Courtesy: Stockdale Filtration systems Ltd., Macclesfield, Cheshire.*

For the handling of edible oils, molten sulphur, effluents and foodstuffs, a Filtra-Matic unit is used in which either the bundle is retracted from the shell as a unit, or the filter tank is retracted leaving the filter leaves and filter cover in position. Such units are available in cylindrical, conical, or trough shell configurations, and cleaning may be either wet or dry, manual or automatic. In the latter case, for dry discharge, vibration systems are used, and for wet removal, spray jets mounted in an overhead manifold sweep the entire leaf surface in an oscillating motion. In this design, the heavy-duty leaves covered with cloth or screen are all interchangeable, and, whether round or rectangular, are all the same size to give uniform precoat, cake buildup, and filtration. In horizontal tray pressure filters, used in batch processes and intermittent flows, the trays are mounted horizontally with connections to a vertical filtrate manifold at the rear, and such units are ideally suited where cake washing and positive cake drying are required. In many cases, the accumulated cake may be sluiced off without removing trays from the filter, and a special recovery leaf is provided where heel filtration is required in which a thin layer of cake is left semi-permanently in contact with the filter medium to improve the clarity of the filtrate. This system is used in various clarification processes and is ideal for handling high flows of liquids with a low solids concentration. In most designs, the tubes are mounted vertically from a tube sheet at the top of the tank, and cleaning is provided with a self-contained internal "air-pump" backwash, thus, avoiding the use of large volumes of sluicing liquid or separate pumps to provide fast and complete removal of the filter cake. The heavy-gauge perforated tube cores are covered with a seamless cloth sleeve sealed at either end by a clamping device. As an alternative, heavy-gauge wire is wound around the centre core, with controlled spacing to give reliable filtration and easy cake release. Tubular element units of this type are available in standard sizes up to 40 m^2.

Cartridge filters

One particular design of pressure filter is the *filter cartridge*, typified by the Metafilter which employs a filter bed deposited on a base of rings mounted on a fluted rod, and is extensively used for clarifying liquids containing small quantities of very fine suspended solids. The rings are accurately pressed from sheet metal of very uniform thickness and are made in a large number of corrosion-resistant metals, though stainless steels are usually employed. The standard rings are 22 mm in external diameter, 16 mm in internal diameter, and 0.8 mm thick, and are scalloped on one side, as shown in Fig. 10.14, so that the edges of the discs are separated by a distance of 0.025–0.25 mm according to requirements. The pack is formed by mounting the rings, all the same way up, on the drainage rod and tightening them together by a nut at one end against a boss at the other, as shown in Fig. 10.15. The packs are mounted in the body of the filter, which operates under either positive or reduced pressure.

The bed is formed by feeding a dilute suspension of material, to the filter usually a form of kieselguhr, which is strained by the packs to form a bed about 3 mm thick. Kieselguhr is available in a number of grades and forms a bed of loose structure which is capable of trapping

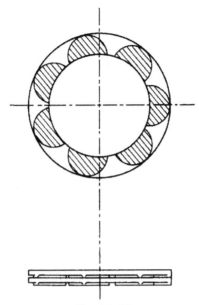

Fig. 10.14

Rings for metafilter. *Image Courtesy: Stella Meta Filters Ltd., Whitchurch, Hants.*

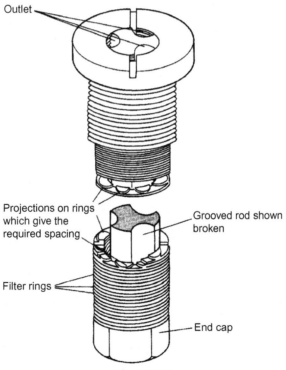

Fig. 10.15

Metafilter pack. *Image Courtesy: Stella Meta Filters Ltd., Whitchurch, Hants.*

particles much smaller than the channels. During filtration, the solids build up mainly on the surface and do not generally penetrate >0.5 mm into the bed. The filtrate passes between the discs and leaves through the fluted drainage rod, and operation is continued until the resistance becomes too high. The filter is then cleaned by backflushing, which causes the filter cake to crack and peel away. In some cases, the cleaning may be incomplete as a result of channelling. If for any reason, the spaces between the rings become blocked, the rings may be quickly removed and washed.

The Metafilter is widely used for filtering domestic water, beer, organic solvents, and oils. The filtration characteristics of clay-like materials can often be improved by the continuous introduction of a small quantity of filter aid to the slurry as it enters the filter. On the other hand, when the suspended solid is relatively coarse, the Metafilter will operate successfully as a strainer, without the use of a filter bed.

The Metafilter is very robust and is economical in use because there is no filter cloth, and the bed is easily replaced, and hence, labour costs are low. Mono pumps or diaphragm pumps are most commonly used for feeding the filter. These are discussed in Volume 1A, Chapter 7.

Example 10.3

The relation between flow and head for a certain slurry pump may be represented approximately by a straight line, the maximum flow at zero head being 0.0015 m^3/s, and the maximum head at zero flow 760 m of liquid.

Using this pump to feed a particular slurry to a pressure leaf filter:

(a) How long will it take to produce 1 m^3 of filtrate?
(b) What will be the pressure across the filter after this time?

A sample of the slurry was filtered at a constant rate of 0.00015 m^3/s through a leaf filter covered with a similar filter cloth but of one-tenth the area of the full-scale unit, and after 625 s, the pressure across the filter was 360 m of liquid. After a further 480 s, the pressure was 600 m of liquid.

Solution

For constant rate filtration through the filter leaf:

$$V^2 + \frac{LA}{v}V = \frac{A^2(-\Delta P)t}{r\mu v}$$

(Eq.10.17)

At a constant rate of 0.00015 m^3/s, then, when the time $= 625$ s:

$V = 0.00015 \times 625 = 0.094\,m^3, (-\Delta P) = 360 \times 9.81 \times 1000 = 3530\,kN/m^2$
and, at $t = 1105\,s$: $V = 0.00015 \times 1105 = 0.166\,m^3$ and $(-\Delta P) = 600 \times 9.81 \times 1000 = 5890\,kN/m^2$

Substituting these values into Eq. (10.17) gives:

$$(0.094)^2 + LA/v \times 0.094 = (A^2/r\mu v) \times 3530 \times 625$$

or:
$$0.0088 + 0.094LA/v = 2.21 \times 10^6 A^2/r\mu v \qquad \text{(i)}$$

and:
$$(0.166)^2 + LA/v \times 0.166 = (A^2/r\mu v) \times 5890 \times 1105$$

or:
$$0.0276 + 0.166LA/v = 6.51 \times 10^6 A^2/r\mu v \qquad \text{(ii)}$$

Eqs (i) and (ii) may be solved simultaneously to give:

$$LA/v = 0.0154 \text{ and } A^2/r\mu v = 4.64 \times 10^{-9}$$

As the full-size plant is 10 times that of the leaf filter, then:

$$LA/v = 0.154 \text{ and } A^2/r\mu v = 4.64 \times 10^{-7}$$

If the pump develops 760 m (7460 kN/m^2) at zero flow and has zero head at $Q = 0.0015$ m^3/s, its performance may be expressed as:

$$-\Delta P = 7460 - (7460/0.0015)Q$$

or:
$$-\Delta P = 7460 - 4.97 \times 10^6 Q (\text{kN/m}^2)$$

$$\frac{dV}{dt} = \frac{A^2(-\Delta P)}{r\mu v(V + LA/v)} \qquad \text{(Eq.10.16)}$$

Substituting for $(-\Delta P)$ and the filtration constants gives:

$$\frac{dV}{dt} = \frac{A^2}{r\mu v} \frac{(7460 - 4.97 \times 10^6 dV/dt)}{(V + 0.154)}$$

Because $Q = dV/dt$, then:

$$\frac{dV}{dt} = \frac{4.67 \times 10^{-7}[7460 - 4.97 \times 10^6 (dV/dt)]}{(V + 0.154)}$$

$$(V + 2.464)\frac{dV}{dt} = 3.46 \times 10^{-3}$$

The time to collect 1 m^3 is then given by:

$$(V + 0.154 + 2.31)\frac{dV}{dt} = \int_0^0 (3.46 \times 10^{-3})dt$$

and:
$$\underline{t = 857\,s}$$

The pressure at this time is found by substituting in Eq. (10.17) with $V = 1$ m^3 and $t = 857$ s to give:

$$1^2 + 0.154 \times 1 = 4.64 \times 10^{-7} \times 857(-\Delta P)$$

and:
$$-\Delta P = 2902 \text{ kN/m}^2$$

10.4.6 Vacuum Filters

If, in a horizontal filter, both sides are open to atmosphere, then, as slurry is introduced above the filter medium, filtration will occur at an adequate rate as long as the hydrostatic head of slurry above the filter medium is sufficient. The unit then operates as a gravity filter. If, however, the top of the vessel is enclosed, and the slurry is introduced under pressure, then the unit operates as a pressure filter, and the driving force is the amount by which the applied pressure exceeds atmospheric pressure below the filter medium. Greatly enhanced filtration rates may then be achieved. Similarly, if the pressure beneath the filter is reduced below atmospheric, the unit operates as a vacuum filter, and the driving force may again be appreciably greater than that available with a gravity filter. The maximum theoretical driving force is then 101 kN/m^2 less the vapour pressure of the filtrate, and much higher flow rates may be obtained than with gravity filtration. With, for example, a head of liquor of 0.3 m in a gravity filter, the maximum driving force is about 3 kN/m^2, whereas, in theory, the driving force of a vacuum filter could be some 30 times greater.

Vacuum filtration has many advantages, not the least being the fact that the feed slurry, often containing abrasive solids in a corrosive liquid, can be delivered by gravity flow or by low-pressure pumps, which need only to overcome the resistance in the feed pipework. In addition, vacuum filtration equipment does not have to withstand high pressures and can, therefore, be manufactured in a wider range of materials. When, because of corrosive conditions or the need to prevent product contamination, expensive materials of construction are required, the low pressures involved allow considerable economy in manufacturing costs. A further advantage is that, after the bulk of the mother liquor has been filtered from the solids, the cake is accessible to mechanical dewatering, flood washing, and sampling, and cake removal can be carried out quickly as it is easily automated. Because of the ease of solids removal, a vacuum filter may be operated with very short cycle times of around 60 s, and the resulting thin filter cakes give very high flow rates per unit area, typically six times as great as those obtained with a pressure filter operating at the same differential pressure. Where delicate solids are handled, very low differential pressures may be used, and, because the feed and the filtrate are not handled under positive pressure, there is no possibility of liquor leakage from the system as might be the case with high pressure differentials. This is an important advantage, especially where corrosive or hazardous materials are involved.

Batch type vacuum filters

Batch vacuum filters were developed from gravity filters, and, in essence, Buchner funnels as used in laboratories and Nutsche type filters as used in industry are similar to gravity filters, except that they feature a vacuum pump or some other vacuum-generating equipment to reduce the pressure under the filter medium, thereby increasing the driving force across the filter medium.

Simple filters operating in this manner are often referred to as pan filters, and they incorporate certain features which are now being incorporated into continuously operating machines. These filters enable the initial cake formation to be made, either under gravity or at very low differential pressures, so that uniform distribution of the solids over the filter medium is obtained. Preferential settling occurs so that the heavier, and usually coarser, solids contact the filter medium first, thereby protecting it from the slower settling fine solids. Following this, the filtrate drainage section may be subjected to higher vacuum to give greater differential pressures, higher flow rates per unit area, and greater total flows. A typical cycle for a batch type vacuum filter might be:

(a) Feeding of slurry,
(b) Cake formation and filtration, initially under gravity if required, followed by vacuum operation,
(c) Initial drying for maximum removal of mother liquor,
(d) Cake washing and drying,
(e) Manual discharge of solids, and
(f) Cloth washing.

The advantages of such units include simplicity in both design and operation, low cost per unit area, and flexibility with regard to filtration, washing and drying periods. In addition, very sharp separation of the mother liquor and various wash liquors can be obtained due to the wash liquors being completely contained within the pan. The filter cakes remain in position between batches without the necessity for power to be used for their retention. In addition, the flow through the filter medium can be stopped or be restricted to a very low rate in order to obtain leaching from the solids in the filter bed.

Against this, it may be necessary with a batch vacuum filter to provide holding vessels for feed, and possibly filtrate and wash. The discharge of solids is intermittent, and this may lead to complications in feeding the solids to a continuous dryer or to some other continuous process, and cake hoppers and a feeding arrangement may have to be provided.

Because the liquor flow rate decreases as filtration proceeds, facilities to cope with this must be incorporated in the feed and filtrate handling sections of the system, and the discharge of solids must be a manual operation if the solids cannot be sluiced off.

Equipment has been developed to overcome the disadvantages of manual discharge of solids from the simple batch pan type vacuum filter, and such equipment incorporates scrapers or rakes for the removal of the solids. Other pan-type vacuum filters are arranged so that the pan is tipped over, when the solids are discharged, either under their own weight or sometimes with the assistance of blowback air admitted to the underside of the filter cloth when the pan is inverted.

Further developments eliminate, or reduce to acceptable proportions, the cyclic variations noted previously, and one simple development which achieves this to a limited extent involves a number of individual tipping pans arranged in parallel. For example, with three pans discharging into a common slurry tank, the operation is arranged so that while one pan is being charged with feed material, and, while solids–mother liquor separation is being carried out in it, the second pan is at the washing stage, and the third pan is at the drying and discharging stage. This arrangement obviously reduces cyclic variations, though it necessitates diversion of the feed and wash liquor from one pan to another. The operation can, however, be carried out automatically by the use of a process timer or similar control unit.

Classification of batch and continuous vacuum filters

The disadvantages of cyclic operation, attributable to the need to remove the cake from batch filters, have been largely overcome by the development of a wide range of continuous vacuum filters, as summarised in Table 10.2, which includes data on the selection for a given duty. This is further detailed in Table 10.3 and is based on the work of G.H. Duffield of Stockdale Engineering and E. Davies of EDACS Ltd.

Horizontal continuous filters

One way of limiting still further cyclic variations is to use vacuum filters of the horizontal continuous type, and many designs have been developed, which in the main, consist of pans arranged in line or in a circle so that each in turn goes through the same cycle of operations, namely: feeding, dewatering, washing, drying, automatic solids discharge, and cloth washing. The pans move past the stationary feed and wash areas, where the feed and wash liquors are continuously introduced to the filter. Alternatively, a plain filter surface may be arranged as a linear belt or as a circular table, in place of pans.

All filters of this type have the distinct advantage of continuous operation, and, as a result, the feed and wash liquors may be fed to the equipment at steady rates. Dense, quick-settling solids can be handled, and the cake thickness—and the washing and drying times—can be varied independently between quite wide limits. The cake can be flooded with wash liquor, which can be taken away separately and reused to obtain countercurrent washing of the cake if desired. Disadvantages of this equipment are the high cost per unit area, mechanical complexity, and the fact that some effectively use only 50% of the total filtration area. With the horizontal linear pan and the rotary tipping pan filters, a truly continuous discharge of solids is not obtained because these units operate as short time-cycle batch-discharge machines.

The *horizontal pan filter* incorporates a number of open top pans which move in a horizontal plane through the feed, filtration, washing, and drying. At the discharge point, the pans pass over a roller and then return to the feed end of the machine in the inverted position. The design of this type of filter varies greatly with different manufacturers, though one frequently used type has the walls of the pans built up on an open type of horizontal linear conveyor. The conveyor,

Table 10.2 Classification of vacuum filters

Type of Vacuum Filter	Note	Usual Max. Area (m²)	Characteristics of Suitable Slurry					Performance Index		
			A	B	C	D	E	Cake Dryness	Cake Washing	Filtrate Clarity
			See Notes Below					See Notes Below		
(A) Cake filters										
Single tipping pan	1	3	X	X				4–7	8–9	8
Multi tipping pan	2	12	X	X				4–7	8–9	8
Horizontal linear tipping pan	3	25	X	X				4–7	8–9	7
Horizontal linear belt	4	20	X	X				5–8	7–8	6
Rotary tipping pan	3	200	X	X				4–7	8–9	7
Horizontal rotary table	4	15	X	X				4–7	7–8	7
Rotary drum—string discharge	5	80		X	X			5–8	6	8
Rotary drum—knife discharge	5	80		X	X			5–8	6	8
Rotary drum—roller discharge	5	80			X	X		5–6	5	8
Rotary drum—belt discharge	5	80			X	X	X	5–8	6	7
Top feed drum	6	10	X	X				1–2	1	6
Hopper/dewaterer	6	10	X					1–2	1	6
Internal drum	7	12		X	X			3–4	1	6
Single comp. Drum	8	15			X			2–3	4	5
Disc	9	300		X	X			2–3	1	6
(B) Media filters										
Rotary drum precoat	10	80				X	X	–	6	9

Notes:

Filters

1. For small batch production. Has very wide application, is very adaptable, and can be automated.
2. Usually 2–4 pans, for medium size batch production. Very wide application, very adaptable, can be automated.
3. For free-draining materials where very good washing is required with sharp separation between mother liquor and wash liquors.
4. For free-draining materials where very good washing is required.
5. Wide range of types and size available. Generally suitable for most slurries in categories B and C. Can usually be fitted with various mechanical devices to improve the washing and drying.
6. Restricted to very free-draining materials not requiring washing.
7. Restricted to very free-draining materials not requiring washing, but where the solids can be retained by vacuum alone.
8. Allows use of high drum speed and is capable of very high flow rates.
9. Large throughputs for small floor space.
10. Suitable for almost any clarification and for handling materials which blind normal filter media.

Slurries

A. High solids concentration, normally >20%, having solids which are free-draining and fast settling, giving difficulty in mechanical agitation and giving high filtration rates.
B. Rapid cake formation with reasonably fast settling solids which can be kept in suspension by mechanical agitation.
C. Lower solids concentration with solids giving slow cake formation and thin filter cakes which can be difficult to discharge.
D. Low solids concentration with solids giving slow cake formation and a filter cake having very poor mechanical strength.
E. Very low solids concentration (i.e., clarification duty), or containing solids which blind normal filter media. Filtrate usually required.
X. Slurries handled.

Performance index

9 = the highest possible performance.
1 = very poor or negligible performance.

(Based on the information provided by Stockdale Filtration systems Ltd., Macclesfield, Cheshire).

Table 10.3 Operating data for some vacuum filter applications

Application	Type of Vacuum Filter Frequently Used (See Note 2)	Solids Content of Feed (Percent by Mass)	Solids Handling Rate kg Dry Solids/s m² Filter Surface (See Note 3)	Moisture Content of Cake (Percent by Mass)	Air Flow	
					Filter Surface (See Note 4) (m³/s m²)	Vacuum (kN/ m² Pressure, Below Atmospheric)
Chemicals						
Aluminium hydrate	Top feed drum	25–50	0.12–0.2	10–17	0.025	84
Barium nitrate	"	80	0.34	5	0.13	67
Barium sulphate	Drum	40	0.013	30	0.005	33
Sodium bicarbonate	"	50	0.5	12	0.15	61
Calcium carbonate	"	50	0.03	22	0.010	33
Calcium (pptd)	"	30	0.04	40	0.010	26
Calcium sulphate	Tipping pan	35	0.16	30	0.025	41
Caustic lime mud	Drum	30	0.20	50	0.03	51
Sodium hypochlorite	Belt disch. Drum	12	0.04	30	0.015	33
Titanium dioxide	Drum	30	0.03	40	0.010	33
Zinc stearate	"	5	0.007	65	0.015	33
Minerals						
Frothed coal (coarse)	Top feed drum	30	0.20	18	0.020	61
Frothed coal (fine)	Drum or disc	35	0.10	22	0.015	51
Frothed coal tailings	Drum	40	0.05	30	0.010	26
Copper concentrates	"	50	0.07	10	0.010	30
Lead concentrates	"	70	0.27	12	0.015	26
Zinc concentrates	"	70	0.20	10	0.010	33
Flue dust (blast furnace)	"	40	0.04	20	0.015	33
Fluorspar	"	50	0.27	12	0.025	51
Paper						
Kraft pulp	Deep sub. Drum or disc	1.5	0.05	90	Barometric leg	
Effluents	Drum	3	0.005	80	0.015	41
Bleach washer	"	1.0	0.20 (filtrate)	90	0.025	67

Table 10.3 Operating data for some vacuum filter applications—cont'd

Application	Type of Vacuum Filter Frequently Used (See Note 2)	Solids Content of Feed (Percent by Mass)	Solids Handling Rate kg Dry Solids/s m² Filter Surface (See Note 3)	Moisture Content of Cake (Percent by Mass)	Air Flow	
					Filter Surface (See Note 4) (m³/s m²)	Vacuum (kN/ m² Pressure, Below Atmospheric)
Foodstuffs						
Starch	Drum	50	0.05	40	0.015	16
Gluten	Belt discharge drum	15	0.007	48	0.010	33
Salt	Top feed drum	30	0.5	6	0.20	91
Sugar cane mud	Drum	10	0.03	30	0.005	33
Glucose (spent carbon)	"	15	0.02	50	0.0025	41
Glucose (44%)	Drum— Precoat	2	0.14 (filtrate)	50	0.0025	33
Effluents						
Primary sewage	Drum	8	0.01	65	0.008	33
Primary sewage, digested	"	6	0.005	65	0.008	33
Neutralised H_2SO_4 pickle	"	8	0.013	50	0.010	41
Plating shop effluent	Drum— Precoat	2	0.10 (filtrate)	80	0.013	33

Notes:
1. The information given should only be used as a general guide, for slight differences in the nature, size range, and concentration of solids, and in the nature and temperature of the liquor in which they are suspended can significantly affect the performance of any filter.
2. It should not be assumed the type of filter stated is the only suitable unit for each application. Other types may be suitable, and the ultimate selection will normally be a compromise based on consideration of many factors regarding the process and the design features of the filter.
3. The handling rate (in kg/m²s) generally refers to dry solids except where specifically referred to as filtrate.
4. The air volumes stated are measured at the operating vacuum (i.e., they refer to attenuated air).
(Based on the information provided by Stockdale Filtration systems Ltd., Macclesfield, Cheshire).

within the walls of the pans, serves as the support for the filter media and for the liquor drainage system, and elsewhere, the conveyor serves as the mechanical linkage between adjacent pans. The conveyor passes over vacuum boxes in the feed, filtration, washing, and drying zones.

The *horizontal rotary tipping pan filter* has the pans located in a circle moving in a horizontal plane, and, in effect, the pans are supported between the spokes of a wheel turning in a vertical shaft. Each pan is connected to a single, centrally-mounted rotary valve which controls the vacuum applied to the feed, filtration, washing, and drying zones and from which the mother liquor and the various wash liquors can be diverted to individual vessels. At the discharge position, the pans are inverted through 180 degrees, and, in some designs, cake removal is assisted by the application of compressed air to the underside of the filter medium via the rotary valve. One example of this type of filter is the Prayon continuous filter, developed particularly for the vacuum filtration of the highly corrosive liquids formed in the production of phosphoric acid and phosphates. It is made in five sizes providing filtering surfaces from about 2–40 m^2, and it consists of a number of horizontal cells held in a rotating frame, and each cell passes in turn under a slurry feed pipe and a series of wash liquor inlets. Connections are arranged so as to provide three countercurrent washes. The timing of the cycle is regulated to give the required times for filtration, washing, and cake drying. The discharge of the cake is achieved by gravity as the cell is automatically turned upside-down, and can be facilitated if necessary by washing or by the application of back pressure. The cloth is cleaned by spraying with wash liquid and is allowed to drain before it reaches the feed point again. The filter has a number of advantages including:

(a)   A large filtration area, up to 85% of which is effective at any instant.
(b)   Low operating and maintenance costs and simple cloth replacement.
(c)   Clean discharge of solids.
(d)   Fabrication in a wide range of corrosion-resistant materials.
(e)   Countercurrent washing.
(f)   Operation at a pressure down to 15 kN/m^2.

The *horizontal rotary table filter* incorporates, as its name suggests, a horizontal rotary circular table divided into segments or drainage compartments, each of which is "ported" to a rotary valve underneath the table. The top surface of the table is fitted with a filter medium, and walls are provided around the periphery of the table and around the hub. The feed slurry is distributed across the table through a feed box, and vacuum is applied to the segments in the feed position to extract the mother liquor. On rotation of the table, the filter cake then passes through the washing and drying zones. At the discharge position, the vacuum on the segments can be broken to atmospheric pressure, and the filter cake removed by a horizontal rotary scroll, which lifts the solids over the periphery wall of the table. The scroll cannot be in contact with the filter medium, and, therefore, a "heel" of solids remains on the table after passing the discharge point. This heel can either be disturbed by blowback air and reslurried with the incoming feed, or, in certain applications, the heel can be retained intact on the filter medium in order to obtain better retention of solids from the incoming feed than would be achieved with a clean filter cloth.

Horizontal belt or band filters

In the horizontal belt or band filter, shown in Fig. 10.16, an endless belt arranged in the horizontal plane and running over pulleys at about 0.05 m/s at the feed and discharge ends (similar to the horizontal band conveyor). The principle is applied in the Landskröna band filter, as described by Parrish and Ogilvie.[32] In the filtration, washing, and drying zones, where the

Fig. 10.16
Rigid belt filter.

filter medium passes over vacuum boxes, it may be supported on a separate endless belt which also serves as the drainage member and as the valve. This belt can be provided with side walls to contain the feed slurry and wash liquors, or a flat belt can be used in conjunction with rigid static walls, against which the belt slides. Rubber (or similar) wiper blades which drag against the cake surface can be used to isolate the filtration and washing zones from each other. In some designs, the belts move continuously; in others, the belts are moved along in stages.

The applications of horizontal belt filters are discussed by Blendulf and Bond.[33] A typical filter for dewatering concentrates is shown in Fig. 10.17, and it may be noted that this type of equipment is the most expensive per unit area.

An interesting development is the Adpec filter described by Bosley[34] that has an intermittently moving belt. The slurry is fed to one end and the vacuum applied to the underside boxes, and filtering occurs with the belt stationary. As the discharge roll at the other end moves inwards, it trips the system so that the vacuum is relieved and the belt moved forward, thus avoiding the problem of pulling the belt continuously over the sections under vacuum.

Band filters have several advantages over rotary vacuum filters which are described in the following section. These include:

(a) Some gravitational drainage of liquid occurs because the cake and the belt are vertically separated at all stages of the filtration operation. In addition, a vacuum is not needed to keep the cake in place, and this simply supplements the effect of gravity on the flow of filtrate.

Fig. 10.17
Delkor horizontal belt filter dewatering copper flotation concentrate (filter area 21 m^2).
Image Courtesy: Delfilt Ltd., Bath, Avon.

(b) Because the feed is from the top, it is not necessary to agitate the suspension in the feed trough.

(c) Large particles in the slurry tend to reach the surface of the filter cloth first and to form a layer which, therefore, protects the cloth from the blinding effect of the fines, although this does lead to the formation of a nonuniform cake.

(d) Washing is more effective because of the improved structure of the filter cake.

(e) The fitting of an impervious cover over the cake toward its exit will reduce the loss of vacuum due to air flow through cracks in the cake.

Another development by Pannevis in Holland, shown in Fig. 10.18, incorporates vacuum trays which support the filter cloth as it moves at a constant speed, which may be varied as necessary. The vacuum tray is evacuated, filtration takes place, and the tray is pulled forward by the cloth in such a way that cloth and tray move at the same velocity, thus, offering negligible sealing problems, minimum wear, low driving power requirements, and the possibility of high vacuum operation. At the end of the vacuum stroke, the vacuum is released, the tray is vented and pulled back quickly by pneumatic action to its starting position. Although an intermittent vacuum is required, the Pannevis design is, in effect, a continuous unit because the suspension feed, wash liquor feed, and cake discharge are all fully continuous. A vacuum tray is typically 1.4 m long and 150 mm deep, and the width and length of the belt are 0.25–3.0 m and 2.8–25 m, respectively.

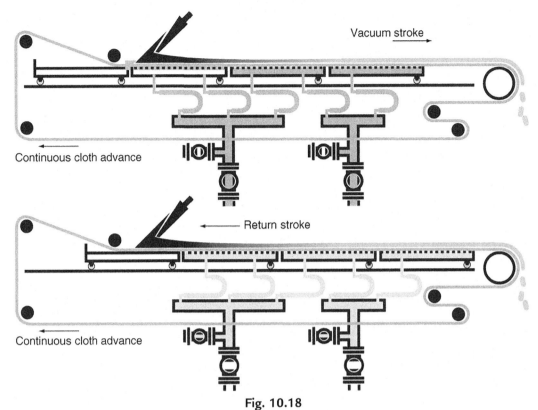

Fig. 10.18
Working principle of the Pannevis vacuum belt filter.

Rotary drum filters

Because of its versatility and simplicity, one of the most widely used vacuum filters is the *rotary drum filter*, and a filter of this type was patented in England in 1872 by William and James Hart. The basic design varies with different manufacturers, although essentially all drum-type vacuum filters may be divided into two categories:

(a) Those where vacuum is created within compartments formed on the periphery of the drum, and

(b) Those where vacuum is applied to the whole of the interior of the drum.

The most frequently used continuous drum type filters fall into the first category. These give maximum versatility and low cost per unit area, and also allow a wide variation of the respective time periods devoted to filtration, washing, and drying.

Essentially, a multicompartment drum-type vacuum filter consists of a drum rotating about a horizontal axis, arranged so that the drum is partially submerged in the trough into which the material to be filtered is fed. The periphery of the drum is divided into compartments, each of which is provided with a number of drain lines. These pass through the inside of the drum and terminate as a ring of ports covered by a rotary valve, through which vacuum is applied. The surface of the drum is covered with a filter fabric, and the drum is arranged to rotate at low speed, usually in the range 0.0016–0.004 Hz (0.1–0.25 rpm) or up to 0.05 Hz (3 rpm) for very free filtering materials.

As the drum rotates, each compartment undergoes the same cycle of operations, and the duration of each of these is determined by the drum speed, the submergence of the drum, and the arrangement of the valve. The normal cycle of operations consists of filtration, drying, and discharge. It is also possible, however, to introduce other operations into the basic cycle, including:

(a)   Separation of initial dirty filtrate—which may be an advantage if a relatively open filter fabric is used.
(b)   Washing of the filter cake.
(c)   Mechanical dewatering of the filter cake.
(d)   Cloth cleaning.

Fig. 10.19A shows a typical layout of a rotary drum installation and Fig. 10.19B shows the sequence of cake formation, washing, and dewatering. A large rotary drum vacuum filter is shown in Fig. 10.20.

In order to achieve consistent performance of a continuous filter, it is necessary to maintain the filter medium in a clean condition. With a drum-type vacuum filter, this requires the complete and continuous removal of the filter cake from the drum surface, and the operating conditions are often influenced by the need to form a fully dischargeable cake. Again, in order to achieve high capacity and good cake washing and/or drying, it is very often desirable to operate with very thin cakes. Therefore, the cake discharge system of most drum-type vacuum filters must be arranged so as to ensure the complete and continuous removal of extremely thin filter cakes. The most effective way of achieving this is determined to a large extent by the physical nature of the solids being handled.

The various discharge systems that are suitable for drum type filters include the string discharge technique, which is effective for an extremely wide range of materials. Essentially, this involves forming the filter cake on an open type of conveyor which is in contact with the filter medium in the filtration, washing, and drying zones. Consequently, the solids which are trapped by the filter medium form a cake on top of the "open" conveyor. From the discharge point on the drum, the conveyor transports the filter cake to a discharge roll, at which point, the cake is dislodged. The conveyor then passes through an aligning mechanism and over a return roll, which guides it back into contact with the filter drum at the commencement of the cycle and just above the level of feed liquor in the filter trough.

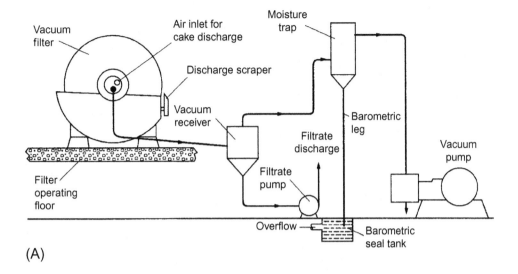

(A)

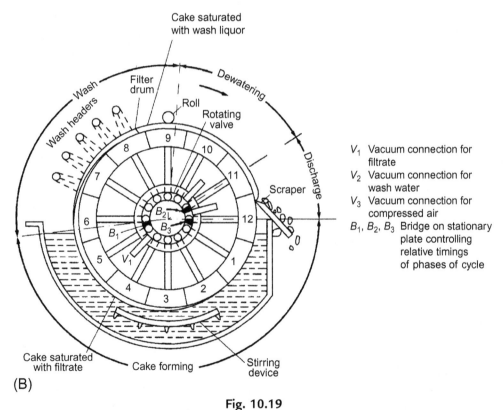

(B)

Fig. 10.19

Typical layout of the rotary drum filter installation (A) typical layout (B) Sequence of cake formation, washing and dewatering.

Fig. 10.20
Rotary vacuum drum filter used in a zinc leaching operation.

In the string discharge system, the conveyor consists of a number of endless strings which are spaced at a pitch of approximately 12 mm over the width of the filter drum. The string spacing may, however, be in the range 6–25 mm, depending on the mechanical properties of the solids.

The advantages of the string discharge system are:

(a) Thin and sticky filter cakes down to about 1.5 mm of materials such as clay may be effectively discharged.

(b) The filter cloth is almost free of mechanical wear and tear so that thin and delicate cloths may be used, and these can be selected almost solely for their filtration properties. Such cloths are usually less prone to plugging than the stronger and thicker cloths required for other discharge systems.

(c) The cloth can be attached to the drum in a simple manner so that fitting and subsequent replacement can be carried out quickly. Normally, the cloth is loosely wrapped around the drum, and it is secured to the drum at the edges and, once across the drum, at the overlap, by a simple caulking system. The use of wire winding, clamping bars, and the necessity of securing the cloth at every panel, which is essential with other discharge systems, is avoided.

(d) The use of compressed air to loosen cake from the drum surface is avoided, and consequently, there is no possibility of blowing back into the filter cake moisture which has previously been removed under vacuum. This is a possibility with knife discharge filters operating with blowback.

(e) If required, the path of the discharge strings can be altered so that the filter cake is conveyed by the strings to a convenient point for feeding a continuous dryer-extruder, or other processing equipment.

A typical string discharge mechanism is shown in Fig. 10.21.

The knife discharge system incorporates a knife which is arranged so that the surface of the drum runs on or near to the knife edge. The cake is dislodged from the cloth either by its own weight, with thick and heavy cakes, or by applying compressed air to the underside of the filter cloth. The blowback air can either be admitted at low pressure for a long period, or at high pressure and instantaneously by means of a mechanical "blow-off timer." With suitable solids, it is possible to operate with the knife spaced from the drum so that not all of the cake is removed. The heel of solids is then retained on the drum, and this acts as the filter media. This discharge system is particularly suitable for friable cakes that do not have the mechanical properties to bridge over the strings of a string discharge system.

With the roller cake discharge system, the cake is transferred from the drum to a discharge roll, from which it is removed by a knife. This is a relatively simple method of removing thin and sticky filter cakes without having a knife rubbing against and wearing the filter cloth. Certain applications require facilities for washing the cloth either continuously or intermittently

Fig. 10.21
String discharge mechanism on a filter handling silica gel.

without dilution of the trough contents, and such a feature is provided by the belt discharge system where the filter cloth also acts as cake conveyor. In such a system, the cake is completely supported between the drum and the discharge roll so that thin cakes and cakes of low mechanical strength can be handled, and higher drum speeds, and hence, higher filtration rates, can be achieved.

The performance of drum-type vacuum filters for given feed conditions can be controlled by three main variables—drum speed, vacuum (if necessary with differential vacuum applied to the filtration, washing, and drying zones), and the percentage of drum surface submerged in the feed slurry. Most drum filters have facilities which allow for easy manual adjustment of these variables, although automatic adjustment of any, or all, of them can be actuated by changes in the quality and/or quantity of the feed or cake. For maximum throughput, a drum filter should be operated at the highest submergence and at the highest possible drum speed. The limiting conditions affecting submergence are:

(a)   Any increase in submergence limits the proportion of the drum area available for washing and/or drying.
(b)   Drum submergence above approximately 40% entails the use of glands where the drum shaft and valve hub pass through the trough.
(c)   High submergence may complicate the geometry of the discharge system.

Under all conditions, it is essential to operate with combinations of drum speed, submergence, and vacuum, which, for the feed conditions that apply, will ensure that a fully dischargeable cake is formed. If this is not done, then progressive deterioration in the effectiveness of the filter medium will occur, and this will adversely affect the performance of the machine. Typical filtration cycles for drum filters are given in Table 10.4.

Drum type vacuum filters can be fitted, if required, with simple hoods to limit the escape of toxic or obnoxious vapours. These may be arranged for complete sealing, and for operation under a nitrogen or similar blanket, although this complicates access to the drum and necessitates a design in which the vacuum system and the cake receiving system are arranged to prevent gas loss. The cake moves through the washing and drying zones in the form of a continuous sheet, and, because the cake and filter medium are adequately supported on the drum shell, it is possible to fit the filter with various devices that will improve the quality of the cake regarding both washing and drying, prior to discharge.

Simple rolls, extending over the full width of the filter, can be so arranged that any irregularities or cracks in the cake are eliminated, and subsequent washing and drying are, therefore, applied to a uniform surface. Otherwise, wash liquors and air tend to short circuit, or "channel," the deposited solids. The cake compression system may also incorporate a compression or wash blanket, which limits still further any tendency for the air or wash liquor to "channel." Wash blankets also avoid disturbance of the cake, which might occur when high pressure sprays are

Table 10.4 Filtration cycles for drum-type rotary vacuum filter (times in seconds)

| Speed | | 25% Submergence | | | | | 37.5% Submergence | | | | | 50% Sub. | | 66.7% Sub. | | 75% Sub. | |
| | | A | B | C | | | A | B | C | | | A | B | A | B | A | B |
(rev/min)	(Hz)	Filter	Dewater	Initial Dewater	Wash	Final Dewater	Filter	Dewater	Initial Dewater	Wash	Final Dewater	Filter	Dewater	Filter	Dewater	Filter	Dewater
3	0.05	5	10	2	5	3	7	8	2	4	2	8	5	12	2	13	1
$2\frac{1}{2}$	0.04	6	12	2	7	3	9	10	2	5	3	11	6.5	16	2.2	17	1.8
2	0.033	7	15	3	8	4	12	14	2	9	3	13	7.5	19	3	21	2
$1\frac{1}{2}$	0.025	10	23	5	12	6	16	18	3	11	4	16	10	24	4	28	3
1	0.017	15	30	7	16	7	22	24	3	16	5	27	15	38	5	42	4
$\frac{2}{3}$	0.011	22	45	10	24	11	33	37	5	24	7	40	22	55	8	63	6
$\frac{1}{2}$	0.008	30	60	13	32	15	44	48	6	32	10	54	30	76	10	84	8
$\frac{1}{3}$	0.0055	45	91	20	48	23	66	72	9	48	15	80	44	110	16	126	12
$\frac{1}{4}$	0.004	60	120	26	64	30	88	96	12	64	20	110	60	150	21	170	17
$\frac{1}{5}$	0.0033	75	150	32	80	38	110	120	15	80	25	135	75	180	27	210	21
$\frac{1}{6}$	0.0028	90	180	39	96	45	132	144	18	96	30	160	90	220	32	280	28
$\frac{1}{7}$	0.0024	105	210	45	112	54	154	168	21	112	35	190	105	260	37	310	31
$\frac{1}{8}$	0.0021	120	240	52	128	60	176	192	24	128	40	215	120	300	42	340	34
$\frac{1}{10}$	0.0017	135	300	65	160	75	220	240	30	160	50	240	135	360	54	420	42

Columns A + B or A + C should be used but *not* all three.

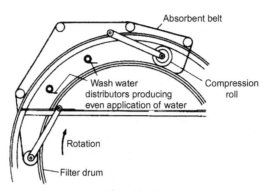

Fig. 10.22
Mechanism of cake compression.

directed onto the cake, and they allow the wash liquor to be applied much nearer to the point where the cake emerges from the slurry. Compression rolls, with or without blankets, can be arranged with pneumatic or hydraulically loaded cylinders, or they can be weighted or fitted with springs to increase further the pressure applied to the cake, as shown in Fig. 10.22. In some applications, the rolls can be arranged to give a controlled cake thickness, and so, give some rearrangement of the solids in the filter cake with the advantages of the elimination of cracks, compaction of the cake, and the liberation of moisture from thixotropic materials, with the liberated moisture finally being drawn into the vacuum system prior to discharge.

Other devices which find varying degrees of success in the treatment of thixotropic materials include rotary beaters and pulsating valves to interrupt the air flow in the drying zone. Both of these result in movement of the cake, giving a reduction in moisture content.

A drum-type vacuum filter can easily be arranged so that the trough and hood, if fitted, are thermally insulated. In addition, the filter can be equipped with heat exchange equipment, such as jackets for heating fluid or refrigerant, and electric resistance wires for heating. These features can be of particular significance in applications such as the dewaxing of lubricating oil which is normally carried out at temperatures below 263 K ($-10°C$), and for salt drying, which may be effected at temperatures approaching 673 K (400°C) on a filter/dryer.

In most applications, bottom feed drum filters, that is, those where the drum is partially suspended in the feed slurry, must be fitted with agitators to keep the solids in suspension. Normally, pendulum-type rakes, which are pivoted on or above the drum axis, are used so as to avoid the use of glands, although rotary paddles fitted at low level in the filter trough also find application. Solids suspension may also be improved by giving careful attention to the flow pattern within the trough such as, for example, by feeding slurry into the trough at low level and at a rate in excess of that at which the material can be filtered. With this technique, a constant overflow is provided so that upward movement within the trough assists in conveying solids toward the filter surface. Air sparge pipes may also be used to prevent solids settlement, although this results in higher operating costs than with a mechanical agitation system.

Top-feed drum filters are used for the treatment of very fast settling solids that cannot easily be kept in suspension. A drum filter may be preferable, rather than a horizontal pan or similar filter, because of its low cost, simplicity, and reliability, and because of the relative ease with which it can be fitted with appropriate accessory features. A top-feed drum filter incorporates multicompartments, drain lines, and a rotary valve identical to those in a conventional drum filter. The feed slurry is introduced at, or just before, top dead centre, and the cake is discharged from 90 degrees to 180 degrees from the feed point. The feed slurry can be distributed across the drum by sprays or a weir box, or it may be contained within a 3-sided head box, which is sealed against the moving drum at each end and across the back on the ascending filter surface. For materials which do not cake together sufficiently to be self-supporting on the drum surface, a hopper-type top-feed filter may be used, which is similar to a conventional top feed filter, except that the walls extend above the drum surface around all sides of each compartment, and these support the filter cake. Because larger particles settle more rapidly, these are deposited near the filter cloth, and smaller particles form the outer portions of the filter cake. A cake of relatively high porosity is thereby obtained, and high filtration rates are achieved.

Modern plants utilise drums with surfaces of 60–100 m^2, as compared with the 20 m^2 of the older cast iron drums. Construction materials such as stainless steel, titanium, epoxy resins, and plastics all give much improved corrosion resistance for many slurries and hence, longer life. The replacement of the knife system by some form of belt has given better cake discharge and permitted the use of thinner filtering media, such as synthetic fibres. The belt provides some support for the cake and materially assists the effect of compressed air for lifting off the cake. Drying can be improved by totally covering the filter with a hood. Improvements have also been made in techniques for reducing cake cracking.

Recent developments in rotary filters include equipment marketed by Dorr–Oliver, illustrated in Fig. 10.23, which combines vacuum filtration with pressure filtration, in which filter cake moisture is reduced by 20%–200%. A combined filtering and drying plant has been developed in which a continuous belt, rather like a bed-spring in construction, passes round the underside of the drum filter, and the filter cake is deposited on the belt to which it adheres. The belt leaves the filter toward the top and is then carried through a cabinet dryer. It then returns to the underside of the drum filter after the cake has been removed by agitation. The metal belt, which assists the drying of the material by virtue of its good heat-conducting properties, is formed into loops which are carried through the dryer on a slat conveyor. With finely divided materials, the total loss of solid from the belt is as little as 1% or 2%.

Fig. 10.23
Dorr–Oliver press-belt drum filter.

Example 10.4

A slurry containing 40% by mass solid is to be filtered on a rotary drum filter 2 m diameter and 2 m long, which normally operates with 40% of its surface immersed in the slurry and under a pressure of 17 kN/m^2. A laboratory test on a sample of the slurry using a leaf filter of area 200 cm^2 and covered with a cloth similar to that on the drum produced 300 cm^3 of filtrate in the first 60 s and 140 cm^3 in the next 60 s, when the leaf was under an absolute pressure of 17 kN/m^2. The bulk density of the dry cake was 1500 kg/m^3, and the density of the filtrate was 1000 kg/m^3. The minimum thickness of cake that could be readily removed from the cloth was 5 mm.

At what speed should the drum rotate for maximum throughput, and what is this throughout in terms of the mass of the slurry fed to the unit per unit time?

Solution

For the leaf filter:

$$A = 0.02 \, \text{m}^2.$$
$$(-\Delta P) = (101.3 - 17) = 84.3 \, \text{kN/m}^2 \text{ or } 84,300 \, \text{N/m}^2$$

When $t = 60 \, \text{s}$: $V = 0.0003 \, \text{m}^3$
When $t = 120 \, \text{s}$: $V = 0.00044 \, \text{m}^3$

These values are substituted into the constant pressure filtration equation:

$$V^2 + \frac{2LAV}{v} = \frac{2(-\Delta P)A^2 t}{r\mu v} \qquad \text{(Eq.10.18)}$$

which enables the filtration constants to be determined as:

$$L/v = 2.19 \times 10^{-3} \text{ and } r\mu v = 3.48 \times 10^{10}$$

For the rotary filter, Eq. (10.18) applies as the whole operation is at a constant pressure. The maximum throughput will be attained when the cake thickness is a minimum, that is 5 mm or 0.005 m.

$$\text{Area of filtering surface} = (2\pi \times 2) = 4\pi \, \text{m}^2.$$

$$\text{Bulk volume of cake deposited} = (4\pi \times 0.005) = 0.063 \, \text{m}^3/\text{revolution}.$$

If the rate of filtrate production $= w$ kg/s, then the volume flow is $0.001 \, w \, \text{m}^3/\text{s}$.

For a 40% slurry : $S/(S + w) = 0.4$, and the mass of solids $= 0.66w$.

Thus:

$$\text{volume of solids deposited} = (0.66w/1500) = 4.4 \times 10^{-4} w \, \text{m}^3/\text{s}.$$

$$\text{If one revolution takes } t \text{ s, then} : 4.4 \times 10^{-4} wt = 0.063.$$

$$\text{and the mass of filtrate produced per revolution} = 143 \, \text{kg}.$$

$$\text{Rate of production of filtrate} = 0.001w \, \text{m}^3/\text{s} = V/t$$

Thus :
$$V^2 = 1 \times 10^{-6} w^2 t^2 = 1 \times 10^{-6}(143)^2 = 0.02 \, \text{m}^6$$
and :
$$V = 0.141 \, \text{m}^3$$

Substituting $V = 0.141 \, \text{m}^3$ and the constants into Eq. (10.18), gives:

$$(0.141)^2 + 2 \times 2.19 \times 10^{-3} \times 0.141 \times 4\pi = 2 \times 84{,}300 \times (4\pi)^2 t/(3.48 \times 10^{10})$$

from which $t = 36.13$ s, which is equal to time of submergence/revolution.

Thus :
$$\text{time for 1 revolution} = (36.13/0.4) = 90.3 \, \text{s}$$
and :
$$\text{speed} = (1/90.3) = \underline{0.011 \, \text{Hz}}$$

$$w = (143/90.3) = 1.58 \, \text{kg/s}$$
$$s = (0.66 \times 1.58) \, \text{kg/s}$$
and :
$$\text{mass of slurry} = (1.66 \times 1.58) = \underline{\underline{2.63 \, \text{kg/s}}}$$

Example 10.5

A plate and frame press with a filtration area of 2.2 m^2 is operated with a pressure drop of 413 kN/m^2 and with a downtime of 21.6 ks (6 h). In a test with a small leaf filter 0.05 m^2 in area, operating with a pressure difference of 70 kN/m^2, 0.00025 m^3 of filtrate was obtained in 300 s and a total of 0.00040 m^3 in 600 s. Estimate the optimum filtration time for maximum throughput.

If the operating cost during filtration is £10/ks, and the cost of a shutdown is £100, what is the optimum filtration time for minimum cost?

Solution

Substituting $V = 0.00025$ m^3 at $t = 300$ s in Eq. (10.39) gives:

$$(300/0.00025) = 0.00025r\mu v/(2 \times 0.05^2 \times 70 \times 10^3) + r\mu L/(0.05 \times 70 \times 10^3)$$

or:
$$1.2 \times 10^6 = 7.14 \times 10^{-6} r\mu v + 2.86 \times 10^{-4} r\mu L \qquad (i)$$

Substituting $V = 0.00040$ m^3 at $t = 400$ s in Eq. (10.39) gives:

$$1.0 \times 10^6 = 11.42 \times 10^{-6} r\mu v + 2.86 \times 10^{-4} r\mu L \qquad (ii)$$

Solving Eqs (i) and (ii) simultaneously gives:

$$r\mu v = 7 \times 10^{12}\,\text{Ns/m}^4 \text{ and } r\mu L = 4.6 \times 10^9\,\text{N/sm}^3$$

Thus, for the plate and frame filter:

$$B_1 = \frac{r\mu v}{2A^2(-\Delta P)} = 7 \times 10^{12}/(2 \times 2.2^2 \times 413 \times 10^3) = 1.75 \times 10^6\,\text{s/m}^6$$

and:
$$B_2 = \frac{r\mu L}{A(-\Delta P)} = 4.6 \times 10^9/(2.2 \times 413 \times 10^3) = 5.06 \times 10^3\,\text{s/m}^3$$

Substituting for V from Eq. (10.43) into Eq. (10.41), the filtration time for maximum throughput is:

$$t = t' + B_2(t'/B_1)^{0.5}$$
$$= 21.6 \times 10^3 + 5.06 \times 10^3[(21.6 \times 10^3)/(1.75 \times 10^6)]^{0.5}$$
$$= 2.216 \times 10^4\,\text{s or } 22.2\,\text{ks}(6.2\,\text{h})$$

From Eq. (10.43):

$$V = [(21.6 \times 10^3)/(1.75 \times 10^6)]^{0.5} = 0.111\,\text{m}^3$$

and the mean rate of filtration is:

$$V/(t + t') = 0.111/(2.216 \times 10^4 + 21.6 \times 10^3) = 2.54 \times 10^{-6}\,\text{m}^3/\text{s}$$

The total cost is:

$$C = 0.01t + 100\,\text{£/cycle}$$

or:
$$C = (0.01t + 100)/V\,\text{£/m}^3$$

Substituting for t from Eq. (10.41) gives:

$$C = \left(0.01B_1V^2 + 0.01B_2V + 100\right)/V$$

Differentiating and putting $dC/dV = 0$:

$$V = \left(100/0.1B_1\right)^{0.5} m^3$$

and from Eq. (10.41), the optimum filtration time for minimum cost is:

$$t = \left(100/0.01\right) + B_2\left(100/0.01\,B_1^{0.5}\right) s$$

Substituting for B_1 and B_2:

$$t = 10^4 + 5.06 \times 10^3 \left(10^4/1.75 \times 10^6\right)^{0.5}$$
$$= 1.03 \times 10^4 \, s \text{ or } \underline{\underline{10.3\,ks\,(2.86h)}}$$

Example 10.6

A slurry containing 0.2 kg of solid per kilogram of water is fed to a rotary drum filter 0.6 m long and 0.6 m in diameter. The drum rotates at one revolution in 360 s, and 20% of the filtering surface is in contact with the slurry at any instant. If filtrate is produced at the rate of 0.125 kg/s, and the cake has a voidage of 0.5, what thickness of cake is produced when filtering with a pressure difference of 65 kN/m²? The density of the solids is 3000 kg/m³.

The rotary filter breaks down and the operation has to be carried out temporarily in a plate and frame press with frames 0.3 m square. The press takes 120 s to dismantle and 120 s to reassemble, and, in addition, 120 s is required to remove the cake from each frame. If filtration is to be carried out at the same overall rate as before, with an operating pressure difference of 175 kN/m², what is the minimum number of frames that need to be used, and what is the thickness of each? It may be assumed that the cakes are incompressible, and that the resistance of the filter medium may be neglected.

Solution

Drum filter

$$\text{Area of filtering surface} = (0.6 \times 0.6\pi) = 0.36\pi\,m^2$$

$$\text{Rate of filtration} = 0.125\,kg/s$$

$$= (0.125/1000) = 1.25 \times 10^{-4} m^3/s \text{ of filtrate}$$

1 kg or 10^{-3} m³ water is associated with 0.2 kg of solids $= 0.2/(3 \times 10^3) = 6.67 \times 10^{-5}$ m³ of solids in the slurry.

Because the cake porosity is 0.5, 6.67×10^{-5} m³ of water is held in the filter cake and $(10^{-3} - 6.67 \times 10^{-5}) = 9.33 \times 10^{-4}$ m³ appears as filtrate, per kg of total water in the slurry.

Volume of cake deposited by unit volume of filtrate, $v = (6.67 \times 10^{-5} \times 2)/(9.33 \times 10^{-4}) = 0.143$.

Volumetric rate of deposition of solids $= (1.25 \times 10^{-4} \times 0.143) = 1.79 \times 10^{-5}$ m^3/s. One revolution takes 360 s. Therefore, the given piece of filtering surface is immersed for $(360 \times 0.2) = 72$ s.

The bulk volume of cake deposited per revolution $= (1.79 \times 10^{-5} \times 360) = 6.44 \times 10^{-3}$ m^3. Thickness of cake produced $= (6.44 \times 10^{-3})/(0.36\pi) = 5.7 \times 10^{-3}$ m or 5.7 mm.

Properties of filter cake

$$\frac{dV}{dt} = \frac{(-\Delta P)A}{r\mu l} = \frac{(-\Delta P)A^2}{r\mu Vv} \qquad \text{(from Eqs 10.2 and 10.8)}$$

At constant pressure:

$$V^2 = \frac{2}{r\mu v}(-\Delta P)A^2 t = K(-\Delta P)A^2 t \,(\text{say}) \qquad \text{(from Eq.10.11)}$$

Expressing pressures, areas, times, and volumes in N/m^2, m^2, s, and m^3 respectively, then for one revolution of the drum:

$$\left(1.25 \times 10^{-4} \times 360\right)^2 = K\left(6.5 \times 10^4\right)(0.36\pi)^2 \times 72$$

because each element of area is immersed for one-fifth of a cycle,

and $$K = 3.38 \times 10^{-10}$$

Filter press

Using a filter press with n frames of thickness b m, the total time, for one complete cycle of the press $= (t_f + 120n + 240)$ s, where t_f is the time during which filtration is occurring.

$$\text{Overall rate of filtration} = \frac{V_f}{t_f + 120n + 240} = 1.25 \times 10^{-4} \text{m}^3/\text{s}$$

where V_f is the total volume of filtrate per cycle.

The volume of frames/volume of cake deposited by unit volume of filtrate, v, is given by:

$$V_f = 0.3^2 nb/0.143 = 0.629nb$$

But : $V_f^2 = (3.38 \times 10^{-10}) \times (1.75 \times 10^5)(2n \times 0.3 \times 0.3)^2 t_f$
    $= 0.629nb^2$ (from Eq.10.11)
        and : $t_f = 2.064 \times 10^5 b^2$

Thus : $$1.25 \times 10^{-4} = \frac{0.629nb}{2.064 \times 10^5 b^2 + 120n + 240}$$

That is : $$25.8b^2 + 0.015n + 0.030 = 0.629nb$$

or : $$n = \frac{0.030 + 25.8b^2}{0.629b - 0.015}$$

n is a minimum when $dn/db = 0$, that is when:

$$(0.629b - 0.015) \times 51.6b - (0.030 + 25.8b^2) \times 0.629 = 0$$
$$b^2 - 0.0458b - 0.001162 = 0$$

Thus : $b = 0.0229 \pm \sqrt{(0.000525 + 0.001162)}$ and taking the positive root :

$$d = 0.0640 \text{ m or } \underline{64 \text{ mm}}$$

Hence:
$$n = \frac{(0.030 + 25.8 \times 0.0640^2)}{(0.629 \times 0.0640 - 0.015)}$$
$$= 5.4$$

Thus, a minimum of 6 frames must be used.

The sizes of frames which will give exactly the required rate of filtration when 6 frames used are given by:

or:
$$0.030 + 25.8b^2 = 3.774b - 0.090$$
$$b^2 - 0.146b + 0.00465 = 0$$

and:
$$b = 0.073 \pm \sqrt{(0.005329 - 0.00465)}$$
$$= 0.047 \, or \, 0.099 \, m$$

Thus, 6 frames of thickness either 47 mm or 99 mm will give exactly the required filtration rate; intermediate sizes give higher rates.

Thus, any frame thickness between 47 and 99 mm will be satisfactory. In practice, 50 mm (2 in) frames would probably be used.

Rotary disc filters

In essence, the *disc filter* operates in a manner similar to a bottom-feed drum filter. The principal differences to be noted are the compartments which are formed on both faces of vertical discs. These comprise eight or more segments, each of which is connected to the horizontal shaft of the machine. The filtrate and air are drawn through the filter medium, into the drainage system of the segments, and finally, through passages in the rotary shaft to a valve located at one or both ends.

The discs are arranged on the shaft about 0.3 m apart as shown in Fig. 10.24, resulting in economy of space, and consequently, a far greater filtration area can be accommodated in a given floor space than is possible with a drum type filter. A disc filter area of approximately 80 m² can be achieved with eight discs, each approximately 2.5 m in diameter, in the same space as a 2.5 m diameter × 2.5 m drum filter, which would have a filtration area of 20 m². Because of the large areas, and consequently, the large liquor and air flows, two rotary valves, one at each end of the main shaft, are frequently fitted. Cake discharge is usually achieved using a knife or wire, and discharge may be assisted by blowback air. The cake falls through vertical openings in the filter trough.

In general, disc filters have the advantages that, not only is the cost per unit filtration area low and large filtration areas can be accommodated in a small floor space, but also, the trough can be divided into two sections so that the different slurries can be handled at the same time in the same machine. The disadvantages of these units are that heavy filter cloths are necessary, with a great tendency to "blind," and cake washing is virtually impossible due to the ease with which

Fig. 10.24
Rotary disc filter.

deposited solids can be disturbed from the vertical faces. In contrast to a drum type filter, in which all parts of the filter medium take the same path through the slurry, each part of a disc segment takes a different path, depending on its radial distance from the hub. Consequently, if homogeneous conditions cannot be maintained in the filter trough, then very uneven cakes may be formed, a situation which favours preferential air flow and results in a widely variable moisture content across each segment. Disc filters also have the limitation that, because every segment must be fully submerged in slurry prior to the application of vacuum, very little variation in submergence level can be accommodated.

Precoat filters

All the filters described previously have been in the category of cake filters, for they all rely on the solids present in the feed slurry having properties which are suitable for them to act as the actual filter medium and to form cakes which are relatively easily discharged. Applications arise where extremely thin filter cakes, perhaps only a fraction of a millimetre thick, have to be discharged, where extremely good filtrate clarity is essential, or where a particularly blinding substance is present, and filtration through a permanent filter medium alone, such as a filter fabric, may not be satisfactory. A much more efficient performance is achieved by utilising a bed of easily filtered material which is *precoated* onto what is otherwise a standard drum type filter.

After the precoat is established, the solids to be removed from the filter feed are trapped on the surface of the precoated bed. This thin layer of slime is removed by a knife, which is caused to advance slowly toward the drum. The knife also removes a thin layer of the precoated bed so that a new surface of the filter medium is exposed. This procedure allows steady filtration rates to be achieved.

A *precoated rotary drum vacuum filter* is the only filtration equipment from which extremely thin filter cakes can be positively and continuously removed in a semidry state.

The usual precoat materials for most production processes are conventional filter aids, such as diatomaceous earth and expanded Perlite, although frothed coal, calcium sulphate, and other solids, which form very permeable filter cakes, may be used where such materials are compatible with the materials being processed. Wood flour and fly ash have also been used in some applications, particularly in effluent treatment plants, where some contamination of the filtered liquor is of no great consequence. The establishment of the precoat bed should preferably be carried out using a precoat slurry of low solids concentration, at high drum speed and with low submergence, so that the bed is built up in the form of a thin layer with each revolution, thereby ensuring a uniform and compacted bed which will not easily be disturbed from the drum. For long periods of operation, the maximum bed thickness should be used. This is often determined by the mechanical design of the filter and by the cake formation properties of the precoat material. Normally, the bed thickness is 75–100 mm, and, with most materials, this can be established on the drum in a period of about 1 h. The type and grade of precoat material used, in addition to process considerations, can influence the rate at which the bed must be removed and the cost of precoat vacuum filter operation. Generally, for maximum economy, a precoat filter should operate at the highest possible submergence, consistent with the required degree of cake washing and drying, when required, and with the lowest possible knife advance rate. Some precoat filters are capable of operation at up to 70% drum submergence, though this, particularly with the facility to accommodate 100 mm thick beds, necessitates special design features to give adequate clearance between the drum and the trough, while at the same time giving acceptable angles for the discharge knife. Knife advance rates are 0.013–0.13 mm per revolution, and drum speeds are frequently 0.002–0.03 Hz (0.1–2 rev/min). By careful control of all the variables, and by correct sizing of the equipment, it is possible to obtain precoat beds which last 60–240 h.

To obtain maximum economy in operation, a precoat type vacuum filter should have:

(a) Easy variation of the knife advance rate.
(b) Very accurate control of the knife advance rate.
(c) A very rigid knife assembly (due to the need to advance at extremely low rates over a face width up to approximately 6 m).
(d) Rapid advance and retraction of the knife.

Mechanical and hydraulic systems for control of the knife are now used. The latter allows easy selection of the rate and direction of knife movement, and the system is particularly suited to remote control or automatic operation.

Example 10.7

A sludge is filtered in a plate and frame press fitted with 25 mm frames. For the first 600 s, the slurry pump runs at maximum capacity. During this period, the pressure rises to 415 kN/m², and 25% of the total filtrate is obtained. The filtration takes a further 3600 s to complete at constant pressure, and 900 s is required for emptying and resetting the press.

It is found that if the cloths are precoated with filter aid to a depth of 1.6 mm, the cloth resistance is reduced to 25% of its former value. What will be the increase in the overall throughput of the press, if the precoat can be applied in 180 s?

Solution

Case 1

$$\frac{dV}{dt} = \frac{A^2(-\Delta P)}{vr\mu\left(V + \frac{AL}{v}\right)} = \frac{a}{V + b} \quad \text{(Eq.10.16)}$$

For constant rate filtration:

or:

$$\frac{V_0}{t_0} = \frac{a}{V_0 + b}$$

$$V_0^2 + bV_0 = at_0$$

For constant pressure filtration:

$$\frac{1}{2}(V^2 - V_0^2) + b(V - V_0) = a(t - t_0)$$

$$t_0 = 600\,\text{s}, \quad t - t_0 = 3600\,\text{s}, \quad V_0 = V/4$$

$$\frac{V^2}{16} + b\frac{V}{4} = 600a$$

and:

$$\frac{1}{2}(V^2 - V^2/16) + b(V - V/4) = 3600a$$

Thus:

$$3600a = \frac{15}{32}V^2 + \frac{3}{4}bV = \frac{3}{8}V^2 + \frac{3}{2}bV$$

and:

$$b = \frac{V}{8}$$

Thus:

$$a = \frac{1}{600}\left(\frac{V^2}{16} + \frac{V^2}{32}\right) = \frac{3}{19,200}V^2$$

Total cycle time $= (900 + 4200) = 5100\,\text{s}.$

Filtration rate $= V/5100 = 0.000196\,V.$

Case 2

$$\frac{V_1}{t_1} = \frac{a}{V_1 + \frac{b}{4}} = \frac{V_0}{t_0} = \frac{a}{V_0 + b}$$

$$\frac{1}{2}\left(\frac{49}{64}V^2 - V_1^2\right) + \frac{b}{4}\left(\frac{7}{8}V - V_1\right) = a(t - t_1)$$

Thus:

$$\frac{V}{2400} = \frac{\frac{3}{19,200}V^2}{V_1 + \frac{V}{32}}$$

and:

$$t_1 = \frac{t_0}{V_0}V_1 = \frac{600}{V/4} \times \frac{11}{32}V = \frac{3300}{4}s = 825\,s$$

Substituting gives:

$$\frac{1}{2}\left(\frac{49}{64}V^2 - \frac{121}{1024}V^2\right) + \left(\frac{1}{4}\right)\frac{V}{8}\left(\frac{7}{8}V - \frac{11}{32}V\right) = \frac{3}{19,200}V^2(t - t_1)$$

or:

$$\frac{49}{128} - \frac{121}{2048} + \frac{17}{1024} = \frac{3}{19,200}(t - t_1)$$

$$t - t_1 = \left(\frac{19,200}{3}\right)\left(\frac{784 - 121 + 34}{2048}\right)$$

$$= 2178\,s$$

$$\text{Cycle time} = (180 + 900 + 825 + 2178) = 4083\,s$$

$$\text{Filtration rate} = \left(\frac{7}{8} \times \frac{V}{4083}\right) = 0.000214V$$

$$\text{Increase} = \frac{(0.000214 - 0.000196)V}{0.000196V} \times 100 = \underline{\underline{9.1\%}}$$

10.4.7 The Tube Press

One of the major problems in coal preparation is the dewatering of fine coal to a moisture content sufficiently low enough both to meet market requirements and to ease the problems of handleability encountered at some collieries, and there is an interest in different types of dewatering equipment in addition to the conventional rotary vacuum filter. As a result, a tube press built by the former English China Clays Limited, now Imerys, was installed at a colliery, where it was used to treat raw slurry which was blended into the final product. As described by Gwilliam[35] and Brown.[36] It was found that the product from the tube press was very low in moisture content. In the china clay industry, for which the tube press was first developed, the formation of a very dry cake reduces the load on the drying plant, where the cost of removing a given quantity of water is an order of magnitude greater than in filtration.

In essence, the tube press is an automatic membrane filter press operating at pressures up to 10 MN/m^2 (100 bar) and consists of two concentric cylindrical tubes, the inner filter candle and the outer hydraulic casing. Cylindrical tubes are used as these have the ability to withstand high pressure without a reinforcing structure and are available commercially. Between the filter candle and the hydraulic casing is a flexible membrane which is fastened to both ends of the hydraulic casing. Hydraulic pressure is exerted onto the flexible membrane, applying filtration pressure to the slurry which is contained between the flexible membrane and filter candle. The membrane, being flexible, allows for the slurry feed volume to be varied to suit each individual application, and this permits near-optimum filter cake thicknesses to be achieved. The filter candle is perforated, and around its outer circumference is fitted a drain mesh, felt backing cloth, and top filter cloth, forming the filter media. The backing cloth protects the filter cloth against the hydraulic high pressure, while the mesh allows the filtrate to run into the centre of the filter candle and be drained away. Under high pressure conditions, the resistance of the filter cloth is low compared with the resistance of the filter cake. This allows a tightly woven filter cloth to be used, and any imperfections in this cloth are compensated for by the felt backing cloth. In this way, clear filtrate is obtained from the tube press.

The sequence of operation of the tube press, illustrated in Fig. 10.25, is as follows. The empty tube filter press is closed, and hydraulic vacuum is applied to the flexible membrane, dilating it against the outer hydraulic casing (A). Slurry is fed into the machine, partially filling the volume contained within the flexible membrane, the filter media, and the end pieces (B). The exact volume of slurry fed into the tube press depends upon the characteristics of each individual slurry and corresponds to the amount required to give optimum cake thickness. The machine will automatically discharge a cake of 16–19 mm maximum thickness. The minimum thickness is dependent on the cake discharge characteristics, although is normally 4–5 mm. Low pressure hydraulic fluid is fed into the tube press between the outer hydraulic casing and the flexible membrane. The membrane contracts, reducing the volume between the membrane and filter cloth and expelling the entrained air. The low pressure hydraulic flow is stopped, and high pressure hydraulic fluid is applied. Filtration commences, and the high pressure hydraulic fluid continues contracting the membrane, reducing the suspension volume and expelling the filtrate through the filter medium into the centre of the filter candle and away through the filtrate drain line (C). Filter cake is formed around the cloth, and high pressure is maintained until filtration is completed (D). Hydraulic vacuum is applied to the flexible membrane, dilating it against the outer hydraulic casing. When this is accomplished, the filter candle is lowered to open the machine (E). A pulse of air admitted into the centre of the filter candle expands the filter cloth, dislodging the cake, which breaks and falls out of the machine (F). This action, together with the candle movement, may be repeated to ensure complete discharge. The filter candle is then raised, and the machine closes, leaving it ready to commence the next cycle (A).

The unique design of the tube press allows for this cycle to be amended, however, to include air pressing and/or cake washing. With air pressing, once the initial filtration is complete, air is

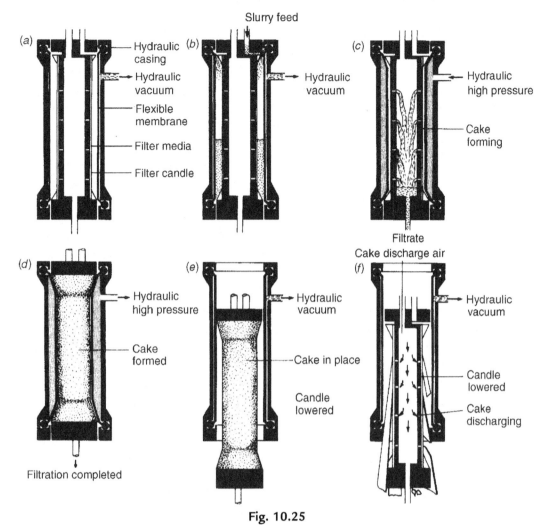

Fig. 10.25

Sequence of operation of the tube press. *Based on the information provided by Charlestown Engineering, St. Austell, Cornwall.*

introduced between the membrane and the cake. The pressure cycle is then repeated. Typically, an air press will further reduce the moisture content of china clay by 2.5%–8%. The final moisture contents with other materials are shown in Fig. 10.26. Water washing, which is used for the removal of soluble salts, is similar to air pressing, except that it is water that is introduced between membrane and cake.

Finally, detailed discussions concerning the relative merits and demerits, mode of operation, maintenance and troubleshooting, etc., of a range of filtration equipment in use are available in the literature.[37–41]

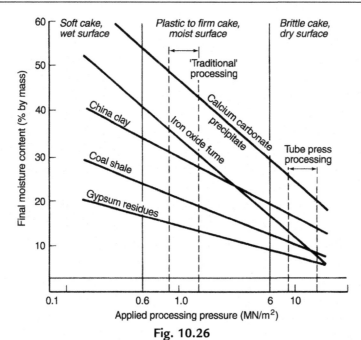

Fig. 10.26

Final moisture content as a function of applied pressure. *Image Courtesy: Charlestown Engineering, St. Austell, Cornwall.*

10.5 Nomenclature

		Units in SI System	Dimensions in M, L, T
A	Cross-sectional area of bed or filtration area	m^2	L^2
B_1	Coefficient	s/m^6	$L^{-6}T$
B_2	Coefficient	s/m^3	$L^{-3}T$
b	Frame thickness	m	L
C	Volume concentration of solids in the filter	–	–
C'	Total cost of filtration per unit volume	£/m^3	L^{-3}
C_0	Value of C at filter surface	–	–
e	Voidage of bed or filter cake	–	–
e_0	Liquid fraction in feed slurry	–	–
e_V	Liquid fraction in slurry in vessel	–	–
e_z	Voidage at distance z from surface	–	–
J	Mass fraction of solids in slurry	–	–
L	Thickness of filter cake with same resistance as cloth	m	L
l	Thickness of filter cake or bed	m	L

n'	Compressibility index	—	—
P_1	Pressure at downstream face of cake	N/m^2	$\mathbf{ML}^{-1}\mathbf{T}^{-2}$
P_2	Pressure at upstream face of cake	N/m^2	$\mathbf{ML}^{-1}\mathbf{T}^{-2}$
P_z	Pressure at distance z from surface	N/m^2	$\mathbf{ML}^{-1}\mathbf{T}^{-2}$
$-\Delta P$	Total drop in pressure	N/m^2	$\mathbf{ML}^{-1}\mathbf{T}^{-2}$
$-\Delta P'$	Drop in pressure across cake	N/m^2	$\mathbf{ML}^{-1}\mathbf{T}^{-2}$
$-\Delta P''$	Pressure drop across cloth	N/m^2	$\mathbf{ML}^{-1}\mathbf{T}^{-2}$
R	Rate of cake production	m/s	$\mathbf{LT}^{-1}$
$\mathbf{r}$	Specific resistance of filter cake	m^{-2}	$\mathbf{L}^{-2}$
$\bar{\mathbf{r}}$	Mean value of $\mathbf{r}_z$	m^{-2}	$\mathbf{L}^{-2}$
$\mathbf{r}_z$	Specific resistance of compressible cake at distance z from surface (Eq. 10.25)	m^{-2}	$\mathbf{L}^{-2}$
$\bar{\mathbf{r}}', \bar{\mathbf{r}}''$	Functions of $\mathbf{r}_z$ independent of ΔP	m^{-2}/(N/m^2)$^{n'}$	$\mathbf{M}^{-n'}\mathbf{L}^{n'-2}\mathbf{T}^{2n'}$
S	Specific surface	m^{-1}	$\mathbf{L}^{-1}$
t	Time	s	$\mathbf{T}$
t'	Time of dismantling filter press	s	$\mathbf{T}$
t_1	Time at beginning of operation	s	$\mathbf{T}$
u_c	Mean velocity of flow calculated over the whole area	m/s	$\mathbf{LT}^{-1}$
V	Volume of liquid flowing in time t	m^3	$\mathbf{L}^3$
V_1	Volume of liquid passing in time t_1	m^3	$\mathbf{L}^3$
$\mathbf{V}$	Volume of vessel	m^3	$\mathbf{L}^3$
v	Volume of cake deposited by unit volume of filtrate	—	—
v'	Volume of solids deposited by unit volume of filtrate	—	—
w	Mass rate of production of filtrate	kg/s	$\mathbf{MT}^{-1}$
W	Overall volumetric rate of filtration	m^3/s	$\mathbf{L}^3\mathbf{T}^{-1}$
z	Distance from surface of filter cake	m	$\mathbf{L}$
λ	Filter coefficient	m^{-1}	$\mathbf{L}^{-1}$
μ	Viscosity of fluid	Ns/m^2	$\mathbf{ML}^{-1}\mathbf{T}^{-1}$
ρ	Density of fluid	kg/m^3	$\mathbf{ML}^{-3}$
ρ_s	Density of solids	kg/m^3	$\mathbf{ML}^{-3}$
σ	Volume of solids deposited per unit volume of filter	—	—

References

1. Ives KJ, editor. *The scientific basis of filtration.* Leyden: Noordhoff; 1975.
2. Ives KJ. Advances in deep-bed filtration. *Trans Inst Chem E* 1970;**48**:T94.
3. Suttle HK. Development of industrial filtration. *Chem Eng* 1976;**314**:675.
4. Ruth BF, Montillon GH, Montonna RE. Studies in filtration. I. Critical analysis of filtration theory. II. Fundamental axiom of constant-pressure filtration. *Ind Eng Chem* 1933;**25**:76 and 153.
5. Ruth BF. Studies in filtration. III. Derivation of general filtration equations. IV. Nature of fluid flow through filter septa and its importance in the filtration equation. *Ind Eng Chem* 1935;**27**:708 and 806.
6. Ruth BF, Kempe L. An extension of the testing methods and equations of batch filtration practice to the field of continuous filtration. *Trans Am Inst Chem Eng* 1937;**33**:34.
7. Ruth BF. Correlating filtration theory with practice. *Ind Eng Chem* 1946;**38**:564.
8. Grace HP. Resistance and compressibility of filter cakes. *Chem Eng Prog* 1953;**49**:303, 367, and 427.
9. Hoffing EH, Lockhart FJ. Resistance to filtration. *Chem Eng Prog* 1951;**47**:3.
10. Carman PC. Fundamental principles of industrial filtration. *Trans Inst Chem Eng* 1938;**16**:168.
11. Heertjes PM. Studies in filtration. *Chem Eng Sci* 1957;**6**:190 and 269.
12. Valleroy VV, Maloney JO. Comparison of the specific resistances of cakes formed in filters and centrifuges. *AIChE J* 1960;**6**:382.
13. Tiller FM, Huang CJ. Filtration equipment. Theory. *Ind Eng Chem* 1961;**53**:529.
14. Tiller FM, Shirato M. The role of porosity in filtration: Part VI. New definition of filtration resistance. *AIChE J* 1964;**10**:61.
15. Tiller FM, Yeh CS. The role of porosity in filtration: Part X. Deposition of compressible cakes on external radial surfaces. *AIChE J* 1985;**31**:1241.
16. Rushton A, Hameed MS. The effect of concentration in rotary vacuum filtration. *Filtr Sep* 1969;**6**:136.
17. Ehlers S. The selection of filter fabrics re-examined. *Ind Eng Chem* 1961;**53**:552.
18. Wrotnowski AC. Nonwoven filter media. *Chem Eng Prog* 1962;**58**(12):61.
19. Purchas DB. *Industrial filtration of liquids.* 2nd ed. London: Leonard Hill; 1971
20. Matteson MS, Orr C. *Filtration principles and practice.* 2nd ed. New York: Marcel-Dekker; 1987.
21. Tarleton S, Wakeman RJ. *Solid-liquid separation: equipment selection and design.* Oxford: Butterworth-Heinemann; 2007.
22. Purchas D, Sutherland K. *Handbook of filter media.* 2nd ed. Oxford: Elsevier; 2002.
23. Tiller FM, Cheng KS. Delayed cake filtration. *Filtr Sep* 1977;**14**:13.
24. Mackley MR, Sherman NE. Cross-flow cake filtration mechanics and kinetics. *Chem Eng Sci* 1997;**47**:3067.
25. Holdich RG, Cumming IW, Ismail B. The variation of cross-flow filtration rate with wall shear stress and the effect of deposit thickness. *Chem Eng Res Des* 1995;**73**:20.
26. Cleasby JL. Filtration with granular beds. *Chem Eng* 1976;**314**:663.
27. Ives KJ. *Theory of filtration. Proc. Int. Water Supply Assn. Eighth Congress, Vienna*, Vol. 1. London: International Water Supply Association; 1969 Special Subject No. 7.
28. Iwasaki T. Some notes on sand filtration. *J Am Water Works Assn* 1937;**29**:1591.
29. Spielman LA, Friedlander SK. Role of the electrical double layer in particle deposition by convective diffusion. *J Colloid Interface Sci* 1974;**46**:22.
30. Harker JH. Getting the best out of batch filtration. *Processing* 1979;**2**:35.
31. Cherry GB. New developments in filter plates and filter presses. *Filtr Sep* 1974;**11**:181.
32. Parrish P, Ogilvie H. *Calcium superphosphates and compound fertilisers. Their chemistry and manufacture.* Hutchison; 1939.
33. Blendulf KAG, Bond AP. The development and application of horizontal belt filters in the mineral, processing and chemical industries. *I Chem E Symp Ser* 1980;**59**:1. 5/1.
34. Bosley R. Vacuum filtration equipment innovations. *Filtr Sep* 1974;**11**:138.
35. Gwilliam RD. The E.C.C. tube filter press. *Filtr Sep* 1971;**8**:173.
36. Brown A. The tube press. *Filtr Sep* 1979;**16**:468.

37. Cheremisinoff NP. *Liquid filtration*. 2nd ed. Oxford: Butterworth-Heinemann; 1998.
38. Perlmutter BA. *Solid-liquid filtration*. Oxford: Butterworth-Heinemann; 2015.
39. Fernando Concha A. *Solid-liquid separation in the mining industry*. Cham, Switzerland: Springer; 2014.
40. Sparks T, Chase G. *Filters and filtration handbook*. 6th ed. Oxford: Butterworth-Heinemann; 2015.
41. Gabelman A. An overview of filtration. *Chem Eng* 2015;**122**(#11):50–8.

Further Reading

Suttle HK, editor. *Process engineering technique evaluation–filtration*. London: Morgan–Grampian (Publishers) Ltd.; 1969.
Svarovsky L, editor. *Solid–liquid separation*. 4th ed. Oxford: Butterworth-Heinemann; 2000.

CHAPTER 11

Centrifugal Separations

11.1 Introduction

There is now a wide range of situations in which centrifugal force is used in place of the gravitational force in order to effect separations in a broad range of settings. Typical examples include (in addition to the traditional applications in the process, chemical, and mining areas) fractionation of lysed cells, separation of cells for the purpose of cloning, biotechnological applications, dairy and food processing, separation of polymer molecules depending upon their molecular weight, etc. The resulting accelerations may be several thousand times those attributable to gravity. Some of the benefits include far greater rates of separation, the possibility of achieving separations which are either not practically feasible (or are actually impossible) in the gravitational field, and a substantial reduction of the size of the equipment. Recent developments in the use of centrifugal fields in *process intensification* are discussed in Chapter 12.

Centrifugal fields can be generated in two distinctly different ways:

(a) By introducing a fluid with a high tangential velocity into a cylindrical or conical vessel, as in the *hydrocyclone* and in the *cyclone separator* (Chapter 4). In this case, the flow pattern in the body of the separator approximates to a *free vortex* in which the tangential velocity varies *inversely* with the radius (see Volume 1A, Chapter 2). Generally, the larger and heavier particles will collect and be removed near the walls of the separator, and the smaller and lighter particles will be taken off through an outlet near the axis of the vessel.

(b) By the use of the *centrifuge*. In this case, the fluid is introduced into some form of rotating bowl and is rapidly accelerated. Because the frictional drag within the fluid ensures that there is very little *rotational slip* or relative motion between fluid layers within the bowl, all the fluid tends to rotate at a constant angular velocity ω and a *forced vortex* is established. Under these conditions, the tangential velocity will be *directly* proportional to the radius at which the fluid is rotating.

In this chapter, attention is focused on the operation of the centrifuge. Some of the areas where it is extensively used are as follows:

Coulson and Richardson's Chemical Engineering. https://doi.org/10.1016/B978-0-08-101098-3.00012-3

(a) *For separating particles on the basis of their size or density.* This is effectively using a centrifugal field to achieve a higher rate of sedimentation than could be achieved under gravity.

(b) *For separating immiscible liquids of different densities,* which may be in the form of dispersions or even emulsions in the feed stream. This is the equivalent of a gravitational decantation process.

(c) *For filtration of a suspension.* In this case, centrifugal force replaces the force of gravity or the force attributable to an applied pressure difference across the filter.

(d) *For the drying of solids and, in particular, crystals.* Liquid may be adhering to the surface of solid particles and may be trapped between groups of particles. Drainage may be slow in the gravitational field, especially if the liquid has a high viscosity. Furthermore, liquid is held in place by surface tension forces which must be exceeded before liquid can be freed. This is particularly important with fine particles. Thus, processes which are not possible in the gravitational field can be carried out in the centrifuge.

(e) *For breaking down of emulsions and colloidal suspensions.* A colloid or emulsion may be quite stable in the gravitational field where the dispersive forces, such as those due to Brownian motion, are large compared with the gravitational forces acting on the fine particles or droplets. In a centrifugal field, which may be several thousand times more powerful than the gravity, however, the dispersive forces are no longer sufficient to maintain the particles in suspension, and separation is effected.

(f) *For the separation of gases.* In the nuclear industry, isotopes are separated in the gas centrifuge, in which the accelerating forces are sufficiently great to overcome the dispersive effects of molecular motion. Because of the very small difference in density between isotopes and between compounds of different isotopes, fields of very high intensity are needed.

(g) *For mass transfer processes.* Because far greater efficiencies and higher throughputs can be obtained before flooding occurs, centrifugal packed-bed contactors are finding favor and are replacing ordinary packed columns in situations where compactness is important, or where it is desirable to reduce the hold-up of materials undergoing processing because of their hazardous properties. An important application is the use of inert gases in the desorption of oxygen from sea water in order to reduce its corrosiveness; in North Sea oil rigs, the sea water is used as a coolant in heat exchangers. In addition, centrifugal contactors for liquid–liquid extraction processes now have important applications, as discussed in Volume 2B. These are additional areas where centrifugal fields, and the employment of centrifuges, are gaining in importance.

11.2 Shape of the Free Surface of the Liquid

For an element of liquid in a centrifuge bowl which is rotating at an angular velocity of ω, the centrifugal acceleration is $r\omega^2$, compared with the gravitational acceleration of g. The ratio

$r\omega^2/g$ is one measure of the separating effect obtained in a centrifuge relative to that arising from the gravitational field; values of $r\omega^2/g$ may be very high (up to 10^4) in some industrial centrifuges and more than an order of magnitude greater in the ultracentrifuge, as discussed in Section 11.8. In practice, the axis of rotation may be vertical, horizontal, or intermediate, and the orientation is usually determined by the means adopted for introducing feed and removing product streams from the centrifuge.

Fig. 11.1 shows an element of the free surface of the liquid in a bowl which is rotating at a radius r_0 about a vertical axis at a very low speed; the centrifugal and gravitational fields will then be of the same order of magnitude. The centrifugal force per unit mass is $r_0\omega^2$, and the corresponding gravitational force is g. These two forces are perpendicular to one another and may be combined as shown to give the resultant force, which must, at equilibrium, be at right angles to the free surface. Thus, the slope at this point is given by:

$$\frac{dz_0}{dr_0} = \frac{\text{radial component of force}}{\text{axial component of force}} = \frac{r_0\omega^2}{g} \tag{11.1}$$

where z_0 is the axial coordinate of the free surface of the liquid.

Eq. (11.1) may be integrated to give:

$$z_0 = \frac{\omega^2}{2g}r_0^2 + \text{constant}$$

If z_a is the value of z_0 which corresponds to the position where the free surface is at the axis of rotation ($r_0 = 0$), then:

$$z_0 - z_a = \frac{\omega^2}{2g}r_0^2 \tag{11.2}$$

Eqs. (11.1) and (11.2) correspond with Eqs. (2.80) and (2.79) in Volume 1A, Chapter 2. Taking the base of the bowl as the origin for the measurement of z_0, positive values of z_a

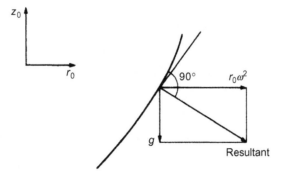

Fig. 11.1
Element of surface of liquid.

correspond to conditions where the whole of the bottom of the bowl is covered by liquid. Negative values of z_a imply that the paraboloid of revolution describing the free surface would cut the axis of rotation below the bottom, and therefore the central portion of the bowl will be dry.

Normally $r_0\omega^2 \gg g$, the surface is nearly vertical, and z_0 has a very large negative value. Thus, in practice, the free surface of the liquid will be effectively concentric with the walls of the bowl. It is seen, therefore, that the operation of a high speed centrifuge is independent of the orientation of the axis of rotation.

11.3 Centrifugal Pressure

A force balance on a sector of fluid in the rotating bowl, carried out as in Volume 1A, Chapter 2, gives the pressure gradient at a radius r:

$$\frac{\partial P}{\partial r} = \rho\omega^2 r \tag{11.3}$$

Unlike the vertical pressure gradient in a column of liquid which is constant at all heights, the centrifugal pressure gradient is a function of radius of rotation r, and increases toward the wall of the basket. Integration of Eq. (11.3) at a given height gives the pressure P exerted by the liquid on the walls of the bowl of radius R when the radius of the inner surface of the liquid is r_0 as:

$$P = \frac{1}{2}\rho\omega^2\left(R^2 - r_0^2\right) \tag{11.4}$$

11.4 Separation of Immiscible Liquids of Different Densities

The problem of the continuous separation of a mixture of two liquids of different densities is most readily understood by first considering the operation of a gravity settler, as shown in Fig. 11.2. For equilibrium, the hydrostatic pressure exerted by a height z of the denser liquid must equal that due to a height z_2 of the heavier liquid and a height z_1 of the lighter liquid in the separator.

$$\text{Thus}: \quad z\rho_2 g = z_2\rho_2 g + z_1\rho_1 g$$
$$\text{or}: \quad z = z_2 + z_1\frac{\rho_1}{\rho_2} \tag{11.5}$$

For the centrifuge, it is necessary to position the overflow on the same principle, as shown in Fig. 11.3. In this case, the radius r_i of the weir for the less dense liquid will correspond approximately to the radius of the inner surface of the liquid in the bowl. That of the outer weir r_w must be such that the pressure developed at the wall of the bowl of radius R by

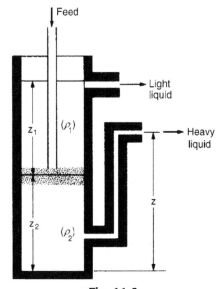

Fig. 11.2

Gravity separation of two immiscible liquids.

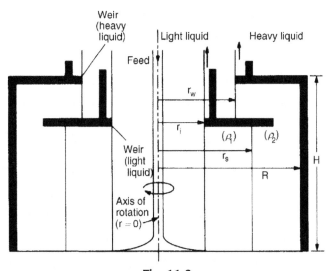

Fig. 11.3

Separation of two immiscible liquids in a centrifuge.

the heavy liquid alone as it flows over the weir is equal to that due to the two liquids within the bowl. Thus, applying Eq. (11.4) and denoting the densities of the light and heavy liquids by ρ_1 and ρ_2, respectively, and the radius of the interface between the two liquids in the bowl as r_s:

$$\frac{1}{2}\rho_2\omega^2\left(R^2 - r_\omega^2\right) = \frac{1}{2}\rho_2\omega^2\left(R^2 - r_s^2\right) + \frac{1}{2}\rho_1\omega^2\left(r_s^2 - r_i^2\right)$$

$$\text{or}: \quad \frac{r_s^2 - r_i^2}{r_s^2 - r_\omega^2} = \frac{\rho_2}{\rho_1} \tag{11.6}$$

If Q_1 and Q_2 are the volumetric rates of feed of the light and heavy liquids, respectively, on the assumption that there is no slip between the liquids in the bowl, and that the same residence time is required for the two phases, then:

$$\frac{Q_1}{Q_2} = \frac{r_s^2 - r_i^2}{R^2 - r_s^2} \tag{11.7}$$

Eq. (11.7) enables the value of r_s to be calculated for a given operating condition.

The retention time t_R necessary to give adequate separation of the liquids will depend on their densities and interfacial tension, and on the form of the dispersion, and can only be determined experimentally for that mixture. The retention time is given by:

$$t_R = \frac{V'}{Q_1 + Q_2} = \frac{V'}{Q} \tag{11.8}$$

where Q is the total feed rate of liquid, Q_1 and Q_2 refer to the light and heavy liquids, respectively, and V' is the volumetric hold-up of liquid in the bowl.

$$\text{Approximately}: \quad V' \approx \pi(R^2 - r_i^2)H \tag{11.9}$$

where H is the axial length (or clarifying length) of the bowl.

$$t_R = \frac{Q}{\pi(R^2 - r_i^2)H} \tag{11.10}$$

Eq. (11.10) gives the relation between Q and r_i for a given retention time, and determines the required setting of the weir. In practice, the relative values of r_i and r_w are adjusted to modify the relative residence times of the individual phases to give extra separating time for the more difficult phase. In the decanter centrifuge, r_i is adjusted to influence either discharged cake dryness or the conveying efficiency.

In Eqs. (11.6) and (11.7), ρ_2/ρ_1 and Q_1/Q_2 are determined by the properties and composition of the mixture to be separated, and R is fixed for a given bowl; r_i and r_w are governed by the settings of the weirs. The radius r_s of the interface between the two liquids is then the only unknown, and this may be eliminated between the two equations. Substitution of the value r_i from Eq. (11.10) then permits calculation of the required radius r_w of the weir for the heavy liquid.

This treatment gives only a simplified approach to the design of a system for the separation of two liquids. It will need modification to take account of the geometric configuration of the centrifuge—a topic which is discussed in Section 11.8.

11.5 Sedimentation in a Centrifugal Field

Centrifuges are extensively used for separating fine solids from suspension in a liquid. As a result of the far greater separating power, compared with that available using gravity, fine solids, colloids, and even molecules differing in their molecular weights may be separated. Furthermore, it is possible to break down emulsions and to separate dispersions of fine liquid droplets, though in this case, the suspended phase is in the form of liquid droplets which will coalesce following separation. Centrifuges may be used for batch operation when dealing with small quantities of suspension although, on the large scale, arrangements must sometimes be incorporated for the continuous removal of the separated constituents. Some of the methods of achieving this are discussed along with the various types of equipment in Section 11.8. When centrifuges are used for *polishing*, the removal of the very small quantities of finely divided solids needs to be carried out only infrequently, and manual techniques are then often used.

Because centrifuges are normally used for separating fine particles and droplets, it is necessary to consider only the Stokes' law region in calculating the drag between the particle and the liquid.

It is seen in Chapter 6 that, as a particle moves outward toward the walls of the bowl of a centrifuge, the accelerating force progressively increases, and, therefore, the particle never reaches an equilibrium velocity, as is the case in the gravitational field. Neglecting the inertia of the particle, then:

$$\frac{dr}{dt} = \frac{d^2(\rho_s - \rho)r\omega^2}{18\mu} \tag{11.11}$$

$$= u_0\frac{r\omega^2}{g} \tag{11.12}$$

At the walls of the bowl of radius R, dr/dt is given by:

$$\left(\frac{dr}{dt}\right)_{r=R} = \frac{d^2(\rho_s - \rho)R\omega^2}{18\mu} \tag{11.13}$$

The time taken to settle through a liquid layer of thickness h at the walls of the bowl is given by integration of Eq. (11.11) between the limits $r=r_0$ (the radius of the inner surface of the liquid), and $r=R$ to give Eq. (6.124). This equation may be simplified where $R - r_0$ ($=h$) is small compared with R, as in Eq. (6.125).

$$\text{Then:} \quad t_R = \frac{18\mu h}{d^2(\rho_s - \rho)R\omega^2} \tag{11.14}$$

It may be seen that Eq. (11.14) gives the time taken to settle through the distance h at a velocity given by Eq. (11.13). t_R is then the minimum retention time required for all particles of size greater than d to be deposited on the walls of the bowl. Thus, the maximum throughput Q at which all particles larger than d will be retained is given by substitution for t_R from Eq. (11.8) to give:

$$Q = \frac{d^2(\rho_s - \rho)R\omega^2 V'}{18\mu h} \tag{11.15}$$

$$\text{or}: \quad Q = \frac{d^2(\rho_s - \rho)g}{18\mu} \frac{R\omega^2 V'}{hg} \tag{11.16}$$

$$\text{From Eq.(6.24)}: \quad \frac{d^2(\rho_s - \rho)g}{18\mu} = u_0$$

where u_0 is the terminal falling velocity of the particle in the gravitational field, and hence:

$$Q = u_0 \frac{R\omega^2 V'}{hg} \tag{11.17}$$

Writing the capacity term as:

$$\Sigma = \frac{R\omega^2 V'}{hg}$$

$$= \frac{\pi R(R^2 - r_0^2)H\omega^2}{hg} \tag{11.18}$$

$$= \pi R(R + r_0)H\frac{\omega^2}{g}$$

$$\text{Then}: \quad Q = u_0\Sigma \tag{11.19}$$

Eq. (11.18) implies that the greater the depth h of liquid in the bowl, that is the lower the value of r_0, the smaller will be the value of Σ, but this is seldom borne out in practice in decanter centrifuges. This is probably due to high turbulence experienced and the effects of the scrolling mechanism. Σ is independent of the properties of the fluid and the particles and depends only on the dimensions of the centrifuge, the location of the overflow weir, and the speed of rotation. It is equal to the cross-sectional area of a gravity settling tank with the same clarifying capacity as the centrifuge. Thus, Σ is a measure of the capacity of the centrifuge and gives a quantification of its performance for clarification. This treatment is attributable to the work of Ambler.[1]

For cases where the thickness h of the liquid layer at the walls is comparable in order of magnitude with the radius R of the bowl, it is necessary to use Eq. (6.124) in place of Eq. (11.14) for the required residence time in the centrifuge or:

$$t_R = \frac{18\mu}{d^2(\rho_s - \rho)\omega^2} \ln\frac{R}{r_0} \tag{11.20}$$

(from Eq. 6.124)

$$\text{Then}: \quad Q = \frac{d^2(\rho_s - \rho)\omega^2 V'}{18\mu \ln(R/r_0)} \tag{11.21}$$

$$= \frac{d^2(\rho_s - \rho)g}{18\mu} \frac{\omega^2 V'}{g \ln(R/r_0)} \tag{11.22}$$

$$= u_0 \Sigma \tag{11.23}$$

$$\text{In this case}: \quad \Sigma = \frac{\omega^2 V'}{g \ln(R/r_0)} \\ = \frac{\pi(R^2 - r_i^2)H\omega^2}{\ln(R/r_0)} \frac{1}{g} \tag{11.24}$$

A similar analysis can be carried out with various geometrical arrangements of the bowl of the centrifuge. Thus, for example, for a disc machine (described later), the value of Σ is very much greater than that for a cylindrical bowl of the same size. Values of Σ for different arrangements are quoted by Hayter,[2] Trowbridge,[3] and others.[4, 5]

This treatment leads to the calculation of the condition where all particles larger than a certain size are retained in the centrifuge. Other definitions are sometimes used, such as, for example, the size of the particle which will just move half the radial distance from the surface of the liquid to the wall, or the condition when just half of the particles of the specified size will be removed from the suspension.

Example 11.1

In a test on a centrifuge, all particles of a mineral of density $2800 \, \text{kg/m}^3$ and of size $5 \, \mu\text{m}$, equivalent spherical diameter, were separated from suspension in water fed at a volumetric throughput rate of $0.25 \, \text{m}^3/\text{s}$. Calculate the value of the capacity factor Σ.

What will be the corresponding size cut for a suspension of coal particles in oil fed at the rate of $0.04 \, \text{m}^3/\text{s}$? The density of coal is $1300 \, \text{kg/m}^3$, the density of the oil is $850 \, \text{kg/m}^3$, and its viscosity is $0.01 \, \text{Ns/m}^2$.

It may be assumed that Stokes' law is applicable.

Solution

The terminal falling velocity of particles of diameter $5 \, \mu\text{m}$ in water, of density $\rho = 1000 \, \text{kg/m}^3$, and of viscosity $\mu = 10^{-3} \, \text{Ns/m}^2$, is given by:

$$u_0 = \frac{d^2(\rho_s - \rho)g}{18\mu} = \frac{25 \times 10^{-12} \times (2800 - 1000) \times 9.81}{18 \times 10^{-3}}$$

$$= 2.45 \times 10^{-5} \, \text{m/s} \quad (\text{Eq.6.24})$$

From the definition of Σ:

$$Q = u_0 \Sigma \quad (\text{Eq.11.19})$$

$$\text{and}: \Sigma = \frac{0.25}{(2.45 \times 10^{-5})} = 1.02 \times 10^4 \text{m}^2.$$

For the coal-in-oil mixture:

$$u_0 = \frac{Q}{\Sigma} = \frac{0.04}{(1.02 \times 10^4)} = 3.92 \times 10^{-6} \text{m/s}.$$

From Eq. (6.24):

$$d^2 = \frac{18\mu u_0}{(\rho_s - \rho)g}$$

$$= \frac{18 \times 10^{-2} \times 3.92 \times 10^{-6}}{(1300 - 850) \times 9.81}.$$

$$\text{and}: d = 4.0 \times 10^{-6}\text{m or } \underline{\underline{4\mu m}}$$

Example 11.2

A centrifuge is fitted with a conical disc stack with an included angle of 2θ, and there are n flow passages between the discs. A suspension enters at radius r_1 and leaves at radius r_2. Obtain an expression for the separating power Σ of the centrifuge. It may be assumed that the resistance force acting on the particles is given by Stokes' law.

Solution

For two discs AA′ and BB′, as shown in Fig. 11.4, the particle which is most unfavourably placed for collection will enter at point A at radius r_1, and be deposited on the upper plate at point B′ at radius r_2. It is assumed that the suspension is evenly divided between the discs. The particles will move not quite in a straight line because both velocity components are a function of r.

At radius r, the velocity of the liquid in the flow channel is:

$$\frac{Q}{2\pi r a n} = \frac{dx}{dt} \tag{i}$$

where x is the distance parallel to the discs and a is the spacing.

At radius r, the centrifugal sedimentation velocity of a particle whose diameter is d is given by:

$$\frac{dr}{dt} = \frac{d^2 r \omega^2 (\rho_s - \rho)}{18\mu} \quad (\text{Eq.11.11})$$

$$= u_0 \frac{r\omega^2}{g} \quad (\text{Eq.11.12}) \tag{ii}$$

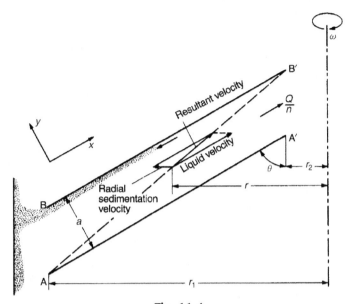

Fig. 11.4
Path of limit particle through separation channel.

From the geometry of the system:

$$-\frac{dr}{dx} = \sin\theta$$

Thus from Eq. (i):

$$-\frac{dr}{dt} = \frac{Q}{2\pi \, ran} \sin\theta \tag{iii}$$

$$\text{and}: \quad \frac{dy}{dr} = \cos\theta$$

Thus from Eq. (ii):

$$\frac{dy}{dt} = u_0 \frac{r\omega^2}{g} \cos\theta \tag{iv}$$

Dividing Eq. (iv) by Eq. (iii) gives:

$$-\frac{dy}{dr} = \left(\frac{u_0 r\omega^2 \cos\theta}{g}\right) \cdot \left(\frac{2\pi \, ran}{Q\sin\theta}\right)$$

$$= \frac{2\pi \, nau_0\omega^2 \cot\theta}{Qg} r^2$$

The particle must move through distance a in the y direction as its radial position changes from r_1 to r_2.

$$\text{Thus:}\quad -\int_0^a dy = \frac{2\pi n a u_0 \omega^2 \cot\theta}{Qg}\int_{r_1}^{r_2} r^2 dr$$

$$a = \frac{2\pi n a u_0 \omega^2 \cot\theta}{3Qg}(r_1^3 - r_2^3)$$

$$Q = u_0\frac{2\pi n \omega^2 \cot\theta}{3g}(r_1^3 - r_2^3)$$

$$= u_0\Sigma \quad \text{(from the definition of } \Sigma, \text{ Eq.11.19)}$$

$$\text{or:}\quad \Sigma = \frac{2\pi\omega^2 n \cot\theta\,(r_1^3 - r_2^3)}{3g}$$

11.6 Filtration in a Centrifuge

When filtration is carried out in a centrifuge, it is necessary to use a perforated bowl to permit removal of the filtrate. The driving force is the centrifugal pressure due to the liquid and suspended solids, and this will not be affected by the presence of solid particles deposited on the walls. The resulting force must overcome the friction caused by the flow of liquid through the filter cake, the cloth, and the supporting gauze and perforations. The resistance of the filter cake will increase as solids are deposited, although the other resistances will remain approximately constant throughout the process. Considering filtration in a bowl of radius R, and supposing that the suspension is introduced at such a rate that the inner radius of the liquid surface remains constant as shown in Fig. 11.5, then at some time t after the

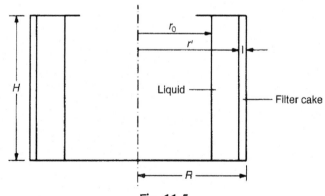

Fig. 11.5
Filtration in a centrifuge.

commencement of filtration, a filter cake of thickness l will have been built up, and the radius of the interface between the cake and the suspension will be r'.

If dP' is the pressure difference across a small thickness dl of cake, the velocity of flow of the filtrate is given by Eq. (9.2), and:

$$\frac{1}{A}\frac{dV}{dt} = u_c = \frac{1}{\mathbf{r}\mu}\left(\frac{-dP'}{dl}\right) \tag{11.25}$$

where $\mathbf{r}$ is the specific resistance of the filter cake and μ is the viscosity of the filtrate.

If the centrifugal force is large compared with the gravitational force, the filtrate will flow in an approximately radial direction, and will be evenly distributed over the axial length of the bowl. The area available for flow will increase toward the walls of the bowl. If dV is the volume of filtrate flowing through the filter cake in time dt, then:

$$u_c = \frac{1}{2\pi r'H}\frac{dV}{dt}$$

$$\text{Thus}: \frac{1}{\mathbf{r}\mu}\left(\frac{-dP'}{dl}\right) = \frac{1}{2\pi r'H}\frac{dV}{dt} \tag{11.26}$$

$$-dP' = \frac{\mathbf{r}\mu dl}{2\pi r'H}\frac{dV}{dt}$$

and thus the total pressure drop through the cake at time t is given by:

$$-\Delta P' = \frac{\mathbf{r}\mu}{2\pi H}\frac{dV}{dt}\int_0^l \frac{dl}{r'}$$

$$= \frac{\mathbf{r}\mu}{2\pi H}\frac{dV}{dt}\int_{r'}^R \frac{dr'}{r'} \tag{11.27}$$

$$= \frac{\mathbf{r}\mu}{2\pi}\frac{dV}{dt}\ln\frac{R}{r'}$$

If the resistance of the cloth is negligible, $-\Delta P'$ is equal to the centrifugal pressure. More generally, if the cloth, considered together with the supporting wall of the basket, is equivalent in resistance to a cake of thickness L, situated at the walls of the basket, the pressure drop $-\Delta P''$ across the cloth is given by:

$$\frac{-\Delta p''}{\mathbf{r}\mu L} = \frac{1}{2\pi HR}\frac{dV}{dt}$$

$$-\Delta P'' = \frac{\mathbf{r}\mu}{2\pi H}\frac{dV}{dt}\frac{L}{R} \tag{11.28}$$

Thus, the total pressure drop across the filter cake and the cloth $(-\Delta P)$, say, is given by:

$$(-\Delta P) = (-\Delta P') + (-\Delta P'')$$

$$\text{Thus}: -\Delta P = \frac{\mathbf{r}\mu}{2\pi H} \frac{dV}{dt} \left(\ln\frac{R}{r'} + \frac{L}{R} \right) \tag{11.29}$$

Before this equation can be integrated, it is necessary to establish the relation between r' and V. If v is the bulk volume of the incompressible cake deposited by the passage of unit volume of filtrate, then:

$$v\,dV = -2\pi r' H\,dr'$$

$$\frac{dV}{dt} = -\frac{2\pi H r'}{v} \frac{dr'}{dt} \tag{11.30}$$

and substituting for dV/dt in the previous equation gives:

$$-\Delta P = \frac{\mathbf{r}\mu}{2\pi H} \frac{2\pi H r'}{v} \frac{dr'}{dt} \left(\ln\frac{R}{r'} + \frac{L}{R} \right) \tag{11.31}$$

$$\text{Thus}: \quad \frac{v(-\Delta P)}{\mathbf{r}\mu}\,dt = \left(\ln\frac{r'}{R} - \frac{L}{R} \right) r'\,dr'$$

This may be integrated between the limits $r' = R$ and $r' = r'$ as t goes from 0 to t. $-\Delta P$ is constant because the inner radius r_0 of the liquid is maintained constant:

$$\frac{(-\Delta P)vt}{\mathbf{r}\mu} = \int_R^{r'} \left\{ \left(\ln\frac{r'}{R} - \frac{L}{R} \right) r' \right\} dr'$$

$$= \frac{1}{4}\left(R^2 - r'^2 \right) + \frac{L}{2R}\left(R^2 - r'^2 \right) + \frac{1}{2}r'^2 \ln\frac{r'}{R} \tag{11.32}$$

$$\text{and}: \quad \left(R^2 - r'^2 \right)\left(1 + 2\frac{L}{R} \right) + 2r'^2 \ln\frac{r'}{R} = \frac{4(-\Delta P)vt}{\mathbf{r}\mu}$$

$$= \frac{2vt\rho\omega^2}{\mathbf{r}\mu}\left(R^2 - r_0^2 \right) \tag{11.33}$$

$$\text{since}: \quad -\Delta P = \frac{1}{2}\rho\omega^2\left(R^2 - r_0^2 \right) \quad \text{(from Eq.11.4)}$$

From this equation, the time t taken to build up the cake to a given thickness r' may be calculated. The corresponding volume of cake is given by:

$$Vv = \pi\left(R^2 - r'^2 \right) H \tag{11.34}$$

and the volume of filtrate is:

$$V = \frac{\pi}{v}\left(R^2 - r'^2\right)H \tag{11.35}$$

Haruni and Storrow[6] have carried out an extensive investigation into the flow of liquid through a cake formed in a centrifuge and have concluded that, although the results of tests on a filtration plant and a centrifuge are often difficult to compare because of the effects of the compressibility of the cake, it is frequently possible to predict the flowrate in a centrifuge to within 20%. They have also shown that, when the thickness varies with height in the basket, the flowrate can be calculated on the assumption that the cake has a uniform thickness equal to the mean value; this gives a slightly high value in most cases.

Example 11.3

When an aqueous slurry is filtered in a plate and frame press, fitted with two 50 mm thick frames each 150 mm square, operating with a pressure difference of 350 kN/m², the frames are filled in 3600 s (1 h). How long will it take to produce the same volume of filtrate as is obtained from a single cycle when using a centrifuge with a perforated basket, 300 mm diameter and 200 mm deep? The radius of the inner surface of the slurry is maintained constant at 75 mm, and the speed of rotation is 65 Hz (3900 rpm).

It may be assumed that the filter cake is incompressible, that the resistance of the cloth is equivalent to 3 mm of cake in both cases, and that the liquid in the slurry has the same density as water.

Solution

In the filter press

Noting that $V = 0$ when $t = 0$, then:

$$V^2 + 2\frac{AL}{v}V = \frac{2(-\Delta P)A^2 t}{r\mu v} \quad \text{(from Eq.10.18)}$$

$$\text{and:} \quad V = \frac{lA}{v} \quad \text{(from Eq.10.6)}$$

$$\text{Thus:} \quad \frac{l^2 A^2}{v^2} + \left(\frac{2AL}{v}\right)\left(\frac{lA}{v}\right) = \frac{2(-\Delta P)A^2 t}{r\mu v}$$

$$\text{or:} \quad l^2 + 2Ll = \frac{2(-\Delta P)vt}{r\mu}$$

For one cycle

$$l = 25\,\text{mm} = 0.025\,\text{m}; \quad L = 3\,\text{mm} = 0.003\,\text{m}$$

$$-\Delta P = 350\,\text{kN/m}^2 = 3.5 \times 10^5\,\text{N/m}^2$$

$$t = 3600\,\text{s}$$

$$\therefore \quad 0.025^2 + (2 \times 0.003 \times 0.025) = 2 \times 3.5 \times 10^5 \times 3600 \times \left(\frac{v}{r\mu}\right)$$

$$\text{and}: \quad \left(\frac{r\mu}{v}\right) = 3.25 \times 10^{12}$$

In the centrifuge

$$\left(R^2 - r'^2\right)\left(1 + 2\frac{L}{R}\right) + 2r'^2 \ln\frac{r'}{R} = \frac{2vt\rho\omega^2}{r\mu}\left(R^2 - r_0^2\right) \quad \text{(Eq.11.33)}$$

$R = 0.15\,\text{m}$, $H = 0.20\,\text{m}$, and the volume of cake $= 2 \times 0.050 \times 0.15^2 = 0.00225\,\text{m}^3$

$$\text{Thus}: \pi\left(R^2 - r'^2\right) \times 0.20 = 0.00225$$

$$\text{and}: \left(R^2 - r'^2\right) = 0.00358$$

$$\text{Thus}: r'^2 = \left(0.15^2 - 0.00358\right) = 0.0189\,\text{m}^2$$

$$\text{and}: r' = 0.138\,\text{m}$$

$$r_0 = 75\,\text{mm} = 0.075\,\text{m}$$

$$\text{and}: \omega = 65 \times 2\pi = 408.4\,\text{rad/s}$$

The time taken to produce the same volume of filtrate or cake as in one cycle of the filter press is, therefore, given by:

$$\left(0.15^2 - 0.138^2\right)\left(1 + 2 \times 0.003/0.15\right) + 2(0.0189)\ln\left(0.138/0.15\right)$$

$$= \frac{2 \times t \times 1000 \times 408.4^2}{3.25 \times 10^{12}}\left(0.15^2 - 0.075^2\right)$$

$$\text{or}: 0.00359 - 0.00315 = 1.732 \times 10^{-6}t$$

$$\text{from which}: t = \frac{4.4 \times 10^{-4}}{1.732 \times 10^{-6}}$$

$$= \underline{254\text{s or }4.25\,\text{min}}.$$

11.7 Mechanical Design

Features of the mechanical design of centrifuges are discussed in Volume 6, where, in particular, the following items are considered:

(a) The mechanical strength of the bowl, which will be determined by the dimensions of the bowl, the material of construction, and the speed of operation.
(b) The implications of the critical speed of rotation, which can cause large deflections of the shaft and vibration.
(c) The slow gyratory motion, known as *precession*, which can occur when the bowl or basket is tilted.

11.8 Centrifugal Equipment

11.8.1 Classification of Centrifuges

Centrifuges may be grouped into two distinct categories—those that utilise the principle of filtration and those that utilise the principle of sedimentation, both enhanced by the use of a centrifugal field. These two classes may be further subdivided according to the method of discharge, particularly the solids discharge. This may be batch, continuous, or a combination of both. Further subdivisions may be made according to source of manufacture or mechanical features, such as the method of bearing suspension, axis orientation, or containment. In size, centrifuges vary from the small batch laboratory tube spinner to the large continuous machines with several tons of rotating mass, developing a few thousands of *g* relative centrifugal force. Records[7] has suggested the following classification.

Filtration centrifuges

(a) Batch discharge, vertical axis, perforate basket.
(b) Knife discharge, pendulum suspension, vertical axis, perforate basket.
(c) Peeler: horizontal axis, knife discharge at speed.
(d) Pusher.
(e) Scroll discharge.

Sedimentation centrifuges

(a) Bottle spinner.
(b) Tubular bowl.
(c) Decanter — scroll discharge.
(d) Imperforate bowl — skimmer pipe discharge, sometimes also a knife.
(e) Disc machine.
    (i) Batch.
    (ii) Nozzle discharge.
    (iii) Opening bowl (solids ejecting).
    (iv) Valve discharge.

Liquid–liquid separation centrifuges

(a) Tubular bowl.
(b) Three-phase scroll discharge decanter.
(c) Disc machine.

Clearly, with such a wide range of equipment, it is not realistic to describe examples of machinery in each category. Discussion will, therefore, be limited to the more important types of machine and the conditions under which they may be used.

11.8.2 Simple Bowl Centrifuges

Most batch centrifuges are mounted with their axes vertical and, because of the possibility of uneven loading of the machine, the bowl is normally supported in bearings, and the centrifuge itself is supported by resilient mountings. In this way, the inevitable out-of-balance forces on the supporting structure are reduced to, typically, 5% or less.

In the case of the underdriven batch machine, where the drive and bearings are underneath, as shown in Fig. 11.6, access to the bowl is direct, while product is discharged from the top in manual machines and from the bottom in automatically controlled machines. In the case of the overdriven batch centrifuge, where the bowl is suspended from above, a valve is often incorporated in the bottom of the bowl for easy discharge of the solids. With a purpose designed drive, the unit can handle higher throughputs of freely filtering feed slurry by cycling more quickly than the underdriven machine.

Batch centrifuges with imperforate bowls are used either for producing an accelerated separation of solid particles in a liquid, or for separating mixtures of two liquids. In the former case, the solids are deposited on the wall of the bowl, and the liquid is removed through an overflow or skimming tube. The suspension is continuously fed in until a suitable depth of solids has been built up on the wall; this deposit is then removed either by hand or by a

Fig. 11.6
Underdriven batch centrifuge.

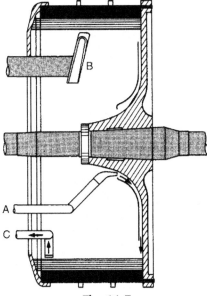

Fig. 11.7

Horizontally mounted bowl with automatic discharge of solids. A—feed, B—cutter, C—skimming tube.

mechanical scraper. With the bowl mounted about a horizontal axis, solids are more readily discharged because they can be allowed to fall directly into a chute.

In the centrifuge shown in Fig. 11.7, the liquid is taken off through a skimming tube and the solids, which may be washed before discharge if desired, are removed by a cutter. This machine is often mounted vertically, and the cutter knife sometimes extends over the full depth of the bowl.

Perforated bowls are used when relatively large particles are to be separated from a liquid, as, for example, in the separation of crystals from a mother liquor. The mother liquor passes through the bed of particles and then through the perforations in the bowl. When the centrifuge is used for filtration, a coarse gauze is laid over the inner surface of the bowl, and the filter cloth rests on the gauze. Space is, thus, provided behind the cloth for the filtrate to flow to the perforations.

When a mixture of liquids is to be separated, the denser liquid collects at the imperforate bowl wall, and the less dense liquid forms an inner layer. Overflow weirs are arranged so that the two constituents are continuously removed. The design of the weirs has been considered in a previous section.

11.8.3 Disc Centrifuges—General

For a given rate of feed to the centrifuge, the degree of separation obtained will depend on the thickness of the liquid layer formed at the wall of the bowl and on the total depth of the bowl, because both these factors control the time the mixture remains in the machine.

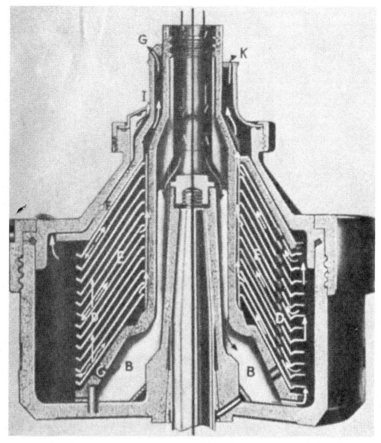

Fig. 11.8

Bowl with conical discs (left-hand side for separating liquids, right-hand side for separating solid from liquid). *Image Courtesy: Alfa-Laval Sharples Ltd., Camberley, Surrey.*

A high degree of separation could, therefore, be obtained with a long bowl of small diameter, although the speed required would then need to be very high. By comparison, the introduction of conical discs into the bowl, as illustrated in Fig. 11.8, enables the liquid stream to be split into a large number of very thin layers and permits a bowl of greater diameter. The separation of a mixture of water and dirt from a relatively low density oil takes place, as shown in Fig. 11.9, with the dirt and water collecting close to the undersides of the discs and moving radially outward, and with the oil moving inward along the top sides.

A disc bowl, although still a high speed machine, can thus be run at lower speeds relative to the excessive speeds required for a long bowl of small diameter. Also, its size is very much smaller relative to a bowl without discs, as may be seen from Fig. 11.10. The separation of two liquids in a disc-type bowl is illustrated in the left-hand side of Fig. 11.8. Liquid enters through the distributor AB, passes through C, and is distributed between the discs E through the holes D. The denser liquid is taken off through F and I, and the less dense liquid through G.

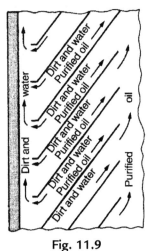

Fig. 11.9
Separation of water and dirt from oil in disc bowl.

Fig. 11.10
Two bowls of equal capacity; with discs (left) and without discs (right). *Image Courtesy: Alfa-Laval Sharples Ltd., Camberley, Surrey.*

A disc-type bowl is often used for the separation of fine solids from a liquid and its construction is shown in the right-hand side of Fig. 11.8. Here, there is only one liquid outlet K, and the solids are retained in the space between the ends of the discs and the wall of the bowl.

11.8.4 Disc Centrifuges—Various Types

Disc centrifuges vary considerably according to the type of discharge. The liquid(s) can discharge freely or through centrifugal pumps. The solids may be allowed to accumulate within the bowl and then be discharged manually (batch bowl). A large number of disc machines (opening bowl) discharge solids intermittently when ports are opened automatically at the periphery of the bowl on a timed basis or are actuated by a signal from a liquid clarity meter. The opening on sophisticated machines may be adjusted to discharge only the compacted solids.

Another type of disc centrifuge has nozzles or orifices at the periphery where thickened solids are discharged continuously, and sometimes, a fraction of this discharge is recycled to ensure that the nozzles are able to prevent breakthrough of the clarified supernatant liquid. Sometimes, these nozzles are internally fitted to the bowl. Others are opened and closed electrically or hydrostatically from a build up of the sludge within the bowl.

Centrifuges of these types are used in the processing of yeast, starch, meat, fish products, and fruit juices. They form essential components of the process of rendering in the extraction of oils and fats from cellular materials. The raw material, consisting of bones, animal fat, fish offal, or vegetable seeds, is first disintegrated, and then, after a preliminary gravitational separation, the final separation of water, oil, and suspended solids is carried out in a number of valve nozzle centrifuges.

11.8.5 Decanting Centrifuges

The widely used decanting type continuous centrifuge, shown in Fig. 11.11, is mounted on a horizontal axis. The mixture of solids and liquids is fed to the machine through a stationary pipe which passes through one of the support bearings, shown in Fig. 11.12. The feed pipe discharges the mixture near the centre of the machine. The heavier solids settle on to the wall of

Fig. 11.11
Solid-bowl decanter centrifuge. *Image Courtesy: Alfa-Laval Sharples Ltd., Camberley, Surrey.*

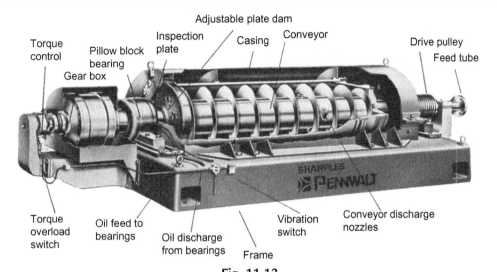

Fig. 11.12

Solid-bowl decanter centrifuge—principle of operation. *Image Courtesy: Alfa-Laval Sharples Ltd., Camberley, Surrey.*

the imperforate or solid bowl under the influence of the centrifugal force. Inside the bowl is a close fitting helical scroll which rotates at a slightly different speed from the solid bowl. Typically, the speed differential is in the range of 0.5 rpm to 100 rpm (0.01–2 Hz).

At one end of the bowl, a conical section is attached, giving a smaller diameter. The liquid runs round the helical scroll and is discharged over weir plates fitted at the parallel end of the bowl. The solids are moved by the conveying action of the helical scroll up the gentle slope of the conical section, out of the liquid, and finally, out of the machine. Decanters of this type are known as solid bowl decanters.

A variant of the decanting centrifuge is the screen bowl decanter, shown in Figs. 11.13 and 11.14. In this unit, a further perforated section is attached to the smaller diameter end of the

Fig. 11.13

Screen bowl decanter centrifuge (cover removed). *Image Courtesy: Thomas Broadbent & Sons Ltd., Huddersfield, West Yorkshire.*

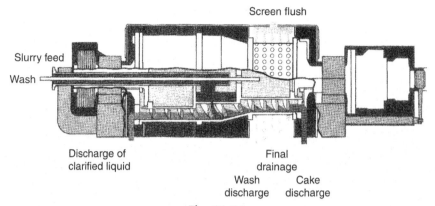

Fig. 11.14

Screen bowl decanter centrifuge—principle of operation. *Image Courtesy: Thomas Broadbent & Sons Ltd., Huddersfield, West Yorkshire.*

conical section. This is known as the screen and allows further drying and/or washing of the solids to take place.

Machines may be tailored to meet specific process requirement by altering the diameter of the liquid discharge, the differential speed of the helical scroll, the position at which the feed enters the machine, and the rotational speed of the bowl. Decanting centrifuges are available in a wide range of diameters and lengths. Diameters are 0.2–1.5 m, and lengths are typically 1.5–5 times the diameter. The longer machines are used when clear liquids are required. Shorter bowl designs are used to produce the driest solids. Continuous throughputs in excess of 250 tons/h (70 kg/s) of feed can be handled by a single machine.

11.8.6 Pusher-Type Centrifuges

This type of centrifuge is used for the separation of suspensions and is fitted either with a perforated or imperforate bowl. The feed is introduced through a conical funnel, and the cake is formed in the space between the flange and the end of the bowl. The solids are intermittently moved along the surface of the bowl by means of a reciprocating pusher. The pusher comes forward and returns immediately and then waits until a further layer of solids has been built up before advancing again. In this machine, the thickness of filter cake cannot exceed the distance between the surface of the bowl and the flange of the funnel. The liquid either passes through the holes in the bowl or, in the case of an imperforate bowl, is taken away through an overflow. The solids are washed by means of a spray, as shown in Fig. 11.15.

A form of pusher-type centrifuge which is particularly suitable for filtering slurries of low concentrations is shown in Fig. 11.16. A perforated pusher cone gently accelerates the feed and secures a large amount of preliminary drainage near the apex of the cone. The solids from the

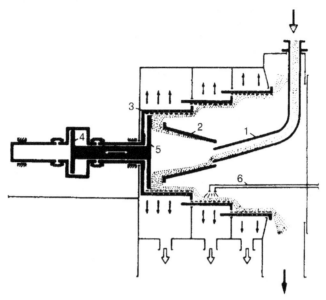

Fig. 11.15

Pusher-type centrifuge. 1: Inlet; 2: inlet funnel; 3: bowl; 4: piston; 5: pusher disc; 6: washing spray.

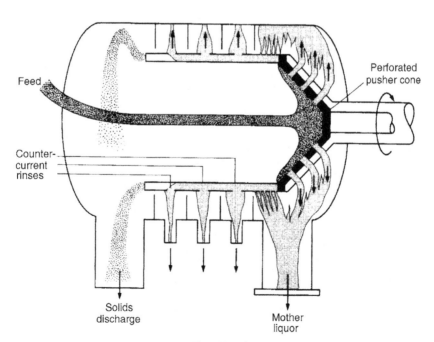

Fig. 11.16

Pusher-type centrifuge. *Image Courtesy: Alfa-Laval Sharples Ltd., Camberley, Surrey.*

partially concentrated suspension are then evenly laid on the cylindrical surface, and, in this way, the risk of the solids being washed out of the bowl is minimised.

11.8.7 Tubular-Bowl Centrifuge

Because, for a given separating power, the stress in the wall is a minimum for machines of small radius, machines with high separating powers generally use very tall bowls of small diameters. A typical centrifuge, shown in Fig. 11.17, would consist of a bowl about 100 mm diameter and 1 m long incorporating longitudinal plates to act as accelerator blades to bring the liquid rapidly up to speed. In laboratory machines, speeds up to 50,000 rpm (1000 Hz) are used to give accelerations 60,000 times the gravitational acceleration. A wide range of materials of construction can be used.

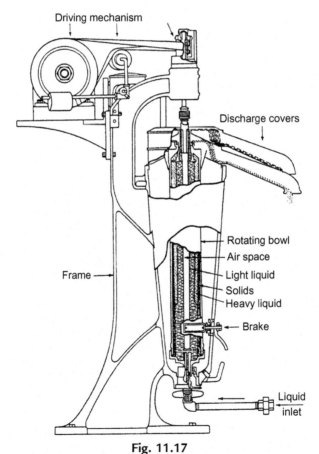

Fig. 11.17
Sectional view of the supercentrifuge. *Image Courtesy: Alfa-Laval Sharples Ltd., Camberley, Surrey.*

The position of the liquid interface is determined by balancing centrifugal forces as in Fig. 11.3. The lip, of radius r_w, over which the denser liquid leaves the bowl is part of a removable ring. Various sizes maybe fitted to provide for the separation of liquids of various relative densities.

Often, the material fed to these machines contains traces of denser solids in addition to the two liquid phases. These solids are deposited on the inner wall of the bowl, and the machine is dismantled periodically to remove them. A common application is the removal of water and suspended solids from lubricating oil.

The supercentrifuge is used for clarification of oils and fruit juices and for the removal of oversize and undersize particles from pigmented liquids. The liquid is continuously discharged, but the solids are retained in the bowl and must be removed periodically.

11.8.8 The Ultracentrifuge

For separation of colloidal particles and for breaking down emulsions, the ultracentrifuge is used. This operates at speeds up to 100,000 rpm (1600 Hz) and produces a force 100,000 times the force of gravity for a continuous liquid flow machine, and as high as 500,000 times for gas phase separations, although these machines are very small. The bowl is usually driven by means of a small air turbine. The ultracentrifuge is often run either at low pressures or in an atmosphere of hydrogen in order to reduce frictional losses, and a fivefold increase in the maximum speed can be attained by this means.

11.8.9 Multistage Centrifuges

Among many specialist types of centrifuge is the multistage machine, which consists of a series of bowls mounted concentrically on a vertical axis. The feed suspension is introduced into the innermost bowl, and the overflow then passes successively to each larger bowl in turn. As the separating force is directly proportional to the radius of rotation, the largest particles are separated out at the first stage, and progressively finer particles are recovered at each subsequent stage. The finest particles are collected in the outermost vessel from which the remaining liquid or suspension is discharged. The multistage system may also incorporate a series of concentric vertical baffles in a single bowl, with the suspension flowing upward and downward through successive annular channels. Both the design and operation are complex, and the machines are somewhat inflexible.

11.8.10 The Gas Centrifuge

A specialised, though nevertheless highly important, function for which the centrifuge has been developed is the separation of radioactive isotopes as described by Fishman.[8] The concentration of uranium-235 from less than 1% up to about 5% is achieved by subjecting uranium hexafluoride (UF_6) to an intense centrifugal field. The small differences in density of the components of the mixture necessitate the use of very high accelerations in order to obtain the desired separation. Mechanical considerations dictate the use of small diameter rotors (0.1–0.2 m) rotating at speeds up to 2000 Hz (10^5 rpm), giving velocities of up to 700 m/s and accelerations of up to $10^6 g$ at the periphery. Under these conditions, the gas flow can vary from free molecular flow in the low pressure region at the centre, to a high Mach number flow at the periphery. Furthermore, pressure will change by a factor of 10 over a distance of about 2 mm, as compared with a distance of about 20 km in the earth's gravitational field.

For use in a radioactive environment, the gas centrifuge must be completely maintenance free. It has been used for the separation of xenon isotopes, and consideration has been given to its application for separation of fluorohydrocarbons. Worldwide, about a quarter of a million gas centrifuges have been manufactured. As an order of magnitude figure, an investment of £1000 is necessary to obtain 0.3 g/s (10 g/h) of product.

Further information is given in the papers by Whitley.[9, 10]

11.9 Nomenclature

		Units in SI System	Dimensions in M, L, T
A	Cross-sectional area of filter	m^2	$\mathbf{L}^2$
a	Distance between discs in stack	m	$\mathbf{L}$
d	Diameter (or equivalent diameter of particle)	m	$\mathbf{L}$
g	Acceleration due to gravity	m/s^2	$\mathbf{LT}^{-2}$
H	Length (axial) of centrifuge bowl	m	$\mathbf{L}$
h	Depth of liquid at wall of bowl	m	$\mathbf{L}$
L	Thickness of filter cake with same resistance as cloth	m	$\mathbf{L}$
l	Thickness of filter cake	m	$\mathbf{L}$
n	Number of passages between discs in bowl	–	–
P	Pressure	N/m^2	$\mathbf{ML}^{-1}\mathbf{T}^{-2}$
P'	Pressure in cake	N/m^2	$\mathbf{ML}^{-1}\mathbf{T}^{-2}$
P''	Pressure in cloth	N/m^2	$\mathbf{ML}^{-1}\mathbf{T}^{-2}$
$-\Delta P$	Pressure drop (total)	N/m^2	$\mathbf{ML}^{-1}\mathbf{T}^{-2}$
$-\Delta P'$	Pressure drop over filter cake	N/m^2	$\mathbf{ML}^{-1}\mathbf{T}^{-2}$
$-\Delta P''$	Pressure drop over filter cloth	N/m^2	$\mathbf{ML}^{-1}\mathbf{T}^{-2}$
Q	Volumetric feed rate to centrifuge	m^3/s	$\mathbf{L}^3\mathbf{T}^{-1}$
R	Radius of bowl of centrifuge	m	$\mathbf{L}$
$\mathbf{r}$	Specific resistance of filter cake	m^{-2}	$\mathbf{L}^{-2}$

r	Radius of rotation	m	**L**
r_0	Radius of inner surface of liquid in bowl	m	**L**
r_1	Radius at inlet to disc bowl centrifuge	m	**L**
r_2	Radius at outlet of disc bowl centrifuge	m	**L**
r_i	Radius of weir for lighter liquid	m	**L**
r_s	Radius of interface between liquids in bowl	m	**L**
r_w	Radius of overflow weir for heavier liquid	m	**L**
r'	Radius at interface between liquid and filter cake	m	**L**
t	Time	s	**T**
t_R	Retention time in centrifuge	s	**T**
u_0	Terminal falling velocity of particle	m/s	$\mathbf{LT^{-1}}$
u_c	Superficial filtration velocity $\left[\frac{1}{A}\frac{dV}{dt}\right]$	m/s	$\mathbf{LT^{-1}}$
V	Volume of filtrate passing in time t	m³	$\mathbf{L^3}$
V'	Volumetric capacity of centrifuge bowl	m³	$\mathbf{L^3}$
v	Volume of cake deposited by passage of unit volume of filtrate	–	–
x	Distance parallel to discs in disc bowl centrifuge	m	**L**
y	Distance perpendicular to discs in disc bowl centrifuge	m	**L**
z	Vertical height	m	**L**
z_0	Vertical height of free surface of liquid (at radius r_0)	m	**L**
z_a	Value of z_0 at vertical axis of rotation ($r_0=0$)	m	**L**
μ	Viscosity of liquid	Ns/m²	$\mathbf{ML^{-1}T^{-1}}$
ρ	Density of liquid	kg/m³	$\mathbf{ML^{-3}}$
ρ_s	Density of particles	kg/m³	$\mathbf{ML^{-3}}$
θ	Half included angle between discs	–	–
ω	Angular velocity	rad/s	$\mathbf{s^{-1}}$
Σ	Capacity term for centrifuge	m²	$\mathbf{L^2}$

References

1. Ambler CA. The evaluation of centrifuge performance. *Chem Eng Prog* 1952;**48**:150.
2. Hayter AJ. Progress in centrifugal separations. *J Soc Cosmet Chem* 1962;**13**:152.
3. Trowbridge MEO'K. Problems in the scaling-up of centrifugal separation equipment. *Chem Eng* 1962;**162**:A73.
4. Sparks T, Chase G. *Filters and filtration handbook*. 6th ed. Oxford: Butterworth-Heinemann; 2015.
5. Leung WW-F. *Industrial centrifugation technology*. New York: McGraw Hill; 1998.

6. Haruni MM, Storrow JA. Hydroextraction. *Ind Eng Chem* 1952;**44**:2751. Chem. Eng. Sci. **1** (1952) 154: **2** (1953) 97, 108, 164 and 203.

7. Records FA. *Alfa Laval Sharples Ltd.* Private communication, March 1990.

8. Fishman AM. Developments in uranium enrichment, 43. The centar gas centrifuge enrichment project: economics and engineering considerations. *AIChemE Symp Ser* 1977;**169**:73.

9. Whitley S. *A summary of the development of the gas centrifuge.* Capenhurst: British Nuclear Fuels PLC, Enrichment Division; June 1988.

10. Whitley S. Review of the gas centrifuges until 1962. Part I: principles of separation physics. Part II: principles of high-speed rotation. *Rev Mod Phys* 1984;**56**:41 & 67.

Further Reading

Ambler CM. In: McKetta JJ, editor. *Encyclopaedia of chemical processing and design.* Vol. 7. New York: Marcel Dekker; 1978.

Hsu H-W. Separations by liquid centrifugation. *Ind Eng Chem Fundam* 1986;**25**:588.

Lavanchy AC, Keith FW. Centrifugal separation. In: Kirk RE, Ohmer DF, editors. *Encyclopaedia of chemical technology.* Vol. 4. Wiley-Interscience; 1979. p. 710.

Mullin JW. Centrifuging. In: Cremer HW, Davies T, editors. *Chemical engineering practice.* Vol. 6. Butterworths; 1958. p. 528.

Zeitsch K. *Centrifugal filtration.* Butterworth & Co; 1981.

CHAPTER 12

Product Design and Process Intensification

12.1 Product Design

12.1.1 Introduction

Because, in its earlier years, chemical engineering was overshadowed by the requirements of the bulk chemical and petroleum industries, it was concerned with operations for the large-volume manufacture of relatively low-value materials of simple structures. Initially, chemical engineers made a major contribution in the development of separation processes—an area that was largely neglected and little understood by chemists. The study of the design and operation of chemical reactors came to the fore only in the early 1950s, at a time when the importance of flow patterns and residence time distributions was only just being appreciated. Pioneering work in this field was carried out, among others, by Danckwerts,[1, 2] whose classic papers form the foundation for much of the later work. On reflection, it seems incredible that so little attention had been given to optimising the whole system, that is, the reactor and the downstream processing. Many of the problems inherent in the separation of reactor products were attributable to the absence of any real attempt to design the reactor in such a way so as to maximise the yield of the desired product. In many ways, chemical engineers were the victims of their own success, in that they concentrated overmuch on the design of processes and paid little attention to the design of products for developing markets.

At this time, the turn of the millennium, interest has rapidly turned to meeting the needs and aspirations of an ever-demanding consumer industry that needs to supply products directly to the end user. Bulk chemicals, as such, have always been predominantly intermediates, forming the feedstock for the production of the final products to be used by the consumer industry, or by a proxy consumer in the case of healthcare products. For bulk, or "commodity" chemicals, price competition is severe and the tendency is for production to be located in those parts of the world where costs of labor, raw materials, energy, and so on are relatively low. Economy of scale has been an important factor, with the result that production is tending to take place in a very few plants of high throughputs. The plants themselves are viable only if they operate with a very efficient utilisation of resources of all kinds, and, at the same time, satisfy the requirements for safe operation and for being "environmentally friendly." Thus, to work in this field is, in no sense, a "soft option." With bulk chemical operation, there is also a "squeeze" that is becoming increasingly more severe. Customers are insisting on tighter specifications for final products, while, at the same time, raw material quality is deteriorating as a result of the tendency to use up the best resources of raw materials first. The continuing challenge is therefore, to make a better product from a lower-quality starting material.

One feature of many commodity chemicals is that they are essentially intermediates used in the production of a wide range of consumer products. It was once suggested that the per capita rate of production of sulfuric acid was a measure of a country's prosperity, although the consumer demand for sulfuric acid itself is almost zero. Similarly styrene, most of which is converted into polystyrene, is hardly a saleable product in the retail market. The exception, on the other hand, is the market for materials that are finally utilised as fuels, many of which are distributed to the final consumer following various degrees of "polishing." Even so, the greatest demand for fuels comes from the electrical supply industry, which dominates the market, and not from individual consumers.

The shift toward synthetic specialty chemicals and high-value products has been marked in recent years. There have been several driving forces at work. The first is the realisation, even by companies sitting on very large amounts of raw materials, that reserves are limited, and that, in the long term anyway, it is in their interest to upgrade at least part of their reserves into higher value products. Thus, for instance, in the china clay industry, the vast bulk of whose product finds its application as a filler or surface-coating agent in the paper industry, kaolin is used to form lightweight structural materials on the one hand, and support materials in chromatographic columns on the other. The importance of added value is rapidly being appreciated. Many of the more sophisticated products are replacing materials from natural sources, and there is a large expansion in the sale of factory-made products both in the food and in the cosmetics industries. Ice creams, now often of highly complex formations, are structured materials consisting of tiny air bubbles and fats dispersed in an essentially aqueous continuous phase. Selling air dispersed in water has always been a very attractive proposition! The structure involving, as it does, the size distribution, concentration, and stability of air bubbles, the size and form of the ice crystals is of critical importance, as it is in scores of other products in the food, personal care, and pharmaceutical sectors, for example; see Fuller[3] and Litster.[4] These factors all contribute to give the desired rheology and structure, which are at least as important as the material composition. Toothpaste must have the correct rheology — it must not run out of the tube of its own volition, and yet, it must be easy to extrude it onto the brush, where it will remain in place until it is sheared when used on the teeth. In other words, it needs to be a shear-thinning material with a yield stress. These characteristics are discussed in Volume 1A, Chapter 3. Face creams must, above all, have an appropriate texture, must be easy to apply, must stay in place, and must "feel right" — a the most difficult condition to define! Consumers are seldom concerned with the chemical make-up, except to satisfy themselves that it is not injurious, though they need to be satisfied that it is 'fit for purpose'. Similarly, it is a fair expectation (often taken for granted!) that table sugar crystals should be white (pleasant appearance), nonsticky, and free flowing, and should resist caking tendency. At the other extreme are the herbicide products (which are of granular form) are expected to be free flowing, attrition resistant, and nondusty, yet these must readily disperse in water for ultimate applications via spraying. In all these cases, providing the appropriate feel, whether it be for a foodstuff or cosmetic, depends on getting the correct microstructure, which, in turn, exerts a strong influence on the rheology of the product, both of which themselves are influenced by the processes used to manufacture these products. This three way link — structure-properties-processing—has been amply illustrated in the case of chocolates.[5] Understanding the nature of the microstructure of the material is, therefore, of paramount important in these circumstance.

In general, the required production rates of these more sophisticated materials are orders of magnitude less than those of the commodity materials referred to earlier. Some will be made in dedicated plants, but many others may be manufactured in multiproduct plant, which presents new problems in the scheduling of efficient production. These include:

- the need to provide buffer to storage to cover demand when the equipment is being used for making other products;
  the need for the plant to be designed so as to facilitate cleaning between runs;
- the need for batch times to be optimized — short runs mean that the downtime becomes unacceptably long, and long runs mean that the buffer storage needs to be greater;
- sequencing of the runs on the various products can sometimes be arranged so that the first product in a sequence is the one which is the least tolerant to cross-run contamination, and so on. This can mean that the intermediate cleaning operations within a sequence may need to be less thorough than those between sequences.

To quote from Cussler and Moggridge,[6] "Product design is the procedure by which customer needs are translated into commercial products." This involves assessing the essential requirements in a material which will satisfy the customer, and then designing a material with the requisite physical and sometimes chemical and structural properties. There may well be a very large number of ways in which the needs can be met, possibly by using starting materials with widely different chemical compositions, and the final judgement will be based on a complex synthesis of considerations of competing attributes and costs of both raw materials and processing. Finally, the product must be "safe" to use and must not have harmful environmental features.[4]

In general, there is a good correlation between the selling price per unit mass of the material, C, and its production rate, P. Dunnill[7] has produced a logarithmic plot of unit price against production rate for biochemical products, and this covered several orders of magnitude. At the high-cost end are expensive pharmaceutical products, and at the bottom cost limit is water (off the scale), with a wide range of products of intermediate complexity in between. The plot, which is reproduced in Fig. 12.1, is seen to be represented by a straight line, of negative

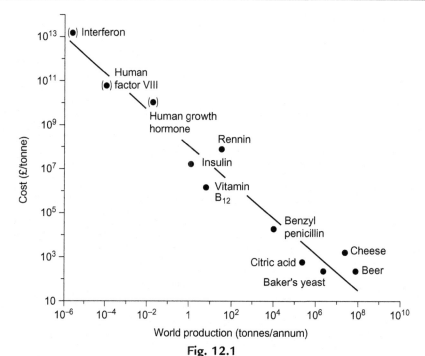

Fig. 12.1

World production tonnages and prices per ton of some products of biotechnology[7] (1983 data).

slope $-n$, which passes approximately through the majority of points giving an expression of the form:

$$C = \text{constant}\, P^{-n} \tag{12.1}$$

If n were equal to unity, it would imply that the annual production or utilisation value of all the components considered was approximately constant, a somewhat striking situation!

Although Fig. 12.1 is based on information relating only to the biochemical industry and the absolute level of prices and throughputs may be significantly out of date, the general trend is also applicable to other industrial areas as well — even to the production of motor vehicles!

12.1.2 Design of Particulate Products

Edwards and Instone,[8] have reviewed the manufacture and use of the particulate products, which, as they describe it, are made by the "fast moving consumer goods" industry and used by consumers around the world. It is claimed that all the products of this sector have the following common features:

(a) The products are created from a range of raw materials to yield complex multiphase mixtures which include emulsions, suspensions, gels, agglomerates, solid mixtures like breakfast cereals, and so on.

(b) During the processing of the product, a microstructure is created on the scale of $1-100\,\mu m$.

(c) This microstructure is usually required to be stable through the supply chain until the product is used by the consumer.

(d) It is the microstructure which determines the appearance of the product and its efficacy in use.

(e) The microstructure assembled during processing is destroyed during the use of the product.

Edwards and Instone[8] discuss the production of particulate products by spray drying and by binder granulation because these two processes are widely used because of their capability to produce multicomponent, designed granules aimed at meeting particular consumer needs. It is shown that the product microstructure resulting from such processing depends upon a complex dynamic interaction between the ingredients of the formulation and the processing conditions used during the manufacture. It is this microstructure which is generally destroyed when the product is used by the consumer, and this is also a complex, dynamic interaction between the applications of conditions and the formulated microstructure. Edwards and Instone[8] and Litster[4] argue that the understanding of the control of microstructure in particular products is far from complete, and that interdisciplinary research, ranging from measurement science, phase equilibrium, and microstructure kinetics to process engineering is needed to advance knowledge in this demanding area. It seems fairly clear that, if such knowledge of formulation and processing can lead to new and improved products, then the potential commercial returns are very high indeed.

12.1.3 The Role of the Chemical Engineer

In a further paper, Edwards[9] classified the processes and operations which occur in the manufacture and supply of products using an appropriate length scale as follows:

	Length (m)
Molecular level	$10^{-10} - 10^{-7}$
Microlevel	$10^{-6} - 10^{-3}$
Unit operations	$1-10$
Factory	$10^2 - 10^3$
Supply chain	$10^3 - 10^6$

Chemical engineers are well-versed in the design and sizing of unit operations such as reactors, mixing vessels, heat exchangers, and separation units, operating on a length scale appropriate to the equipment of around $1\,m$. Chemical engineers are also able to integrate individual operations to create an entire plant or factory which is on a scale of around 100 m. The supply chain, which includes raw material supply, manufacturing, and distribution to the consumer, involves a much larger scale, often in excess of 100 km. In the microstructural scale of around 1–100 μm, small gas bubbles, liquid droplets, and suspended fine particles in multiphase products are encountered together with microstructures created by surfactants, polymers, clays, and so on. The molecular reactions and interactions that create this microstructure take place within an even finer level of scrutiny. The microstuctures of products can be very complex as, for example, when a product contains more than ten components and where processing can involve flows that create wide residence time distributions and varying stress and temperature levels. Edwards[9] and others[10] have argued that chemical engineers can provide a key role in producing optimum microstructures provided they can link the physical and chemical sciences of microstructure formation with processing conditions. This indeed necessitates an inter-disciplinary approach involving solid state physics and chemistry, biological engineering, and even graphics and packaging, etc., as shown schematically in Fig. 12.2. In this way, chemical engineers must, in addition to dealing with flow and heat transfer in complex equipment, be able to determine the associated product structure. In the supply chain, the total system from the raw material supply, through manufacturing to distribution must be optimised, ensuring that the desired microstructure is delivered intact to the consumer. Cost effective solutions within this supply chain require a systems engineering approach which, Edwards[9] claims, chemical engineers are well placed to tackle if, in addition to the heartland of processing, the challenges of purchasing, material supply, packing activities, and distribution can also be met.

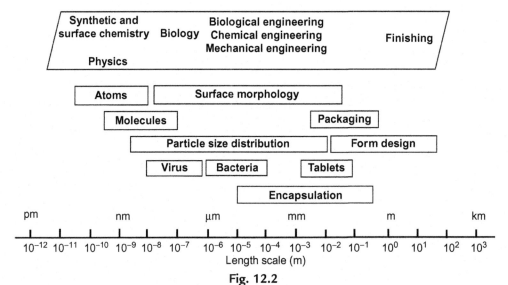

Fig. 12.2

Length scales and domain knowledge for product design.

12.1.4 Green Chemistry

Increasing concern for the need to conserve and to use effectively world reserves of raw materials and, at the same time, to reduce the quantities of waste materials which are likely to have an adverse effect on the environment has led to pressure for the increased use of renewable resources and so-called 'green chemistry'. The principles of green chemistry and green engineering have been enunciated, among others, by Hamley and Poliakoff,[11] Allen and Shonnard,[12] Marteel-Parrish and Abraham,[13] and Allen et al.,[14] as follows:

(a) It is better to prevent waste than to treat or clean up waste after it is formed.

(b) Synthetic methods should be designed to maximise the incorporation of all materials used in the process into the final product.

(c) Wherever practicable, synthetic methodologies should be designed to use and generate substances that possess little or no toxicity to human health and the environment.

(d) Chemical products should be designed to preserve efficacy of function while reducing toxicity.

(e) The use of auxiliary substances, such as solvents and separation agents, should be made unnecessary wherever possible, and innocuous when used.

(f) Energy requirements should be recognised for their environmental and economic impacts and should be minimised. Synthetic methods should be conducted at as close as possible to ambient temperature and pressure conditions.

(g) A raw material or feedstock should be renewable, rather than depleting, wherever this is technically and economically practicable.

(h) Unnecessary derivatisation (blocking group, protection/deprotection, temporary modification of physical/chemical processes) should be avoided wherever possible.

(i) Catalytic reagents (as selective as possible) are superior to stoichiometric reagents.

(j) Chemical products should be designed so that at the end of their function, they do not persist in the environment, but break down into innocuous degradation products.

(k) Analytical methodologies need to be developed further to allow for real-time in-process monitoring to minimise the potential for chemical accidents, including releases, explosions, and fires.

The current emphasis on the production of very high-value products has led to a complete rethinking of the way in which processes are carried out. There are considerable gains to be achieved by carefully controlling the conditions in a chemical reactor in order to minimise the formation of unwanted products. In many cases, the product itself has an inhibitory effect on the progress of the reaction, and considerable gains in productivity can often be achieved in combining the reaction and separation stages into a single unit. The concept is not new in the sense that reactive distillation, in which, in effect, a chemical reaction takes place within a column with continuous separation of the products, has been practiced for many years. The technique is now being applied over a far wider range of conditions. Similar ideas have been recently reinforced and expanded by Anastas and Zimmerman.[15]

12.1.5 New Processing Techniques

Although the general principles of separation processes are applicable widely across the process industries, more specialised techniques are now being developed. Reference is made in the chapter on liquid–liquid extraction in Volume 2B to the use of supercritical fluids, such as carbon dioxide, for the extraction of components from naturally produced materials in the food industry, and to the applications of aqueous two-phase systems of low interfacial tensions for the separation of the products from bioreactors, many of which will be degraded by the action of harsh organic solvents. In many cases, biochemical separations may involve separation processes of up to 10 stages, possibly with each utilising a different technique. Very often, differences in both physical and chemical properties are utilised. Frequently the materials are stereo-isomers, differing only in the relative spatial orientations of the functional groups. With pharmaceutical products, near complete separation of the isomers is essential, one having the desired therapeutic properties and the other possibly being highly toxic, or even worse, as in the case of phthalidamide. Separations are then only possible by using 'molecular recognition' techniques and high resolution columns with the packing treated with a suitable ligand, to which one of the isomers will selectively attach itself. This is sometimes referred to as a 'lock and key' situation.

For many of the new highly specialised products, both reactors and separation stages need to be designed for low hold-ups and rapid processing, and these conditions may frequently be achieved by carrying out the processes under conditions of 'accelerated gravity' by the deployment of centrifugal in place of the normal gravitational field. Under such conditions, retention times and holdups are low—the latter being of particular importance when multipurpose plant is used and where shorter cleaning times minimise the loss of material when changing from one product to the next. Furthermore, reduction of hold-up may contribute to safer operation when the material has hazardous properties. The use of intensified fields for separation processes is not confined solely to gravitational fields. In many cases, differences in, say, electrical or magnetic properties may form the essential driving force for separation. Where significant differences in any single property are not sufficiently great, the use of combined fields (gravitational plus magnetic or electrical, for example) may provide the most satisfactory basis for separation.

In the following section, the use of intensified (gravitational) fields is given as an example of the way in which the operation of both reactors and separation units may be intensified, and smaller, more efficient, units may be developed.

12.2 Process Intensification

12.2.1 Introduction

In many unit operations such as distillation, absorption, and liquid–liquid extraction, fluids pass down columns solely under the force of gravity, and this limits not only the flow rates that may be attained, but also the rates of mass and heat transfer. The fluid dynamics (and the efficacy in terms of separation, rates of heat, and mass transfer) of these processes is influenced to a large extent by the magnitude of the term (gΔs), where Δs is the density difference between the two phases. In the limiting condition of when either $\Delta s \approx 0$ or $g \approx 0$, the behaviour of the system is governed by the surface forces. In this case, not only is bubble/drop coalescence suppressed, but also the countercurrent mode of contacting is not possible, and the size of bubbles/droplets may also become large. This lowers the rate of heat/mass transfer on both counts, namely, lowering the specific surface area as well as the interfacial shear. Thus, when a force other than gravity is utilised (such as a centrifugal force, for example), then, in theory, there is no limit to the force which can be applied nor indeed to the increase in heat and mass transfer rates that may be achieved. A reduction in residence times is also possible, and this leads to a decrease in the physical size of the plant itself. One fairly simple example of this reduction in plant size, or *process intensification* as it is termed, may be seen by comparing the relatively vast size of a settling tank with the very modest size of a centrifuge accomplishing the same task. The advantage of reducing plant size in this way has been prompted by the fairly recent requirement for processing units which are suitable for confined spaces, such as, for example, on oil-drilling rigs. There are also important benefits to be gained, however, from reducing the size of land-based units and not least in minimising intrusions on the environment. Rotating devices probably provide the most important way of achieving this processing intensification, and, where very thin films are produced on spinning discs, for example, these have the advantage that mass and heat transfer take place in this thin film. This means that the resistance to diffusion from the film to the bulk of the fluid is low. Similar considerations apply not only to physical operations, but also to those systems involving a chemical reaction.

It is the aim of this section to offer but a brief introduction to this relatively new field of endeavour that will surely have a not inconsiderable role in shaping the whole future of both physical and chemical processing. It may be noted at the outset, however, that it is not the use of centrifugal forces which is a relatively new development, but rather their application in spinning discs, and, indeed, centrifugal devices are already widely used in processing, as indicated by the following examples:

(a) Centrifugal fluidised beds are being used much more widely mainly due to the fact that the centrifugal field increases the minimum fluidising velocity and consequently increases the flow rates which can be handled. Much smoother fluidisation may also be achieved.

(b) In the chapter on liquid–liquid extraction in Volume 2B, various rotating devices are described which include the Podbielniak extractor, the Alfa-Laval centrifugal extractor, and the Scheibel column.

(c) The whole of Chapter 11 is devoted to a discussion of the use of centrifugal forces for carrying out the processes of settling, thickening, and filtration.

It is for this reason that the present chapter is, in essence, concerned with recent developments in which spinning discs are the dominant feature. Lest the impression be gained that process intensification (PI) is restricted only to reactions and separations, indeed, PI includes reduction in energy usage, capital investment, and raw material usage, on one hand, and increased process flexibility, process safety, improved product quality, not to mention better environmental performance, on the other. Thus, the philosophy of PI applies to all segments and aspects of chemical plants.[16, 17]

12.2.2 Principles and Advantages of Process Intensification

Process intensification, pioneered by Ramshaw[17, 18] in the 1980s, may be defined as a strategy which aims to achieve process miniaturisation, reduction in capital cost, improved inherent safety and energy efficiency, and often, improved product quality. In recent years, process intensification has been seen to provide processing flexibility, just-in-time manufacturing capabilities, and opportunities for distributed manufacturing. In order to develop a fully intensified process plant, it is essential that all the unit operation systems, that is, reactors, heat exchangers, distillation columns, separators, and so on, should be intensified. Wherever possible, the aim should be to develop and use multifunctional modules for performing heat transfer, mass transfer, and separation duties. As discussed by Stankiewicz and Moulijn,[19, 20] process intensification encompasses not only the development of novel, more compact equipment, but also intensified methods of processing which may involve the use of ultrasonic and radiation energy sources.

Additional benefits of process intensification include improved intrinsic safety, simpler scale-up procedures, and increased energy efficiency. Adopting a process intensification approach can substantially improve the intrinsic safety of a process by significantly reducing the inventory of potentially hazardous chemicals in the processing unit. A further advantage of process intensification is that it allows the replacement of batch processing by small continuous reactors, which frequently give more efficient overall operation, especially in the case of highly exothermic reactions where heat can be rapidly removed continuously as it is being released. The inherent safety aspect of process intensification, and its role in minimising hazards in the chemical and process industries, has been discussed in a survey article by Hendershot,[21] who has suggested that the design considerations to be taken into account in intensifying a process include:

(a)   Is the process based on batch or continuous technology?
(b)   What is the rate-limiting step — heat transfer, mass transfer, mixing, and so on?
(c)   What are the appropriate intensification tools, modules, and concepts?
(d)   Is it possible to eliminate solvents?
(e)   Is it possible to use supported catalysts?
(f)   Can pressure and temperature gradients be reduced?
(g)   Can the number of processing steps be reduced by using multifunctional modules?
(h)   Does it achieve the ultimate aim of enhancing transport rates by orders of magnitude?

Process intensification may also be seen as an ideal vehicle for performing chemical reactions based on what is known as "green chemistry" as discussed in Section 12.1.4. Intensification can provide appropriate reactor technologies for utilising heterogeneous catalysis, phase transfer catalysis, supercritical fluid chemistry, and ionic liquids. Process intensification allows chemical processes to be accelerated by using reactors such as spinning disc reactors, heat exchanger (HEX) reactors, oscillatory baffle reactors, microwave reactors, microreactors, cross-corrugated membrane reactors, and catalytic plate reactors.[17] For example, intensification may permit the use of higher reactant concentrations, giving significantly beneficial effects with regard to the kinetics, selectivity, and inventory. Often, due to limitations attributable to the heat and mass transfer resistances and inadequate mixing in the reactor, the effective concentrations of reactants are reduced, resulting in slow rates, poor selectivity, and the need for extensive downstream separation processing. Process intensification, thus, presents a range of exciting processing tools and opportunities in processing, which have seldom been explored in the past. It now offers the following characteristics:

(a)   It gives every molecule approximately the same processing experience.
(b)   It matches the mixing and transport rates to the reaction rate.
(c)   Significant enhancements are offered in heat and mass transfer rates.
(d)   The reaction rate is limited only by the design and performance of the equipment.
(e)   Selectivity and yield are both improved.
(f)   Product quality and specification are both improved.
(g)   A rapid grade change is possible because of the low hold-up and ease of cleaning the equipment.
(h)   A rapid response to set-points is possible.
(i)   For certain processes, the laboratory-scale equipment may constitute the full-scale unit.

On the aesthetic side, it is likely that intensified process plants will be less intrusive on the environment, making them far less of an eyesore than the unsightly and massive constructions that are characteristic of present processing units making up the entire chemical plant. In some cases, the plant may be mobile, thereby offering the opportunity for distributed manufacturing of chemicals close to the point of utilisation. This may reduce the quantities of hazardous products currently being transported by road and rail, thereby improving safety. The improved energy efficiency obtainable in intensified unit operations constitutes yet another highly attractive benefit in a world where there is overwhelming concern over the ever-growing demand on nonrenewable energy resources, and also over the release of greenhouse gases such as carbon dioxide. In this respect, there is a great and urgent need for the development of new process technologies, which will utilise energy in an efficient manner, and process intensification may represent a positive step in this direction. Large enhancements in the rates of heat and mass transfer, two of the most fundamental and frequently encountered in process engineering, can be achieved in intensified units. Such improvements could permit processing times and the associated energy consumption to be dramatically reduced for a given operation. An additional advantage in the case of packed bed units is that the operating range is considerably increased by the use of a centrifugal force. This may be seen by an examination of Fig. 7.17, where it is quite clear that an increase in "g," which appears in the ordinate, will bring many operating points well into the region below the flooding curve.

Green et al.[22] have discussed and detailed a series of possible intensified processes that include nitration in a compact heat exchanger, a gas–liquid reaction using static mixers, and hypochlorous acid production in a rotating packed bed. All of these clearly illustrate the benefits of process intensification that may be achieved in real processing situations. Keller and Bryan[23] have suggested that process intensification will dictate the future advancement of the chemicals and process industries. The process benefits that may be achieved based on green chemistry are shown in Fig. 12.3.

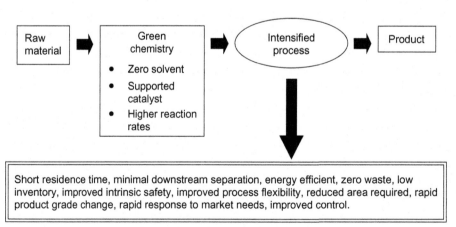

Fig. 12.3
Process characteristics of an intensified plant using green chemistry.

12.2.3 Heat Transfer Characteristics on Rotating Surfaces

Introduction

Heat transfer has been identified by Reay[24] as an important area in which process intensification is expected to offer major benefits in terms of energy efficiency, pollution control and plant operating costs. So-called *passive* techniques, including modifying the walls of a plant unit, for example, are routinely used to improve heat transfer coefficients in evaporation and condensation and to raise critical heat fluxes. The use of *active* methods, which offer high potential rewards in terms of efficiency and compactness, has been less well explored and less extensively applied. Of the several active methods in operation, the use of high gravity fields created by rotation is potentially the most rewarding because this offers the following advantages over other active techniques such as stirring, scraping, or vibration:

(a) Variable rotation speed offers a further degree of freedom in exchanger design and operation.
(b) The increased '*g*' coupled with built-in surface roughness factors, enhances film processes.
(c) Because there is a self-cleaning action, rotating devices can handle liquids containing solids.
(d) Reduced fluid residence times in the heating zone permit the processing of heat-sensitive fluids.

Heat transfer studies on smooth rotating surfaces have shown that, with thin films, heat transfer rates may be significantly enhanced, although Brauner and Maron[25] have shown that, where a fluid film flows over a surface, ripples may develop, and these may be responsible for a marked improvement in the rates of both the heat and mass transfer. Jachuck and Ramshaw[26] have suggested that surface irregularities might enhance the heat transfer characteristics of a rotating surface still further and have investigated this proposal.

Experimental tests and results

Jachuck and Ramshaw[26] have carried out tests on the rotating disc heat exchanger, shown in Fig. 12.4, which included one base disc and four top discs, thus, allowing a degree of flexibility in studying various surfaces. These included normal groove (shown in Fig. 12.4A), reentry groove (shown in Fig. 12.4B), metal-sprayed, and smooth discs. The normal groove disc had seven concentric grooves of a geometry that promoted and also created instabilities in the flow by generating surface waves at each of the groove sites. The metal-sprayed disc was coated with an aluminium-bronze composite powder, and the smooth surface disc was identical to that of the other top discs, except that the upper surface had no grooves or metal coating.

Thermocouples were connected to a data acquisition system by way of a slip-ring assembly incorporating a protective shroud, as shown in Fig. 12.4, which was cooled by compressed air.

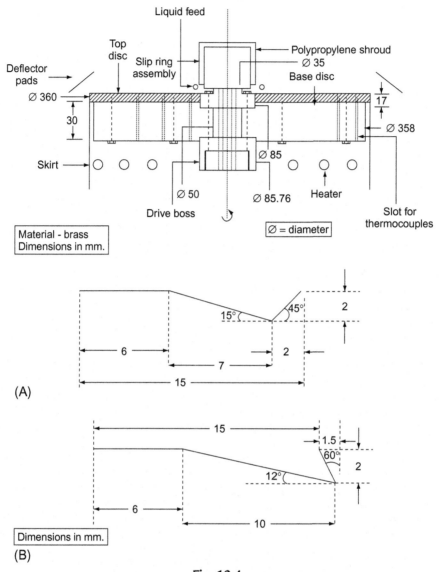

Fig. 12.4
Rotating disc heat exchanger.[26] (A) Normal groove. (B) Reentry groove.

The disc speed was recorded by an analogue tachometer, and the bulk liquid temperature was measured by pressing the edge of a thin rubber strip against the surface of the disc. Due to its high velocity, the liquid was forced up the side of the rubber strip at the point of contact where there was a build-up of the liquid, large enough for its temperature to be measured by a thermocouple. This method was used because the film was too thin for a thermocouple to be inserted directly into the liquid without it touching the disc surface and thereby giving an incorrect reading. It is estimated that this technique was accurate to within 0.1 deg. K.

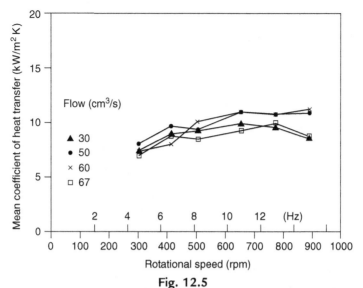

Fig. 12.5
Heat transfer characteristics of the metal-sprayed disc.[26]

Knowing the heat flux, the disc surface temperature, and the liquid film temperature, it was possible to calculate the average and local heat transfer coefficients for a variety of flow conditions. For the *metal-sprayed disc*, the results obtained are summarised in Fig. 12.5. For a given rotational speed, increasing the feed flow rate from 30 to 67 cm^3/s resulted in an increase in the mean heat transfer coefficient, as shown in Fig. 12.5. Jachuck and Ramshaw[26] suggested that the heat transfer coefficient is dependent on both the liquid film thickness and the surface waves, or instabilities in the liquid film. It was considered that the best results would be achieved for conditions where thin films with large instabilities were formed to give high shear mixing. For a flow of 30 cm^3/s, the value of the coefficient decreased at rotational speeds in excess of 10.8 Hz (650 rpm), probably because extremely thin films flowed smoothly over the disc surface without generating any surface waves. At low rotational speeds, it was thought that a combination of reasonably thin films and large surface waves were responsible for a steady increase in the heat transfer rates as the rotational speed was increased. For a flow rate of 50 cm^3/s, it was noted that the heat transfer coefficient increased linearly up to 10.8 Hz (650 rpm) and then remained almost constant with further increase in the rotational speed. This effect may be due to a combination of thin films and surface waves increasing the heat transfer rate at lower rotational speeds, while, at higher rotational speeds, the surface waves decreased, and, therefore, the average heat transfer coefficient, thought to be dependent on both the surface wave and film thickness, increased only very slightly with an increase in the rotational speed. It was expected that beyond a particular rotational speed, the heat transfer coefficients would drop, and this occurred at a flow rate of 30 cm^3/s. It is important to note that this decrease in the average heat transfer coefficient was observed not to be due to dry spots. At a flow of 67 cm^3/s, the average heat transfer coefficient increased linearly with increased rotational speed, suggesting that the surface waves play an important role in the heat transfer performance of thin films on rotating discs. It was expected, however, that for a flow rate of 67 cm³/s, the average heat transfer coefficient would drop at very high rotational speeds, when the surface waves ceased to exist. Results for a flow of 70 cm³/s suggested that, for a given rotational speed, the average heat transfer coefficient obtained was less than that for 67 cm³/s. This suggests that there is a cut-off point in the feed flow rate, and, therefore, a compromise between the film thickness and the formation of surface waves should be made in order to achieve the best results.

Currently, there are no correlations available between the surface wave function, the film thickness, and the average heat transfer coefficient that successfully describe the experimental results. The increase in the average heat transfer coefficient for an increasing rotational speed may be due to better shear mixing, resulting from thinner films, and smaller and more concentrated surface waves. Similar phenomena have been observed by both Elsaadi[27] and Lim,[28] who explained the existence of the maximum value of the mass transfer coefficient by suggesting that, by increasing the flow rate, the film thickness would increase, thereby creating waves that would induce progressively more efficient mixing in the film. When the film thickness was increased beyond some optimum value, the waves would be unable to exert the levels of mixing required for the higher mass transfer rates.

The results from tests with the *normal grooved discs*, shown in Fig. 12.6, indicate that the average heat transfer coefficient increased with an increase in the liquid flow rate, although only up to a certain point, above which the films were too thick, and, therefore, the heat transfer

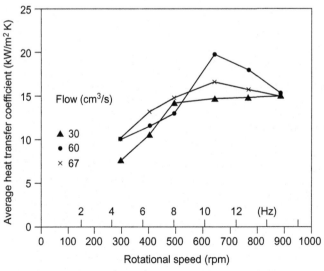

Fig. 12.6
Heat transfer characteristics of the normal grooved disc.[26]

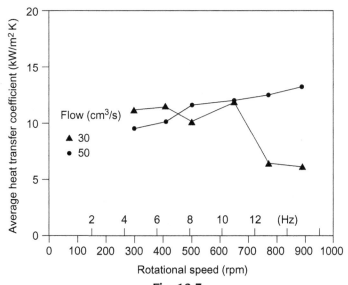

Fig. 12.7
Heat transfer characteristics of the reentry disc.[26]

performance decreased. The coefficients with the grooved disc were higher than those obtained with the metal-sprayed disc, and this may be due to better mixing and the creation of surface waves by continual creation and breakdown of the boundary layer. The peak in the heat transfer profile may be due to the forward-mixing effect, although, ideally, a grooved disc should be operated under conditions where forward mixing does not take place. It may be that, for a viscous liquid melt, such as a polymer, higher rotational speeds could be used before the peak in the profile was experienced. Test results with the reentry type of grooved disc are shown in Fig. 12.7. It was observed that a considerable proportion of the liquid was being thrown off the disc, due to the reentry effect of the disc. A proportion of the liquid that experienced a hydraulic jump, discussed in Volume 1A, Chapter 3, at the grooves caused a forward-mixing effect, though most of the liquid was thrown off the disc. It may be concluded that the reentry groove design will be suitable for denser liquids, as it can provide effective mixing and create surface waves. Test results for the *smooth disc*, shown in Fig. 12.8, again confirm that the heat transfer coefficient was dependent on both the surface waves and the film thickness of the liquid, and that there was clearly an optimum flow rate above which there would be an adverse effect on the mean heat transfer coefficient. At higher rotational speeds, there was a sharp decrease in the performance of the metal-sprayed and the reentry discs, although the normal grooved disc continued to perform well. Even though the performance of the metal-sprayed and the reentry discs decreased at higher rotational speeds, they performed considerably better than the smooth disc.

A stroboscope operating at twice the rotational speed of the disc was used to obtain a clear view of the flow pattern on the spinning disc. Photographs were taken to study the flow behaviour.

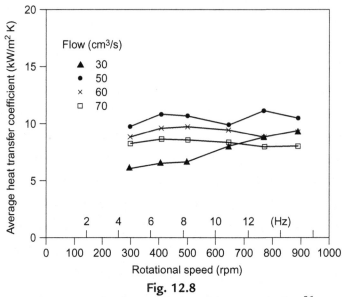

Fig. 12.8

Heat transfer characteristics of the smooth disc.[26]

On increasing the speed of rotation, the films became thinner, and, in addition, the number of surface waves increased. At lower rotational speeds, the surface waves were larger, whereas at higher speeds, the wavelength of the waves was reduced, and several smaller waves were generated. This effect created larger instabilities and, therefore, enhanced the heat transfer performance. For a given flow rate and speed of rotation, surface instabilities were a minimum for the smooth disc and a maximum for the grooved discs. It was difficult to differentiate visually between the performance of the normal and of the reentry grooved disc, although, at lower rotational speeds, the reentry disc generated more waves than the normal grooved disc. This was probably the reason why the reentry grooved disc performed better than the normal disc at very low rotational speeds. At higher speeds, the advantages of the reentry disc were undermined, due to the forward-mixing effect. It was suggested that the presence of surface imperfections clearly promoted the formation of a large number of small waves, which created instabilities in the film and enhanced the heat transfer rate.

In summary, this investigation showed that:

(a) Heat transfer coefficients were dependent on both the liquid-film thickness and the nature of the surface waves, or the instabilities, in the liquid film. Typically, the best results were achieved for conditions that resulted in thin films with large instabilities, which ensured high-shear mixing.

(b) Mean heat transfer coefficients as high as $18\,kW/m^2\,K$ were achieved by using a grooved rotating surface.

(c) For a given flow rate, an increase in the rotational speed increased the average heat transfer coefficient, although at very high rotational speeds, the heat transfer coefficient decreased for the grooved and coated surfaces and showed very little change compared with that for a smooth disc.

(d) In general, a normal grooved disc performed better than the other tailored surfaces, although, at low rotational speeds, the reentry groove disc seemed to perform better. At higher speeds, the performance of a reentry disc dropped because the liquid was thrown off the disc surface due to its extremely high velocity, as a result of which, cold liquid from the center of the disc mixed with that at the edge.

(e) Jachuck and Ramshaw[26] suggested that, with viscous liquids, a reentry disc would perform better than a normal grooved disc, even at higher rotational speeds.

12.2.4 Condensation in Rotating Devices

Introduction

Certain sectors of industry are seeking to use lightweight and corrosion-resistant compact heat exchangers for condensation as well as for convection duties. This requirement is of particular interest in the aviation, automobile, and domestic heating and ventilation industries. Also, recent interest in the concept of mobile chemical plants necessitates the use of lightweight compact heat exchangers. Such plants may well have an important role to play in the future of processing as they provide flexibility, improved inherent safety, and tighter process control. One way of reducing the weight of a heat exchanger is to use polymeric materials in their construction. Although many polymers are poor thermal conductors, and, with a wall thickness of 0.5–1 mm, the heat transfer performance can be adversely affected, a relatively new material, poly-ether-ether-ketone (PEEK) has many advantages in that it is a thermoplastic with a working temperature of up to 500 K, which is capable of being formed into a 100 µm thick film, can be easily corrugated, and, depending on the geometry, can withstand differential pressures of up to 1 MN/m^2. In addition, PEEK has attractive chemical resistance properties, making it suitable for application in chemically aggressive environments. Burns and Jachuck[29] have explored the performance of this material, in the form of cross-corrugated films, for the condensation of water in the presence of a noncondensable gas.

Experimental tests and results

Jachuck[30] describes a compact heat exchanger formed from corrugated sheets of PEEK, which provided a total surface area for heat transfer of approximately 0.125 m^2. The heat exchanger was mounted inside a thick Perspex box to minimise heat losses from the PEEK sheets. The gas stream was initially heated by use of heating tape wrapped around the piping and moisture was added to the gas by bubbling some or all of it through a bath of thermostatically controlled water. The pressure drop over the gas side was measured using a water-filled U-tube connected to the entrance and exit chambers of the heat exchanger to give a resolution of 10 kN/m^2 and a range of up to 8000 kN/m^2.

Heat transfer characteristics of the system were expressed in terms of an overall heat transfer coefficient and it was found that condensation of water vapor within the system was the major component in the transfer of energy from the gas side. Thermal effectiveness, as defined in Volume 1B, Chapter 1, was used to gauge the efficiency of the process. It was suggested that standard quantities, such as Nusselt number and number of transfer units, were meaningless for this system due to the heat generation from the condensation process in the presence of a noncondensable gas. The tests carried out focused on the effects of gas and liquid flow rates on heat transfer performance for two different packing orientations. The humidity of the input gas was close to saturation for all the tests, and the outlet air stream was also saturated with water vapor. The packing orientations used differed only in the angle at which the flow met the corrugation. The primary purpose of the tests was to examine the influence of gas flow rate on the heat exchanger performance. In conventional systems, increased gas flow rates result in well-defined increases in heat transfer performance through the improved forced convective mixing occurring at high Reynolds numbers. In the case of condensation processes, however, much lower gas flow rates are encountered, and hence, there are lower Reynolds numbers and less turbulent mixing. The results are reported by Burns and Jachuck[29] and are shown in Figs 12.9 and 12.10. Overall heat transfer coefficients in the range of $70\text{–}370 \text{ W/m}^2 \text{ K}$ were observed for the first packing configuration shown in Fig. 12.9 and $60\text{–}300 \text{ W/m}^2 \text{ K}$ for the second, shown in Fig. 12.10, with little significant difference in performance for the different orientations. The greatest influence observed for both packing configurations was the liquid flow rate, and, with the exception of the highest liquid flow rate in the first set of tests, little change was observed in the heat transfer coefficient as gas flow rate was increased. It was felt that, although changes in mixing patterns on the liquid side should be negligible at the low Reynolds numbers prevailing (<33) for this flow, variations in the temperature

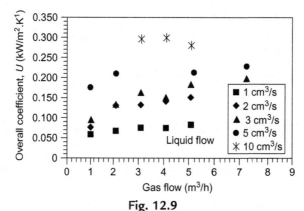

Fig. 12.9
Overall heat transfer coefficients in the first configuration.[29]

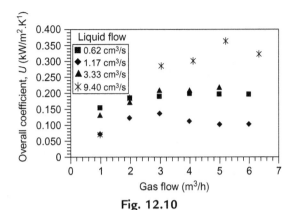

Fig. 12.10
Overall heat transfer coefficients in the second configuration.[29]

distribution over the packing surface caused by changing the liquid flow rate might, however, have influenced the condensation rate, and this would be the most likely explanation for the observed results. In summary, it may be concluded that heat transfer performance was not significantly influenced by corrugation orientation or gas flow rate, but was more strongly influenced by liquid flow rate over the range of conditions examined.

Thermal effectiveness values for the condensation process in both series of tests were in excess of 0.9 for most of these tests where the liquid/gas mass flow ratios were >2. These high values were partly due to the small change in liquid temperature when the gas temperature was allowed to approach that of the liquid input, but more importantly, due to the high heat transfer capability of the compact heat exchanger. For lower mass flow ratios, the effectiveness was only 0.7–0.9, due mainly to the greater warming of the cooling water. Values of the gas-side heat transfer coefficient which were calculated from the overall heat transfer coefficients were of the order of 63–520 W/m² K. Pressure drops for the gas flow across the unit were measured for normal operating conditions and compared with measurements taken when using low-humidity 'dry' air. This comparison showed a much higher pressure drop when condensation from the air was occurring, and this suggests that there was a significant hold-up of condensate within the gas flow passages. Visual analysis of flow in the upper gas layer of the heat exchanger was carried out, and photographs taken of the upper semitransparent PEEK sheet revealed the presence of condensate within this layer. Dark areas on the photographs showed where liquid had collected, and it was confirmed that drop-wise (rather than film-wise) condensation was taking place within the heat exchanger. It was found, in this respect, that a significant volume of liquid condensate was retained within the gas layer and that this decreased as the gas flow rate was increased.

In summary, although the influence of gas flow rate on heat transfer performance was shown to be small, it did exert a strong effect on the quantity of condensate remaining inside the heat exchanger. Liquid flow rate appeared to have a stronger influence on heat transfer performance, and it was suggested that the temperature distribution over the PEEK film might have an additional effect on the condensation process.

12.2.5 Two-Phase Flow in a Centrifugal Field

Introduction

As described by Bisschops et al.,[31, 32] centrifugal technology, using countercurrent contact of the process liquid with micrometre-range adsorbent particles, constitutes a new technique for carrying out adsorption and ion-exchange processes in a centrifugal field. Because the use of very small particles results in large interfacial areas and short distances for diffusion, the mass-transfer rates are extremely fast, and centrifugal adsorption equipment is usually much more compact than that used for conventional countercurrent processes, which operate as either as fluidised beds under gravity or as packed beds. A more detailed description of the advantages of centrifugal adsorption technology over existing techniques has also been published by Bisschops et al.[33, 34] A critical aspect of this technology is that very small adsorbent particles are moving countercurrent to the liquid flow. Because centrifugal force increases with radius, the critical condition for countercurrent flow occurs at the solids feed position in the contact zone where the centrifugal force is lowest. In order to evaluate the hydrodynamic capacity of the rotor, the relation between the centrifugal force and the flow rates of both the liquid and solid phases must be known. If the liquid flow exceeds a certain maximum, solids will be rejected at the entrance. This phenomenon has been called 'flooding' by Elgin and Foust.[35] As the adsorbent particles move through the countercurrent flow region toward the rim of the rotor, they will experience an increase in the centrifugal force, and, as Bisschops points out, this may affect the two-phase flow in either of the following two ways:

(a) The increase in the terminal settling velocity with increasing centrifugal force leads to an increase in the slip velocity between particles and liquid. Because the flow rates of the two phases do not change over the contact zone, this will lead to an increase of the void fraction toward the periphery of the rotor.

(b) The increase in the centrifugal force acting upon the particles results in a denser fluidised bed, in which the particles experience more hindrance due to the interparticle interactions. This leads to a decrease in slip velocity and an increase of particle concentration toward the periphery of the rotor.

The interfacial area in the contactor, which is directly related to the solids hold-up, strongly influences the mass transfer rate. To maximise the overall mass-transfer rate per unit volume of equipment, a high solids hold-up is necessary. On the other hand, the solids hold-up also influences the pressure drop over the contactor. The pressure drop has a hydrostatic and a dynamic component, both of which rise with increased solids hold-up. Because the adsorbent consists of extremely small particles, fluid friction between liquid and solids may lead to a relatively high dynamic pressure drop. The hydrostatic pressure drop is attributable to the density difference between the suspension in the contact zone and in the liquid.

di Felice[36] points out that liquid fluidised systems with large density differences between the particle and liquid phases may give rise to heterogeneous behaviour similar to that exhibited by gas-fluidised systems. According to Bisschops et al., similar heterogeneous fluidisation phenomena can occur in centrifugal systems, where the action of the centrifugal force is similar to that of increasing the apparent density differences between particles and liquid, although this phenomenon does not seem to have been reported for centrifugal systems. Because the occurrence of heterogeneous flow may be detrimental to the desired countercurrent flow in centrifugal adsorption technology, experiments must be carefully checked for fluidisation regimes.

Bisschops et al.[33, 34] have investigated the hydrodynamic capacity and solids holdup in countercurrent two-phase flow in the centrifugal field, as well as the relation between the pressure drop and void fraction. Moreover, the analysis included a check as to whether the two-phase flow in the centrifugal field was homogeneous or heterogeneous.

Experimental studies

Bisschops et al.[31, 32] adopted the following strategy in this investigation:

(a) Verification of the homogeneous two-phase flow model under gravity with two solid phases—relatively large particles (>1 mm) in water with a small density difference and relatively small particles (100 μm) in water with a large density difference.
(b) Verification of the model in the centrifuge, using solid phases in water with different density differences and particle diameters.

Tests under gravity were carried out with Maxazyme Gl Immob. biocatalyst particles in water, and small ballotini glass beads in tap water at ambient temperature. The expansion behaviour of the systems was investigated by fluidisation experiments in a Perspex column, shown in Fig. 12.11, which is equipped with water manometers at several heights. Tests with centrifugal force were carried out with two devices: (a) the low-*g* rotor and (b) the high-*g* rotor.

The low-*g* rotor, which is shown in Fig. 12.12, consisted of a rotating disc, fitted with two straight horizontal columns. The rotor, fitted with a rotary seal, was contained in a transparent Perspex drum to prevent leakage, and the liquid feed, liquid discharge, and solids feed were directed along the axis of rotation. The maximum speed of rotation was 4.2 Hz (250 rpm). As shown in Fig. 12.13, the solid phase was fed to the rotor as a slurry from a fluidised bed, and the solids left the rotor by way of the solids discharge, either in the form of a nozzle or as a backflow system from which they flowed to a separate settler.

For tests under higher centrifugal accelerations, the high-*g* rotor which was constructed, as shown in Fig. 12.14, from stainless steel which allowed high speeds of rotation to be used, although the maximum was limited to 42 Hz (2500 rpm). Solids were fed to the rotor in a

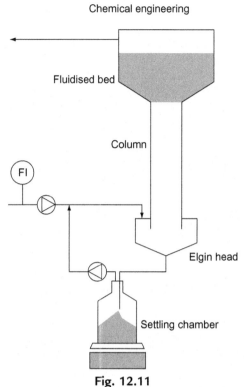

Fig. 12.11
Experimental system for measuring flooding characteristics under gravity.[31]

suspension containing approximately 30% solids by volume, and two geometries for the solids discharge and liquid distributor head at the rim of the columns were tested, as shown in Fig. 12.15. The first distributor was designed such that the solids passed the liquid inlet and were collected and discharged at the periphery of the rotor. In order to obtain a stable solids discharge without clogging, the liquid fraction in the solids discharge flow rate had to exceed a certain minimum, usually about 10% by volume. The second distributor was constructed with the solids discharge located closer to the axis of rotation than the liquid feed. This allowed a fluidised bed to accumulate in the contactor, from which a dense slurry could be extracted from the rotor. Initially, the rotor was operated with a very high solids feed rate, and the surplus solids were continuously rejected in the liquid effluent.

Results

All the model systems investigated by Bisschops et al.[31] fell within the completely homogeneous fluidisation regime modelled by Gibilaro et al.[37] With the low-g rotor and the nozzle discharge, the results for flooding conditions obtained from the tests are shown in Fig. 12.16, in which u_L is the superficial liquid velocity, u_S is the superficial solids velocity, and u_0 is the terminal falling velocity of the particles in the gravitational field. For glass ballotini

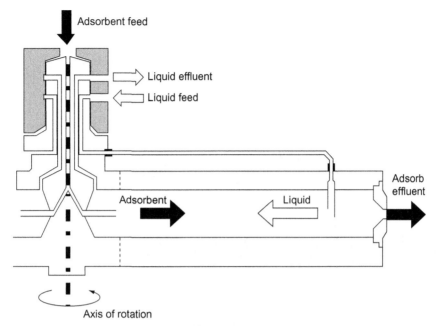

Fig. 12.12
Layout of the low-*g* rotor.[31]

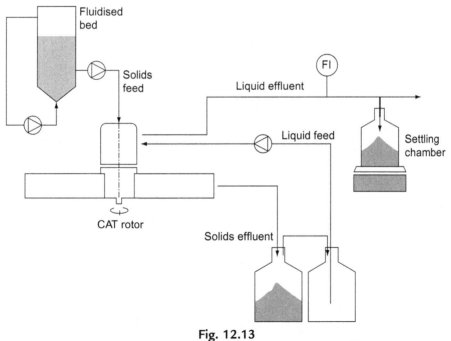

Fig. 12.13
Equipment for experiments with low-*g* rotor.[31]

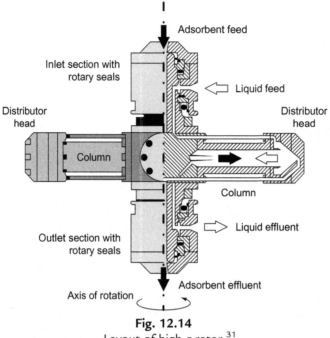

Fig. 12.14
Layout of high-*g* rotor.[31]

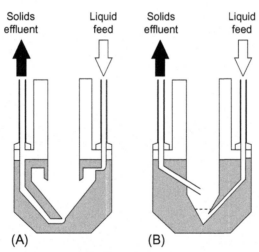

Fig. 12.15
Layout of distributor heads applied in experiments with high-*g* rotor.[31] (A) The Elgin-head distributor. (B) The fluidised-bed distributor.

beads, much higher capacities were achieved than those predicted for a single particle at the solids inlet position. The capacity of the low-*g* rotor was also tested with backflow of the solids discharge, and the results are compared with values predicted by the homogeneous flow model, as shown in Fig. 12.17. Again, the capacity of the rotor was higher than was expected. The

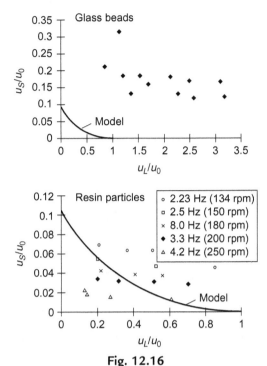

Fig. 12.16
Experiment flooding data obtained in the low-*g* rotor with nozzle discharge.[31]

capacity data, shown in Fig. 12.18, show that the predicted maximum flow would be exceeded in practice.

Bisschops et al.[31] developed a model for describing countercurrent two-phase flow in a centrifugal field, based on homogeneous flow conditions. When the model was tested under normal gravity conditions, it was found that, for systems with relatively low-density differences between dispersed and continuous phase, it described the observations very accurately. For small glass ballotini beads with a larger density, however, the model failed to describe the countercurrent flow pattern accurately. Large inhomogeneties were observed in this system, although the stability criterion of Gibilaro et al.[37] was met. In addition, the model did not accurately describe the hydrodynamic capacity of the centrifugal rotor. When the rotor was equipped with a nozzle discharge, however, the capacity for ion-exchange resin of relatively low density difference compared with water, was predicted to the correct order of magnitude, although the influence of the superficial liquid velocity was not taken into account, probably because the solids throughput depends entirely on the nozzle diameter, and the pressure drop acts across the solids discharge port. The low-*g* rotor also yielded much higher capacities than expected, possibly due to the formation of vortices, although deviations from the model became less as the centrifugal force was increased.

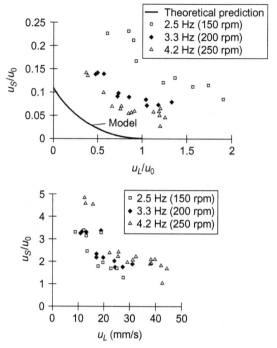

Fig. 12.17

Experimental flooding data for the ion-exchange resin ($d = 190\,\mu m$, $\rho_s = 1$, $250\,kg/m^3$) in the low-g rotor with backflow of the solids discharge.[31]

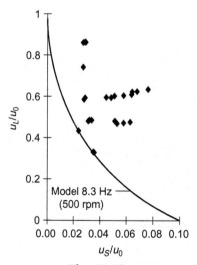

Fig. 12.18

Experimental results obtained in the high-g rotor in fluidized state under overload conditions at 500 rpm (8.3 Hz).[31]

Bisschops et al.[31] concluded that the stability criterion for fluidised beds, as derived by Gibilaro et al.,[37] cannot be applied to countercurrent systems, and that it is uncertain which mode of transport dominates in countercurrent flow in a centrifugal field. It was found that vortices may not appear if the flow is unrestricted in a tangential direction, and for scale-up of the centrifugal adsorption technology, a contact zone that covers the entire periphery is preferred. Although this might lead to a decrease in the capacity of the rotor, it could reduce the degree of back-mixing, and thereby improve the overall separation performance of the process.

12.2.6 Spinning Disc Reactors (SDR)

Performance characteristics

A further example of the benefits to be gained from process intensification is the spinning disc reactor, illustrated in Fig. 12.19. The effect of centrifugal force is to produce highly sheared thin films, as shown in Fig. 12.20, on the surfaces of rotating discs or cones. Extensive heat and mass transfer studies carried out by Jachuck and coworkers[38, 39] using spinning disc devices have shown that convective heat transfer coefficients as high as $14 \, kW/m^2 K$ and values of overall mass transfer coefficients $-K_L$, for the liquid phase, as high as $30 \times 10^{-5} \, m/s$ and K_G, for the gas phase, as high as $12 \times 10^{-8} \, m/s$, can be achieved while simultaneously providing micromixing and an appropriate fluid dynamic environment for achieving faster reaction kinetics. The disc may be 60–500 mm in diameter, and its surface may be smooth, grooved, or meshed, depending on the application and the required throughput. Rotational speeds may be

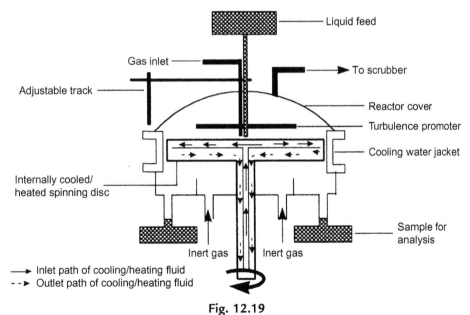

Fig. 12.19
Schematic diagram of a spinning disc reactor.[38]

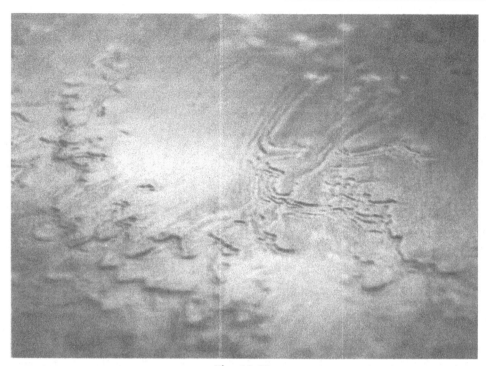

Fig. 12.20
View of sheared thin films on a spinning disc reactor.[38]

2–100 Hz (120–6000 rpm), and typically around 25 Hz (1500 rpm). The spinning disc reactor has been successfully used by Boodhoo and Jachuck[38, 40] to perform both free radical and condensation polymerisations, fast precipitation reactions for the production of monodispersed particles and catalysed organic reactions. As an example, Fig. 12.21 highlights the time saving that may be achieved by carrying out the polymerisation of styrene on a spinning-disc reactor. The characteristics of the device may be summarised as follows:

(a)   intense mixing in the thin liquid film,
(b)   short liquid residence time allowing the use of higher processing temperatures,
(c)   plug-flow characteristics,
(d)   high solid–liquid heat and mass transfer rates, and
(e)   high liquid–vapour heat and mass transfer rates.

Similarly, spinning disc reactors have been used by Wilson et al.[41] to carry out catalytic reactions using supported zinc triflate catalyst for the rearrangement of α-pinene oxide to yield campholenic aldehyde. The results of this study, presented in Table 12.1, suggest that, by using a supported catalyst on a spinning disc reactor, it is possible to achieve faster reaction rates, improved yield, and the elimination of the need for any downstream separation process for catalyst recovery.

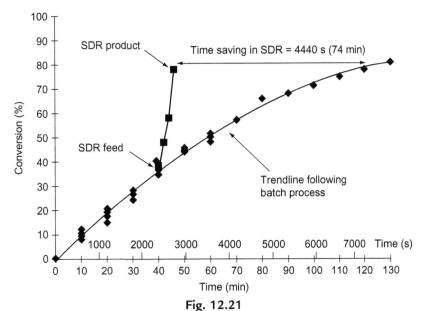

Fig. 12.21

Process time saving for the polymerisation of styrene in a spinning disc reactor with a disc of 1.08 m in diameter rotating at 6.7 Hz (400 rpm).[38]

Table 12.1   Performance of batch, catalysed spinning-disc reactors[41]

Rearrangement of α-Pinene Oxide to Campholenic Aldehyde	Batch Reactor	Catalysed SDR
Feed or total throughput (cm^3)	100	100
Conversion (%)	50	95
Yield (%)	42	71
Processing time (s)	900	17

Microreactors

Improved methods of manufacturing at the microscale are opening up new avenues for the development of compact devices for performing a range of chemical processes from reactions to extraction and separation. Equipment achieving rapid heat and mass transfer within the submillimetre scale channels offers the possibility of a low inventory environment with a high degree of control over the chemical process. Microreactors based on this concept, as described by Burns and Ramshaw,[42] can provide intrinsically safe environments for catalysed and noncatalysed fluid processing. One such example of this is the nitration of organic compounds in a PTFE capillary reactor, where rapid heat transfer rates allow higher acid concentrations to be used at lower temperatures resulting in the formation of less organic oxidisation byproducts (Fig. 12.22).

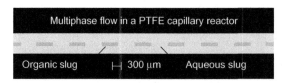

Fig. 12.22

PTFE capillary reactor used for nitration of organic compounds.[42]

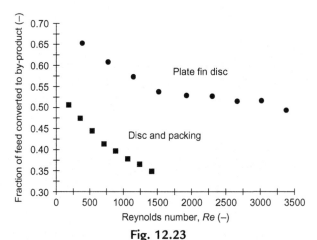

Fig. 12.23

Use of HEX reactors for reducing byproduct formation.[43] *Courtesy BHR Solutions.*

Phillips and Edge[43] describe the HEX reactors developed by BHR Solutions as another example of an intensified process unit which operates under steady-state conditions. This technology has been successfully used to demonstrate the reduction in byproduct formation for an exothermic organic reaction as a function of Reynolds number, as shown in Fig. 12.23, where, in calculating the Reynolds number, the length is taken as the disc diameter, and the velocity as the tip velocity of the disc. The use of high intensity gas–liquid mixers as described by Zhu and Green,[44] rotor-stator mixers, as described by Sparks et al.,[45] and tubular reactors with static mixers, as described by Schutz,[46] for performing chemical reactions with improved selectivity have also been successfully demonstrated. The catalysed plate reactor concept described by Charlesworth[47] is an example of yet another innovative process intensification unit in which effective heat transfer is achieved by performing exothermic and endothermic reactions on opposite sides of a catalysed plate. A schematic diagram representing the coupling of methane steam reforming and combustion of methane, is shown in Fig. 12.24. This technology enables the size of equipment to be reduced by several orders of magnitude, and eliminates NO_X emissions as the process is operated at lower temperatures.

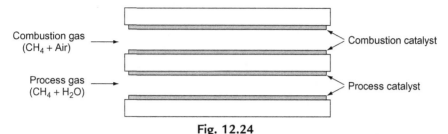

Fig. 12.24
Exothermic and endothermic reactions using a catalysed plate reactor.[44]

Cross-corrugated multi-functional membranes

Hall et al.[48] discuss the use of cross-corrugated membrane modules, illustrated in Fig. 12.25, as offering the potential for developing multifunctional units for performing both reactions and product separation in one miniaturised module. The use of microporous membranes in a reaction system is a relatively new method of solving various reaction–separation problems. Miscible fluids must be kept apart in order to control the overall reaction rate. Alternatively, the membrane may be designed to be permselective and to allow only a desired species to pass through it, such as the product of an organic synthesis, while holding back another, such as a byproduct, and unreacted feed. This type of application may be successfully applied for organic reactions in the manufacture of pharmaceuticals, cosmetics, agricultural chemicals, dyes and flavoring ingredients by using the process of phase-transfer catalysis which is discussed by Reuben and Sjoberg.[49] Fig. 12.26 shows a representation of the early phase-transfer work of

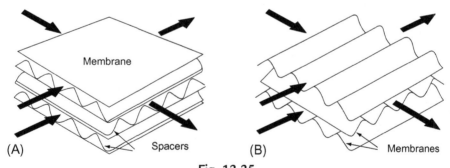

Fig. 12.25
Membrane flow cell (A) flat-sheet (B) cross-corrugated form.[48]

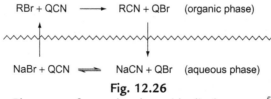

Fig. 12.26
Phase-transfer catalysed cyanide displacement.[50]

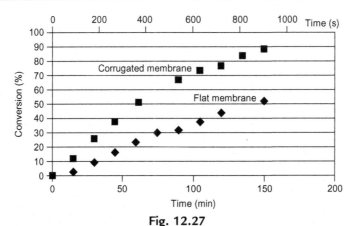

Fig. 12.27

Effect of corrugated membranes on oxidation reaction rates.[50]

Starks,[50] in which 1-chloro-octane was reacted with aqueous cyanide to produce 1-cyano-octane. In a batch reactor, with no phase-transfer agent present, no reaction occurs, although with the addition of a small amount of quaternary ammonium salt, the reaction goes to completion in just a few hours.

In industry, many of these reactions are currently carried out in stirred tanks, and the separation of the resulting mixture for product and catalyst recovery requires a significant energy input. Tests carried out in a cross-corrugated membrane module using phase transfer catalysts, such as tetra-butyl ammonium salts for the oxidation of benzyl alcohol, have produced encouraging results, as shown in Fig. 12.27. It may be that similar results can be obtained for other reactions such as alkylations, esterifications, oxidations and reductions, epoxidations, condensation reactions, polymerisations, and so on.

References

1. Danckwerts PV. Continuous flow systems. Distribution of residence times. *Chem Eng Sci* 1953;**2**:1. Measurement of molecular homogeneity in a mixture. *Chem. Eng. Sci.* **7** (1958) 116. The effect of incomplete mixing on homogeneous reactions. *Chem. Eng. Sci.* 8 (1958) 93. Gas absorption accompanied by chemical reaction. *AIChE J.* **1** (1955) 456.
2. Danckwerts PV. *Gas–liquid reactions.* New York: McGraw-Hill; 1970.
3. Fuller GW. *New food product development.* 2nd ed. Boca Raton, FL: CRC Press; 2005.
4. Litster J. *Design and processing of particulate products.* Cambridge: Cambridge University Press; 2016.
5. Windhab EJ. What makes for smooth, creamy chocolate? *Phys Today* 2006;**82**:81–3.
6. Cussler EL, Moggridge GD. *Chemical product design.* Cambridge: Cambridge University Press; 2001.
7. Dunnill P. The future of biotechnology. *Biochem Soc Symp* 1983;**48**:9–23.
8. Edwards MF, Instone T. *Particulate products—their manufacture and use.* Port Sunlight: Unilever Research; 2000.
9. Edwards MF. The importance of chemical engineering in delivering products with controlled microstructure to customers. *Inst Chem Eng* 1998;**8**: North Western Branch Paper.
10. Brockel U, Meier W. *Product design and engineering: formulation of gels and pastes.* New York: Wiley; 2013.

11. Hamley P, Poliakoff M. Green chemistry: the next industrial revolution? *Chem Eng* 2001;**721**:27.

12. Allen DT, Shonnard DR. *Green engineering: environmentally conscious design of chemical processes.* New Jersey: Prentice Hall; 2001.

13. Marteel-Parrrish AE, Abraham MA. *Green chemistry and engineering: a pathway to sustainability.* New York: Wiley; 2013.

14. Allen DT, Shonnard DR, Huang Y, Schuster D. Green engineering education in chemical engineering curricula: a quarter century of progress and prospects for future transformations. *ACS Sustain Chem Eng* 2016;**4**:5850–4.

15. Anastas PT, Zimmerman JB. Design through the twelve principles of green engineering. *Environ Sci Technol* 2003;**37**:94A–101A.

16. Semel J, editor. *Process intensification in practice: applications and opportunities.* London, UK: Mechanical Engineering Publications; 1997.

17. Reay D, Ramshaw C, Harvey A. *Process intensification.* 2nd ed. Oxford, UK: Butterworth-Heinemann; 2013.

18. Ramshaw C. Process intensification: a game for n-players. *Chem Eng* 1985;**416**:30–2.

19. Stankiewicz AI, Moulijn JA. Process intensification: transforming chemical engineering. *Chem Eng Prog* 2000;**96**(2):8.

20. Stankiewicz A, Moulijn JA. *Re-engineering the chemical processing plant.* New York: Marcel Dekker; 2004.

21. Hendershot DC. Process minimisation: making plants safer. *Chem Eng Prog* 2000;**96**(1):35–40.

22. Green A, Johnson B, John A. Process intensification magnifies profits. *Chem Eng (Albany)* 1999;**106** (13):66–73.

23. Keller GE, Bryan PF. Process engineering: moving in new directions. *Chem Eng Prog* 2000;**96**(1):41–50.

24. Reay DA. Heat transfer enhancement—a review of techniques and their possible impact on energy efficiency in the U.K. *Heat Recovery Syst CHP* 1991;**11**(1):1–4.

25. Brauner N, Maron DM. Modeling of wavy flow in inclined thin-films. *Chem Eng Sci* 1983;**38**(5):775–88.

26. Jachuck RJJ, Ramshaw C. Process intensification: heat transfer characteristics of tailored rotating surfaces. *Heat Recovery Syst CHP* 1994;**14**(5):475.

27. Elsaadi MS. *Heat transfer to thin film liquids on closed rotating discs systems.* M. Phil. Thesis, University of Newcastle upon Tyne, UK; 1992.

28. Lim ST. *Hydrodynamics and mass transfer processes associated with the absorption of oxygen in liquid films flowing across a rotating disc.* Ph.D. thesis, University of Newcastle upon Tyne, UK; 1980.

29. Burns JR, Jachuck RJJ. Condensation studies using cross-corrugated polymer film compact heat exchanger. *App Thermal Eng* 2001;**21**:495–510.

30. Jachuck RJJ. In: *Opportunities presented by cross-corrugated polymer film compact heat exchangers.* Proceedings of the international conference on compact heat exchangers and enhancement technology for the process industries, Banff, Canada; 1999. p. 243–50.

31. Bisschops MAT, van der Wielen LAM, Luyben KCAM. Hydrodynamics of countercurrent two-phase flow in a centrifugal field. *AIChE J* 2001;**47**(6):1263–77.

32. Bisschops MAT, van der Wielen LAM, Luyben KCAM. PCT patent application, PCT/NL 97/ 00121; 1997.

33. Bisschops MAT, van der Wielen LAM, Luyben KCAM. Operating conditions of centrifugal ion exchange. In: Grieg JA, editor. *Ion exchange developments and applications, IEX'96, SCI, London*; 1996. p. 297.

34. Bisschops MAT, van der Wielen LAM, Luyben KCAM. Centrifugal adsorption technology for the removal of volatile organic compounds from water. In: Semel J, editor. *Process intensification in practice, applications and opportunities.* London: BHR Group; 1999. p. 229.

35. Elgin JC, Foust HC. Countercurrent flow of particles through moving continuous liquid; pressure drop and flooding in spray-type liquid towers. *Ind Eng Chem* 1950;**42**:1127.

36. di Felice R. Hydrodynamics of liquid fluidization. *Chem Eng Sci* 1995;**50**:1213.

37. Gibilaro LG, Hossian I, Foscolo PU. Aggregate behavior of liquid fluidized beds. *Can J Chem Eng* 1986;**64**:931.

38. Boodhoo KVK, Jachuck RJJ. Process intensification: spinning disc reactor for styrene polymerisation. *Appl Therm Eng* 2000;**20**:1127.

39. Jachuck RJJ, Ramshaw C. Process intensification: spinning disc polymeriser (IChemE research event, first European conference, 1995).

40. Boodhoo KVK, Jachuck RJJ. Process intensification: Spinning disc reactor for condensation polymerisation. *Green Chem* 2000;**4**:235–44.

41. Wilson K, Renson A, Clark JH. Novel heterogeneous zinc triflate catalysts for the rearrangement of alpha-pinene oxide. *Catal Lett* 1999;**61**(1–2):51–5.

42. Burns JR, Ramshaw C. In: *A microreactor for the nitration of benzene and toluene.* Proceedings of the 4th international conference on microreaction technology, Atlanta, GA; 2000. p. 133–40.

43. Phillips CH, Edge AM. *The reactor–heat exchanger. 6th. Heat exchanger action group (HEXAG) meeting, Leatherhead, UK*; October 1996.

44. Zhu ZM, Green A. Use of high intensity gas–liquid mixers as reactors. *Chem Eng Sci* 1992;**47**:2847–52.

45. Sparks TG, Brown DE, Green A. Assessing rotor/stator mixers for rapid chemical reactions using overall power characteristics. BHR conference series. Publication 18. London: Mechanical Engineering Publications Ltd; 1995.

46. Schutz J. Agitated thin film reactors and tubular reactors with static mixers for rapid exothermic multiple reactions. *Chem Eng Sci* 1988;**43**(8):1975.

47. Charlesworth RJ. *The steam reforming and combustion of methane on micro-thin catalysts for use in a catalytic plate reactor.* Ph.D. Thesis, University of Newcastle upon Tyne; 1996.

48. Hall K, Scott K, Jachuck RJJ. In: *Mass transfer characteristics of cross-corrugated membranes.* Presented at ICoM '99, Toronto; June 1999.

49. Reuben B, Sjoberg K. Phase transfer catalysis in industry. *Chem Tech* May 1981;313–5.

50. Starks CM. Phase transfer catalysis I. Heterogeneous reactions involving anion transfer by quaternary phosphonium salts. *J Am Chem Soc* 1971;**93**:195.

Further Reading

Bisschops MAT, van der Wielen LAM, Luyben KCAM. In: Semel J, editor. *Process intensification in practice, applications and opportunities.* London: BHR Group; 1999.

Bisschops MAT, van der Wielen LAM, Luyben KCAM. In: Grieg JA, editor. *Ion exchange developments and applications, IEX'96.* London: SCI; 1996.

Cussler EL, Moggridge GD. *Chemical product design.* Cambridge: Cambridge University Press; 2001.

Favre E, Marchal-Heusler L, Kind M. Chemical product engineering: Research and educational challenges. *Chem Eng Res Design* 2002;**80**:65.

Ulrich KT, Eppinger SD. *Product design and development.* New York: McGraw Hill; 2017.

CHAPTER 13

Colloidal Dispersions

13.1 Introduction

As discussed in several chapters of this book, the particulate solids which are omnipresent, are handled in dry form, wet form, or as their dispersion in a fluid, depending on the requirements of the unit operations involved in their processing. Some common everyday examples of formulations and essential fluids that are particulate dispersions are paint, ink, blood, milk, etc. The discussion in this chapter is restricted to dispersions of solid particles, i.e., hard particles that are not deformable. In particular, the focus will be on very fine particulate dispersions, i.e., particles larger than molecules, but much smaller than macroscopic objects. If the sizes of the particles in the dispersions are in few nanometers to one-micrometre range, such fine particle dispersions are called colloids. If the particles are characterised by two or more physical dimensions (which is the case for nonspherical particles discussed in Chapter 15), at least one dimension must fall in the nm to μm size range to classify them as colloidal particles. Detailed discussions of the fundamentals of colloidal dispersions, such as interparticle interactions, stability, thermodynamic aspects, microstructure (spatial arrangement of particles), and transport properties, are available in several excellent books that serve both beginners and advanced readers.[1–7] This chapter deals with the basic knowledge and essential features of the science of colloidal dispersions and some aspects relevant to the fluid-particle processes discussed in this book.

13.2 Sources of Model Colloidal Particles

Particles of colloidal dimensions generated in industrial processes, vehicular emissions, and activities such as combustion of camphor have complex characteristics, i.e., different sizes and morphologies, as the conditions for their formation are not very well controlled. While there is a lot of interest and scope for studying fate and transport of aerosols (dispersions of particles in gaseous medium), design of hydrophobic and hydrophilic surfaces by soot particle coating, so on and so forth, they cannot be used as model particles to investigate fundamental principles of colloids. The term "model colloids" refers to the colloids of well-defined sizes, shapes, and surface properties formed under controlled conditions such that the particles in the dispersion are nearly identical in almost all aspects. As shown schematically in Fig. 13.1A, when all the particles are of identical size, a colloidal dispersion is said to contain *monodisperse* particles. The synthesis protocols can be tuned to

Coulson and Richardson's Chemical Engineering. https://doi.org/10.1016/B978-0-08-101098-3.00014-7

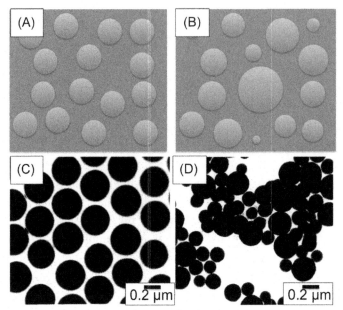

Fig. 13.1

Schematic representation (A, B) and experimental realisation (C, D) of monodisperse (A, C) and polydisperse (B, D) particles. The particles shown in (C) and (D) are silica particles synthesised by the Stöber method. *C and D: Reproduced with permission from Bogush GH, Tracy MA, Zukoski CF. Preparation of monodisperse silica particles: control of size and mass fraction. J Non-Cryst Solids 1988;**104** (1):95–106.*

obtain particles of varying size, or *polydisperse* particles, as shown in Fig. 13.1B and D, respectively. The electron microscopy images of monodisperse and polydisperse spherical silica particles prepared by a well-established Stöber method,[8,9] are shown respectively in Fig. 13.1C and D, which will be discussed further. Though the silica colloids shown in Fig. 13.1C appear to be monodisperse, the particle population shows a very narrow size distribution. In that sense, there are no true monodisperse particles; i.e., the size distribution can either be narrow or broad. Both monodisperse and polydisperse particles can be considered model colloids depending on the colloidal phenomena under investigation. For example, the study of crystallisation of colloids typically involves particles that are highly monodisperse, whereas the investigation of the effect of polydispersity on the packing of solids involves particles with tailored polydispersity.

13.2.1 Commercial Availability

Model colloidal particles of a wide range of sizes, shapes, and surface functionality are commercially available either as dispersions or in powder form. A series of silica nanoparticle dispersions which are sold as LUDOX HS-40, LUDOX HS-30, LUDOX TM-50, LUDOX AS-40, LUDOX AS-30, and LUDOX CL-30 can be purchased (Sigma Aldrich). These are

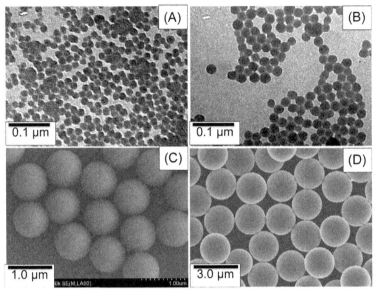

Fig. 13.2

The particles shown in (A) and (B) are transmission electron microscopy images of the silica particles in Ludox LS and Ludox TM nanoparticle dispersions which are commercially available. The scanning electron micrographs of spherical silica (procured from Fibre Optic Centre Inc.) and polystyrene (procured from Thermo Fischer) are shown in (C) and (D), respectively. *The transmission electron microscopy images are adapted with permission from J Phys Chem B **2009**;113 (11):3423–30, Copyright (2009) American Chemical Society; Images courtesy of Thriveni G. Anjali, IIT Madras.*

available as charge-stabilised aqueous dispersions. The number at the end of each product name corresponds to the weight % of particles in the dispersion; for example, LUDOX TM-50 contains 50% particles by weight. These dispersions are highly stable and have a narrow particle size distribution. The transmission electron microscopy images of silica nanoparticles in Ludox LS and TM shown in Fig. 13.2A and B reveal an average radius of 12.1 ± 1 and 22.4 ± 1.7 nm. Fig. 13.2C shows 890 ± 30-nm-diameter, commercially available silica spheres in dry power form (Fibre Optic Centre Inc.). The scanning electron microscopy image in Fig. 13.2D shows positively charged latex particles (amidine-functionalized polystyrene particles) of 2.2 ± 0.05 μm diameter. A large variety of polystyrene particles of size ranging from sub 100 nm to tens of microns are now commercially available (Thermo Fisher, Sigma Aldrich), including particles that are magnetically functionalized, fluorescently labelled, and differently surface-functionalized with carboxyl groups, sulfate groups, etc. In addition to the particles shown in Fig. 13.2, spherical and nonspherical particles of various sizes and shapes are also readily available, example include, titanium dioxide (TiO_2) spheres and nanotubes, single and multiwalled carbon nanotubes, graphene nanoplatelets, graphene oxide, quantum dots, aluminium oxide (Al_2O_3), iron oxide (Fe_3O_4), gold nanospheres and nanorods, silver nanospheres and nanowires. Some of the commercially available particles are highly expensive. Due to the contributions of several colloid chemists, most notably, Egon Matijević,[10] particles of uniform size, shape, and surface chemistry can easily be prepared using the established synthesis protocols at much lower costs; some of these well known procedures are briefly discussed in the following sections.

13.2.2 Synthesis of Colloids

Colloidal scale particles can be prepared using several techniques that can be broadly classified as top-down and bottom-up approaches. In the top-down approach, a bulk solid is converted into smaller particles by high-energy ball milling, lithography, etching, and the electric synthesis route,[11] etc. On the other hand, in the bottom-up approach, the constituent molecules or atoms are assembled for example, by processes such as nucleation and growth, to give rise to particles that are one or more orders larger in size than the atomic length scale. Detailed procedures for the bottom-up approach for the synthesis of model inorganic, metallic, metal oxide, polymeric, and other types of colloids are well documented.[12–15] The methods described in the literature and their variations are far too numerous, and this is clearly beyond the scope of this chapter; therefore, we restrict our discussion here to a select few well-established methods.

Stöber process for silica (SiO_2) colloids

The chemical reagents used for the Stöber synthesis of silica colloids are tetraethyl orthosilicate (TEOS), ethyl alcohol (C_2H_5OH), water (H_2O), and ammonia (NH_3).[8,9] The first step involves the reaction of TEOS with H_2O in ethanolic aqueous solution, which is a hydrolysis reaction catalysed by NH_3. Following this, a condensation reaction occurs, leading to the conversion of silanol molecules to monodisperse silica (SiO_2) colloids. For a detailed synthesis procedure and discussion of the underlying mechanism, the reader is referred to the works of Pontoni et al.[16] and Han et al.[17] Bogush et al.[9] carried out more than one hundred 25°C syntheses by considering reagents in the concentration range of 0.1–0.5 M TEOS, 0.5–17.0 M H_2O, and 0.5–3 M NH_3, and reported the following correlation for the prediction of final particle diameter, D_p, in nanometers[9]:

$$D_P = A[H_2O]^2 e^{-B[H_2O]^{\frac{1}{2}}} \tag{13.1}$$

$$A = [TEOS]^{\frac{1}{2}}(85 - 151[NH_3]) + 1200[NH_3]^2 - 266[NH_3]^3 \tag{13.2}$$

$$B = 1.05 + 0.523[NH_3] - 0.128[NH_3]^2 \tag{13.3}$$

In Eqs. (13.1)–(13.3), the units of the concentration of the reaction species and the reaction media is mol/L. A transmission electron microscopy image of nearly monodisperse silica nanoparticles of diameter 253 ± 11 nm synthesised by the Stöber method is shown in Fig. 13.3A.

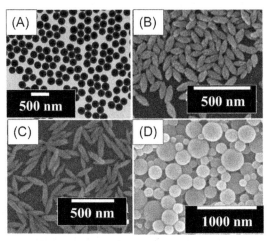

Fig. 13.3

(A) Transmission electron microscopy images of silica nanoparticles of diameter 253 ± 11 nm synthesised using the Stöber method. Scanning electron microscopy images in (B) and (C) show haematite ellipsoids of aspect ratio 2.1 and 4.2, respectively, synthesised by forced hydrolysis. (D) Scanning electron microscopy image shows polydisperse polystyrene particles synthesised by nanoprecipitation method. *A: Image courtesy—Dr. S Manigandan, NIT Warangal; B, C: Reprinted with permission from Lama H, Basavaraj MG, Satapathy DK. Desiccation cracks in dispersion of ellipsoids: effect of aspect ratio and applied fields. Phys Rev Mater 2018;**2**:085602; D: Image courtesy—Ashna Rajeev, IIT Madras.*

Recently, the Stöber process for silica growth occurring inside emulsion droplets has been exploited to synthesise uniform rod-like silica colloids.[18] This modified synthesis procedure offers several advantages such as large yield and tunability of particle aspect ratio (defined as the ratio of the length of the major axis of particles to the length of the minor axis).

Metal oxide particles by forced hydrolysis

Highly uniform metal oxide particles with very narrow size distribution can be prepared by aging aqueous solutions of the respective metal salts at elevated temperature. This method is called the forced hydrolysis method. Metal oxide particles of different sizes, morphologies (shape), and surface chemistries have been synthesised using this technique. Matijević pioneered the application of this method for the synthesis of uniform colloids and investigation of their colloidal behaviour.[12,13] One of the simple forced hydrolysis procedures for the synthesis of spindle or needle-like haematite (α-Fe$_2$O$_3$) particles involves aging a known volume (50 mL or 500 mL) of 0.020 M aqueous ferric chloride (FeCl$_3$) solution containing sodium dihydrogen phosphate (NaH$_2$PO$_4$). The aspect ratio of the particles can be tuned by changing the concentration of NaH$_2$PO$_4$ in the range of 1.0×10^{-4} to 4.5×10^{-4} M. Typically, the aqueous reaction mixture containing FeCl$_3$ and NaH$_2$PO$_4$ is introduced into a Pyrex bottle, tightly stoppered, and kept in an oven preheated to 100°C for 2 days or 7 days.[19] This procedure yields haematite spindles of aspect ratio >1 to $\sim$6.

A slightly modified procedure in which the forced hydrolysis of iron (III) perchlorate salt is carried out in the presence of urea is shown to yield haematite spindles of aspect ratios up to 10.[20] The formation of nonspherical particles is ascribed to the adsorption of phosphate ions on the haematite nuclei formed in the initial stages of the reaction. The adsorption of phosphate ions prohibits the isotropic growth of the particle and preferentially directs the particle growth in only one direction. The scanning electron microscopy images in Fig. 13.3B and C show haematite ellipsoids of aspect ratios 2.1 and 4.2, respectively, synthesised by this technique.

Nanoprecipitation

Nanoprecipitation is a simple one-step process that has been established for the synthesis of polymeric particles and is extensively used for encapsulation of active ingredients in pharmaceutical formulations. In general, particle synthesis by nanoprecipitation requires three components: a polymer and two liquids that are completely miscible. One of the liquids must be a good solvent for the polymer, and the other must be a nonsolvent. The polymer micro- and nanoparticles precipitate when a small quantity of polymer dissolved in a good solvent is added to a nonsolvent under the conditions of continuous agitation or mixing. When a drop of polymer solution is added to the nonsolvent, the polymeric particles are formed due to the interfacial deposition of the polymer as the solvent is displaced with the nonsolvent.[21–23] Solvents used for the precipitation of polystyrene particles include tetrahydrofuran (THF), 1, 4-Dioxane, acetone, and 2,6-lutidine; water is used as a nonsolvent.[23] One of the disadvantages of conventional nanoprecipitation is the high degree of size polydispersity of the particles formed. Several modifications such as coupling dialysis with nanoprecipitation[21] and flash nanoprecipitation[22] have been suggested to enable the formation of particles with narrow size distribution. The scanning electron microscopy image in Fig. 13.3D shows polydisperse polystyrene particles synthesised by the nanoprecipitation method.

Polystyrene (PS) and poly(methyl methacrylate) (PMMA) particles by polymerisation

PS and PMMA particles can be prepared by the polymerisation of the respective monomers, i.e., styrene and methylmethacrylate. Dispersion and emulsion polymerisations are widely used for this purpose, and the polymerisation reaction involves the use of monomer, initiator, solvent (both aqueous and organic solvents have been used), and suitable stabilisers.[24–29]

Almog et al.[24] reported a simple method for the formation of PS and PMMA spheres in which the polymerisation of the monomers is carried out in an alcohol using a combination of steric and charge stabilisers. A series of alcohols such as methanol, ethanol, isopropanol, t-butanol, etc., and vinyl alcohol/vinyl acetate copolymer stabilisers such as polyvinylpyrrolidone (PVP), poly(methyl vinyl ether) and poly (vinylpyrrolidone vinylacetate) copolymer and other stabilisers are used. A known quantity of styrene containing a small initiator azobisisobutironitrile (AIBN) is added to a known volume of alcohol in which a small quantity of stabiliser is present. An inert gas such as argon or nitrogen is bubbled through the mixture to remove air to ensure that the reaction occurs in an inert atmosphere. The polymerisation reactions are carried out in a constant temperature bath in a three-neck round-bottom (RB) flask with a provision to stir the reaction mixture, add reagents and purge the inert gas. The mean diameter of the particles and their size distribution can be controlled depending on the nature of the alcohol and the concentration of the other reagents. A seeded growth procedure for the synthesis of larger particles is also proposed by Almog et al.[24]

13.3 Consequence of Small Size

For colloidal size particles, the inertial forces are invariably negligible. The collision of the particles with the molecules of the surrounding fluid in which they are dispersed results in a chaotic motion of the particles. This random motion, ubiquitous to tiny particles dispersed in a fluid, is called the Brownian motion. Such particles are also called Brownian particles. The Brownian-motion-driven fluctuations in the position of the particle are due to the thermal energy, $k_B T$, of solvent molecules, where k_B is the Boltzmann's constant ($=1.38 \times 10^{-23}$ J/K) and T is the absolute temperature. When the particle size becomes larger, the thermal fluctuations become negligible, especially when the diameter of the particle is larger than about 5 μm, and such suspensions are said to be non-Brownian.

13.3.1 Settling of Particles

In Chapter 6, the motion of a single particle through a fluid under the action of an external force is discussed. Due to the density difference between the particle and the fluid, the settling of solid particles (or rising of drops or bubbles) is expected under the influence of gravitational force. However, when the particle under consideration is in the colloidal size range, the Brownian force (F_B) given by[3]:

$$F_B = \frac{k_B T}{D_p} \tag{13.4}$$

where D_p is the characteristic dimension of the particle (for spherical particle, D_p is the diameter of the particle), becomes comparable to the net gravitational force (F_g) acting on the particle, which is given by[3]:

$$F_g = V_p (\rho_p - \rho_f) g \tag{13.5}$$

where V_p is the volume of the particle, ρ_p is the density of the particle, ρ_f is the density of the fluid, and g is the acceleration due to gravity (9.81 m/s²). Consider a spherical silica nanoparticle of diameter $D_p = 100$ nm and density $\rho_p = 2000$ kg/m³ dispersed in water of density $\rho_f = 997$ kg/m³ (at a temperature of 298 K); the ratio of the Brownian force to gravitational force, F_B/F_g is approximately 8000. Therefore, the Brownian force dominates over gravity, and the particles exhibit extraordinary stability against gravity sedimentation.

13.3.2 Brownian Motion and Stokes-Einstein Equation

Perrin carried out the very first quantitative study of Brownian motion.[30,31] By careful experimental protocols, he developed a system of model colloids of approximately 1.0 μm diameter of equal size, and ensured that the experiments were carried out under controlled conditions such that (1) the influence of gravity and other external forces are absent, and (2) particle-particle and particle-wall interactions were negligible. The trajectories of three equal sized particles dispersed in water from Perrin's experiments are shown in Fig. 13.4. The results from the experiments of Jean Perrin elucidated that the chaotic motion of particles, as seen in Fig. 13.4, is solely due to the Brownian motion, which originates from the continuous bombardment of fluid molecules with the particles. In each particle trajectory, the dots correspond to the location of the particle at a particular time instant, and successive dots are separated by a time interval of 30 s. The line that connects the successive dots is the displacement of the particle in 30 s. The use of well-characterised particles and the analysis of trajectories of the particles enabled the quantitative estimation of diffusivity of particles from the theory of Brownian motion developed by Einstein and Smoluchowski (Eq. 13.10). These experiments further enabled the calculation of Avogadro's number and proved the very existence of atoms. The mean square displacement of the particle is related to the particle diffusivity, D as:

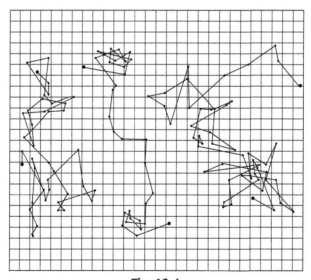

Fig. 13.4
Typical trajectories of three individual particles of approximately 1.0 μm diameter dispersed in water. The size of each square in the graph is 3.125 μm by 3.125 μm. *Reprinted from Perrin J. Atoms (D.L. Hammick, Trans.) London: Constable & Company Ltd; 1916.*

where

$$\langle \Delta x^2 (\Delta t)\rangle = 2dDt \tag{13.6}$$

$$\langle \Delta x^2 (\Delta t)\rangle = \langle |x(t+\Delta t) - x(t)|^2 \rangle \tag{13.7}$$

is the mean square displacement of the particle, $x(t+\Delta t)$ and $x(t)$ correspond to the position of the particle at times $t+\Delta t$ and t, respectively, d is the dimensionality ($n=3$ for Brownian motion in three dimensions), and D is the diffusivity of the particles in the fluid. The Brownian motion of particles results in the hydrodynamic (Stokes) drag force, F_h, on the particle given by:

$$F_h = 3\pi \eta v D_p \tag{13.8}$$

where $v = \bar{x}/t$ is the average distance travelled by the particles in time interval t, D_p is the diameter of the particle, and η is the viscosity of the fluid in which the particles are dispersed. Equating the Brownian energy that causes the motion of the particles and the hydrodynamic energy,[6]

$$k_B T = 3\pi \eta v D_p \bar{x} \tag{13.9}$$

Therefore,

$$\frac{\bar{x}^2}{t} = D = \frac{k_B T}{3\pi \eta D_p} \tag{13.10}$$

where D is the diffusivity of the particle in the fluid. Eq. (13.10) is known as the Stokes-Einstein equation extensively used for the characterisation of dimension of the colloidal particles by optical microscopy and dynamic light-scattering methods.

13.4 Stability of Colloidal Dispersions

A dispersion of colloids is said to be stable if the particles in the dispersion continue to exist as individual units, that is, if they do not cluster together or form aggregates. The stabilisation of colloids is all about how to prevent particles from aggregating or flocculating. In general, colloids may be stable either thermodynamically or kinetically.

An example of a thermodynamically stable colloidal dispersions is the microemulsions which form when oil, water, and surfactant are mixed in certain proportions. This is widely studied in a mixture of decane, water, and an anionic surfactant, aerosol-OT (AOT). A micellar dispersion that results when a small quantity of surfactant above a critical concentration called the critical micellar concentration is mixed with water is another example of thermodynamically stable colloidal dispersion. Such colloidal dispersions form spontaneously upon mixing, and their formation is thermodynamically favoured due to a reduction in the overall free energy of the mixture. Thermodynamically stable colloids exhibit extraordinary long-term stability. In several cases, the presence of colloidal particles in a well-dispersed state is not thermodynamically favoured; for example, due to high surface area and attractive interactions, metallic colloids often cluster together and precipitate out of the solution. A common strategy used to stabilise colloids against aggregation is to impart kinetic stability, such that a large energy barrier for the processes that lead to destabilisation of colloids exists, or to completely eliminate attractive forces between the particles. If the particles are so small that the gravity effects are negligible, and if not thermodynamically stable, the aggregation of particles can be prevented by means of refractive index matching, electrostatic stabilisation, steric stabilisation, and depletion stabilisation; some of these are described in the following sections.[1,10,32,33]

13.4.1 Electrostatic Stabilisation and Electrical Double Layer

In electrostatic stabilisation, an energy barrier (for destabilisation) that is much larger than the thermal energy, $k_B T$, is achieved by imparting interparticle repulsion as a consequence of charges on the particle surface. Various mechanisms have been proposed to impart surface charge on the colloidal particles and to explain the origin of electrostatic stabilisation.[1,10,32,33]

Colloidal particles can be engineered so that one or more chemical moieties are covalently linked or attached to the molecules/atoms on the particle surface. When such colloids are dispersed in a fluid, the chemical groups on the particle surface may dissociate, leading to a charged group on the particle surface and releasing counter ions into the solution. For example, in the case of a particle with COOH group dispersed in aqueous solution, dissociation leads to COO^- group on the particle surface and H^+ ion, leading to a particle with net negative charge. In many commercially available charge stabilised dispersions, the surface of the particle carries charge due to this mechanism. Schematics in Fig. 13.5A and B show negatively charged and positively charged surface groups, respectively, on the commercially available sulfate and amidine polystyrene latex particles. The particles may also acquire surface charge by the adsorption of ions on the particle surface, as shown in Fig. 13.5C and D. The adsorption of H^+ ions or OH^- ion on the surface of haematite particles (α-Fe_2O_3) that leaves the particle positive or negatively charged is a common example of the origin of surface charge by this mechanism. Charged molecules added to a dispersion of colloids may adsorb on particle surfaces, leading to improved stability. In Fig. 13.5E–H, colloidal particles are shown to acquire charge as a result of adsorption of charged surfactant molecules and charged polymers, respectively.

Irrespective of the mechanism that leads to charges on colloidal particles, a charge-stabilised colloidal dispersion will have (1) charged particles and charges that, for all practical purposes, can be assumed to be attached to the particles, and (2) counter ions, which are the ions that loosely said are released into the solution when the particle acquires surface charge, i.e., the charges that belong to the continuous medium. The total charge on all the colloidal particles in the dispersion must be equal but opposite in magnitude to the total charge of all the

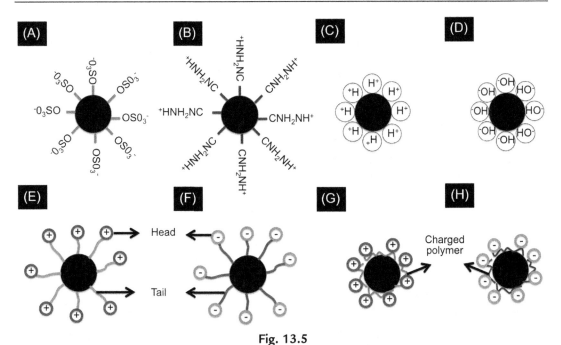

Fig. 13.5

Schematic representation of different schemes or mechanisms that result in charges on particle surface: dissociation of covalently attached chemical species on particle surface (A, B), adsorption of ions (C, D), surfactant molecules (E, F), and charged polymers or polyelectrolytes (G, H).

counter ions in the dispersion such that the dispersion as a whole is electroneutral. In addition, there may also be co-ions (whose sign is the same as the charge on the particle surface) and counter ions from externally added electrolyte or due to the dissolved ionic species.

As a result of the electric field generated due to a charged particle, the counter ions/co-ions in the solution are attracted/repelled from the particle surface. The thermal energy of the ions contributes to negate the electrostatic force, resulting in a configuration shown schematically in Fig. 13.6. The thermal motion acts to drive the ions away from the charged surface, leading to a diffused distribution of ions. The configuration of a charged particle in solution, shown in Fig. 13.6, is comprised of a layer of charges on the particle surface and a layer of counter-ions in the immediate neighbourhood of the charged surface. This arrangement is called electrical double layer (EDL). When a charged particle is subjected to a perturbation, such as an externally applied electric field, or an electrolyte is added to a dispersion of charged colloids, or charged particles are transferred into an electrolyte solution, the free ions redistribute, resulting in the restructuring of the electrical double layer.

For mathematical treatment of charged surfaces in solution, the particle surface charge is expressed in terms of a potential difference. Because the potential at a very far distance, i.e., outside the double layer, is zero (due to electroneutrality), a charged surface is typically

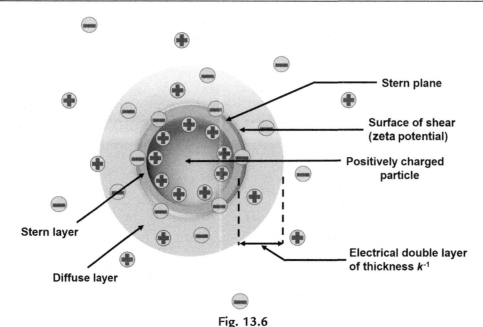

Fig. 13.6
Schematic representation of a charged spherical particle depicting the formation of electrical double layer. *Figure courtesy—Preetika Rastogi, IIT Madras.*

characterised by a surface potential, ψ_0. The layer of counter-ions immediately next to the charged surface shown in Fig. 13.6 is called the stern layer. These ions are strongly bound to the charged surface due to the electric field generated due to the charged surface and are expected to move with the particle when it is in motion. The charges on the particle surface and the ions in the stern layer resemble the arrangement of a capacitor, proposed first by Helmholtz. As per the capacitor model, the surface potential in this region varies linearly. Beyond the stern layer, the surface potential decreases more gradually due to the diffused distribution of ions, until reaching a distance κ^{-1} from the charged surface. The length scale κ^{-1} is called the thickness of the double layer (typically expressed in nanometers) and is the inverse of the Debye-Hückel parameter, κ. The Debye-Hückel parameter, κ, is given by[1,3,34]:

$$\kappa = \left[\frac{1000e^2 N_{AV}}{\epsilon_0 \epsilon_r k_B T} \sum_i z_i^2 M_i \right]^{\frac{1}{2}} \tag{13.11}$$

where $e = 1.60 \times 10^{-19}$ C is the elementary charge, N_{AV} is the Avogadro's number, ϵ_r is the relative permittivity, ϵ_0 is the permittivity of free space, k_B is the Boltzmann's constant $(=1.38 \times 10^{-23}$ J K$^{-1})$, T is the absolute temperature, z is the valency of the ions, z_i is the valency, and M is the molar concentration and i refers to ith specie in the solution. Repulsive interactions between the colloidal particles arise when the electrical double layers of neighbouring particles overlap. The dielectric constant of the medium, the temperature, and the concentration of ions in the solution can be used to manipulate the thickness of the electrical double layer. For example, according to Eq. (13.11), addition of an electrolyte to a colloidal dispersion increases the ionic strength and κ; hence, the thickness of the double layer, κ^{-1} decreases, thereby reducing the overall repulsion, which will be discussed in the following sections. A plane in the immediate vicinity of the stern layer, which demarks whether the counter-ions in the solution are strongly bound to the particle surface or free to move is called the plane of shear, and the potential at this location, which is accessible experimentally, is the zeta potential (ζ). The zeta potential is an important parameter that needs to be measured to assess the istability of charge-stabilised colloids. The zeta potential is positive if the particle is positively charged, and it is negative if the particles are negatively charged. The higher the zeta potential, the higher the surface charge and the more stable is the dispersion. In a charge-stabilised dispersion, a zeta potential of zero is called the isoelectric point or the point of zero charge, which corresponds to the conditions of maximum aggregation.

13.4.2 Steric Stabilisation

In steric stabilisation of colloids, long chain molecules are anchored on the surface of the particle such that each particle is decorated with a thin shell or envelope of long-chain molecules,[35] as shown schematically in Fig. 13.7. Thus, particles can be thought of as having bristles (very much like a paintbrush or a toothbrush) giving a brush-like appearance to the particle surface. The anchoring of long-chain molecules can be achieved by chemical synthesis (also called chemical grafting), as shown schematically in Fig. 13.7A, or through adsorption. Fig. 13.7B and C show the steric stabilisation of colloids due to the adsorbed polymer molecules and surfactant molecules, respectively. When such particles are dispersed in a solvent, the molecules that form brush-like architecture may assume an extended conformation or a collapse configuration. When the dimensions of the brush-like layers exceed a few nanometers, the attractive van der Waals forces are completely masked and, thus, provide stability to the colloids.

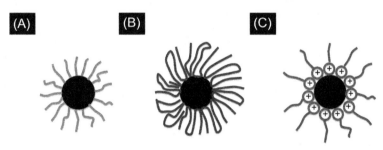

Fig. 13.7
Schematic representation of particles with brushes on their surface due to (A) covalently attached long-chain molecules, (B) adsorption of long chain molecules such as polymers, and (C) adsorption of surfactant molecules such that the tail group is exposed to continuous medium in which the particles are dispersed.

13.4.3 Refractive Index Matching

In this approach to the stabilisation of colloids, the particles are dispersed in a solvent whose refractive index is same as that of the particles. This ensures that the omnipresent van der Waals forces of attraction between the particles become negligible. Silica particles dispersed in ethoxylated trimethylolpropane triacrylate,[36] poly (methyl methacrylate) (PMMA) particles grafted with a thin layer of poly-12-hydrostearic acid dispersed in a mixture of cis-decaline and tetraline (cis-decaline mass fraction ~0.2)[37] or a mixture of solvents cyclohexylbromide and cis-decalin (3:1 wt./wt.) saturated with tetrabutylammoniumbromide are some examples of index matched colloidal dispersions.[36]

13.4.4 Depletion Stabilisation

The depletion stabilisation involves the addition of nonionic polymers to colloidal dispersion.[38] Unlike steric stabilisation of particles, in depletion stabilisation, polymer molecules that are added to a colloidal dispersion are not adsorbed onto the particle surface and are free to move. When particles that are in a pool of polymer solution approach each other, in general, the polymer molecules between the particles are expected to leave the gap i.e., the region between the particles. That is, the polymer is supposed to deplete from the region between the particles. However, when the interaction between the polymer and the solvent is preferred, i.e., when the polymer is in a good solvent, polymer remains in the region between the particles, as this is energetically more favoured. In such cases, the polymer molecules continue to exist between the particles and this provides excellent stability to colloidal dispersions via what is called the depletion stabilisation.[1,39,40]

13.5 Characterisation of Colloidal Dispersions

General methods to characterise the size of solid particles are discussed in Chapter 2. In this section, several quantities that must be measured for a comprehensive characterisation of colloidal dispersions are discussed. This is important both for framing a scientifically correct problem as well as for the analysis and interpretation of the experimental results. Typically, the colloidal particles in the dispersion are characterised by their dimension(s), density, and surface properties such as charge density, grafting density, roughness, etc. The size of particles can be measured by one of the techniques discussed in Chapter 2.

13.5.1 Particle Density and Specific Surface Area

The particle density (ρ_p) is defined as the mass of a particle divided by its volume. The assumption that the density of a particle is same as the density of the corresponding bulk material breaks down as the particle size becomes comparable to the molecular dimensions. Moreover, the density of particles may depend on the protocol used for their synthesis, the morphology of particles (surface roughness, porous nature), and on the properties of the dispersion medium. Particle density is a useful quantity for estimating the specific surface area of the particles. Under the assumption of ideal mixing,[41] the total volume V_d of a colloidal dispersion consisting of solid particles of total volume V_p and the suspending medium of total volume V_s can be written as:

$$V_d = V_p + V_s \tag{13.12}$$

Dividing both sides of the equation by the total mass M_d of the dispersion and multiplying first and second term on the right hand side by M_p/M_p and M_s/M_s, respectively, where M_p is the total mass of the particles, and M_s is the total mass of the suspending medium, we obtain:

$$\frac{1}{\rho_d} = \frac{1}{\rho_p}\frac{M_p}{M_d} + \frac{1}{\rho_s}\frac{M_s}{M_d} \tag{13.13}$$

Substituting $M_p/M_d = X$ and $M_s/M_d = 1 - X$, where X is the mass fraction of particles in the dispersion,

$$\frac{1}{\rho_d} = \frac{1}{\rho_p}X + \frac{1}{\rho_s}(1 - X) \tag{13.14}$$

Eq. (13.14) can be recast as:

$$\frac{1}{\rho_d} = \frac{1}{\rho_s} + \left(\frac{1}{\rho_p} - \frac{1}{\rho_s}\right)X \tag{13.15}$$

In order to obtain the particle density, a series of homogeneous dispersions containing particles of known weight are prepared, and the solution densitometry is used to measure the density of the dispersions. To obtain the weight fraction of particles in the dispersion, a fixed quantity of dispersion is weighed and dried under controlled conditions to ensure complete evaporation of the suspending medium. The particle density is obtained from the slope and intercept of $1/\rho_p$ vs. X plot, in accordance with Eq. (13.15).

If the particles in the dispersion are of a well-defined shape, the particle density can be used to estimate the specific surface area of the particle, if the dimensions of the particles are known. The specific surface area of the particle (A_{sp}), defined as the surface area of a particle (A_p) divided by the mass of that particle (M_p), is given by:

$$A_{sp} = \frac{A_p}{M_p} \tag{13.16}$$

For a spherical particle of diameter D_p,

$$A_{sp} = \frac{6}{\rho_p D_p} \tag{13.17}$$

Therefore, the particle density can be used to obtain the specific surface area, which is one of the quantities exemplifying the fact that the surface area of the particle increases with decreasing particle size.

13.5.2 Volume Fraction of Particles in the Dispersion

The volume fraction of the particles in the dispersion (ϕ) is an important quantity that will help compare several colloidal scale phenomena, including transport properties such as viscosity, across dispersions of different types of particles. The volume fraction of particles in the dispersion (ϕ) is defined as the volume of the particles in the dispersion (V_p) divided by the total volume of the dispersion.

$$\phi = \frac{V_p}{V_d} \tag{13.18}$$

If the dispersion contains N spherical particles each of volume v_p and diameter D_p, then,

$$\phi = \frac{Nv_p}{V_d} = \frac{N\left(\pi D_p^3/6\right)}{V_d} \tag{13.19}$$

In terms of easily measurable quantities such as the total mass of particles in the dispersion (M_p) and the particle density (ρ_p), this relation can be recast as:

$$\phi = \frac{M_p}{V_d \rho_p} \tag{13.20}$$

Though accurate measurement of volume fraction is very important, detailed procedures adopted in determining this quantity are seldom discussed. Analysis of different methods typically used for the measurement of volume fraction and the uncertainties involved in its estimation can be found in a review by Poon et al.[42]

13.5.3 Grafting Density

As discussed in earlier sections, in steric stabilisation, either polymer molecules or any long-chain molecules are physically or chemically attached to the surface of the colloidal particles. Such particles are characterised, among other parameters, by the surface grafting density or grafting density (σ_g), which is defined as the number of molecules or chains grafted (or tethered) onto the surface of a particle (N_g) divided by the surface area of the particle (A_p).

$$\sigma_g = \frac{N_g}{A_p} \tag{13.21}$$

Typically, σ_g is expressed in units of number of molecules (or chains) per nm^2. A convenient way to manipulate the grafting density, especially if a chemical synthesis route is used for grafting, is to vary the reactant concentration or reaction time. A simple method for the measurement of grafting density uses the principle of thermogravimetry analysis (TGA). In a typical TGA experiment, a sample of known weight is subjected to a thermal field (heating or cooling) under controlled conditions (heating rate, inert atmosphere, etc.), and the change in the mass of the sample as a function of time is noted. In order to obtain the grafting density, the sample is heated so that all the grafted molecules in the particles in the sample decompose. From the knowledge of the onset and completion of thermal decomposition and the weight loss vs. temperature data, the total mass, m_g of the grafted molecules is estimated. The total number of grafted molecules on the surface of all the particles in the sample, N_g can be obtained by,

$$N_g = \left(\frac{m_g}{MW}\right) N_{AV} \tag{13.22}$$

where MW is the molecular weight of the grafted molecule, and N_{AV} is the Avogadro's number. The grafting density (σ_g) per particle is then calculated as:

$$\sigma_g = \left(\frac{N_g}{A_p N_p}\right) \tag{13.23}$$

where N_p is the number of particles in the sample.

13.5.4 Zeta Potential and Surface Charge Density

The ability of charged particles or ionic species to respond to externally applied electric field, a phenomenon called electrokinetics, can be exploited to measure the zeta potential of charged colloidal particles, which is the potential at the plane of shear. Depending on whether the charged particle is forced to move or the ions are made to flow past a charged surface, the electrokinetic techniques can be classified as electrophoresis, electroosmosis, streaming potential and sedimentation potential.[27] Electrophoresis is by far the most widely used technique for the measurement of zeta potential. In electrophoresis, a dilute dispersion of colloidal particles is placed between two electrodes and the movement of the colloidal particle with respect to a stationary dispersion medium is tracked either by a microscope or by means of light scattering. The movement of particles measured by electrophoresis is expressed in terms of the electrophoretic mobility, u, which is defined as the velocity, v with which the particle moves under the influence of an external field of magnitude E.[1,5,34]

$$u = \frac{v}{E} \tag{13.24}$$

For a dilute dispersion of nonconducting colloids, the Henry's equation can be used to obtain the zeta potential (ζ) of particles from the measured electrophoretic mobility (u) data:[1,5]

$$u = \frac{2}{3}\frac{\zeta \epsilon_0 \epsilon_r}{\eta} f(\alpha) \tag{13.25}$$

where η is the viscosity of the dispersion medium, ε_r is the relative permittivity, ε_0 is the permittivity of the free space, and $\alpha = \kappa R_p$, where R_p is the radius of the particle, and κ is the Debye-Hückel parameter. The function $f(\alpha)$ when $\kappa R_p < 1$ is given[1,5,34]:

$$f(\alpha) = \left(1 + \frac{1}{16}\alpha^2 - \frac{5}{48}\alpha^3 - \frac{1}{96}\alpha^4 - \frac{1}{96}\alpha^5 - \left[\frac{1}{8}\alpha^4 - \frac{1}{96}\alpha^6 \right] \exp(\alpha) \int_{\infty}^{\alpha} \frac{e^{-t}}{t} dt \right) \tag{13.26}$$

and when $\kappa R_p > 1$,

$$f(\alpha) = \left(\frac{3}{2} - \frac{9}{2}\alpha^{-1} + \frac{75}{2}\alpha^{-2} - 330\alpha^{-3} \right) \tag{13.27}$$

For the limiting case of $\kappa R_p < 0.1$, the Henry's equation simplifies to the Hückel equation, which is valid for cases where the particles are dispersed in extremely dilute electrolyte solutions, i.e., when the electrical double layer is thick:

$$u = \frac{2}{3}\frac{\varsigma\epsilon_0\epsilon_r}{\eta} \tag{13.28}$$

For the case of thin electrical double layers, which corresponds to $\kappa R_p > 100$, the Henry's equation simplifies to the Helmholtz-Smoluchowski equation[1]:

$$u = \frac{\varsigma\epsilon_0\epsilon_r}{\eta} \tag{13.29}$$

The zeta potential of particles can further be used to estimate the surface charge density of particles. Consider a particle of surface area A_p with q charge on its surface. The surface charge density (σ_s) is defined as the total charge on the particle surface (q) divided by the total surface area of the particle (A_p). σ_s is typically reported in coulombs per square meter (Cm^{-2}). If the surface area of the particle is known, then the number of charges on the surface can be calculated. For a spherical particle of diameter D_p,

$$\sigma_s = \frac{q}{\pi D_p^2} \tag{13.30}$$

The zeta potential, obtained from electrokinetic measurements can be used to estimate the surface charge density using a semi-empirical method proposed by Loeb et al.[43]:

$$\sigma_s = \epsilon_r\epsilon_0 \frac{k_B T}{ze}\kappa \left[2\sinh\frac{ze\zeta}{2k_B T} + \frac{8}{\kappa D_p}\tanh\frac{ze\zeta}{4k_B T} \right] \tag{13.31}$$

where k_B is the Boltzman's constant ($= 1.38 \times 10^{-23}$ J/K)) and T is the absolute temperature, z is the valency of the counterions, $e = 1.60 \times 10^{-19}$C is the elementary charge, ζ is the zeta potential, R_p is the radius of the particle, and κ is the Debye-Hückel parameter.

The conductometric and potentiometric titration methods are also widely used for the measurement of the surface charge density of colloidal particles. In these methods, a chemical reagent such as an electrolyte, an acid or a base is added to a dispersion containing a known quantity of colloids and the change in the concentration of ionic species is measured by monitoring the conductivity (conductometric titration) or pH (potentiometric titration). The sudden change in the conductivity or pH, which are associated with marked changes in the concentration of ions during the titration, is used for the estimation of surface charge density.[44–47]

13.5.5 Surface Heterogeneity

The colloidal particles may have one or more types of surface heterogeneities, which are either physical or chemical in nature. For example, imagine a particle with a uniform surface charge density such that the zeta potential (ζ) of the particle is −50 mV as shown schematically in Fig. 13.8A. However, if the dissociable surface groups that render charge to colloidal particles are not uniformly distributed across the particle surface, different regions on the particle surface may have different zeta potential[48] as shown schematically in Fig. 13.8B. Moreover, individual particles can be selectively functionalized such that domains that are - positively and negatively charged or hydrophilic or hydrophobic as shown respectively in Fig. 13.8C and D coexist on the particle surface. The schematic of particles

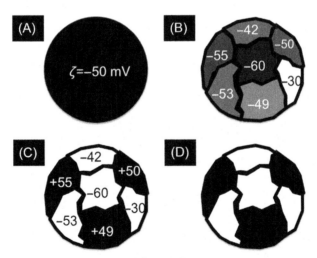

Fig. 13.8

Schematic illustration of surface charge heterogeneity: (A) A homogeneously charged spherical particles (B) particle with nonuniform surface charge distribution (C) particle with oppositely charged domains—positively charged domains are shown in *black*, and negatively charged domains are shown in *white* (D) particle with hydrophilic *(black)* and hydrophobic *(white)* regions. *A and B: Redrawn with permission from Feick JD, Chukwumah N, Noel AE, Velegol D. Altering surface charge nonuniformity on individual colloidal particles. Langmuir 2004;20(8):3090–3095.*

shown in Fig. 13.8B–D exemplifies those which exhibit chemical heterogeneity. The characterisation of such chemical heterogeneity requires the use of advanced techniques such as rotational electrophoresis.[48] It is possible to measure force-displacement curves using atomic force microscopy or a surface force apparatus and estimate the extent of surface heterogeneity using appropriate models to fit the experimental data.

Surface roughness, an example of physical heterogeneity, which can be easily measured, is shown to significantly influence a number of colloidal phenomena. A class of techniques based on "etching" can be used to fabricate particles with rough surfaces. In this method, the molecules or atoms on the surface of a smooth particle are dissolved by controlled addition of a chemical reagent called "etchant" to a colloidal dispersion, leaving behind a rough particle. It is possible to exploit the aggregation between oppositely charged particles to fabricate colloidal particles with controlled surface roughness. Electron microscopy images of rough particles obtained by the electrostatic aggregation of micron-sized, positively charged particles and negatively charged sub-100 nm particles are shown in Fig. 13.9.

The first step in the characterisation of surface roughness of particles involves the measurement of roughness profile using a suitable technique such as optical surface profiler; X-ray computed tomography, atomic force microscopy, etc., depending on the size of the asperities on the particle surface. With these techniques, the three-dimensional surface topography that shows peaks and valleys that are representative of the asperities on the particle surface is measured. From the surface profiles, line profiles that show the height variations along a line of interest are extracted. Such roughness profiles discretized into n equally spaced intervals with y_i representing the vertical height of the peak or valley at the ith interval with reference to a mean line. The parameters most commonly used to characterise the surface roughness are the

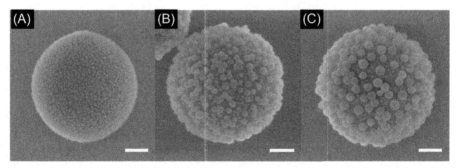

Fig. 13.9

Colloidal scale particles with surface roughness fabricated by exploiting aggregation between oppositely charged particles: Positively charged silica particles of 960 nm ± 36 nm diameter are shown to be deposited with (A) 12 nm, (B) 39 nm and 12 nm, and (C) 72 nm and 12 nm negatively charged silica nanoparticles. The scale bar in each electron microscopy image corresponds to 200 nm. *Reprinted with permission from Zanini M, Hsu C-P, Magrini T, Marini E, Isa L. Fabrication of rough colloids by heteroaggregation. Colloids Surf A; 2017;**532**:116–24.*

arithmetic mean roughness (R_{Mean}) and the root mean square roughness (R_{RMS}), which are calculated as[49]:

$$R_{Mean} = \frac{1}{n}\sum_{1}^{n}|y_i|$$
(13.32)

$$R_{RMS} = \frac{1}{n}\sum_{1}^{n}y_i^2$$
(13.33)

13.6 Colloidal Interactions

The ability to tune the interaction forces or the energy of interaction between colloidal particles in a dispersion is crucial for the design of optimal solution processing routes for the fabrication of novel materials. The structural and functional properties of the final materials derived following solution processing can be modulated by a precise control of the colloidal interactions. The tailoring of colloidal interactions has been exploited to create diverse assembly of colloids and nanoparticles that have applications in photonics, display devices, etc. A basic and clear understanding of the colloidal interactions is useful to predict stability, induce de-stabilisation of particles in dispersion and also to tailor transport properties of dispersions. For example, several unit operations, such as slurry transport, demand that individual particles in a colloidal dispersion are in a well-dispersed state so that there is no sedimentation or aggregation; however, certain operations, such as water treatment, rely on the formation of particle aggregates or flocs such that the unwanted particles are effectively and efficiently removed by gravity sedimentation. The interactions between colloidal particles in the bulk develop when the concentration of particles in the colloidal dispersion is sufficiently large. When two or more particles in the dispersion are appreciably close, they start to "interact" with each other or "feel" the presence of each other. The interaction between colloidal particles, analogous to atomic and molecular interactions, may be purely attractive or purely repulsive or a combination of both. Typically, colloidal interactions are quantified by plotting the interaction energy (or force) as a function of the distance of separation between the two particles. The origin of different types of interactions relevant to colloidal systems, their magnitude, and functional form depends on several factors as discussed further in this section.

13.6.1 Molecular/Atomic Interactions

From elementary material science and through the discussions in the first chapter of this volume, we know that the atoms or molecules that constitute solids are held together via interatomic or intermolecular forces. Some common examples of the different types of interactions in solids are covalent interactions, ionic interactions, and metallic bonding, etc. The properties (such as hardness, conductivity, melting point, etc.) that are unique to a given material are a consequence of these interatomic or intermolecular interactions. The interaction potential between any pair of objects (atoms, molecules, or particles) refers to the potential energy of interaction between the two objects as a function of distance of separation. In general, the overall interaction or total interaction is the sum of the contributions of attractive and repulsive components:

$$\Phi(r) = \Phi_A(r) + \Phi_R(r) \tag{13.34}$$

where Φ refers to the interaction potential, subscripts A and R refer to attraction and repulsion, respectively, and r is the separation distance. Depending on the nature of the objects (atoms, molecules, or particles), Φ_A, Φ_B, or both may contribute to the overall interaction potential. The functional form of the variation of Φ_A and Φ_B with separation distance depends on the specific type of pairs of objects considered. It is a general convention that positive potential refers to repulsion and negative potential refers to attraction. If the interaction potential (Φ) is known, the functional form of the interaction force can easily be obtained as force is the negative gradient of the potential, i.e., $F(r) = -\frac{d\Phi(r)}{dr}$. One of the most celebrated functional forms of such an interaction potential between two neutral molecules is the Lenard-Jones (LJ) potential, given by[1]:

$$\Phi_{LJ}(r) = \frac{\alpha}{r^{12}} - \frac{\beta}{r^6} \tag{13.35}$$

A plot of the interaction energy versus separation distance obtained by substituting $\alpha = 1.58 \times 10^{-134} \, \mathrm{J \, m^{12}}$ and $\beta = 1.02 \times 10^{-77} \, \mathrm{J \, m^6}$ for a pair of argon molecules is shown in Fig. 13.10. The interaction energy is scaled with thermal energy, $k_B T$ ($T = 298 \, \mathrm{K}$). The dotted and dashed lines respectively correspond to the repulsion and attraction, and the overall interaction is shown as a continuous line. Because the attractive interactions counterbalance the repulsive interactions, the overall interaction represented by the continuous line in Fig. 13.10 exhibits a minimum. The location of the minimum provides an estimate of the equilibrium separation distance and the attraction energy that a pair of argon molecules experience at this separation. In the context of the colloidal dispersions, a control of the behaviour of colloids in many formulations and processing steps rely on tuning the interparticle interactions and therefore require a basic understanding of the range of interactions. The range of interactions refers to the distance of separation over which the interactions become significant. In general, interactions can be either short-ranged or long-ranged. With respect to the interaction potential shown in Fig. 13.10, the repulsive interactions are short ranged (becomes 0 at a smaller r) and in comparison, the attractive interactions are long-ranged.

13.6.2 Hard Sphere Interactions

Poly-methylmethacrylate (PMMA) grafted with poly-12-hydroxystearic acid chains dispersed in dodecane is an example of a colloidal dispersion that has been widely used for experimental investigations of phase behaviour and flow properties of hard sphere colloids.[50] Typically, in

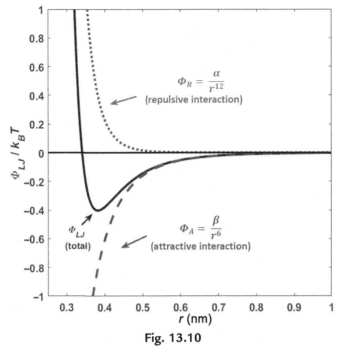

Fig. 13.10

Interaction energy versus separation distance for a pair of argon molecules. The *dotted line* corresponds to repulsion, the *dashed line* corresponds to attraction, and the overall interaction is represented by the *continuous line*. Note that the interaction energy is scaled with the thermal energy, $k_B T$. *Figure courtesy—Preetika Rastogi, IIT Madras.*

hard sphere colloids (i) there are small hair-like structures on the particle surface, and (ii) dispersion medium is a solvent whose refractive index matches that of the particles. This ensures that there is effectively no interparticle attraction. Therefore, the overall interaction between the particles as shown schematically in Fig. 13.11 is purely repulsive and short-ranged, given by:

$$\Phi(r) = \infty \quad \text{when } r < d$$
$$= 0 \quad \text{when } r > d$$

(13.36)

13.6.3 DLVO Interactions

The celebrated DLVO theory attributed to four scientists, Derjaguin, Landau, Verwey, and Overbeek, is a mean field approach proposed to describe the behaviour of charged colloids in solutions. Developed in the 1940s, the DLVO theory is widely used to explain the experimental results for charged colloids, for example, self-assembly of particles, deposition of particles on solid surfaces, aggregation, etc. According to the DLVO theory, in a dispersion of charged colloids, the net or overall interaction between a pair of particles is due to the

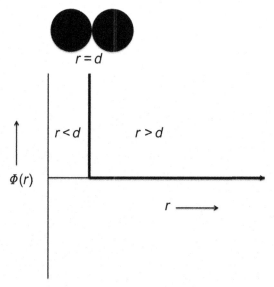

Fig. 13.11
Schematic representation of hard sphere interaction potential.

attractive van der Waals (vdW) interactions and the repulsive interactions originate due to the overlap of the electrical double layers (EDL), that is:

$$\Phi_{DLVO}(r) = \Phi_{vdW}(r) + \Phi_{EDL}(r) \tag{13.37}$$

The van der Waals interactions

The term, $-\beta/r^6$, in the Lenard-Jones (LJ) interaction potential (Eq. 13.35) corresponds to the van der Waals attraction. The columbic interactions due to the permanent and induced dipole result in the van der Waals interactions. The induced dipole in a molecule arises due to the instantaneous fluctuations in the distribution of electrons. The contribution to the overall van der Waals interactions includes interaction between the permanent dipole (PD) of one molecule with permanent dipole of the other molecule, interaction between the permanent dipole of one molecule with the induced dipole (ID) of other molecule, and interaction between the induced dipole of one molecule with the induced dipole of the other molecule. The van der Waals interactions between two colloidal particles are obtained by summing the contributions of all pairs of atoms or molecules that the colloidal particles are made of. The functional form of the van der Waals interaction between colloidal particles depends on the geometry of the particles, and their orientation if the particles are nonspherical. For two spherical particles of radius R_1 and R_2, separated by a surface-to-surface distance d such that both R_1 and R_2 are much larger than d, i.e., $R_1, R_2 \gg d$, the van der Waals interaction potential is given by[1]:

$$\Phi_{vdW}(d) = -\frac{AR_1R_2}{6d(R_1+R_2)} \tag{13.38}$$

where A is a constant proportional to β (in Eq. 13.35). The constant A takes into account the contributions of PD-PD, PD-ID and ID-ID interactions and is called the Hamaker constant. From Eq. (13.38), it is apparent that A has units of energy and for most material takes values ranging from 10^{-20} to 10^{-19} J. Metallic colloids typically exhibit very high values of the Hamaker constant and therefore are prone to aggregation due to the van der Waals attraction. For example, the Hamaker constant of gold particles interacting via vacuum is about 4.5 times higher than that of polystyrene particles interacting via the same medium, $A_{\text{polystyrene}} \sim 8 \times 10^{-20}$ J. The presence of any medium is known to reduce the value of the Hamaker constant and hence the van der Waals force of attraction. For example, the Hamaker constant for silica particles in water is[51] $\sim 2.0 \times 10^{-20}$ compared to $\sim 8.0 \times 10^{-20}$ J in vacuum.[52] For spherical particles of two equal radii R_p, Eq. (13.38) reduces to

$$\Phi_{vdW}(d) = -\frac{AR_p}{12d} \tag{13.39}$$

Electrical double layer interaction

The repulsive interactions in charge stabilised colloids arise when the electrical double layers (EDL) of the two neighbouring particles overlap. If the distance between the two particles is much larger than κ^{-1}, there is no overlap of their respective electrical double layers and hence no interparticle repulsion occurs under these conditions. However, as the particles approach each other, either due to increase in the concentration of particles or due to the Brownian motion, if the separation distance is less than κ^{-1}, repulsive interactions are triggered. Due to the electrical double layer overlap, the concentration of ions in the overlapping region increases which manifests in the form of the osmotic force and the electrostatic force arising due to potential around the charge particles eventually repel the particles away from each other. For two spherical particles of low surface charge densities such that the zeta potential (ς) is approximately 25 mV or less, the electrical double layer repulsion is given by the expression[1]:

$$\Phi_{EDL}(d) = 4\pi R^2 \epsilon_0 \epsilon_r \psi_0^2 \exp\frac{-\kappa d}{d+2R_p} \tag{13.40}$$

where R_p is the radius of the spherical particle, ψ_0 is the surface potential, ε_r is the relative permittivity, ε_0 is the permittivity of free space, κ is the inverse screening length, and the surface-to-surface particle separation distance, d is related to the center-to-center separation, r as $d = r - 2R_p$. In the calculation of electrical double layer repulsion, surface potential ψ_0 is replaced with zeta potential ζ. A general expression applicable when the charge-stabilised dispersion contains particles of any surface potential and symmetric electrolyte of valency $z{:}z$ is given by[5]

$$\Phi_{EDL}(d) = 32\varepsilon_0\epsilon_r R_p \left(\frac{k_BT}{ze}\right)^2 \tanh^2 \frac{\psi_0 ez}{4k_BT} \exp(-kd) \qquad (13.41)$$

where $e = 1.60 \times 10^{-19}$C is the elementary charge on a particle, k_B is the Boltzmann's constant ($= 1.38 \times 10^{-23}$ JK^{-1}), T is the absolute temperature, z is the valency of the ions, and $d = r - 2R_p$ is the surface to surface particle separation distance.

Eqs. (13.40) and (13.41) reveal that the electrical double layer interactions decrease exponentially with an increase in the separation distance. The interaction potential can be modulated by changing the solution conditions such as temperature, dielectric constant, electrolyte concentration, valency of ions and properties of particles (particle size and surface charge). A schematic of the interaction potential versus separation distance for a pair of particles in a charge-stabilised dispersion is shown in Fig. 13.12.

13.6.4 Depletion Interactions

Depletion interactions arise when a colloidal dispersion contains at least one more dispersed component whose size is much smaller than the dimension of the colloidal particle.[33,38,53,54] Colloidal dispersions are often multicomponent and contain additives such as polymers, surfactants or other particulate matter. While some additives are used as viscosity modifiers, i.e., to tune the flow behaviour, the addition of a small quantity of high molecular weight polymers can impart exceptional stability against gravity settling, and certain additives can also tailor colloidal interactions. Typical examples of multicomponent colloidal mixtures in which depletion interactions may occur are (1) colloid-polymer mixtures, i.e., a dispersion containing colloidal particles dispersed in a polymer solution, (2) dispersions

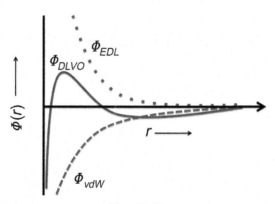

Fig. 13.12
A typical plot of interaction energy versus separation distance for a pair of particles in a charge-stabilised dispersion. The *dotted line* corresponds to electrical double layer repulsion (Φ_{EDL}), the *dashed line* corresponds to the van der Waals attraction (Φ_{vdW}) and the overall interaction represented by the *continuous line* is the total interaction (Φ_{DLVO}).

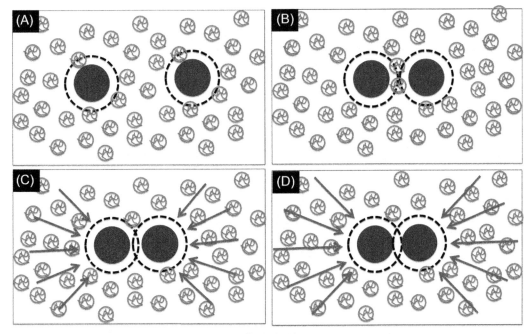

Fig. 13.13

Schematic illustration of depletion interaction between two large spherical particles *(filled large circles)* dispersed in a fluid containing smaller dispersed species (represented by *smaller circles*) at different separation distance: (A) the particles in the polymer solution are sufficiently far (B) the depletion layer around each particles touch. In (A) and (B), the osmotic pressure acting across the particle surface is same in all the directions. (C) and (D) the over lap of depletion layers leads to the formation of a lens. The volume of the depletion zone (or the lens) is larger in (D) than that in (C).

containing particles and surfactants at a concentration above critical micellar concentration (which is the concentration above which the surfactant molecules self-assemble to form micelles which are small nanosized objects) and (3) a dispersion containing two or more colloidal particles of significantly different dimensions.

The concept of depletion interactions is shown schematically in Fig. 13.13. The scheme shows two large spherical particles (filled large circles) in a pool of smaller dispersed species (represented by smaller circles) at different separation distances. For the ease of understanding, a polymer molecule is shown inside the smaller circle. This picture refers to a classic case of the depletion interactions in the colloid-polymer mixtures. However, the smaller circles can be nanoparticles or micelles or any other species that is small, and does not adsorb onto the surface of larger particles. For simplicity, the configuration of polymer is shown to be identical, however, due to polymer-solvent interactions and thermal fluctuations, the configuration of each polymer molecule may vary slightly. Polymers when dispersed in a good solvent can be assumed to be in a coil-like state and take the shape of a sphere. The configuration of polymer molecule in solutions is characterised by the radius of gyration (R_g), which is the root-mean-square radius of all the elements that the polymer is made of from its centre of mass. When close to a solid surface, the loss of conformational entropy of polymer molecules leads to the formation of a depletion zone around each particle whose thickness is approximately equal to R_g of the polymer.[33,54] The dashed circles in Fig. 13.13 indicate the depletion zone.

When the colloidal particles are sufficiently far from each other, the distribution of polymer molecules around the particles will be similar and so would be the concentration of polymer, therefore, the osmotic pressure acting on the surface of the particle given by,[54]

$$P = n_P k_B T \tag{13.42}$$

is isotropic, i.e., same across the surface of spherical particles. In Eq. (13.42), n_P is the concentration of polymer molecules in the solution expressed in terms of the number density, i.e., the number of polymer molecules per unit volume and $k_B T$ is the thermal energy. In Fig. 13.13B, the depletion layer around each sphere just touch each other and the polymer molecules still have access to the region between the particles. When the separation distance, as shown in Fig. 13.13C and D, is less than $2R_g$, the polymer molecules are expelled from the gap between the particles leading to the 'depletion of polymer molecules' in the gap. This creates an imbalance in the osmotic pressure i.e., the osmotic pressure in the region between the particles is zero, whereas, that in the region outside the depletion zone is nonzero. Thus, an attractive force results between the large colloidal particles. The strength of the depletion attraction is proportional to the volume of the "lens" that is formed as a result of the overlap of the depletion layers. Note that the size of the arrows that represent the magnitude of the force due to such osmotic pressure imbalance is smaller in Fig. 13.13C compared to that in Fig. 13.13D.

Asakura and Oosawa[54] developed a theoretical treatment for the analysis of depletion phenomena shown schematically in Fig. 13.13. The depletion potential, also called the AO potential, between a pair of spherical particles is given by the expression[54,55]:

$$\begin{aligned}\Phi_{Dep}(d) &= -PV_{ov} \quad \text{when } 0 \le d \le 2R_g \\ &= 0 \quad \text{when } d \ge 2R_g\end{aligned} \tag{13.43}$$

where $d = r - 2R_p$ is the surface to surface particle separation distance, P is the osmotic pressure given by Eq. (13.42), V_{ov} is the overlap volume. The overlap volume depends on the surface-to-surface separation distance and geometry of the colloidal particles. For the case of rectangular plates, the overlap volume is given by the area of the plate times the distance between the plates. For the case of two spherical particles of radius R_p, the volume of the lens formed (See Fig. 13.13) as the two particles approach each other is given by[33,54]:

$$V_{ov} = \frac{\pi}{6}(2R_g - d)^2 \left(3R + 2R_g + \frac{d}{2}\right) \tag{13.44}$$

A schematic representation of the depletion attraction between two spheres dispersed in a solution of polymer of radius of gyration, R_g, is shown in Fig. 13.14.

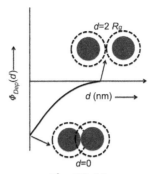

Fig. 13.14
A schematic of depletion interaction potential.

13.6.5 Steric Interactions

In Section 13.4, we discussed that the colloidal particles anchored with long chain molecules at their surface can stabilise dispersions. Typically, the anchoring of the long chain molecules on the particle surface is achieved by physical adsorption, chemisorption or via a covalent attachment of long chain molecules through a controlled chemical reaction. A schematic representation of the sterically stabilised colloids at different separation distance is shown in Fig. 13.15.

Fig. 13.15 shows two particles (smaller filled circles) grafted with long chain molecules that are attached onto the particle surface. The larger circles are drawn as guide to eye. The configuration of the surface anchored molecules (such as polymers) are known to depend on the nature of the polymer-solvent interactions. When the particles are sufficiently close to each other, there will be a competition between the polymer molecules to occupy the same space, i.e., the region represented in the form of lens in Fig. 13.15B. Therefore, steric repulsion can be viewed as arising due to the osmotic pressure imbalance, which in this case is due to an increase in the concentration of polymer chains in the lens-like region. In general, the total interaction potential for sterically stabilised colloids interacting via a fluid medium is given by:

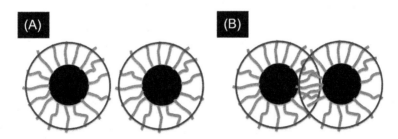

Fig. 13.15
A schematic of sterically stabilised colloids at (A) large (B) close separation.

$$\Phi_{total}(r) = \Phi_{vdW}(r) + \Phi_{steric}(r) \qquad (13.45)$$

The functional form of Φ_{steric}, depends on several factors such as length of the polymer chains, the grafting density, configuration of the polymer chains, the separation distance, temperature, etc.

13.7 Destabilisation of Colloids

While stabilisation of colloids is important in the context of increasing the shelf life of products such as inks, paints, liquid oral dosage forms such as syrups and others, the process of destabilisation of particles suspended in a fluid is also sought-after during several unit operations. One of the most important processes where destabilisation of colloids is exploited is in the field of the wastewater treatment. In order to remove colloidal scale suspended particles, which cannot be separated by gravity settling, the rate of sedimentation is often increased by the addition of flocculants such that the large aggregates that form can be separated by gravity settling. The destabilisation of colloids occurs due to interparticle attraction, achieved typically by the addition of a third component such as an electrolyte, acid/base, polymer, and particles, or by changing the solvent quality (when the particles are stabilised by surface-grafted polymers). Some of the most common methods used for the controlled destabilisation of colloids are discussed here.

13.7.1 Destabilisation of Charge-Stabilised Colloids

Charged stabilised colloids can be destabilised by the addition of suitable electrolytes, oppositely charged particles, or oppositely charged polymers. A dispersion of charge-stabilised colloids is kinetically stable due to the existence of a large energy barrier for the destabilisation of colloids. This energy barrier can be manipulated by the addition of symmetric or asymmetric electrolytes such as sodium chloride (NaCl), potassium chloride (KCl), calcium chloride ($CaCl_2$), etc. A schematic that shows the effect of the addition of an electrolyte on the DLVO interaction energy between a pair of particles in a charge-stabilised dispersion plotted versus separation distance is shown in Fig. 13.16. With an increase in the electrolyte concentration, the energy barrier progressively decreases and eventually vanishes completely. Assuming that the addition of electrolyte does not alter the dielectric constant, the van der Waals interactions remain unaffected. Therefore, the contribution of the electrical double layer interaction to the DLVO interaction decreases with the increasing electrolyte concentration.

Several types of colloidal particles, when dispersed in water, exhibit pH dependent stability behaviour. Typical examples include aqueous dispersions of silica (SiO_2), titanium dioxide (TiO_2), alumina (Al_2O_3), haematite (α-Fe_2O_3), amidine-functionalized polystyrene, carboxylated polystyrene, etc. The stability of these colloids is typically regulated by exploiting the fact that it is possible to modulate the surface charge density or zeta potential of particles.

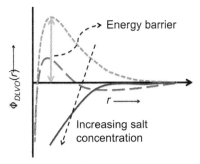

Fig. 13.16

A schematic showing the effect of addition of electrolyte on the DLVO interaction energy (Φ_{DLVO}) versus separation distance for a pair of particles in a charge-stabilised dispersion. With increase in the concentration of electrolyte, the energy barrier progressively decreases and vanishes completely.

For example, the zeta potential of haematite particles in pH 2 aqueous dispersion is +40 mV, i.e., these particles are positively charged. However, in pH 12 aqueous dispersion, the haematite particles are highly negatively charged with a zeta potential of −75 mV.[56] As the pH of the dispersion is either increased from 2 or decreased from 12, the zeta potential decreases, reaching a value close to 0 at a pH of ∼ 9. In order to use an appropriate equation for the conversion of electrophoretic mobility to zeta potential, it is essential to carry out electrophoretic measurements in the presence of electrolytes of known concentration. The zeta potential of haematite particles in aqueous dispersions at different pH has been measured in the presence of 0.0001 M NaCl.[56] The pH that corresponds to a zeta potential of zero is called the isoelectric point (IEP) or the point of zero charge (PZC). A colloidal dispersion at IEP exhibits least stability, and it progressively increases as the pH of the dispersion is driven sufficiently away from the IEP by the addition of appropriate quantity of an acid or a base. Similar to the effect of addition of electrolytes, the energy barrier for aggregation drastically reduces in the vicinity of IEP. However, a decrease in the magnitude of the energy barrier is a consequence of the reduction in the zeta potential and not due to the reduced double layer thickness caused by the electrolyte addition.

A dispersion of charged colloids can also be destabilised by the addition of oppositely charged particles. Because the aggregation is due to the electrostatic attraction and involve particles of opposite charge (different type), the destabilisation occurs due to the electrostatic heteroaggregation.[57–59] The destabilisation by this mechanism depends on the size ratio, surface charge ratio and the number ratio of colloidal particles in the mixture. For particles of identical size (size ratio = 1) and surface charges (charge ratio = 1), complete destabilisation is expected to occur when the dispersion mixture contains equal number of oppositely charged particles, i.e., when the number ratio is 1. In general, for particles of same size, the condition for maximum aggregation is given by,[58]

$$N_n|q_n| = N_p|q_p| \qquad (13.46)$$

where N_n and q_n, respectively, are the number and effective surface charge of the negatively charged particles in the mixture: N_p and q_p are the number and effective surface charge, respectively, of the positively charged particles in the mixture. Eq. (13.46) simply enforces the electroneutrality of a mixture of the oppositely charged colloids. A dispersion of charged colloids can also be destabilised by the addition of oppositely polyelectrolytes. The stability and aggregation behaviour of dispersions of charged colloidals in the presence of oppositely charged polyelectrolytes is qualitatively similar to that of the dispersions of large charged colloids in the presence of much smaller oppositely charged colloids.

13.7.2 Destabilisation by Polymer Addition

The addition of polymer to a colloidal dispersion can lead to attraction or repulsion between the particles and therefore there can be either destabilisation of colloids or stabilisation of particles against aggregation. This depends on a number of parameters: polymer-particle interactions, i.e., whether the polymer adsorbs onto the particle surface or not, and polymer-solvent interactions which influence the configuration of polymer molecules in solutions, polymer-particle size ratio, polymer concentration, particle concentration, etc. The polymer added to a colloidal dispersion can cause aggregation of particles and eventually destabilisation due to depletion effects or particle-polymer bridging. As discussed in Section 13.6, depletion effects occur when the size of a nonadsorbing polymer added to the colloidal dispersion is much smaller than the size of the particle. When the surface-to-surface distance between the colloidal particles is $<2R_g$, the particles experience a net attractive force. Thus, the configuration of a homogeneous polymer-colloid mixture shown schematically in Fig. 13.17A may eventually transform to a destabilised dispersion consisting of polymer rich and colloid rich regions. However, whether the particles cluster together or remain as individual units, as shown schematically in Fig. 13.17B and C, depends on the surface properties of the colloids, strength of van der Waals interactions, depletion attraction, etc. If the particle-polymer interactions are favoured, the polymer molecules can adsorb onto the particles, creating polymer bridges between the particles. This process is called bridging flocculation and is shown schematically in Fig. 13.17D. This leads to a destabilised dispersion wherein particles and polymers coexist. Note the significant change in the configuration of the polymer molecules that are adsorbed onto the particle surface.

13.8 Settling of Aggregates

In Chapters 6 and 8 of this volume, the settling behaviour of single particles and sedimentation of particulate dispersions are discussed. In these instances, the sedimentation process involves primary particles that are adequately characterised for their size and shape. Even if the particles are of complex geometry or irregular shape, settling behaviour can be analysed by invoking the concept of sphericity or by approximating the particle geometry to be close to one of the regular

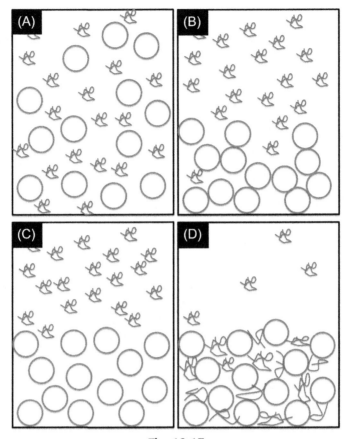

Fig. 13.17

Destabilisation of colloids by the addition of polymers: (A) Initially homogeneous colloid-polymer mixture (B) and (C) destabilisation evidenced by the formation of polymer-rich and colloid-rich regions due to depletion effects (D) destabilisation due to bridging flocculation. *Adapted by permission from Lekkerkerker HNW, Tuinier R. Stability of colloid–polymer mixtures. COPYRIGHT - Springer Nature; 2011.*

shapes. However, when a colloidal dispersion containing individual particles, shown in Fig. 13.18A, is destabilised by any of the methods discussed in the preceding sections, due to the attractive interactions, the primary particles cluster together to form multiparticle aggregates shown schematically in Fig. 13.18B. Therefore, the fundamental units that respond to any externally applied force field (be it gravity, electric, or magnetic) are the aggregates. As a result of aggregation, the particles lose kinetic independence and all the particles in an aggregate move as a collective single unit. Also, the characteristics of such aggregates may change or evolve depending on the destabilisation process at play and the strength of the attraction between the particles. If the particles are weakly attractive, any perturbation, be it physical or chemical, may lead to some rearrangement of the individual particles in the aggregates, thus changing their characteristics. In general, the gravity settling behaviour of

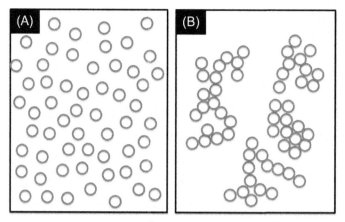

Fig. 13.18
(A) A stable colloidal dispersion containing individual particles (B) a destabilised colloidal dispersion containing multiparticle aggregates.

aggregates depends on several factors such as the number of particles in the aggregate, size distribution, the spatial arrangement of particles within the aggregate, their shape, morphology (such as open particle network versus compact nature), etc. Therefore, structural characterisation of colloidal aggregates is essential to analyse the results of the settling experiments and also to develop theories that predict the settling behaviour of such aggregates.

13.8.1 Structure of Aggregates

Fig. 13.19A shows a schematic of a spherical aggregate that is completely solid. Such aggregates are not real, and typically most aggregates are porous in nature as shown in Fig. 13.19B–D. The pores in the aggregates dispersed or settling in any solution are often filled with liquid. Therefore, for the aggregates shown in Fig. 13.19B–D, certain material properties such a density will be intermediate between the density of the solid particles and the suspending liquid. The density of an aggregate (ρ_{agg}) can be approximated as:

$$\rho_{agg} = \rho_p(1 - \epsilon) + \rho_l \epsilon \tag{13.47}$$

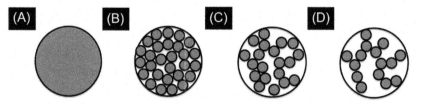

Fig. 13.19
Schematic representation of (A) hypothetical solid spherical aggregate, (B–D) corresponding porous spherical aggregates that may result following the destabilisation of colloidal dispersions.

where ρ_p is the density of the primary particles, ρ_l is the density of the suspending liquid, and ϵ is the porosity, which is the total volume of pores in the aggregate (total volume of liquid in the pores) divided by the volume of the aggregate.

The concept of fractals has been widely used to characterise colloidal aggregates. In doing so, it is assumed that the size of the aggregates is much larger than the size of individual colloidal particles, and that the aggregates are self-similar structures over the length scale considered for their analysis.[60] For the hypothetical solid spherical aggregate without any pores, the mass of the particle, m_p, is

$$m_p = \rho_p V_p \tag{13.48}$$

The mass of the particle scales with radius to the power 3,[1] i.e.,

$$m_p(r) = \rho_p \frac{4}{3}\pi r^3 \tag{13.49}$$

However, for a porous spherical aggregates shown in Fig. 13.19B–D, the mass scales with an exponent <3. In general, the variation of the mass of the porous aggregate with radius is expressed as,[1,60]

$$m_p(r) \sim \rho_p \frac{4}{3}\pi r^{D_f} \tag{13.50}$$

where $D_f < 3$ is the fractal dimension for a three-dimensional aggregate and $D_f < 2$ for a two-dimensional aggregate. For the aggregates shown in Fig. 13.20, the fractal dimension of the aggregate shown in (B) is greater than the fractal dimension of aggregate shown in (C), and the aggregate in (D) exhibits the lowest fractal dimension.

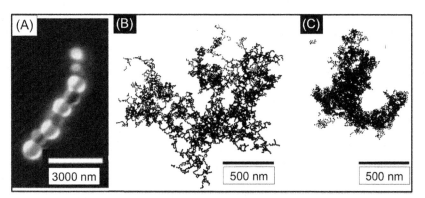

Fig. 13.20

Experimental visualisation of fractal aggregates: (A) linear chain-like aggregate of $D_f = 1.21 \pm 0.15$ composed of oppositely charged spherical particles, (B) aggregate of $D_f = 1.75 \pm 0.05$ formed due to diffusion limited aggregation, and (C) aggregate of $D_f = 2.05 \pm 0.05$ formed due to reaction limited aggregation. *B and C: Reprinted with permission from Weitz DA, Huang JS, Lin MY, Sung J. Limits of the fractal dimension for irreversible kinetic aggregation of gold colloids. Phys Rev Lett 1985;54:1416–9.*

Fig. 13.20 shows micrographs of the structure of fractal aggregates observed in experiments. A linear, chain-like aggregate of positive spheres of 1 μm diameter and negative spheres of 1.1 μm formed in the studies on heteroaggregation is shown in Fig. 13.20A.[59] Note that one of the particle appears bigger due to the different microscopy imaging modes used. The fractal dimensions of these aggregates measured by the static light scattering is $D_f = 1.21 \pm 0.15$. Fig. 13.20B shows aggregate of $D_f = 1.75 \pm 0.05$ formed due to the diffusion limited aggregation. A stable dispersion of 14.5 nm diameter gold nanoparticles in which the particles are about 870 nm apart (60 particle diameter) is considered and destabilised by the addition of pyridine. When the pyrene molecules replace the citrate ions that electrostatically stabilise gold nanoparticles, aggregation occurs due to the van der Waals forces. When sufficient quantity of pyridine is used, every encounter between the particles is effective in contributing to aggregation, i.e., the sticking probability is close to 1. Such a process is referred to as diffusion-limited aggregation.[60,61] This process leads to an open network of aggregated particles as shown in Fig. 13.20B, and the aggregate shown here consists of approximately 5000 primary particles.[62] However, at low concentrations of pyridine, the aggregation phenomenon depends on the interplay of both the electrical double layer repulsion and the van der Waals attraction. Therefore, sticking probability upon particle contact is significantly reduced, leading to reaction-limited aggregation.[60,61] This results in a more compact, denser aggregate with $D_f = 2.05 \pm 0.05$[62] shown in Fig. 13.20C. The fractal dimensions of the aggregates are typically obtained from the real space images of the aggregates via the analysis of the mass of the aggregate as a function of its radius, or by the analysis of the light intensity scattered from the aggregates measured in light-scattering experiments.

13.8.2 Settling of Spherical Aggregates

The motion of aggregates under the influence of gravity and the prediction of their terminal velocity is a complex and vexing problem. This is due to their intricate morphology, size, shape, deformation, or particle rearrangement when aggregates move under the influence of an external force field. The aggregates that settle under gravity in a quiescent fluid can be assumed to retain their size and shape if the attractive strength that holds the particles together is sufficiently large. If the aggregates are dense and impermeable (i.e., the fractal dimension of the particles is close to 3) without any fluid exchange between the aggregate and the surrounding, their settling behaviour can be analysed by extending the approach followed for single particle settling, discussed in Chapter 6 of this volume. The first step in such an analysis involves identifying the settling regime. Consider a dispersion of 15 nm radius particles such that individual particles do not settle under standard gravitational conditions. Let us assume that the controlled destabilisation leads to the formation of a 500 nm spherical impermeable aggregate. If we assume that the particles in this aggregate are closely packed, assuming a packing density of 0.74 for hexagonal close packing, the aggregate is expected to contain approximately 27,400 particles. Yet, due to their small size, the motion of such an aggregate under the influence of gravity is expected to be in the Stokes settling regime. Therefore, the expression for the terminal velocity for the settling of spherical particles can be modified to predict the settling behaviour of spherical impermeable aggregates.[63]

$$U_{t,agg} = \frac{g D_{agg}^2 \left(\rho_{agg} - \rho_l \right)}{18\eta} \tag{13.51}$$

where $U_{t,\,agg}$ is the terminal settling velocity of the aggregate, g is the acceleration due to gravity, D_{agg} is the diameter of the aggregate, ρ_{agg} is the density of the aggregate, ρ_l and η respectively are the density and viscosity of the fluid in which the aggregate is settling.

Substituting for ρ_{agg} from Eq. (13.47),

$$U_{t,agg} = \frac{g D_{agg}^2 \left(\rho_p - \rho_l \right)(1 - \epsilon)}{18\eta} \tag{13.52}$$

where ϵ is the porosity of the aggregate. The volume of the aggregate, V_{agg} is related to the volume of primary particles, V_p and the number of particles in the aggregate, N_p as:

$$V_{agg} = N_p V_p + V_l = N_p V_p + \epsilon V_{agg} \tag{13.53}$$

where V_l is the volume of the liquid in the aggregate. Eq. (13.53) in terms of diameter of the aggregate, D_{agg} and the diameter of the primary particle D_P is written as follows:

$$D_{agg}^3 = \frac{N_p D_P^3}{(1 - \epsilon)} \tag{13.54}$$

Combining Eqs. (13.52) and (13.54)

$$U_{t,agg} = \frac{g \left(D_P^2 N_P^{\frac{2}{3}} (1 - \epsilon)^{-2/3} \right) \left(\rho_p - \rho_l \right)(1 - \epsilon)}{18\eta} \tag{13.55}$$

which can be further rearranged as,

$$U_{t,agg} = N_P^{\frac{2}{3}} (1 - \epsilon)^{1/3} \frac{g D_P^2 \left(\rho_p - \rho_l \right)}{18\eta} \tag{13.56}$$

Therefore,

$$U_{t,agg} = N_P^{\frac{2}{3}} (1 - \epsilon)^{1/3} U_t \tag{13.57}$$

where U_t is the terminal settling velocity of the primary particle under free settling condition. Therefore, the settling velocity of aggregate depends on both the number of particles in the aggregate and the porosity of the aggregate.

In a similar way, if the attractive strength that holds the particles together is sufficiently large, i.e., there are no changes in the morphology and number of particles in the aggregate, the analysis of settling of linear aggregates of fractal dimension D_f close to 1 can be considered analogous to the settling of cylindrical or rod-like particles.[64] When the aggregates are of intermediate D_f, the estimation of their projected area required for the drag force calculations is nontrivial. The measurement and prediction of the terminal settling velocities of such fractal aggregates is a subject of ongoing research.[63]

13.9 Viscosity of Colloidal Dispersions

Several colloidal processing operations, which may involve transportation and pumping of dispersions, require the characterisation of response of colloidal dispersions to controlled flow fields. The study of flow and deformation of colloidal dispersions, a field referred to as rheology of colloidal dispersions, deals with response of stable and aggregated dispersions to externally applied flow fields. There are books dedicated to this field that discuss both basic elements and advanced aspects.[3] In this section, the effect of presence of particles on the viscosity of liquids is briefly discussed. The viscosity of colloidal dispersion is a useful quantity for predicting the settling behaviour of dilute and concentrated particle filled fluids. For a colloidal dispersion in which particle settling occurs in the Stokes' law regime, it is possible to account for hindered settling effects by applying Stokes' law developed for single particle settling, however, by replacing the properties of fluid with properties of the dispersion[65]:

$$U_{t,d} = \frac{gD_p^2(\rho_p - \rho_d)}{18\eta_d} \tag{13.58}$$

where $U_{t,\,d}$ is the relative terminal settling velocity of particles in the dispersion, D_p is the diameter of the particles, ρ_p is the density of the particle, ρ_d and η_d are the density and viscosity of the dispersion, respectively, and g is the acceleration due to gravity. This approach appears reasonable, as effectively, each particle is settling in a fluid whose density and viscosity are modified due to the presence of particles. Both ρ_d and η_d are a function of volume fraction of particles in the dispersion (or the volume fraction of liquid in the dispersion). The density of dispersion ρ_d is expressed in terms of density of particle (ρ_p), density of the dispersion medium (ρ_l) and the volume fraction of particles (ϕ) as:

$$\rho_d(\phi) = \rho_p\phi + \rho_l(1 - \phi) \tag{13.59}$$

Eq. (13.59) written in terms of the volume fraction of liquid in the dispersion, ϵ is:

$$\rho_d(\epsilon) = \rho_p(1 - \epsilon) + \rho_l\epsilon \tag{13.60}$$

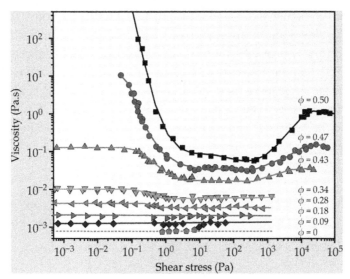

Fig. 13.21

The variation of viscosity of colloidal dispersions of different volume fraction ϕ with applied shear stress. The plot shows Newtonian behaviour (viscosity independent of shear stress), shear thinning behaviour (a decrease in viscosity with increase in shear stress) and shear thickening behaviour (an increase in viscosity with increase in shear stress). *Reprinted with permission from Wagner NJ, Brady JF. Shear thickening in colloidal dispersions. Phys Today 2009;**62**:27–32.*

A plot of the variation of viscosity of colloidal dispersions of latex particles at different volume fraction ϕ as a function of shear stress is shown in Fig. 13.21.[66] The viscosity at $\phi = 0, 0.09$, and 0.18 is independent of shear stress (and shear rate). For colloidal dispersions up to $\phi \leq =0.20$, shear stress (τ) varies linearly with shear rate ($\dot{\gamma}$), i.e.,

$$\tau = \eta\dot{\gamma} \tag{13.61}$$

where the proportionality constant η is the viscosity of fluid. Eq. (13.61) is the well-known Newton's law of viscosity. Therefore, colloidal dispersions up to $\phi \leq 0.20$ behave as Newtonian fluids.[67] When the concentration of particles in the dispersion is increased, the viscosity vs. shear stress plots show Newtonian plateau at low as well as high shear stresses, similar to dispersions with $\phi \leq 0.20$. However, at intermediate shear stresses, a marginal decrease in viscosity is observed with increase in shear stress. This behaviour associated with a decrease in viscosity as the fluid sample is subjected to higher shear stress is called shear thinning.[3,66,67] In Fig. 13.21, this behaviour is observed for dispersions at $\phi = 0.28$ and 0.34. As shown in Fig. 13.21, with further increase in the concentration of particles in the dispersion, that is, $\phi > 0.43$, the shear thinning becomes more prominent. Interestingly, at very high shear stresses, dispersions at $\phi = 0.43, 0.47$, and 0.50 exhibit an increase in viscosity with increase in shear stress. This behaviour is called shear thickening.[3,66,67] The simultaneous measurement of the microstructure of colloidal dispersions during flow has shown that the shear thinning is due to the flow-induced organisation of particles. The shear thickening observed in concentrated colloidal dispersions at higher shear stresses is attributed to the hydrodynamic interactions, i.e., the flow-induced formation of hydrodynamic clusters (which completely disappear when the flow is stopped) and the local change in particle concentration. It must be noted from Fig. 13.21 that the viscosity of colloidal dispersion for $\phi > 0$ is always higher than the viscosity of the suspending fluid ($\phi = 0$). That is, in general, the presence of particles in a liquid increases its viscosity. This is due to the perturbation of the flow field due to the presence of particles in the liquid. The dependence of viscosity of dispersion, η_d on the volume fraction of particles (ϕ) is typically expressed as a power series[3]:

$$\eta_d(\phi) = A + B\phi + C\phi^2 + D\phi^3 + \dots \tag{13.62}$$

As the concentration of particles in the dispersion decreases, the viscosity of dispersion decreases and approaches that of the suspending liquid at $\phi = 0$; therefore, Eq. (13.64) takes the form:

$$\eta_d(\phi) = 1 + B\phi + C\phi^2 + D\phi^3 + \dots \tag{13.63}$$

where B, C, and D are the constants that depend on the characteristics of dispersed species such as rigidity (hard, soft, nondeformable), size distribution, shape, etc. The use of Eq. (13.63) in conjunction with Eq. (13.58) for predicting the relative terminal settling velocity of particles in the dispersion requires the estimation of various constants. Einstein developed an equation for predicting the viscosity of dilute dispersions containing rigid nondeformable spherical particles[3,4]:

$$\eta_d(\phi) = \eta(1 + 2.5\phi) \tag{13.64}$$

Defining η_d/η as the relative viscosity η_r

$$\eta_r = 1 + 2.5\phi \tag{13.65}$$

Experiments have shown that Eq. (13.65) is applicable for $\phi \leq 0.05$, which is dilute enough that particle-particle interactions do not influence the viscosity of dispersions. At higher particle concentrations, the particles experience hydrodynamic interactions due to reduced interparticle separation, which can be accounted for by retaining terms up to ϕ^2 in Eq. (13.63). The relative viscosity of dispersions up to $\phi \sim 0.10$ can be calculated from Batchelor's equation[3,4,68]:

$$\eta_r = 1 + 2.5\phi + 7.6\phi^2 \tag{13.66}$$

A useful empirical expression for determining the viscosity of dispersions of high ϕ is the Krieger-Dougherty equation[3,4]:

$$\eta_r = \left(1 - \frac{\phi}{\phi_{max}}\right)^{-[\eta]\phi_{max}} \tag{13.67}$$

where ϕ_{max} is the maximum possible volume fraction of particle in the dispersion. ϕ_{max} is approximately 0.63–0.64 for random packing of hard spherical particles and is a function of particle shape. $[\eta]$ is the intrinsic viscosity, which for hard spheres is equal to $2.5/\rho_p$.[3]

13.10 Nomenclature

		Units in SI System	Dimensions in $\mathbf{M, N, L, T, \theta,}$
A_p	surface area of the particle	m^2	L^2
A_{sp}	specific surface area of the particle	m^2/kg	$M^{-1} L^2$
d	dimensionality	–	–
D	diffusivity of the particle in the fluid	m^2/s	$L^2 T^{-1}$
D_{agg}	diameter of the aggregate	m	L
D_f	fractal dimension for a three-dimensional aggregate	–	–
D_P	diameter of the particle	m	L
Δt	time interval	s	T
Δx^2	mean square displacement	m^2	L^2
e	elementary charge	C	$M^{1/2} L^{3/2} T^{-1}$
E	strength of the applied electric field	V/m	$M^{1/2} L^{-1/2} T^{-1}$
F_B	Brownian force on the particle	N	$M L T^{-2}$
F_g	force of gravity on the particle	N	$M L T^{-2}$
F_h	hydrodynamic drag force	N	$M L T^{-2}$
g	acceleration due to gravity	m/s^2	$L T^{-2}$
k_B	Boltzmann's constant	J/K	$M L^2 T^{-2} \theta^{-1}$
m_g	mass of the grafted molecules	kg	M
M	molar concentration	mol/m^3	$N L^{-3}$
M_d	mass of the colloidal dispersion	kg	M
M_p	mass of the particle	kg	M
M_s	mass of the suspending medium	kg	M
MW	molecular weight	kg/mol	$M N^{-1}$
N	number of particles	–	–
N_{AV}	Avogadro's number	–	–
N_g	number of grafted molecules or chains	–	–
N_n	number negatively charged particles	–	–
N_p	number positively charged particles	–	–
q	particle surface charge	C	$M^{1/2} L^{3/2} T^{-1}$
q_n	effective surface charge on negatively charged particle	C	$M^{1/2} L^{3/2} T^{-1}$

q_p	effective surface charge on positively charged particle	C	$M^{1/2} L^{3/2} T^{-1}$
R_p	radius of the particle	m	L
t	time	s	T
T	absolute temperature	K	θ
u	electrophoretic mobility	$m^2 s^{-1} V^{-1}$	$M^{-1/2} L^{3/2}$
U_t	terminal settling velocity of the particle	$m s^{-1}$	$L T^{-1}$
$U_{t,agg}$	terminal settling velocity of the aggregate	$m s^{-1}$	$L T^{-1}$
$U_{t,d}$	relative terminal settling velocity of particles in the dispersion	$m s^{-1}$	$L T^{-1}$
v	velocity of the particle	$m s^{-1}$	$L T^{-1}$
v_p	volume of single particle	m^3	L^3
V_d	volume of the colloidal dispersion	m^3	L^3
V_p	volume of the particles	m^3	L^3
V_s	volume of the suspending medium	m^3	L^3
x	position of the particle	m	L
$\bar{x}$	average distance travelled by the particles	m	L
X	mass fraction of particles in the dispersion	–	–
z	valency of ions	–	–

Greek Letters

ε	permittivity	$C^2 N^{-1} m^{-2}$	–
ε_0	permittivity of free space	$C^2 N^{-1} m^{-2}$	–
ε_r	relative permittivity	–	–
η	viscosity	kg/m s	$M^1 L^{-1} T^{-1}$
η_r	relative viscosity	–	–
η_d	viscosity of the dispersion	kg/m s	$M^1 L^{-1} T^{-1}$
$[\eta]$	intrinsic viscosity	m^3/kg	$M^{-1} L^3$
γ	shear rate	s^{-1}	T^{-1}
κ	Debye-Hückel parameter	m^{-1}	L^{-1}
κ^{-1}	Debye screening length	m	L
ϕ	volume fraction of particles	–	–
ϕ_{max}	maximum possible volume fraction of particles	–	–
ϕ_p	packing density of particles in the aggregate	–	–
ρ	density	kg/m^3	$M L^{-3}$
ρ_{agg}	density of the aggregate	kg/m^3	$M L^{-3}$
σ_g	grafting density	m^{-2}	L^{-2}
σ_s	surface charge density	C/m^2	$M^{1/2} L^{-1/2} T^{-1}$
τ	shear stress	Pa	$ML^{-1} T^{-2}$
ζ	zeta potential	V	$M^{1/2} L^{1/2} T^{-1}$

Subscripts

d	dispersion	–	–
f	fluid	–	–
i	*i*th species	–	–
n	negative	–	–
p	positive	–	–
P	particle	–	–
r	relative	–	–
s	suspending medium	–	–

References

1. Hiemenz PC, Rajagopalan R. *Principles of colloid and surface chemistry*. M. Dekker; 1997.
2. Israelacvili JN. *Intermolecular and surface forces*. Academic Press; 2011.
3. Mewis J, Wagner NJ. *Colloidal suspension rheology*. Cambridge University Press; 2011.
4. Larson RG. *The structure and rheology of complex fluids*. Oxford University Press; 1999.
5. Russel WB, Saville DA, Schowalter WR. *Colloidal dispersions*. Cambridge University Press; 1989.
6. Goodwin J. *Colloids and interfaces with surfactants and polymers*. London: John Wiley & Sons Ltd; 2004.
7. Hunter RJ. *Foundations of colloid science*. Oxford University Press; 1989.
8. Stöber W, Fink A, Bohn E. Controlled growth of monodisperse silica spheres in the micron size range. *J Colloid Interface Sci* 1968;**26**(1):62–9.
9. Bogush GH, Tracy MA, Zukoski CF. Preparation of monodisperse silica particles: control of size and mass fraction. *J Non-Cryst Solids* 1988;**104**(1):95–106.
10. Kallay N. Egon Matijević, his personality and achievements. *J Colloid Interface Sci* 2013;**392**:1–6.
11. Mukhopâdhyâya J. Electric synthesis of colloids. *J Am Chem Soc* 1915;**37**(2):292–7.
12. Matijevic E. Monodispersed metal (hydrous) oxides—a fascinating field of colloid science. *Accounts of Chemical Research* 1981;**14**(1):22–9.
13. Matijevic E. Production of monodispersed colloidal particles. *Annu Rev Mater Sci* 1985;**15**:483–516.
14. Sugimoto T. Preparation of monodispersed colloidal particles. *Adv Colloid Interface Sci* 1987;**28**:65–108.
15. Dugyala VR, Daware SV, Basavaraj MG. Shape anisotropic colloids: synthesis, packing behavior, evaporation driven assembly, and their application in emulsion stabilization. *Soft Matter* 2013;**9**(29):6711–25.
16. Pontoni D, Narayanan T, Rennie AR. Time-resolved SAXS study of nucleation and growth of silica colloids. *Langmuir* 2002;**18**(1):56–9.
17. Han Y, Lu Z, Teng Z, Liang J, Guo Z, Wang D, Han M-Y, Yang W. Unraveling the growth mechanism of silica particles in the Stöber method: in situ seeded growth model. *Langmuir* 2017;**33**(23):5879–90.
18. Kuijk A, van Blaaderen A, Imhof A. Synthesis of monodisperse, rodlike silica colloids with tunable aspect ratio. *J Am Chem Soc* 2011;**133**(8):2346–9.
19. Ozaki M, Kratohvil S, Matijević E. Formation of monodispersed spindle-type hematite particles. *J Colloid Interface Sci* 1984;**102**:146–51.
20. Ocaña M, Morales MP, Serna CJ. Homogeneous precipitation of uniform α-Fe2O3 particles from iron salts solutions in the presence of urea. *J Colloid Interface Sci* 1999;**212**:317–23.
21. Hornig S, Heinze T, Becer CR, Schubert US. Synthetic polymeric nanoparticles by nanoprecipitation. *J Mater Chem* 2009;**19**:3838–40.
22. Zhang C, Pansare VJ, Prud'homme RK, Priestley RD. Flash nanoprecipitation of polystyrene nanoparticles. *Soft Matter* 2012;**8**:86–93.
23. Rajeev A, Erapalapati V, Madhavan N, Basavaraj MG. Conversion of expanded polystyrene waste to nanoparticles via nanoprecipitation. *J Appl Polym Sci* 2016;**133**(4):42904.
24. Almog Y, Reich S, Levy M. Monodisperse polymeric spheres in the micron size range by a single step process. *Br Polym J* 1982;**14**:131–6.

25. Okubo M, Ikegami K, Yamamoto Y. Preparation of micron-size monodisperse polymer microspheres having chloromethyl group. *Colloid & Polymer Sci* 1989;**267**:193–200.

26. Richez AP, Yow HN, Biggs S, Cayre OJ. Dispersion polymerization in non-polar solvent: evolution toward emerging applications. *Prog Polym Sci* 2013;**38**(6):897–931.

27. Barrett KEJ. *Dispersion polymerization in organic media*. Wiley; 1974.

28. Klein SM, Manoharan VN, Pine DJ, Lange FF. Preparation of monodisperse PMMA microspheres in nonpolar solvents by dispersion polymerization with a macromonomeric stabilizer. *Colloid Polym Sci* 2003;**282**:7–13.

29. Peng B, van der Wee E, Imhof A, van Blaaderen A. Synthesis of monodisperse, highly cross-linked, fluorescent PMMA particles by dispersion polymerization. *Langmuir* 2012;**28**(17):6776–85.

30. Perrin J. Mouvement Brownien et Realite Moleculaire. *Ann Chim Phys* 1909;**18**:5–104.

31. Perrin J. *Atoms* [D.L. Hammick, Trans.]. London: Constable & Company Ltd; 1916.

32. Napper DH. Colloid stability. *Ind Eng Chem Prod Res Dev* 1970;**9**:467–77.

33. Lekkerkerker HNW, Tuinier R. *Colloids and the depletion interaction*. Springer; 2011.

34. Masliyah JH, Bhattacharjee S. *Electrokinetic and colloid transport phenomena*. New Jersey: Wiley; 2006.

35. Napper DH. Steric stabilization. *J Colloid Interface Sci* 1977;**58**:390–407.

36. Wu YL, Derks D, van Blaaderen A, Imhof A. Melting and crystallization of colloidal hard-sphere. *Proc Natl Acad Sci USA* 2009;**106**:10564–9.

37. Liu Y, Bławdziewicz J, Cichocki B, Dhont JKG, Lisicki M, Wajnryb E, Young Y-N, Lang PR. Near-wall dynamics of concentrated hard-sphere suspensions: comparison of evanescent wave DLS experiments, virial approximation and simulations. *Soft Matter* 2015;**11**:7316–27.

38. Feigin RI, Napper DH. Depletion stabilization and depletion flocculation. *J Colloid Interface Sci* 1980;**75**:525–41.

39. Kim S, Hyun K, Moon JY, Clasen C, Ahn KH. Depletion stabilization in nanoparticle-polymer suspensions: multi-length-scale analysis of microstructure. *Langmuir* 2015;**31**(6):1892–900.

40. Semenov AN, Shvets AA. Theory of colloid depletion stabilization by unattached and adsorbed polymers. *Soft Matter* 2015;**11**(45):8863–78.

41. Lee YS, Wagner NJ. Rheological properties and small-angle neutron scattering of a shear thickening, nanoparticle dispersion at high shear rates. *Ind Eng Chem Res* 2006;**45**(21):7015–24.

42. Poon WCK, Weeks ER, Royall CP. On measuring colloidal volume fractions. *Soft Matter* 2012;**8**:21–30.

43. Hunter RJ. *Zeta potential in colloid science: principles and applications*. New York/London: Academic Press; 1981.

44. Labib ME, Robertson AA. The conductometric titration of latices. *J Colloid Interface Sci* 1980;**77**:151–61.

45. Lutterbach N, Versmold H, Reus V, Belloni L, Zemb T. Charge-stabilized liquidlike ordered binary colloidal suspensions. 1. Ultra-small-angle X-ray scattering characterization. *Langmuir* 1999;**15**(2):337–44.

46. Stone-Masui J, Watillon A. Characterization of surface charge on polystyrene latices. *J Colloid Interface Sci* 1975;**52**:479–503.

47. Sprycha R, Jablonski J, Matijevic E. Zeta potential and surface charge of monodispersed colloidal yttrium (III) oxide and basic carbonate. *J Colloid Interface Sci* 1992;**149**:561–8.

48. Feick JD, Chukwumah N, Noel AE, Velegol D. Altering surface charge nonuniformity on individual colloidal particles. *Langmuir* 2004;**20**(8):3090–5.

49. Kerckhofs G, Pyka G, Moesen M, Van Bael S, Schrooten J, Wevers M. High-resolution microfocus x-ray computed tomography for 3D surface roughness measurements of additive manufactured porous materials. *Adv Eng Mater* 2013;**15**:153–8.

50. Petekidis G, Moussaïd A, Pusey PN. Rearrangements in hard-sphere glasses under oscillatory shear strain. *Phys Rev E* 2002;**66**:051402.

51. Valmacco V, Elzbieciak-Wodka M, Besnard C, Maroni P, Trefalt G, Borkovec M. Dispersion forces acting between silica particles across water: influence of nanoscale roughness. *Nanoscale Horizons* 2016;**1**(4):325–30.

52. Bergström L. Hamaker constants of inorganic materials. *Adv Colloid Interface Sci* 1997;**70**:125–69.

53. Xing X, Hua L, Ngai T. Depletion versus stabilization induced by polymers and nanoparticles: the state of the art. *Current Opinion in Colloid & Interface Science* 2015;**20**:54–9.

54. Tuinier R, Fan T-H, Taniguchi T. Depletion and the dynamics in colloid–polymer mixtures. *Current Opinion in Colloid & Interface Science* 2015;**20**:66–70.

55. Asakura S, Oosawa F. On interaction between two bodies immersed in a solution of macromolecules. *J Chem Phys* 1954;**22**:1255–6.

56. Dugyala VR, Basavaraj MG. Control over coffee-ring formation in evaporating liquid drops containing ellipsoids. *Langmuir* 2014;**30**(29):8680–6.

57. Islam AM, Chowdhry BZ, Snowden M. Heteroaggregation in colloidal dispersions. *J Adv Colloid Interface Sci* 1995;**62**:109–36.

58. Bansal P, Deshpande AP, Basavaraj MG. Hetero-aggregation of oppositely charged nanoparticles. *J Colloid Interface Sci* 2017;**492**:92–100.

59. Kim AY, Hauch KD, Berg JC, Martin JE, Anderson RA. Linear chains and chain-like fractals from electrostatic heteroaggregation. *J Colloid Interface Sci* 2003;**260**(1):149–59.

60. Bushell GC, Yan YD, Woodfield D, Raper J, Amal R. On techniques for the measurement of the mass fractal dimension of aggregates. *Adv Colloid Interface* 2002;**95**:1–50.

61. Lazzari S, Nicoud L, Jaquet B, Lattuada M, Morbidelli M. Fractal-like structures in colloid science. *Adv Colloid Interface Sci* 2016;**235**:1–13.

62. Weitz DA, Huang JS, Lin MY, Sung J. Limits of the fractal dimension for irreversible kinetic aggregation of gold colloids. *Phys Rev Lett* 1985;**54**:1416–9.

63. Johnson CP, Li X, Logan BE. Settling velocities of fractal aggregates. *Environ Sci Technol* 1996;**30**:1911–8.

64. Komar PD. Settling velocities of circular cylinders at low Reynolds numbers. *The Journal of Geology* 1980;**88**(3):327–36.

65. Rhodes M. *Introduction to particle technology*. Chichester: Wiley; 2008.

66. Wagner NJ, Brady JF. Shear thickening in colloidal dispersions. *Phys Today* 2009;**62**:27–32.

67. Laun HM. Rheological properties of aqueous polymer dispersions. *Die Angewandte Makromolekulare Chemie* 1984;**123**(1):335–59.

68. Batchelor G, Green J. The determination of the bulk stress in a suspension of spherical particles to order c2. *J Fluid Mech* 1972;**56**:401–27.

Further Reading

Kovalchuk NM, Starov VM. Aggregation in colloidal suspensions: effect of colloidal forces and hydrodynamic interactions. *Adv Colloid Interface Sci* 2012;**179–182**:99–106.

Gambinossia F, Mylonb SE, Ferria JK. Aggregation kinetics and colloidal stability of functionalized nanoparticles. *Adv Colloid Interface Sci* 2015;**222**:332–49.

Wu H, Lattuada M, Morbidelli M. Dependence of fractal dimension of DLCA clusters on size of primary particles. *Adv Colloid Interface Sci* 2013;**195–196**:41–9.

Zanini M, Hsu C-P, Magrini T, Marini E, Isa L. Fabrication of rough colloids by heteroaggregation. *Colloids Surf A* 2017;**2017**(532):116–24.

Lama H, Basavaraj MG, Satapathy DK. Desiccation cracks in dispersion of ellipsoids: effect of aspect ratio and applied fields. *Phys Rev Materials* 2018;**2**:085602.

CHAPTER 14

Health and Explosion Hazards

14.1 Introduction

Handling and processing (conveying, metering, mixing, heating, or cooling, etc.) of fine particles—powders and dust—in wide-ranging industrial settings, including agricultural, chemical, cement, coal and minerals, foodstuffs, metals, pharmaceutics, plastics, woodworking, etc., is much more difficult than that of gases and liquids. These difficulties stem from a variety of sources (particle shape and size, free flowing or not, chemical nature like toxicity, explosive, mechanical aspects like hardness, abrasive nature, combustion characteristics, etc.). Thus, the severity of the problem varies both with the type of unit operation and the specific material at hand. While the first aspect related to the mixing, conveying, size reduction or enlargement, liquid–solid separation, etc., have been discussed in detail in Chapters 3–5 of this volume, consideration is given here to the corresponding health, fire, and explosion hazards of powders, which are, indeed, quite acute in a range of situations including granaries, flour mills, sand blasting, grinding, machining, and milling, and a host of other processing operations. Indeed, the so-called dust (comprising solid and liquid constituents) is generated during the course of machining, grinding of particulate minerals and ores, metals, charging and discharging of grain silos, filter bags, flour mills, pouring of powders into empty vessels and those containing liquids, sanding, rapid impact, detonation and decrepitation (by heat) of limestone and nitrates of inorganic and organic substances, woodworking, mining and dredging operations, combustion processes, etc., to name a few. While it is difficult to define precisely the size of the so-called dust, it does clearly span a range of particle size, typically in micron range, and it also depends upon the process of its generation. In order to develop a feel for particle size, it is useful to mention here that a typical cotton fibre is on the order of 15–20 μm in diameter, human hair ranges from 50 μm to 500 μm, and normal red blood cells are in 8–10 μm range. Based on our everyday experience, coupled with the fact that the human eye can see particles as small as 35–40 μm, while we have no problem in seeing a human hair, a microscope is needed to examine the red blood cells. However, the phrase 'particle size' itself is ambiguous, as seen in Chapter 2, in the context of a nonspherical shape. The definition of dust varies from one application to another; in the context of industrial hygiene, the National Safety Council (United States) simply defines dust as the 'fine' solid or fluid particles generated during the course of handling (charging or discharging of grain silos, pneumatic conveying, powdering, sieving, for instance), grinding, crushing, sanding, etc.

Coulson and Richardson's Chemical Engineering. https://doi.org/10.1016/B978-0-08-101098-3.00015-9

Notwithstanding this uncertainty, Table 14.1 lists typical particle size (or range) for scores of materials. This encompasses six orders of magnitude, with viruses and atmospheric dust being at the lower end of this spectrum, and beach sand, pollens, etc., lying at the other end. Naturally, smaller the particle, the larger is its specific surface area, thereby resulting in advantages in terms of high rates of heat and mass transfer, chemical reactions, etc. On the other hand, such small particles settle very slowly and, thus, remain suspended in air

Table 14.1 Typical particle size[1]

Substance	Particle Size (μm)
Anthrax (solid)	1–5
Asbestos	0.7–90
Atmospheric dust	0.001–40
Car emissions	1–150
Bacteria	0.3–60
Beach sand	100–10^4
Bone dust	3–300
Burning wood	0.2–3
Ca-Zn dust	0.7–20
Carbon black dust	0.2–10
Cement dust	3–100
Clay	0.1–50
Coal dust	1–100
Coal flue gas	0.08–0.2
Copier toner	0.3–15
Dust mites	100–300
Face (Talc) powder	0.1–30
Fibreglass insulation	1–10^3
Fertiliser	10–10^3
Flour, milled corn	1–100
Fly ash	1–10^3
Ginger powder	25–40
Grain dust	5–10^3
Ground limestone	10–10^3
Insecticides dust	0.5–10
Iron dust	4–20
Lead dust	2
Metallurgical dusts and fumes	0.1–10^3
Mist	70–350
Mould	3–12
Mould spores	10–30
Oil smoke	0.03–1
Paints and pigments	0.1–5
Pollens	10–10^3
Radioactive fallout	0.1–10
Rosin smoke	0.01–1
Saw dust	30–600
Smoke from synthetic materials	1–50

Table 14.1   Typical particle size—cont'd

Substance	Particle Size (μm)
Smouldering or flaming cooking oil	0.03–0.9
Spider web	2–3
Spores	3–40
Talcum dust	0.5–50
Tea dust	8–300
Textile dust	6–20
Viruses	0.005–0.3
Yeast cell	1–50

and water for significantly long periods. Depending upon their concentration and nature (toxic, carcinogenic, abrasive, etc.) and the duration of their exposure to such atmospheres, there are varying levels of a range of health hazards. These are discussed in detail in Section 14.2. By the same reasoning, when the particles are prone to ignition, combustion, and burning in the presence of oxygen and/or sources of ignition or at high temperatures, this can lead to fires and explosions with devastating consequences. This is discussed in Section 14.3. Suffice it to add here that excellent books are available on health hazards and industrial hygiene,[2–4] as well as on dust fires and explosions,[1, 4–7] and reference should be made to these sources for more details and guidelines of regulatory bodies such as the National Safety Council of the United States, the Occupational Safety and Health Administration (OSHA) in the United Kingdom, etc.

14.2 Health Hazards and Risks of Dust and Fine Powders

Perhaps the single most important characteristic of fine particles and powders that directly impinge upon human health is their ability to become airborne and to form dust clouds capable of being inhaled through the mouth and nose, although some particles also act as irritants and/or can enter the human body through skin. This is generally less of a problem than that caused by inhalation. In industrial settings, there is a broad spectrum of dusts and airborne particles (varying in concentration, toxicity, hazard index, size) in the environment in which people work and live. For instance, imagine living next to a coal-based thermal power station or a flour mill! The health hazards stemming from a regular exposure over long periods of time often have short-term and long-term ill effects on the well-being of people. It is generally believed that, in addition to their chemical characteristics, the particle size and specific surface area are the other two important factors in determining their potential in inflicting inflammatory injury, oxidative damage, and other biological damage depending upon where the inhaled particles end up inside the human body. For instance, very fine and ultrafine particles can penetrate deep into the air passages of the respiratory tract, and these can even reach alveoli, in which up to 50% of the inhaled particles can be retained in the lung

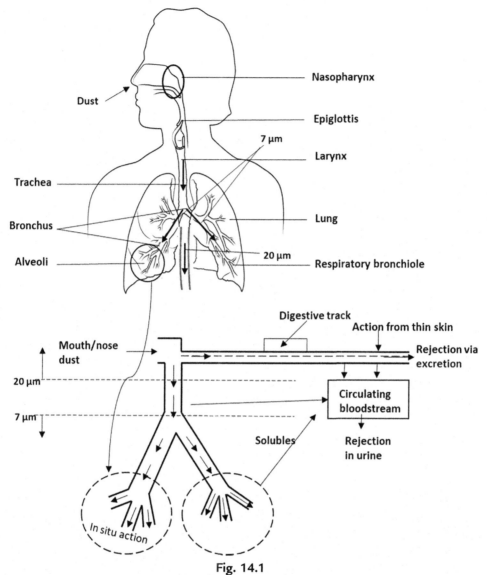

Fig. 14.1

Schematics of the respiratory tract system in humans.

parenchyma. In this context, the respiratory tract system, shown schematically in Fig. 14.1, can be visualised as a tube progressively bifurcating into smaller and smaller diameter branches. Thus, the so-called coarse ($>\sim10$–$15\,\mu$m) particles are filtered out by nose and upper air passages. The so-called fine ($<2.5\,\mu$m) and ultrafine ($<0.1\,\mu$m) particles penetrate deep into the respiratory system, reaching all the way to the alveoli.[8, 9] Furthermore, while some dusts exist in the form of primary particles, agglomeration is not uncommon in these systems. For instance, metal fumes start with a base particle size of about $\sim0.01\,\mu$m, which almost instantaneously form agglomerates as big as 0.2–0.5 μm in size. In addition to size, of course, chemical composition of particles also plays a significant role in determining the severity of potential health hazard in a given situation.

14.2.1 The Respiratory System and Health Effects

In its simplest form, our respiratory tract system is shown in Fig. 14.1, which can be visualised as a tube progressively bifurcating into narrow passages. Therefore, when we inhale a particle, the health risk it poses very much depends upon exactly where it ends up in the system. For instance, very fine particles can really penetrate deep, all the way up to the alveoli, and can eventually pass into the blood stream. As noted earlier, the course of action depends principally on the particle size. For instance, the diameter of the main bronchi and terminal bronchioles are ~10,000 and 500 μm, respectively. These values together with the particle size influence the deposition and passages of particles entering the lungs. Terminal bronchioles connect to the respiratory bronchioles, ultimately leading to the respiratory space composed of a number of alveoli (~100 μm in diameter). The surface of the alveoli is covered by a thin blanket (~0.5–1 μm) made up of a network of capillaries which facilitates gas exchange involving inflow of oxygen into the blood and that of carbon dioxide in the opposite direction.

On the other hand, the cilia—hair-like structures or organelles—covering the surface of trachea and the bronchial capillaries create a mechanism whereby the particles deposited can be dislodged. Beyond the terminal bronchioles, the surface lining of the air passages does not have this action, and, therefore, the particles cannot be dislodged once deposited.

For particles entering the respiratory tree through nasal (pharyngeal regions) via the mouth rather than through the nose, breathing being normal for moderately demanding physical work, soluble solids dissolve on a wet surface, or if insoluble in water, they remain as a solid. Such particles can be rejected by the body via spitting or nose blowing, or they may be swallowed, making their way to the digestive tract. Coughing and sneezing also carry the particles (>10 μm) toward the upper parts of the respiratory system. On the other hand, soluble solids may pass into the bloodstream, leading to serious health conditions.

Obviously, the inhalation of particles is a problem associated mainly with dusty processes and working environments, but there are other routes through which fine particles can enter the human body. Two such mechanisms are briefly mentioned here. The dust particles deposited on hands, clothes, and other personal articles like cell phones, computers, etc., contribute to the total intake of particles into the body by so-called ingestion. Even with good housekeeping practices in place to control the concentration of airborne dust, a casual working culture can lead to a greater intake of solids by ingestion through smoking (cigarettes, pipes), eating, and drinking activities at the workplace.

Similarly, some industrial dusts directly attack the skin, causing irritation, itching, and even dermatitis. Moreover, some organo-metal and organo-phosphorous compounds may also be absorbed into the body via the intact skin or open wounds, etc.

Currently available research results suggest the following health effects caused by the inhalation of particles.[4, 8, 9]

(1) When the particles are lodged in the pulmonary region, it leads to the malfunction of the lungs. This, in turn, leads to the loss of elasticity of the lung walls, which impedes the diffusion of O_2 and CO_2. This can be caused by a number of chemical species, free crystalline silica (encountered in mines or grinding wheels, for instance) present in the inhaled dust being the main one. Other potential species leading to pneumoconiosis include asbestos, beryllium, china clay, iron oxides, talc powder, etc.

(2) If the toxic particles (Mn, Cd, Pb, etc.) penetrate into the circulation network and internal organs after dissolution, the human body fails through so-called system poisoning.

(3) The relationship between long-term exposure to working atmospheres with arsenic, chromate, nickel, and radioactive particles, and lung cancer has been studied extensively for a long time.[9] While the debate continues, soluble cancer-causing agents are regarded as a risk to the lungs as well as to the other organs.

(4) Less severe hazards include irritation and allergies. The presence of particles like fumes of acids, cadmium, beryllium, vanadium, plastics, pesticides, acid mists, and fluorides in the work atmosphere is a common source of irritation, which may manifest in the form of bronchitis, pneumonitis, and pulmonary oedema, etc. On the other hand, vegetable dusts from bagasse, corn, cotton, flour, sawdust, pollens, metallic dusts of nickel and chromium, etc., result in allergic reactions in the form of asthma, hay fever, metal fume fever, etc. Also, airborne particles carrying fungal, viral, or bacterial pathogens result in the transmission and spread of various types of infections.

More details concerning other health hazards, threshold limiting values, and underlying mechanisms leading to the aforementioned conditions are available in the literature.[4, 8, 9] Some remedial measures in the form of respiratory protective devices have been described by Tanaka and Hori.[10]

14.3 Dust Explosions

A dust explosion is caused by a quick release of a large amount of energy due to the combustion of a particulate material of small size. Given the fact that >50% of the solids processed in chemical and allied process engineering applications are combustible to some extent, accompanied by a rapid increase in pressure due to confinement, dust explosions pose a serious process safety hazard in a wide range of settings entailing a broad range of obvious and not so obvious materials. For instance, coal dust, flour, sulphur, etc., immediately come to mind as

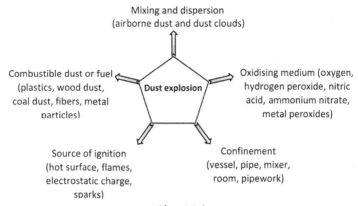

Fig. 14.2
Dust explosion pentagon.

posing a risk of dust explosion, whereas one would normally not consider metallic dust (aluminium, copper, iron, etc.) as prone to dust explosions. Generally, a dust explosion occurs when the following five conditions are met: fuel (combustible particulate), mixing (turbulence), oxidising medium (generally air), ignition source (hot surfaces, open flames, sparks, electrostatic charging), and confinement (partial or complete). These are often shown schematically in the form of the so-called dust explosion pentagon; a term introduced by Kauffman possibly for the first time,[11] shown schematically in Fig. 14.2. The process can simply be described as follows: when a combustible particulate material forms a dust cloud of certain concentration in air when ignited, the combustion process initiates, releasing a large amount of heat which increases the pressure in a confined space very rapidly, eventually leading to the so-called dust explosion.

Indeed, in their extensive review, Abbasi and Abbasi[12] have documented the available historical data on dust explosion dating back to the 18th century. A quick glance at this review reveals the following three aspects of dust explosions:

(a) *Diversity of materials*
   Dust explosions have occurred in the handling and processing of wheat flour, grains and grain dust, corn starch powder, sulphide dust, silicone, aluminium, magnesium-aluminium alloys, fish meat, coal, rape seed flour pellets, textile dust, potassium chlorate, cotton waste, rubber waste, resins and gums, polymeric materials, agricultural dusts, dyes, pesticides, sugar dust, etc. Undoubtedly, this range of materials posing the risk of a dust explosion is beyond the imagination of an average process engineer!

(b) *Processing operations and equipment*
   Bearing in mind that one side of the explosion pentagon is confinement, Abbasi and Abbasi[12] have also included the setting of dust explosions. These include bakeries, ships, storage and shipping bins, hammer mills, elevators, flour mills, grain silos, milling

stations, mixing tanks, mines, cement plants, bucket elevators, coal dust burners, dust removal and collection systems, power houses, ovens, etc. Additional possibilities of dust explosions also exist in cyclones, scrubbers, electrostatic precipitators, different types of dryers, packaging facilities, etc. This list reinforces the fact that even a partial confinement can lead to a dust explosion.

(c) *Consequences*

The number of workers injured and/or deaths resulting from such explosions, not to mention the financial loss and down time, are also troubling and discomforting.

Therefore, this chapter aims to provide a brief overview of dust explosions including the identification and characterisation of factors contributing to such a risk, and its mitigation. However, more detailed discussions can be found in excellent books[1, 5–7] and other publications.[11–15]

Types of Dust

It is rather difficult to offer a precise definition of dust itself. For instance, according to the National Fire Protection Association (of the United States),[16] any finely divided solid of diameter $\leq 420 \mu m$ is classified as dust. In contrast, according to the British Standards,[17] any fine particulate material of particle size smaller than $76 \mu m$ (200 BS mesh size) is regarded as dust. Notwithstanding the 6-fold range, it is perhaps justified to accept the conservative proposal of Palmer,[18] which does not exclude particles as big as 1 mm from such a discussion. Because one of the key factors in initiating and sustaining a dust explosion is the amount of energy released by the combustion reaction, the particle size, together with the heat of combustion of the dust material, are important factors in this regard. As a particle progressively becomes smaller, its specific surface area increases, which accelerates its combustion, and hence, the quantum of energy liberated. Second, it is much easier to form a cloud of fine particles with air, and this also facilitates their combustion and flame propagation. Representative values of heat of combustion (for complete oxidation) of a few common dusts are listed in Table 14.2.

Evidently, the dust explosion of metallic dusts can be much more violent and energetic than that of nonmetallic dusts. Because combustion reactions lead to the formation of stable oxides, dusts comprising such oxides (like silicates and carbonates, for instance) are generally not explosive. Instead, such substances are used as suppressants for dust explosions, e.g. the use of limestone in coal mines. Such inert materials act as heat sinks thereby regulating the temperature rise. Similarly, not all combustible dusts explode, but all explosible dusts must be combustible in nature. For instance, both graphite and anthracite have very high values of heat of combustion but are not easily explosible. Similarly, fly ash, a product of combustion reaction, is regarded to be nonexplosible. If, however, if it becomes contaminated with unburned fuel, (such as pulverised coal or petroleum coke) or the fly ash is a reaction

Table 14.2 Typical values of heat of combustion

Dust	Heat of Combustion (kJ/mol Oxygen)
Aluminium	1100
Calcium	1270
Carbon	400
Chromium	750
Coal	400
Iron	530
Magnesium	1240
Polyethylene	390
Silicon	830
Starch	470
Sucrose	470
Sulphur	300

*Modified from Abbasi T, Abbasi SA. Dust explosions—cases, causes, consequences, and control. J Hazard Mater 2007;**140**:7–44.*

product of incomplete combustion of coal, nonexplosible fly ash can transition to an explosible mixture.

Numerous criteria are used to characterise a dust for its explosion potential. These include the so-called Group A (capable of sustaining flame propagation following its ignition) or Group B (flame does not propagate and thus explosion does not occur).[19] However, this rather crude classification is based on the observations when the ignition occurs about 25°C. Thus, it is quite possible that a dust classed as Group B can migrate to Group A if it is ignited at temperatures >25°C. The second approach is based on the nature of combustion[20, 21] and this scheme classifies dusts into six types, simply labelled as CC1 to CC6. The salient characteristics of each type are summarised in Table 14.3.

This classification is, however, based on the behaviour of a well defined dust heap when exposed to a gas flame or hot platinum wire as the source of ignition. Thus, it is more of a measure of the ignitability of a dust layer and intensity of burning of a dust layer. This is of

Table 14.3 Combustion-based classification of dusts[21]

	Type	Key Characteristics
	CC1	No ignition and no self-sustained combustion
	CC2	Short ignition, quick extinguishing; local combustion of short duration
	CC3	Localised burning and glowing, but without spreading and propagation.
↓ Increasing Explosibility	CC4	Spreading of a glowing fire and propagation of smouldering combustion
	CC5	Spreading of an open fire; propagating open flame
	CC6	Explosive burning and combustion

direct relevance to the applications wherein powders or dusts are accumulated on hot surfaces such as that on dryers, kilns, heat exchangers, etc.

Both preceding schemes are direct measures of the intrinsic fire and explosion risk associated with a given dust. Returning to Fig. 14.2, confinement is a key element for a dust explosion to occur. This is related to the maximum pressure and the rate of pressure rise in the confined region or inside process equipment. Neither of the preceding classification schemes account for this aspect, and this deficiency is rectified in the scheme proposed by the US Bureau of Mines,[22] which categorises the hazard potential of a material using the ignition and explosion characteristics of Pittsburgh seam coal (PC) as the reference.

This approach relies on the use of two individual parameters, both of which are dimensionless: ignition sensitivity (IS) and explosion severity (ES), defined as follows:

$$IS = \frac{\{MIT \times MIE \times MEC\}_{PC}}{\{MIT \times MIE \times MEC\}_{dust\ sample}} \tag{14.1}$$

where MIT = minimum ignition temperature, MIE = minimum ignition energy, MEC = minimum explosive concentration

$$ES = \frac{\{MEP \times MRPR\}_{dust\ sample}}{\{MEP \times MRPR\}_{PC}} \tag{14.2}$$

where MEP = maximum explosion pressure and MRPR = maximum rate of pressure rise.

And finally, the index of explosibility (IE) is given simply by the product of these two parameters, i.e.

$$IE = IS \times ES \tag{14.3}$$

Because IE is a relative measure, it is less dependent on the experimental methods employed to measure the values of MIT, MIE, MEC, MEP, and MRPR, etc., as long as the same procedure is employed for both materials. Its main disadvantage, however, is the fact that the evaluation of IE necessitates the full range of tests. Based on extensive experimental data for scores of powders of pure metals, alloys, copper, nickel, and lead-bearing ores, Jacobsen et al.[22] reported values of IE ranging from IE ~ 0 to IE > 10. Based on the values of IE, they also classified various dusts posing no (~ 0), weak (IE < 0.1), moderate ($0.1 \leq$ IE ≤ 1), strong ($1 \leq$ IE ≤ 10), and severe (IE > 10) risks of dust explosion. To put this classification in context, atomised aluminium dust poses a severe risk, as do the magnesium, uranium, and thorium powders. This fits in qualitatively with the fact that their heats of combustion are rather high, Table 14.2. Next, the hydrides of titanium and zirconium, for instance, fall within the strong risk category. Obviously, some of the individual quantities appearing in Eqs (14.1) and (14.2) are strongly influenced by the particle size and cloud concentration, source of ignition, size of apparatus, etc.; these aspects are discussed in a later section in this chapter.

Finally, perhaps the most widely used classification scheme is based on the so-called K_{st} value, which is a measure of the MRPR (maximum rate of pressure rise) in $1\,m^3$ vessel upon the ignition of the dust. This concept was initially introduced by Bartknecht [as cited in Ref. 6] in the following form:

$$\left(\frac{dP}{dt}\right)_{max} \cdot V^{1/3} = K_{st} \quad \text{(constant for a dust)} \tag{14.4}$$

Eq. (14.4) has been shown to be applicable for test vessel volumes greater than $\sim 0.04\,m^3$. In order to make a sensible comparison between the K_{st} values from various sources, it is tacitly assumed that (i) geometrically similar test vessels yield geometrically similar flame characteristics, (ii) the flame thickness is negligible compared to the radius of the vessel so that the lateral spreading of flame is not influenced, and (iii) the burning velocity as a function of temperature and pressure is identical in vessels of different volumes. Deviations from these assumptions can lead to wide-ranging values of K_{st} for a given dust. For instance, maize starch dust clouds tested in the range of $0.0012 \leq V \leq 13.4\,m^3$ vessels, the reported values of K_{st} vary from ~ 3 to $\sim 209\,bar\,m/s$! While some of the deviations can be ascribed to the differences in the dust samples themselves (such as particle size and moisture content, for instance), both degree of turbulence in the dust clouds and flame characteristics (especially flame thickness) are regarded to be the main contributing factors here,[7] both of which are difficult to control and quantify. This indicates a degree of arbitrariness in the value of K_{st} obtained from nonstandard tests.[23] On the other hand, the values of K_{st} obtained using the standardised protocols seldom vary by $>10\%–15\%$. Generally, materials with $K_{st} \sim 0$ are classified as nonexplosible, $K_{st} < 200\,bar\,m/s$ as being weakly explosible, and $200 < K_{st} < 300$ as very strongly (severely) explosible.

14.3.2 Experimental Methods

In essence, two systems—the Hartmann vertical tube[17] and the so-called 20-L sphere – have been used extensively in the literature. Most of the experimental results prior to the 1980s were obtained using the Hartmann vertical tube method, shown schematically in Fig. 14.3. It consists of a 300 mm long and 64 mm inner diameter tube (approximate volume of $\sim 1.2\,L$) in which dust is dispersed in the form of a cloud using air blast, which is then ignited using a hot wire or a spark device. Flame propagation is monitored as a function of particle size and shape, concentration, ignition energy, and temperature, all of which are known to influence the degree of potential risk, irrespective of the classification one uses.

Notwithstanding the fact that it is virtually impossible to achieve a uniform dust cloud and/or to regulate the level of turbulence, additional effects arise from confinement (wall effects) and from the fact that the flame travels in both directions—upward and downward – following its initial spherical expansion. Naturally, all these factors influence (generally tend to reduce) the rate of combustion, and hence, the value of MRPR. Therefore, these values must be

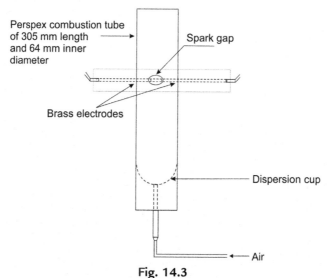

Perspex combustion tube
of 305 mm length
and 64 mm inner
diameter

Spark gap

Brass electrodes

Dispersion cup

Air

Fig. 14.3
Schematics of the Hartmann vertical tube.

treated with reserve and used for indicative purposes only, rather than using them as a basis for designing appropriate mitigation systems.

Over the years, many improved versions of this apparatus have been developed to overcome some of these deficiencies. Perhaps the best of all such designs is the one which employs a large spherical vessel rather than a vertical tube. Two designs – the so-called 20-L vessel[24-26] and the ISO standard $1\,m^3$ vessel[23] – have gained wide acceptance in the literature. In this method, the dust is introduced via a pressurised container or through a "ring sparger" type device, as shown schematically in Fig. 14.4, for both 20 L and $1\,m^3$ versions. The source of ignition is located at the centre of the sphere. Detailed experimental protocols for the $1\,m^3$ ISO vessel are available in the relevant ASTM standard.[6] While the results obtained in the 20 L and $1\,m^3$ vessels are not always in perfect agreement, one important factor to bridge this gap between the two values is to use an ignition source smaller than 5 kJ in a 20-L vessel, whereas it is fixed at a value of 10 kJ in the case of the $1\,m^3$ vessel. Of course, the mechanical details of the methods of dust injection and dispersion also play a role in the case of the 20-L apparatus.

14.3.3 Influencing Factors

From the foregoing discussion, it is abundantly clear that the severity and consequences of a dust explosion (irrespective of what classification is used) are determined by an intricate interplay between numerous factors. Probably, the most significant of these are particle size, dust concentration, oxidant concentration, ignition energy and temperature, turbulence level in the cloud, value of MRPR, etc. These are summarised in Table 14.4 for a quick reference.

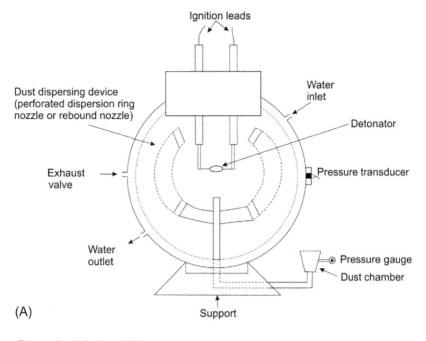

(A)

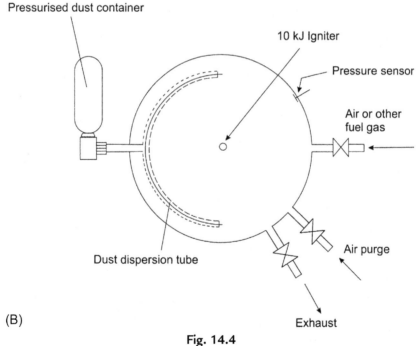

(B)

Fig. 14.4

Schematics of spherical test vessels for measuring the maximum pressure (P_{max}) and K_{st} values (A) 20 L vessel (B) 1 m³ (150) vessel.

Table 14.4 Parameters used to characterise dust explosion risk[7]

Factor	Description
P_{max}	Maximum explosion pressure in a constant volume device
MRPR	Maximum rate of pressure rise in a constant volume explosion
K_{st}	The value of MRPR in a $1\,m^3$ constant volume explosion and accounts for other volumes via Eq. (14.4)
MEC	Minimum explosible dust concentration
MIE	Minimum energy needed to ignite a dust cloud
MIT	Minimum ignition temperature of a dust cloud
LIT	Minimum ignition temperature of a deposit/layer of dust
LOC	Limiting oxygen concentration in the atmosphere for flame propagation in a dust cloud

Because the key underlying processes in dust explosion are the combustion and flame propagation, both of which are also influenced by these very factors. In this section, the role of each of these factors is discussed in brief, and more detailed discussions can be found in numerous references, e.g. see Refs. 1, 6, 7. Table 14.5 provides a quick guide to the variation of these factors with the dust characteristics, and these are described in brief in the following section.

(a) *Particle Size*

As mentioned earlier, the specific surface area of a particle available for combustion increases with the decreasing particle size, and, thus, the smaller the particle size, the greater is the hazard associated with it. On the other hand, some fine particles tend to agglomerate, and if the resulting lumps are $>\sim 500\,\mu m$, the substance may even become nonexplosible. Fig. 14.5 shows the effect of particle size on the value of K_{st} reported for HDPE when burned with pure air and blended with various hydrocarbons.[27] Clearly, the severity of the dust explosion increases (as reflected by the increasing values of K_{st}) with the decreasing particle size and/or the addition of hydrocarbons to air.

(b) *Particle Concentration in dust cloud*

Naturally, a dust cloud explodes only within a certain range of concentrations. Typical values of the so-called minimum concentration required to initiate an explosion are $50–100\,g/m^3$, and that of the maximum concentration are $2–3\,kg/m^3$. For many combustible dusts, a dust concentration of $\sim 500\,g/m^3$ produces the most devastating overpressures and the values of MRPR. The lower limit of the concentration is set by the fact that the combustion should lead to moderate values of K_{st} to pose a threat. On the other hand, the maximum concentration is limited by the availability of sufficient oxygen required for a sustained combustion, which may not always be possible. Also, flame propagation is somewhat impeded at high concentration thereby lowering the probability of dust explosion. Thus, it is virtually impossible to disperse and efficiently burn excessively thick dust deposits.

Table 14.5 Influence of dust properties/characteristics on dust explosion parameters[12, 17]

Parameter	Increases With	Decreases With
Explosibility of the dust	1. Lower explosible concentration 2. Minimum ignition temperature 3. Lower minimum ignition energy 4. Burning velocity 5. Maximum rate of pressure increase 6. Presence of chemical groups such as COOH, OH, NH_2, NO_2, $C\equiv N$, $C=N$, $N=N$ 7. Presence of volatile matter in the dust at levels above 10% 8. Relatively small proportion of fines 9. Increasing oxygen concentration	1. Presence of chemical groups such as Cl, Br, F 2. Presence of inert material at concentrations above 10%–20% 3. Dust moisture content above 30%
Effect of particle size on the likelihood of explosion initiation	50–70 μm < particles size (μm) <500 μm (with decreasing particle size)	500 μm < particles size (μm) <50–70 μm (with increasing particle size)
Minimum explosive concentration	1. Increasing moisture content 2. Increasing concentration of inertant	1. Decreasing particle size 2. Increasing volatile matter 3. Increasing oxygen concentration
Minimum ignition temperature	1. Increasing moisture content 2. Increasing concentration of inertant	1. Decreasing particle size 2. Increasing volatile matter content 3. Increasing oxygen concentration 4. Increasing thickness of the dust layer
Maximum permissible oxygen concentration	Decreasing dust temperature	Increasing dust temperature
Maximum explosion pressure	Decreasing particle size, though weakly	–
Maximum rate of pressure rise	1. Decreasing particle size 2. Increasing volatile matter content 3. Increasing oxygen concentration	1. Increasing moisture content 2. Increasing concentration of inertant

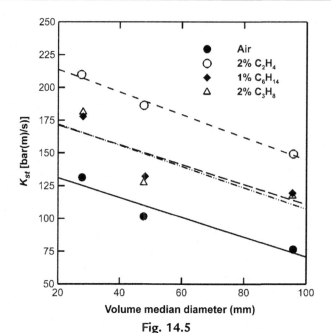

Fig. 14.5

Dependence of K_{st} on particle size and type of fuel for polyethylene obtained using 20 L test vessel.[27]

From another vantage point, one can appreciate the role of dust layer thickness (h) through the following equation:

$$C = \rho_{bulk} \left(\frac{h}{H} \right) \tag{14.5}$$

Where ρ_{bulk} is the bulk density of dust layer (typical values for coal dust, corn flour, and iron powder are $560 \, kg/m^3$, $820 \, kg/m^3$, and $2800 \, kg/m^3$, respectively). In Eq. (14.5), H is the height of the dust cloud and h is the thickness of dust layer. For the purpose of illustration, here, let $\rho_{bulk} = 500 \, kg/m^3$ and $h = 1 \, mm$. In a 5 m high room, this dust layer will create a dust cloud of $100 \, g/m^3$ concentration, which will rise to $500 \, g/m^3$ if the height of the room is reduced to 1 m. Thus, it stands to reason that even a small amount of dust, under suitable conditions, can lead to explosions. Finally, dust concentration influences both the value of P_{max} and MRPR, as shown in Fig. 14.6 for two coal dust samples. Typical values of the minimum explosive concentration are listed in Table 14.6.

(c) *Concentration of Oxidising Medium*

Most common oxidising medium is oxygen present in the atmospheric air. Naturally, artificially oxygen enriched air increases the rate of combustion, and vice versa. Therefore, the air participating in combustion is progressively depleted in oxygen, thereby slowing down the burning rate with the passage of time. Under such conditions, either the combustion may cease to occur, or if an explosion occurs, it may not be very severe.

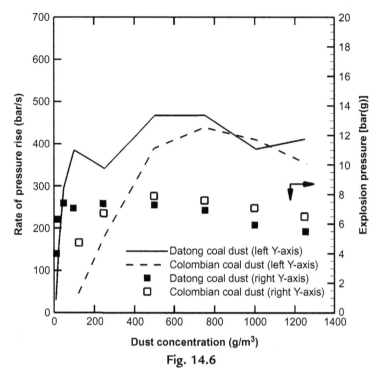

Fig. 14.6

Maximum pressure (right scale, ■: Datong coal dust, □: Colombian coal dust) and rate of pressure rise (left scale, Datong coal dust, Colombian coal dust) in two coal dust. *Repotted from Amyotte PR, An introduction to dust explosions. Oxford, UK: Elsevier; 2013.*

Generally, fires cannot be sustained when the oxygen concentration in the air falls below 10%.

In this context, it is customary to introduce the so-called limiting oxygen concentration (LOC). It is defined as the highest O_2 concentration in the dust/air/inert gas mixture which does not lead to an explosion. Naturally, such values vary from one inertant to another. Typical LOC values of various dusts with N_2 as the inertant are presented in Table 14.7. These values are also influenced by the temperature and pressure conditions.

(d) *Sources of Ignition and Minimum Ignition Energy (MIE)*

One of the pillars of Fig. 14.2 is the source of ignition. Both Abbasi and Abbasi[12] and Taveau[29] offer long and somewhat similar lists of potential sources of ignition in processing industries. These include open flames and direct heat, mechanical sparks, welding and cutting, self-heating, electrostatic discharges, self-ignition, hot surfaces, fire, shock waves, lighting, etc. Indeed, dusts can be ignited by low-energy as well as high-energy sources.

Over the years, standardised equipment and procedures have been developed,[7,12] namely, the so-called MIKE 3 apparatus (Kuhner A.G., Basel, Switzerland) and the ASTM

Table 14.6 Typical MEC and MIT values[1]

Dust	MEC (g/m^3)	MIT (°C)
Aluminium	45	650
Al-Mg alloy	–	550
Chromium	230	–
Coal	55	610
Copper	–	900
Corn starch	–	430
Epoxy resin	20	530
Flax shive	–	430
Grain dust	–	430
Iron dust	100	420
Wheat flour	–	430
Wheat straw	–	470
Rye dust	–	430–500
Magnesium dust	20	520
Rice	–	440
Silicone	110	–
Soy flour	–	540
Tin	190	630
Titanium	45	460
Zinc	480	600

Table 14.7 Typical LOC (with N_2 as inert medium) values[28]

Type of Dust	LOC (%v/v)
Aluminium	5
Cadmium stearate	12
High density polyethylene	10
Organic pigment	12
Pea flour	15.5
Sulphur	7
Wheat flour	11

standard[30] for measuring the minimum ignition energy for a given dust. The MIKE 3 setup consists of a 1.2L cylindrical glass chamber, similar to the Hartmann tube shown in Fig. 14.3, into which dust is dispersed and ignited by an electrical spark (≤ 1 J), and the ignition criterion is that the resulting flame must propagate at least up to 60 mm away from the electrode. Because the MIKE3 device uses 1 mH inductance, it results in a longer duration spark, and, thus, one tends to measure a somewhat smaller value of MIE than what would be obtained in the absence of any inductance in the spark circuit. The available practical experience indicates that most, if not all, electrostatic discharges in chemical and

Table 14.8 Typical values of MIE[31]

| Material | MIE (mJ) | |
	With (1 mH) Inductance	Without Inductance
Aluminium granulate	50	500
Benzanthron	0.9	1.0
Epoxy coating powder	1.7	2.5
Epoxy polyester coating powder	2.3	9.0
Flock	~70–100	1300–1600
Lycopodium	5	50
Magnesium granulate	25	200
Organic stabiliser	0.4	0.4
Polyamide coating powder	4	19
Polyester coating powder	2.9	15
Polyurethane coating powder	2	8

processing plants entail negligible inductance. Table 14.8 presents representative values of MIE (with and without) inductance.[31] The values of MIE with inductance are seen to be appreciably lower than that in purely capacitive cases. In either event, these values are of the order of 1.5 J only.

However, the values of MIE (with inductance) generally decrease with the decreasing particle size, as does the corresponding minimum ignition temperature.[32]

(e) *Ignition Temperature*

When an ignitable dust and air mixture is heated, at some temperature, it catches fire. The lowest temperature at which such a mixture ignites is called the minimum ignition temperature (MIT). However, this is generally different form the ignition point of a dust layer (LIT). For a given dust, MIT decreases with the decreasing particle size (due to increased surface area) and increasing proportion of volatile matter present in the dust cloud and/or in oxygen-rich atmosphere. On the other hand, concentration of water (moisture) and other inert molecules (like carbonates, slilicates) raise the minimum ignition temperature.

The so-called BAM oven along with ASTM E1491 standard[7, 33] is used to experimentally measure the minimum ignition temperature of a dust cloud. In this method, a dust cloud is generated by squeezing a rubber bulb (in which the sample to be tested is loaded), and the resulting cloud impinges on a circular concave metal plate heated to a known temperature ($\leq 600°C$). The ignition is ascertained by visual observation of the flame exiting from the rear of the oven covered by a metal flap. Typical ignition temperatures of common dusts in air are summarised in Table 14.6.

(f) *Effect of Turbulence*

While it is virtually impossible to measure and quantify the level of turbulence in the context of dust explosions, its signatures can be seen in terms of the high rate of mixing, heat and mass transfer, and the extent of chemical reaction, all of which facilitate the release of a large amount of energy and the temperature and pressure rise in a confined system. Generally, it is customary to distinguish between so-called pre- and post-ignition turbulence. Pre-ignition turbulence is related to the process of converting a heap or layer of dust into a dust cloud. Thus, it is influenced by the type of unit operation, equipment, etc., used to create a dust cloud. This turbulence seems to have a much greater effect on the rate of pressure rise than on over-pressurisation due to the explosion.

As the name implies, the post-ignition turbulence is caused by flame propagation in the unburned part of the dust cloud. Because the rate of heat removal from the ignition zone under turbulent conditions is intensified, the ignition temperature (MIT) and ignition energy (MIE) requirements generally increase with the post-ignition turbulence levels. It has been argued in the literature that the degree of mixing is increased, leading to the formation of dynamic, three-dimensional structures comprising burned, burning, and unburned fuel particles. This, in turn, further creates combustion sites, thus, leading to rapid combustion and pressure rise at higher turbulence levels.

It must be borne in mind that the values of MIT included in Table 14.6 are only indicative because some of these dusts (grain and flour dusts) can be ignited (in the form of smouldering) even at temperatures as low as 200 °C under appropriate conditions, e.g. fine particles, high turbulence and low relative humidity, etc.

14.4 Prevention of Dust Explosions

Recalling the explosion pentagon (Fig. 14.2), an explosion will not occur if one of the five elements is missing in a given application. Therefore, most of the tools developed to minimise the risk of an explosion can be traced back to Fig. 14.2. The available explosion prevention strategies fall into the following four main categories: process modification, elimination of the generation of dust cloud, elimination or minimisation of the occurrence of ignition sources, and inertion of dust and/or of the oxidising medium. Each of these are discussed here in some detail.

14.4.1 Process Modifications

This idea is similar to the philosophy of inherently safer process design and process intensification. While it is not always possible or feasible to replace the existing processes by new ones eliminating the presence of combustible dusts, one can always modify the process (or part thereof) so that the tendency for the formation of dust cloud is reduced, e.g. using mass flow silos and hoppers rather than the common practice of funnel flow to charge a vessel. Similarly, the use of nitrogen as the transporting medium and as the sealing medium in silos as opposed to air will lower the risk. Other possibilities, including maintaining the dust concentration below the MEC value, or controlling moisture in pipes and storage silos, and installing isolation valves between silos to minimise the possibility of secondary or sequential explosions, etc., have been discussed in detail by Amyotte et al.[32] and others.[1, 12] In line with the idea of inherently safer design, the use of smaller amounts of hazardous materials, or replacement with less hazardous ones, or use of a material in its least hazardous form by suitable modifications to the process and process conditions, etc., are some of the tools employed in industry to reduce the risk of a potential hazard.

14.4.2 Formation of Explosible Dust Clouds

While it is desirable to keep the dust concentration below MEC, it is always not possible due to process requirements and operating conditions which are often dictated by economic and other considerations. Notwithstanding this aspect, the following preventive steps may be helpful in reducing the possibilities of dust formation:

(i)   Free fall of dust should be avoided as much as possible.
(ii) Work with smaller piles and heaps of dust rather than with one large heap or layer.
(iii) Dust removal from a gaseous process stream may be carried out at an early stage if the process permits it.
(iv) Good housekeeping practices like proper ventilation and cleaning help reduce the accumulation of dust on the floor, external surfaces, etc. For instance, NFPA standard[34] recommends that dust layers of a thickness of 0.8 mm warrant immediate cleaning of the region; a dust layer this thick is prone to create hazardous conditions if it covers >5% of the floor area or ~90 m^2, whichever is smaller; such calculations should also include the dust on walls, overhead beams, ductwork, conduit cabling, piping, etc. Similarly, emission of dusts should also be reduced by proper design of process equipment. On the other hand, housekeeping practices must be selected based on the explosion characteristics of a dust, e.g. vigorous sweeping or the use of compressed air to blow down equipment in dusty areas may themselves lead to the formation of explosible dust clouds. Thus, it may be desirable to use a suitable vacuum cleaner first before sweeping or blowing down the plant area.

14.4.3 Management of Ignition Sources

While it is generally not possible to precisely identify the source of ignition in a dust explosion, some general observations are made here. There are two types of ignition sources present in a plant: the types which originate from routine conditions and/or working styles of individuals such as smoking, unprotected light bulbs, open flames, welding, cutting and grinding, etc. These can be addressed by staff training and good working practices and discipline.

However, more serious are those which are inherent to the processing operations itself. Typical examples include hot surfaces (dryers, kilns), open flames, smouldering nests and exothermic decompositions and oxidation reactions, electric sparks, electrostatic discharges, etc. In this case, the hazard cannot be completely eliminated, but it can certainly be reduced by periodic removal of the dust layers deposited on such hot surfaces, or by proper earthing of equipment and paying attention to the unusual operation of the equipment.

14.4.4 Inertion

This refers to the ways and means of lowering the oxygen concentration in a plant area or in a process equipment by adding an inert gas to the extent that the dust cloud can no longer sustain a propagating flame upon ignition, i.e. the fuel is deprived of the oxygen it requires to burn. Inertion is also used in another way by adulterating the dust cloud using a noncombustible entity. This is, however, not used often because it is not always possible to add an impurity to a product.

Typical inerting gases include nitrogen, carbon dioxide, water vapor, and rare gases. However, the final choice of a particular gas depends upon its chemical affinity with the dust. For example, carbon dioxide is an excellent inerting gas in many situations, but it cannot be used with aluminium dust due to its severe reactivity in this case. Similarly, nitrogen, a very effective inertant for many combustible dusts, reacts with magnesium dust, thereby rendering it unsuitable for such an application. In addition, economic considerations also influence the final choice of an inerting material in a given situation. In view of the cost and problems in achieving complete inertion, lowering the risk of explosion through partial inertion is quite common in practical situations. Admittedly, numerous studies are available in the literature, e.g. see,[12] but it is not yet possible to put forward definitive guidelines in this regard. For instance, it is generally believed that the value of K_{st} varies linearly with the percentage of oxygen in the gas phase. Another factor to be borne in mind is that the values of the maximum permissible oxygen concentration to prevent ignition are generally measured at ambient temperatures. Therefore, caution must be exercised in using such values at temperatures above 100°C or so. Thus, additional tests must be carried out in such cases or in cases of hybrid vapor-dust mixtures. It is thus not uncommon to use 2% as a safety margin, i.e. if the maximum permissible oxygen concentration is known to be 11%, the oxygen concentration must be maintained below 9%. Of course, reducing the risk of explosion by inertion increases the risk of suffocation.

Liquid/solid inerting materials are used with two objectives: prevention or mitigation/control of a dust explosion. As noted earlier, in the context of the first objective, such inertants are added to an otherwise explosible dust in sufficient amounts to render it nonexplosible. One outstanding example of this approach is to spray rock dust (mainly $CaCO_3$ with or without $MgCO_3$) in coal mine galleries. The rock dust simply acts as a heat sink and absorbs heat from the flame front, thereby inhibiting its propagation. Naturally, smaller particles are more effective than coarse ones due to their higher specific surface areas, but very fine particles lose

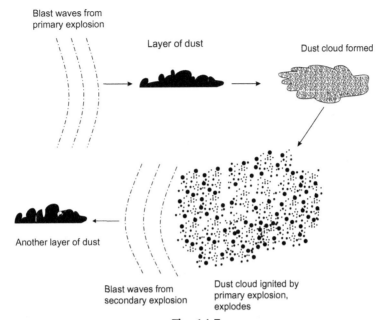

Fig. 14.7

Domino effect leading to secondary explosion. *Replotted from Abbasi T, Abbasi SA, Dust explosions—cases, causes, consequences, and control.* J Hazard Mater 2007;*140*:7–44.

some of this effectiveness due to their tendency to form agglomerates. This approach is, however, generally limited to coal mines only. The second idea behind the use of solid inertants is to control an explosion by dispersing such a solid in adequate quantity to the vessel on fire. The success of this approach also hinges on quenching the flame or inhibiting flame propagation.

The discussion in this chapter has, thus far, been limited to isolated dust explosions. In practice, however, a dust explosion can lead to several secondary explosions, depending upon the proximity of dust heaps to the site of the primary explosion. This is known as the 'domino effect' and is shown schematically in Fig. 14.7. Similarly, when a layer of dust is ignited, the celerity of the flame depends upon the heat of combustion. Such a fire in an unconfined environment generally does not lead to an explosion because the heat is dissipated to the atmosphere by convection and radiation, thereby raising the temperature of atmospheric air in a small region of flame (Fig. 14.8). Also, the resulting flame can act as an ignition source directly or indirectly by heating nearby process equipment.

Many other issues relating to dust explosions, including mitigation, relevant standards, strategies for inherently safe design to reduce the potential risk, expert systems, etc., are discussed in several excellent reviews[12] and books.[1, 6, 7] However, this chapter is expected to provide a useful starting point to the vast field of dust hazards and explosions.

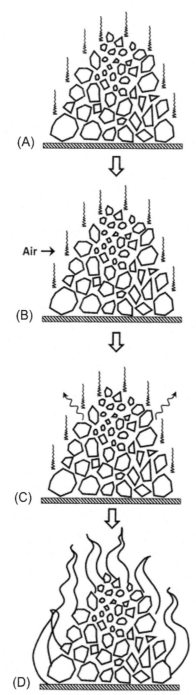

Fig. 14.8

Dust layer fire: (A) dust layer with various particle sizes; (B) air ingress and initial self-heating; (C) rate of removal of heat by convection and radiation lower than the rate of heat generation; (D) self-ignition at some critical temperature. *Redrawn from Amyotte PR, An introduction to dust explosions. Oxford, UK: Elsevier; 2013.*

14.5 Nomenclature

		SI Unit	Dimensions
C	Concentration	g/m^3	$\mathbf{ML^{-3}}$
h	Height of dust layer	m	$\mathbf{L}$
H	Height of room	m	$\mathbf{L}$
ES	Explosion severity	–	$\mathbf{M^0L^0T^0}$
IE	Index of explosibility	–	$\mathbf{M^0L^0T^0}$
IS	Ignition sensivity	–	$\mathbf{M^0L^0T^0}$
K_{st}	Parameter, Eq. (14.4)	bar.m/s	$\mathbf{MT^{-3}}$
LIT	Minimum ignition temperature for a dust layer	K	$\boldsymbol{\theta}$
LOC	Limiting oxygen concentration (volume % basis)	–	–
MEC	Minimum explosion concentration on volume basis	m^3/m^3	$\mathbf{M^0L^0T^0}$
MEP	Maximum explosion pressure	Pa	$\mathbf{ML^{-1}\,T^{-2}}$
MIE	Minimum ignition energy	J	$\mathbf{ML^{-2}\,T^{-2}}$
MIT	Minimum ignition temperature of a dust cloud	K	$\boldsymbol{\theta}$
MRPR	Maximum rate of pressure rise	Pa/s	$\mathbf{ML^{-1}\,T^{-3}}$
P	Pressure	Pa	$\mathbf{ML^{-1}T^{-2}}$
t	Time	s	$\mathbf{M^0L^0T}$
V	Volume	m^3	$\mathbf{M^0L^3T^0}$
ρ_{bulk}	Bulk density of solid	kg/m^3	$\mathbf{ML^{-3}T^0}$

References

1. Cheremisinoff NP. *Dust explosion and fire prevention handbook: a guide to good industry practices*. Beverly, MA: Scrivener/Wiley; 2014.
2. Plog BA, Quinlan PJ. *Fundamentals of industrial hygiene*. 6th ed. Washington, DC: National Safety Council; 2012.
3. Fuller TP. *Essentials of industrial hygiene*. Washington, DC: National Safety Council; 2015.
4. Masuda H, Higashitani K, Yoshida H, editors. *Powder technology handbook*. 3rd ed. Boca Raton, FL: Taylor and Francis; 2006.
5. Baker WE, Tang MJ. *Gas, dust and hybrid explosions*. Amsterdam: Elsevier; 1991.
6. Eckhoff RK. *Dust explosion in the process industries*. 3rd ed. New York: Gulf Publishing/Elsevier; 2003.
7. Amyotte PR. *An introduction to dust explosions*. Oxford, UK: Elsevier; 2013.
8. Anderson JO, Thundiyil JG, Stolbach A. Clearing the air: a review of the effects of particulate matter air pollution on human health. *J Med Toxicol* 2005;**8**:166–75.
9. Valavanidis A, Fiotakis K, Vlachogianni T. Airborne particulate matter and human health: toxicological assessment and importance of size and composition of particles for oxidative damage and carcinogenic mechanisms. *J Environ Sci Health* 2008;**26C**:339–62.
10. Tanaka I, Hori H. Respiratory protective devices for particulate matter. In: *Powder technology handbook*. 3rd ed. Boca Raton, FL: CRC Press; 2006.
11. Kauffman CW. Agricultural dust explosions in grain handling facilities. In: Lee JHS, Guirao CM, editors. *Fuel-air explosions*. Waterloo, ON, Canada: University of Waterloo Press; 1982. p. 305–47.

12. Abbasi T, Abbasi SA. Dust explosions—cases, causes, consequences, and control. *J Hazard Mater* 2007;**140**:7–44.

13. Cartwright P, Pilkington G. *Explosions–Part 1, The Chemical Engineer (U.K.);* 24 November 1994. p. 15–7. Part 2, ibid, 12 January 1995, pp. 16–18; Part 3, ibid, 23 February, 1995, pp. 17–18.

14. Agarwal A. Dust explosions, prevention and protection. *Chem Eng* 2012;**115**:26–30.

15. Glor M. A synopsis of explosion hazards during the transfer of powders into flammable solvents and explosion preventative measures. *Pharm Eng* 2010;**30**(1):1–8.

16. National Fire Protection Association, NFPA-68. *Guide for venting of deflagrations*, Quincy, MA, 2002.

17. Lees FP. *Lees' loss prevention in the process industries: hazard identification, assessment and control.* 3rd ed. vols. 1–3. Oxford: Butterworth–Heinemann; 2005.

18. Palmer KN. *Dust explosions and fire.* London, UK: Chapman and Hall; 1973.

19. Amyotte PR, Basu A, Khan FI. Dust explosion hazard of pulverized fuel carry-over. *J Hazard Mater* 2005;**122**:23–30.

20. Gummer J, Lunn GA. Ignitions of explosive dust clouds by smouldering and flaming agglomerates. *J Loss Prev Process Ind* 2003;**16**:27–32.

21. ISSA. *Determination of the combustion and explosion characteristics of dusts.* Mannheim, Germany: International Social Security Agency; 1998.

22. Jacobsen M, Cooper AR, Nagy J. *Explosibility of metal powders.* New York: Bureau of Mines Report # 6516; 1964.

23. International Standards Organization. *Explosion protection systems part 1. Determination of explosion indices of combustible dusts in air, ISO 6184/1.* Geneva: ISO; 1985.

24. Siwek R. *20-L laboratory apparatus for the determination of the explosion characteristics of flammable dusts.* [Dissertation], Winterthur, Switzerland: Winterthur Engineering College; 1977. p. 109.

25. Cashdollar KL. Flammability of metals and other elemental dust cloud. *Process Saf Prog* 1994;**13**:139–45.

26. Denkevits A, Dorofeev S. Explosibility of fine graphite and tungsten dusts and their mixtures. *J Loss Prev Process Ind* 2005;**19**:174–80.

27. Amyotte PR, Lindsay M, Domaratzki R, Marchand N, Di Benedetto A, Russo P. Prevention and mitigation of dust and hybrid mixture explosions. *Process Saf Prog* 2010;**29**:17–21.

28. Hoppe T, Jaeger N. Reliable and effective inerting methods to prevent explosions. *Process Saf Prog* 2005;**24**:266–72.

29. Taveau J. Secondary dust explosions: how to prevent them or mitigate their effects? *Process Saf Prog* 2012;**31**:36–50.

30. ASTM E 2019-03. *Standard test method for minimum ignition energy of a dust cloud in air.* West Conshohocken, PA, 2003.

31. von Pidoll U. The ignition of clouds of sprays, powders and fibers by flames and electric sparks. *J Loss Prev Process Ind* 2001;**14**:103–9.

32. Amyotte PR, Khan F, Boilard S, Iarossi I, Cloney C, Dastidar A, Eckhoff R, Marmo L, Ripley R. In: *Explosibility of non- traditional dusts: experimental and modelling challenges.* Hazards–XXIII, I Chem E Symp. Ser No. 158, November 13–15; 2012. p. 83–90.

33. ASTM E1491-06. *Standard test method for minimum autoignition temperature of dust clouds.* West Conshohocken, PA, 2006.

34. NFPA 654. *Standard for the prevention of fire and dust explosions from the manufacturing, processing and handling of combustible particulate solids;* 2000.

Further Reading

Barton J, editor. *Dust explosion prevention and protection: a practical guide.* IChem E (U.K.): Warwickshire; 2002.

CHAPTER 15

Advanced Topics in Particle Technology

15.1 Introduction

In this chapter some advanced topics in particle technology are introduced. Although, not discussed in traditional particle technology books, these topics are relevant in the context of current research developments in the field and the application of these concepts in several areas of industrial relevance. The topics covered in this chapter include anisotropic particles, use of particles for emulsion stabilisation, soft and deformable particles, separating particles using microfluidics, and response of particles to external fields. The basic foundation to these topics provided are useful to understand widely researched and rapidly growing literature in these fields. Over the last few decades, the advances in the field of particle science and technology grew tremendously. The references to specialised books and reviews that provide comprehensive overview of these topics are included.

15.2 Anisotropic Particles

15.2.1 Types and Classification

There is a very large body of literature dealing with several fundamental principles developed about the aspects of particle science and technology that deal with idealised case of 'spherical' or 'isotropic' particles. However, most of the particles that occur in nature such as clay and minerals, those of biological origin such as blood cells, bacteria and viruses including tobacco mosaic virus and others, as well as pollen grains from plants are all examples of anisotropic or asymmetric particles. Cutting-edge developments in synthetic chemistry and innovative particle synthesis approaches have been instrumental in the development of protocols for the fabrication of several types of novel particles.[1-3] Fig. 15.1 illustrates typical particles of different geometric shapes and complex morphologies that have been realised to date.[1] Intense research in the last two decades focused on the investigation of anisotropic particles have revealed several unusual behavior associated with such particles.

Shape anisotropic particles

These are particles of well-defined geometric shapes, typically described by more than one characteristic dimension. For such particles, an accurate estimate of the surface area and volume can be obtained from a geometrical considerations. A particle's shape effects are key

Coulson and Richardson's Chemical Engineering. https://doi.org/10.1016/B978-0-08-101098-3.00016-0

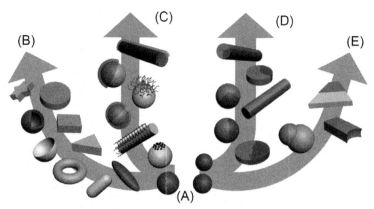

Fig. 15.1

A graphic of typical types of anisotropic particles represented by four branches are shown in (B) to (E): (A) spherical or isotropic particles, (B) shape anisotropic particles, (C) surface anisotropic particles, (D) and (E) particles designed to be anisotropic in volume, and surface (D) and (E), respectively, correspond to particles of regular shape and complex architecture. *Reproduced with permission from Lee KJ, Yoon J, Lahann J. Recent advances with anisotropic particles.* Curr Opin Colloid Interface Sci 2011;*16*(3):195–202.

attributes of anisotropic particles. Particle shape effects on various fluid particle operations and particle characterisation are taken into account in this book by describing the deviation of particle shape from spherical shape in terms of *sphericity*. As it is defined, the use of sphericity is appropriate only when the particles are anisotropic by shape. Examples of shape anisotropic particles include acorn-like, spherical cap-like, cuboids, donuts, disks, peanut or dumbbell-like, ellipsoids, rods, spherocylinders, hollow tubes, stars, and others.[1–3] A schematic of such particles is shown in Fig. 15.1B. The scanning electron microscopy images in Fig. 15.2 display representative anisotropic particles of different shapes: (a) peanut, (b) spherocylinder, (c) cuboid, and (d) rod. These particles were prepared by chemical synthesis route, one of the many techniques available for the synthesis of anisotropic particles.[4] A discussion of different methods for the synthesis of such particles in presented in Chapter 13.

Surface anisotropic particles

As shown schematically in Figs 15.1C and 15.3, a portion of the surface of the particles is modified by either physical or chemical means to render selective functionality to the particles.[5] Such particles, which are typically synthesised, starting from the isotropic or shape anisotropic particles of homogeneous surface properties, are called surface anisotropic particles. When two equal halves of the particle surface area are made of different surface chemistry (or composition), the surface anisotropic particles are called *Janus particles*, named after the two-faced Roman god, Janus. In the case of such particles, half the surface area can be hydrophilic and the other half could be hydrophobic, one half of the surface can be positively charged and the other could be negatively charged, etc. When one or multiple small portions of the surface

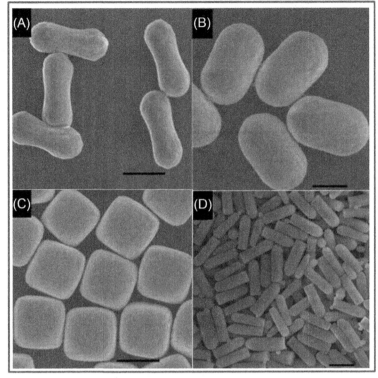

Fig. 15.2

The scanning electron microscopy images of (A) peanut, (B) spherocylindrical, (C) cuboid, and (D) rod-shaped particles. The scale bar in the images correspond to 1 µm. *Reproduced with permission from Anjali TG, Basavaraj MG. General destabilization mechanism of pH-responsive Pickering emulsions.* Phys Chem Chem Phys 2017;**19**:30790–7.

area are functionalised, the particles are referred to, respectively, as *single or multipatchy* particles. Additional parameters that are typically used to characterise such particles are the number of patches, size of the patches, the patch coverage, which is defined as the area occupied by the patch divided by total surface area of the particles, and the location of the patch. The scanning electron microscopy images in Fig. 15.3 show the patchy particles with a single patch of varying patch coverage. The spherical polystyrene particles in each image are of 2.2 µm diameter and their surface is partially coated with gold nanoparticles. The region of the polystyrene particle coated with gold appears shiny.[5]

Volume and surface anisotropic particles

These are particles derived from multiple components and are anisotropic, both by volume and by surface.[1] That is, the particles are fabricated by joining two or more segments of the particles, which are made of different materials, so that the particles possess both volume anisotropy and surface anisotropy. The final particle may assume a regular shape or complex

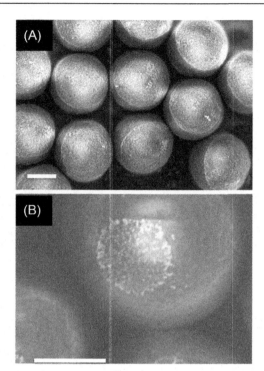

Fig. 15.3
Surface anisotropic patchy particles: (A) and (B) correspond to the scanning electron microscopy images of the polystyrene particles with a single patch of gold. The shiny surface of polystyrene particle is caused by a thin gold layer, which is a few nanometres thick. The scale bar corresponds to 1 µm. *Reproduced with permission from Sabapathy M, Kollabattula V, Basavaraj MG, Mani E. Visualization of the equilibrium position of colloidal particles at fluid–water interfaces by deposition of nanoparticles. Nanoscale 2015;7:13868–76.*

shape as shown, respectively, in Fig. 15.1D and E. Multifaceted particles shown in Fig. 15.1 have innumerable applications and exhibit fascinating behaviour, which are beyond the scope of this chapter. Refer to books and review articles devoted to these topics. In here, the role of particle shape on the packing characteristics is briefly discussed.

15.2.2 Effect of Particle Shape on Packing Characteristics

Besides being a subject of intense theoretical research for several centuries, how solids pack has several practical applications. For example, in operations that deal with the conversion of a collection of particles into multiparticle aggregates; a process called granulation or in packed and fluidised bed operations; it is the geometrical aspects of packing of solids that dictate the efficacy of the fluid-particle processes. The packing of solids is also relevant to the processes dealing with the flow of slurries, especially when the concentration of solids in the slurry is high. Also, it is relevant to the processing of particle-filled composite materials.

When spherical particles are used as packing material to create a randomly packed bed, the maximum random packing fraction (MRPF) that is typically achieved is ~0.64, such that the bed porosity is 0.36. The maximum random packing of spheres and the resulting packing fraction have been confirmed via experiments and computer simulations. Similar effort on prediction and measurement of MRPF of a nonspherical object has been a subject of more recent work. The MRPF of ellipsoids of different aspect ratio (defined as the ratio of the length of major axis to the length of the minor axis) investigated by simulations and experiments show a nonmonotonic evolution as show in Fig. 15.4.[6, 7] The graph in Fig. 15.4 shows that the MRPF increases as the particle shape deviates slightly from spherical shape. In other words, when the particle aspect ratio becomes slightly greater than one, it reaches a maximum, and then decreases continuously as the aspect ratio further increases. The data shown with black squares are results of computer simulations,[6] and the red circles are from experiments on colloidal ellipsoids.[7] The continuous lines are guides for the eye to emphasise the nonmonotonic variation of MRPF with increase in the aspect ratio. The MRPF for packing of silica-coated ellipsoids (an example of a core-shell particle) of 1.6 aspect ratio is measured to be ~0.694, higher than MRPF of spherical particles. Besides packing density, the elongation of particles from sphere to ellipsoids increases the average number of immediate neighbours of particles in the packing (coordination number), and it also leads to reduction in the percolation threshold. These geometric factors are crucial for the development of ultralightweight materials that are cost effective.[3]

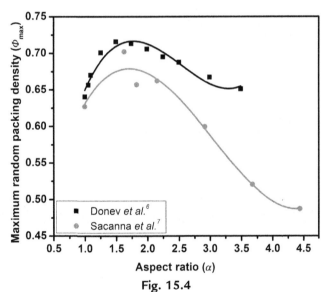

Fig. 15.4

The variation of maximum random packing density with aspect ratio for prolate ellipsoids. The results represented by *squares* and *circles*, respectively, are from simulations of Donev et al.[6] and experiments from Sacanna et al.[7] The *continuous lines* in the graph demonstrate that the maximum random packing density is highest at intermediate aspect ratio. *Courtesy of Preetika Rastogi, IIT Madras.*

15.3 Particle Stabilised Emulsions

15.3.1 Particles as Surfactants

Emulsions are an example of liquid-in-liquid dispersions. The emulsions formulated by considering two immiscible liquids consist of drops of one liquid dispersed in the other. During the creation of liquid droplets by the vigorously mixing of immiscible liquids, typically, a third component in the from of surface-active agent is added to achieve emulsification. Traditionally, surfactants are used for the stabilisation of the large interface area created during the emulsification process. A schematic representation of surfactant stabilised emulsions is shown in Fig. 15.5A and B. Surfactants are amphiphilic molecules, which consist of a bulky head group, which is hydrophilic and a slender tail group, typically a hydrocarbon chain, which is hydrophobic as shown schematically in Fig. 15.5A and B. In any emulsion, the hydrophilic part of the surfactant molecule resides predominantly in the aqueous phase and the hydrophobic part in the oil phase. Depending on whether the surfactant is water soluble or oil soluble, the surfactants can stabilise respectively oil-in-water or water-in-oil emulsions as shown in Fig. 15.5A and B.

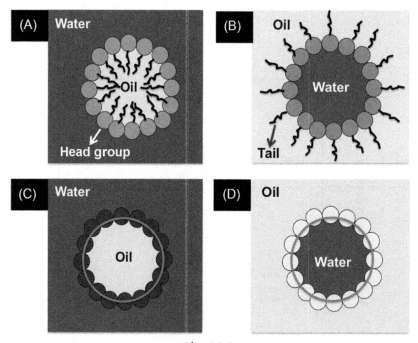

Fig. 15.5

Schematic exhibits emulsions that are surfactant stabilised (A, B) and particle stabilised (C, D). Examples (A) and (C) illustrate oil-in-water emulsions, and (B) and (D) depict water-in-oil emulsions. The particle-stabilised emulsions shown in (C) and (D) are also called Pickering emulsions, named after British scientist S.U. Pickering.

As an alternative to conventional surfactants, the idea of using solid particles for the stabilisation of emulsions demonstrated by Ramsden in 1904[8] and Pickering in 1907[9] is more than a century old. A schematic representation of particle stabilised emulsions oil-in-water and water-in-oil is shown in Fig. 15.5C and D. The smaller circles in (C) and (D) represent the spherical particles that surround the droplet, and the line represented as larger circle corresponds to the water–oil interface. Similar to the surfactant stabilised emulsions, the liquid that is wetted more by the particles becomes the continuous phase. That is, when the majority of the particle surface is in contact the water phase, i.e. when the particles are hydrophilic, the emulsion type is oil-in-water which is illustrated in Fig. 15.5C. As depicted in Fig. 15.5D, the majority of the particle resides in the oil phase, i.e. the particles are hydrophobic, and therefore, the oil forms the continuous phase, and hence the emulsion type is water-in-oil.

In Fig. 15.6, the use of 1 μm diameter spherical silica particles for emulsion stabilisation is demonstrated. The image in Fig. 15.6A shows a vial containing emulsion obtained by manual mixing of aqueous dispersion of 0.5 wt. % particles and decane in the 2:1 volume ratio. The figure reveals that the emulsion droplets cream to the top, which is typical of oil-in-water emulsions.[4] Also, as the particles used to create emulsions are unmodified silica spheres, which are hydrophilic ($\theta \sim 33° < 90°$) particles, the formation of oil-in-water emulsions is expected. A magnified view confirms the presence of droplets that undergo no further coalescence is shown in Fig. 15.6B. Remarkable long-term stability of the particle-stabilised emulsions is typically due to the formation of a close packed particle monolayer around each droplet as show in Fig. 15.6C. The high magnification microscopy image confirms the dense close packed

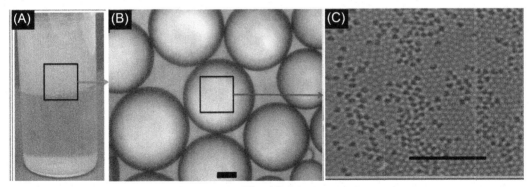

Fig. 15.6

Silica particle stabilised oil-in-water emulsions. (A) The photographs of the emulsions obtained after mixing a dispersion containing 0.5 wt.% spherical silica particles and decane in 2:1 volume ratio. (B) The bright field optical microscopy images of silica particle stabilised decane droplets (scale corresponds to 250 μm). (C) The high magnification microscopy image confirms the dense close packed arrangement of 1 μm diameter particles on the drop surface (scale bar corresponds to 10 μm). *Reproduced with permission from Anjali TG, Basavaraj MG. General destabilization mechanism of pH-responsive Pickering emulsions. Phys Chem Chem Phys 2017;**19**:30790–7.*

arrangement of 1 μm diameter particles on the drop surface that inhibits the coalescence of droplets when two or more such drops come into contact.

The factors that influence the formation and stability of particle stabilised emulsions are the three-phase contact angle of particle at the interface and the energy required to detach a particle that is already adsorbed at the drop surface,[10] which are discussed further in the next sections.

15.3.2 Wettability of Particles at Interfaces

The wettability of a particle adsorbed at an interface is a measure of its equilibrium position with respect to the interface. In other words, it is a quantitative measure of the fraction of the particle surface that is in contact with the fluids used to create the fluid-fluid interface. When the particles are adsorbed at the interface of two immiscible liquids, the wettability is expressed in terms of the three-contact angle (θ). Consider a schematic, Fig. 15.7, showing a solid particle (P) located at liquid (L1) and liquid (L2) interface, which is the case for the particle trapped at the interface encountered in the particle stabilised emulsions. As displayed, three distinct interfaces coexist: (i) Particle—Liquid 1 interface with the surface tension γ_{PL1}, (ii) Particle—Liquid 2 with the surface tension γ_{PL2}, and (iii) Liquid 1 and Liquid 2 with the interfacial tension γ_{L1L2}. The position of a particle located at liquid-liquid interface is characterised by means of three-phase contact angle (θ), which is related to the surface tension of the three interfaces given by the Young's equation[10]:

$$\cos\theta = \frac{\gamma_{PL1} - \gamma_{PL2}}{\gamma_{L1L2}} \tag{15.1}$$

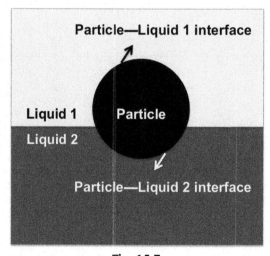

Fig. 15.7
Schematic of a particle (P) at the interface of two immiscible liquids L1 and L2.

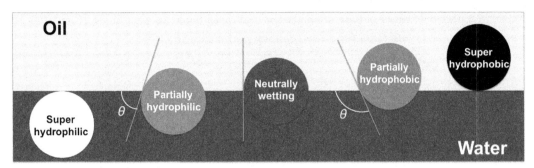

Fig. 15.8

Schematic representation of the particles of different wettability adsorbed at oil–water interface: super hydrophilic *(first sphere from the left)*, partially hydrophilic *(second sphere from the left)*, neutrally wetting *(third sphere from the left)*, partially hydrophobic *(third sphere from the left)*, and super hydrophobic *(fifth sphere from the left)*.

Typically, Liquid 1, i.e. the upper phase is less dense than the bottom liquid (L2). In the case of emulsions created using oil and water as the immiscible fluids, L1 is the oil phase and L2 is the aqueous phase. The three-phase contact angle, θ, is a crucial parameter to assess the type of emulsion resulting from the use of colloidal particles for emulsion stabilisation. The particles can either be completely wetted by one of the fluids, more wetted by either of the fluids, or equally wetted by both the fluids that constitute the interface depending on the combination of particle and the immiscible fluids as shown schematically in Fig. 15.8. Typically, the measurement of the three-phase contact angle of a solid particle located at an interface involves (1) identification of the three-phase contact point (where the solid and the two fluid phases meet) (2) drawing a tangent to the particle surface at the three-phase contact point (3) estimation of the angle that the tangent makes with the interface, measured through the aqueous phase.

Based on the value of the three-phase contact angle, all particles, i.e. those synthesised using various routes and those that occur naturally, can be classified as: super hydrophilic ($\theta=0°$), partially hydrophilic ($0° < \theta < 90°$), neutrally wetting ($\theta=90°$), partially hydrophobic ($90° < \theta < 180°$), and super hydrophobic ($\theta=180°$). When oil and water are emulsified, if the particles used are adsorbed to the interface such that $\theta < 90°$, the oil-in-water emulsion is formed and if $\theta > 90°$, the emulsion formed is water-in-oil type. In general, the particles used for emulsification must be partially wetted by both oil and water for efficient emulsion stabilisation, i.e. $20° < \theta < 160°$.[10] From the three-phase contact angle of the particles at the interfaces and the type of emulsion formed, tabulated in Table 15.1, it is clear that partially hydrophobic and partially hydrophilic particles, respectively, form water-in-oil (W/O) and oil-in-water (O/W) emulsions.

Table 15.1 Relation between type of emulsions and three-phase contact angle of particles at interfaces for a few common particulate emulsifiers

Serial No.	Particles Used for Emulsification	Type of Emulsion	Wettability	Contact Angle	Reference
1	Silica	O/W	Partially hydrophilic	~81°	4, 11
2	Polystyrene (latex)	W/O	Partially hydrophobic	~128°	12
3	Haematite (iron oxide)	O/W	Partially hydrophilic	~47°	4, 13–15
4	Graphene oxide	O/W	Partially hydrophilic	~62°	16, 17
5	Carbon nanotubes	W/O	Partially hydrophobic	~144°	18, 19
6	Starch	O/W	Neutrally wetting	~90°	20
7	Wax	W/O	Partially hydrophobic	~135°	21

15.3.3 Energy to Detach Particles From Interfaces

The use of particles for emulsion stabilisation and the remarkable long-term stability that the Pickering emulsions exhibit are attributed to irreversible adsorption of particles to the interfaces. That is, once a particle is adsorbed to the interface, newly created during emulsification, it is practically impossible to detach particles from the interface, especially if the particles are in the micron size range. For a spherical particle of radius R adsorbed at the liquid–liquid interface of the interfacial tension γ_{L1L2} such that the contact angle is θ, the free energy ΔG to detach the particle from the interface is[10]:

$$\Delta G = \pi \gamma_{L1L2} (1 - |\cos\theta|)^2 \tag{15.2}$$

Eq. (15.2) can be derived from a simple surface energy balance approach.[22] It must be noted that the estimation of particle wettability is crucial to determine the detachment energy of particles from interface. Moreover, the energy of adsorption scales with the square of the particle radius. Therefore, with the average size of the order of ~1 nm, the energy required to detach the surfactant molecule from the interface is of the order of thermal energy ($k_B T$), and therefore, small surface active molecules readily desorb from the interface leading to the destabilisation of emulsions. In marked contrast, for micron-sized particles, the energy of detachment is of the order of $10^7 \, k_B T$. Hence, the particles adsorbed on the surface of liquid drops or gas bubbles provide a steric barrier against coalescence due to the irreversible adsorption of solid particles to interfaces.

15.4 Soft Particles

The particles that do not undergo any morphological change, i.e. those that remain intact during a given particle or fluid-particle processing operation, can be treated as 'hard' particles. For example, organic (polystyrene poly(methyl methacrylate), etc.), inorganic (silica, haematite, boron nitride), or metallic (gold, silver, etc.) particles that have elastic modulus ranging from few MPa to GPa[23, 24] dispersed in a suitable fluid and subjected to operations such as mixing, transfer from one phase to other, addition of electrolytes, moderate change in temperature retain their size and shape, and therefore, they can be regarded as hard particles. The classification of whether a particle is hard or soft depends on a number of parameters. For example, 'hard' polymeric particles when heated to a temperature close to the glass transition temperature become 'soft'. Similarly, polystyrene particles dispersed in water can be either made soft or completely dissolved (giving rise to a solution of polystyrene molecules) when adequate quantity of solvents such as tetrahydrofuran (THF) is added.

The particles that change size or shape, or that may completely disintegrate when subjected to external stimuli, for example, different solution conditions (pH, electrolyte, any chemical reagent), or applied fields (thermal, gravitational, electric, or magnetic) are termed 'soft' particles. A common example of soft and deformable particles are red blood cells (RBCs), which are known to undergo extensive shape changes as the blood flows through smaller capillaries, i.e. when these cells flow through confined channels.[23–26] It is well accepted that the deformability of RBCs is important since it points to the health of the cells, disease conditions, and it also affects the viscosity of the blood.[23, 24, 26] A brief introduction to the growing interest in the study of soft and deformable particles is presented in this section.

15.4.1 Microgels

In commonly used conventional polymer particles such as polystyrene (PS) or poly(methyl methacrylate) (PMMA), the number density or the concentration of polymer molecules across different regions of the particle is almost identical. In contrast, in a class of polymeric particles called 'microgels', the crosslinking density of polymer molecules form which the particles are made of, varies across the particle volume. The crosslinked polymer network can be made to collapse or swell using stimuli such as pH, electrolytes, thermal fields, electric fields, etc.[27–30] The collapse of the polymer network involves the segregation of the polymer and expulsion of the solvent from the particle. The microgel particles made of thermoresponsive polymer poly (N-isopropylacrylamide) (PNIPAM) are examples of thermoresponsive microgels that undergo a change in size when the temperature of the dispersion medium is changed.[27–30]

The hydrodynamic diameter of the PNIPAM microgel particles dispersed in water measured as a function of temperature is shown in Fig. 15.9. The hydrodynamic diameter of the colloidal scale particles dispersed in a fluid is typically larger than the diameter of the particles measured

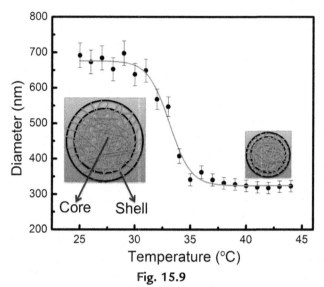

Fig. 15.9

The hydrodynamic diameter of PNIPAM microgel particles as a function of temperature. The inset shows the swollen and collapsed state of PNIPAM, respectively, at lower and higher temperature. The inner and outer *circles* that surround the crosslinked polymer network respectively represent the high and low crosslink density regions. *Courtesy of M. Mayarani and Dillip K. Satapathy, IIT Madras.*

in dry state. The schematic in the inset in Fig. 15.9 displays the microgel particles that can be thought of as the core-shell particles with both the core and the shell made of an entangled polymer network in which the polymer molecules are crosslinked by a covalent bond. However, the crosslinking densities in the core and the shell regions are significantly different. In particular, the crosslinking density in the core is higher and the crosslinking density in the shell is low, leading to free polymer chains in the shell. Therefore, the degree of swelling and deswelling of the core and shell regions are expected to be different. At 25°C, the particles are approximately 700 ± 40 nm in diameter. As the microgel dispersion is heated above 30°C, the diameter decreases drastically to 330 ± 15 nm at temperatures >37°C, beyond which, the microgels do not shrink further. The volume phase transition of PNIPAM particles displayed by the sigmoidal curve in Fig. 15.9 is smooth and occurs at ~32°C, which is the lower critical solution temperature (LCST). This volume phase transition is thermoreversible.

15.4.2 Drops and Bubbles

Smaller drops or bubbles assume a spherical shape as a result of dominant surface tension forces. In such drops or bubbles of radius R, the pressure in the interior is greater than that outside and the pressure differential is related to the surface tension γ as[31]:

$$\Delta P = \frac{2\gamma}{R} \qquad (15.3)$$

Eq. (15.3) is the well-known Young-Laplace equation. However, the bigger drops or bubbles residing on a substrate or falling under the influence of gravity in a quiescent fluid, can assume different shapes. The non-dimensional Bond number B_0 given by,[31]

$$B_0 = \frac{\Delta \rho R^2 g}{\gamma} \tag{15.4}$$

can be used as a measure of relative importance of gravity and surface tension forces in asserting the deformation of the drops and the bubbles. In Eq. (15.4), $\Delta \rho$ is the density difference between the drop (or bubble) and the outer fluid and g is the acceleration due to gravity. For raindrops of 10 mm diameter, the Bond number, $B_0 \sim 3.4$ ($\Delta \rho \sim 1000$ kg/m^3, $\gamma = 72$ mN/m and $g = 9.81$ m/s^2). Therefore, the larger raindrops in the absence of other external factors are expected to deform due to the action of gravity and flatten in the direction of the fall. When the magnitude of $B_0 \ll 1$, this gravity-driven deformation can be completely neglected. As discussed in Chapter 6, the drops and bubbles that are in motion can also experience deformation due to differences in the pressures acting on various parts of their surface. This deformation enhances the resistance to their motion and, therefore, the terminal velocity of the larger drops and bubbles, in general, is lowered. In addition, the drops can also deform when subjected to a shear or electric field.[32–35]

Coating the drops or bubbles with field responsive materials is another strategy to induce the morphological changes. For example, particles, polymers, or other molecules can be coated on the surface of drops or bubbles such that they respond to externally applied magnetic, electric, acoustic, optical or thermal fields. An example where the thermoresponsive behaviour of PMIPAM microgels (discussed in Section 15.4.2) is exploited to obtain thermoresponse PNIPAM coated hollow shell-like structures is shown in Fig. 15.10.[36] In Fig. 15.10A, the relative diameter, defined as ratio of diameter at a given temperature divided by the diameter at 20°C is plotted. The filled square symbols in Fig. 15.10A show the temperature response of PNIPAM microgels dispersed in water when the temperature is varied from 20 to 55°C and the diameter of the microgels at 55°C decreases by a factor of 0.5 compared to the initial size. The PNIPAM microgel coated shells are prepared from the PNIPAM stabilised emulsion droplets as templates (discussed in Section 15.3). The PNIPAM particles are amphiphilic and can readily adsorb at the oil–water interface leading to the PNIPAM stabilised Pickering emulsions. The particles at the drop surface can be crosslinked using suitable procedures. The removal of the inner and outer fluid results in the hollow shells made of colloidal particles, and therefore, these structures are called 'colloidosomes', as they structurally resemble 'liposomes' (discussed in Section 15.4.3). The filled circles in Fig. 15.10A show the temperature response of the PNIPAM microgel coated shells redispersed in water. As evident, when the temperature is increased from 20 to 55°C, the diameter of the PNIPAM colloidosomes decreases by a factor of 0.58, which corresponds to approximately 80% decrease in volume. These structures are so robust that upon cooling the dispersion to 20°C, the colloidosome once again reaches the initial

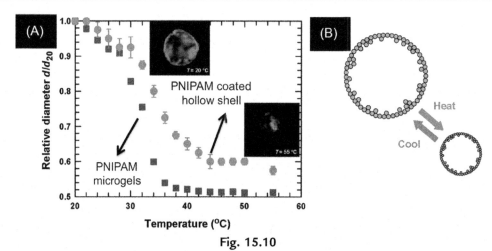

Fig. 15.10

(A) The variation of relative diameter as a function of temperature for aqueous dispersions of PNIPAM microgels and hollow shells coated with PNIPAM microgels. (B) Schematic representation of the reversible temperature responsive behaviour of PNIPAM coated shells in water. *Reprinted with permission from Shah RK, Kim J-W, Weitz DA. Monodisperse stimuli-responsive colloidosomes by self-assembly of microgels in droplets.* Langmuir **2010**;26:1561–5.

size. The insets in Fig. 15.10A show the microscopy images of the larger colloidosomes at $T = 20°C$ and the colloidosomes in the shrunken state at $T = 55°C$. Fig. 15.10B is a schematic, which shows that the temperature driven size change in the PNIPAM colloidosomes is reversible.

15.4.3 Micelles and Vesicles

In Section 15.2, the use of surfactants for the stabilisation of water-in-oil and oil-in-water emulsions is briefly discussed. When added to certain liquids, surfactant molecules can spontaneously form molecular aggregates shown schematically in Fig. 15.11. These structures formed due to the self-assembly of surfactants in solution are spherical micelles, reverse or inverted micelles, rod-like micelles, vesicles, or planar bi-layers (not shown in the schematic). The self-assembly of surfactant molecules into one of the geometric shapes, depends on the packing of surfactant molecules and can be predicted from the packing factor, P, which depends on the length, volume and area associated with surfactant molecules. In general, the tail or the hydrocarbon chain may be fully extended, stiff, or fluid like depending on the surfactant-liquid combination, and therefore, can be characterised by volume (v) and a critical chain length (l_c). Similarly, the head group of the surfactant molecules occupies an optimal area (a), which again depends on the geometry of the head group, configuration of surfactant molecules, van der Waals, electrostatic, and other interactions. It is possible to estimate the area that the head group occupies by combining the measurement of surface tension of surfactant solutions as a function of concentration and the Gibb's adsorption isotherm formalism.[37] It is possible to estimate the

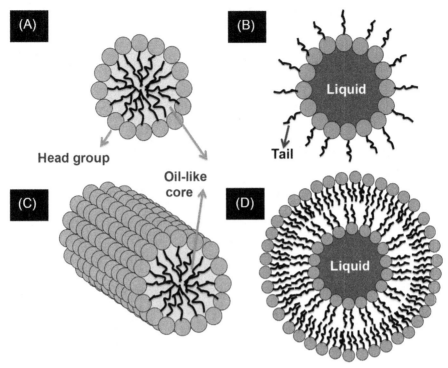

Fig. 15.11

Schematic shows the self-assembly of surfactant molecules into (A) micelle, (B) reverse micelle, (C) rod- or worm-like micelle, and (D) vesicles. These structures form when surfactant molecules are added to one or more liquids. The drawings shown in (A), (B), and (D) are the two-dimensional representation of the three-dimensional spherical molecular aggregates.

critical chain length and volume of a hydrocarbon chain containing n carbon atoms using the relations[37, 38] $l_c \sim (0.154 + 0.1265n)$ nm and $v \sim (0.0274 + 0.0269n)$ nm^3. The packing factor P is defined as[37, 38]:

$$P = \frac{v}{al_c} \tag{15.5}$$

The spherical and inverted micelles shown in Fig. 15.11A and B result when the packing of the surfactant molecules in the self-assembled state is such that $P < 1/3$ and $P > 1$, respectively. While $P < 1/3$ corresponds to surfactant molecules taking the shape of a cone, at $P > 1$, the configuration resembles an inverted truncated cone. The rod-like micelles and vesicles shown in Fig. 15.11C and D form when the surfactant molecules assume the shape of truncated cones of different base areas such that the packing factor satisfies the condition, $\frac{1}{3} < P < \frac{1}{2}$ and $\frac{1}{2} < P < 1$, respectively. The vesicles can be thought of as core-shell spheres consisting of a fluid-core and a bi-layer shell formed from curved surfactant bilayers. Typically, surfactant bi-layers from when $P \sim 1$. When the vesicles are formed with lipids, which are naturally

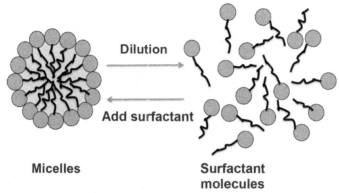

Fig. 15.12

Disintegration of micelles to surfactant molecules upon dilution and reformation of micelles upon further addition of surfactant molecules.

occurring surfactants, which form lipid membranes that surround cells, the vesicle are often called 'liposomes'. The structures that resemble liposomes formed using colloids and polymers are respectively called 'colloidosomes' and 'polymerosomes'.

One of the characteristics of micelles or vesicles is that the interaction forces that hold the molecules together in these aggregates are much weaker than the forces that hold the molecules or atoms in hard particles of metallic or inorganic origin such as gold, titanium dioxide, or iron oxide particles. Therefore, these molecular aggregates are soft, flexible, and deformable.[37] It is well known that surfactant molecules form micelles when the concentration of surfactant added to a liquid is above a critical concentration called the 'critical micelle concentration (CMC)' or 'critical aggregation concentration (CAC)'.[37] For example, sodium dodecyl sulphate (SDS) forms spherical micelles when added to water, when the concentration of SDS is above 8.1–8.4 mM (at 25°C), depending on the purity of the surfactant. However, when a large quantity of water is added to the surfactant solution that contains spherical micelles, the surfactant molecules disintegrate to form monomers as shown schematically in Fig. 15.12. This process is reversible. Even simple dilution can change the shape or size of these molecular aggregates. These structures can also fuse or rupture due to changes in solution conditions such as pH, addition of electrolyte, temperature, external electric field, etc.

15.5 Particle Separation Using Microfluidic Devices

Over the last two decades, microfluidic devices are widely used in many chemical and biomedical applications such as separating target particles, diagnostics, chemical and biological analyses, automotive and electronic industries, food and chemical processing, etc.[39, 40] Microfluidics has major implications in the development of Lab-On-Chip devices.

The reason for the sudden shift from conventional techniques to microfluidic devices is their enhanced efficiency and accuracy at a lower cost and high portability. Such devices require less sample volume, and hence, low reagent cost and analysis time. The smaller size of the device and the low fabrication costs allow parallelisation, and hence, increases the throughput.[39] Particle separation techniques have immense applications in many small and large-scale processes. Recent advances in the separation techniques suggest that the separation of particles is easy to achieve by using geometries similar in size with the particles, and hence, it is efficient to separate the particles of microsize using microfabricated geometries.[40] Therefore, in the field of continuous separation techniques, microfluidic devices show promise. The separation can be done on the basis of size, shape, and deformability by either externally applied fields such as electrical, optical, acoustic, and magnetic field or internally applied fields. The techniques that use externally applied fields are known as active separation techniques. In addition, various passive separation methods such as insulator-based dielectrophoresis (iDEP), deterministic lateral displacement (DLD), hydrodynamic filtration, hydrophoresis, split-flow thin-cell fractionation (SPLITT), pinched flow fractionation (PFF), and inertial microfluidics, etc. are based on the interaction between the particles, the flow field, and the channel structure. The continuous sorting of micron-sized particles based on size offers many applications including chemical syntheses, mineral processing, and biological analyses.[41–43] For example, harmful bacterial activity can be hampered by such separation technologies, whereas in defence applications these techniques can be used to detect threatening agents, etc. In diagnostics, microfluidic separation techniques are used to separate living cells from dead cells, cancer cells from normal cells, infected cells from healthy cells.[44] The separation of cells based on the physical properties is relevant to healthcare applications. The aforementioned techniques of continuous particle separation methods are applicable for Newtonian fluids.[45] However, most of the fluids encountered in industrial or natural systems such as polymeric solutions, colloidal solutions, bodily fluids (e.g. blood and saliva) are non-Newtonian that exhibit either shear thinning, or viscoelastic behaviour.[46–48] Recently, a particle separation technique developed by Lu et al., known as elasto-inertial pinched flow fractionation (eiPFF) exploited elastic and inertial forces to separate particles from viscoelastic solutions.[45]

Diseases are known to alter the physical properties of cells. For instance, epithelial cancer cells are larger in size compared to the healthy cells,[44] whereas malarial parasite-infected red blood cells are 50 times more rigid than healthy cells that make it difficult to pass through capillaries.[49] In digital microfluidics where the droplets are used for encapsulation or microreactors for applications in pharmaceuticals, cosmetic, and food and materials industries, require the separation of the droplets.[50] Recent advances in particle separation using microfluidic devices have enabled the development of several novel separation methods for both Newtonian and non-Newtonian fluids. In the following discussion, a few representative active and passive separation techniques are described.

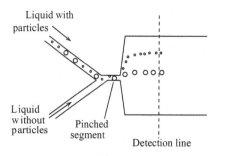

Fig. 15.13
Schematic of particle separation using pinched flow fractionation.

15.5.1 Passive Techniques

Pinched flow fractionation (PFF)

This technique is based upon the assumption that the system of incompressible fluid is in steady state, the flow corresponds to low Reynolds number and no-slip condition is applicable at the walls. PFF uses the characteristics of laminar flow for sizing of particles in microchannels. A secondary fluid without any particles is injected in the microchannel with a pinched segment. The output stream of the pinched segment contains a mixture of fluid and particles, which later get separated by spreading streamlines based on their size. A schematic illustration is shown in Fig. 15.13. The mechanism of separation is based on the tendency of particles to flow along the streamlines passing through their centre of mass which separates at the exit of the pinched segment. The centre of mass is closer to one of the walls for the smaller particles and towards the centre for the larger particles. Separation efficiency mainly depends upon the precise distribution of the flow rates, whereas the fractionation quality depends upon the shape of the pinch segment. Efficiency decreases with the width of the pinched segment, which further depends upon the size of the particles to be separated. The particles are small enough that they do not perturb the base flow and interact with the walls. The empirical relation for the dependence of the pinched segment on the particle size is given by

$$Y_0 = \left(w_p - \frac{D}{2}\right)\frac{w_0}{w_p}. \tag{15.6}$$

Here w_0 is the width of the outlet, w_p is the width of the pinched segment, Y_0 is the effluent position of the particle centre at the outlet, D is the particle diameter. PPF is mainly applicable for particles whose size range magnitude is equal or lower than the order of roughness of the sidewall.[39]

Inertia and dean flow fractionation

The particles of different size moving along the flow in a microchannel get distributed based on the size of the particles, and hence, can be separated. A schematic is shown in Fig. 15.14.

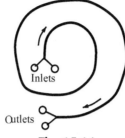

Fig. 15.14
Schematic illustration of particle separation using Dean flow fractionation.

The phenomenon of self-sorting of the particles is based on the existence of two lift forces acting in the microchannel. The first lift force is induced by shear gradient and the second results from wall effects. The velocity gradient is higher near the walls than that existing near the centreline as a result of which the particles move towards the walls. Due to the presence of the wall, the flow field around the particle gets disturbed, which further enables the particle to move away from the wall. The superposition of these forces acting in opposite directions causes the lateral movement of the particles, and hence, the particles self-sort based on their size relative to the microchannel width. This method of separation is based on the strength of the lift forces. The microchannel is fabricated with sudden expansions and contraction. When the flow stream containing the well-organised particles is subjected to sudden expansion, the shear gradient-induced lift force dominates the wall-induced lift force and the particles move towards the wall. It is observed that the lateral migration of the particles is directly proportional to the square of the particle diameter and the flow rate of the fluid. Thus bigger particles are subjected to a higher lateral shift compared to the particles with a small diameter, and thus, separate.[39]

Microvortex manipulation

Artificial vortices are produced to separate the particles using the microvortex manipulation technique. Several herringbone grooves are constructed at the bottom of the microchannel. When the fluid flows past these groves, vortices are produced in the direction of the streams, which results in a helical motion. The drag force applied by the microvortices causes the particles to move along the interface of the vortices. If a particle is lighter than the medium it is subjected to the upward lift and it moves towards the channel wall and attains an equilibrium position due to the balance between the buoyancy force, the drag forces and the downward gravitational force. The heavier particle settles at the bottom of the channel, and thus, separation occurs.[39]

Deterministic lateral displacement (DLD)

Deterministic lateral displacement method is based on the phenomenon of the asymmetric bifurcation of the laminar flow passing around micro-scale obstacles, which was first discovered by Huang et al.[51] Cylindrically shaped posts (obstacles) are placed in the form of an array in the flow direction. The direction of migration of particles is characterized by θ, the angle of migration. The posts can be structured in two configurations, rotated square and rhombic. These obstacles break the symmetry of the particle trajectory, which induces a lateral displacement of the particle. The array of the obstacles is periodic which leads to accumulation of small lateral displacement of the particles, which results in a significantly large migration angle. The successive lateral displacement, which accumulates along the flow direction, separates the large and small-sized particles. Theoretically, the movement of the particles can be studied by Stokesian dynamics simulations to predict the movement of the particles.

Several other methods, which fall into the category of passive separation techniques in microfluidic devices, are the Zweifach-Fung effect, filtration, hydrodynamic filtration, and micro hydrocyclone, which are studied by several authors to separate the particles based on the size, shape, and deformability of the particles and have different areas of applications. In the next section, the techniques that belong to the active separation techniques in microfluidic devices will be discussed.

15.5.2 Active Separation Techniques

Dielectrophoresis

Dielectrophoresis (DEP) is the phenomenon that occurs when a neutral particle become polarised when placed in a nonuniform electric field (see Fig. 15.15). The phenomenon has wide applications such as in fractionation and characterisation of particles, chemical, and biological analysis, electrically controlled trapping, etc. Dielectrophoresis is widely used to separate particles from a mixture using the frequency dependent behaviour of the dielectric property of the particle. DEP force acting on a particle will be different as the force is proportional to the third power of the radius and dielectric properties of the particle. However, the application of the DEP is limited to the particles, which have significantly different dielectric properties. In some applications where the particles to be separated have similar

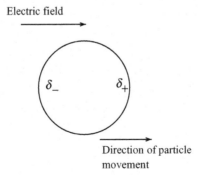

Fig. 15.15
Schematic illustration of dielectrophoresis.

intrinsic physical properties, the DEP method is modified for the separation. The target particles are labelled with polymeric beads to achieve a significant difference in the intrinsic physical properties, and hence, the DEP force will deflect only the labelled particles. In many cases, where multiple particles are to be separated, the multitarget dielectrophoresis activated cell sorter chip should be used.[52]

Magnetic separation

Magnetic separation of cells exploits their magnetic properties. It offers several advantages, such as high specificity, fewer chances of cell damage, and shorter sorting time. The high-gradient magnetic field and the fluorescent labelling on the cells make the process more efficient. More than one target cell can be separated using the multitarget magnetic-activated cell sorter developed by Adams et al.[53] The target cells are separated through a combination of hydrodynamic force produced by the flow, and magnetophoretic force induced by ferromagnetic strips. Different target cells are tagged differently based on the magnetic properties and depending on the criterion of the dominance of either drag or magnetic forces, the particles get separated.

A few more techniques which exploit optical and acoustic fields are useful based on specific application. A detailed discussion of these techniques can be found in the latest literature.

15.6 Phoretic Motion: Response of Particles to External Fields

Phoretic motion refers to the flux of colloidal particles that results when the colloidal suspension is subjected to nonuniform 'field' such as concentration gradient of one of the solutes other than the colloid, temperature, electric potential, etc.[54] Phoretic motion is very different from sedimentation, wherein the body force due to gravity leads to a net flux of colloidal particles in the direction of gravity. In that case, there is a net force acting on each colloidal particle. However, in phoretic motion, the net force acting on each colloidal particle is zero, and there is interaction between the nonuniform field and the interfacial region near the surface of the particle. A well known example of phoretic motion is the 'electrophoresis' wherein the applied electric field acts on the charged electrical double layer surrounding the particle.[55,56] More generally, the inhomogeneous applied fields (in concentration, temperature, or electric potential) alters the interfacial region near the surface of the particle, which then is set in motion. Depending on the nature of the externally imposed gradients, phoretic motion can be classified as follows:

1. electrophoresis: the particle responds to an external electric field gradient.
2. thermophoresis: the particle responds to an external temperature gradient.
3. diffusiophoresis: the colloidal particle interacts with smaller solute molecules in the surrounding bulk, which themselves are inhomogeneously distributed. Thus, the concentration gradient of the solute molecules provides the driving force.

The specific case of phoretic motion of a liquid drop due to an imposed temperature gradient on the surrounding liquid can be understood as follows.[57] The interfacial tension of the drop-surrounding liquid interface is usually a strong function of temperature, with regions of higher temperature of the interface having lower surface tension. In other words, the hotter 'pole' of the drop will have the lowest interfacial tension and the colder pole of the drop will have the highest interfacial tension. This gradient in interfacial tension across results in a discontinuity in the tangential stress across the interface, leading to a flow in the two bulk fluids. Thus, there will be flows (often termed 'Marangoni flows') induced by the nonuniform surface tension along the droplet surface, which will drag the neighbouring fluid toward the colder pole. The drop, as a consequence, propels itself in the opposite direction. Thus, droplets under the influence of temperature gradients tend to move toward the hotter region. It is also possible to envisage that a similar effect can be achieved by inhomogeneous distribution of surfactants at the interface, leading to a solutal Marangoni flow. Such phoretic motion of drops under the influence of inhomogeneous temperature fields is called 'thermocapillary' motion.

During the phenomenon of thermophoresis, small colloidal particles are driven from high- to low-temperature regions.[58] Thermophoresis is closely related to the phenomenon of thermal diffusion, which is transport of a given species due to temperature gradient in a multicomponent system. The mechanism underlying thermophoresis can be understood by realizing that the average kinetic energy of molecules is directly proportional to the temperature, and hence a colloidal particle in a temperature gradient will be subjected to asymmetric bombardment by the surrounding molecules, leading to a systematic motion of the colloidal particle down the temperature gradient. A similar process occurs in diffusiophoresis, wherein the concentration gradient of a solute leads to asymmetric collisions of the solute with the colloidal particle, leading to the motion of the colloidal particle. In all phoretic motions, the net force acting on the particle is zero. A key unifying feature among all phoretic phenomena is that there is an interplay between fluid dynamics, surface phenomena, and transport of mass/charge/internal energy.

In recent years, another type of phoretic motion (called 'self phoresis') has also garnered much attention.[57] This type of motion is exhibited by a 'janus'-shaped spherical solid particle in which only one-half is coated with a solid catalyst (see Section 15.2). The solid particle is placed in a reactant solution, and only the hemispherical surface coated with the catalyst transforms the reactant to a product. This leads to an asymmetric distribution of product molecules around the spherical particle, with more product molecules close to the coated part of the sphere. This leads to asymmetric interaction of the product molecules with the janus particle, leading to a phoretic motion of the particle. In contrast to the previous case of thermocapillary motion, the asymmetry is not imposed externally, but is created by the particle itself.

15.6.1 A Simple Model for Diffusiophoresis

Consider a colloidal particle in a solution in which there is a dissolved solute, which are molecules of significantly smaller dimensions than the colloidal particle.[54] The solute molecules are considered to be electrically neutral. These solute molecules interact with the surface of the colloidal particle via van der Walls forces or via excluded volume effects. In the following analysis, we are dealing with a thin region close to the particle surface, such that the curvature of the spherical particle is negligible, and the particle locally appears flat, and hence a locally Cartesian coordinate frame can be used with y being the direction perpendicular to the surface, and x being the direction along the imposed concentration gradient. The net interaction between the solute molecules and the colloidal particle is denoted by $\Phi(y)$ (called the 'potential of mean force'), and the force experienced by a colloidal particle is $-\nabla\Phi$. If the concentration of solute molecules at a given point is C, then the net force (per volume) is $-C\nabla\Phi$. The variation of C in the direction normal to the colloid is given by the Boltzmann distribution $C(y) = C^s \exp\left[-\Phi/k_BT\right]$, where C^s is the concentration at the particle surface. The momentum equations in the direction normal (y) and parallel (x) to the particle surface are:

$$\frac{\partial p}{\partial y} + C\frac{\partial \Phi}{\partial y} = 0, \tag{15.7}$$

$$\eta\frac{\partial^2 v_x}{\partial y^2} - \frac{\partial p}{\partial x} = 0. \tag{15.8}$$

By using the Boltzmann distribution for C, the pressure field can be found from the y-momentum equation. This is then substituted in the x-momentum equation to obtain the velocity v_x, which when evaluated for $y \rightarrow \infty$, yields the slip velocity:

$$v^s = -\frac{k_BT}{\eta}\int_0^\infty dy\left[y\exp[(-\Phi/k_BT) - 1]\frac{dC^s}{dx}\right]. \tag{15.9}$$

15.6.2 Basic Theory of Electrophoresis

A charged colloidal particle in an electrolyte moves under the influence of an externally imposed electric field. Colloidal particles, when dispersed in aqueous solutions, often acquire a surface charge due to a variety of physicochemical mechanisms, as discussed in Section 13.4.1 (also see Ref.[55]). Such charged particles, when dispersed in an electrolytic solution, acquire the so-called electrical double layer, wherein the ions of sign opposite to the particle charge (called 'counterions') tend to accumulate near the particle due to electrostatic interactions. The length scale of this region where counterions are in excess concentration compared to their bulk values

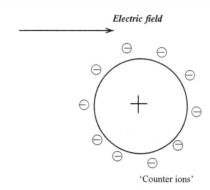

Electric field

'Counter ions'

Fig. 15.16
A spherical positively charged particle is surrounded by negative counterions in the solution. This particle-counter ion arrangement is subjected to an externally imposed electric field.

is often very small compared to the particle size itself. A schematic of such a particle is shown in Fig. 15.16. When electric field is applied on such a colloidal particle with an electrical double layer, there is a force on the excess charge in the liquid part (i.e. on the ions outside the stern layer) of the electrical double layer, which is balanced by the viscous stresses. This balance results in a steady motion of the particle. It is first useful to also consider the closely related phenomenon of electro-osmosis, which is the motion of the fluid with respect to the solid channel walls, due to the imposition of an electric field.

Consider a solid surface (e.g., the walls of a channel) with a charge density (number of charges per unit area of the channel) of q, which is surrounded by a cloud of ions in the liquid (of opposite sign to those in the channel wall), with charge density $-q$. When an electric field is applied, then, the wall and the fluid will experience equal and opposite forces given by qE_{ext} and $-qE_{ext}$, respectively. If we assume the channel walls to be stationary, then this means that the net force on the solid side (due to the imposed electric field) is balanced by some mechanical force, which keeps the surface stationary. Thus, the net force (per volume) on a charged surface and the quiescent electrical double layer is $-qE_{ext}$. This net force must be balanced by the forces of magnitude qE_x generated by a viscous flow. Thus, an electric field applied to an initially net neutral system induces a net fluid motion, if the (charged) solid surface is maintained stationary by an external mechanical force. This electric-field induced flow is usually termed as 'electro-osmosis'.[55]

If, instead of the wall being stationary, we imagine a charged particle in an electrolyte, with an imposition of external electric field, but we impose the far-field condition that the fluid velocity is zero at distances far away from the charged particle. If there is no restraining force on the solid particle (as was assumed in electro-osmosis), then the charged particle will tend to move under the influence of applied electric field. Such a motion of a charged particle under the influence of an external electric field is what is conventionally referred to as electrophoresis.

We can illustrate this using a simple one-dimensional mathematical model. The momentum balance equation for the fluid in the direction of the motion is given by

$$0 = \mu \frac{\partial^2 u}{\partial y^2} + \rho_E E_{ext,wall}, \tag{15.10}$$

where the inertial terms in the left side are neglected owing to the very small Reynolds numbers (estimated based on the particle size). Using the Poisson equation for expressing the charge density ρ_E in terms of the electrostatic potential ϕ

$$0 = \mu \frac{\partial^2 u}{\partial y^2} - \varepsilon \frac{\partial^2 \phi}{\partial y^2} E_{ext,wall} \tag{15.11}$$

We now solve these equations with the boundary condition that at $y \gg \lambda_D$ (which is the inverse Debye-Hückel parameter or the thickness of the electrical double layer or the Debye layer thickness), $u \to 0$, and at $y = 0$, $u = u_{wall}$. A simple integration of the previous equation, along with these boundary conditions, yields the solution to the fluid velocity

$$u(y) = \varepsilon E_{ext,wall} \phi(y) / \mu \tag{15.12}$$

When this equation is applied at $y = 0$, then we obtain

$$u_{wall} = \varepsilon E_{ext,wall} \phi_{wall} / \mu, \tag{15.13}$$

where ϕ_{wall} is the electrostatic potential at the wall. If we substitute the potential variation as given by the Debye-Huckel theory in Eq. (15.12), i.e., $\phi(y) = \phi_{wall} \exp[-y/\lambda_D]$. Thus, the fluid velocity profile is given by

$$u(y) = \frac{\varepsilon E_{ext,wall}}{\mu} \phi_{wall} \exp[-y/\lambda_D] = \frac{q\lambda_D E_{ext}}{\mu} \exp[-y/\lambda_D]. \tag{15.14}$$

In the last step, the relation between surface potential ϕ_{wall} and surface charge density q as given by the Debye-Huckel theory has been used.

While the electrophoretic velocity (Eq. 15.13) is specifically derived for a flat charged surface of infinite extent, it can be shown that when the Debye layer thickness is small compared to the particle dimensions, then the result can be used for an arbitrary shaped particle as well. However, when the particle size is not very large compared to the Debye layer thickness, then Eq. (15.13) must be modified by a correction factor.[55]

15.6.3 Applications of Phoretic Motion

Electrophoresis has been widely used in the separation and identification of DNA (a negatively charged biopolymer) in biotechnological applications. In addition, electrophoresis is frequently used in the analysis of proteins, testing of antibiotics, testing and production of vaccines, etc. The fundamental principle by which separation is achieved in electrophoresis is the variation in electrophoretic mobilities of various particles. Further, electrophoresis of colloidal particles is used for particle manipulation in microfluidic devices.[55] Thermophoresis is used for particle deposition in the preparation of 'preforms' for optical fibre manufacture. It is also occasionally used to manipulate biomolecules. Diffusiophoresis is used in the formation of rubber gloves and in the deposition of paint onto a steel surface.

15.7 Nomenclature

		Units in SI System	Dimensions in **M, L, T**
a	optimal head group area	m^2	L^2
B_0	Bond number	–	–
ΔG	detachment energy	J	$M\,L^2\,T^{-2}$
g	acceleration due to gravity	m/s^2	$L\,T^{-2}$
γ	surface tension	N/m	$M\,T^{-2}$
l_c	critical chain length	m	L
L1	liquid 1	–	–
L2	liquid 2	–	–
p	particle	–	–
P	packing factor	–	–
ΔP	pressure difference	N/m^2	$M\,L^{-1}\,T^{-2}$
ϕ_{wall}	electrostatic potential at the wall	V	$M^{1/2}\,L^{1/2}\,T^{-1}$
$\Phi(y)$	potential of mean force	J	$M\,L^2\,T^{-2}$
R	radius of the drop or bubble	m	L
θ	three-phase contact angle	radians	–
V	volume	m^3	L^3

References

1. Lee KJ, Yoon J, Lahann J. Recent advances with anisotropic particles. *Curr Opin Colloid Interface Sci* 2011;**16**(3):195–202.
2. Sacanna S, Pine DJ. Shape-anisotropic colloids: building blocks for complex assemblies. *Curr Opin Colloid Interface Sci* 2011;**16**:96–105.
3. Dugyala VR, Daware SV, Basavaraj MG. Shape anisotropic colloids: synthesis, packing behavior, evaporation driven assembly, and their application in emulsion stabilization. *Soft Matter* 2013;**9**(29):6711–25.
4. Anjali TG, Basavaraj MG. General destabilization mechanism of pH-responsive Pickering emulsions. *Phys Chem Chem Phys* 2017;**19**:30790–7.
5. Sabapathy M, Kollabattula V, Basavaraj MG, Mani E. Visualization of the equilibrium position of colloidal particles at fluid–water interfaces by deposition of nanoparticles. *Nanoscale* 2015;**7**:13868–76.
6. Donev A, Cisse I, Sachs D, Variano EA, Stillinger FH, Connelly R, Torquato S, Chaikin PM. Improving the density of jammed disordered packings using ellipsoids. *Science* 2004;**303**:990–3.

7. Sacanna S, Rossi L, Wouterse A, Philipse AP. Observation of a shape-dependent density maximum in random packings and glasses of colloidal silica ellipsoids. *J Phys Condens Matter* 2007;**19**:376108.

8. Ramsden W. Separation of solids in the surface-layers of solutions and 'suspensions' (observations on surface-membranes, bubbles, emulsions, and mechanical coagulation)—preliminary account. *Proc R Soc Lond* 1903;**72**:156–64.

9. Pickering SU. Emulsions. *J Chem Soc* 1907;**91**:2001–21.

10. Binks BP. Particles as surfactants-similarities and differences. *Curr Opin Colloid Interface Sci* 2002;**7**: 21–41.

11. Frelichowska J, Bolzinger MA, Chevalier Y. Pickering emulsions with bare silica. *Colloids Surfaces A Physicochem Eng Asp* 2009;**343**(1–3):70–4.

12. Paunov VN. Novel method for determining the three-phase contact angle of colloid particles adsorbed at air-water and oil-water interfaces. *Langmuir* 2003;**19**(19):7970–6.

13. Shang J, Flury M, Harsh JB, Zollars RL. Comparison of different methods to measure contact angles of soil colloids. *J Colloid Interface Sci* 2008;**328**(2):299–307.

14. De Folter JWJ, Hutter EM, Castillo SIR, Klop KE, Philipse AP, Kegel WK. Particle shape anisotropy in Pickering emulsions: cubes and peanuts. *Langmuir* 2014;**30**(4):955–64.

15. Madivala B, Vandebril S, Fransaer J, Vermant J. Exploiting particle shape in solid stabilized emulsions. *Soft Matter* 2009;**5**:1717–27.

16. Moon IK, Lee J, Ruoff RS, Lee H. Reduced graphene oxide by chemical graphitization. *Nat Commun* 2010;**1** (6):73.

17. He Y, Wu F, Sun X, Li R, Guo Y, Li C, Zhang L, Xing F, Wang W, Gao J. Factors that affect Pickering emulsions stabilized by graphene oxide. *ACS Appl Mater Interfaces* 2013;**5**(11):4843–55.

18. Wang H, Hobbie EK. Amphiphobic carbon nanotubes as macroemulsion surfactants. *Langmuir* 2003;**19** (8):3091–3.

19. Pavese M, Musso S, Bianco S, Giorcelli M, Pugno N. An analysis of carbon nanotube structure wettability before and after oxidation treatment. *J Phys Condens Matter* 2008;**20**(47).

20. Ge S, Xiong L, Li M, Liu J, Yang J, Chang R, Liang C, Sun Q. Characterizations of Pickering emulsions stabilized by starch nanoparticles: influence of starch variety and particle size. *Food Chem* 2017;**234**:339–47.

21. Binks BP, Rocher A. Effects of temperature on water-in-oil emulsions stabilised solely by wax microparticles. *J Colloid Interface Sci* 2009;**335**(1):94–104.

22. Pieranski P. Two-dimensional interfacial colloidal crystals. *Phys Rev Lett* 1980;**45**:569–72.

23. Guo D, Xie G, Luo J. Mechanical properties of nanoparticles: basics and applications. *J Phys D Appl Phys* 2014;**47**:013001.

24. Guo Q, Duffya SP, Matthews K, Santoso AT, Scottc MD, Ma H. Microfluidic analysis of red blood cell deformability. *J Biomech* 2014;**47**:1767–76.

25. Shin S, Ku Y, Park M, Suh J. Deformability of red blood cells: a determinant of blood viscosity. *J Mech Sci Technol* 2005;**19**:216–23.

26. Chien S. Red cell deformability and its relevance to blood flow. *Annu Rev Physiol* 1987;**49**:177–92.

27. Hashmi SM, Dufresne ER. Mechanical properties of individual microgel particles through the deswelling transition. *Soft Matter* 2009;**5**(19):3682–8.

28. Lai H, Chen Q, Wu P. The core–shell structure of PNIPAM collapsed chain conformation induces a bimodal transition on cooling. *Soft Matter* 2013;**9**:3985–93.

29. Plamper A, Richtering W. Functional microgels and microgel SystemsFelix. *Acc Chem Res* 2017;**50**(2):131–40.

30. Karthickeyan D, Gupta DK, Tata BVR. Identification of volume phase transition of a single microgel particle using optical tweezers. *J Opt* 2016;**18**:105401.

31. de Gennes P, Brochard-Wyart F, Quéré D. *Capillarity and wetting phenomena*. New York: Springer; 2004.

32. Stone HA. Dynamics of drop deformation and breakup in viscous fluids. *Annu Rev Fluid Mech* 1994;**26**:65–102.

33. Grace HP. Dispersion phenomena in high viscosity immiscible fluid systems and application of static mixers as dispersion devices in such systems. *Chem Eng Commun* 1982;**14**:225–77.

34. Sherwood JD. The deformation of a fluid drop in an electric field: a slender-body analysis. *J Phys A Math Gen* 1991;**24**:4047–53.

35. Corson LT, Tsakonas C, Duffy BR, Mottram NJ, Sage IC, Brown CV, Wilson SK. Deformation of a nearly hemispherical conducting drop due to an electricfield: theory and experiment. *Phys Fluids* 2014;**26**:122106.

36. Shah RK, Kim J-W, Weitz DA. Monodisperse stimuli-responsive colloidosomes by self-assembly of microgels in droplets. *Langmuir* 2010;**26**:1561–5.

37. Israelachvili JN. *Intermolecular and surface forces.* Boston: Academic Press; 1992.

38. Larson RG. *The structure and rheology of complex fluids.* Oxford University Press; 1999.

39. Sajeesh P, Sen AK. Particle separation and sorting in microfluidic devices: a review. *Microfluid Nanofluid* 2014;**17**:152.

40. Sai Y, Yamada M, Yasuda M, Seki M. Continuous separation of particles using a microfluidic device equipped with flow rate control valves. *J Chromatogr A* 2006;**1127**:214–20.

41. Manz A, Harrison D, Verpoorte EMJ, Fettinger JC, Paulus A, Ludi H, Widmer HM. Planar chips technology for miniaturization and integration of separation techniques into monitoring system capillary electrophoresis on a chip. *J Chromotogr* 1992;**593**:253–8.

42. Reyes DR, Lossifidis D, Auroux PA, Manz A. Micro total analysis systems: introduction, theory, and technology. *Anal Chem* 2002;**74**:2623–36.

43. Toner M, Irimia D. Blood on a chip. *Annu Rev Biomed Eng* 2005;**7**:77–103.

44. Suresh S, Spatz J, Mills JP, Micoulet A, Dao M, Lim CT, Beil M, Seufferlein T. Connections between single-cell biomechanics and human disease states: gastrointestinal cancer and malaria. *Acta Biomater* 2005;**1**(1):15–30.

45. Lu X, Xuan X. Continuous microfluidic particle separation via elasto-inertial pinched flow fractionation. *Anal Chem* 2015;**87**:6389–96 [1977/78].

46. Pipe CJ, McKinley GH. Microfluidic rheometry. *Mech Res Commun* 2009;**36**:110–20.

47. Berli CLA. The apparent hydrodynamic slip of polymer solutions and its implications in electrokinetics. *Electrophoresis* 2013;**34**:622–30.

48. Zhao C, Yang C. Electrokinetics of non-Newtonian fluids: a review. *Adv Colloid Interf Sci* 2013;**201–202**:94–108.

49. Cranston HA, Boylan CW, Caroll GL, Sutera SP, Williamson JR, Gluzman IY, Krogstad DJ. Plasmodium falciparum maturation abolishes physiologic red cell deformability. *Science* 1984;**223**(4634):400–3.

50. Link DR, Grasland-Mongrain E, Duri A, Sarrazin F, Cheng Z, Cristobal G, Marquez M, Witz DA. Electric control of droplets in microfluidic devices. *Angew Chem* 2006;**118**(16):2618–22.

51. Huang LR, Cox EC, Austin RH, Sturm JC. Continuous particle separation through deterministic lateral displacement. *Science* 2004;**304**(5673):987–90.

52. Kim U, Qian J, Kenrick SK, Daugherty PS, Soh HT. Multi target dielectrophoresis activated cell sorter. *Anal Chem* 2008;**80**:8656–61.

53. Adams JD, Kim U, Soh HT. Multi target magnetic activated cell sorter. *Proc Natl Acad Sci U S A* 2008;**105**(47):18165–70.

54. Anderson JL. Colloid transport by interfacial forces. *Ann Rev Fluid Mech* 1989;**21**(1):61–99.

55. Kirby B. *Micro- and nanoscale fluid mechanics: transport in microfluidic devices.* Cambridge: Cambridge University Press; 2010.

56. Bruus H, *Theoretical microfluidics.* Oxford, UK: Oxford University Press; 2008.

57. Illien P, Golestanian R, Sen A. Fuelled motion: phoretic motility and collective behaviour of active colloids. *Chem Soc Rev* 2017;**46**:5508–18.

58. Friedlander SK. *Smoke, dust and haze: fundamentals of aerosol behaviour.* 2nd ed Oxford: Oxford University Press; 2000.

Problems

(Several of these questions have been taken from examination papers)

2.1. The size analysis of a powdered material in terms of its mass fraction is represented by a straight line from 0% at 1 μm particle size to 100% by mass at 101 μm particle size. Calculate the surface mean diameter of the particles constituting the system.

2.2. The equations giving the number distribution curve for a powdered material are $dn/dd = d$ for the size range 0–10 μm, and $dn/dd = 100,000/d^4$ for the size range 10–100 μm, where d is in μm. Sketch the number, surface, and mass distribution curves and calculate the surface mean diameter for the powder.

Explain briefly how the data for the construction of these curves may be obtained experimentally.

2.3. The fineness characteristic of a powder on a cumulative basis is represented by a straight line from the origin to 100% undersize at particle size 50 μm. If the powder is initially dispersed uniformly in a column of liquid, calculate the proportion by mass which remains in suspension in the time from commencement of settling to that at which a 40-μm particle falls the total height of the column.

2.4. The size distribution of a dust as measured by a microscope is as follows. Convert these data to obtain the distribution on a mass basis, and calculate the specific surface, assuming spherical particles of density 2650 kg/m³.

Size Range (μm)	Number of Particles in Range (–)
0–2	2000
2–4	600
4–8	140
8–12	40
12–16	15
16–20	5
20–24	2

3.1. The performance of a solids mixer was assessed by calculating the variance occurring in the mass fraction of a component among a selection of samples withdrawn from the mixture. The quality was tested at intervals of 30 s, and the data obtained are:

Sample variance (–)	0.025	0.006	0.015	0.018	0.019
Mixing time (s)	30	60	90	120	150

If the component analysed represents 20% of the mixture by mass, and each of the samples removed contains approximately 100 particles, comment on the quality of the mixture produced, and present the data in graphical form showing the variation of mixing index with time.

4.1. In a mixture of quartz of density 2650 kg/m^3 and galena of density 7500 kg/m^3, the sizes of the particles range from 0.0052 to 0.025 mm.

On separation in a hydraulic classifier under free settling conditions, three fractions are obtained, one consisting of quartz only, one a mixture of quartz and galena, and one of galena only. What are the ranges of sizes of particles of the two substances in the original mixture?

4.2. A mixture of quartz and galena of a size range from 0.015 to 0.065 mm is to be separated into two pure fractions using a hindered settling process. What is the minimum apparent density of the fluid that will give this separation? How will the viscosity of the bed affect the minimum required density? The density of galena is 7500 kg/m^3, and the density of quartz is 2650 kg/m^3.

4.3. The size distribution by mass of the dust carried in a gas, together with the efficiency of collection over each size range, is as follows:

Size range (µm)	0–5	5–10	10–20	20–40	40–80	80–160
Mass (%)	10	15	35	20	10	10
Efficiency (%)	20	40	80	90	95	100

Calculate the overall efficiency of the collector and the percentage by mass of the emitted dust that is smaller than 20 µm in diameter. If the dust burden is 18 g/m^3 at entry, and the gas flow is 0.3 m^3/s, calculate the mass flow of dust emitted.

4.4. The collection efficiency of a cyclone is 45% over the size range 0–5 µm, 80% over the size range 5–10 µm, and 96% for particles exceeding 10 µm. Calculate the efficiency of collection for a dust with a mass distribution of 50% 0–5 µm, 30% 5–10 µm, and 20% above 10 µm.

4.5. A sample of dust from the air in a factory is collected on a glass slide. If dust on the slide was deposited from 1 cm^3 of air, estimate the mass of dust in g/m^3 of air in the factory, given the number of particles in the various size ranges to be as follows:

Size range (μm)	0–1	1–2	2–4	4–6	6–10	10–14
Number of particles (−)	2000	1000	500	200	100	40

It may be assumed that the density of the dust is $2600\,kg/m^3$, and an appropriate allowance should be made for particle shape.

4.6. A cyclone separator 0.3 m in diameter and 1.2 m long has a circular inlet 75 mm in diameter and an outlet of the same size. If the gas enters at a velocity of 1.5 m/s, at what particle size will the theoretical cut occur?

The viscosity of air is $0.018\,mN\,s/m^2$, the density of air is $1.3\,kg/m^3$, and the density of the particles is $2700\,kg/m^3$.

5.1. A material is crushed in a Blake jaw crusher such that the average size of particle is reduced from 50 to 10 mm, with the consumption of energy of 13.0 kW/(kg/s). What will be the consumption of energy needed to crush the same material of average size 75 mm to average size of 25 mm:

(a) assuming Rittinger's Law applies,
(b) assuming Kick's Law applies?

Which of these results would be regarded as being more reliable and why?

5.2. A crusher was used to crush a material with a compressive strength of $22.5\,MN/m^2$. The size of the feed was *minus* 50 mm, *plus* 40 mm, and the power required was 13.0 kW/(kg/s). The screen analysis of the product was:

Size of Aperture (mm)	Amount of Product (%)
Through 6.0	All
On 4.0	26
On 2.0	18
On 0.75	23
On 0.50	8
On 0.25	17
On 0.125	3
Through 0.125	5

What power would be required to crush 1 kg/s of a material of compressive strength $45\,MN/m^2$ from a feed of *minus* 45 mm, *plus* 40 mm, to a product of 0.50 mm average size?

5.3. A crusher reducing limestone of crushing strength $70\,MN/m^2$ from 6 mm diameter average size to 0.1 mm diameter average size, requires 9 kW. The same machine is used to crush dolomite at the same output from 6 mm diameter average size to a product consisting of 20% with an average diameter of 0.25 mm, 60% with an average diameter of 0.125 mm, and a balance having an average diameter of 0.085 mm. Estimate the power required, assuming that the crushing strength of the dolomite is 100 MN/m², and that crushing follows Rittinger's Law.

5.4. If crushing rolls 1 m diameter are set so that the crushing surfaces are 12.5 mm apart, and the angle of nip is 31 degrees, what is the maximum size of particle which should be fed to the rolls?

If the actual capacity of the machine is 12% of the theoretical, calculate the throughput in kg/s when running at 2.0 Hz, if the working face of the rolls is 0.4 m long, and the feed density is 2500 kg/m³.

5.5. A crushing mill which reduces limestone from a mean particle size of 45 mm to the following product

Size (mm)	Amount of Product (%)
12.5	0.5
7.5	7.5
5.0	45.0
2.5	19.0
1.5	16.0
0.75	8.0
0.40	3.0
0.20	1.0

requires 21 kJ/kg of material crushed.

Calculate the power required to crush the same material at the same rate, from a feed having a mean size of 25 mm to a product with a mean size of 1 mm.

5.6. A ball mill 1.2 m in diameter is run at 0.8 Hz, and it is found that the mill is not working satisfactorily. Should any modification in the condition of operation be suggested?

5.7. 3 kW is supplied to a machine crushing material at the rate of 0.3 kg/s from 12.5 mm cubes to a product having the following sizes: 80% 3.175 mm, 10% 2.5 mm, and 10% 2.25 mm.

What power should be supplied to this machine to crush 0.3 kg/s of the same material from 7.5 mm cube to 2.0 mm cube?

6.1. A finely ground mixture of galena and limestone, in the proportion of 1–4 by mass, is subjected to elutriation by an upwardly flowing stream of water flowing at a velocity of 5 mm/s. Assuming that the size distribution for each material is the same, and is as shown in the following table, estimate the percentage of galena in the material carried away and in the material left behind. The viscosity of water is $1\,mN\,s/m^2$, and Stokes' equation may be used.

Diameter (μm)	20	30	40	50	60	70	80	100
Undersize (% mass)	15	28	48	54	64	72	78	88

The densities of galena and limestone are $7500\,kg/m^3$ and $2700\,kg/m^3$, respectively.

6.2. Calculate the terminal velocity of a steel ball, 2 mm diameter and of density $7870\,kg/m^3$, in an oil of density $900\,kg/m^3$ and viscosity $50\,mN\,s/m^2$.

6.3. What is the terminal settling velocity of a spherical steel particle of 0.40 mm diameter, in an oil of density $820\,kg/m^3$ and viscosity $10\,mN\,s/m^2$? The density of steel is $7870\,kg/m^3$.

6.4. What will be the terminal velocities of mica plates, 1 mm thick and ranging in area from 6 to $600\,mm^2$, settling in an oil of density $820\,kg/m^3$ and viscosity $10\,mN\,s/m^2$? The density of mica is $3000\,kg/m^3$.

6.5. A material of density $2500\,kg/m^3$ is fed to a size separation plant where the separating fluid is water, which rises with a velocity of 1.2 m/s. The upward vertical component of the velocity of the particles is 6 m/s. How far will an approximately spherical particle, 6 mm diameter, rise relative to the walls of the plant before it comes to rest in the fluid?

6.6. A spherical glass particle is allowed to settle freely in water. If the particle starts initially from rest, and if the value of the Reynolds number with respect to the particle is 0.1 when it has attained its terminal falling velocity, calculate:

(a) the distance travelled before the particle reaches 90% of its terminal falling velocity,
(b) the time elapsed when the acceleration of the particle is one hundredth of its initial value.

6.7. In a hydraulic jig, a mixture of two solids is separated into its components by subjecting an aqueous slurry of the material to a pulsating motion, and allowing the particles to settle for a series of short time intervals such that their terminal falling velocities are not attained. Materials of densities 1800 and $2500\,kg/m^3$, whose particle size ranges from 0.3 to 3 mm diameter, are to be separated. It may be assumed that the particles are approximately spherical, and that Stokes' law is applicable. Calculate the approximate maximum time interval for which the particles may be allowed to settle so that no particle of the less dense material falls a greater distance than any particle of the denser material. The viscosity of water is $1\,mN\,s/m^2$.

6.8. Two spheres of equal terminal falling velocities settle in water starting from rest starting at the same horizontal level. How far apart vertically will the particles be when they have both reached their terminal falling velocities? It may be assumed that Stokes' law is valid, and this assumption should be checked.

The diameter of one sphere is 40 μm, and its density is 1500 kg/m³, and the density of the second sphere is 3000 kg/m³. The density and viscosity of water are 1000 kg/m³ and 1 mN s/m², respectively.

6.9. The size distribution of a powder is measured by sedimentation in a vessel having the sampling point 180 mm below the liquid surface. If the viscosity of the liquid is 1.2 mN s/m², and the densities of the powder and liquid are 2650 and 1000 kg/m³, respectively, determine the time which must elapse before any sample will exclude particles larger than 20 μm.

If Stokes' law does not apply when the Reynolds number is greater than 0.2, what is the approximate maximum size of particle to which Stokes' law may be applied under these conditions?

6.10. Calculate the distance a spherical lead shot of diameter 0.1 mm settles in a glycerol/water mixture before it reaches 99% of its terminal falling velocity

The density of lead is 11,400 kg/m³, and the density of liquid is 1000 kg/m³. The viscosity of liquid is 10 mN s/m².

It may be assumed that the resistance force may be calculated from Stokes' law and is equal to $3\pi\mu du$, where u is the velocity of the particle relative to the liquid.

6.11. What is the mass of a sphere of material of density 7500 kg/m³ which falls with a steady velocity of 0.6 m/s in a large, deep tank of water?

6.12. Two ores, of densities 3700 and 9800 kg/m³, are to be separated in water by a hydraulic classification method. If the particles are all of approximately the same shape, and each is sufficiently large for the drag force to be proportional to the square of its velocity in the fluid, calculate the maximum ratio of sizes which can be separated if the particles attain their terminal falling velocities. Explain why a wider range of sizes can be separated if the time of settling is so small that the particles do not reach their terminal velocities.

An explicit expression should be obtained for the distance through which a particle will settle in a given time if it starts from rest, and if the resistance force is proportional to the square of the velocity. The acceleration period should be taken into account.

6.13. Salt, of density 2350 kg/m³, is charged to the top of a reactor containing a 3 m depth of aqueous liquid of density 1100 kg/m³ and viscosity 2 mN s/m², and the crystals must dissolve completely before reaching the bottom of the reactor. If the rate of dissolution of the crystals is given by:

$$-\frac{dd}{dt} = 3 \times 10^{-6} + 2 \times 10^{-4}u$$

where d is the size of the crystal (m) at time t (s), and u is its velocity in the fluid (m/s), calculate the maximum size of crystal which should be charged. The inertia of the particles may be neglected, and the resistance force may be taken as that given by Stokes' law ($3\pi\mu du$), where d is taken as the equivalent spherical diameter of the particle.

6.14. A balloon of mass 7 g is charged with hydrogen to a pressure of $104\,kN/m^2$. The balloon is released from ground level, and, as it rises, hydrogen escapes in order to maintain a constant differential pressure of $2.7\,kN/m^2$, under which condition the diameter of the balloon is 0.3 m. If conditions are assumed to remain isothermal at 273 K as the balloon rises, what is the ultimate height reached, and how long does it take to rise through the first 3000 m?

It may be assumed that the value of the Reynolds number with respect to the balloon exceeds 500 throughout, and that the resistance coefficient is constant at 0.22. The inertia of the balloon may be neglected, and at any moment, it may be assumed that it is rising at its equilibrium velocity.

6.15. A mixture of quartz and galena of densities 3700 and $9800\,kg/m^3$, respectively with a size range is 0.3–1 mm is to be separated by a sedimentation process. If Stokes' law is applicable, what is the minimum density required for the liquid if the particles all settle at their terminal velocities?

A separating system using water as the liquid is considered in which the particles were to be allowed to settle for a series of short time intervals so that the smallest particle of galena settled a larger distance than the largest particle of quartz. What is the approximate maximum permissible settling period?

According to Stokes' law, the resistance force F acting on a particle of diameter d, settling at a velocity u in a fluid of viscosity μ is given by:

$$F = 3\pi\mu du$$

The viscosity of water is $1\,mN\,s/m^2$.

6.16. A glass sphere, of diameter 6 mm and density $2600\,kg/m^3$, falls through a layer of oil of density $900\,kg/m^3$ into water. If the oil layer is sufficiently deep for the particle to have reached its free falling velocity in the oil, how far will it have penetrated into the water before its velocity is only 1% above its free falling velocity in water? It may be assumed that the force on the particle is given by Newton's law, and that the particle drag coefficient, $R'/\rho u^2 = 0.22$.

6.17. Two spherical particles, one of density $3000\,kg/m^3$ and diameter $20\,\mu m$, and the other of density $2000\,kg/m^3$ and diameter $30\,\mu m$, start settling from rest at the same horizontal level in a liquid of density $900\,kg/m^3$ and of viscosity $3\,mN\,s/m^2$. After what period of settling will the particles be again at the same horizontal level? It may be assumed that Stokes' law is applicable, and the effect of mass acceleration of the liquid moved with each sphere may be ignored.

6.18. What will be the terminal velocity of a glass sphere 1 mm in diameter in water if the density of glass is 2500 kg/m^3?

6.19. What is the mass of a sphere of density 7500 kg/m^3 which has a terminal velocity of 0.7 m/s in a large tank of water?

7.1. In a contact sulphuric acid plant, the secondary converter is a tray-type converter, 2.3 m in diameter, with the catalyst arranged in three layers, each 0.45 m thick. The catalyst is in the form of cylindrical pellets 9.5 mm in diameter and 9.5 mm long. The void fraction is 0.35. The gas enters the converter at 675 K and leaves at 720 K. Its inlet composition is:

$$SO_3 \; 6.6, \quad SO_2 \; 1.7, \quad O_2 \; 10.0, \quad N_2 \; 81.7 \; \text{mole\%}$$

and its exit composition is:

$$SO_3 \; 8.2, \quad SO_2 \; 0.2, \quad O_2 \; 9.3, \quad N_2 \; 82.3 \; \text{mole\%}$$

The gas flow rate is $0.68 \text{ kg/m}^2 \text{ s}$. Calculate the pressure drop through the converter. The viscosity of the gas is 0.032 mN s/m^2.

7.2. Two heat-sensitive organic liquids of average molecular weight of 155 kg/kmol are to be separated by vacuum distillation in a 100 mm diameter column packed with 6-mm stoneware Raschig rings. The number of theoretical plates required is 16, and it has been found that the HETP is 150 mm. If the product rate is 5 g/s at a reflux ratio of 8, calculate the pressure in the condenser so that the temperature in the still does not exceed 395 K, equivalent to a pressure of 8 kN/m^2. It may be assumed that $a = 800 \text{ m}^2/\text{m}^3$, $\mu = 0.02 \text{ mN s/m}^2$, $e = 0.72$, and that the temperature changes, and the correction for liquid flow may be neglected.

7.3. A column 0.6 m diameter and 4 m high is packed with 25-mm ceramic Raschig rings and used in a gas absorption process carried out at 101.3 kN/m^2 and 293 K. If the liquid and gas approximate to those of water and air, respectively, and their flow rates are 2.5 and $0.6 \text{ kg/m}^2 \text{ s}$, what is the pressure drop across the column? By how much may the liquid flow rate be increased before the column floods?

7.4. A packed column, 1.2 m in diameter and 9 m tall, is packed with 25-mm Raschig rings and used for the vacuum distillation of a mixture of isomers of molecular weight 155 kg/kmol. The mean temperature is 373 K, the pressure at the top of the column is maintained at 0.13 kN/m^2, and the still pressure is 1.3–3.3 kN/m^2. Obtain an expression for the pressure drop on the assumption that this is not appreciably affected by the liquid flow and may be calculated using a modified form of Carman's equation. Show that, over the range of operating pressures used, the pressure drop is approximately directly proportional to the mass rate of flow of vapour, and calculate the pressure drop at a vapour rate of 0.125 kg/m^2. The specific surface of packing, $S = 190 \text{ m}^2/\text{m}^3$, the mean voidage of bed, $e = 0.71$, the viscosity of vapour, $\mu = 0.018 \text{ mN s/m}^2$, and the molecular volume $= 22.4 \text{ m}^3/\text{kmol}$.

8.1. A slurry containing 5 kg of water/kg of solids is to be thickened to a sludge containing 1.5 kg of water/kg of solids in a continuous operation. Laboratory tests using five different concentrations of the slurry yielded the following results

Concentration (kg water/kg solid)	5.0	4.2	3.7	3.1	2.5
Rate of sedimentation (mm/s)	0.17	0.10	0.08	0.06	0.042

Calculate the minimum area of a thickener to effect the separation of 0.6 kg/s of solids.

8.2. A slurry containing 5 kg of water/kg of solids is to be thickened to a sludge containing 1.5 kg of water/kg of solids in a continuous operation

Laboratory tests using five different concentrations of the slurry yielded the following data:

Concentration (kg water/kg solid)	5.0	4.2	3.7	3.1	2.5
Rate of sedimentation (mm/s)	0.20	0.12	0.094	0.070	0.050

Calculate the minimum area of a thickener to effect the separation of 1.33 kg/s of solids.

8.3. When a suspension of uniform coarse particles settles under the action of gravity, the relation between the sedimentation velocity u_c and the fractional volumetric concentration C are given by

$$\frac{u_c}{u_0} = (1 - C)^n,$$

where $n = 2.3$, and u_0 is the free-falling velocity of the particles. Draw the curve of solids flux ψ against concentration and determine the value of C at which ψ is a maximum and where the curve has a point of inflexion. What is implied about the settling characteristics of such a suspension from the Kynch theory? Comment on the validity of the Kynch theory for such a suspension.

8.4. For the sedimentation of a suspension of uniform fine particles in a liquid, the relation between observed sedimentation velocity u_c and fractional volumetric concentration C are given by

$$\frac{u_c}{u_0} = (1 - C)^{4.8}$$

where u_0 is the free falling velocity of an individual particle.

Calculate the concentration at which the rate of deposition of particles per unit area is a maximum and determine this maximum flux for 0.1 mm spheres of glass of density 2600 kg/m^3 settling in water of density 1000 kg/m^3 and viscosity 1 mN s/m^2.

It may be assumed that the resistance force F on an isolated sphere is given by Stokes' law.

8.5. A binary suspension consists of equal masses of spherical particles whose free falling velocities in the liquid are 1 and 2 mm/s, respectively. The system is initially well mixed, and the total volumetric concentration of solids is 20%. As sedimentation proceeds, a sharp interface forms between the clear liquid and suspension consisting only of small particles, and a second interface separates the suspension of fines from the mixed suspension. Choose a suitable model for the behaviour of the system and estimate the falling rates of the two interfaces. It may be assumed that the sedimentation velocity, u_c, in a concentrated suspension of voidage e is related to the free-falling velocity u_0 of the particles by:

$$(u_c/u_0) = e^{2.3}.$$

9.1. Oil, of density 900 kg/m³ and viscosity 3 mN s/m², is passed vertically upwards through a bed of catalyst consisting of approximately spherical particles of diameter 0.1 mm and density 2600 kg/m³. At approximately what mass rate of flow per unit area of bed will (a) fluidisation, and (b) transport of particles occur?

9.2. Calculate the minimum velocity at which spherical particles of density 1600 kg/m³ and of diameter 1.5 mm will be fluidised by water in a tube of diameter 10 mm. Discuss the uncertainties in this calculation. The viscosity of water is 1 mN s/m², and Kozeny's constant is 5.

9.3. In a fluidised bed, isooctane vapour is adsorbed from an air stream onto the surface of alumina microspheres. The mole fraction of isooctane in the inlet gas is 1.442×10^{-2}, and the mole fraction in the outlet gas is found to vary with time as follows:

Time From Start (s)	Mole Fraction in Outlet Gas ($\times 10^2$)
250	0.223
500	0.601
750	0.857
1000	1.062
1250	1.207
1500	1.287
1750	1.338
2000	1.373

Show that the results may be interpreted on the assumptions that the solids are completely mixed, that the gas leaves in equilibrium with the solids, and that the adsorption isotherm is linear over the range considered. If the flow rate of gas is 0.679×10^{-6} kmol/s, and the mass of solids in the bed is 4.66 g, calculate the slope of the adsorption isotherm. What evidence do the results provide concerning the flow pattern of the gas?

9.4. Cold particles of glass ballotini are fluidised with heated air in a bed in which a constant flow of particles is maintained in a horizontal direction. When steady conditions have been reached, the temperatures recorded by a bare thermocouple immersed in the bed are:

Distance Above Bed Support (mm)	Temperature (K)
0	339.5
0.64	337.7
1.27	335.0
1.91	333.6
2.54	333.3
3.81	333.2

Calculate the coefficient for heat transfer between the gas and the particles, and the corresponding values of the particle Reynolds and Nusselt numbers. Comment on the results and on any assumptions made. The gas flow rate is $0.2 \, \text{kg/m}^2$ s, the specific heat capacity of air is $0.88 \, \text{kJ/kg K}$, the viscosity of air is $0.015 \, \text{mN s/m}^2$, the particle diameter is $0.25 \, \text{mm}$, and the thermal conductivity of air $0.03 \, \text{W/mK}$.

9.5. The relation between bed voidage e and fluid velocity u_c for particulate fluidisation of uniform particles which are small compared with the diameter of the containing vessel is given by

$$\frac{u_c}{u_0} = e^n$$

where u_0 is the free-falling velocity.

Discuss the variation of the index n with flow conditions, indicating why this is independent of the Reynolds number Re with respect to the particle at very low and very high values of Re. When are appreciable deviations from this relation observed with liquid fluidised systems?

For particles of glass ballotini with free-falling velocities of 10 and 20 mm/s, the index n has a value of 2.39. If a mixture of equal volumes of the two particles is fluidised, what is the relation between the voidage and fluid velocity if it is assumed that complete segregation is obtained?

9.6. Obtain a relationship for the ratio of the terminal falling velocity of a particle to the minimum fluidising velocity for a bed of similar particles. It may be assumed that Stokes' law and the Carman–Kozeny equation are applicable. What is the value of the ratio if the bed voidage at the minimum fluidising velocity is 0.4?

9.7. A bed consists of uniform spherical particles of diameter 3 mm and density $4200 \, \text{kg/m}^3$. What will be the minimum fluidising velocity in a liquid of viscosity $3 \, \text{mN s/m}^2$ and density $1100 \, \text{kg/m}^3$?

9.8. Ballotini particles 0.25 mm in diameter are fluidised by hot air flowing at the rate of 0.2 kg/ m^2 cross-section of bed to give a bed of voidage 0.5, and a cross-flow of particles is maintained to remove the heat. Under steady-state conditions, a small bare thermocouple immersed in the bed gives the following data:

Distance Above Bed Support (mm)	Temperature (°C)
0	66.3
0.625	64.5
1.25	61.8
1.875	60.4
2.5	60.1
3.75	60.0

Assuming plug flow of the gas and complete mixing of the solids, calculate the coefficient for heat transfer between the particles and the gas. The specific heat capacity of air is 0.85 kJ/kg K.

A fluidised bed of total volume 0.1 m^3 containing the same particles is maintained at an approximately uniform temperature of 425 K by external heating, and a dilute aqueous solution at 375 K is fed to the bed at the rate of 0.1 kg/s so that the water is completely evaporated at atmospheric pressure. If the heat transfer coefficient is the same as that previously determined, what volumetric fraction of the bed is effectively carrying out the evaporation? The latent heat of vaporisation of water is 2.6 MJ/kg.

9.9. An electrically heated element of surface area 12 cm^2 is immersed so that it is in direct contact with a fluidised bed. The resistance of the element is measured as a function of the voltage applied to it giving the following data:

Potential (V)	1	2	3	4	5	6
Resistance (ohms)	15.47	15.63	15.91	16.32	16.83	17.48

The relation between resistance R_w and temperature T_w is:

$$\frac{R_w}{R_0} = 0.004T_w - 0.092$$

where R_0, the resistance of the wire at 273 K, is 14 Ω and T_w is in K. Estimate the bed temperature and the value of the heat transfer coefficient between the surface and the bed.

9.10.

(a) Explain why the sedimentation velocity of uniform coarse particles in a suspension decreases as the concentration is increased. Identify and, where possible, quantify the various factors involved.

(b) Discuss the similarities and differences in the hydrodynamics of a sedimenting suspension of uniform particles and of an evenly fluidised bed of the same particles in the liquid.

(c) A liquid fluidised bed consists of equal volumes of spherical particles 0.5 and 1.0mm in diameter. The bed is fluidised, and complete segregation of the two species occurs. When the liquid flow is stopped, the particles settle to form a segregated two-layer bed. The liquid flow is then started again. When the velocity is such that the larger particles are at their incipient fluidisation point, what, approximately, will be the voidage of the fluidised bed composed of the smaller particles?

It may be assumed that the drag force F of the fluid on the particles under the free falling conditions is given by Stokes' law, and that the relation between the fluidisation velocity u_c and voidage, e, for particles of terminal velocity, u_0, is given by:

$$u_c/u_0 = e^{4.8}$$

For Stokes' law, the force F on the particles is given by $F = 3\pi\mu d u_0$, where d is the particle diameter, and μ is the viscosity of the liquid.

9.11. The relation between the concentration of a suspension and its sedimentation velocity is of the same form as that between velocity and concentration in a fluidised bed. Explain this in terms of the hydrodynamics of the two systems.

A suspension of uniform spherical particles in a liquid is allowed to settle, and, when the sedimentation velocity is measured as a function of concentration, the following results are obtained:

Fractional Volumetric Concentration (C)	Sedimentation Velocity (u_c m/s)
0.35	1.10
0.25	2.19
0.15	3.99
0.05	6.82

Estimate the terminal falling velocity u_0 of the particles at infinite dilution. On the assumption that Stokes' law is applicable, calculate the particle diameter d.

The particle density, $\rho_s = 2600\,kg/m^3$, the liquid density, $\rho = 1000\,kg/m^3$ and the liquid viscosity, $\mu = 0.1\,Ns/m^2$.

What will be the minimum fluidising velocity of the system? Stokes' law states that the force on a spherical particle $= 3\pi\mu d u_0$.

9.12. A mixture of two sizes of glass spheres of diameters 0.75 and 1.5 mm is fluidised by a liquid, and complete segregation of the two species of particles occurs, with the smaller particles constituting the upper portion of the bed and the larger particles in the lower portion. When the voidage of the lower bed is 0.6, what will be the voidage of the upper bed?

The liquid velocity is increased until the smaller particles are completely transported from the bed. What is the minimum voidage of the lower bed at which this phenomenon will occur?

It may be assumed that the terminal falling velocities of both particles may be calculated from Stokes' law, and that the relationship between the fluidisation velocity u and the bed voidage e is given by:

$$(u_c/u_0) = e^{4.6}$$

9.13.

(a) Calculate the terminal falling velocities in water of glass particles of diameter 12 mm and density 2500 kg/m^3, and of metal particles of diameter 1.5 mm and density 7500 kg/m^3

It may be assumed that the particles are spherical, and that, in both cases, the friction factor $R'/\rho u^2$ is constant at 0.22, where R' is the force on the particle per unit of projected area of the particle, ρ is the fluid density, and u the velocity of the particle relative to the fluid.

(b) Why is the sedimentation velocity lower when the particle concentration in the suspension is high? Compare the behaviour of the concentrated suspension of particles settling under gravity in a liquid with that of a fluidised bed of the same particles.
    At what water velocity will fluidised beds of the glass and metal particles have the same densities? The relation between the fluidisation velocity u_c terminal velocity u_0 and bed voidage e is given for both particles by:

$$(u_c/u_0) = e^{2.30}$$

9.14. Glass spheres are fluidised by water at a velocity equal to one half of their terminal falling velocities. Calculate:

(a) the density of the fluidised bed,
(b) the pressure gradient in the bed attributable to the presence of the particles.

The particles are 2 mm in diameter and have a density of 2500 kg/m^3. The density and viscosity of water are 1000 kg/m^3 and 1 mN s/m^2, respectively.

10.1. A slurry containing 0.2 kg of solid/kg of water is fed to a rotary drum filter 0.6 m in diameter and 0.6 m long. The drum rotates at one revolution in 360 s, and 20% of the filtering surface is in contact with the slurry at any given instant. If filtrate is produced at the rate of 0.125 kg/s, and the cake has a voidage of 0.5, what thickness of cake is formed when filtering at a pressure difference of 65 kN/m^2? The density of the solid is 3000 kg/m^3.

The rotary filter breaks down, and the operation has to be carried out temporarily in a plate and frame press with frames 0.3 m × 0.3 m. The press takes 120 s to dismantle and 120 s to reassemble, and, in addition, 120 s is required to remove the cake from each frame. If filtration is to be carried out at the same overall rate as before, with an operating pressure difference of $275 \, \text{kN/m}^2$, what is the minimum number of frames that must be used, and what is the thickness of each? It may be assumed that the cakes are incompressible, and the resistance of the filter media may be neglected.

10.2. A slurry containing $100 \, \text{kg}$ of whiting/m^3 of water is filtered in a plate and frame press, which takes $900 \, \text{s}$ to dismantle, clean, and reassemble. If the filter cake is incompressible and has a voidage of 0.4, what is the optimum thickness of cake for a filtration pressure of $1000 \, \text{kN/m}^2$? The density of the whiting is $3000 \, \text{kg/m}^3$. If the cake is washed at $500 \, \text{kN/m}^2$, and the total volume of wash water employed is 25% of that of the filtrate, how is the optimum thickness of cake affected? The resistance of the filter medium may be neglected, and the viscosity of water is $1 \, \text{mN s/m}^2$. In an experiment, a pressure of $165 \, \text{kN/m}^2$ produced a flow of water of $0.02 \, \text{cm}^3/\text{s}$ though a centimetre cube of filter cake.

10.3. A plate and frame press, gave a total of $8 \, \text{m}^3$ of filtrate in $1800 \, \text{s}$ and $11.3 \, \text{m}^3$ in $3600 \, \text{s}$ when filtration was stopped. Estimate the washing time if $3 \, \text{m}^3$ of wash water is used. The resistance of the cloth may be neglected, and a constant pressure is used throughout.

10.4. In the filtration of a sludge, the initial period is effected at a constant rate with the feed pump at full capacity, until the pressure differences reaches $400 \, \text{kN/m}^2$. The pressure is then maintained at this value for a remainder of the filtration. The constant rate operation requires $900 \, \text{s}$, and one-third of the total filtrate is obtained during this period.

Neglecting the resistance of the filter medium, determine (a) the total filtration time and (b) the filtration cycle with the existing pump for a maximum daily capacity, if the time for removing the cake and reassembling the press is $1200 \, \text{s}$. The cake is hot washed.

10.5. A rotary filter operating at $0.03 \, \text{Hz}$ filters at the rate of $0.0075 \, \text{m}^3/\text{s}$. Operating under the same vacuum and neglecting the resistance of the filter cloth, at what speed must the filter be operated to give a filtration rate of $0.0160 \, \text{m}^3/\text{s}$?

10.6. A slurry is filtered in a plate and frame press containing 12 frames, each $0.3 \, \text{m} \times 0.3 \, \text{m}$ and $25 \, \text{mm}$ thick. During the first $180 \, \text{s}$, the filtration pressure is slowly raised to the final value of $400 \, \text{kN/m}^2$ and, during this period, the rate of filtration is maintained constant. After the initial period, filtration is carried out at constant pressure and the cakes are completely formed in a further $900 \, \text{s}$. The cakes are then washed with a pressure difference of $275 \, \text{kN/m}^2$ for $600 \, \text{s}$, using *thorough washing*. What is the volume of filtrate collected per cycle, and how much wash water is used?

A sample of the slurry was tested, using a vacuum leaf filter of $0.05 \, \text{m}^2$ filtering surface and a vacuum, giving a pressure difference of $71.3 \, \text{kN/m}^2$. The volume of filtrate collected in the first $300 \, \text{s}$ was $250 \, \text{cm}^3$, and, after a further $300 \, \text{s}$, an additional $150 \, \text{cm}^3$ was collected. It may be assumed that cake is incompressible, and the cloth resistance is the same in the leaf as in the filter press.

10.7. A sludge is filtered in a plate and frame press fitted with 25 mm frames. For the first 600 s, the slurry pump runs at maximum capacity. During this period, the pressure difference rises to 415 kN/m^2, and 25% of the total filtrate is obtained. The filtration takes a further 3600 s to complete at constant pressure, and 900 s is required for emptying and resetting the press.

It is found that, if the cloths are precoated with filter aid to a depth of 1.6 mm, the cloth resistance is reduced to 25% of its former value. What will be the increase in the overall throughput of the press if the precoat can be applied in 180 s?

10.8. Filtration is carried out in a plate and frame filter press, with 20 frames 0.3 m × 0.3 m and 50 mm thick, and the rate of filtration is maintained constant for the first 300 s. During this period, the pressure is raised to 350 kN/m^2, and one-quarter of the total filtrate per cycle is obtained. At the end of the constant rate period, filtration is continued at a constant pressure of 350 kN/m^2 for a further 1800 s, after which the frames are full. The total volume of filtrate per cycle is 0.7 m^3, and dismantling and refitting of the press takes 500 s.

It is decided to use a rotary drum filter, 1.5 m long and 2.2 m in diameter, in place of the filter press. Assuming that the resistance of the cloth is the same in the two plants, and that the filter cake is incompressible, calculate the speed of rotation of the drum which will result in the same overall rate of filtration as was obtained with the filter press. The filtration in the rotary filter is carried out at a constant pressure difference of 70 kN/m^2, and the filter operates with 25% of the drum submerged in the slurry at any instant.

10.9. It is required to filter a slurry to produce 2.25 m^3 of filtrate per working day of 8 h. The process is carried out in a plate and frame filter press with 0.45 m × 0.45 m frames and a working pressure of 450 kN/m^2. The pressure is built up slowly over a period of 300 s, and, during this period, the rate of filtration is maintained constant.

When a sample of the slurry is filtered using a pressure of 35 kN/m^2 on a single leaf filter of filtering area 0.05 m^2, 400 cm^3 of filtrate is collected in the first 300 s of filtration, and a further 400 cm^3 is collected during the following 600 s. Assuming that the dismantling of the filter press, the removal of the cakes, and the resetting of the press takes an overall time of 300 s, plus an additional 180 s for each cake produced, what is the minimum number of frames that need be employed? The resistance of the filter cloth may be taken as the same in the laboratory tests as on the plant.

10.10. The relation between flow and head for a slurry pump may be represented approximately by a straight line, the maximum flow at zero head being 0.0015 m^3/s, and the maximum head at zero flow 760 m of liquid. Using this pump to feed a slurry to a pressure leaf filter:

(a) how long will it take to produce 1 m^3 of filtrate, and

(b) what will be the pressure difference across the filter after this time?

A sample of the slurry was filtered at a constant rate of $0.00015\,\text{m}^3/\text{s}$ through a leaf filter covered with a similar filter cloth, but of one-tenth the area of the full-scale unit, and after $625\,\text{s}$, the pressure drop across the filter was $360\,\text{m}$ of liquid. After a further $480\,\text{s}$, the pressure drop was $600\,\text{m}$ of liquid.

10.11. A slurry containing 40% by mass solid is to be filtered on a rotary drum filter $2\,\text{m}$ diameter and $2\,\text{m}$ long, which normally operates with 40% of its surface immersed in the slurry and under a pressure of $17\,\text{kN/m}^2$. A laboratory test on a sample of the slurry using a leaf filter of area $200\,\text{cm}^2$ and covered with a similar cloth to that on the drum produced $300\,\text{cm}^3$ of filtrate in the first $60\,\text{s}$ and $140\,\text{cm}^3$ in the next $60\,\text{s}$, when the leaf was under an absolute pressure of $17\,\text{kN/m}^2$. The bulk density of the dry cake was $1500\,\text{kg/m}^3$, and the density of the filtrate was $1000\,\text{kg/m}^3$. The minimum thickness of cake which could be readily removed from the cloth was $5\,\text{mm}$.

At what speed should the drum rotate for maximum throughput, and what is this throughput in terms of the mass of the slurry fed to the unit per unit time?

10.12. A continuous rotary filter is required for an industrial process for the filtration of a suspension to produce $0.002\,\text{m}^3/\text{s}$ of filtrate. A sample was tested on a small laboratory filter of area $0.023\,\text{m}^2$ to which it was fed by means of a slurry pump to give filtrate at a constant rate of $12.5\,\text{cm}^3/\text{s}$. The pressure difference across the test filter increased from $14\,\text{kN/m}^2$ after $300\,\text{s}$ filtration to $28\,\text{kN/m}^2$ after $900\,\text{s}$, at which time, the cake thickness had reached $38\,\text{mm}$. What are suitable dimensions and operating conditions for the rotary filter, assuming that the resistance of the cloth used is one-half that on the test filter, and that the vacuum system is capable of maintaining a constant pressure difference of $70\,\text{kN/m}^2$ across the filter?

10.13. A rotary drum filter $1.2\,\text{m}$ diameter and $1.2\,\text{m}$ long handles $6.0\,\text{kg/s}$ of slurry containing 10% of solids when rotated at $0.005\,\text{Hz}$. By increasing the speed to $0.008\,\text{Hz}$, it is found that it can then handle $7.2\,\text{kg/s}$ of slurry. What will be the percentage change in the amount of wash water which may be applied to each kilogram of cake caused by the increased speed of rotation of the drum, and what is the theoretical maximum quantity of slurry which can be handled?

10.14. A rotary drum with a filter area of $3\,\text{m}^2$ operates with an internal pressure of $70\,\text{kN/m}^2$ below atmospheric, and with 30% of its surface submerged in the slurry. Calculate the rate of production of filtrate and the thickness of cake when it rotates at $0.0083\,\text{Hz}$, if the filter cake is incompressible, and the filter cloth has a resistance equal to that of $1\,\text{mm}$ of cake.

It is desired to increase the rate of filtration by raising the speed of rotation of the drum. If the thinnest cake that can be removed from the drum has a thickness of $5\,\text{mm}$, what is the maximum rate of filtration which can be achieved, and what speed of rotation of the drum is required? The voidage of cake $=0.4$, the specific resistance of cake $=2\times 10^{12}\,\text{m}^{-2}$, the density of solids $=2000\,\text{kg/m}^3$, the density of filtrate $=1000\,\text{kg/m}^3$, the viscosity of filtrate $=10^{-3}\,\text{Ns/m}^2$, and the slurry concentration $=20\%$ by mass of solids.

10.15. A slurry containing 50% by mass of solids of density $2600 \, kg/m^3$ is to be filtered on a rotary drum filter 2.25 m in diameter and 2.5 m long, which operates with 35% of its surface immersed in the slurry and under a vacuum of 600 mmHg. A laboratory test on a sample of the slurry, using a leaf filter with an area of $100 \, cm^2$ and covered with a cloth similar to that used on the drum, produced $220 \, cm^3$ of filtrate in the first minute, and $120 \, cm^3$ of filtrate in the next minute when the leaf was under a vacuum of 550 mmHg. The bulk density of the wet cake was $1600 \, kg/m^3$, and the density of the filtrate was $1000 \, kg/m^3$.

On the assumption that the cake is incompressible, and that 5 mm of cake is left behind on the drum, determine the theoretical maximum flow rate of filtrate obtainable. What drum speed will give a filtration rate of 80% of the maximum?

10.16. A rotary filter that operates at a fixed vacuum gives a desired rate of filtration of a slurry when rotating at 0.033 Hz. By suitable treatment of the filter cloth with a filter aid, its effective resistance is halved, and the required filtration rate is now achieved at a rotational speed of 0.0167 Hz (1 rpm). If, by further treatment, it is possible to reduce the effective cloth resistance to a quarter of the original value, what rotational speed is required? If the filter is now operated again at its original speed of 0.033 Hz (2 rpm), by what factor will the filtration rate be increased?

11.1. If a centrifuge is 0.9 m diameter and rotates at 20 Hz, at what speed should a laboratory centrifuge of 150 mm diameter be run if it is to duplicate the performance of the large unit?

11.2. An aqueous suspension consisting of particles of density $2500 \, kg/m^3$ in the size range 1–10 μm is introduced into a centrifuge with a basket 450 mm diameter rotating at 80 Hz. If the suspension forms a layer 75 mm thick in the basket, approximately how long will it take for the smallest particle to settle out?

11.3. A centrifuge basket 600 mm long and 100 mm internal diameter has a discharge weir 25 mm diameter. What is the maximum volumetric flow of liquid through the centrifuge such that when the basket is rotated at 200 Hz, all particles of diameter greater than 1 μm are retained on the centrifuge wall? The retarding force on a particle moving liquid may be taken as $3\pi\mu du$, where u is the particle velocity relative to the liquid μ is the liquid viscosity, and d is the particle diameter. The density of the liquid is $1000 \, kg/m^3$, the density of the solid is $2000 \, kg/m^3$, and the viscosity of the liquid is $1.0 \, mN \, s/m^2$. The inertia of the particle may be neglected.

11.4. When an aqueous slurry is filtered in a plate, and frame press, fitted with two 50-mm-thick frames, each 150 mm × 150 mm, at a pressure difference of $350 \, kN/m^2$, the frames are filled in 3600 s. The liquid in the slurry has the same density as water.

How long will it take to produce the same volume of filtrate as is obtained from a single cycle when using a centrifuge with a perforated basket 300 mm in diameter and 200 mm deep? The radius of the inner surface of the slurry is maintained constant at 75 mm, and the speed of rotation is 65 Hz (3900 rpm).

It may be assumed that the filter cake is incompressible, that the resistance of the cloth is equivalent to 3 mm of cake in both cases, and that the liquid in the slurry has the same density as water.

11.5. A centrifuge with a phosphor bronze basket, 380 mm in diameter, is to be run at 67 Hz with a 75 mm layer of liquid of density 1200 kg/m^3 in the basket. What wall thickness is required in the basket? The density of phosphor bronze is 8900 kg/m^3, and the maximum safe stress for phosphor bronze is 87.6 MN/m^2.

Index

Note: Page numbers followed by *f* indicate figures and *t* indicate tables.

Printed in the United States
by Baker & Taylor Publisher Services